AF767799

Operational Research Tools for Solving Sustainable Engineering Problems

Operational Research Tools for Solving Sustainable Engineering Problems

Editors

Dragan Pamucar
Željko Stević

MDPI • Basel • Beijing • Wuhan • Barcelona • Belgrade • Manchester • Tokyo • Cluj • Tianjin

Editors

Dragan Pamucar
Department of Logistics
Military Academy
University of Defence
in Belgrade
Belgrade
Serbia

Željko Stević
Faculty of Transport and Traffic
Engineering
University of East Sarajevo
Doboj
Bosnia & Herzegovina

Editorial Office

MDPI
St. Alban-Anlage 66
4052 Basel, Switzerland

This is a reprint of articles from the Special Issue published online in the open access journal *Sustainability* (ISSN 2071-1050) (available at: www.mdpi.com/journal/sustainability/special_issues/Operational_Research_Tools).

For citation purposes, cite each article independently as indicated on the article page online and as indicated below:

LastName, A.A.; LastName, B.B.; LastName, C.C. Article Title. *Journal Name* **Year**, *Volume Number*, Page Range.

ISBN 978-3-0365-1796-4 (Hbk)
ISBN 978-3-0365-1795-7 (PDF)

© 2022 by the authors. Articles in this book are Open Access and distributed under the Creative Commons Attribution (CC BY) license, which allows users to download, copy and build upon published articles, as long as the author and publisher are properly credited, which ensures maximum dissemination and a wider impact of our publications.
The book as a whole is distributed by MDPI under the terms and conditions of the Creative Commons license CC BY-NC-ND.

Contents

About the Editors

Dragan Pamucar

Dr. Dragan Pamucar is an associate professor at the University of Defence in Belgrade, Department of Logistics, Serbia. Dr. Pamucar received his PhD in Applied Mathematics with a specialization in multi-criteria modeling and soft computing techniques from the University of Defence in Belgrade, Serbia, in 2013 and his MSc degree from the Faculty of Transport and Traffic Engineering in Belgrade in 2009. His research interests are in the field of computational intelligence; multi-criteria decision-making problems; neuro-fuzzy systems; fuzzy, rough, and intuitionistic fuzzy set theory; and neutrosophic theory. His research can be applied to areas including a wide range of logistics, management, and engineering problems.

Željko Stević

Dr. Željko Stević is an assistant professor of the Faculty of Transport and Traffic Engineering Doboj, University of East Sarajevo. He received his PhD in Transport and Traffic Engineering from the University of Novi Sad, Faculty of Technical Sciences, in 2018. His interests include logistics, supply chain management, transport, traffic engineering, multi-criteria decision-making problems, rough set theory, sustainability, fuzzy set theory, and neutrosophic set theory. He has published over 150 papers from these areas of interest. He has contributed outstanding research in the mentioned fields. In all his research, he has provided very good application studies and practical contributions to solving different problems in transportation, logistics, supply chain management, traffic engineering, etc. Dr. Stević has authored/co-authored papers published in refereed international journals including *Applied Soft Computing, Expert systems with applications, Neural Computing and Applications, Sustainability, Symmetry, Engineering Economics, Soft Computing, Transport, Scientometrics, Information, ECECSR, Technical gazette, SIC, Mathematics* (MDPI), and others. He is the editor in chief of the journal *Operational Research in Engineering Sciences: Theory and Applications*.

Preface to "Operational Research Tools for Solving Sustainable Engineering Problems"

The need for operational research (OR) tools and techniques is manifested in its ability to balance conflicting objectives (goals or interests) where there are many alternative courses of action available to the decision-maker when resources, time, and funds are limited. Therefore, beyond the domain of theoretical knowledge and model-building activities, OR has all of the pervasive applications in decision making regarding problems in sustainable engineering, sustainable society, and business. OR tools and techniques find applications in all aspects of sustainability operations, such as sustainable supply chain planning, sustainable distribution, sustainable traffic flow optimization, industrial waste reduction, energy conservation, sustainable city planning, etc. Over the years, the applications of OR have been extended to solving the problems of communication of information and socioeconomic fields. In this context, the academic community needs to take the lead in the design, development, and demonstration of sustainable operational research models, endowed and supported by organizations.

Engineering is the application of scientific principles for practical objectives such as the processes, manufacture, design, and operation of products while accounting for constraints invoked by sustainable factors. There are various factors that need to be considered in order to address engineering sustainability, which is critical for the overall sustainability of human development and activity.

This Special Issue publishes high-quality papers that develop OR optimization models in the field of sustainable engineering.

Dragan Pamucar, Željko Stević
Editors

sustainability

MDPI

Article

Development of Modified SERVQUAL–MCDM Model for Quality Determination in Reverse Logistics

Željko Stević [1,*], Ilija Tanackov [2], Adis Puška [3], Goran Jovanov [4], Jovica Vasiljević [5] and Darko Lojaničić [6]

1 Faculty of Transport and Traffic Engineering Doboj, University of East Sarajevo, Vojvode Mišića 52, 74000 Doboj, Bosnia and Herzegovina
2 Faculty of Technical Sciences, University of Novi Sad, Trg Dositeja Obradovića 6, 21000 Novi Sad, Serbia; ilijat@uns.ac.rs
3 Government of Brčko District B&H, Bulevara mira 1, 76100 Brčko, Bosnia and Herzegovina; adispuska@yahoo.com
4 Department of forensics, University of Criminal Investigation and Police Studies, Cara Dušana 196, 11080 Belgrade, Serbia; goran.jovanov@kpu.edu.rs
5 State Secretary for Public Transport, City of Belgrade, 27. Marta 43-45, 11000 Belgrade, Serbia; jovica.vasiljevic@gmail.com
6 Faculty of Business and Law, MB University, Knez Mihajlova 33, 11000 Belgrade, Serbia; dlojanicic@yahoo.com
* Correspondence: zeljko.stevic@sf.ues.rs.ba or zeljkostevic88@yahoo.com

Abstract: To run a business successfully, quality determination and customer relations are very important factors. Therefore, it is necessary to measure quality and identify critical points of business. In this paper, an original integrated model for measuring the service quality of reverse logistics (RL) was developed for the company Komunalac Teslić, which was used as an example. The Delphi and Full Consistency Method (FUCOM) was applied to determine the significance of the quality dimensions, while a modified SERVQUAL (SQ) model was used to measure the service quality of the logistics. An original SQ questionnaire was formed with a total of 21 statements that were arranged in five standard dimensions. Examining the reliability of the questionnaire for quality dimensions using the Cronbach Alpha coefficient, it was found that the measurement scales for dimensions are appropriate in terms of user expectations, while in terms of quality perception there is no measurement scale for the empathy dimension. An extensive statistical analysis was then performed to verify the results. A Signum test was applied to identify the relationship between the responses in terms of expectations and perceptions, i.e., to examine their differences. The findings obtained by this research show that the expectations were higher than the perceived quality of the services and that there was a significant statistical difference for 12 of the SQ statements. For two statements, there was a significant statistical difference in favor of perceived quality compared to expectations. Based on the results obtained, the company must improve its services in order for service quality to be at a satisfactory level.

Keywords: quality; reverse logistics; sustainability; SERVQUAL model; waste management; Signum test; FUCOM

check for updates

Citation: Stević, Ž.; Tanackov, I.; Puška, A.; Jovanov, G.; Vasiljević, J.; Lojaničić, D. Development of Modified SERVQUAL–MCDM Model for Quality Determination in Reverse Logistics. *Sustainability* **2021**, *13*, 5734. https://doi.org/10.3390/su13105734

Academic Editor: António Abreu

Received: 14 April 2021
Accepted: 17 May 2021
Published: 20 May 2021

Publisher's Note: MDPI stays neutral with regard to jurisdictional claims in published maps and institutional affiliations.

Copyright: © 2021 by the authors. Licensee MDPI, Basel, Switzerland. This article is an open access article distributed under the terms and conditions of the Creative Commons Attribution (CC BY) license (https://creativecommons.org/licenses/by/4.0/).

1. Introduction

The globalization of business together with information technology development has influenced the changes that are happening in the market. The world market is available to all organizations and they can participate in it, which leads to increased competition in the market. Hoping to improve market competitiveness and ensure better long-term development, companies are devoting more and more attention to logistics. Strong, healthy, and well-operated logistics can be an efficient means to reduce costs and increase profit margins [1]. Social pressures, environmental legislation, and economic opportunities have put pressure on companies to increasingly advocate for sustainable development policies [2].

All of this has influenced companies to increasingly address the issue of how they affect the environment. This has motivated companies to establish the concepts of circular and green economies, as well as sustainable and environmentally friendly logistics development, with reverse logistics (RL) emerging as a means of strengthening their competitive position in the market and mitigating their environmental impact [3]. RL includes recycling and the reuse of materials and goods [4]. RL is part of logistics and its main task is to enable the return of products from the customer to the manufacturer in order to fully recycle the product or to separate the components that could be reused. The remanufacturing process is widely used in RL [5]. The main applications of remanufacturing are forecasting problems, production scheduling, capacity planning, production planning, and inventory management [2]. The focus of RL is to maximize the value of end-of-life products through reuse, refabrication, remanufacturing, recycling, and energy recovery of the products [6]. RL represents an important segment of sustainability due to aspects of the recycling process and green supply chain issues [7]. The proper application of RL not only creates a cleaner environment and allows for proper waste management, but it can also have a significant impact on a country's economic development [8].

Adequate waste management, including the implementation of RL, is one of the main challenges facing all countries. It is characteristic of Balkan countries to only recycle a small percentage of waste and to use RL rarely [9]. In addition, there is a generally negative perception of quality among consumers regarding the application of RL, and it is necessary to enhance efforts to raise awareness among consumers in the Balkans [10]. It is necessary to strengthen the citizens' awareness of recycling and introduce RL in Bosnia and Herzegovina (B&H) wherever possible in order to manage waste.

This research aims to determine the quality of services in RL for the utility company Komunalac Teslić by its service users using an original integrated SERVQUAL–MCDM model. The research, conducted at the company Komunalac Teslić, aims to identify and possibly eliminate certain shortcomings. Using the SERVQUAL model and its five dimensions: responsiveness, empathy, assurance, reliability, and tangibles, the service quality of the utility company Komunalac Teslić will be determined. The most important issues, i.e., goals, that this study addresses relate to the following:

Forming an original SERVQUAL questionnaire of 21 statements used for the first time in the literature of reverse logistics.

Forming an integrated model for determining the quality of service, using Delphi and FUCOM methods to identify the significance of dimensions.

Identifying where the biggest gap is in terms of expectations and perceptions of the quality of RL services in this company.

The application of this approach will enable the company Komunalac Teslić to improve the quality of RL services. In this way, the amount of waste in the city of Teslić will be reduced and the environment will be protected. Therefore, it is necessary to obtain feedback from users in order for RL to have better results in waste management.

Apart from the clear motivation and significance of this field of research, in this paper the literature related to RL is reviewed and the use of the SERVQUAL model in measuring the quality of services in logistics as support to create a good model is also reviewed. The main parts of the paper are research methodology, case study, and results. In Section 3, the research process is clarified, while the part related to the case study explains how the survey data were collected using the original SERVQUAL–MCDM model. Section 5 summarizes and explains the research findings with extensive statistical analysis, while the next part of the paper provides a discussion of the results obtained. Moreover, the most important conclusions reached by this research and the limitations and guidelines for future research are presented.

2. Literature Review

Reverse logistics is a term commonly used to describe the management of end-of-life products, and mostly refers to the terms reduce, reuse, remanufacture, and recycle [11].

Reduce is a term that refers to waste reduction in manufacturing and the packaging of products. The term reuse refers to the return of an unused product to the manufacturer in order to put the product back into use. The term remanufacture refers to a process of repairing, restoring, or overhauling products to extend their lifespan. Recycle refers to a process in which any component of a product that contains a certain value is returned to the manufacturer. RL should be designed outside the company and should not be limited by waste collection and recycling actions [12], but other activities should also be included to preserve the value and usefulness of materials for the longest possible period, which would make significant gains for the company's value chain [13]. Implementing RL helps reduce production waste and helps companies make a profit [14].

There are many reasons why business professionals and scientists turn their attention to RL, including the following: growing concern for the environment, competitive advantage, financial potential, legal reasons, and social responsibility [15]. RL is closely related to the elements of sustainability within supply chains [7]. Creating a sustainable supply chain and sustainability in business itself are the main conditions for competing in the global world market [16]. RL plays a significant role in many traditional efforts related to the sustainability of enterprises [17]. However, RL is not always required for a supply chain to be sustainable or environmentally friendly, but it is linked to the environmental awareness of enterprises [7]. Therefore, it is necessary to observe RL separately from a sustainable supply chain, since it is not an element of sustainability but plays a significant role in reducing the negative effects that a company has on ecology and the environment.

In contrast to the supply chain, i.e., logistics, RL starts from the final destination (customers) and ends at the place of origin (suppliers) [18]. Based on that, it can be stated that the user is a key participant in RL. Therefore, it is necessary to develop RL based on user expectations or increase existing customer satisfaction [19]. Increasing customer satisfaction is achieved by improving services. The improvement of RL services is achieved by improving the system of quality in companies. Quality management does not directly affect competitiveness, but it does affect certain dimensions, such as increasing customer satisfaction, attracting new customers, improving the image of companies and various other factors that lead to an improved competitiveness within companies and their market survival in times of crisis [20]. Quality management seeks not only to meet or exceed customer expectations but also to meet the expectations of other interested parties important to the company, e.g., the public, regulatory bodies, and suppliers [21].

In order to manage quality in RL, it is necessary to examine customer satisfaction, as this is key to RL. To achieve this, different models are used, and the most prominent is SERVQUAL. Wang et al. [22] showed that the SERVQUAL model was used the most and cited by researchers, and thanks to that the model significantly contributed to service quality research. Apart from that, many organizations have improved their quality after the application of the SERVQUAL model after obtaining poor results in the initial stage. In addition, SERVQUAL is a very useful tool for recognizing customer requirements [23]. The SERVQUAL model was known as the Gap Model, and it is used to measure quality in relation to expectations and the evaluation of performance [24]. The difference between expectations and the evaluation of performance is quality. If expectations are higher than the evaluation of performance, then the level of quality is low and vice versa.

Meidutė-Kavaliauskienė et al. [25] have proved that the SERVQUAL method is suitable for identifying sectoral value gaps in logistics and its application ensures competitive advantages. Prentkovskis et al. [26] proved the applicability of the SQ model in combination with a MCDM method using the example of a logistics service in an express post company. In their research applying SERVQUAL, Kilibarda et al. [27] proved that the quality of logistics services was not at a satisfactory level in Serbia. Using SERVQUAL, Knop [28] showed that the quality of the service provided by transport and logistics operators in the pharmaceutical industry was such that the expectations regarding the quality of services provided by these operators were higher than the actual quality level obtained for all dimensions of the service quality being evaluated. Limbourg et al. [29] examined the quality of logistics services using SERVQUAL on a sample of 200 logistics service users in the city of Da Nang and showed that the customer support programs needed to be improved. Using the SERVQUAL model, Memić et al. [30] showed that the users were not satisfied with the logistics services of a passenger transport company since all the dimensions had negative values regarding the difference between the observations and the expectations. Czajkowska and Stasiak-Betlejewska [31] used the SERVQUAL method to measure the expectations and the perceptions of the quality of logistics services in companies operating in Eastern Europe and showed that the quality of the services in the areas of "Empathy" and "Materiality" should be improved. Roslan et al. [32] proposed a SERVQUAL-based model to measure the differences between customer satisfaction and desire in terms of the quality of logistics services provided by manufacturers in Iskandar, Malaysia. Parmata et al. [33] used SERVQUAL to measure the quality of the service of three major pharmaceutical distributors in India and to show how service quality affects service satisfaction. These studies have shown the effectiveness of the SERVQUAL model in testing the quality of services in logistics. Therefore, in this research, it was decided to use the SERVQUAL model to measure the quality of RL services using the example of the company Komunalac Teslić.

3. Methodology

In order to determine the service quality of Komunalac Teslić in terms of the application of RL, a methodology consisting of three phases (Table 1) was used. Phase 1 of this research was data collection. When measuring the quality of RL at the company Komunalac Teslić, the users of these services were surveyed. First, the SERVQUAL questionnaire was adapted to measure the quality of RL. Then, this questionnaire was set up online using a template of Google forms. After the survey was completed, the data were further processed and prepared for analysis.

Table 1. Research methodology.

Phase	Steps
Phase 1. Data collection	Forming an original SERVQUAL questionnaire Sending the SERVQUAL questionnaire to users Data processing and preparation for analysis
Phase 2. Determining dimension weights	Surveying users about the dimension importance percentages they have for them Synthesis of dimension weight ratings obtained by users Implementing the Delphi method Collecting data from decision-makers Calculating the weights of quality dimensions using the FUCOM method
Phase 3. Research results	Descriptive analysis of results Comparison of expectations and perceptions in respondents' answers Determining insignificant and significant differences in expectations and quality perception Conducting a Signum test
Phase 4. Discussion of results	Analysis of the results obtained Comparison of results in terms of expectations and quality perception Proposed guidelines for quality improvement

Phase 2 of the research was applied by determining the weights of the quality dimensions. This weight determination was performed in two ways. The first way collected users' opinions about the importance that a certain dimension had for them, by applying the Delphi method [34]. The second way applied the FUCOM method [35–37], where five decision-makers were selected for the evaluation of the quality dimensions.

Phase 3 included the results of the research. After the data were processed and prepared for analysis, the distribution of the obtained results was presented. Then, the obtained results were compared in terms of the service users' expectations and perceptions, thus comparing their agreement or disagreement with the given statements in terms of expected and perceived quality. After that, the insignificant and significant differences were determined, first for the users' expectations, and then for the users' quality perception. The last part of the analysis of the research results was the implementation of a Signum test where the significance between responses for expectations and perceptions were determined. Because the collected results had non-parametric characteristics that deviated from expected binomial distributions, a Signum test was used [23].

Phase 4 of the research involved conducting a discussion of the results obtained. First, an analysis of the obtained results was completed, and it was determined where there was the biggest difference, i.e., for which statements the expectations were higher than the perceptions and vice versa. Based on the comparison of the results, it was identified why such results were obtained. Then, the guidelines on how the company Komunalac Teslić will improve the quality in the implementation of RL were provided.

4. Case Study

At the very beginning of its business, the enterprise for utility services consisted of a utility company and a water supply system in the municipality of Teslić, and it was not until 2001 that these two businesses separated. Since 2001, the utility company has been operating independently under the name Komunalac Teslić. So far, the company has about 7000 registered users; 6500 are natural persons and 500 are legal entities. In the beginning, waste collection and transport were only undertaken in urban and suburban areas, while in the last few years the business has changed and expanded into the rural areas of the municipality of Teslić, thus increasing the number of users. Waste collection charges are fixed and are taken on a monthly basis. The basic function of the utility company Komunalac Teslić is waste management in the territory of the municipality of Teslić. Waste collection in the urban zone is carried out twice a day, except on weekends when waste collection is carried out once a day. Waste collection in rural areas is carried out once a week. Each rural settlement has a particular day when waste is collected. In this case, waste management includes waste collection and its disposal at a landfill under the supervision of the utility company. In the last few years, the company Komunalac Teslić has initiated activities to open a recycling center. In 2020, the company started collecting waste that will be recycled. The company Komunalac Teslić decided to set up containers for sorting waste intended for recycling. The containers are at accessible locations, so that users can dispose of waste in a very easy and fast way. The number of containers and their volume at a location depends on the number of users in that area. Service users have been informed by the media about the provision of a new service to collect waste that will be recycled. The aim of this company is to increase the percentage of waste for recycling. In order to improve the quality of this service, users were surveyed about their expectations before setting up the containers, and their quality perceptions after setting up the containers. In this way, the company Komunalac Teslić will receive the necessary information on how to improve its RL services through waste recycling.

Users were surveyed through the application of the SERVQUAL model. The customized questionnaire consisted of 21 questions related to expectations and 21 questions related to quality perception. The questions in the questionnaire were divided into five dimensions: reliability, assurance, empathy, responsiveness, and tangibles (Table 2). Users were offered answers for the rating of expectations and quality perceptions in the form of linguistic values, which ranged from "I completely disagree with this statement" assigned grade one, and "I completely agree with it" assigned grade five. A Likert scale of five levels of disagreement or agreement with given statements was used. At the end of the questionnaire, users had to rate each quality dimension in the form of percentages and thus needed to determine which of the dimensions was the most important to them. The total sum of the evaluated dimensions should be 100%. During the evaluation, the users were guided by which of the given dimensions they thought had the greatest impact on the quality of the provided service of the utility company. The questionnaire was posted online using Google forms. The questionnaire was active from March to May 2020. The links were forwarded by social networks and the e-mail services of the company Komunalac Teslić. The questionnaire was accessed by 170 service users, and it was correctly filled in by 112 users.

Table 2. SERVQUAL questionnaire.

Dimensions		Statements	Grades				
Reliability	Q1	Services will be provided at the expected time	1	2	3	4	5
	Q2	Waste will be collected regularly	1	2	3	4	5
	Q3	Waste collection will be performed without difficulties	1	2	3	4	5
Assurance	Q4	Workers will be careful when performing work tasks	1	2	3	4	5
	Q5	The user will be informed in a timely manner	1	2	3	4	5
	Q6	The cost of the waste collection service will be fixed	1	2	3	4	5
Tangibles	Q7	The cost of the service will be acceptable	1	2	3	4	5
	Q8	No noise will be generated during waste collection	1	2	3	4	5
	Q9	Invoices will be clear and delivered to home addresses	1	2	3	4	5
	Q10	The streets will be clean and tidy	1	2	3	4	5
	Q11	The containers will be placed close to the household	1	2	3	4	5
	Q12	There will be no unpleasant odors at waste disposal sites	1	2	3	4	5
	Q13	Waste collection vehicles will be modern	1	2	3	4	5
Empathy	Q14	Services will be flexible and customized	1	2	3	4	5
	Q15	The time of waste collection and transport will be appropriate	1	2	3	4	5
	Q16	When charging, population categories will be taken into account	1	2	3	4	5
Responsiveness	Q17	Workers will be professional during the waste collection process	1	2	3	4	5
	Q18	Novelties will be accepted quickly	1	2	3	4	5
	Q19	Users' needs will be adequately responded to	1	2	3	4	5
	Q20	Waste collection will be fast and adequate	1	2	3	4	5
	Q21	Traffic will not be disturbed	1	2	3	4	5

As we mentioned in the introduction, the original SERVQUAL questionnaire contained 21 statements that were used for the first time in the literature. The questionnaire was formed based on the experiences of managers from the field of reverse logistics and the need of the utility company Komunalac Teslić, for which this research was performed. It is important to note that this questionnaire contained statements mostly related to waste collection management. Bosnia and Herzegovina is a very poor country from the aspect of the full application of reverse logistics (reduce, reuse, recycle), so we were forced to only consider waste disposal as one of the channels of reverse logistics.

In addition to user surveys, the FUCOM method was applied to determine the weights of the quality dimensions by decision-makers. Five decision-makers who are regional experts in the field were selected. They are informed on a daily basis and they make decisions related to waste management activities in reverse logistics. Decision-makers based their preferences on the results of the Delphi method, i.e., the initial ranking for the FUCOM method.

5. Results

Before showing the results collected from the users in terms of expectations and quality perceptions, the findings related to the weights of the quality dimensions are presented first. At the end of the questionnaire, each service user evaluated individual quality dimensions with a percentage of how important a certain dimension was to them. Each dimension was summed up and divided by the total rating, i.e., the sum of the percentage values of one dimension was divided by the sum of the percentage values for all dimensions. In this way, the results of the weights of the quality dimensions were obtained (Table 3). Based on the results obtained, it can be concluded that the most important dimension for users was responsiveness (w = 0.216), followed by the reliability dimension (w = 0.204) which was slightly smaller than the responsiveness dimension. The assurance dimension (w = 0.199) and the tangibles dimension (0.191) had approximately the same results, while the empathy dimension (0.190) was in the last position.

Table 3. The weight values of the quality dimensions obtained by users applying the Delphi method.

Weight Coefficients	Values
Reliability	0.204
Assurance	0.199
Tangibles	0.191
Empathy	0.190
Responsiveness	0.216
Total	1.000

Final weight values were determined using the FUCOM method by five decision-makers. Based on the ranking defined by users applying the Delphi method, the decision-makers compared the criteria. After applying the other steps of the FUCOM method, their final values were calculated by the Lingo 17 software, as shown in Table 4. The same procedure was applied for all decision-makers and the weights were calculated for each decision-maker. The final weight values, related to the decision-makers' opinions, were obtained by applying the average value for the quality dimensions. The highest value was given to the responsiveness dimension (w = 0.231), followed by the reliability dimension (w = 0.211), assurance (w = 0.197), tangibles (w = 0.189), while the empathy dimension gained the least weight (w = 0.172).

Table 4. Final weight values of the quality dimensions obtained by the decision-makers using the FUCOM method.

	Reliability	Assurance	Tangibles	Empathy	Responsiveness
DM1	0.203	0.203	0.194	0.167	0.233
DM2	0.214	0.196	0.181	0.174	0.235
DM3	0.214	0.195	0.195	0.172	0.224
DM4	0.217	0.198	0.192	0.175	0.217
DM5	0.207	0.194	0.182	0.172	0.244
Wj	0.211	0.197	0.189	0.172	0.231

After the final weights of the quality dimensions were determined by users and decision-makers, descriptive statistics of the data received from the users were performed (Table 5). The highest value for user expectations was given to the assurance dimension (mean = 3.915), while the lowest value was given to the empathy dimension (mean = 3.342). When observing the results obtained for quality perception, the highest value was given to the reliability dimension (mean = 3.783), while the lowest evaluated dimension was empathy (mean = 3.247). When determining the dispersion of the service users' responses, it was identified that the tangibles dimension (ST = 1.348) had the largest dispersion in responses in terms of user expectations, while the reliability dimension had the lowest dispersion (ST = 1.230).

When observing the results for quality perception, the dimension tangibles (ST = 1.321) had the largest dispersion in users' responses, while the dimension reliability (ST = 1.189) had the smallest dispersion. The internal consistency of the measurement scales was measured using the Cronbach Alpha (CA) coefficient. The results obtained using the CA coefficient are such that there was an internal consistency in the dimensions for measuring user expectations, as all CA values are higher than 0.6 [38]. However, in the responses related to perception in the empathy dimension there was no consistency of the measurement scales, and certain segments of the SQ questionnaire have been deleted.

Table 5. Descriptive analysis of quality dimensions, Cranbanch alpha indicator and quality identified.

Dimension	Expectations			Perception			SQ Gap
	AV	ST	CA	AV	ST	CA	
Reliability	3.848	1.230	0.845	3.783	1.189	0.813	−0.014
Assurance	3.915	1.234	0.746	3.385	1.269	0.600	−0.009
Tangibles	3.712	1.348	0.822	3.655	1.321	0.618	−0.017
Empathy	3.342	1.281	0.942	3.247	1.229	0.338	−0.018
Responsiveness	3.833	1.287	0.810	3.699	1.242	0.600	−0.029
SERVQUAL	3.730	1.276	0.833	3.554	1.250	0.600	−0.017

When filling out the questionnaire, the respondents filled in their expectations and perceptions regarding the RL service quality of the company Komunalac Teslić. The distribution of ratings in terms of expectations and quality perception are shown in Table 6. These results show that the highest expectations of service users were related to statement Q9 (Invoices will be clear and delivered to home addresses) where there is the highest rating by all users (mean = 4.1786), while the lowest expectations were for statement Q12 (There will be no unpleasant odors at waste disposal sites), which received the lowest rating by users (mean = 3.3750). The ratings of RL service quality perception are such that the highest rating was given to statement Q5 (The user is informed in a timely manner) (mean = 3.9107), while the lowest rating was given to statement Q16 (When charging, population categories are taken into account) (mean = 3.2411).

Table 6. Distribution of ratings in terms of expectations and perceptions.

Ordinals	Expectation Ratings					Mean	Perception Ratings					Mean
	E(1)	E(2)	E(3)	E(4)	E(5)		P(1)	P(2)	P(3)	P(4)	P(5)	
Q1	11	11	9	36	45	3.8304	11	15	7	43	36	3.6964
Q2	10	12	10	40	40	3.7857	7	17	7	40	41	3.8125
Q3	6	9	13	43	41	3.9286	5	15	12	41	39	3.8393
Q4	7	15	8	39	43	3.8571	8	16	6	43	39	3.7946
Q5	5	14	10	33	50	3.9732	5	12	12	42	41	3.9107
Q6	6	13	12	32	49	3.9375	11	11	11	44	35	3.7232
Q7	6	18	8	38	42	3.8214	7	12	9	45	39	3.8661
Q8	9	13	10	41	39	3.7857	9	15	17	39	32	3.6250
Q9	6	5	15	23	63	4.1786	13	13	8	40	38	3.6875
Q10	14	14	9	35	40	3.6518	9	20	12	35	36	3.6161
Q11	9	14	9	40	40	3.7857	18	13	15	29	37	3.4821
Q12	21	16	8	34	33	3.3750	9	18	12	40	33	3.6250
Q13	15	24	11	27	35	3.3839	11	14	27	33	27	3.4554
Q14	9	16	18	42	27	3.5536	5	16	12	39	40	3.8304
Q15	10	12	15	30	45	3.7857	11	14	27	33	27	3.4554
Q16	8	15	4	38	47	3.9107	18	23	15	26	30	3.2411
Q17	11	13	11	34	43	3.7589	8	13	13	45	33	3.7321
Q18	6	14	10	46	36	3.8214	12	14	24	33	29	3.4732
Q19	6	16	21	38	31	3.6429	5	14	30	34	29	3.6071
Q20	10	13	14	31	44	3.7679	6	20	12	46	28	3.6250
Q21	4	14	11	35	48	3.9732	7	19	9	38	39	3.7411
sum	189	291	236	755	881		195	324	297	808	728	

Considering the rating of expectations and quality perceptions, it is obvious that the number of ratings received increased: for grade one, from 189 to 195 (+6); for grade two, from 291 to 324 (+33); for grade three from 236 to 297 (+61); and for grade four, from 755 to 808 (+53), while the number of grades five given by the users in terms of quality perception decreased from 881 to 728 (−153). In this way, it was shown that expectations were higher than the perceived RL quality.

This fluctuation in ratings indicates the exclusive disappointment of the respondents who had the highest expectations (Figure 1). However, the presentation of expectations and perceptions of 112 respondents (Figure 2) by expectation and perception indicates uncharacteristic fluctuations. This fluctuation has a dominant tendency to decrease ratings for user perceptions, but in some cases, it also has a tendency to increase ratings for user perceptions. Therefore, it is necessary to consider this relationship through a linear correlation between the same ratings of expectations and perceptions.

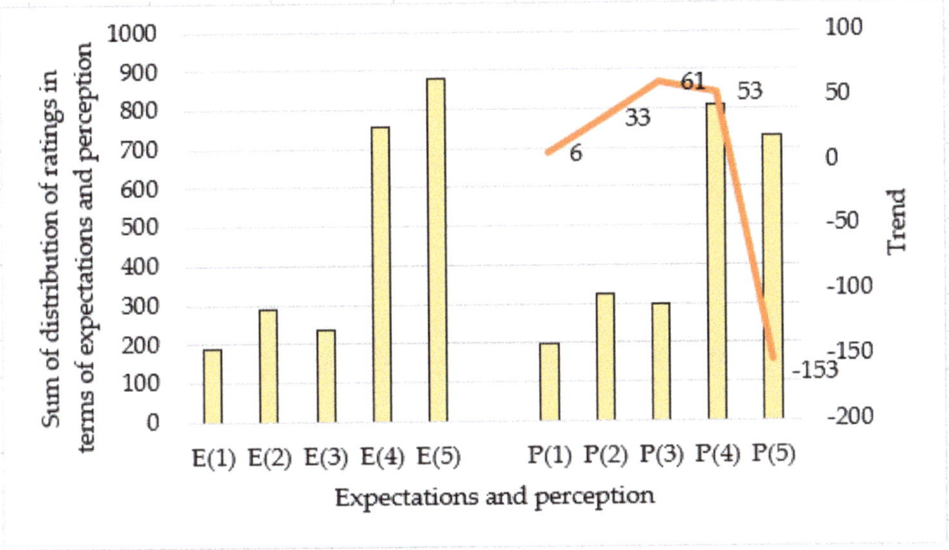

Figure 1. The difference in the sum of distribution in terms of expectation and perception and their trend.

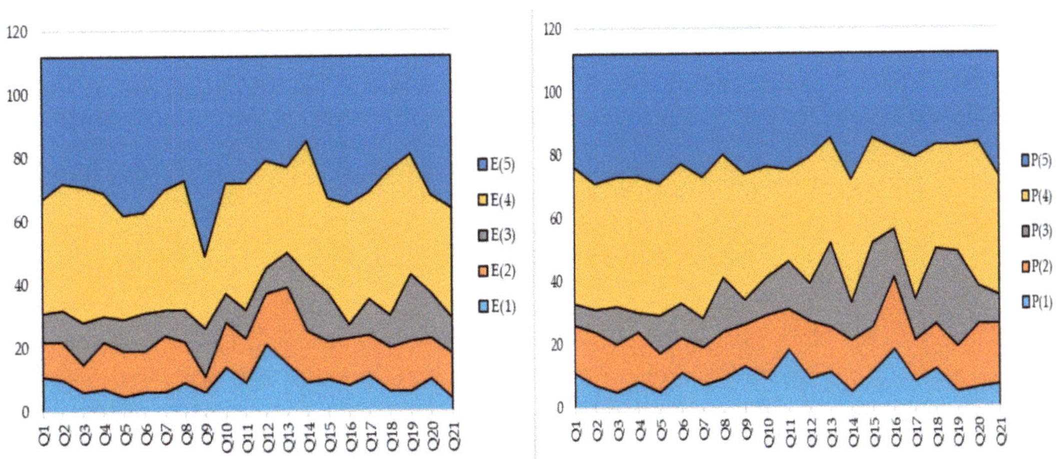

Figure 2. Stacked plots of all variables of expectations and perception.

The results of the correlation coefficient show that there is a small correlation between the ratings observed (Table 7). Only observing the same ratings shows that a small number of respondents in their perception rating repeated their rating from expectations. At grade four, there was a negative correlation between these responses (r = −0.149). However, the weak correlation between the ratings does not allow a relevant conclusion to be drawn about the relation between expectations and responses, but at the same time they point to analytical dynamics resulting from the obvious consistent cooperation and interest of users in participating in the survey.

Table 7. Results of a linear correlation between the ratings of expectations and perception.

	E(1)	E(2)	E(3)	E(4)	E(5)
P(1)	**+0.109**	+0.293	+0.090	−0.075	−0.318
P(2)	−0.056	**+0.051**	+0.404	−0.237	−0.307
P(3)	−0.465	−0.296	**+0.396**	+0.102	−0.138
P(4)	−0.119	+0.110	−0.014	**−0.149**	+0.214
P(5)	+0.284	−0.109	−0.421	+0.207	**+0.223**

Research into the nonparametric characteristics of the distribution of expectations in responses has not established a specific type of distribution in any case. Binomial distribution, established in some previous research [23] was potentially the closest, but did not meet the verification requirements with a given significance threshold of $p > 0.05$ in the case of expectation and perception. Therefore, the basis for further analysis is a nonparametric Signum test.

In accordance with the Signum test results, insignificant and significant differences in the values of each of the expectations $E_m(n)$ $n \in [1,21]$, $m \in [1,112]$ are given in Table A1, while the Signum test results of insignificant and significant differences in the values of each of the answers $P_m(n)$ $n \in [1,21]$, $m \in [1,112]$ are given in Table A2.

We can conclude that the tests of expectations are dominantly different from each other and that they represent a reference basis for the estimation of expectations. Expectation E12 had the strongest logical differentiation, and expectation E18 had the weakest logical differentiation. In accordance with the Signum test results, it should be noted that expectations E01 were absolutely in compliance with E07 ($p = 1.0000$) and that expectations E01 were absolutely in compliance with E18 ($p = 1.0000$). The relationship between expectations E07 and E18 is significantly consistent, but not absolute ($p = 0.7728$). Moreover, we can conclude that the results of the response tests are mostly different from each other and that they represent a reference basis for estimating expectations.

Perception P16 to expectation E16 had the strongest logical differentiation, and perception P09 to expectation E09 had the weakest logical differentiation. In accordance with the Signum test results, P10 had the highest number of four absolute compliances with the following: P08, P12, P19, and P20. It should be emphasized that expectations E10 and E19 were absolutely in compliance. The coefficients of linear correlation between these expectations and answers are as follows:

- Between E10 and E19 $r = +0.792$
- Between P10 and P19 $r = +0.654$
- Between E10 and P10 $r = +0.945$
- Between E19 and P19 $r = +0.915$

The high compliance of the rating distribution of expectations between E10 and E19 was based on two subgroups of respondents that were not in compliance and therefore the results were such that there was a moderate correlation ($r = 0.7921$), and their compliance differentiation was also expressed in perceptions P10 and P19 ($r = 0.6538$). However, the consistency of the relationship between their expectations and responses was evident in the correlations for E10 and P10 ($r = 0.9450$) and E19 with P19 ($r = 0.9145$), so we conclude that there are no significant quantitative differences between E10/E19 expectations and P10/P19 perceptions, but there are significant qualitative differences in the relationship between E10/P10 and E19/P19 expectations due to the inversion of the groups with opposite attitudes to expectations and perceptions.

In accordance with the Signum test results, absolute compliance was also established between the following perceptions: P01/P09, P03/P14, P06/P17, P11/P18, P13/P15, and P17/P21. The specific determination of differences between individual expectations and responses is given in Table A3, and based on the results it can be noted that:

- A total of 12 responses had significantly lower ratings than expected, such as: E01/P01, E03/P03, E04/P04, E06/P06, E08/P08, E09/P09, E11/P11, E15/P15, E16/P16, E18/P18, E20/P20, E21/P21;
- A total of 7 responses remained at the level of expectations: E02/P02, E05/P05, E07/P07, E10/P10, E13/P13, E17/P17, E19/P19;
- A total of 2 responses had significantly higher ratings than expected: E12/P12, E14/P14.

The largest drop in ratings was found for expectation E16 to perception P16, which is extremely significant (p = 0.0000) and in absolute value is $\Delta(16) = -0.6696$, and in relative terms it represents a loss of 17.12% of the expectation. The E16/P16 relation is specific because the perception is absolutely heterogeneous (there is no compliance by the Signum test with another perception). The second drop in the ratings value was found for expectation E09 to perceptions P09, which is also absolutely significant (p = 0.0000) and in absolute value is $\Delta(09) = -0.4910$, and in relative terms it represents a loss of 11.75% of the expectation.

6. Discussion

When assessing the quality of RL services at the company Komunalac Teslić, the original SERVQUAL–MCDM model was used. Using this model, users' opinions about the quality of services were examined. A total of 21 statements were used and were divided into five dimensions: reliability, assurance, empathy, responsiveness, and tangibles. RL is specific because the initiation of activities is by the customer, ending them at the supplier [18], which distinguishes it from classical logistics.

The questionnaire containing these statements was completed by 170 users, and 112 of them completed it correctly. First, the correlation between ratings for expectations and perceptions was examined. The results have shown that there is a weak correlation between the ratings, which has proved that the collected results have non-parametric characteristics that deviate from the expected binomial distributions.

By comparing the results obtained by measuring expectations and perceptions, it was shown that user expectations differ significantly from the perception of quality in 14 statements. The results obtained using this model showed that in statements Q16 (When charging, population categories will be taken into account) and Q9 (Invoices will be clear and delivered to home addresses) there is the highest degree of discrepancy between expectations and perceptions by users. As can be seen, both of these statements are not related to RL services but to invoices and charging. The company Komunalac Teslić should first work on improving the service of distributing invoices, and when charging, take into account the categories of the population and adjust the cost of services to the categories. However, the results showed that there was also a significant statistical difference in responses for the other 12 statements. Based on that, the company Komunalac Teslić must first introduce novelties in its business, then harmonize the time of waste collection and transport with the requirements of households and set up containers close to the households. This is only a part of the service that the company Komunalac Teslić must improve in order to improve the quality of RL services. The reason for this is that the service users had high expectations regarding the introduction of the new RL service.

What is particularly significant for the results obtained is that there is a higher rating of quality perception than user expectation for two statements, and those statements are Q12 and Q14. There is also a significant statistical difference between these statements. Based on that, it can be concluded that the company Komunalac Teslić has additionally adjusted its services to the users' needs, improving flexibility and reducing unpleasant odors at waste disposal sites. In addition, there is a higher rating of perceived quality for three other statements than of expected quality, and those statements are Q2, Q7 and Q13. However, there is no significant statistical difference between expectations and perceptions of quality for these statements.

In order for Komunalac Teslić to advance its RL services, it is necessary to improve the services aiming to improve the perceived quality of services. This needs to be carried out particularly for those services where there is a significant statistical difference between user expectations and their perception. It is necessary to re-examine the perception of users after a certain time in order to determine whether the quality has improved. Only in this way is it possible to constantly improve the quality of the services since there is feedback from service users.

7. Conclusions

Every company strives to improve the quality of its services and products in order to be more competitive in the market. When a new service is introduced, it is necessary to determine how the service has been accepted by users. In this paper, the quality of RL services was examined using the example of the company Komunalac Teslić, using the original SERVQUAL–MCDM model for measuring the quality of services. It was used to measure the expectations and the perceptions of the quality of the user services when introducing RL services. The research findings have shown that there are higher user expectations than quality perceptions for most of the statements. This especially refers to statements Q9 and Q16. However, the results have shown that there is a significant statistical difference between the perception of quality and user expectations for statements Q12 and Q14. In order to improve the quality of its services, the company Komunalac Teslić must continuously examine the user perception of the quality of the services. Only in this way will timely information will be obtained.

The main disadvantage of this research can be seen through the fact that many users of the services of the company Komunalac Teslić were not included. However, not all customers use RL services because they are not familiar with the importance of RL for preserving the environment. Therefore, it is necessary to increase people's awareness of the importance of RL in order for more of them to use these services. The research methodology has shown great flexibility in work and can be used in measurement and other services that occur in logistics.

Author Contributions: Conceptualization, Ž.S. and A.P.; methodology, Ž.S., D.L. and G.J.; formal analysis, I.T. and J.V.; investigation, Ž.S.; writing—original draft preparation, Ž.S., A.P. and D.L.; writing—review and editing, I.T. and G.J.; supervision, I.T.; project administration, J.V. All authors have read and agreed to the published version of the manuscript.

Funding: This research received no external funding.

Institutional Review Board Statement: Not applicable.

Informed Consent Statement: Not applicable.

Data Availability Statement: Not applicable.

Conflicts of Interest: The authors declare no conflict of interest.

Appendix A. Statistical Analysis

Table A1. Signum test results of insignificant and significant differences in the values of statements in expectations.

	E01	E02	E03	E04	E05	E06	E07	E08	E09	E10	E11	E12	E13	E14	E15	E16	E17	E18	E19	E20	E21
E01		0.131	0.022	0.450	0.000	0.002	**1.000**	0.182	0.000	0.000	0.182	0.000	0.000	0.000	0.131	0.008	0.013	**1.000**	0.000	0.046	0.000
E02	0.131		0.000	0.013	0.000	0.000	0.289	0.480	0.000	0.000	0.480	0.000	0.000	0.000	0.752	0.001	0.505	0.387	0.002	0.752	0.000
E03	0.022	0.000		0.043	0.302	**1.000**	0.003	0.000	0.000	0.000	0.000	0.000	0.000	0.000	0.002	0.803	0.000	0.002	0.000	0.001	0.267
E04	0.450	0.013	0.043		0.001	0.016	0.221	0.013	0.000	0.000	0.013	0.000	0.000	0.000	0.043	0.077	0.003	0.343	0.000	0.009	0.001
E05	0.000	0.000	0.302	0.001		0.134	0.000	0.000	0.000	0.000	0.000	0.000	0.000	0.000	0.000	0.070	0.000	0.000	0.000	0.000	0.617
E06	0.002	0.000	**1.000**	0.016	0.134		0.001	0.000	0.000	0.000	0.000	0.000	0.000	0.000	0.000	0.547	0.000	0.002	0.000	0.000	0.221
E07	**1.000**	0.289	0.003	0.221	0.000	0.001		0.289	0.000	0.000	0.221	0.000	0.000	0.000	0.421	0.016	0.046	0.773	0.000	0.145	0.000
E08	0.182	0.480	0.000	0.013	0.000	0.000	0.289		0.000	0.001	0.480	0.000	0.000	0.000	0.773	0.001	0.547	0.343	0.001	0.773	0.000
E09	0.000	0.000	0.000	0.000	0.000	0.000	0.000	0.000		0.000	0.000	0.000	0.000	0.000	0.000	0.000	0.000	0.000	0.000	0.000	0.000
E10	0.000	0.000	0.000	0.000	0.000	0.000	0.000	0.001	0.000		0.000	0.000	0.000	0.054	0.000	0.000	0.002	0.001	**1.000**	0.001	0.000
E11	0.182	0.480	0.000	0.013	0.000	0.000	0.221	0.480	0.000	0.000		0.000	0.000	0.000	0.773	0.001	0.505	0.387	0.002	0.752	0.000
E12	0.000	0.000	0.000	0.000	0.000	0.000	0.000	0.000	0.000	0.000	0.000		**1.000**	0.001	0.000	0.000	0.000	0.000	0.000	0.000	0.000
E13	0.000	0.000	0.000	0.000	0.000	0.000	0.000	0.000	0.000	0.000	0.000	**1.000**		0.002	0.000	0.000	0.000	0.000	0.000	0.000	0.000
E14	0.000	0.000	0.000	0.000	0.000	0.000	0.000	0.000	0.000	0.054	0.000	0.001	0.002		0.000	0.000	0.000	0.000	0.004	0.000	0.000
E15	0.131	0.752	0.002	0.043	0.000	0.000	0.423	0.773	0.000	0.000	0.773	0.000	0.000	0.000		0.001	0.450	0.522	0.002	0.480	0.000
E16	0.008	0.001	0.803	0.077	0.070	0.547	0.016	0.001	0.000	0.000	0.001	0.000	0.000	0.000	0.001		0.000	0.034	0.000	0.000	0.070
E17	0.013	0.505	0.000	0.003	0.000	0.000	0.046	0.547	0.000	0.002	0.505	0.000	0.000	0.000	0.450	0.000		0.190	0.021	**1.000**	0.000
E18	**1.000**	0.387	0.002	0.343	0.000	0.002	0.773	0.343	0.000	0.001	0.387	0.000	0.000	0.000	0.522	0.034	0.190		0.000	0.286	0.000
E19	0.000	0.002	0.000	0.000	0.000	0.000	0.000	0.001	0.000	**1.000**	0.002	0.000	0.000	0.004	0.002	0.000	0.021	0.000		0.008	0.000
E20	0.046	0.752	0.001	0.009	0.000	0.000	0.149	0.773	0.000	0.001	0.752	0.000	0.000	0.000	0.480	0.000	**1.000**	0.286	0.008		0.000
E21	0.000	0.000	0.267	0.001	0.617	0.221	0.000	0.000	0.000	0.000	0.000	0.000	0.000	0.000	0.000	0.070	0.000	0.000	0.000	0.000	

Table A2. Signum test results of insignificant and significant differences in the values of statements in perception.

	P01	P02	P03	P04	P05	P06	P07	P08	P09	P10	P11	P12	P13	P14	P15	P16	P17	P18	P19	P20	P21
P01		0.001	0.000	0.003	0.000	0.371	0.000	0.080	**1.000**	0.027	0.000	0.043	0.000	0.000	0.000	0.000	0.387	0.000	0.134	0.099	0.182
P02	0.001		0.505	0.617	0.003	0.016	0.114	0.000	0.001	0.000	0.000	0.000	0.000	0.724	0.000	0.000	0.039	0.000	0.000	0.000	0.013
P03	0.000	0.505		0.182	0.013	0.001	0.450	0.000	0.000	0.000	0.000	0.000	0.000	**1.000**	0.000	0.000	0.002	0.000	0.000	0.000	0.003
P04	0.003	0.617	0.182		0.001	0.043	0.013	0.000	0.002	0.000	0.000	0.000	0.000	0.343	0.000	0.000	0.096	0.000	0.001	0.000	0.077
P05	0.000	0.003	0.013	0.001		0.000	0.131	0.000	0.000	0.000	0.000	0.000	0.000	0.008	0.000	0.000	0.000	0.000	0.000	0.000	0.000
P06	0.371	0.016	0.001	0.043	0.000		0.000	0.010	0.343	0.010	0.000	0.010	0.000	0.002	0.000	0.000	**1.000**	0.000	0.031	0.029	0.789
P07	0.000	0.114	0.450	0.013	0.131	0.000		0.000	0.000	0.000	0.000	0.000	0.000	0.343	0.000	0.000	0.000	0.000	0.000	0.000	0.001
P08	0.080	0.000	0.000	0.000	0.000	0.010	0.000		0.169	**1.000**	0.003	0.683	0.000	0.000	0.000	0.000	0.002	0.000	0.823	0.773	0.004
P09	**1.000**	0.001	0.000	0.002	0.000	0.343	0.000	0.169		0.080	0.000	0.121	0.000	0.000	0.000	0.000	0.302	0.000	0.200	0.190	0.077
P10	0.027	0.000	0.000	0.000	0.000	0.010	0.000	**1.000**	0.080		0.001	**1.000**	0.001	0.000	0.001	0.000	0.006	0.001	**1.000**	**1.000**	0.001
P11	0.000	0.000	0.000	0.000	0.000	0.000	0.000	0.003	0.000	0.001		0.002	0.710	0.000	0.710	0.000	0.000	**1.000**	0.030	0.010	0.000
P12	0.043	0.000	0.000	0.000	0.000	0.010	0.000	0.683	0.121	**1.000**	0.002		0.000	0.000	0.000	0.000	0.002	0.000	0.845	0.752	0.001
P13	0.000	0.000	0.000	0.000	0.000	0.000	0.000	0.000	0.000	0.001	0.710	0.000		0.000	**1.000**	0.000	0.000	0.683	0.000	0.000	0.000
P14	0.000	0.724	**1.000**	0.343	0.008	0.002	0.343	0.000	0.002	0.000	0.000	0.000	0.000		0.000	0.000	0.003	0.000	0.000	0.000	0.004
P15	0.000	0.000	0.000	0.000	0.000	0.000	0.000	0.000	0.000	0.001	0.713	0.002	**1.000**	0.000		0.000	0.000	0.681	0.000	0.001	0.000
P16	0.000	0.000	0.000	0.000	0.000	0.000	0.000	0.000	0.000	0.000	0.000	0.000	0.000	0.000	0.000		0.000	0.000	0.000	0.000	0.000
P17	0.385	0.039	0.002	0.091	0.000	**1.000**	0.003	0.002	0.302	0.006	0.000	0.002	0.000	0.003	0.000	0.000		0.000	0.008	0.006	**1.000**
P18	0.000	0.000	0.000	0.000	0.000	0.000	0.000	0.000	0.000	0.001	**1.000**	0.000	0.683	0.000	0.683	0.000	0.000		0.000	0.000	0.000
P19	0.134	0.000	0.000	0.001	0.000	0.031	0.000	0.823	0.200	**1.000**	0.030	0.845	0.000	0.000	0.000	0.000	0.008	0.000		0.823	0.015
P20	0.099	0.000	0.000	0.000	0.000	0.029	0.000	0.773	0.190	**1.000**	0.010	0.752	0.000	0.000	0.000	0.000	0.006	0.000	0.823		0.002
P21	0.182	0.013	0.003	0.077	0.000	0.789	0.001	0.004	0.077	0.001	0.000	0.001	0.000	0.004	0.000	0.000	**1.000**	0.000	0.015	0.002	

Table A3. Signum test results.

	Mean Expectations	Relation	Mean Perception	Difference Δ	Zeta (from Signum)	p
Q1	3.830357	>	3.696429	−0.133928	3.614784	0.000301
Q2	3.785714	≈	3.812500	+0.026786	0.755929	0.449692
Q3	3.928571	>	3.839286	−0.089285	2.598076	0.009375
Q4	3.857143	>	3.794643	−0.062500	2.267787	0.023342
Q5	3.973214	≈	3.910714	−0.062500	1.809068	0.070440
Q6	3.937500	>	3.723214	−0.214286	4.694855	0.000003
Q7	3.821429	≈	3.866071	+0.044642	1.109400	0.267258
Q8	3.785714	>	3.625000	−0.160714	4.006938	0.000062
Q9	4.178571	>	3.687500	**−0.491071**	7.142749	0.000000
Q10	3.651786	≈	3.616071	−0.035715	0.801784	0.422678
Q11	3.785714	>	3.482143	−0.303571	5.659453	0.000000
Q12	3.375000	<	3.625000	+0.250000	5.102520	0.000000
Q13	3.383929	≈	3.455357	+0.071428	1.322876	0.185877
Q14	3.553571	<	3.830357	+0.276786	5.388159	0.000000
Q15	3.785714	>	3.455357	−0.330357	5.918364	0.000000
Q16	3.910714	>	3.241071	**−0.669643**	7.682213	0.000000
Q17	3.758929	≈	3.732143	−0.026786	0.485071	0.627626
Q18	3.821429	>	3.473214	−0.348215	6.084870	0.000000
Q19	3.642857	≈	3.607143	−0.035714	0.866025	0.386476
Q20	3.767857	>	3.625000	−0.142857	3.061862	0.002200
Q21	3.973214	>	3.741071	−0.232143	4.902903	0.000001

References

1. Chen, Z.-S.; Zhang, X.; Govindan, K.; Wang, X.-J.; Chin, K.-S. Third-party reverse logistics provider selection: A computational semantic analysis-based multi-perspective multi-attribute decision-making approach. *Expert Syst. Appl.* **2021**, *166*, 114051. [CrossRef]
2. Tighazoui, A.; Turki, S.; Sauvey, C.; Sauer, N. Optimal design of a manufacturing-remanufacturing-transport system within a reverse logistics chain. *Int. J. Adv. Manuf. Technol.* **2019**, *101*, 1773–1791. [CrossRef]
3. Govindan, K.; Bouzon, M. From a literature review to a multi-perspective framework for reverse logistics barriers and drivers. *J. Clean. Prod.* **2018**, *187*, 318–337. [CrossRef]
4. Awan, M.A.; Ali, Y. Sustainable modeling in reverse logistics strategies using fuzzy MCDM: Ase of China Pakistan Economic Corridor. *Manag. Environ. Qual. Int. J.* **2019**, *30*, 1132–1151. [CrossRef]
5. Ardeshirilajimi, A.; Azadivar, F. Reverse supply chain plan for remanufacturing commercial returns. *Int. J. Adv. Manuf. Technol.* **2014**, *77*, 1767–1779. [CrossRef]
6. Yu, H.; Sun, X.; Solvang, W.D.; Zhao, X. Reverse Logistics Network Design for Effective Management of Medical Waste in Epidemic Outbreaks: Insights from the Coronavirus Disease 2019 (COVID-19) Outbreak in Wuhan (China). *Int. J. Environ. Res. Public Health* **2020**, *17*, 1770. [CrossRef]
7. Morgan, T.R.; Tokman, M.; Richey, R.G.; Defee, C. Resource commitment and sustainability: A reverse logistics performance process model. *Int. J. Phys. Distrib. Logist. Manag.* **2018**, *48*, 164–182. [CrossRef]
8. Ayilara, M.S.; Olanrewaju, O.S.; Babalola, O.O.; Odeyemi, O. Waste Management through Composting: Challenges and Potentials. *Sustainability* **2020**, *12*, 4456. [CrossRef]
9. Galijašević, E.; Nunić, Z.; Alihodžić, A. Waste management system and reverse logistics of Tešanj. *HORIZONS B* **2018**, *5*, 41–52. [CrossRef]
10. Aćimović, S.; Mijušković, V.; Rajić, V. The impact of reverse logistics onto green supply chain competitiveness evidence from Serbian consumers. *Int. J. Retail. Distrib. Manag.* **2020**, *48*, 1003–1021. [CrossRef]
11. Hazen, B.; Cegielski, C.; Hanna, J.B. Diffusion of green supply chain management. *Int. J. Logist. Manag.* **2011**, *22*, 373–389. [CrossRef]
12. Homrich, A.S.; Galvão, G.; Abadia, L.G.; Carvalho, M.M. The circular economy umbrella: Trends and gaps on integrating pathways. *J. Clean. Prod.* **2018**, *175*, 525–543. [CrossRef]
13. Julianelli, V.; Caiado, R.G.G.; Scavarda, L.F.; Cruz, S.P.D.M.F. Interplay between reverse logistics and circular economy: Critical success factors-based taxonomy and framework. *Resour. Conserv. Recycl.* **2020**, *158*, 104784. [CrossRef]
14. Moghaddam, K.S. Fuzzy multi-objective model for supplier selection and order allocation in reverse logistics systems under supply and demand uncertainty. *Expert Syst. Appl.* **2015**, *42*, 6237–6254. [CrossRef]
15. Marić, J.; Opazo-Basáez, M. Green Servitization for Flexible and Sustainable Supply Chain Operations: A Review of Reverse Logistics Services in Manufacturing. *Glob. J. Flex. Syst. Manag.* **2019**, *20*, 65–80. [CrossRef]
16. Agrawal, S.; Singh, R.K. Analyzing disposition decisions for sustainable reverse logistics: Triple Bottom Line approach. *Resour. Conserv. Recycl.* **2019**, *150*, 104448. [CrossRef]
17. Kumar, V.; Brady, M.; Garza-Reyes, J.A.; Simpson, M. Resolving forward-reverse logistics multi-period model using evolutionary algorithms. *Int. J. Prod. Econ.* **2017**, *183*, 458–469. [CrossRef]
18. Henninger, C.E.; Bürklin, N.; Niinimäki, K. The clothes swapping phenomenon—when consumers become suppliers. *J. Fash. Mark. Manag. Int. J.* **2019**, *23*, 327–344. [CrossRef]
19. Sajjanit, C.; Rompho, N. Measuring customer-oriented product returns service performance. *Int. J. Logist. Manag.* **2019**, *30*, 772–796. [CrossRef]
20. Bagur-Femenías, L.; Perramon, J.; Amat, O. Impact of quality and environmental investment on business competitiveness and profitability in small service business: The case of travel agencies. *Total. Qual. Manag. Bus. Excel.* **2014**, *26*, 1–14. [CrossRef]
21. Bastas, A.; Liyanage, K. Sustainable supply chain quality management: A systematic review. *J. Clean. Prod.* **2018**, *181*, 726–744. [CrossRef]
22. Wang, Y.L.; Luor, T.; Luarn, P.; Lu, H.-P. Contribution and Trend to Quality Research—A literature review of SERVQUAL model from 1998 to 2013. *Inform. Econ.* **2015**, *19*, 34–45. [CrossRef]
23. Stević, Ž.; Đalić, I.; Pamučar, D.; Nunić, Z.; Vesković, S.; Vasiljević, M.; Tanackov, I. A new hybrid model for quality assessment of scientific conferences based on Rough BWM and SERVQUAL. *Scientometrics* **2019**, *119*, 1–30. [CrossRef]
24. Rezaei, J.; Kothadiya, O.; Tavasszy, L.; Kroesen, M. Quality assessment of airline baggage handling systems using SERVQUAL and BWM. *Tour. Manag.* **2018**, *66*, 85–93. [CrossRef]
25. Meidutė-Kavaliauskienė, I.; Vasilienė-Vasiliauskienė, V.; Vasiliauskas, A.V. Identification of sectoral logistics service quality gaps by applying servqual method. *Transport* **2020**, *35*, 419–434. [CrossRef]
26. Prentkovskis, O.; Erceg, Ž.; Stević, Ž.; Tanackov, I.; Vasiljević, M.; Gavranović, M. A new methodology for im-proving service quality measurement: Delphi-FUCOM-SERVQUAL model. *Symmetry* **2018**, *10*, 757. [CrossRef]
27. Kilibarda, M.; Nikolicic, S.; Andrejic, M. Measurement of logistics service quality in freight forwarding companies. *Int. J. Logist. Manag.* **2016**, *27*, 770–794. [CrossRef]
28. Knop, K. Evaluation of quality of services provided by transport & logistics operator from pharmaceutical industry for improvement purposes. *Transp. Res. Procedia* **2019**, *40*, 1080–1087. [CrossRef]

29. Limbourg, S.; Giang, H.T.Q.; Cools, M. Logistics Service Quality: The Case of Da Nang City. *Procedia Eng.* **2016**, *142*, 124–130. [CrossRef]
30. Memić, Z.; Vasiljević, M.; Tanackov, I.; Stević, Ž. Measuring the quality of logistics services in the transport company using the servqual model. In Proceedings of the 2nd International Conference on Management, Engineering and Environment, ICMNEE 2018, Belgrade, Serbia, 11–12 October 2018; pp. 119–129.
31. Czajkowska, A.; Stasiak-Betlejewska, R.; Agnieszka, C.; Renata, S.-B. Quality management tools applying in the strategy of logistics services quality improvement. *Serb. J. Manag.* **2015**, *10*, 225–234. [CrossRef]
32. Roslan, N.A.A.; Wahab, E.; Abdullah, N.H. Service Quality: A Case Study of Logistics Sector in Iskandar Malaysia Using SERVQUAL Model. *Procedia Soc. Behav. Sci.* **2015**, *172*, 457–462. [CrossRef]
33. Parmata, U.M.D.; Sankara Rao, B.; Rajashekhar, B. Measuring service quality in pharmaceutical supply chain—distributor's perspective. *Int. J. Pharm. Healthc. Mark.* **2016**, *10*, 258–284. [CrossRef]
34. Vesković, S.; Stević, Ž.; Stojić, G.; Vasiljević, M.; Milinković, S. Evaluation of the railway management model by using a new integrated model DELPHI-SWARA-MABAC. *Decis. Mak. Appl. Manag. Eng.* **2018**, *1*, 34–50. [CrossRef]
35. Pamučar, D.; Stević, Ž.; Sremac, S. A New Model for Determining Weight Coefficients of Criteria in MCDM Models: Full Consistency Method (FUCOM). *Symmetry* **2018**, *10*, 393. [CrossRef]
36. Ali, Y.; Mehmood, B.; Huzaifa, M.; Yasir, U.; Khan, A.U. Development of a new hybrid multi criteria deci-sion-making method for a car selection scenario. *Facta Univ. Ser. Mech. Eng.* **2020**, *18*, 357–373.
37. Badi, I.; Abdulshahed, A. Ranking the Libyan airlines by using Full Consistency Method (FUCOM) and Analytical Hierarchy Process (AHP). *Oper. Res. Eng. Sci. Theory Appl.* **2019**, *2*, 1–14. [CrossRef]
38. Puška, A.; Stević, Ž.; Šadić, S. Uticaj razmjene informacija sa dobavljačem i kupcem na organizacione performanse prehrambenih preduzeća u Bosni i Hercegovini. *EMC-Review* **2019**, *17*, 33–52. [CrossRef]

sustainability

MDPI

Article

A Study on Fiscal Risk of China's Employees Basic Pension System under Longevity Risk

Min Le [1,*], Xinrong Xiao [2], Dragan Pamučar [3] and Qianling Liang [4]

1 School of Economics and Management, Zhejiang Business Technology Institute, Ningbo 315012, China
2 School of Banking and Finance, University of International Business and Economics, Beijing 100029, China; xiaoxinrong@uibe.edu.cn
3 Department of Logistics, University of Defence, 11000 Belgrade, Serbia; dpamucar@gmail.com
4 School of Insurance, Central University of Finance and Economics, Beijing 102206, China; lql111_222@163.com
* Correspondence: nculemin@163.com; Tel.: +86-135-6664-6872

Abstract: It is generally accepted that China's Employees Basic Pension System (CEBPS) cannot cover its expenses. The government needs to fill the gap in income and expenditure with fiscal revenue to ensure sustainability of the system, which may cause it to take fiscal risk caused by the volatility of the fund gap. In this article, through the establishment of a prediction model for the income and expenditure of CEBPS with dynamic mortality, we aimed to measure the fiscal risk caused by longevity risk and provide policy basis for the government. We found that longevity risk leads to serious fiscal risk. The income and expenditure gap of CEBPS fluctuates greatly, and the 2.5% and 97.5% quantiles of fund balance in 2067 are 1.52 and 0.44 times the expected value, respectively. The knock-on effect of fiscal risk, measured by value-at-risk (VaR), is 1.15 times gross domestic product and 4.75 times state fiscal expenditure in 2020. In this article, we not only calculate the expected value like the other literatures but also discuss the volatility of the CEBPS fund gap.

Keywords: China's Employees Basic Pension System; dynamic stochastic model; the fund gap; fiscal risk; longevity risk

check for updates

Citation: Le, M.; Xiao, X.; Pamučar, D.; Liang, Q. A Study on Fiscal Risk of China's Employees Basic Pension System under Longevity Risk. *Sustainability* **2021**, *13*, 5526. https://doi.org/10.3390/su13105526

Academic Editor: Michael McAleer

Received: 12 April 2021
Accepted: 13 May 2021
Published: 15 May 2021

Publisher's Note: MDPI stays neutral with regard to jurisdictional claims in published maps and institutional affiliations.

Copyright: © 2021 by the authors. Licensee MDPI, Basel, Switzerland. This article is an open access article distributed under the terms and conditions of the Creative Commons Attribution (CC BY) license (https://creativecommons.org/licenses/by/4.0/).

1. Introduction

With the advancement of medical technology and improvement in living standards, the life expectancy of the population continues to increase. The resulting longevity risk threatens the sustainable operation of public pension systems in various countries [1–5]. In China, according to "World Health Statistics 2014" and "World Health Statistics 2018", the average life expectancy of the population increased by 6 years from 1990 to 2012 and by 1.4 years from 2012 to 2016. Life expectancy increases rapidly. At the same time, the family planning policy of the 1980s caused a sharp drop in the proportion of the current working population aged 15 to 65 years. The proportion of the working population reached 73.05% in 2009 and dropped to 71.02% in 2018, a decrease of 2.03 percentage points in the past decade. With the increase in life expectancy and the overlapping effect of China's family planning policy, the dependency ratio of the elderly has increased sharply. According to the National Bureau of Statistics, China's old-age dependency ratio was 11.6% in 2009 but increased to 16.8% in 2018. The number of people over 65 years of age has increased by 53.51 million from 2009 to 2018. The Development Research Center of the State Council predicts that the proportion of people over 60 years will exceed 20% of the total population by 2025. The sharp increase in the old-age dependency ratio has placed tremendous pressure on the income and expenditures of China's basic pension insurance fund. By 2018, the fund income of China's Employees Basic Pension System (CEBPS) had increased 3.08 times, and fund expenditure increased 3.83 times, in the previous decade. The rapid increase in the old-age dependency ratio caused the growth rate of fund expenditure to exceed that of fund income, resulting in a fund gap. Without considering central fiscal subsidies, the current fund gap in 2018 reached RMB 414.74 billion. According to the "China Pension Actuarial Report 2019–2050", with the change from nearly two payers supporting a retiree in 2019 to almost one payer supporting a retiree in 2050, the sustainability of the CEBPS fund will be greatly impacted.

Because of its public attribute, it is inherently reasonable for public finance to subsidize the fund gap of CEBPS. The government's social security responsibility is the product of historical changes and social progress. The basic pension insurance system aims to share the risks of the whole population and guarantee a basic living standard for the people; the government has an inescapable responsibility for this. A. C. Pigou, the founder of welfare economics, believed that socio-economic welfare depends not only on the total national income, but also on the distribution of social income. The more equal the distribution of social income, the greater the socio-economic welfare. A basic pension insurance system is one means of government redistribution. The government's intervention in the operation and investments of the basic pension insurance fund, including the use of state revenue to subsidize the fund gap, can give full play to the redistribution function of pension insurance and increase socio-economic welfare. Therefore, the consensus of researchers is that the state should subsidize the pension insurance fund with fiscal revenue [6–8]. In fact, the National Development and Reform Commission (NDRC) (2000) No. 8 document issued specific instructions on the timely and full payment of basic pension and fiscal subsidies for retired employees. In action, according to "Statistical Bulletin on the Development of Human Resources and Social Security", local governments subsidized the fund of CEBPS by RMB 164.6 billion in 2009; in 2017, the figure reached RMB 800.4 billion, an average annual increase of 54%.

"Fiscal risk" is defined as a source of financial stress that could face a government in the future [9]. As an important means of macro-control, changes in fiscal revenue and expenditure and their risks are important issues for the government to consider when it regulates the economy and formulates policies. If the income and expenditure gap of China's basic pension insurance fund can be accurately predicted and measured, then the annual fiscal subsidy can be determined, so the government can prepare in advance and avoid fiscal risk. However, many risks lead to fluctuations in the income and expenditure gap. Many researchers have used the absolute index and scenario analysis methods to analyze the fiscal risk of CEBPS in China, but these methods do not consider the adverse impact of risk factors and the probability of adverse changes. Risk factors (for example, longevity risk) lead to the fluctuation of the fund gap. If the government withdraws only the expected gap, when the worst situation occurs within a given confidence level, it will lead to a financial crisis due to insufficient provision to resist fund imbalance. That is, fiscal risk results when the actual gap is greater than the expected gap. Therefore, in this paper, we focus on the most important longevity risk and discuss the fiscal risk of the basic pension insurance system under longevity risk.

With regard to the fiscal risk of the basic pension insurance system, there are two frequently used indicators: the income and expenditure gap of the pension insurance fund and implicit debt. From the perspective of financial balance, it is more important to calculate the flow-based annual pension gap. The cash-flow model based on demographics, economics, and pension systems is commonly used in quantitative measurement. According to the different assignment methods of input variables, cash-flow models are generally divided into three types: deterministic models, stochastic models, and micro-simulation models [10]. It is widely known that the United States is at the forefront of the world in the use of stochastic models in projecting the balance of social security funds. The Office of the Chief Actuary (OCA) has introduced a stochastic prediction model (the OCA Stochastic Model, OSM for short). Meanwhile, the Congressional Budget Office has built a long-term micro-simulation model of social security funds based on individual longitudinal historical data since 2002 [11]. Most studies in other countries have adopted deterministic prediction cash-flow models, while some have begun to establish stochastic forecasting models. Blake and Mayhew [1] focused on the sustainability of the UK state pension system in light of population ageing and declining fertility by using a simple model and assessed the consequences of six scenarios involving different possible assumptions about productivity growth, activity rates, and different policy responses. Metzger [12] analyzed the medium-term sustainability of the Swiss old-age pension scheme (AHV) by estimating a "Swedish" actuarial balance sheet, which compared pension liabilities with the explicit and implicit assets of the pension scheme. Belolipetskii and Lepskaya [13] considered the probability of ruin of a pension fund on a finite time interval based on the standard Cramer–Lundberg model, which was modified by specifying enrolment and contribution parameters in the form of random variables. Sonsbeek [3] used a dynamic micro-simulation model for calculating the financial and economic implications of the ageing problem and the policy measures considered based on micro-datasets of all Dutch pensions and pension entitlements in the Netherlands. With population ageing becoming a world-wide trend, longevity risk has been a concern. Mortality models play a basic role in the evaluation of longevity risk. The classic model is the Lee–Carter model [14,15], which is very influential because of its simplicity and high accuracy. In addition, the age-period-cohort model (APC model) and the Cairns–Blake–Dowd model (CBD model) are often used in related research [16,17]. Their performance strongly depends on the different patterns shown by mortality data in different countries [18].

Regarding the pension gap of the Chinese system, most researchers have adopted deterministic prediction cash-flow models. For example, Jing et al. [19] constructed an actuarial model to analyze the financial imbalance risk of contribution rate reduction in several scenarios and to investigate the possibility of further reducing the contribution rate. Many studies have used actuarial models to evaluate the financial sustainability of CEBPS under the existing policy scenario and several sets of hypothetical policy scenarios [20–22]. Very few studies have built stochastic prediction models. Tian and Zhao [23] adopted time series modeling techniques to estimate the mortality rate and the fertility rate, focused on the stochastic forecast of the financial sustainability of basic pension. The results showed that an imbalance of the basic pension will occur in 2026. Chen and Yang [24] explored the financial self-balancing ability of the individual accounts of CEBPS. In the particularly serious scenario that the individual accounts' previous accumulated funds are zero, the bookkeeping rate and the investment return rate are considered as stochastic variables. Some studies have considered the impact of longevity risk. Xie et al. [25] aimed to investigate the impact of China's new fertility policy on the actuarial balance of its CEBPS fund, with a stochastic mortality model included to address longevity risk. Zhao et al. [26] built a comprehensive risk assessment system to evaluate the solvency sustainability of CEBPS by integrating the Lee-Carter model into their population projection.

The existing literature includes a considerable amount of research on the operation and fiscal risk of CEBPS in China, but there is room for improvement. First, when considering the fiscal risk of CEBPS, the existing literature does not describe all cash flows according to the social security system. For example, when calculating the expenditure of the basic pension insurance fund, in addition to the basic pension, funeral expenses and individual account balances should also be considered. The payment and distribution methods of the insured have policy provisions, and cash flow should be calculated according to the policy. Moreover, the income and expenditure model does not take a person as the basic unit for more accurate calculation. Second, most researchers still use the income and expenditure gap to measure and analyze the fiscal risk of China's basic pension insurance system. To introduce more general risk measurement methods (e.g., the unconditional variance method or the value-at-risk (VaR) method) into the research, it is important to put forward more precise and targeted measures to prevent fiscal risk in the pension insurance system.

Based on the existing problems, in this paper, we first predicted the dynamic mortality rate by age and gender based on the Lee–Carter model and the data from the "China Statistical Yearbook". Second, we established a model for the income and expenditure of CEBPS according to the actuarial balance principle in combination with the actual payment and distribution method in China. Then we predicted the dynamic and random future changes in the income and expenditure gap of CEBPS by using actual data from China and analyzed the sustainability of the pension insurance. Finally, we used risk measurement methods (including the unconditional variance method, the value-at-risk (VaR) method, and others) to discuss fiscal risk and conduct a sensitivity analysis.

The paper is organized as follows: Section 2 describes an actuarial model of basic pension insurance; Section 3 includes the dynamic mortality prediction and parameter calibration; Section 4 provides a fiscal risk assessment of CEBPS under longevity risk; and Section 5 summarizes the research.

2. Actuarial Model of Basic Pension Insurance

We established a prediction model for the income and expenditure of CEBPS with dynamic mortality to measure the fiscal risk caused by longevity risk and provide policy basis for the government.

2.1. Dynamic Mortality Model

By introducing time and age factors and a random disturbance term into the Lee–Carter model [14,15], it has been shown possible to achieve high goodness-of-fit for mortality prediction in many countries. Although the prediction effect of the model on limited data needs to be improved, considering the complexity of the improved Lee–Carter model and the sufficient data in our study, and after comparing multiple dynamic mortality models, we considered the Lee–Carter model to be effective for fitting the data of China's population mortality. Therefore, we used the classic Lee–Carter model to build a dynamic model of China's mortality by gender and age. The expression of the Lee–Carter model based on single-age mortality data is as follows:

$$\ln(m_{x,t}) = \alpha_x + \beta_x k_t + \varepsilon_{x,t} \tag{1}$$

where $m_{x,t}$ is the central mortality rate at x years old in year t; α_x is the average level of logarithmic central mortality rate at x years old, β_x is the age factor (these two parameters are only related to age); k_t is the time factor, which indicates the change in mortality rate with time; and $\varepsilon_{x,t}$ is the random error term of normal distribution. The least square method and weighted least square method were used to estimate the parameters. First, the ordinary least squares method (OLS) was used to obtain the following results:

$$\min \sum_{x=0}^{N} \sum_{t=1}^{T} (\ln(m_{x,t}) - \alpha_x - \beta_x k_t)^2 \tag{2}$$

Under the boundary condition $\sum\limits_{x=0}^{N} \beta_x = 1$, $\sum\limits_{t=1}^{T} k_t = 0$, we obtained the unique solution of the parameter α_x:

$$\hat{\alpha}_x = \frac{\sum\limits_{t=1}^{T} \ln(m_{x,t})}{T}, x = 0, 1, 2, \ldots\ldots, N \tag{3}$$

The estimation methods for Lee–Carter model parameters include matrix singular value decomposition (SVD), OLS, and weighted least square (WLS). The SVD and OLS methods give the same weight to the mortality of different age groups. Many researchers have shown that different age groups have large differences in population size and numbers of deaths [27], so the WLS method was better than the SVD and OLS methods. The WLS method estimates the parameters k_t and β_x of the Lee–Carter model through the following two steps.

The first step is to sum the left and right sides of Equation (1) for all x under condition $\sum_{x=0}^{N} \varepsilon_{x,t} = 0$:

$$\hat{k}_t = \sum_{x=0}^{N} (\ln(m_{x,t}) - \hat{\alpha}_x) \tag{4}$$

In the second step, the variance proved by Wilmoth [28] is approximately equal to the reciprocal of the death toll, so it can be used as the weight of the sum of squares of the residuals. By minimizing the sum of the squares of the weighted residuals, the following results are obtained:

$$\min \sum_{x=0}^{N} \sum_{t=1}^{T} d_{x,t} (\ln(m_{x,t}) - \hat{\alpha}_x - \beta_x \hat{k}_t)^2 \tag{5}$$

Taking the derivative and setting the derivative to zero, we have

$$\hat{\beta}_x = \sum_{t=1}^{T} d_{x,t} k_t (\ln(m_{x,t}) - \hat{\alpha}_x) / \sum_{t=1}^{T} d_{x,t} k_t^2 \tag{6}$$

To predict the dynamic change of mortality, it was necessary to predict the time factor k_t. It could be estimated using the differential autoregressive moving average model ARIMA (p, d, q). Generally, establishing the ARIMA (p, d, q) model requires several steps, including data stabilization, ARIMA model parameter selection, parameter estimation, model applicability test and optimization, and sequence prediction, to predict the future k_{T+t}. Central mortality can be calculated according to $m_{T+t} = \exp(\hat{\alpha}_x + \hat{\beta}_x \hat{k}_t + \varepsilon_{x,T+t})$. Assuming that the mortality rate follows an exponential distribution, we have $m_{x,T+t} = \mu_{x,T+t} = -\ln(1 - q_{x,T+t})$, where $q_{x,T+t} = 1 - \exp(-m_{x,T+t})$. The disturbance term $\varepsilon_{x,T+t}$ follows a normal distribution with a mean value of zero and variance of σ^2. The estimated variance value can be estimated by the sample variance. The confidence interval of mortality can be obtained by Monte Carlo simulation.

2.2. Income Model

Referring to Liao [29], and taking the current moment as the zero moment, the income cash flow of the public account SA and individual account PA of CEBPS in period $t (t \geq 1)$ are

$$SA^t = \sum_{x=\underline{x}}^{\overline{x}} \eta B_{x+t}^t \cdot LW_x \cdot {}_t p_x \cdot g_{x+t}^t + \sum_{s=1}^{t} \eta B_{c+t-s}^t \cdot NLW_c(s) \cdot {}_{t-s} p_c \cdot g_{c+t-s}^t \tag{7}$$

$$PA^t = \sum_{x=\underline{x}}^{\overline{x}} \tau B_{x+t}^t \cdot LW_x \cdot {}_t p_x \cdot g_{x+t}^t + \sum_{s=1}^{t} \tau B_{c+t-s}^t \cdot NLW_c(s) \cdot {}_{t-s} p_c \cdot g_{c+t-s}^t \tag{8}$$

where $\underline{x}$ and $\overline{x}$, respectively, indicate the minimum age and maximum age of employees participating in CEBPS; B_{x+t}^t represents the payment wage base of employees aged $x + t$ in period t; LW_x is the number of employees aged x at zero moment; $NLW_c(s)$ is the number of new insured persons aged c in period s; η and τ, respectively, represent the enterprise contribution rate and the individual contribution rate; and g_{x+t}^t is the payment persistence rate of employees aged $x + t$ in period t. The survival probability ${}_t p_x$ and ${}_{t-s} p_c$ are determined using Formula (1).

2.3. Expenditure Model

The fund expenditure of CEBPS is divided into three parts: expenditure of retired staff, expenditure of on-the-job employees, and expenditure of the newly insured among on-the-job employees. The survival probability and death probability used in the three parts of the expenditure were determined by Formula (1).

2.3.1. Expenditure Model of Retired Staff

At zero moment, the expenditures of the public account BPR and individual account PPR of CEBPS to retired staff in period $t(t \geq 1)$ are, respectively,

$$BPR^t = \sum_{y=\underline{y}}^{\overline{y}} LR_y\{{}_tp_y{\cdot}BP_y{\cdot}\prod_{k=1}^{t}(1+\alpha_k) + {}_{t-1}p_y{\cdot}q_{y+t-1}{\cdot}D{\cdot}\prod_{k=1}^{t}(1+\beta_k) + {}_tp_y{\cdot}PP_y I_{(t+y-t^r \geq h)}\} \quad (9)$$

$$PPR^t = \sum_{y=\underline{y}}^{\overline{y}} LR_y\{{}_tp_y{\cdot}PP_y[1 - I_{(t+y-t^r \geq h)}] + {}_{t-1}p_y{\cdot}q_{y+t-1}{\cdot}PP_y\max[h-(t+y-t^r),0]\} \quad (10)$$

where $\underline{y}$ and $\overline{y}$ represent the minimum and maximum age of the employees in the retirement pool, respectively; LR_y is the number of insured retirees aged y at zero moment; h is the divisor factor of the retirees aged t^r (the divisor factor is defined as the number of planned payment months during the payment period of pension insurance benefits, and is measured in years here); BP_y is the average basic pension level of retirees aged y at zero moment; α_t is the average growth rate of basic pension in year t; D is the funeral expenses in period 0; β_t is the average growth rate of funeral expenses in year t; and PP_y is the average pension level of the personal account of (y) at zero moment. When the number of receiving months of pension in individual account exceeds the number of planned payment months, the value of the indicative function $I_{(t+y-t^r \geq h)}$ is 1; otherwise, it is 0.

2.3.2. Expenditure Model of On-the-Job Employees

Taking the current moment as the zero moment, we have

$$BPW^t = \sum_{x=\underline{x}}^{\overline{x}} LW_x{\cdot}\{{}_tp_x \cdot [1 - I_{(t<t^r-x)}]BPW_x^t + {}_{t-1}p_x q_{x+t-1} \cdot D\prod_{k=1}^{t}(1+\beta_k)\} \quad (11)$$

$$PPW^t = \sum_{x=\underline{x}}^{\overline{x}} LW_x\{{}_tp_x \cdot [1 - I_{(t<t^r-x)}] \cdot PPW_x^t + {}_{t-1}p_x q_{x+t-1} \cdot [1 - I_{(t<t^r-x)}] \cdot PAS_x^t + {}_{t-1}p_x q_{x+t-1} \cdot I_{(t<t^r-x)} \cdot PA_x^t\} \quad (12)$$

$$BPW_x^t = BP_x^{t^r} \prod_{k=t^r-x}^{t}(1+\alpha_k) + PP_x^{t^r} I_{(t-t^r+x \geq h)} \quad (13)$$

$$BP_x^{t^r} = \overline{W}^{t^r-x-1} \cdot Z_x^{equ} \cdot N_x^{equ} \cdot 1.2\% + 0.005T'\left(\overline{W}^{t^r-x-1} + \overline{W}^{t^r-x-1} \cdot IX\right) \quad (14)$$

$$PPW_x^t = PP_x^{t^r} \cdot [1 - I_{(t-t^r+x \geq h)}] \quad (15)$$

$$PP_x^{t^r} = \frac{\sum_{k=0}^{t^r-x} \eta B_{x+k}^k \cdot (1+r)^{t^r-k-x} + PAY_x \cdot (1+r)^{t^r-k-x}}{h} \quad (16)$$

$$PAS_x^t = PPW_x^t max\{h-(t+x-t^r),0\} \quad (17)$$

$$PA_x^t = \sum_{k=0}^{t} \eta B_{x+k}^k (1+r)^{t-k} + PAY_x(1+r)^t \quad (18)$$

where BPW^t and PPW^t are, respectively, the expenditures of the public account and the individual account of CEBPS to on-the-job employees in period $t(t \geq 1)$; BPW_x^t and PPW_x^t are, respectively, the pension level of the public account and the pension level of the individual account of on-the-job employees aged x at the time t; t^r is the retirement age of on-the-job employees; $BP_x^{t^r}$ is the initial basic pension level of x-year-old employees after retirement at t^r years old; Z_x^{equ} and N_x^{equ} are the deemed contribution index and the deemed contribution period corresponding to the transitional pension; IX is the average contribution index of on-the-job employees; PAS_x^t is the balance of individual account of on-the-job employees aged x at the time t; PAY_x is the existing accumulated amount of individual account of on-the-job employees aged x at zero moment; and PA_x^t is the existing accumulated amount of individual account of on-the-job employees aged x at the time t.

2.3.3. Expenditure Model of New Insured On-the-Job Employees

Taking the current moment as the zero moment, the expenditures of the public account and individual account of CEBPS to new insured on-the-job employees in period $t(t \geq 1)$ are, respectively,

$$BPN^t = \sum_{s=1}^{t} NLW_c(s)\{_{t-s}p_c \cdot [1 - I_{(t<r)}]BPW_c^t(s) + {}_{t-s-1}p_c q_{c+t-s-1} \cdot \mathrm{D}\prod_{k=1}^{t}(1+\beta_k)\} \tag{19}$$

$$PPN^t = \sum_{s=1}^{t} NLW_c(s) \cdot \{_{t-s}p_c \cdot [1 - I_{(t<r)}] \cdot PPW_c^t(s) + {}_{t-s-1}p_c q_{c+t-s-1} \cdot [1 - I_{(t<r)}] \cdot PAS_c^t(s) + {}_{t-s-1}p_c q_{c+t-s-1} \cdot I_{(t<r)} \cdot PA_c^t(s)\} \tag{20}$$

$$BPW_c^t(s) = BPW_c^r(s)\prod_{k=r}^{t}(1+\alpha_k) + PPW_c^r(s)I_{(t\geq r+h))} \tag{21}$$

$$BPW_c^r = \frac{1}{2}(\overline{W}^{r-1} + \overline{W}^{r-1} \cdot EZ_c^r(s)) \cdot EN_c^r(s) \cdot 1\% \tag{22}$$

$$PPW_c^t(s) = PPW_c^r(s) \cdot [1 - I_{(t\geq r+h)}] \tag{23}$$

$$PPW_c^r = \frac{\sum_{k=s}^{r} \tau B_{c+k}^k \cdot (1+r)^{r-k}}{h} \tag{24}$$

$$PAS_c^t(s) = PPW_c^t(s) max\{h - (t-r), 0\} \tag{25}$$

where $s(0 \leq s \leq t)$ and c are the time and age of the new insured employees entering the insurance system for the first time; the corresponding retirement time of this group is $r = s + t^r - c$; $EZ_c^r(s)$ is the expected contribution index of c participating in insurance at time s and retiring at time r; and $EN_c^r(s)$ is the expected payment period of c participating in insurance at time s and retiring at time r.

3. Mortality Prediction and Parameter Calibration

3.1. Mortality Prediction

3.1.1. Revision and Supplement of Mortality Data

In the "China Population and Employment Statistical Yearbook", except for the data obtained from the census in 1990, 2000, and 2010, all data is sampling data. The sampling data for each year is not fixed, and sampling frequency is random. In 1995, 2005, and 2015, about 1% of the population was sampled, whereas sampling data in other years was about 0.1% of the total population. To facilitate estimation and analysis, we adopted 1% sampling. The small sampling base can result in a lack of death population in some age groups, which is inconsistent with reality. For this type of abnormal data, we can use the linear interpolation method to correct the data. The formula is as follows:

$$\ln(m_{x,t}) = \frac{1}{T_2 - T_1} \times [(T_2 - t)\ln(m_{x,T_1}) + (t - T_1)\ln(m_{x,T_2})] \tag{26}$$

where T_1 and T_2 are relatively reliable adjacent statistical years. For example, the data for 1996 are missing, but we can use data of 1995 and 2000 for linear interpolation.

In addition to the problem of abnormal data, the data provided by the "China Population and Employment Statistical Yearbook" also has the problem of deletion of high-age mortality data, so it is necessary to extrapolate the mortality of the elderly population. Because there are few elderly people over 100 years old, the mortality rate used in this paper is accurate to 100 years old. For the different lengths of statistical data and the lack of high age population data, we used the Coale–Kisker (C–K) model to expand the mortality rate. The expression of the improved C–K model is as follows:

$$\ln(m_x) = a + bx + cx^2 + \varepsilon \tag{27}$$

The values of parameters a, b, and c are estimated by the least squares method. According to the above formula, we can expand the mortality rates of men and women aged 90 to 99 from 1994 to 2017. The results are shown in Figures 1 and 2, where it can be seen that mortality at the same age is decreasing year by year, and the greater the age, the higher the mortality, which is in line with the actual situation.

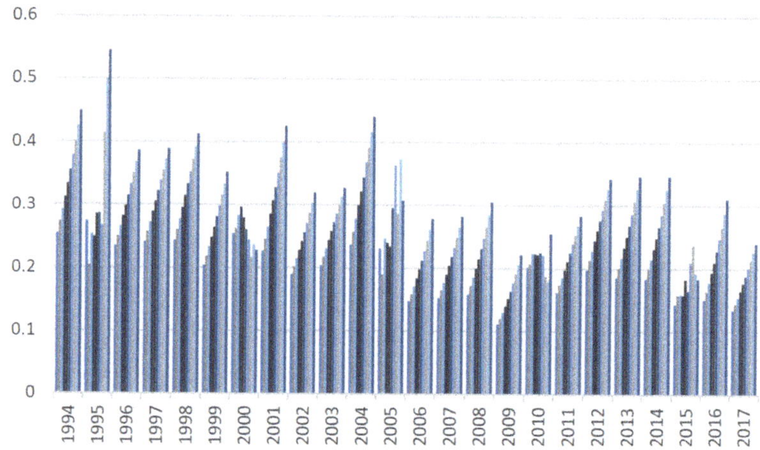

Figure 1. Expanded chart of mortality rates of men aged 90 to 99 from 1994 to 2017 by using the improved Coale–Kisker model.

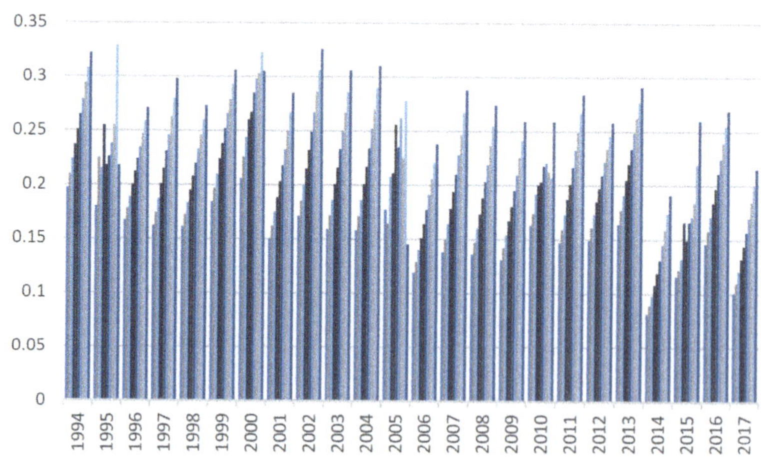

Figure 2. Expanded chart of mortality rates of women aged 90 to 99 from 1994 to 2017 by using the improved Coale–Kisker model.

3.1.2. The Lee–Carter Model and Prediction of Mortality

Based on mortality of the elderly and the correction of abnormal data obtained in the previous section, we obtained relatively complete mortality data for all age groups from 1994 to 2017. Using these data, and with the help of the OLS and WLS methods, we estimated the parameters α_x, β_x, and k_t. The estimated values of parameters α_x and β_x are given in Appendix A, and the estimated value of time factor k_t from 1994 to 2017 is given in Appendix B.

Using the estimated data, we needed to predict parameter k_t, which was the key to using the Lee–Carter model to fit and predict mortality. We used the ARIMA (p, d, q) model to fit and created a trend chart of male and female time factors k_t based on the estimated data. In Figure 3, we can see that both male and female time factors k_t show an obvious downward trend with the passage of time, and that values turn from positive to negative. Reasonable values of the ARIMA model parameters p, d, and q can be obtained through several steps, such as stationary processing, ARMA model selection, parameter estimation, model applicability test and optimization, or sequence prediction.

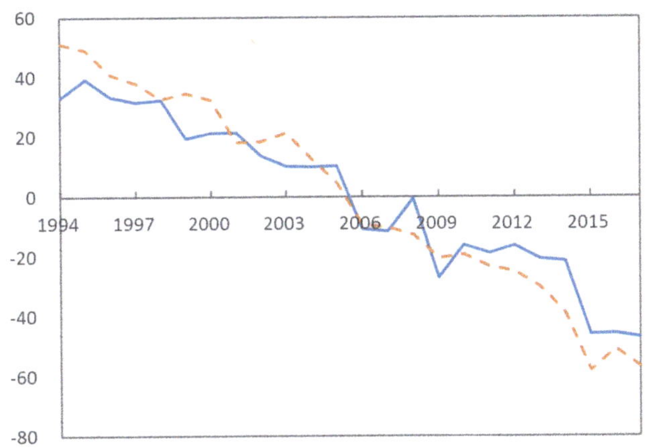

Figure 3. Trend chart of male (solid line) and female (dashed line) time factors k_t from 1994 to 2017.

Because of space limitations, the steps of parameter estimation and applicability test of the ARIMA (p, d, q) model are given in Appendix C. But through the applicability test, we believe that the male sequence k_t belongs to the ARIMA (0, 1, 1) model, which extracts sufficient information. The specific form of the ARIMA (0, 1, 1) model in accordance with the male sequence k_t is as follows:

$$k_t = k_{t-1} - 3.8891 - 0.9643\varepsilon_{t-1} + \varepsilon_t \tag{28}$$

The female sequence k_t also belongs to the ARIMA (0, 1, 1) model. The specific form of ARIMA (0, 1, 1) model in accordance with female sequence k_t is as follows:

$$k_t = k_{t-1} - 4.9166 - 0.9997\varepsilon_{t-1} + \varepsilon_t \tag{29}$$

According to the specific formulas of the ARIMA model, we can predict future values of male and female time factors k_t. The predicted values for the next 50 years are given in Appendix D.

3.1.3. Validity Test of the Lee–Carter Model

We selected the latest available data in 2017 to compare the mortality rates of men and women in all age groups with those predicted by the Lee–Carter model, as shown in Figures 4 and 5. Figure 4 indicates that the curves of actual and predicted values of male mortality in 2017 almost coincide, indicating that the fitting of the Lee–Carter model to predict male mortality in China is close to ideal. Figure 5 indicates that the fitting effect of actual and predicted values of female mortality is very good until the age of 88. After this point, the predicted value of mortality is slightly higher than the real value, but the difference is small and within the acceptable range.

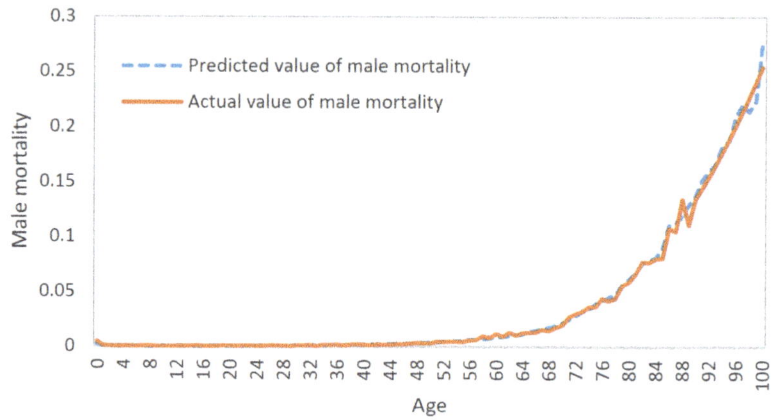

Figure 4. Comparison of actual (solid line) and predicted (dashed line) male mortality in 2017.

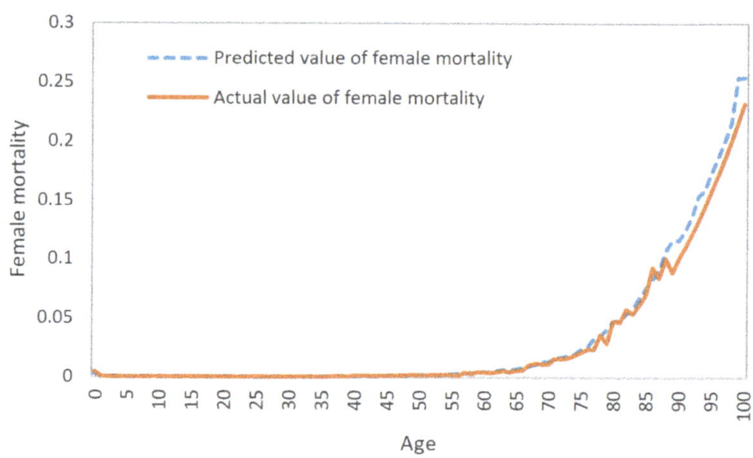

Figure 5. Comparison of actual (solid line) and predicted (dashed line) female mortality in 2017.

3.2. Parameter Setting and Basis

3.2.1. Future Population Structure

The future population structure can be determined by the existing population structure, future mortality, and future birth rate.

- Existing population structure. The proportion of the population of each age was obtained from "China Population and Employment Statistical Yearbook 2010" and was used as the initial population structure of the model.
- Future mortality. According to the Lee–Carter model, we obtained 10,000 samples of mortality in each age every year and used the expected value of the sample to represent mortality at each age.
- Future birth rate. Many researchers believed that implementation of the two-child policy would result in a short-term increase in China's birth rate. However, due to the lag of policy and contemporary youth's fear of marriage and childbirth, China's fertility rate has not increased significantly in recent years. The National Bureau of

Statistics announced that the fertility rate in 2018 was only 1.094, the lowest value on record. In this paper, we assumed that the effect of the two-child policy is not obvious and took the average fertility rate of women of childbearing age in 2014 and 2015 as the corresponding fertility rate (the National Bureau of Statistics has only published the fertility rate of women of childbearing age in 2015 and before) to calculate the birth rate.

3.2.2. Employment and Wage System

- Human capital parameters. In this paper, we took the average value of human capital parameters calculated by the average wages at every age of Beijing employees participating in CEBPS in 2008 and 2009 as the value of human capital level at every age.
- Wage growth rate. We assumed that the future wage growth rate of enterprise employees was 7.7% in 2018–2020, 6.6% in 2021–2025, and 5.7% in 2026–2068.
- Unemployment rate (the introduction of unemployment rate was due to the interruption of contributions caused by unemployment, which affects pension income). According to the data released by the National Bureau of Statistics, the unemployment rate in China has been relatively stable since the beginning of this century. It has fluctuated between 3.8% and 4.3% from 2002 to 2018. The average value of the unemployment rate over 17 years is about 4.09%. We assumed that the unemployment rate would remain unchanged at 4.09% in the future.
- Urbanization rate (the reason for the introduction of urbanization rate is that China's urbanization rate has not yet reached saturation and is growing, which will affect the number of new insured people of CEBPS). According to "National Population Development Plan (2016–2030)", the urbanization rate of China's permanent population will reach 70% in 2030, whereas it was 58.52% in 2018. Therefore, we predicted that China's urbanization rate would increase by 1% per year from 2019 to 2030 and would remain unchanged after 2030.

3.2.3. Basic Pension Insurance System

- Age parameters. We assumed that urban residents begin work at the age of 21, male employees retire at the age of 60, and female employees retire at the age of 55.
- Contribution rate and divisor factor. According to the "Decision of the State Council on Improving the Basic Pension Insurance System for Enterprise Employees" (NDRC (2005) No. 38), the contribution rate of the individual account is set at 8%, the contribution rate of the public account is set at 20%, the divisor factor (defined as the number of planned payment months during the payment period of pension insurance benefits, here measured in years) of male retirees is 12 years, and the divisor factor of female retirees is 14 years. In 2019, the State Council promulgated "Notice of the General Office of the State Council on Printing and Distributing Comprehensive Plans for Reducing Social Insurance Rates", which indicated that the contribution rate of the public account of CEBPS would be reduced to 16% from May 1, 2019. Based on the actual situation, the public account contribution rate was calculated as 20% before May 2019 and 16% after May 2019.
- Return on investment. Referring to "The Annual Report of the National Social Security Fund Council Fund (2013)", the annual return on investment of the national social security fund since its establishment is 8.13%. Therefore, we assumed that the return on investment of both the pension insurance fund and the personal account would be 8.13% after 2017.
- Basic pension growth rate. According to urban basic pension insurance data released by the National Bureau of Statistics, the annual growth rate of the per capita basic pension was 10% from 2006 to 2015. In 2016, the government work report set the growth rate of the basic pension at 6.5%. It was 5.5% in 2017 and 5% in 2018 and 2019. Therefore, the basic pension growth rate was set at 5%.

- Funeral allowance. Because of the different calculation methods of funeral subsidies in different regions, referring to Liao [29], we assumed that the one-time payment of funeral expenses and pension was 60% of the average social wage at the time of death.

3.2.4. Insurance Status

- Composition of existing insured persons. The "China Labor Statistics Yearbook 2018" shows the number of urban on-the-job employees participating in CEBPS and the number of urban retired employees participating in CEBPS in 2017. Assuming that the proportion of insured people at every age in 2017 is the same as the proportion of the insured people at every age in Beijing in 2008, we calculated the number of the insured people of every age of both on-the-job and retired urban employees in 2017.
- Composition of new insured persons. In 2017, the 21-year-old population as a proportion of the total population was 1.22%. Assuming that it will remain unchanged in future, the size of the 21-year-old population in the future can be calculated, and the product of 21-year-old population, urbanization rate, and labor participation rate can represent the number of new insured people.

4. Fiscal Risk Assessment of CEBPS under Longevity Risk

4.1. Model Validation

Using actual data from China and the model in this paper, we calculated the average values of income and expenditure of CEBPS from 2009 to 2018 and compared it with actual income and expenditure. The results are shown in Table 1.

Table 1. Income and expenditure of CEBPS from 2009 to 2018 (unit: RMB 100 million) [1].

Year	Actual Income	Estimated Income	Absolute Percentage Error of Income	Actual Expenditure	Estimated Expenditure	Absolute Percentage Error of Expenditure
2009	9534	10,068.66	5.61%	8894.4	9735.32	9.45%
2010	11,110	11,765.67	5.90%	10,554.9	10,167.91	3.67%
2011	13,956	13,945.07	0.08%	12,765.0	11,174.73	12.46%
2012	16,467	15,740.66	4.41%	15,561.8	15,622.41	0.39%
2013	18,634	17,547.78	5.83%	18,470.4	19,588.41	6.05%
2014	20,434	22,947.38	12.30%	21,754.7	20,762.44	4.56%
2015	23,016	24,922.45	8.28%	25,812.7	26,352.53	2.09%
2016	26,768	28,013.43	4.65%	31,853.8	31,583.61	0.85%
2017	33,403	30,992.80	7.22%	38,051.5	37,338.04	1.87%
2018	38,813	37,198.93	4.16%	42,960.8	41,928.16	2.40%

[1] The data came from "China Labor Statistics Yearbook 2018" and the budget and final accounts of the Ministry of Finance in 2018.

From the perspective of quantitative analysis, we can use the mean absolute percentage error (MAPE) to analyze the effectiveness of the basic pension. The smaller the value of MAPE, the better the fitting effect of income and expenditure. The expression of MAPE is as follows:

$$MAPE(\hat{x}) = \frac{1}{N}\sum_{i=1}^{N} \frac{\left|x_i - \hat{x}_i\right|}{x_i} \tag{30}$$

where N is sample size, x_i is the actual value, and $\hat{x}_i$ is the predicted value. The actual income of the pension insurance fund refers to the actual income collected by the pension insurance fund without the central adjustment fund. According to data in Table 1, the MAPE values of income and expenditure of CEBPS from 2009 to 2018 were 6.03% and 4.60%, respectively. Referring to a goodness-of-fit measurement standard table, we can see that the actuarial model of basic pension had a good fitting effect.

4.2. Fiscal Risk Assessment of CEBPS under Longevity Risk

4.2.1. Fluctuation of the Fund Gap under Longevity Risk

Taking 2017 as the starting year, we calculated the fund operation of CEBPS from 2018 to 2067. Without fiscal subsidies, the results of the average fund gap of CEBPS and its confidence interval are shown in Figure 6.

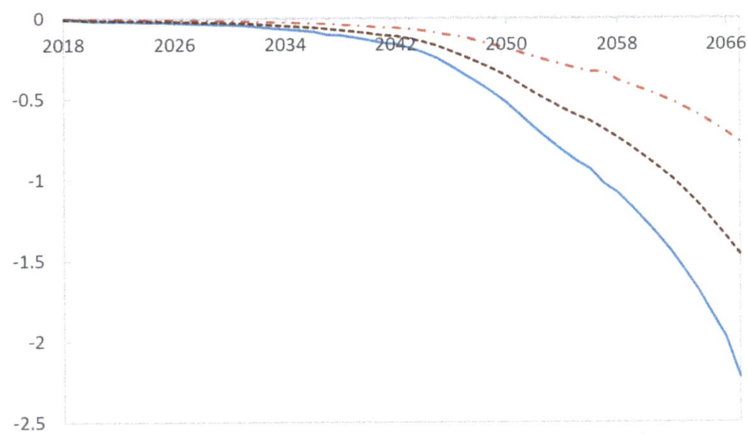

Figure 6. Calculation of the current fund balance of CEBPS (the solid line indicates the current fund balance floor, the dashed line indicates the expected value of the current fund balance, and the dashed-dotted line indicates the current fund balance cap; unit = RMB 100 trillion).

If there is no external financing, there will be a fund gap of CEBPS under longevity risk. According to the budget and final accounts of the pension insurance fund of the Ministry of Finance in 2018, the current balance of the CEBPS fund was negative, and the amount of the fund gap was as high as RMB 414.74 billion, accounting for 2.26% of fiscal revenue in 2018. There is no doubt that the fund gap will increase fiscal pressure. As can be seen in Figure 6, without changing the existing basic pension insurance policy for urban employees, the 95% confidence interval value of the current fund balance has been and continues to be negative, and the fund gap extends outward in a trumpet shape. The expected value of the current fund balance tends toward the lower limit of the fund balance year by year, and the curve shows a negative exponential decline. In 2067, the 2.5% and 97.5% quantiles of fund balance are 1.52 and 0.44 times the expected value, respectively, which indicates a strong likelihood of fiscal risk.

Many researchers have used the absolute index and scenario analysis methods to analyze the fiscal risk of CEBPS, but these methods do not consider the adverse impact of risk factors and the probability of adverse changes. We used more general risk measurement methods to provide a new perspective for evaluating and managing the fiscal risk of CEBPS. Table 2 shows the fiscal risk assessment of CEBPS in the future using various risk measurement methods.

Table 2. Risk measurement results of fund gap (unit: RMB 100 million).

Year	Mean	Standard Deviation	Coefficient of Variation	VaR	ES	TVaR
2027	22,883.7	140,641.7	6.1	35,313.3	34,361.0	722,533.1
2037	66,945.3	755,578.2	11.3	99,814.3	96,610.3	2,032,019.4
2047	240,002.1	6,249,071.7	26.0	350,758.4	337,653.3	7,103,824.8
2057	686,336.9	40,403,533.2	58.9	992,681.7	952,773.2	20,048,146.2
2067	1,464,021.7	196,744,166.8	134.4	2,124,397.7	2,038,943.3	42,903,264.3

Note: VaR, value-at-risk; ES, expected shortfall; TVaR, tail value-at-risk.

From Table 2, it can be seen that the mean value of the fund gap increases in multiples, and the coefficient of variation changes from 6.1 in 2027 to 134.4 in 2067. This volatility in the fund gap increases the possibility that the actual fund gap is higher than the expected fund gap. Under the given 95% confidence level, the maximum risk value of the fund gap in 2067 is RMB 212,439.77 billion; that is, the government must prepare RMB 212,439.77 billion to guard against fiscal risk of CEBPS in 2067. If the government withdraws only RMB 146,402.17 billion and the worst situation occurs within a given confidence level, it will lead to a financial crisis due to insufficient provision to resist fund imbalance. The tail risk cannot be ignored. The ES (expected shortfall) and TVaR (tail value-at-risk) values of the fund gap also increase rapidly. On the premise of maintaining the existing CEBPS, the fiscal risk borne by the government is increasing.

In the calculation of the fund gap of CEBPS, various results are presented based on different model settings and different assumptions of demographic, economic and policy parameters [10,19,21], as shown in Table 3. But the basic conclusions are similar: the gap will inevitably exist and remain huge in the long term without policy adjustment and systematic reform. Moreover, we have tested the validity of the model, but previous studies have not, and our results are relatively more reliable. Our results show that the fund gap extends outward in a trumpet shape, which is similar to other studies that also use the Lee–Carter model [26].

Table 3. Comparisons to the results of the previous studies (unit: RMB 100 million).

Year	This Article Mean Fund Gap	Median Pension Gap under Low Mortality Rates [10]	Current Deficits under the Baseline Scenario [19]	Annual Deficit in Baseline Scenario [21]
2027	22,883.7		54,544.6	86,100
2037	66,945.3		131,313.3	214,400
2047	240,002.1	15,510	249,426.2	420,200
2057	686,336.9		367,640.5	578,900
2067	1,464,021.7	33,820		742,300

Note: The data in the previous studies are for 2030, 2040, 2050, 2060, 2070.

4.2.2. The Knock-On Effect of the Fund Gap under Longevity Risk

The knock-on effect of the fund gap reflects the fiscal pressure faced by the government in the future and the risk of insufficient preparation for fiscal subsidy. Figure 7 shows the knock-on effect of the fund gap of CEBPS from 2018 to 2067 (with the help of the idea of international financial institutions' solvency capital demand VaR, the knock-on effect of fund gap of CEBPS under longevity risk is defined as the difference between the maximum risk value and the expected value of the fund gap in a given confidence interval. The solvency capital demand VaR refers to the maximum loss that an asset or portfolio may suffer under a given level of confidence interval. The expression is $VaR(x;p) = F_X^{-1}(p)$, where $F_X^{-1}(p) = \inf\{x \in R/F_X(x) \geq p\}$.)

In Figure 7, we can see that the knock-on effect of the current fund gap of CEBPS increases rapidly over time. The knock-on effect of the current fund gap in 2018 is only RMB 223.57 billion, but with the increased life expectancy of the population and the lack of potential energy released by the second-child policy, the knock-on effect of the current fund gap will increase to RMB 64,067.13 billion in 2067, with an annual growth rate of 12.68%. To more directly reflect the systematic impact caused by the knock-on effect of the fund gap in the next 50 years, we discounted the total knock-on effect to 2020 with the present value of RMB 116.73 trillion, which is equivalent to 1.15 times GDP and 4.75 times the state fiscal expenditure in 2020. This indicates that the gap of public fiscal subsidy fund is facing great payment pressure in the future, which is likely to lead to fiscal and political risk. It should be dealt with through corresponding policies.

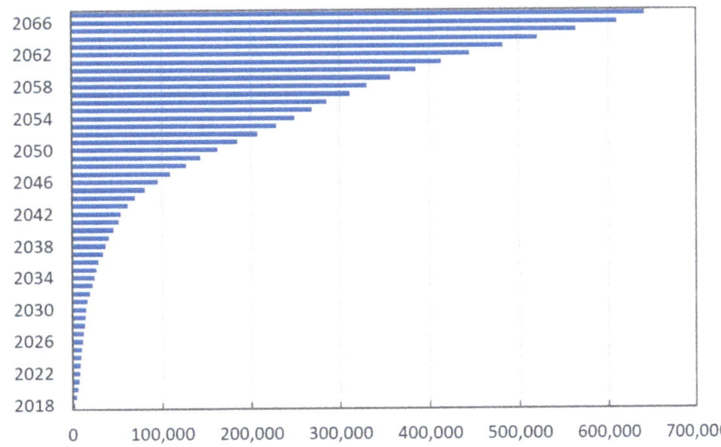

Figure 7. Knock-on effect of current fund gap from 2018 to 2067 (unit: RMB 100 million).

4.2.3. Sensitivity Analysis of Fiscal Risk of CEBPS

Measurement of the benchmark situation was based on reasonable assumptions of actual policy and economic factors, but future institutional and economic factors are uncertain, and changes in these factors will affect the measurement result and fiscal risk. Because the fiscal risk of CEBPS is positively related to the knock-on effect of the fund gap, we took the enterprise contribution rate, wage growth rate, urbanization rate, pension growth rate, and age of receiving pension as sensitive factors to analyze their impact on the present value of the total knock-on effect of CEBPS. In a high-speed change scheme, we increased the enterprise contribution rate, wage growth rate, urbanization rate, and basic pension growth rate by 1% and increased the age of receiving pension by five years. In a low-speed change scheme, we decreased the enterprise contribution rate, wage growth rate, urbanization rate, and basic pension growth rate by 1% and decreased the age of receiving pension by five years. The results of the sensitivity analysis are shown in Table 4.

Table 4. Sensitivity analysis on the knock-on effect of the fund gap (unit: trillion yuan) [1].

Sensitive Factor	Reference Value	High-Speed Value	High Rate of Change	Elasticity	Low-Speed Value	Low Rate of Change	Elasticity
Enterprise contribution rate	116.730	111.554	−0.044	−0.709	120.081	0.029	−0.459
Wage growth rate	116.730	104.758	−0.103	−0.599	148.919	0.276	−1.611
Urbanization rate	116.730	116.325	−0.003	−0.237	117.225	0.004	−0.289
Pension growth rate	116.730	143.236	0.227	1.135	99.059	−0.151	0.757
Age of receiving pension	106.884	95.485	−0.107	−2.091	152.683	0.428	−3.539

[1] Rate of change = (variable value − reference value)/reference value.

The results of the sensitivity analysis in Table 4 show that the fiscal risk of CEBPS under longevity risk is inversely related to the enterprise contribution rate, wage growth rate, urbanization rate, and age of receiving pension, and is positively related to the pension growth rate. Compared with the benchmark case, the elasticity of the wage growth rate is −0.599 under the high-speed scheme and −1.611 under the low-speed scheme, and the reduction of wage growth rate has a more significant impact on financial risk. The elasticity of the basic pension growth rate is 1.135 under the high-speed scheme and 0.757 under the low-speed scheme, and an increase in pension growth rate significantly increases fiscal risk. The elasticity of the age of receiving pension is −2.091 under the high-speed scheme and −3.539 under the low-speed scheme. Thus, the age of receiving pension has the most significant impact on fiscal risk. Therefore, the age growth rate, basic pension growth rate, and age of receiving pension have significant impacts on fiscal risk, whereas the urbanization rate has a relatively limited effect. Increasing the wage growth rate, reducing the basic pension growth rate, and delaying retirement could reduce the fiscal risk of CEBPS.

5. Conclusions

Considering the mortality data of all age groups in China from 1994 to 2017 as historical data and taking into account the lack of static mortality, we used the Lee–Carter model to predict dynamic mortality of all age groups in China from 2018 to 2067 under longevity risk. We then established an income and expenditure prediction model of CEBPS in combination with the actual payment and distribution methods in China. Using relevant actual data of China, we predicted the fund income and expenditure of CEBPS and the confidence interval of the fund balance in the next 50 years, and analyzed the impact of longevity risk on the income and expenditure of the fund and the fund gap. At the same time, we used a more general risk measurement method to analyze the fiscal risk of CEBPS under longevity risk and to analyze the knock-on effect of fiscal risk.

Our results lead to the following conclusions. (1) The dependency ratio of the elderly in China is increasing year by year, the remaining life of retirees is increasing, and some employees are retiring. Women's desire for childbirth is low in China. In this background, fund income and expenditure increase exponentially, the rate of increase of fund expenditure is much greater than that of fund income, the fund gap continues to exist and expand rapidly, the fund gap in later stages is tens or hundreds of times that of the current fund gap, and the sustainability of the pension insurance fund is low under existing policies. (2) Under longevity risk, the income and expenditure gap of CEBPS fluctuates greatly, which is likely to cause fiscal risk. The 2.5% and 97.5% quantiles of fund balance in 2067 are 1.52 and 0.44 times the expected value. The knock-on effect of fiscal risk measured by VaR is 1.15 times GDP and 4.75 times the state fiscal expenditure in 2020. (3) Under longevity risk, the fiscal risk of CEBPS is inversely related to the enterprise contribution rate, wage growth rate, urbanization rate, and age of receiving pension, and is positively related to the pension growth rate. (4) According to the above conclusions, the Chinese government should pay attention to the fiscal risk caused by longevity risk and consider not only the expected value but also the volatility of the CEBPS fund gap. The government can try to take the following three measures to reduce the fiscal risk of CEBPS: increasing the wage growth rate, decreasing the basic pension growth rate, and delaying retirement.

With population ageing becoming a world-wide trend, the fiscal risk of the basic pension insurance system under longevity risk has occurred or is about to occur in most countries. In this article, we attempted to measure the fiscal risk caused by longevity risk and provide policy basis for the government. We took a person as the basic unit for more accurate calculations in the income and expenditure model and introduced more general risk measurement methods into the research, which was important to put forward more precise and targeted measures to prevent fiscal risk in the pension insurance system. Our research has a strong reference significance for researchers facing the same problem. Because this article makes some assumptions about the parameters and does not consider policy changes, subsequent studies can attempt to change some parameter settings or consider policy changes to enrich the research on the fiscal risk of the basic pension insurance system under longevity risk.

Author Contributions: Conceptualization, M.L. and D.P.; resources and writing—original draft preparation, M.L. and X.X.; investigation and data curation, Q.L.; writing—review and editing, M.L. and D.P. All authors have read and agreed to the published version of the manuscript.

Funding: This research was funded by the China National Social Science Fund, grant number 17CSH018, Fundamental Research Funds for the Central Universities in UIBE, grant number 19YB14, and the Program for Innovation Research in Central University of Finance and Economics.

Institutional Review Board Statement: Not applicable.

Informed Consent Statement: Not applicable.

Data Availability Statement: Not applicable.

Conflicts of Interest: The authors declare no conflict of interest.

Appendix A. Estimated Values of Parameters α_x and β_x of Male and Female

Gender	Male		Female	
Age\Parameters	α_x	β_x	α_x	β_x
0	−4.62156	0.03271	−4.38418	0.02688
1	−6.57001	0.02045	−6.69297	0.02041
2	−7.00203	0.02498	−7.01375	0.02014
3	−7.20135	0.02155	−7.30388	0.01379
4	−7.36729	0.02054	−7.68324	0.01845
5	−7.52433	0.02074	−7.95896	0.02046
6	−7.72190	0.01231	−8.12368	0.01075
7	−7.62413	0.01199	−8.18860	0.01375
8	−7.65736	0.01433	−8.17953	0.01186
9	−7.76815	0.00256	−8.23297	0.00833
10	−7.81154	0.01327	−8.05884	0.00919
11	−8.09005	0.01938	−7.96781	0.00664
12	−8.03120	0.00917	−8.19955	0.01490
13	−7.76556	0.00978	−8.22248	0.02053
14	−7.72090	0.01125	−8.16831	0.01040
15	−7.64872	0.01023	−7.99545	0.01770
16	−7.55858	0.01105	−8.05549	0.00548
17	−7.43028	0.02019	−7.99888	0.01434
18	−7.27804	0.01939	−7.67284	0.01495
19	−7.12622	0.01435	−7.61026	0.01545
20	−7.09375	0.01772	−7.84089	0.01710
21	−6.85488	0.01066	−7.84289	0.01658
22	−6.95970	0.01249	−7.70703	0.02261
23	−7.04901	0.01194	−7.48115	0.02173
24	−6.99608	0.01458	−7.69706	0.02197
25	−7.00353	0.01483	−7.57436	0.02047
26	−6.93059	0.01595	−7.52795	0.01453
27	−6.94696	0.01032	−7.63655	0.02301
28	−6.85357	0.01518	−7.50668	0.01774
29	−6.78187	0.01044	−7.50092	0.02005
30	−6.69146	0.01362	−7.31876	0.01801
31	−6.60542	0.01285	−7.47704	0.01452
32	−6.60853	0.00635	−7.36607	0.01439
33	−6.60972	0.01400	−7.19650	0.01105
34	−6.46795	0.00980	−7.24633	0.01135
35	−6.33917	0.00811	−7.30183	0.01783
36	−6.42502	0.00989	−7.23281	0.01244
37	−6.40468	0.00844	−7.04852	0.00960
38	−6.22751	0.00639	−7.02321	0.01133
39	−6.14788	0.00515	−6.75124	0.01060
40	−6.04110	0.00769	−6.80048	0.00792
41	−6.08234	0.00677	−6.66756	0.01047
42	−5.96965	0.00471	−6.63577	0.00521
43	−5.90425	0.00671	−6.55378	0.00840
44	−5.89315	0.00712	−6.49435	0.00855
45	−5.75320	0.00450	−6.43030	0.00979

Gender	Male		Female	
Age\Parameters	α_x	β_x	α_x	β_x
46	−5.64048	0.00633	−6.28376	0.00640
47	−5.61615	0.00852	−6.22656	0.00627
48	−5.45081	0.00626	−6.17575	0.01115
49	−5.39686	0.00629	−5.98499	0.00895
50	−5.31300	0.00450	−5.94701	0.00760
51	−5.20185	0.00620	−5.81607	0.00893
52	−5.18605	0.00417	−5.74637	0.00870
53	−5.09421	0.00531	−5.72730	0.01126
54	−5.00930	0.00517	−5.56433	0.00628
55	−4.88421	0.00757	−5.50715	0.00825
56	−4.79198	0.00561	−5.41774	0.00657
57	−4.75442	0.00600	−5.26577	0.00683
58	−4.59815	0.00754	−5.13659	0.01009
59	−4.48090	0.00847	−5.06904	0.00748
60	−4.37499	0.00657	−4.93217	0.00864
61	−4.30909	0.00854	−4.94986	0.00941
62	−4.15282	0.00951	−4.72151	0.00921
63	−4.10731	0.00923	−4.59292	0.00770
64	−4.02560	0.00934	−4.55246	0.00917
65	−3.93652	0.00719	−4.41630	0.00979
66	−3.79845	0.00881	−4.31074	0.00807
67	−3.71748	0.00815	−4.24819	0.00725
68	−3.62275	0.00895	−4.06680	0.00784
69	−3.50980	0.00932	−3.91371	0.00726
70	−3.37198	0.00975	−3.83106	0.00877
71	−3.25782	0.00906	−3.73532	0.00711
72	−3.18089	0.00824	−3.60911	0.00818
73	−3.07531	0.00733	−3.55004	0.00815
74	−3.03662	0.00715	−3.44068	0.00845
75	−2.90625	0.00678	−3.28499	0.00804
76	−2.79576	0.00780	−3.23901	0.00723
77	−2.67531	0.00853	−3.05641	0.00601
78	−2.61787	0.00868	−2.97738	0.00606
79	−2.51028	0.00728	−2.88624	0.00527
80	−2.42028	0.00726	−2.71385	0.00549
81	−2.25320	0.00894	−2.62181	0.00672
82	−2.22365	0.00676	−2.55495	0.00591
83	−2.11842	0.00828	−2.40772	0.00719
84	−2.03546	0.00918	−2.35070	0.00550
85	−2.00357	0.00768	−2.25595	0.00512
86	−1.88175	0.00552	−2.16679	0.00494
87	−1.75388	0.00791	−2.12763	0.00396
88	−1.69046	0.00788	−1.99767	0.00314
89	−1.58424	0.00835	−1.86394	0.00400
90	−1.53899	0.00797	−1.81861	0.00483
91	−1.49201	0.00656	−1.73445	0.00494
92	−1.39869	0.00752	−1.64709	0.00482
93	−1.33766	0.00773	−1.55747	0.00417
94	−1.26771	0.00698	−1.49454	0.00462
95	−1.20362	0.00788	−1.42283	0.00418
96	−1.13022	0.00670	−1.34921	0.00406
97	−1.06942	0.00684	−1.29018	0.00369
98	−0.99529	0.00915	−1.19799	0.00382
99	−0.93120	0.00941	−1.17475	0.00095

Appendix B. Estimated Values of Parameter k_t of Male and Female

Time\Gender	Male	Female
1994	33.2084	51.3184
1995	39.4429	49.1193
1996	33.5458	40.9681
1997	31.7851	38.0873
1998	32.5188	32.8505
1999	19.4786	34.7427
2000	21.4609	32.5212
2001	21.6200	18.2203
2002	13.7045	18.4637
2003	10.2823	21.4180
2004	10.0631	12.6208
2005	10.3486	4.6885
2006	−10.6914	−9.2847
2007	−11.5040	−10.1855
2008	−0.5283	−12.6329
2009	−27.0173	−20.4423
2010	−16.1270	−19.2904
2011	−18.8028	−23.2862
2012	−16.2574	−24.9482
2013	−20.6983	−30.0364
2014	−21.4034	−38.9570
2015	−46.0050	−58.1271
2016	−45.7643	−51.0063
2017	−47.0511	−56.8212

Appendix C. Stationary Processing, ARMA Model Selection, Parameter Estimation, and Model Applicability Test of the ARIMA (p, d, q) Model

Table A1. ADF test of male k_t.

Original Hypothesis	There Is a Unit Root in K1 Sequence		
Model Form Lag Order	Intercept Term 2 (Based on SiC Test, the Maximum Lag Order Is 5)		
		t Value	*p* Value
ADF test statistics		0.6178	0.9867
Significance level	1%	−3.7880	
	5%	−3.0124	
	10%	−2.6461	

Table A2. ADF test of female k_t.

Original Hypothesis	There Is a Unit Root in K2 Sequence		
Model Form Lag Order	Intercept Term 0 (Based on SiC Test, the Maximum Lag Order Is 5)		
		t Value	*p* Value
ADF test statistics		−0.0954	0.9389
Significance level	1%	−3.7529	
	5%	−2.9981	
	10%	−2.6388	

Table A3. Unit root test of first order difference of male k_t.

Original Hypothesis	The First Order Difference Sequence of K1 has a Unit Root		
Model Form Lag Order	Intercept Term 1 (Based on SiC Test, the Maximum Lag Order Is 5)		
		t Value	*p* Value
ADF test statistics		−5.5445	0.0002
Significance level	1%	−3.7880	
	5%	−3.0124	
	10%	−2.6461	

Table A4. Unit root test of first order difference of female k_t.

Original Hypothesis	The First Order Difference Sequence of K2 has a Unit Root		
Model Form Lag Order	Intercept Term 0 (Based on SiC Test, the Maximum Lag Order Is 5)		
		T Value	*p* Value
ADF test statistics		−5.7388	0.0001
Significance level	1%	−3.7696	
	5%	−3.0049	
	10%	−2.6422	

Table A5. Autocorrelation and partial correlation of first order difference of male k_t.

Sample Years Sample Size		1994–2017 23				
Autocorrelation Graph	**Partial Autocorrelation Graph**	**Lag Phase**	**Autocorrelation Value**	**Partial Autocorrelation Value**	**Q Statistic**	***p* Value**
		1	−0.462	−0.462	5.5693	0.018
		2	−0.071	−0.361	5.7063	0.058
		3	0.178	−0.049	6.6137	0.085
		4	−0.071	−0.009	6.7658	0.149
		5	−0.209	−0.282	8.1599	0.148
		6	0.191	−0.14	9.3913	0.153
		7	0.003	−0.028	9.3916	0.226
		8	−0.123	−0.082	9.9715	0.267
		9	0.101	−0.056	10.389	0.320
		10	0.14	0.153	11.251	0.338
		11	−0.217	0.015	13.507	0.261
		12	0.02	−0.095	13.527	0.332

Table A6. Autocorrelation and partial correlation of first order difference of female k_t.

Sample Years Sample Size		1994–2017 23				
Autocorrelation Graph	**Partial Autocorrelation Graph**	**Lag Phase**	**Autocorrelation Value**	**Partial Autocorrelation Value**	**Q Statistic**	***p* Value**
		1	−0.241	−0.241	1.5158	0.218
		2	−0.215	−0.289	2.7764	0.250
		3	−0.049	−0.215	2.8453	0.416
		4	0.001	−0.176	2.8453	0.584
		5	0.082	−0.051	3.0594	0.691
		6	−0.088	−0.153	3.3206	0.768
		7	−0.206	−0.359	4.839	0.680
		8	0.163	−0.163	5.8517	0.664
		9	0.189	0.008	7.3199	0.604
		10	−0.026	0.003	7.3499	0.692
		11	−0.028	0.069	7.3867	0.767
		12	−0.209	−0.181	9.6739	0.645

Table A7. Regression results of male ARIMA model.

Model	AIC	AC	Log Maximum Likelihood Estimation	Are the Parameters Significant?
ARIMA (0, 1, 0)	6.580521	6.7297	−66.09547	No
ARIMA (0, 1, 1)	6.21557	6.3143	−69.479	Yes
ARIMA (0, 1, 2)	6.198652	6.34676	−68.28449	No
ARIMA (1, 1, 0)	6.546365	6.64555	−70.01001	No
ARIMA (1, 1, 1)	5.58491	5.73368	−58.43401	No
ARIMA (1, 1, 2)	5.53601	5.73439	−56.89617	No
ARIMA (2, 1, 0)	6.58021	6.72934	−66.09547	No
ARIMA (2, 1, 1)	6.35593	6.55489	−62.73731	No
ARIMA (2, 1, 2)	6.435176	6.683872	−62.56935	No

Table A8. Regression results of female ARIMA model.

Model	AIC	AC	Log Maximum Likelihood Estimation	Are the Parameters Significant?
ARIMA (0, 1, 0)	7.4249	7.4743	−83.3867	No
ARIMA (0, 1, 1)	6.8043	6.9030	−72.2491	Yes
ARIMA (0, 1, 2)	6.8909	7.0390	−76.2458	No
ARIMA (1, 1, 0)	7.2607	7.3599	−77.8679	Yes
ARIMA (1, 1, 1)	6.1173	6.2660	−64.2899	No
ARIMA (1, 1, 2)	6.9346	7.1330	−72.2808	No
ARIMA (2, 1, 0)	7.2375	7.38670	−72.9941	No
ARIMA (2, 1, 1)	6.1509	6.3498	−60.5841	No
ARIMA (2, 1, 2)	7.0318	7.2805	−68.8342	No

Table A9. White noise test of male residual sequence.

Sample Years			1994–2017			
Sample Size			23			
Autocorrelation Graph	Partial Autocorrelation Graph	Lag Phase	Autocorrelation Value	Partial Autocorrelation Value	Q Statistic	p Value
		1	−0.055	−0.055	0.0799	
		2	−0.125	−0.128	0.5058	0.477
		3	−0.025	−0.041	0.5241	0.769
		4	−0.197	−0.222	1.6967	0.638
		5	−0.335	−0.401	5.2864	0.259
		6	0.052	−0.123	5.3787	0.371
		7	0.044	−0.142	5.4470	0.488
		8	−0.024	−0.205	5.4693	0.603
		9	0.138	−0.133	6.2533	0.619
		10	0.195	−0.013	7.9296	0.541
		11	−0.055	−0.094	8.0753	0.621
		12	−0.038	−0.103	8.1495	0.700

Table A10. White noise test of female residual sequence.

Sample Years			1994–2017			
Sample Size			23			
Autocorrelation Graph	Partial Autocorrelation Graph	Lag Phase	Autocorrelation Value	Partial Autocorrelation Value	Q Statistic	p Value
		1	0.284	0.284	2.1038	
		2	−0.12	−0.218	2.4987	0.114
		3	−0.206	−0.117	3.7214	0.156
		4	−0.215	−0.159	5.123	0.163
		5	−0.161	−0.12	5.9498	0.203
		6	−0.199	−0.24	7.2929	0.200
		7	−0.206	−0.247	8.8191	0.184
		8	0.127	0.1	9.4399	0.223
		9	0.3	0.072	13.142	0.107
		10	0.171	−0.046	14.429	0.108
		11	0.049	0	14.544	0.150
		12	−0.122	−0.126	15.325	0.168

Appendix D. Predicted Values of Parameter k_t

Time	Predicted Values of Male	Predicted Values of Female
2018	−50.2707	−61.7373
2019	−54.1598	−66.6538
2020	−58.0489	−71.5704
2021	−61.9380	−76.4869
2022	−65.8271	−81.4035
2023	−69.7162	−86.3201
2024	−73.6053	−91.2366
2025	−77.4944	−96.1532
2026	−81.3835	−101.0697
2027	−85.2726	−105.9863
2028	−89.1618	−110.9029
2029	−93.0509	−115.8194
2030	−96.9400	−120.7360
2031	−100.8291	−125.6525
2032	−104.7182	−130.5691
2033	−108.6073	−135.4857
2034	−112.4964	−140.4022
2035	−116.3855	−145.3188
2036	−120.2746	−150.2353
2037	−124.1637	−155.1519
2038	−128.0528	−160.0685
2039	−131.9419	−164.9850
2040	−135.8310	−169.9016
2041	−139.7201	−174.8181
2042	−143.6092	−179.7347
2043	−147.4984	−184.6513
2044	−151.3875	−189.5678
2045	−155.2766	−194.4844
2046	−159.1657	−199.4009
2047	−163.0548	−204.3175
2048	−166.9439	−209.2341
2049	−170.8330	−214.1506
2050	−174.7221	−219.0672
2051	−178.6112	−223.9837
2052	−182.5003	−228.9003
2053	−186.3894	−233.8169
2054	−190.2785	−238.7334
2055	−194.1676	−243.6500
2056	−198.0567	−248.5665
2057	−201.9459	−253.4831
2058	−205.8350	−258.3997
2059	−209.7241	−263.3162
2060	−213.6132	−268.2328
2061	−217.5023	−273.1493
2062	−221.3914	−278.0659
2063	−225.2805	−282.9825
2064	−229.1696	−287.8990
2065	−233.0587	−292.8156
2066	−236.9478	−297.7321
2067	−240.8369	−302.6487

References

1. Blake, D.; Mayhew, L. On the sustainability of the UK state pension system in the light of population ageing and declining fertility. *Econ. J.* **2006**, *116*, F286–F305. [CrossRef]
2. Jimeno, J.F.; Rojas, J.A.; Puente, S. Modelling the impact of aging on social security expenditures. *Econ. Model.* **2008**, *25*, 201–224. [CrossRef]

3. Sonsbeek, J. Micro simulations on the effects of ageing-related policy measures: The social affairs department of the Netherlands ageing and pensions model. *Econ. Model.* **2010**, *27*, 968–979. [CrossRef]
4. Chomik, R.; Piggott, J. Pensions, ageing and retirement in Australia: Long term projections and policies. *Aust. Econ. Rev.* **2012**, *45*, 350–361. [CrossRef]
5. Bonenkamp, J.; Meijdam, L.; Ponds, E.; Westerhout, E. Ageing-driven pension reforms. *J. Popul. Econ.* **2017**, *30*, 953–976. [CrossRef]
6. Barr, N. *Economics of the Welfare State*, 4th ed.; Oxford University Press: New York, NY, USA, 2004; p. 204.
7. Zaidi, A. Population aging and financial and social sustainability challenges of pension systems in Europe: A cross-national perspective. In *The Future of Multi-Pillar Pensions*; Bovenberg, L., Ewijk, C.V., Westerhout, E., Eds.; Cambridge University Press: Cambridge, UK, 2012; pp. 17–45.
8. Liu, B.; Yang, Z. Population aging shock and fiscal sustainability in china: Mechanism analysis and effect simulation. *Singap. Econ. Rev.* **2018**. [CrossRef]
9. Brixi, H.P.; Schick, A. *Government at Risk: Contingent Liabilities and Fiscal Risk*; Oxford University Press: New York, NY, USA, 2002; pp. 2–3.
10. Zhao, Q.; Mi, H. Evaluation on the sustainability of urban public pension system in China. *Sustainability* **2019**, *11*, 1418. [CrossRef]
11. U.S. Congressional Budget Office. *Quantifying Uncertainty in the Analysis of Long-term Social Security Projections*; U.S. Congressional Budget Office: Washington, DC, USA, 2005.
12. Metzger, C. An actuarial balance sheet of the Swiss old-age pension scheme. *Int. Soc. Secur. Rev.* **2018**, *71*, 25–49. [CrossRef]
13. Belolipetskii, A.A.; Lepskaya, M.A. A Mathematical Model of Pension Fund Operation and Methods of Fund Stability Analysis. *Comput. Math. Model.* **2018**, *29*, 233–243. [CrossRef]
14. Lee, D.R.; Carter, R.L. Modeling and forecasting US mortality. *J. Inst. Actuar.* **1860**, *13*, 325–358.
15. Girosi, F.; Kin, G. Understanding the Lee-Carter Mortality Forecasting Method. 2007. Available online: http://gking.harvard.edu/files/abs/lc-abs.shtml (accessed on 20 November 2020).
16. Chernyavskiy, P.; Little, M.P.; Rosenberg, P.S. A unified approach for assessing heterogeneity in age-period-cohort model parameters using random effects. *Stat. Methods Med Res.* **2019**, *28*, 20–34. [CrossRef]
17. Cairns, A.J.G.; Blake, D.P.; Dowd, K. A two-factor model for stochastic mortality with parameter uncertainty: Theory and calibration. *J. Risk Insur.* **2006**, *73*, 687–718. [CrossRef]
18. Carfora, M.F.; Cutillo, L.; Orlando, A. A quantitative comparison of stochastic mortality models on Italian population data. *Comput. Stat. Data Anal.* **2017**, *112*, 198–214. [CrossRef]
19. Jing, P.; Chang, C.; Zhu, H.; Hu, Q. Financial imbalance risk and its control strategy of China's pension insurance contribution rate reduction. *Math. Probl. Eng.* **2021**, *12*. [CrossRef]
20. Zhao, Y.; Bai, M.; Liu, Y.; Hao, J. Quantitative analyses of transition pension liabilities and solvency sustainability in China. *Sustainability* **2017**, *9*, 2252. [CrossRef]
21. Wang, H.; Huang, J.; Yang, Q. Assessing the financial sustainability of the pension plan in china: The role of fertility policy adjustment and retirement delay. *Sustainability* **2019**, *11*, 883. [CrossRef]
22. Sun, L.; Su, C.; Xian, X. Assessing the sustainability of china's basic pension funding for urban and rural residents. *Sustainability* **2020**, *12*, 2833. [CrossRef]
23. Tian, Y.; Zhao, X. Stochastic forecast of the financial sustainability of basic pension in China. *Sustainability* **2016**, *8*, 46. [CrossRef]
24. Chen, X.; Yang, Z. Stochastically assessing the financial sustainability of individual accounts in the urban enterprise employees' pension plan in china. *Sustainability* **2019**, *11*, 3568. [CrossRef]
25. Xie, Y.; Zhang, X.; Lv, H.; Guo, X. The new fertility policy and the actuarial balance of china urban employee basic endowment insurance fund based on stochastic mortality model. *Math. Probl. Eng.* **2020**, *3*, 1–12.
26. Zhao, Y.; Bai, M.; Peng, F.; Zhu, M. Stochastic assessments of urban employees' pension plan of china. *Sustainability* **2018**, *10*, 1028. [CrossRef]
27. Koissi, M.C.; Shapiro, A.F.; Hognas, G. Evaluating and extending the Lee-carter mortality forecasting:Bootstrap confidence interval. *Insur. Math. Econ.* **2006**, *38*, 1–20. [CrossRef]
28. Wilmoth, J.R. Mortality projections for Japan: A comparison of four methods. In *Health and Mortality among Elderly Population*; Oxford University Press: New York, NY, USA, 1996.
29. Liao, P.; Su, H.; Pamučar, D. Will Ending the One-Child Policy and Raising the Retirement Age Enhance the Sustainability of China's Basic Pension System? *Sustainability* **2020**, *12*, 8172. [CrossRef]

sustainability

MDPI

Article

Novel Fuzzy Composite Indicators for Locating a Logistics Platform under Sustainability Perspectives

Hana Ayadi [1,2,3,*], Nadia Hamani [2], Lyes Kermad [1] and Mounir Benaissa [3]

1 IUT de Montreuil, University of Paris 8, 93100 Montreuil, France; l.kermad@iut.univ-paris8.fr
2 INSSET, University of Picardie Jules Verne, 02100 Saint Quentin, France; nadia.hamani@u-picardie.fr
3 ISGIS, University of Sfax, Sfax 3021, Tunisia; mounir.benaissa@isgis.usf.tn
* Correspondence: hana.ayedi@etud.univ-paris8.fr

Abstract: The purpose of this paper is to help decision-makers choose the location of a logistics platform with sustainability perspectives. This study presents a compensatory and partially compensatory approach to build composite indicators, using mainly fuzzy multi-criteria decision-making methods. In the first instance, the fuzzy full consistency method (F-FUCOM) was used to calculate the weight of the criteria and sub-criteria. In the second instance, two aggregation methods, namely the fuzzy multi-attribute ideal-real comparative analysis (F-MAIRCA) and the fuzzy preference ranking organization method for enrichment evaluation (F-PROMETHEE), were used to rank the location of a logistics platform. The novelty of the work lays in studying the impact of limited sustainability and weak sustainability on the location of a logistics platform. In this respect, the aggregation of various sustainability criterion in fuzzy compensatory and partially compensatory composite indicators is an innovative and interesting approach used to locate a logistics platform. The obtained results show that economic sustainability is the most important criterion for the selection of a logistics platform, followed by the environmental criterion. Obviously, the F-MAIRCA and F-PROMETHEE methods provided the same ranking orders. Finally, sensitivity analyses were performed to validate the robustness of the proposed approach.

Keywords: logistics platform; facility location selection; composite indicators; multi-criteria decision-making; F-FUCOM; F-PROMETHEE; F-MAIRCA; sustainability

check for updates

Citation: Ayadi, H.; Hamani, N.; Kermad, L.; Benaissa, M. Novel Fuzzy Composite Indicators for Locating a Logistics Platform under Sustainability Perspectives. *Sustainability* **2021**, *13*, 3891. https://doi.org/10.3390/su13073891

Academic Editors: Dragan Pamucar, Željko Stević and Tamás Bányai

Received: 11 January 2021
Accepted: 23 March 2021
Published: 1 April 2021

Publisher's Note: MDPI stays neutral with regard to jurisdictional claims in published maps and institutional affiliations.

Copyright: © 2021 by the authors. Licensee MDPI, Basel, Switzerland. This article is an open access article distributed under the terms and conditions of the Creative Commons Attribution (CC BY) license (https://creativecommons.org/licenses/by/4.0/).

1. Introduction

In an increasingly competitive environment, the freight transport system is undeniably necessary to ensure the proper functioning of the city economy through its presence in the upstream and downstream of the supply chain [1,2]. Despite its importance, this system has not been given great attention by the local, regional and national authorities. Thus, to make the movement of goods more fluid, it is essential to know the characteristics of the urban space, its stakes and its constraints [3,4]. To this end, the establishment of a logistics platform, located a few kilometers from the city center, can certainly alleviate the severity of the impact of freight transport on the city, and in terms of sustainable development, by making deliveries more fluid.

The logistics platform is an infrastructure that affects sustainable development by reducing traffic congestion, minimizing carbon emissions and ensuring efficient land use. The main mission of this platform is to pool resources and to decrease the concentration of flows to the city or, more precisely, to the city center [5]. It also reduces transport costs, delays and nuisances, and facilitates the flow of goods and the transition from one mode of transport to another, using technologically advanced and efficient equipment to handle containers. These activities accompanied by storage operations require large, equipped warehouses.

However, the implications of building infrastructure and operating logistics facilities are far-reaching and often almost imperceptible. The operation of a logistics platform negatively and positively affects the environment of the zone where it is implemented. On the other hand, there are vertical and horizontal relationships between the interests of the actors associated with the implementation of a logistics platform [6].

In this context, the next section presents the existing work addressing the problem of facility location. Our study of the literature shows that no general or systematic method of localizing a logistics platform, specifically related to the sustainability perspectives, has been proposed until now. In fact, this issue depends on a set of locations (alternatives) assessed against a set of weighted criteria that are independent of each other. In this research work, an innovative and interesting approach of locating a logistics platform, based on fuzzy compensatory and partially compensatory composite indicators, is introduced as a support tool for decision-makers to locate a logistics platform with sustainability perspectives. The following research questions are answered in this paper: What are the important criteria that should be applied to select a logistics platform in order to increase the level of sustainability? What are the most commonly used multi-criteria decision-making (MCDM) methods for weighting and aggregation as reported in the literature? How can they be utilized well in fuzzy compensatory and partially compensatory composite indicators' development? Is the selection of a logistics platform under sustainability perspectives affected by the compensation phenomenon?

To answer these research questions, this study aims to:

- Identify the sustainable evaluation criteria and sub-criteria for logistics platform location;
- Select the most suitable weighting and aggregation methods;
- Propose a composite indicator based on compensatory and partially compensatory multi-criteria decision support methods to identify the location of a logistics platform, responding more adequately to the requirements of sustainability;
- Study the impact of a compensation phenomenon on the decision-making process.

This manuscript is organized as follows. The next section reviews the relevant existing approaches for locating a logistics platform and developing composite indicators. Section 3 presents the proposed approach. Then, Section 4 illustrates the results of analyzing the case study applied in the city of Sfax. The implications of this study are presented in Section 5. Finally, the conclusion and future research directions are provided in Section 6.

2. Literature Review

The choice of a location is one of the problems discussed in the literature. We present, in the first subsection, an overview of the existing approaches to locating logistics platforms. The second subsection describes the steps of the construction of the composite indicators because it is useful tool that is increasingly demanded by policy-makers.

2.1. Existing Approaches of Locating Logistics Platform

Many methods, such as multi-criteria decision-making (MCDM) methods, meta-heuristics for multi-objective decision-making methods and multi-objective combinatorial optimization methods, have been applied to solve localization problems [7]. However, the proposed studies based on combinatorial optimization and meta-heuristics are more complex and do not always represent reality. In this regard, the authors of this study focused on multi-criteria decision support methods because they can provide insight about reality by integrating expert's opinions.

In fact, the facility location problem is one of the major problems discussed in the literature. This section describes the theoretical aspects of the suggested approach. It also presents some approaches proposed to solve the problem of localizing infrastructures using multi-criteria decision support methods. Table 1 describes the existing localization MCDM approaches.

Agrebi et al. [8] proposed a decision support system to select the location of distribution centers. The introduced multi-attribute and multi-actor decision-making method based on the Elimination and Choice Expressing Reality method (ELECTRE-I). Given the inherent uncertainty and imprecision of human decision-making, a fuzzy multi-attribute and multi-actor decision-making method was also applied. To check the sensitivity of the chosen solution to variations in the criteria weights, a sensitivity analysis was carried out. The obtained results prove that the two suggested methods met the desired objective of the selection of the best location in a certain/uncertain context of multi-attribute and multi-actor variables.

Kumar et Anbanandam [9] presented a framework to select the location of the multimodal freight terminal under sustainability perspectives. The authors used the intuitive fuzzy (IF) set to incorporate the importance of the expert's group decision-making process and calculated the priority weight of the criteria and its sub-criteria using the hierarchical analysis process (IF-AHP). Then, they evaluated the performance of location by incorporating the technique for order of preference by similarity to ideal solution (IF-TOPSIS).

Yazdani et al. [10] developed a two-step decision-making model to find the most preferred area for establishing logistics centers. In the first step, to identify the efficient and inefficient alternatives, the authors compared the considered communities through data envelopment analysis (DEA). However, in the second step, an evaluation model was designed to assess the performance of effective communities. Researchers employed the rough full consistency method (R-FUCOM), to obtain the optimal weights of the criteria, and the rough combined compromise solution (R-CoCoSo) method to rank efficient communities. The adopted model allowed for capturing the uncertainty and vagueness in the judgments of decision-makers by applying the rough set theory (RST). In addition, sensitivity analyses were performed to validate the robustness of the obtained results.

Cheng et Zhou [11] proposed a method to evaluate the location of the logistics distribution center. To improve the efficiency of decision-making, they developed a fuzzy approach based on the AHP method. Through a case study of four potential locations, the results prove that the adopted method is effective in selecting the best location with both qualitative and quantitative factors.

Table 1. The existing localization multi-criteria decision-making (MCDM) approaches.

Authors	Weighting	Aggregation	Technical	Country	Extension
[9]	FI-AHP	FI-TOPSIS	Compensatory	India	Intuitionistic fuzzy sets
[10]	DEA, R-FUCOM	R-CoCoSo	Compensatory	-	Rough set theory
[8]		ELECTRE I	Non-compensatory	-	Fuzzy sets
[12]	DEMATEL	MAIRCA	Compensatory	China	Fuzzy sets
[13]	DEMATEL	MAIRCA	Compensatory	China	-
[14]	GIS, Fuzzy SWARA	COCOSO	Compensatory	Turkey	Fuzzy sets
[15]	EW- Fuzzy AHP	Fuzzy TOPSIS	Compensatory	China	Fuzzy sets
[16]	Fuzzy AHP	PROMETHEE	Non-compensatory	Turkey	Fuzzy sets
[17]	DEMATEL, ANP	TOPSIS	Compensatory	Turkey	Intuitionistic fuzzy sets

Pamucar et al. [12] applied a hybrid MCDM approach for the sustainable selection of a site to develop a multimodal logistics center. The decision-making trial and evaluation laboratory (DEMATEL) method was also used to determine the weight coefficients of the criteria. Then, a multi-attributive ideal-real comparative analysis (MAIRCA) was carried out to select a location by comparing the theoretical and empirical alternative assessments.

Muravev et al. [13] introduced a new approach based on DEMATEL-MAIRCA to determine the optimal locations of the China Railway Express international logistics centers and to minimize the number of rail routes. This approach found the best solution, closest to the ideal one. In the first instance, the value of the best alternative was identified in line with the observed criterion. In the second instance, the distances of the other alternatives were measured as a function of the observed criterion of the ideal value. The preliminary results showed that, because of the increase in the container turnover between China and the European Union, the determination of the optimal locations for the logistics centers should be done in a dynamic manner.

Ulutaş et al. [14] suggested a new approach combining the geographic information systems (GIS) with the fuzzy step-wise weight assessment ratio analysis (SWARA) method and the CoCoSo method to select the location of the logistics center in Turkey. In addition, sensitivity analyses were performed to validate the robustness of the suggested approach by varying criteria weights and comparing the ranking of the CoCoSo method with other MCDM techniques (complex proportional assessment of alternatives (COPRAS), VIseKriterijumska Optimizacija I Kompromisno Resenje (VIKOR), additive ratio assessment (ARAS), multi-objective optimization on the basis of ratio analysis (MOORA) and multi-attributive border approximation area comparison (MABAC)).

He et al. [15] developed a new hybrid fuzzy multi-criteria decision-making method to select the location of a joint distribution center by taking sustainability into account. First, the weights of the subjective criteria were calculated with a fuzzy AHP method, while the objective criteria were weighted using a fuzzy entropy method. Subsequently, the authors ranked the alternatives with the improved fuzzy TOPSIS method utilizing weighted criterion distances. Finally, a sensitivity analysis was carried out to illustrate the effectiveness and robustness of the proposed method and its ability to promote the sustainability of companies in China.

Kazançoğlu et al. [16] presented a hybrid multi-criteria decision-making approach to locate the logistics center in terms of benchmarks based on the sustainability concerns. This approach uses the fuzzy AHP method to obtain the weights of the criteria and the preference ranking organization method for enrichment evaluations (PROMETHEE) to select the best alternative. It was applied in Turkey, for the location of a sustainable logistics center in Izmir.

Karaşan et al. [17] adopted a new integrated fuzzy decision-making model to select the location of freight villages. In this model, the DEMATEL method was used to determine the most efficient criteria and their internal and external dependencies. The weight coefficients of the criteria were obtained using the analytic network process (ANP) method. Then, the TOPSIS method was employed to find the best alternative location. It was applied to a case study for the city of Istanbul in Turkey.

2.2. Composite Indicators

The literature of the existing works that proposed to solve the localization problem shows that, until now, no general or systematic method has been introduced to locate the logistics platform, and no methods have been specifically dedicated to the sustainability issue. Indeed, it is quite difficult to evaluate the choice of the platform location with several sustainable criteria. In this respect, the aggregation of these criteria in a composite indicator is an innovative and interesting approach that should be applied to localize logistics platforms. In this section, we present an overview of the composite indicators built, according to the literature, using MCDM methods.

The main procedures of building composite indicators include identifying sustainable criteria, weighting the identified criteria and aggregating these criteria into composite indicators [18–20]. Undoubtedly, the application of the weighting and aggregation methods is a key step in developing the composite indicators [18,19,21]. Controversial issues can arise at any stage of constructing the composite indicators. For this reason, the main challenge faced by the decision-maker is to choose the right weighting and aggregation methods that allow for the constructing of composite indicators [22,23].

In the literature, various weighting and aggregating MCDM methods were introduced to assess sustainability [20,23]. They are rather considered as a means of assisting decision-makers in developing composite indicators [23]. Weighting methods can be applied using several techniques such as AHP, ANP, the strengths, weaknesses, opportunities, threats (SWOT) method, SWARA, the best-worst method (BWM), the method of criterion impact loss (CILOS), the integrated determination of objective criteria weights (IDOCRIW), FUCOM, etc. As instances of aggregation methods, we can cite TOPSIS, VIKOR, DEMATEL, weighted aggregated sum product assessment (WASPAS), MAIRCA, CoCoSo,

PROMETHEE, ELECTRE, etc. Nonetheless, each method has specific characteristics and none of the MCDM methods can be applied to solve all types of problems of localizing infrastructures [24,25]. We present below an overview about the widely used weighting and aggregation methods. Then, we focus on three methods applied in the proposed approach.

2.2.1. Weighting Methods

One of the most important steps of constructing a composite indicator for the location of infrastructure is the weight of each criterion (called also indicator) [26]. More precisely, all the criteria may not have the same level of importance [27]. Thus, the weights can significantly affect the results of the overall composite indicators [28]. In fact, determining the weight of the criteria is one of the key problems that complicate the decision-making process. Weighting methods can be categorized into:

- Equal weighting methods [18];
- Objective data-based methods such as the principal component analysis (PCA) [29] and the data envelopment analysis (DEA) [18];
- Subjective participatory methods where the subjective opinions of experts and/or stakeholders are taken into account, such as the budget allocation (BAL) [19], AHP, FUCOM, etc.

Although there are several weighting methods, the AHP is the most intensively used together with recently-proposed methods such as BMW and FUCOM. Table 2 represents the most cited weighting methods.

Table 2. Analysis of the widely applied weighting methods.

Methods	Characteristic	Simplicity	Comparison
AHP [30]	It defines the relationships between the different levels formed by a framework considered as an objective to be achieved. With AHP, it is almost impossible to make perfectly coherent pairwise comparisons with more than nine criteria.	Very critical	$n (n - 1)/2$
BWM [31]	It is based on a non-linear model used to determine the weights of the decision-making criteria by identifying the most preferable and least preferable criteria for making pairwise comparisons.	Average	$2n - 3$
FUCOM [32]	It allows for calculating weights and comparing criteria in pairs using integer, decimal or predefined scale values for the pairwise comparison of criteria.	Simple	$n - 1$

The FUCOM Method

The full consistency method (FUCOM) was recently developed by Pamucar et al. [32]. Thanks to its high stability, robustness and reliability, this method has quickly been applied in several works [33–43].

In a review done by Stojčić et al., the authors noticed that the AHP weighting method is the most implicated in solving problems related to transport and logistics [44]. However, Pamučar et al. [32] stated that the FUCOM is more consistent and preferable than the AHP method. The application of AHP method is somewhat complicated as it requires $n (n - 1)/2$ pairwise comparisons of the criteria. On the other hand, FUCOM uses far fewer comparisons, which is one of its most important advantages. In addition, it provides the same results as those obtained by the BWM and AHP methods applying an integer or decimal scale. This method allows decision-makers to prioritize criteria utilizing a simple algorithm and applying an acceptable scale. Moreover, it makes it possible to obtain the optimal weight coefficients with the possibility of validating them by showing the consistency of the results.

2.2.2. Aggregation Methods

Aggregation determines the mathematical operation of combining weighted criterion values [45]. In other words, in order to combine the different sustainability criteria, certain aggregation methods are needed [46]. When aggregating, data quality is of primary importance. Methods used to calculate the composite indicators can be classified into the following aggregation techniques [27,28,47]:

- Compensatory technique: It operationalizes the weak sustainability and allows for a high level of substitutability between criteria, which means that a poor performance in a criterion can be compensated by a good performance in another criterion. Otherwise, the weakness of one criterion could be hidden behind the strength of another criterion;
- Partially compensatory technique: This technique operationalizes the limited sustainability. It relies on geometric mean-based methods. In this case, a mutually preferential independence condition of indicators is required, with certain limits;
- Non-compensatory technique: It operationalizes the strong sustainability paradigm that partially or completely prevents the substitutability of criteria. Thus, an unfavorable result of one criterion cannot be compensated for by a favorable result of another criterion.

In the literature, several studies, such as those by the authors of [28,29,48] developed composite indicators with different levels of compensation. Moreover, other authors introduced non-compensatory composite indicators [49,50] and partial composite indicators [21,24,51]. Many studies [28,52] have picked up on this problem by discussing the compensatory issue. In this case, identifying dealt with this problem by discussing the compensatory issue. As such, identifying the appropriate aggregation technique to rank the alternatives is critical. The selection of aggregation techniques is one of the most contestable and scientifically relevant questions in the construction of the composite indicators [49]. After studying different MCDM methods used to aggregate the criteria, a comparative study of some aggregation methods was undertaken and the results are presented in Table 3. After that, the PROMETHEE and MAIRCA methods were chosen for various reasons:

- From a theoretical point of view, these methods use two different aggregation techniques. PROMETHEE and MAIRCA are based on compensatory aggregation and partially compensatory aggregation, respectively. Their objective is to study the impact of two sustainability perspectives: limited sustainability with the partially compensatory technique and weak sustainability with the compensatory technique utilized to choose a sustainable logistics platform;
- From a more practical point of view, these methods are known for their stability and robustness. The PROMETHEE and MAIRCA methods offer consistent solutions that do not change with the variation of the scale of values;
- Finally, because of their popularity [12,34,53–55], they were chosen to locate the logistics platform.

Table 3. Comparison between the MCDM methods.

Methods	Type of Information	Stability	Simplicity	Technical
ELECTRE	Mixed	Medium	Moderately critical	Non-compensatory
PROMETHEE	Mixed	Stable	Moderately critical	Partially compensatory
MAIRCA	Mixed	Stable	Simple	Compensatory
TOPSIS	Quantitative	Medium	Moderately critical	Compensatory
VIKOR	Quantitative	Medium	Medium	Compensatory

In the remainder of this section, we first define the two methods applied in this study. Then, we cite the advantages of each one while showing the reasons behind the choice of these two methods. The objective of the joint use of both techniques is not to compare the two methods or to study the technical difference, but to help decision-makers deal with the main challenges of the sustainability perspectives when building the composite indicators. In other words, it aims to understand the location problem, which ultimately depends on the degree of inter-criteria compensation the decision-maker is willing to accept. The main challenge here is: how can the decision be made if different rankings were produced? In this case, it does not mean that one specific method is better than another. The decision-maker should deeply analyze the impact of sustainability on the location of the logistics platform. The objective of this study is to choose the location from the two-fold point of view of sustainability. The choice of one particular method depends on the type of solution expected by the decision-makers. More generally, the limited sustainability perspective is recommended when certain logistics platform localization projects have a good performance with limited compensatory aggregation that makes the rankings of alternatives robust. Consequently, F-PROMETHEE should be used. Furthermore, compensatory methods, such as F-MAIRCA, may be applied to select a second location if the decision-maker accepts changes in the sustainability strength.

Compensatory Aggregation Method: The MAIRCA Method

Multi Attributive Ideal-Real Comparative Analysis (MAIRCA) was developed in 2014 by the Center for Logistics Research of the University of Defense in Belgrade [56]. The MAIRCA has shown more stability, compared to other popular MCDM methods such as TOPSIS, ELECTRE, MOORA and COPRAS [12]. In fact, it is based on linear aggregation methods, which makes it a compensatory method. The MAIRCA was chosen because of its simple mathematical calculations, the stability of the solution and the possibility of combining this method with others [12]. The MAIRCA model is generally applied to assess the gap between the ideal and empirical assessments. The summation of the gaps in each criterion allows determining the total gap for each alternative. The ranking of the alternatives occurs at the end of the process where the highest ranked alternative is the one with the smallest gap value. The alternative with the lowest total gap value represents the alternative having values closer to the ideal scores (the ideal criteria values) [55].

Partially Compensatory Aggregation Method: The PROMETHEE Method

The PROMETHEE (preference ranking organization method for enrichment evaluations) method was developed to overcome the difficulties encountered in the implementation of the existing prioritization methods by Jean-Pierre Brans in the 1980s [57]. PROMETHEE has some limits in terms of compensation, particularly the absence of indifference and/or preference thresholds [28,54,58]. In this case, we used the usual function, characterized by a limited compensation degree and sustainability [53]. Thus, PROMETHEE is a partially compensatory technique. The PROMETHEE method is one of the most well-known and widely used upgrading methods. It can be extended to multi-decision-maker problems. This method is a potential tool to aid in-conflict resolution in a cooperative problem-solving environment. It is often chosen by decision-makers thanks to its easy implementation. It has the ability to imitate the human mind by making preferences among the alternatives in front of different contradictory criteria. Indeed, this method provides the decision-maker with a comprehensible representation and interpretation of the results.

3. Method

This article proposes composite indicators based on compensatory and partially compensatory multi-criteria decision support methods to identify the location of a logistics platform. The objective of formulating this composite indicator based on individual sustainability criteria and alternatives is to choose the best location by taking into account various sustainability considerations.

The human decision-making process is marked by imprecision and the inherent uncertainty associated with the lack of exact data of the criteria [59]. Indeed, with uncertainty, experts are generally unable to evaluate potential locations and to define the criteria weights, which underline the importance of fuzzy set theory in calculating the composite indicators [60]. Experts' uncertainty can also result when their preferences are often ambiguous and uncertain, which impacts the quality of the data resulting from their observations [8,61]. As decision-makers are unlikely to have full knowledge about all issues related to the choice of the location of a logistics platform, uncertainties need to be taken into account. For this reason, the current paper includes triangular fuzzy numbers given their ability to treat uncertainty about data in the decision-making process and reduce its complexity [61,62]. This form of fuzzy numbers is the most common one due to the simplicity and the rapidity of resolution [60,61].

The aim of this study is not to create a new fuzzy MCDM method, but to propose a methodological approach that allows decision-makers to face the main challenges of choosing the right MCDM method. Thus, it was necessary to study the impact of a compensation phenomenon to make solid decisions and ensure that the best localization alternative was selected. In order to solve this problem, a schematic representation of the suggested approach is presented in Figure 1. It consists of five main phases of calculating the composite indicator.

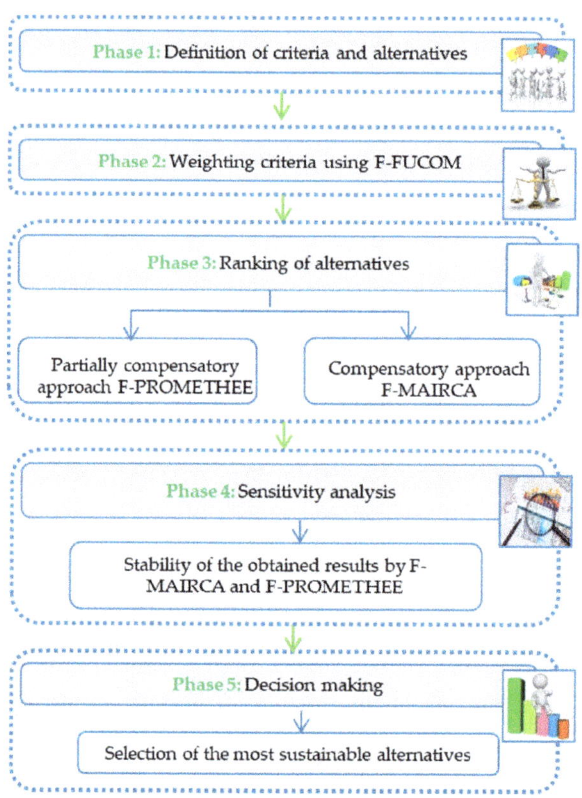

Figure 1. Schematic representation of the proposed approach.

3.1. Phase 1: Definition of Criteria and Alternatives

The first phase consists of identifying the criteria and alternatives. Indeed, the choice of location depends on several sustainability criteria such as social, territorial, economic and environmental criterion, among others. To this regard, it seems important to select the most appropriate criteria.

In the literature, some studies [15,16,63] dealt with the problem of location from the sustainability point of view based on the three traditional criteria, namely, economic, environmental and social. In this context, this paper [15] integrates the economic, environmental and social criteria using 13 sub-criteria. The choice of these criteria was based on the literature and the experts' opinion about their project (City Joint Distribution for Online Shopping) in China. Kumar et Anbanandam [9] examined the choice of the multimodal freight terminal according to five criteria of sustainability: technical, economic, social, environmental and political. Thirteen sub-criteria implied by the authors of [11] were classified according to four criteria (economic, traffic, environmental and government policy). However, Muravev et al. [13] categorized the criteria used to locate the China Railway Express international logistics centers according to three types: social and economic criteria, geographical and infrastructure criteria and transport works criteria. Other researchers, such as the authors of [10,12,14,17], defined the criteria without classifying them according to sustainability criteria.

After analyzing the sustainability criteria discussed in the works presented above, the following section presents the criteria chosen to evaluate the location platform criteria, constructed under the traditional criteria of sustainability (economic, environmental and social) and emerging criteria (political and spatial). A fourth political criterion was introduced, given the interest of political criteria. This criterion represents the awareness and the required impact of local authorities on transport sustainability. A fifth territorial criterion was also suggested in order to choose the location of a logistics platform. Thus, the inclusion of this last criterion and its relationship with the traditional components of sustainability introduces the territorial cohesion perspective of sustainable development, i.e., a spatially equitable, efficient and consistent territory. Table 4 represents the different criteria assessed in this study.

3.1.1. Economic Criteria

Multimodal transport connectivity: Connectivity from the logistics platform to the city center is an important criterion [63]. A logistics platform should connect its location with other transport modes (motorways, rail, seaport, airport, etc.) to facilitate transit. Its proximity to other transport modes helps transport activities operate more efficiently [14].

Cost of land acquisition and construction: This cost must be properly controlled and minimized [64]. The selected location of the logistics platform varies depending on the land price [9], which is an important component of the total cost of the project [14,15]. Obviously, a higher land price will certainly increase the investment costs of building a logistic platform [63]. This cost is an essential component in the selection of the logistics platform [65].

Fiscal policy: Government fiscal support for the development of multimodal transport plays an essential role in stability in policy-making [9]. To attract investors and promote regional development, local and national authorities can offer fiscal advantages to investors [15]. Fiscal policy should be general for the city, imposing no spatial restrictions. In other words, the policy that favors a particular area of the city over other regions influences the chosen location. Fiscal policy should be general without the restriction of certain areas in the city that may influence the chosen location. Besides, governments should facilitate new platform developments via the implementation of new policies and by deregulation [66].

Transport cost: The terminal location must be close to the economic activities and industrial areas, which reduces the cost of short-distance transport and offers a competitive transport cost per ton-kilometer to ensure the easy flow of goods. The right choice of location may minimize the transport costs.

Table 4. The different criteria assessed in this study.

Criteria	[9]	[10]	[11]	[12]	[13]	[14]	[15]	[16]	[17]	[63]	[62]	This Study
						Economic						
Multimodal transport connectivity	*	*		*						*	*	*
Cost of land acquisition and construction	*	*	*		*	*	*	*		*	*	*
Fiscal policy	*								*			*
Transport costs	*		*						*			*
						Environmental						
Conformity with environmental emissions regulations	*			*						*		*
Effect on the natural landscape	*						*		*	*		*
						Social						
Safety and security		*		*	*				*	*	*	*
Noise	*								*			*
Impact on nearby residents							*			*		*
Impact on traffic congestion	*						*	*	*	*		*
						Political						
Current policy	*		*							*		*
Support role for industry	*						*	*				*
						Territorial						
Accessibility to multimodal transport	*	*		*		*		*			*	*
Proximity to the industrial zone	*	*		*		*			*		*	*
Possibility of extending the freight platform	*	*		*			*	*	*		*	*

Abbreviation: * Sustainability criteria used in an existing study.

3.1.2. Environmental Criteria

Conformity with the environmental emissions regulations: The compliance of the logistics platform location with environmental laws and regulations is paramount. The protection of the natural environment and the reduction of urban pollution must also be considered [15]. The selected location should be in line with the spatial structure of the city and land-use planning [63]. Moreover, this platform should reduce the urban air pollution due to the movement of delivery vehicles. However, a bad choice of a location can create several problems of air pollution. This choice can generate transport flows and movement in the wrong place. Such criterion remarkably affects the environment because of its proximity to urban areas. Although the objective of setting up the platform is to encourage multimodality, efforts should nonetheless be directed toward making multimodal transport more sustainable by controlling air pollution and noise [66].

Effect on the natural landscape: The land chosen for the platform must promote harmony with the surrounding landscape. It must also maintain or enhance the original landscape without destroying vegetation and soil [15].

3.1.3. Social/Societal Criteria

Safety and security: the logistics platform has to protect persons against accidents, theft and vandalism [62,63], hence, the importance of the political reform efforts to promote platform security [9].

Noise: The noise generated by the movement of vehicles has a negative impact on the environment. A proposed location should use low-noise equipment to mitigate the noise impact of freight movement on the surrounding area [9]. The noise in a platform and in its surroundings varies from one location to another, where special attention should be paid to reduce noise pollution.

Impact on nearby residents: The choice of location must integrate the social environment. It should not only reduce disruption to urban life, but also relieve pressure on urban congestion and promote healthy development [63]. A location should also be far from densely-populated places to prevent accident occurrence [14].

Impact on traffic congestion: It is important to anticipate the influence of the selected infrastructure on traffic. For this reason, choosing the wrong location of the logistics platform can deteriorate local traffic conditions. To ensure the proper functioning of the logistics platforms, the surrounding traffic environment must be organized [63].

3.1.4. Political Criteria

Current policy: Current policy is included as it is an essential requisite to locate a logistics platform for a solid and sustainable base. A stable government system develops coherent policies for the development of a multimodal system in the whole nation. In fact, political stability plays a crucial role in providing the stability of policy-making [9]. The impact of localization on the city should be examined as part of the current transport policy. In the case where current policy criterion give particular importance to the specific location, regardless of the whole city, this criterion depends on the location compared to another one. Moreover, strong and stable policy is one of the important factors that needs to be considered in locating a logistics platform [66].

Support role for industry: The local government should establish appropriate policies to promote the development of its industry in logistics centers by taking sustainability into account [8,14,15]. Indeed, giving more support to one location by authorities, without considering the stakeholders' views, can influence the choice of location. A fundamental part of sustainable decision-making at the policy level concerns the degree of participation in the locating of a logistics platform [67]. More precisely, the support and the inclusion of relevant stakeholder groups (such as shippers, logistics service providers, receivers, etc.) is important for choosing the best location. Both coordination and cooperation between the government and industry are indispensable to avoid conflicts of interest in selecting the appropriate location [15,66].

3.1.5. Territorial Criteria

Accessibility to multimodal transport: Reduced accessibility weighs on the development of economic centers. Therefore, the location of the logistics platform should be connected to all modes of transport [15,16,63,68].

Proximity to the industrial zone: Proximity to the freight market, to the production area and to freight shippers is considered as the most critical criterion that, in turn, reduces the costs of transportation. A platform should be at the service of companies operating in different sectors. Therefore, an infrastructure location should be close to the industrial area [14,68].

Possibility of extending the freight platform: The availability of suitable land for the development of infrastructure is essential. It requires more land to ensure the possibility of extending the freight platform. The platform size should increase by building a new container yard, warehouses, a parking lot, etc. [62,68].

The sustainability criteria chosen to evaluate the location of the logistics platform are shown in Table 5. They can be classified into cost (C) or benefit (B) criteria. More precisely, cost criteria should be minimized, i.e., the lower the value of the criterion is, the better the alternative will be. However, profit criteria must be maximized, that is to say, the higher the criterion value is, the better the alternative will be.

Table 5. Evaluation criteria and sub-criteria.

Unit	Criteria	Definition	Type
C_1	Economic		
$C_{1.1}$	Multimodal transport connectivity	Connectivity of the location to other modes of transport, e.g., highways, railways, seaport, airport, etc.	Benefit
$C_{1.2}$	Cost of land acquisition and construction	The location of a logistics platform to be selected depends on these costs, which must be properly controlled and minimized.	Cost
$C_{1.3}$	Fiscal policy	The fiscal advantages offered by the authorities to attract investors and promote the development of transport.	Cost
$C_{1.4}$	Transport cost	The location should be close to the source of freight to reduce the cost of transportation.	Cost
C_2	Environmental		
$C_{2.1}$	Conformity with environmental emissions regulations	Choosing the right location can reduce the impact of air pollution on human health and the environment.	Benefit
$C_{2.2}$	Effect on the natural landscape	To promote harmony with the surrounding landscape without destroying the original landscape.	Cost
C_3	Social		
$C_{3.1}$	Safety and security	The platform is protected against accidents, theft and vandalism.	Benefit
$C_{3.2}$	Noise	The noise generated by the movement of vehicles has a negative impact on environments.	Cost
$C_{3.3}$	Impact on nearby residents	A location should promote healthy development for urban residents.	Cost
$C_{3.4}$	Impact on traffic congestion	Traffic environment planning to relieve pressure on urban congestion.	Cost
C_4	Political		
$C_{4.1}$	Current policy	Political stability plays a crucial role in the stability of the development of a multimodal system.	Benefit
$C_{4.2}$	Support role for industry	The local government should establish appropriate policies to promote the development of its industry in platforms.	Benefit
C_5	Territorial		
$C_{5.1}$	Accessibility to multimodal transport	A location should be connected and accessible to all modes of transport.	Benefit
$C_{5.2}$	Proximity to the industrial zone	A platform should be at the service of companies operating at different sectors.	Benefit
$C_{5.3}$	Possibility of extending the freight platform	The infrastructure must have the capacity to increase the size of the platform, to meet the growing freight demands.	Benefit

3.2. Phase 2: Weighting of Criteria Using FUCOM

The second phase was necessary to calculate the weight of the criteria and sub-criteria using the fuzzy FUCOM method (F-FUCOM). In the following section, we present the F-FUCOM algorithm whose application consists in the following five steps.

Step 1: Ranking the criteria according to their level of importance. In the first step, the ranking was evaluated according to the importance of the evaluation criteria $C = \{C_1, C_2, \ldots, C_n\}$. This means the first rank was assigned to the criterion that was expected to have the highest weight coefficient. The last place was occupied by the criterion for which we expected the lowest value of the weight coefficient. The criteria classified according to the expected values of weight coefficients are presented as follows:

$$C_{j(1)} \succ C_{j(2)} \succ \ldots \succ C_{j(n)} \tag{1}$$

where $C_{j(1)}$ and $C_{j(n)}$ respectively stand for the most and the least important criterion among the predefined set of n elements. Notably, if two or more criteria have equal importance, the equality sign "=" is placed between the criteria instead of ">".

Step 2: Determining the comparative priority of the criteria. The mutual comparison of the criteria was carried out using fuzzy linguistic expressions from the defined scale. Mutual comparisons were made by each expert individually {E1, E2, ... , Et} according to his/her preferences. The comparison was performed using the first-ranked (most important) criterion. Thus, for each expert, the fuzzy criteria significance was obtained for all the criteria ranked in Step 1. Based on the defined significances of the criteria and using Equation (2), the fuzzy comparative significance $\varphi_{j/(j+1)}$ was determined.

$$\varphi_{j/(j+1)} = \frac{\varpi_{Cj}}{\varpi_{C(j+1)}} = \left(\frac{w_j^l}{w_{j+1}^u}, \frac{w_j^m}{w_{j+1}^m}, \frac{w_j^u}{w_{j+1}^l} \right) \tag{2}$$

The fuzzy vectors of the comparative significance of the criteria for each individual expert {E1, E2, ... , Et} were obtained applying Equation (3), where $e = (1, 2, \ldots, t)$.

$$\Phi^e = \left(\varphi_{1/2}, \varphi_{2/3}, \ldots, \varphi_{(n-1)/n} \right) \tag{3}$$

Step 3: Defining the limits of the fuzzy model. The final values of weight coefficients should satisfy two constraints:

- **Constraint 1**: the ratio of the weights of the criteria should be the same as their comparative signification between the observed criteria.

$$\varphi_{j/(j+1)} = \frac{w_j}{w_{j+1}} \tag{4}$$

- **Constraint 2**: The final values of weight coefficients should satisfy the transitivity condition, respectively $\varphi_{\frac{j}{j+1}} \otimes \varphi_{\frac{j+1}{j+2}} = \varphi_{\frac{j}{j+2}}$, i.e., $\frac{w_j}{w_{j+1}} \otimes \frac{w_{j+1}}{w_{j+2}} = \frac{w_j}{w_{j+2}}$. This second condition must fulfill the final values of weight coefficients.

$$\frac{w_j}{w_{j+2}} = \varphi_{j/(j+1)} \otimes \varphi_{(j+1)/(j+2)} \tag{5}$$

Step 4: Designing a fuzzy model to calculate the optimal values of the weights of the criteria. In this step, the final values of the fuzzy weight coefficients of the criteria were computed for each expert: $(w_1, w_2, \ldots, w_n)^e$. The conditions defined in the third step should be met with the minimal deviation from the maximal consistency. In other words, the conditions must be met $\frac{w_j}{w_{j+1}} - \varphi_{j/(j+1)} = 0$ and $\frac{w_j}{w_{j+2}} - \varphi_{j/(j+1)} \otimes \varphi_{(j+1)/(j+2)} = 0$. Under these conditions, the maximum coherence amounts to $\chi = 0$.

In order to satisfy the previously presented conditions, the non-linear model was applied to determine the optimal fuzzy values of the weight coefficients for the evaluation criteria, which can be formulated as follows:

$$\begin{gathered}\min \chi \\ \text{s.t.} \\ \left\{\begin{array}{c} \left|\frac{w_j}{w_{j+1}} - \varphi_{j/(j+1)}\right| \leq \chi,\ \forall j \\ \left|\frac{w_j}{w_{j+2}} - \varphi_{j/(j+1)} \otimes \varphi_{(j+1)/(j+2)}\right| \leq \chi,\ \forall j \\ \sum_{j=1}^{n} w_j = 1,\ \forall j \\ w_j^l \leq w_j^m \leq w_j^u \\ w_j^l \geq 0,\ \forall j \\ j = 1, 2, \ldots, n \end{array}\right.\end{gathered} \tag{6}$$

where $w_j = \left(w_j^l, w_j^m, w_j^u\right)$ and $\varphi_{j/(j+1)} = \left(\varphi^l_{j/(j+1)}, \varphi^m_{j/(j+1)}, \varphi^u_{j/(j+1)}\right)$

Considering that the maximum consistency requires fulfilling the condition where $\frac{w_j}{w_{j+1}} - \varphi_{j/(j+1)} = 0$ and $\frac{w_j}{w_{j+2}} - \varphi_{j/(j+1)} \otimes \varphi_{(j+1)/(j+2)} = 0$, model (6) can be transformed into a fuzzy linear model (7) and, by solving it, the optimal fuzzy values of the weight coefficients were obtained.

$$\begin{gathered}\min \chi \\ \text{s.t.} \\ \left\{\begin{array}{c} \left|w_j - w_{j+1} \otimes -\varphi_{j/(j+1)}\right| \leq \chi,\ \forall j \\ \left|w_j - w_{j+2} \otimes \varphi_{j/(j+1)} \otimes \varphi_{(j+1)/(j+2)}\right| \leq \chi,\ \forall j \\ \sum_{j=1}^{n} w_j = 1,\ \forall j \\ w_j^l \leq w_j^m \leq w_j^u \\ w_j^l \geq 0,\ \forall j \\ j = 1, 2, \ldots, n \end{array}\right.\end{gathered} \tag{7}$$

where $w_j = \left(w_j^l, w_j^m, w_j^u\right)$ and $\varphi_{j/(j+1)} = \left(\varphi^l_{j/(j+1)}, \varphi^m_{j/(j+1)}, \varphi^u_{j/(j+1)}\right)$

Step 5: Calculating the final optimal values of the criteria weights. Solving model (6) or (7) allowed us to obtain the weight coefficients of criteria by the experts.

3.3. Phase 3: Ranking of Alternatives

The third phase aims at ranking the different alternatives. We used fuzzy MAIRCA (F-MAIRCA) and fuzzy PROMETHEE (F-PROMETHEE) to evaluate the performance of the location considering a compensatory and partially compensatory reasoning, respectively. The PROMETHEE method is based on the partially compensatory aggregation technique (an unfavorable result of one criterion can be limited, compensated by a favorable result of another). However, the MAIRCA method relies on the compensatory aggregation technique (the weakness of one criterion could be hidden behind the strength of another).

3.3.1. Compensatory Approach: The F-MAIRCA Method

Step 1: Building the initial decision matrix , $\widetilde{D}^t$. On the basis of the linguistic evaluation of alternatives with respect to the considered criteria, the matrix $\widetilde{D}^t = \left[\widetilde{A}_{ij}^{(t)}\right] \forall\, i = 1,2,\ldots,m$ and $j = 1,2,\ldots,n$ was constructed as shown in Equation (8).

$$\widetilde{D}^t = \begin{bmatrix} \widetilde{A}_{11}^{(t)} & \cdots & \widetilde{A}_{1n}^{(t)} \\ \vdots & \ddots & \vdots \\ \widetilde{A}_{m1}^{(t)} & \cdots & \widetilde{A}_{mn}^{(t)} \end{bmatrix} \quad (8)$$

where $\widetilde{A}_{mn}^{(t)}$ demonstrates that the m-th alternative was evaluated linguistically with respect to the n-th criterion by the t-th expert.

Step 2: Building the fuzzy aggregate decision matrix. The fuzzy aggregated decision matrix was constructed using the arithmetic operator, as represented in Equation (9).

$$\widetilde{D} = \begin{bmatrix} \widetilde{A}_{11} & \cdots & \widetilde{A}_{1n} \\ \vdots & \ddots & \vdots \\ \widetilde{A}_{m1} & \cdots & \widetilde{A}_{mn} \end{bmatrix} \text{ where } \widetilde{A}_{11} = \frac{\widetilde{A}_{11}^{(1)} + \widetilde{A}_{11}^{(2)} + \cdots + \widetilde{A}_{11}^{(t)}}{t} \quad (9)$$

Step 3: Determining the preferences of the alternatives. In the third step, we defined preferences in relation to the selection of alternatives. The decision-maker was neutral, i.e., he/she had no preference towards any of the suggested alternatives. Since there was an equal probability between the alternatives, the preferences for each of them can be defined as follows:

$$P_{Ai} = \frac{1}{m}; \quad \sum_{i=1}^{m} P_{Ai} = 1;\ i = 1,2,\ldots,m \quad (10)$$

where m is the number of alternatives.

Step 4: Determining the fuzzy matrix of the theoretical ponder , $\widetilde{T}_{PA}$. This matrix was obtained by multiplying the preferences of alternatives and fuzzy criteria weights provided by FUCOM. The fuzzy matrix of theoretical ponder was calculated as demonstrated in Equation (11).

$$\widetilde{T}_{PA} = \begin{bmatrix} P_{Ai}\widetilde{w}_1 & \cdots & P_{Ai}\widetilde{w}_n \\ \vdots & \ddots & \vdots \\ P_{Ai}\widetilde{w}_1 & \cdots & P_{Ai}\widetilde{w}_n \end{bmatrix} = \begin{bmatrix} \widetilde{t}_{P_{11}} & \cdots & \widetilde{t}_{P_{1n}} \\ \vdots & \ddots & \vdots \\ \widetilde{t}_{P_{m1}} & \cdots & \widetilde{t}_{P_{mn}} \end{bmatrix} \quad (11)$$

Step 5: Constructing the normalized fuzzy aggregated decision matrix. This step consisted of normalizing the fuzzy aggregated decision matrix, as defined in Step 2. This matrix can be expressed as follows: $X = [X_{ij}]_{m,n} \forall i = 1,2,\ldots,m$ and $j = 1,2,\ldots,n$ [69]. If C and B represent the cost and benefit criteria, respectively, then the normalization procedure is as presented in Equations (12) and (13).

$$X_{ij} = \left(\frac{a_j^{L-}}{a_{ij}^N}, \frac{a_j^{L-}}{a_{ij}^M}, \frac{a_j^{L-}}{a_{ij}^L}\right),\ j \in C,\ a_j^{L-} = \min_i a_{ij}^L \quad (12)$$

$$X_{ij} = \left(\frac{a_{ij}^L}{a_j^{N+}}, \frac{a_{ij}^M}{a_j^{N+}}, \frac{a_{ij}^N}{a_j^{N+}}\right),\ j \in B,\ a_j^{N+} = \max_i a_{ij}^N \quad (13)$$

Step 6: Calculating the matrix of fuzzy actual ponder , $\widetilde{T}_{RA}$. This matrix was obtained by multiplying the fuzzy normalized decision matrix, as shown in Step 5, and the fuzzy of theoretical ponders, as revealed in Step 4.

$$\widetilde{T}_{R_A} = \begin{bmatrix} \widetilde{t}_{R_{11}} & \cdots & \widetilde{t}_{R_{1n}} \\ \vdots & \ddots & \vdots \\ \widetilde{t}_{R_{m1}} & \cdots & \widetilde{t}_{R_{mn}} \end{bmatrix} = \begin{bmatrix} \widetilde{n}_{11} \times \widetilde{t}_{P_{11}} & \cdots & \widetilde{n}_{1n} \times \widetilde{t}_{P_{1n}} \\ \vdots & \ddots & \vdots \\ \widetilde{n}_{m1} \times \widetilde{t}_{P_{m1}} & \cdots & \widetilde{n}_{mn} \times \widetilde{t}_{P_{mn}} \end{bmatrix} \quad (14)$$

Step 7: Calculating the total gap matrix. In this step, the Euclidian distance between the matrix of fuzzy theoretical ponder and the actual ponder of each alternative with respect to each criterion was calculated. In other works [13,70,71], the authors suggested subtracting T_{PA} and obtaining the total gap matrix G. It is preferable that the highest rank alternative should have a minimum gap value from each criterion. The calculation was done as shown in Equation (15).

$$g_{ij} = \sqrt{\frac{1}{3}\left[\left(t^{l}_{P_{ij}} - t^{l}_{R_{ij}}\right)^2 + \left(t^{m}_{P_{ij}} - t^{m}_{R_{ij}}\right)^2 + \left(t^{u}_{P_{ij}} - t^{u}_{R_{ij}}\right)^2\right]} \quad (15)$$

Step 8: Summating the gap values and ranking the alternatives. The gap values g_{ij} were summed for each alternative with respect to each criterion by using Equation (16). The final values were then arranged in ascending order, and finally, the preferences were ranked.

$$Q_i = \sum_{j=1}^{n} g_{ij}, \quad i = 1, 2, \ldots, m \quad (16)$$

3.3.2. Partially Compensatory Approach: The F-PROMETHEE Method

Step 1: Creating a decision matrix. Each decision-maker evaluated n criteria according to m alternatives (or actions). Where j = 1,2, ... ,n and i = 1,2, ... ,m.

Step 2: Constructing the standard fuzzy performance matrix. The normalization of the comparison values of the decision matrix was obtained by deploying the expressions given in Equations (12) and (13) to construct a normalized fuzzy performance matrix.

Step 3: Determining the preferred function. The preference function between the two alternatives (a) and (b) was defined by the indifference and preference thresholds. The latter, denoted (p), is the lowest value of $d_j(a,b)$, below which there is indifference between selecting (a or b). The indifference threshold (q) is the lowest of $d_j(a,b)$, and there is strict preference of (a over b). Six different types of preference functions were proposed, namely, Usual (Type I), quasi-criterion (Type II), linear preference criterion (Type III), level criterion (Type IV), linear preference criterion and zone of indifference (Type V), as well as the Gaussian criterion (Type VI) [57]. A preference function was defined for each criterion using Equation (17), where (a) and (b) are two alternatives of the set of (m) alternatives.

$$d_j(a,b) = l_{aj} - u_{bj}; m_{aj} - m_{bj}; u_{aj} - l_{bj} \quad (17)$$

It is preferable that experts define the preference function. However, this step makes the application of the method more cumbersome and complex. To deal with this limitation, we used the first assumption of the strict preference function because it is the most widely used in the literature [69,72,73]. The choice of the usual preference function is based on the need to use an assessment model that can be easily understood by decision-makers. On the other hand, this function does not include any threshold values, which is typically a complex and time-consuming exercise for decision-makers [53]. Regardless of the difference between the alternatives, with this strict preference function, the best evaluated alternative is always the most preferred. However, the effect of the types of preference functions of PROMETHEE on the final preferences is not known. The choice of the usual function could be a limitation of this study. The usual function is without an indifference threshold. With this function, $F_j[d_j(a,b)] = 0$ when $d_j(a,b) \leq 0$ and $F_j[d_j(a,b)] = 1$ when $d_j(a,b) > 0$.

Step 4: Calculating the preference index for each criterion. The preference index, $P_j(a,b)$, which denotes the preference of (a) over (b) for criterion j, was calculated as a function of $d_j(a,b)$ by applying Equation (18). The value of $P_j(a,b)$ varies from 0 to 1 such that [47]:

- $P(a,b) = 0$ means an indifference between (a) and (b) or no preference of (a) over (b);
- $P(a,b) \sim 0$ means a weak preference of (a) over (b);
- $P(a,b) \sim 1$ means a strong preference of (a) over (b);
- $P(a,b) = 1$ means a strict preference of (a) over (b).

$$P_j(a,b) = \left(F_j(l_{aj} - u_{bj}\right); F_j(m_{aj} - m_{bj});\ F_j\left(u_{aj} - l_{bj}\right))P_j(a,b) = (l^j_{ab}, m^j_{ab}, u^j_{ab}) \tag{18}$$

Step 5: Calculating the overall preference index. A multi-criteria degree of preference was then calculated to globally compare each pair of shares. The global preference index represents the degree of preference of alternatives (a) over (b), considering all the criteria simultaneously [57]. The fuzzy global preference index π (a,b) was calculated by applying Equation (19) with the assumption that the relative weights of the criteria were also triangular fuzzy numbers, $W_j = (l''_j, m''_j, u''_j)$.

$$\pi(a,b) = \sum_{j=1}^{n} w_j \otimes P_j(a,b)\,\pi(a,b) = \sum \left(l''_j, m''_j, u''_j\right) \otimes \left(l^j_{ab}, m^j_{ab}, u^j_{ab}\right)\pi(a,b) = (l^{\pi}_{ab}, m^{\pi}_{ab}, u^{\pi}_{ab}) \tag{19}$$

where $\pi(a,b) \approx 0$ denotes a low preference of (a) over (b) and $\pi(a,b) \approx 1$ denotes a strong preference of (a) over (b).

Step 6: Determining the positive outranking flow $(\Phi+)$ and negative outranking flow $(\Phi-)$. The outgoing and incoming flows for each alternative were calculated using the respective Equations (20) and (21) [74]. The leaving flow represents the dominance of one action (a) over the other actions. Thus, the entering flow represents the weakness of an action (a) compared to other actions.

$$\Phi^+(a) = \frac{1}{n-1}\sum \pi(a,x),\ x \in (i = 1\ldots m) \tag{20}$$

$$\Phi^-(a) = \frac{1}{n-1}\sum \pi(x,a),\ x \in (i = 1\ldots m) \tag{21}$$

Step 7: Determining the partially priorities using PROMETHEE I. The partial priorities include three possible outcomes: the preference of one decision point over another, indifference between the decision points and inability to compare the decision points with each other.

$$\begin{cases} aPb & \begin{cases} \Phi^+(a) > \Phi^+(b)\text{and}\Phi^-(a) < \Phi^-(b)\text{or} \\ \Phi^+(a) = \Phi^+(b)\text{and}\Phi^-(a) < \Phi^-(b)\text{or} \\ \Phi^+(a) > \Phi^+(B)\text{and}\Phi^-(a) = \Phi^-(b) \end{cases} \\ aIb & \{\Phi^+(a) = \Phi^+(b)\text{and}\Phi^-(a) = \Phi^-(b) \\ aRb & \begin{cases} \Phi^+(a) > \Phi^+(b)\text{and}\Phi^-(a) > \Phi^-(b)\text{or} \\ \Phi^+(a) < \Phi^+(b)\text{and}\Phi^-(a) < \Phi^-(b) \end{cases} \end{cases} \tag{22}$$

where (P, I, R) respectively denote preference, indifference and incomparability in PROMETHEE.

Step 8: Determining all priorities applying PROMETHEE II. The PROMETHEE II method provides a total pre-order (excludes incomparability and considerably reduces indifference). The net flow is the subtraction of $\Phi^-(a)$ and $\Phi^+(a)$ (Equation (23)). At this step, all the alternatives become comparable, as no incomparability remains. They could thus be classified by full ranking them. The best alternative is the one with the highest net flow. At the end of the calculations, the fuzzy net flow values obtained could be defuzzified to facilitate the comparisons of alternatives.

$$\Phi(a) = \Phi^{+}(a) - \Phi^{-}(a)\begin{cases} aPb \text{ if } \Phi(a) > \Phi(b) \\ aIb \text{ if } \Phi(a) = \Phi(b) \end{cases} \quad (23)$$

3.4. Phase 4: Sensitivity Analysis

In the fourth phase, a sensitivity analysis was performed to rigorously test the robustness and the feasibility of the proposed approach. This analysis allows the decision-maker to check whether the final ranking is sensitive, and depends on the approach adopted to aggregate the scores of the criteria.

3.4.1. Assessment of the Independence of the Aggregation Technique

The objective of this analysis is to evaluate the stability of a solution by changing the applied aggregation technique. Three different methods (F-PROMETHEE, F-MAIRCA and F-TOPSIS) based on compensation techniques were used. Moreover, three sustainability perspectives were considered: limited sustainability with the use of the partial compensation technique, weak sustainability with the use of the compensation technique and strong sustainability with the use of the non-compensation technique. The authors of [21] mentioned that the use of the three aggregation techniques is strongly recommended to adequately assess the obtained results. The type of sustainability is directly related to the aggregation approach and the level of compensability. For this reason, F-TOPSIS was applied using the non-compensation technique (an unfavorable result of one criterion cannot be compensated by a favorable result of another) (Garcia-Bernabeu et al.'s method [21]).

3.4.2. Variation of Criteria Weights

Sensitivity analysis is generally performed to assess the influence of the weights assigned by experts on the ranking of alternatives. The experts' preferences are different. In fact, some give more importance to one criterion over another. In addition, the same criterion may be given different weights by the same expert in different situations. Therefore, the authors of [70] suggested checking the final ranking of the alternatives with a small variation of criteria weights. The sensitivity analysis was carried out in the experiments through 90 scenarios divided into three phases. In phase 1, the weight coefficients of the criteria in 30 scenarios increased or decreased by 45%. In each of the 30 scenarios, one weight coefficient increased by 45%, and it was favored. In the same scenario, the weight coefficients of the remaining criteria decreased by 45%. In phase 2 a similar procedure was applied in the next 30 scenarios, with weight coefficients that increased or decreased by 65%. Finally, in phase 3, the weight coefficients increased or decreased by 85%.

3.5. Phase 5: Decision-Making Process

In the last phase, the potential alternatives were ranked to select the most sustainable location of the logistics platform. In this phase, decision-makers can also identify the impact of criteria on the selection of the platform.

4. Results

4.1. Problem Definition and Alternatives Selection

In the following section, the proposed approach, applied to locate the logistics platform, is described taking sustainability into account. This study was carried out in the city of Sfax. In Tunisia, road transport is the main means of the routing of goods. According to the Ministry of Transport, it concentrates 85% of overland flows of goods. More precisely, freight transport is at the heart of economic and social development. Despite its preponderant place, freight transport generates various problems, especially in cities where the flow of distribution is important. In this study, we chose the city of Sfax because it is known by its significant economic dynamism and rich industrial fabric. In fact, it is the second largest Tunisian city after the capital Tunis, in terms of both its demographic weight and industrial activities. It is located in the center-east of the Republic of Tunisia. Its privileged geographical position, its wide opening to the sea and its port make it a national and international commercial and trading center. In other words, it constitutes a natural corridor for the transport of goods. However, it includes a set of elements and issues that directly or indirectly influence the efficiency of the transport system. The city of Sfax, despite its economic and industrial dynamism, has suffered for decades from serious problems in the transport and circulation system, which affect sustainable development. During peak hours (morning, noon and evening) on the eve of the holidays, the difficulty of transport becomes even more difficult. These congestion and traffic problems are due to the spatial concentration of economic activities and administrative services. They are aggravated by the increase of heavy goods vehicles traffic and the architecture of the road infrastructure. Moreover, there is a strong embarrassment occasioned by flows freight transport that have an origin or destination at the commercial port or in the industrial zones, which are predominantly located in the coastal wings.

For years, this city has suffered greatly from this system, which is unsuited to the country's economic, demographic and urban growth. Further, transport and traffic conditions in Sfax have deteriorated significantly in recent years. Thus, with the intensification of the national and international containerized exchanges, the implementation of a multimodal infrastructure can contribute to the sustainable development of the city and remedy the problems mentioned below. As a solution, a better location of the logistics platform would ensure the interconnection of the available transport infrastructure to improve its performance. Moreover, this infrastructure plays a fundamental role in the development of the regional and extra-regional economic activities by affecting both the prices of production and the sale of goods.

The location of the logistics platform in the city of Sfax has been given great interest for several years. However, the identification of locations depends on the preferences of the decision-makers and the conditions of freight transport. After interviewing several decision-makers, seven potential locations were considered. These locations were defined as alternatives that meet the interests of all the stakeholders in the city. The different alternatives, chosen in Sfax, are as follows (Figure 2): (A1) GARGROUR, (A2) NAKATA, (A3): ELGONNA, (A4) ELHAJEB1, (A5) ELHAJEB2, (A6) LA SIAPE and (A7) SKHIRA.

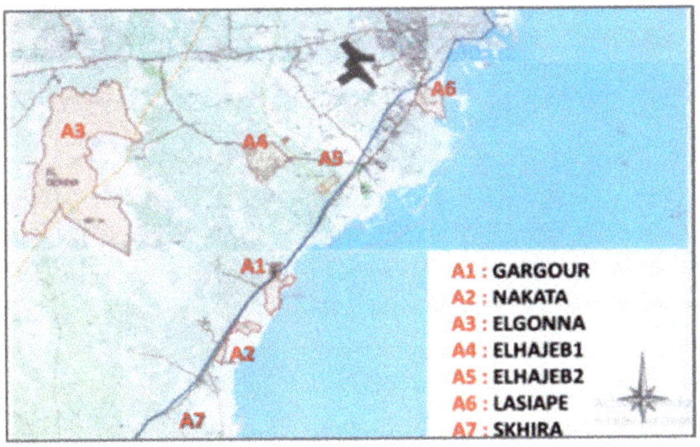

Figure 2. Set of alternatives for logistics platform location in Sfax.

4.2. Weighting of Criteria

4.2.1. Obtaining Linguistic Judgments

In this step, we interviewed seven experts {E1, E2 ... E7} involved in the decision-making process. These experts all had at least 10 years of experience. Table 6 shows the characteristics of the consulted experts. They participated in the process of determining the weights for the criteria and evaluating the alternatives. A given 5-point scale was used in the survey to establish the experts' fuzzy language ratings. Data were collected in the city of Sfax between July and August 2020.

Table 6. Characteristics of the experts.

	Age	Level of Education	Experience	Profession	Organization Name
E1	55 years	PhD degree in electrical engineering	25 years	Regional Director of Transport in Sfax	Ministry of transport
E2	35 years	Industrial engineer	10 years	Logistics Director	SOCOMENIN
E3	49 years	Civil engineer	23 years	Port Director	Office merchant marine and Ports of Sfax
E4	43 years	Master's degree in international trade	18 years	Logistics Director, contractual teacher	Pastry MASMOUDI
E5	58 years	PhD in urban planning	28 years	Municipal civil servant	Municipality of Sfax
E6	49 years	Master in business strategy	23 years	Port Technical Director, temporary teacher	Office merchant marine and Ports of Sfax
E7	36 years	Master's degree in logistics	11 years	Administrator	Governorate of Sfax

4.2.2. The F-FUCOM Results

The C1–C5 criteria constitute the first hierarchical level, while the second hierarchical level consists of the sub-criteria, classified into the five criteria presented in Table 5. Using fuzzy FUCOM, the values of the local weights of the sub-criteria were calculated. After defining the local weights of the sub-criteria, the weights of the criteria were multiplied by the group of weights of the sub-criteria. We obtained the global values, used subsequently in this analysis, to evaluate alternatives with MAIRCA and PROMETHEE. In order to solve this problem, six fuzzy FUCOM models were defined:

- Model 1: Calculation of the values of the weight coefficients of the criteria C1, C2, C3, C4 and C5;
- Model 2: Calculation of the local values of the weight coefficients of the sub-criteria C1.1, C1.2, C1.3 and C1.4;
- Model 3: Calculation of the local values of the weight coefficients of the sub-criteria C2.1 and C2.2;
- Model 4: Calculation of the local values of the weight coefficients of the sub-criteria C3.1, C3.2; C3.3 and C3.4;
- Model 5: Calculation of the local values of the weight coefficients of the sub-criteria C4.1 and C4.2;
- Model 6: Calculation of the local values of the weight coefficients of the sub-criteria C5.1, C5.2 and C5.3.

Step 1: The ranking of the criteria and the sub-criteria was carried out according to the preferences of the experts (E1–E7). The defined ranks are shown in Table 7 using Expression (1).

Step 2: The comparative significance of the criteria and the sub-criteria was determined as shown in Table 7 using Expression (2) (as defined below). This significance was measured by applying triangular fuzzy numbers. At this stage, fuzzy comparisons by pairs of criteria were made according to the preferences of the experts via five linguistic terms, as shown in Table 8 [75].

Table 7. Linguistic evaluations of the criteria and the sub-criteria.

		C1–C5	C1.1–C1.4	C2.1–C2.2	C3.1–C3.4	C4.1–C4.2	C5.1–C5.3
E1	R	C4 > C3 = C2 > C1 > C5	C1.1 > C1.3 > C1.4 > C1.2	C2.1 > C2.2	C3.1 > C3.4 > C3.2 > C3.3	C4.2 > C4.1	C5.1 > C5.2 > C5.3
	C	EI, AI, EI, VI, FI	EI, AI, FI, VI	EI, VI	EI, VI, EI, EI	EI, VI	EI, WI, AI
E2	R	C1 > C5 > C3 > C2 > C4	C1.4 > C1.1 > C1.2 > C1.3	C2.1 > C2.2	C3.4 > C3.1 > C3.3 > C3.2	C4.2 > C4.1	C5.2 > C5.1 > C5.3
	C	EI, AI, FI, WI, WI	EI, FI, FI, WI	EI, WI	EI, AI, FI, WI	EI, WI	EI, VI, FI
E3	R	C2 > C1 > C3 > C4 > C5	C1.1 > C1.3 > C1.4 > C1.2	C2.1 > C2.2	C3.1 > C3.3 > C3.4 > C3.2	C4.2 > C4.1	C5.1 > C5.3 > C5.2
	C	EI, EI, WI, FI, VI	EI, WI, EI, FI	EI, WI	EI, WI, WI, FI	EI, WI	EI, EI, WI
E4	R	C1 > C5 > C4 > C3 > C2	C1.1 > C1.4 > C1.2 > C1.3	C2.2 > C2.1	C3.1 > C3.3 > C3.2 > C3.4	C4.2 > C4.1	C5.1 > C5.2 = C5.3
	C	EI, AI, VI, EI, VI	EI, AI, VI, WI	EI, EI	EI, AI, EI, WI	EI, AI	EI, AI, FI
E5	R	C1 > C4 > C5 > C2 = C3	C1.1 > C1.4 > C1.2 > C1.3	C2.1 > C2.2	C3.1 > C3.4 > C3.2 > C3.3	C4.2 > C4.1	C5.1 > C5.2 > C5.3
	C	EI, VI, VI, WI, EI	EI, FI, VI, AI	EI, WI	EI, VI, WI, VI	EI, WI	EI, FI, AI
E6	R	C4 > C1 > C3 > C2 > C5	C1.1 > C1.4 > C1.3 > C1.2	C2.1 > C2.2	C3.1 > C3.4 > C3.2 > C3.3	C4.1 > C4.2	C5.1 > C5.3 > C5.2
	C	EI, WI, FI, WI, FI	EI, WI, EI, WI	EI, FI	EI, WI, FI, EI	EI, WI	EI, FI, WI
E7	R	C1 > C2 > C5 > C3 > C4	C1.3 > C1.2 > C1.1 > C1.4	C2.1 > C2.2	C3.1 > C3.3 > C3.4 > C3.2	C4.1 > C4.2	C5.3 > C5.2 > C5.1
	C	EI, VI, WI, WI, WI	EI, FI, FI, WI	EI, WI	EI, VI, WI, WI	EI, VI	EI, WI, EI

R: Rank; C: Comparisons.

Table 8. Fuzzy language scale for criteria assessment [75].

Linguistic Terms	Abbreviation	Fuzzy Number in a Triangular Style
Equally important	(EI)	(1, 1, 1)
Weakly important	(WI)	(2/3, 1, 3/2)
Fairly Important	(FI)	(3/2, 2, 5/2)
Very important	(VI)	(5/2, 3, 7/2)
Absolutely important	(AI)	(7/2, 4, 9/2)

The vectors of the comparative significance were defined by calculating the comparative significance of the criteria and the sub-criteria (Appendix A). For example, the comparative significance of the criteria C1–C5, for expert E1, was obtained by applying Expression (2) as follows:

- $\varphi_{C4/C3} = \frac{\varpi_{C4}}{\varpi_{C3}} = \frac{AI}{EI} = (3.5; 4; 4.5)$;
- $\varphi_{C3/C2} = \frac{\varpi_{C3}}{\varpi_{C2}} = \frac{EI}{AI} = (0.22; 0.25; 0.29)$;
- $\varphi_{C2/C1} = \frac{\varpi_{C2}}{\varpi_{C1}} = \frac{VI}{EI} = (2.5; 3; 3.5)$;
- $\varphi_{C1/C5} = \frac{\varpi_{C1}}{\varpi_{C5}} = \frac{FI}{VI} = (0.43; 0.67; 1)$.

Thus, the vector of comparative significance was obtained by applying Expression (3):

$$\Phi^1 = ((3.5,\ 4,\ 4.5); (0.22,\ 0.25,\ 0.29); (2.5,\ 3,\ 3.5); (0.43,\ 0.67,\ 1)).$$

Step 3: In this step, the constraints of the fuzzy model were defined by applying Expression (4) and Expression (5).

The first group of the constraints of criteria C1–C5, for expert E1, was defined as follows: $\frac{\varpi_{C4}}{\varpi_{C3}} = (3.5; 4; 4.5)$, $\frac{\varpi_{C3}}{\varpi_{C2}} = (0.22; 0.25; 0.29)$, $\frac{\varpi_{C2}}{\varpi_{C1}} = (2.5; 3; 3.5)$ and $\frac{\varpi_{C1}}{\varpi_{C5}} = (0.43; 0.67; 1)$. The second group of the constraints resulting from the condition of transitivity of relations was defined as follows: $\frac{W_4}{W_2} = (3.5; 4.00; 4.5) * (0.1; 0.25; 0.28) =$

$(0.35; 1.00; 1.26)$; $\frac{w_3}{w_1} = (0.10; 0.25; 0.28) * (2.50; 3.00; 3.50) = (0.25; 0.75; 0.98)$ and $\frac{w_2}{w_5} = (2.50; 3.00; 3.50) * (0.43; 0.67; 1.00)$. The constraints of the other models were defined in the same way.

Step 4: On the basis of the constraints defined in the previous step, Model (6) was formed to determine the optimal fuzzy values of the weight coefficients of the criteria and the sub-criteria (Appendix B). It was used to specify the weight coefficients of the criteria and the sub-criteria, presented below.

$$\text{Expert 1 (C1} - \text{C5) } \min \chi \quad \text{s.t.}$$
$$\begin{cases}
\left|\frac{w_4^l}{w_3^u} - 3.5\right| \le \chi; \left|\frac{w_4^m}{w_3^m} - 4\right| \le \chi; \left|\frac{w_4^u}{w_3^{ul}} - 4.5\right| \le \chi; \\
\left|\frac{w_3^l}{w_2^u} - 0.22\right| \le \chi; \left|\frac{w_3^m}{w_2^m} - 0.25\right| \le \chi; \left|\frac{w_3^u}{w_2^l} - 0.29\right| \le \chi; \\
\left|\frac{w_2^l}{w_1^u} - 2.5\right| \le \chi; \left|\frac{w_2^m}{w_1^m} - 3\right| \le \chi; \left|\frac{w_2^u}{w_1^l} - 3.5\right| \le \chi; \\
\left|\frac{w_1^l}{w_5^u} - 0.43\right| \le \chi; \left|\frac{w_1^m}{w_3^m} - 0.67\right| \le \chi; \left|\frac{w_1^u}{w_3^l} - 1\right| \le \chi; \\
\left|\frac{w_4^l}{w_2^u} - 0.78\right| \le \chi; \left|\frac{w_2^m}{w_4^m} - 1\right| \le \chi; \left|\frac{w_2^u}{w_4^l} - 1.29\right| \le \chi; \\
\left|\frac{w_3^l}{w_1^u} - 0.56\right| \le \chi; \left|\frac{w_3^m}{w_1^m} - 0.75\right| \le \chi; \left|\frac{w_3^u}{w_1^l} - 1.00\right| \le \chi; \\
\left|\frac{w_2^l}{w_5^u} - 1.07\right| \le \chi; \left|\frac{w_2^m}{w_5^m} - 2.00\right| \le \chi; \left|\frac{w_2^u}{w_5^l} - 3.5\right| \le \chi; \\
\sum_{j=1}^{5} (w_j^l + 4 \cdot w_j^m + w_j^u)/6 = 1, \forall j = 1,2,3,4,5 \\
w_j^l \le w_j^m \le w_j^u, \forall j = 1,2,3,4,5 \\
w_j^l, w_j^m, w_j^u \ge 0, \forall j = 1,2,3,4,5
\end{cases} \quad \cdots$$

$$\text{Expert 1 (C5.1} - \text{C5.3) } \min \chi \quad \text{s.t.}$$
$$\begin{cases}
\left|\frac{w_{5.1}^l}{w_{5.2}^u} - 0.67\right| \le \chi; \left|\frac{w_{5.1}^m}{w_{5.2}^m} - 1\right| \le \chi; \left|\frac{w_{5.1}^u}{w_{5.2}^l} - 1.5\right| \le \chi; \\
\left|\frac{w_{5.2}^l}{w_{5.3}^u} - 2.33\right| \le \chi; \left|\frac{w_{5.2}^m}{w_{5.3}^m} - 4\right| \le \chi; \left|\frac{w_{5.2}^u}{w_{5.3}^l} - 6.7\right| \le \chi; \\
\left|\frac{w_{5.1}^l}{w_{5.3}^u} - 1.56\right| \le \chi; \left|\frac{w_{5.1}^m}{w_{5.3}^m} - 4.00\right| \le \chi; \left|\frac{w_{5.1}^u}{w_{5.3}^l} - 10.07\right| \le \chi; \\
\sum_{j=1}^{3} (w_j^l + 4 \cdot w_j^m + w_j^u)/6 = 1, \forall j = 1,2,3, \\
w_j^l \le w_j^m \le w_j^u, \forall j = 1,2,3 \\
w_j^l, w_j^m, w_j^u \ge 0, \forall j = 1,2,3
\end{cases}$$

Step 5: After solving the model, we obtained the optimal local values of the expert weight coefficients (Appendix C). The global sub-criteria values for each expert were obtained, as shown in Table 9, by multiplying the local sub-criteria values by the criteria weight coefficients. Lingo 17.0 software was utilized to solve the non-linear fuzzy models, which allowed for obtaining the mean value $\chi \approx 0.0$, showing the high consistency of the provided values of the criteria weights.

Table 9. Global values of the fuzzy weight coefficients of the sub-criteria.

Criteria	Weight	Sub-Criteria	Local Weight	Global Weight
C1	(0.2, 0.27, 0.27)	C1.1	(0.2, 0.33, 0.35)	(0.04, 0.09, 0.1)
		C1.2	(0.1, 0.18, 0.19)	0.02, 0.05, 0.05)
		C1.3	(0.15, 0.29, 0.34)	(0.03, 0.08, 0.09)
		C1.4	(0.17, 0.25, 0.27)	(0.03, 0.07, 0.07)
C2	(0.16, 0.22, 0.23)	C2.1	(0.051, 0.53, 0.73)	(0.08, 0.12, 0.17)
		C2.2	(0.41, 0.41, 0.58)	0.07, 0.09, 0.13)
C3	(0.13, 0.21, 0.23)	C3.1	(0.18, 0.26, 0.37)	(0.02, 0.05, 0.08)
		C3.2	(0.17, 0.28, 0.33)	(0.02, 0.06, 0.08)
		C3.3	(0.16, 0.22, 0.23)	(0.02, 0.05, 0.05)
		C3.4	(0.17, 0.25, 0.28)	(0.02, 0.05, 0.06)
C4	(0.11, 0.21, 0.26)	C4.1	(0.43, 0.43, 0.59)	(0.05, 0.09, 0.15)
		C4.2	(0.52, 0.52, 0.67)	(0.06, 0.11, 0.18)
C5	(0.08, 0.14, 0.17)	C5.1	(0.35,0.4, 0.51)	(0.03, 0.05, 0.08)
		C5.2	(0.28, 0.31, 0.5)	(0.02, 0.04, 0.08)
		C5.3	(0.2, 0.24, 0.34)	(0.02, 0.03, 0.06)

4.3. *Ranking of Alternatives*

After determining the weight of the criteria, compensatory (F-MAIRCA) and partially compensatory (F-PROMETHEE) approaches were used to select the best location of the logistics platform.

4.3.1. The F-MAIRCA Results

Alternatives were ranked using Equations (8)–(16). The different steps of F-MAIRCA are described below.

Step 1: Individual evaluations of the alternatives were performed by experts (E1, E2, E3, E4, E5, E6 and E7) using the linguistic terms presented in Table 10 [76]. These evaluations are shown in Appendix D using Equation (8).

Table 10. Alternative language scale [76].

Linguistic Term	Abbreviation	Fuzzy Number in a Triangular Style
Very Low	(VL)	(0, 1, 2)
Low	(L)	(1, 2, 3)
Medium	(M)	(2, 3, 4)
High	(H)	(3, 4, 5)
Very High	(VH)	(4, 5, 6)

Step 2: the aggregated decision values of the triangular fuzzy numbers were calculated by Equation (9) to construct the fuzzy aggregated decision matrix, as shown in Table 11.

Table 11. Fuzzy aggregated decision matrix.

	A1	A2			A6	A7
C1.1	(3.43, 4.43, 5.43)	(3.14, 4.14, 5.14)	…	…	(2.43, 3.43, 4.43)	(1.71, 2.71, 3.71)
C1.2	(2.14, 3.00, 3.86)	(2.57, 3.57, 4.57)	…	…	(3.57, 4.57, 5.57)	(2.86, 3.86, 4.86)
	⋮	⋮	⋱	⋱	⋮	⋮
C5.2	(3.57, 4.57, 5.57)	(2.00, 3.00, 4.00)	…	…	(2.00, 3.00, 4.00)	(2.43, 3.43, 4.43)
C5.3	(2.00, 3.00,4.00)	(2.14, 3.14, 4.14)	…	…	(1.29, 2.29, 3.29)	(2.00, 3.00, 4.00)

Step 3: In this step, each alternative received equal preferences. In this study, we had seven alternatives. Thus, the preferences PAi were calculated as follows: $PAi = 1/7 = 0.143$ by applying Equation (10).

Step 4: Using Equation (11), the fuzzy matrix of theoretical ponder $\widetilde{T}_{PA}$ was obtained as demonstrated in Table 12. Each row of this table indicates the theoretical evaluation of the alternatives for this particular criterion.

Table 12. Fuzzy matrix of theoretical ponder.

	A1	A2			A6	A7
C1.1	(0.01, 0.01, 0.01)	(0.01, 0.01, 0.01)	…	…	(0.01, 0.01, 0.01)	(0.01, 0.01, 0.01)
C1.2	(0.00, 0.01, 0.01)	(0.00, 0.01, 0.01)	…	…	(0.00, 0.01, 0.01)	(0.00, 0.01, 0.01)
	⋮	⋮	⋱	⋱	⋮	⋮
C5.2	(0.00 0.01, 0.01)	(0.00 0.01, 0.01)	…	…	(0.00 0.01, 0.01)	(0.00 0.01, 0.01)
C5.3	(0.00, 0.00 0.01)	(0.00, 0.00 0.01)	…	…	(0.00, 0.00 0.01)	(0.00, 0.00 0.01)

Step 5: To make the fuzzy aggregated decision matrix dimensionless, Equation (12) and Equation (13) were used. We then obtained the fuzzy normalized decision matrix, as shown in Table 13.

Table 13. Fuzzy Normalized Decision Matrix.

	A1	A2			A6	A7
C1.1	(0.63, 0.82, 1.00)	(0.58, 0.76, 0.95)	…	…	(0.45, 0.63, 0.82)	(0.32, 0.50, 0.68)
C1.2	(0.13, 0.38, 0.52)	(0.28, 0.48, 0.59)	…	…	(0.48, 0.59, 0.67)	(0.35, 0.52, 0.62)
	⋮	⋮	⋱	⋱	⋮	⋮
C5.2	(0.64, 0.82, 1.00)	(0.36, 0.54, 0.72)	…	…	(0.36, 0.54, 0.72)	(0.44, 0.62, 0.79)
C5.3	(0.34, 0.59, 0.83)	(0.52, 0.76, 1.00)	…	…	(0.31, 0.55, 0.79)	(0.48, 0.72, 0.97)

Step 6: Using Equation (14), the matrix of fuzzy actual ponder, also known as the real evaluation matrix, was provided, as shown in Table 14.

Table 14. Matrix of fuzzy actual ponder.

	A1	A2			A6	A7
C1.1	(0.004, 0.010, 0.01)	(0.00, 0.010, 0.013)	…	…	(0.003, 0.008, 0.011)	(0.002, 0.006, 0.009)
C1.2	(0.00, 0.004, 0.004)	(0.00, 0.003, 0.003)	…	…	(0.002, 0.003, 0.002)	(0.002, 0.003, 0.003)
	⋮	⋮	⋱	⋱	⋮	⋮
C5.2	(0.002, 0.005, 0.01)	(0.00, 0.003, 0.009)	…	…	(0.001, 0.003, 0.009)	(0.001, 0.004, 0.009)
C5.3	(0.00, 0.003, 0.007)	(0.00, 0.004, 0.008)	…	…	(0.001, 0.003, 0.006)	(0.001, 0.003, 0.008)

Step 7: In this step, the total gap matrix was calculated using Equation (15), as shown in Table 15.

Table 15. Total gap matrix.

	A1	A2	A3	A4	A5	A6	A7
C1.1	0.002	0.002	0.002	0.004	0.004	0.004	0.005
C1.2	0.003	0.003	0.003	0.003	0.003	0.004	0.003
C1.3	0.004	0.005	0.005	0.004	0.005	0.005	0.005
C1.4	0.006	0.008	0.008	0.007	0.007	0.007	0.008
C2.1	0.003	0.004	0.004	0.004	0.004	0.005	0.003
C2.2	0.006	0.006	0.006	0.007	0.007	0.007	0.008
C3.1	0.001	0.001	0.001	0.002	0.002	0.002	0.001
C3.2	0.004	0.005	0.005	0.005	0.005	0.006	0.005
C3.3	0.004	0.004	0.004	0.004	0.004	0.004	0.004
C3.4	0.006	0.006	0.006	0.007	0.007	0.007	0.007
C4.1	0.003	0.005	0.005	0.005	0.006	0.006	0.005
C4.2	0.003	0.007	0.007	0.007	0.007	0.007	0.005
C5.1	0.001	0.003	0.003	0.003	0.003	0.003	0.006
C5.2	0.001	0.003	0.003	0.003	0.003	0.003	0.002
C5.3	0.002	0.001	0.001	0.002	0.002	0.002	0.001

Step 8: Table 16 presents the gap values as given in Equation (16). The aim of the decision-makers is to maintain the smallest possible value between the theoretical and actual evaluation for the best alternative, according to the FMAIRCA method. The final ranking of the gap values in ascending order is as follows: A1 > A3 > A2 > A6 > A4 > A7 > A5.

Table 16. Ranking of alternatives.

	Alternatives						
	A1	A2	A3	A4	A5	A6	A7
Gap values	0.048	0.064	0.063	0.068	0.067	0.070	0.068
Rank	1	3	2	6	4	7	5

4.3.2. The F-PROMETHEE Results

Steps 1 and 2: The individual evaluations of the alternatives according to different criteria and the aggregated decision matrix are shown in Tables 11 and 12.

Steps 3, 4 and 5: The distances between the two alternatives with respect to each criterion were calculated by applying Equation (17). After that, the distances were expressed as a preference function for every pair of alternatives. In this step, the usual criterion preference function was used, and the results were summarized, as shown in Appendix E. The preference index for each pair of alternatives was determined using Equation (18). For example, the preference of alternative A1 over A2 was represented by the index PC1.1 (A1, A2) = (0.2, 0.33, 0.35). The fuzzy preferences index was obtained, as shown in Appendix E, using the value preference index for each criterion and the weights of the criteria. For simplicity, Appendix E presents only the results corresponding to the values of the usual criterion preference function and the overall preference index.

Steps 6, 7 and 8: The defuzzified values of leaving flows, entering flows and net flows were respectively calculated, as shown in Table 17. The ranking results indicated that alternative A1 was the most durable location (with scores of 0.981, 1.30 and 0.317 for Ø net, Ø + and Ø, respectively), while alternative A3 was the second-best sustainable location and alternative A6 was the least sustainable location.

Table 17. Defuzzified upgrading flows.

	A1	A2	A3	A4	A5	A6	A7
Φ^+	1.30	1.30	1.75	0.91	1.37	0.60	1.22
$\Phi-$	0.317	0.792	0.899	1.55	1.385	2.074	1.43
Φnet	0.981	0.506	0.851	−0.64	−0.01	−1.47	−0.21
Rank	1	3	2	6	4	7	5

4.4. Stability of the Obtained Results

4.4.1. Assessment of the Independence of the Aggregation Technique

In Figure 3, the ranking of alternatives was sorted by the F-PROMETHEE, F-MAIRCA and F-TOPSIS methods with the partial compensation technique, compensation technique and non-compensation technique, respectively. The obtained results prove that the use of the non-compensation technique produced a different ranking. As proposed in this paper, the compensatory and partially compensatory composite indicators can be used by decision-makers to locate the logistics platforms.

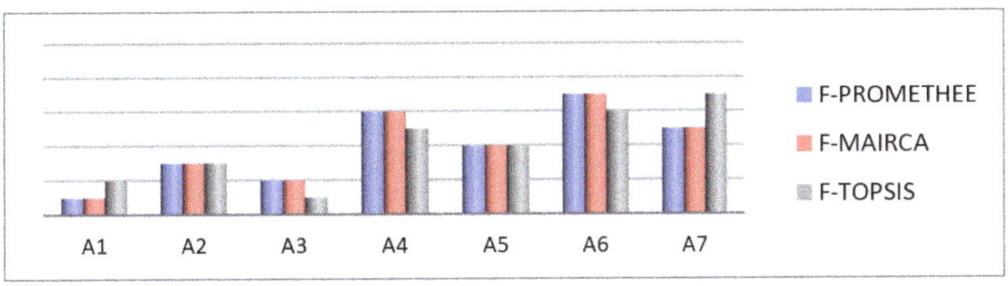

Figure 3. Ranking of alternatives for the three aggregation technique.

4.4.2. Variation of Criteria Weights

The illustrative ranking of F-MAIRCA and F-PROMETHEE changed by varying the applied scenarios of weight sensitivity analysis, as shown in Figure 4. Obviously, the variation of the weights of the criteria through the scenarios changed the ranking of the alternatives in both methods. Despite these changes, a logical ranking was obtained. The best and worse alternative's positions were almost identical to the initial ranking. The obtained results prove that the proposed approach is slightly sensitive to the variations in the criteria weights.

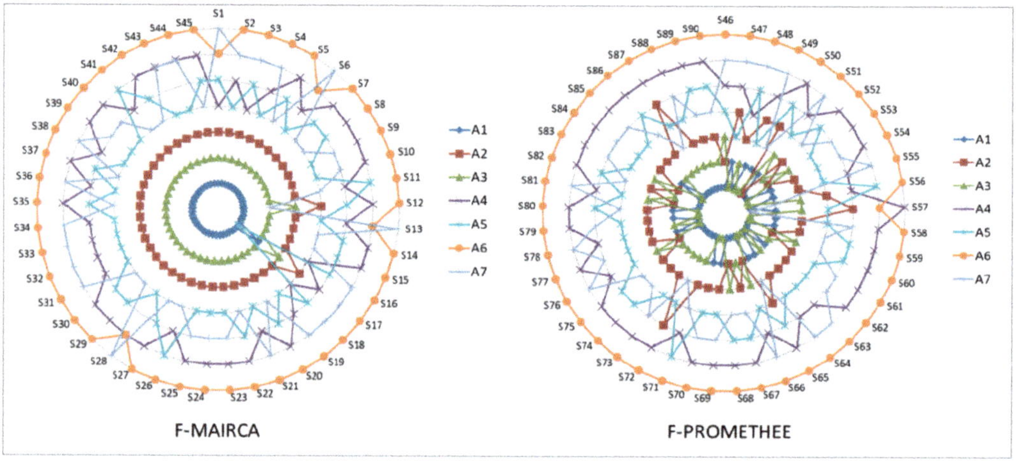

Figure 4. The illustrative ranking of the fuzzy multi-attribute ideal-real comparative analysis (F-MAIRCA) and the fuzzy preference ranking organization method for enrichment evaluation (F-PROMETHEE) obtained by sensitivity analysis.

4.5. Results and Decision-Making Process

The F-FUCOM results revealed the following order: economic criteria > environmental criteria > political criteria > social criteria > territorial criteria. From this order, we noticed that economic sustainability was the most important aspect to be taken into account to evaluate the logistics platform. The next two most important aspects were environmental sustainability and political sustainability.

Figure 5 represents the ranking of the sustainability sub-criteria (indicator) of each criterion (dimension). This figure shows that the conformity with the environmental emissions regulations is ranked one, and were respectively followed by (C4.2) the role of support to the industry and (4.1) the current policy. The effect on the natural landscape, connectivity to multimodal transport and fiscal policies were the most important sub-criteria to locate the logistics platform in the fourth, fifth and sixth place, respectively.

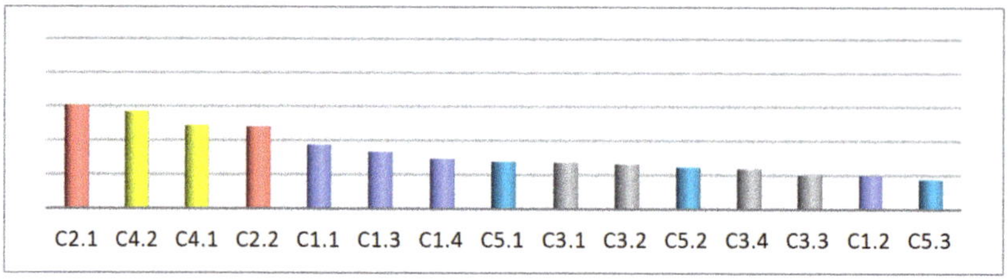

Figure 5. Results obtained by applying the full consistency method (FUCOM)—weighting of the criteria.

- The ranking of the economic criteria was as follows: C1.1 > C1.3 > C1.4 > C1.2. The above results show that a location should ensure connectivity to multimodal transport and offer fiscal policies to attract investors and promote the development of multimodal transport.
- The ranking of the environmental criteria was in the following order: C2.1> C2.2. The conformity with environmental emissions regulations was at the top of the list, which was expected because the improvement of environmental criteria is important in the process of the logistics platform localization.
- The ranking of the social criteria was as follows: C3.1 > C3.2 > C3.4 > C3.3. The results presented above reveal that the logistics platform should ensure the safety and security of the site and the workers, while minimizing the generated noise.
- The results of ranking the political criteria showed the following order: C4.2 > C4.1, which proves the vital role of support and cooperation between both government and industry in choosing the platform location, as locations are often not finalized due to government instability.
- The ranking of the territorial criteria was as follows: C5.1 > C5.2 > C5.3. The above ranking order demonstrates the importance of a location being connected to and accessible by all transport modes. Second, a logistics platform should be close to all industrial areas.

The objective of this study was to develop a multi-criteria approach based on compensation phenomenon to locate the logistics platform by taking sustainability into account. The developed approach was applied to evaluate the influence of both aggregation techniques (compensatory and partially compensatory) on the final results. The findings of this case study showed that the two fuzzy MCDM (F-PROMETHEE and F-MAIRCA) methods provided similar location ranking results (A1 > A3 > A2 > A6 > A4 > A7 > A5) (Figure 6).

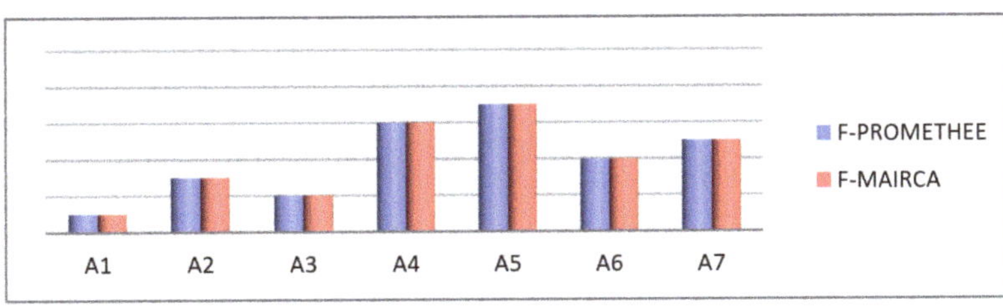

Figure 6. Final ranking of F-PROMETHEE and F-MAIRCA.

The authors can thus conclude that the choice of one of these methods seems appropriate for the selection of the most sustainable location. The purpose of this case study was not to compare the F-PROMETHEE and F-MAIRCA methods, but to highlight the capacity of the proposed approach to allow for the use of both aggregation techniques.

5. Implications

This study provides fuzzy compensatory and partially compensatory composite indicators to locate a logistics platform under sustainability perspectives. In this context, the main contributions of this study can be summed up in four points:

- First, in contrast to the existing localization approaches, in addition to the classic dimensions of sustainability (economic, environmental and social), this study included two other dimensions (political and territorial) identified in the literature as being relevant to urban logistics;
- Second, the proposed approach is characterized by the choice of methods that are most suitable to our study context. Although several MCDM methods were proposed, the decision-maker faces many the challenges when selecting the appropriate method to use to avoid the subjective choice. Thus, this study was carried out to manage the complexity of the decision-making process in situations of uncertainty. From a

methodological point of view, the present work integrates the set of fuzzy numbers with FUCOM, MAIRCA and PROMETHEE to locate the logistics platform. These methods were chosen because of their popularity and stability;

- Third, in the literature, there is no general or systematic localization method specifically related to the sustainability perspectives. This study proposes an innovative and interesting approach as a support tool for decision-makers with sustainability perspectives. The novelty of this approach lays in developing fuzzy compensatory and partially compensatory composite indicators by considering weak sustainability and limited sustainability;
- Fourth, to validate the robustness of the proposed approach, a sensitivity analysis was performed. In the first phase, the independence of the aggregation technique was assessed. However, in the second phase, the effect of sensitivity on the variation of the criteria weight was evaluated.

From a practical perspective, this research provides several results for sustainable facility location problems. The obtained findings provide valuable insights for decision-makers to select a logistic platform from the calculated composite indicators.

6. Conclusions

An approach for locating the logistics platforms with sustainability perspectives was introduced in this paper. It uses an integrated MCDM method with the set of fuzzy numbers. A composite indicator based on compensatory and partially compensatory multi-criteria decision-making methods was also proposed. The composite indicator computation model involves two stages. Firstly, important weights of sustainability criteria were computed using the F-FUCOM method, relying on the experts' linguistic responses. Secondly, alternatives were classified by two aggregation methods: F-MAIRCA and F-PROMETHEE. The suggested approach was applied in the city of Sfax, where it is necessary to construct a logistics platform to reduce the impact of freight transport in the city.

The experimental results reveal that the economic and environmental criteria considerably affect the selection of the logistics platform location. A good connection and accessibility to multimodal transport is essential for a platform's success. The chosen location must ensure conformity with the environmental emissions regulations. Furthermore, the logistics platform should ensure the safety and security of the site and the workers, while minimizing the generated noise. A vital role of support and cooperation between both the government and industry is needed to choose the adequate platform location.

From a practical perspective, this research proposed a solution to sustainable facility location problems. In fact, one of our main contributions is to present a comprehensive list of sustainable evaluation sub-criteria, involving economic, environmental, social, political and territorial criterion, to assess the sustainability of the logistics platform location. The novelty of this approach lays in developing fuzzy compensatory and partially compensatory composite indicators to solve the facility location problem. Then, the stability and the robustness of the proposed approach were demonstrated.

However, this approach has some limitations. Firstly, it used only a limited number of experts. Secondly, the MCDM methods applied in this research work rely on the experts' opinions, which can deteriorate their performances. Finally, this study only considers uncertainty with the fuzzy set theory. Limited efforts were made to simultaneously relate two types of uncertainty with MCDMs to solve decision-making problems [77].

The authors suggest several future research directions. In further research, it is recommended to integrate fuzzy theory with other types of uncertainty (such as stochastic). We expect to build ontologies from the sustainability criteria and the stakeholders of the logistics platform to support the proposed approach. Furthermore, we will try to improve our sensitivity analysis by using other models, methods and tools.

Author Contributions: Conceptualization, H.A. and N.H.; methodology, H.A., N.H., M.B. and L.K.; software, H.A.; validation, H.A., N.H., M.B. and L.K.; formal analysis, H.A.; investigation, H.A.; data curation, H.A.; writing—original draft preparation, H.A.; writing—review and editing, H.A., N.H., M.B. and L.K. All authors have read and agreed to the published version of the manuscript.

Funding: This research received no external funding.

Institutional Review Board Statement: Not applicable.

Informed Consent Statement: Not applicable.

Data Availability Statement: Not applicable.

Acknowledgments: This work could have not been carried out without the generous collaboration of anonymous experts from different fields and disciplines.

Conflicts of Interest: The authors declare no conflict of interest.

Appendix A

Table A1. Vectors of comparative meanings of criteria and sub-criteria.

	Expert 1 (E1)	**Expert 2 (E2)**
C1–C5	(3.5, 4, 4.5) (0.22, 0.25, 0.29) (2.5, 3, 3.5) (0.43, 0.67, 1)	(3.5, 4, 4.5) (0.33, 0.5, 0.71) (0.27, 0.5, 1) (0.45, 1, 2.24)
C1.1–C1.4	(3.5, 4, 4.5) (0.33, 0.5, 0.71) (1, 1.5, 2.33)	(1.5, 2, 2.5) (0.6, 1, 1.67) (0.27, 0.5, 1)
C2.1–C2.2	(0.67, 1, 1.5)	(0.67, 1, 1.5)
C3.1–C3.4	(2.5, 3, 3.5) (0.29, 0.33, 0.4) (1, 1, 1)	(3.5, 4, 4.5) (0.33, 0.5, 0.71) (0.27, 0.5, 1)
C4.1–C4.2	(2.5, 3, 3.5)	(0.67, 1, 1.5)
C5.1–C5.3	(0.67, 1, 1.5) (2.33, 4, 6.72)	(2.5, 3, 3.5) (0.43, 0.67, 1)
	Expert 5 (E5)	**Expert 6 (E6)**
C1–C5	(1, 1, 1) (0.67, 1, 1.5) (1, 2, 3.73) (1, 1.5, 2.33)	(3.5, 4, 4.5) (0.56, 0.75, 1) (0.29, 0.33, 0.4) (2.5, 3, 3.5)
C1.1–C1.4	(0.67, 1, 1.5) (0.67, 1, 1.49) (1.5, 2, 2.5)	(3.5, 4, 4.5) (0.56, 0.75, 1) (0.19, 0.33, 0.6)
C2.1–C2.2	(0.67, 1, 1.5)	(1, 1, 1)
C3.1–C3.4	(0.67, 1, 1.5) (0.45, 1, 2.24) (1, 2, 3.73)	(3.5, 4, 4.5) (0.22, 0.25, 0.29) (0.67, 1, 1.5)
C4.1–C4.2	(0.67, 1, 1.5)	(3.5, 4, 4.5)
C5.1–C5.3	(1, 1, 1) (0.67, 1, 1.5)	(3.5, 4, 4.5) (0.33, 0.5, 0.71)
	Expert 5 (E5)	**Expert 6 (E6)**
C1–C5	(2.5, 3, 3.5) (0.71, 1, 1.4) (0.19, 0.33, 0.6) (0.67, 1, 1.49)	(0.67, 1, 1.5) (1, 2, 3.73) (0.27, 0.5, 1) (1, 2, 3.73)
C1.1–C1.4	(1.5, 2, 2.5) (1, 1.5, 2.33) (1, 1.33, 1.8)	(0.67, 1, 1.5) (0.67, 1, 1.49) (0.67, 1, 1.5)
C2.1–C2.2	(0.67, 1, 1.5)	(1.5, 2, 2.5)
C3.1–C3.4	(2.5, 3, 3.5) (0.19, 0.33, 0.6) (1.67, 3, 5.22)	(0.67, 1, 1.5) (1, 2, 3.73) (0.4, 0.5 0.67)
C4.1–C4.2	(0.67, 1, 1.5)	(0.67, 1, 1.5)
C5.1–C5.3	(1.5, 2, 2.5) (1.4, 2, 3)	(1.5, 2, 2.5) (0.27, 0.5, 1)
	Expert 7 (E7)	
C1–C5	(2.5, 3, 3.5) (0.19, 0.33, 0.6) (0.45, 1, 2.24) (0.45, 1, 2.24)	
C1.1–C1.4	(1.5, 2, 2.5) (0.6, 1, 1.67) (0.27, 0.5, 1)	
C2.1–C2.2	(0.67, 1, 1.5)	
C3.1–C3.4	(2.5, 3, 3.5) (0.19, 0.33, 0.6) (0.45, 1, 2.24)	
C4.1–C4.2	(2.5, 3, 3.5)	
C5.1–C5.3	(0.67, 1, 1.5) (0.67, 1, 1.49)	

Appendix B

Expert 1 (C1 − C5) min χ

s.t.

$$\begin{cases}
\left|\frac{w_4^l}{w_3^u}-3.5\right|\le\chi;\ \left|\frac{w_4^m}{w_3^m}-4\right|\le\chi;\ \left|\frac{w_4^u}{w_3^{ul}}-4.5\right|\le\chi;\\
\left|\frac{w_3^l}{w_2^u}-0.22\right|\le\chi;\ \left|\frac{w_3^m}{w_2^m}-0.25\right|\le\chi;\ \left|\frac{w_3^u}{w_2^l}-0.29\right|\le\chi;\\
\left|\frac{w_2^l}{w_1^u}-2.5\right|\le\chi;\ \left|\frac{w_2^m}{w_1^m}-3\right|\le\chi;\ \left|\frac{w_2^u}{w_1^l}-3.5\right|\le\chi;\\
\left|\frac{w_1^l}{w_5^u}-0.43\right|\le\chi;\ \left|\frac{w_1^m}{w_3^m}-0.67\right|\le\chi;\ \left|\frac{w_1^u}{w_3^l}-1\right|\le\chi;\\
\left|\frac{w_4^l}{w_2^u}-0.78\right|\le\chi;\ \left|\frac{w_2^m}{w_4^m}-1\right|\le\chi;\ \left|\frac{w_2^u}{w_4^l}-1.29\right|\le\chi;\\
\left|\frac{w_3^l}{w_1^u}-0.56\right|\le\chi;\ \left|\frac{w_3^m}{w_1^m}-0.75\right|\le\chi;\ \left|\frac{w_3^u}{w_1^l}-1.00\right|\le\chi;\\
\left|\frac{w_2^l}{w_5^u}-1.07\right|\le\chi;\ \left|\frac{w_2^m}{w_5^m}-2.00\right|\le\chi;\ \left|\frac{w_2^u}{w_5^l}-3.5\right|\le\chi;\\
\sum_{j=1}^{5}(w_j^l+4\cdot w_j^m+w_j^u)/6=1,\ \forall j=1,2,3,4,5\\
w_j^l\le w_j^m\le w_j^u,\ \forall j=1,2,3,4,5\\
w_j^l,w_j^m,w_j^u\ge 0,\ \forall j=1,2,3,4,5
\end{cases}$$

……

Expert 7 (C1 − C5) min χ

s.t.

$$\begin{cases}
\left|\frac{w_1^l}{w_2^u}-2.5\right|\le\chi;\ \left|\frac{w_1^m}{w_2^m}-3\right|\le\chi;\ \left|\frac{w_1^u}{w_2^{ul}}-3.5\right|\le\chi;\\
\left|\frac{w_2^l}{w_5^u}-0.19\right|\le\chi;\ \left|\frac{w_2^m}{w_5^m}-0.33\right|\le\chi;\ \left|\frac{w_2^u}{w_5^l}-0.6\right|\le\chi;\\
\left|\frac{w_5^l}{w_3^u}-0.45\right|\le\chi;\ \left|\frac{w_5^m}{w_3^m}-1\right|\le\chi;\ \left|\frac{w_5^u}{w_3^l}-2.24\right|\le\chi;\\
\left|\frac{w_3^l}{w_4^u}-0.45\right|\le\chi;\ \left|\frac{w_3^m}{w_4^m}-1\right|\le\chi;\ \left|\frac{w_3^u}{w_4^l}-2.24\right|\le\chi;\\
\left|\frac{w_1^l}{w_5^u}-0.48\right|\le\chi;\ \left|\frac{w_1^m}{w_5^m}-1\right|\le\chi;\ \left|\frac{w_1^u}{w_5^l}-2.1\right|\le\chi;\\
\left|\frac{w_2^l}{w_3^u}-0.09\right|\le\chi;\ \left|\frac{w_2^m}{w_3^m}-0.33\right|\le\chi;\ \left|\frac{w_2^u}{w_3^l}-1.34\right|\le\chi;\\
\left|\frac{w_5^l}{w_4^u}-0.2\right|\le\chi;\ \left|\frac{w_5^m}{w_4^m}-1\right|\le\chi;\ \left|\frac{w_5^u}{w_4^l}-5.01\right|\le\chi;\\
\sum_{j=1}^{5}(w_j^l+4\cdot w_j^m+w_j^u)/6=1,\ \forall j=1,2,3,4,5\\
w_j^l\le w_j^m\le w_j^u,\ \forall j=1,2,3,4,5\\
w_j^l,w_j^m,w_j^u\ge 0,\ \forall j=1,2,3,4,5
\end{cases}$$

Expert 1 (C1.1 − C1.4) min χ

s.t.

$$\begin{cases}
\left|\frac{w_{1.1}^l}{w_{1.3}^u}-3.5\right|\le\chi;\ \left|\frac{w_{1.1}^m}{w_{1.3}^m}-4\right|\le\chi;\ \left|\frac{w_{1.1}^u}{w_{1.3}^l}-4.5\right|\le\chi;\\
\left|\frac{w_{1.3}^l}{w_{1.4}^u}-0.33\right|\le\chi;\ \left|\frac{w_{1.3}^m}{w_{1.4}^m}-0.5\right|\le\chi;\ \left|\frac{w_{1.3}^u}{w_{1.4}^l}-0.71\right|\le\chi;\\
\left|\frac{w_{1.4}^l}{w_{1.2}^u}-1.00\right|\le\chi;\ \left|\frac{w_{1.4}^m}{w_{1.2}^m}-1.5\right|\le\chi;\ \left|\frac{w_{1.4}^u}{w_{1.2}^l}-2.33\right|\le\chi;\\
\left|\frac{w_{1.1}^l}{w_{1.4}^u}-1.17\right|\le\chi;\ \left|\frac{w_{1.1}^m}{w_{1.4}^m}-2.00\right|\le\chi;\ \left|\frac{w_{1.1}^u}{w_{1.4}^l}-3.21\right|\le\chi;\\
\left|\frac{w_{1.3}^l}{w_{1.2}^u}-0.33\right|\le\chi;\ \left|\frac{w_{1.3}^m}{w_{1.2}^m}-0.75\right|\le\chi;\ \left|\frac{w_{1.3}^u}{w_{1.3}^l}-1.67\right|\le\chi;\\
\sum_{j=1}^{4}(w_j^l+4\cdot w_j^m+w_j^u)/6=1,\ \forall j=1,2,3,4\\
w_j^l\le w_j^m\le w_j^u,\ \forall j=1,2,3,4\\
w_j^l,w_j^m,w_j^u\ge 0,\ \forall j=1,2,3,4
\end{cases}$$

……

Expert 7 (C1.1 − C1.4) min χ

s.t.

$$\begin{cases}
\left|\frac{w_{1.3}^l}{w_{1.2}^u}-1.5\right|\le\chi;\ \left|\frac{w_{1.3}^m}{w_{1.2}^m}-2\right|\le\chi;\ \left|\frac{w_{1.3}^u}{w_{1.2}^l}-2.5\right|\le\chi;\\
\left|\frac{w_{1.2}^l}{w_{1.1}^u}-0.6\right|\le\chi;\ \left|\frac{w_{1.2}^m}{w_{1.1}^m}-1\right|\le\chi;\ \left|\frac{w_{1.2}^u}{w_{1.1}^l}-1.67\right|\le\chi;\\
\left|\frac{w_{1.1}^l}{w_{1.4}^u}-0.27\right|\le\chi;\ \left|\frac{w_{1.1}^m}{w_{1.4}^m}-0.5\right|\le\chi;\ \left|\frac{w_{1.1}^u}{w_{1.4}^l}-1\right|\le\chi;\\
\left|\frac{w_{1.3}^l}{w_{1.1}^u}-0.9\right|\le\chi;\ \left|\frac{w_{1.3}^m}{w_{1.1}^m}-2\right|\le\chi;\ \left|\frac{w_{1.3}^u}{w_{1.1}^l}-4.17\right|\le\chi;\\
\left|\frac{w_{1.2}^l}{w_{1.4}^u}-0.16\right|\le\chi;\ \left|\frac{w_{1.2}^m}{w_{1.4}^m}-0.5\right|\le\chi;\ \left|\frac{w_{1.2}^u}{w_{1.4}^l}-1.67\right|\le\chi;\\
\sum_{j=1}^{4}(w_j^l+4\cdot w_j^m+w_j^u)/6=1,\ \forall j=1,2,3,4\\
w_j^l\le w_j^m\le w_j^u,\ \forall j=1,2,3,4\\
w_j^l,w_j^m,w_j^u\ge 0,\ \forall j=1,2,3,4
\end{cases}$$

Expert 1 (C2.1 − C2.2) min χ

s.t.

$$\begin{cases}
\left|\frac{w_{2.1}^l}{w_{2.2}^u}-2.5\right|\le\chi;\ \left|\frac{w_{2.1}^m}{w_{2.2}^m}-3\right|\le\chi;\ \left|\frac{w_{2.1}^u}{w_{2.2}^l}-3.5\right|\le\chi;\\
(w_{2.1}^l+4\cdot w_{2.1}^m+w_{2.1}^u)/6+(w_{2.2}^l+4\cdot w_{2.2}^m+w_{2.2}^u)/6=1\\
w_j^l\le w_j^m\le w_j^u,\ \forall j=1,2\\
w_j^l,w_j^m,w_j^u\ge 0,\ \forall j=1,2
\end{cases}$$

……

Expert 7 (C2.1 − C2.2) min χ

s.t.

$$\begin{cases}
\left|\frac{w_{2.1}^l}{w_{2.2}^u}-0.67\right|\le\chi;\ \left|\frac{w_{2.1}^m}{w_{2.2}^m}-1\right|\le\chi;\ \left|\frac{w_{2.1}^u}{w_{2.2}^l}-1.5\right|\le\chi;\\
(w_{2.1}^l+4\cdot w_{2.1}^m+w_{2.1}^u)/6+(w_{2.2}^l+4\cdot w_{2.2}^m+w_{2.2}^u)/6=\\
w_j^l\le w_j^m\le w_j^u,\ \forall j=1,2\\
w_j^l,w_j^m,w_j^u\ge 0,\ \forall j=1,2
\end{cases}$$

Figure A1. *Cont.*

Expert 1 (C3.1 − C3.4) $\min \chi$

s.t.

$$\begin{cases} \left|\frac{w^l_{3.1}}{w^u_{3.4}} - 2.5\right| \le \chi \,;\, \left|\frac{w^m_{3.1}}{w^m_{3.4}} - 3\right| \le \chi \,;\, \left|\frac{w^u_{3.1}}{w^l_{3.4}} - 3.5\right| \le \chi; \\ \left|\frac{w^l_{3.4}}{w^u_{3.2}} - 0.29\right| \le \chi \,;\, \left|\frac{w^m_{3.4}}{w^m_{3.2}} - 0.33\right| \le \chi \,;\, \left|\frac{w^u_{3.4}}{w^l_{3.2}} - 0.4\right| \le \chi; \\ \left|\frac{w^l_{3.2}}{w^u_{3.3}} - 1\right| \le \chi \,;\, \left|\frac{w^m_{3.2}}{w^m_{3.3}} - 1\right| \le \chi \,;\, \left|\frac{w^u_{3.2}}{w^l_{3.3}} - 1\right| \le \chi; \\ \left|\frac{w^l_{3.1}}{w^u_{3.2}} - 0.7\right| \le \chi \,;\, \left|\frac{w^m_{3.1}}{w^m_{3.2}} - 0.1\right| \le \chi \,;\, \left|\frac{w^u_{3.1}}{w^l_{3.2}} - 1.4\right| \le \chi; \\ \left|\frac{w^l_{3.4}}{w^u_{3.3}} - 0.29\right| \le \chi \,;\, \left|\frac{w^m_{3.4}}{w^m_{3.3}} - 0.33\right| \le \chi \,;\, \left|\frac{w^u_{3.4}}{w^l_{3.3}} - 0.4\right| \le \chi; \\ \sum_{j=1}^{4} (w^l_j + 4 \cdot w^m_j + w^u_j)/6 = 1 \,,\, \forall j = 1,2,3,4 \\ w^l_j \le w^m_j \le w^u_j,\ \forall j = 1,2,3,4 \\ w^l_j, w^m_j, w^u_j \ge 0 \,,\, \forall j = 1,2,3,4 \end{cases}$$

……

Expert 7 (C3.1 − C3.4) $\min \chi$

s.t.

$$\begin{cases} \left|\frac{w^l_{3.1}}{w^u_{3.3}} - 2.5\right| \le \chi \,;\, \left|\frac{w^m_{3.1}}{w^m_{3.3}} - 3\right| \le \chi \,;\, \left|\frac{w^u_{3.1}}{w^l_{3.3}} - 3.5\right| \le \chi; \\ \left|\frac{w^l_{3.3}}{w^u_{3.4}} - 0.19\right| \le \chi \,;\, \left|\frac{w^m_{3.3}}{w^m_{3.4}} - 0.33\right| \le \chi \,;\, \left|\frac{w^u_{3.3}}{w^l_{3.4}} - 0.6\right| \le \chi; \\ \left|\frac{w^l_{3.4}}{w^u_{3.2}} - 0.45\right| \le \chi \,;\, \left|\frac{w^m_{3.4}}{w^m_{3.2}} - 1\right| \le \chi \,;\, \left|\frac{w^u_{3.4}}{w^l_{3.2}} - 2.24\right| \le \chi; \\ \left|\frac{w^l_{3.1}}{w^u_{3.4}} - 0.48\right| \le \chi \,;\, \left|\frac{w^m_{3.1}}{w^m_{3.4}} - 1\right| \le \chi \,;\, \left|\frac{w^u_{3.1}}{w^l_{3.4}} - 2.1\right| \le \chi; \\ \left|\frac{w^l_{3.3}}{w^u_{3.2}} - 0.09\right| \le \chi \,;\, \left|\frac{w^m_{3.3}}{w^m_{3.2}} - 0.33\right| \le \chi \,;\, \left|\frac{w^u_{3.3}}{w^l_{3.2}} - 1.34\right| \le \chi; \\ \sum_{j=1}^{4} (w^l_j + 4 \cdot w^m_j + w^u_j)/6 = 1 \,,\, \forall j = 1,2,3,4 \\ w^l_j \le w^m_j \le w^u_j,\ \forall j = 1,2,3,4 \\ w^l_j, w^m_j, w^u_j \ge 0 \,,\, \forall j = 1,2,3,4 \end{cases}$$

Expert 1 (C4.1 − C4.2) $\min \chi$

s.t.

$$\begin{cases} \left|\frac{w^l_{4.2}}{w^u_{4.1}} - 2.5\right| \le \chi \,;\, \left|\frac{w^m_{4.2}}{w^m_{4.1}} - 3\right| \le \chi \,;\, \left|\frac{w^u_{4.2}}{w^l_{4.1}} - 3.5\right| \le \chi \\ (w^l_{2.1} + 4 \cdot w^m_{2.1} + w^u_{2.1})/6 + (w^l_{2.2} + 4 \cdot w^m_{2.2} + w^u_{2.2})/6 = 1 \\ w^l_j \le w^m_j \le w^u_j,\ \forall j = 1,2 \\ w^l_j, w^m_j, w^u_j \ge 0 \,,\, \forall j = 1,2 \end{cases}$$

……

Expert 7 (C4.1 − C4.2) $\min \chi$

s.t.

$$\begin{cases} \left|\frac{w^l_{4.1}}{w^u_{4.2}} - 2.5\right| \le \chi \,;\, \left|\frac{w^m_{4.1}}{w^m_{4.2}} - 3\right| \le \chi \,;\, \left|\frac{w^u_{4.2}}{w^l_{4.1}} - 3.5\right| \le \chi \\ (w^l_{2.1} + 4 \cdot w^m_{2.1} + w^u_{2.1})/6 + (w^l_{2.2} + 4 \cdot w^m_{2.2} + w^u_{2.2})/6 = \\ w^l_j \le w^m_j \le w^u_j,\ \forall j = 1,2 \\ w^l_j, w^m_j, w^u_j \ge 0 \,,\, \forall j = 1,2 \end{cases}$$

Expert 1 (C5.1 − C5.3) $\min \chi$

s.t.

$$\begin{cases} \left|\frac{w^l_{5.1}}{w^u_{5.2}} - 0.67\right| \le \chi \,;\, \left|\frac{w^m_{5.1}}{w^m_{5.2}} - 1\right| \le \chi \,;\, \left|\frac{w^u_{5.1}}{w^l_{5.2}} - 1.5\right| \le \chi; \\ \left|\frac{w^l_{5.2}}{w^u_{5.3}} - 2.33\right| \le \chi \,;\, \left|\frac{w^m_{5.2}}{w^m_{5.3}} - 4\right| \le \chi \,;\, \left|\frac{w^u_{5.2}}{w^l_{5.3}} - 6.7\right| \le \chi; \\ \left|\frac{w^l_{5.1}}{w^u_{5.3}} - 1.56\right| \le \chi \,;\, \left|\frac{w^m_{5.1}}{w^m_{5.3}} - 4.00\right| \le \chi \,;\, \left|\frac{w^u_{5.1}}{w^l_{5.3}} - 10.07\right| \le \chi; \\ \sum_{j=1}^{3} (w^l_j + 4 \cdot w^m_j + w^u_j)/6 = 1 \,,\, \forall j = 1,2,3 \\ w^l_j \le w^m_j \le w^u_j,\ \forall j = 1,2,3 \\ w^l_j, w^m_j, w^u_j \ge 0 \,,\, \forall j = 1,2,3 \end{cases}$$

……

Expert 7 (C5.1 − C5.3) $\min \chi$

s.t.

$$\begin{cases} \left|\frac{w^l_{5.3}}{w^u_{5.2}} - 0.67\right| \le \chi \,;\, \left|\frac{w^m_{5.3}}{w^m_{5.2}} - 1\right| \le \chi \,;\, \left|\frac{w^u_{5.3}}{w^l_{5.2}} - 1.5\right| \le \chi; \\ \left|\frac{w^l_{5.2}}{w^u_{5.1}} - 0.67\right| \le \chi \,;\, \left|\frac{w^m_{5.2}}{w^m_{5.1}} - 1\right| \le \chi \,;\, \left|\frac{w^u_{5.2}}{w^l_{5.1}} - 1.49\right| \le \chi; \\ \left|\frac{w^l_{5.3}}{w^u_{5.1}} - 0.45\right| \le \chi \,;\, \left|\frac{w^m_{5.3}}{w^m_{5.1}} - 2\right| \le \chi \,;\, \left|\frac{w^u_{5.3}}{w^l_{5.1}} - 2.24\right| \le \chi; \\ \sum_{j=1}^{3} (w^l_j + 4 \cdot w^m_j + w^u_j)/6 = 1 \,,\, \forall j = 1,2,3 \\ w^l_j \le w^m_j \le w^u_j,\ \forall j = 1,2,3 \\ w^l_j, w^m_j, w^u_j \ge 0 \,,\, \forall j = 1,2,3 \end{cases}$$

Figure A1

Appendix C

Table A2. Local values of the fuzzy weight coefficients of the sub-criteria.

	E1-C1-C5			**E2-C1-C5**			**E3-C1-C5**			**E4-C1-C5**			**E5-C1-C5**			**E6-C1-C5**			**E7-C1-C5**			**Sum**		
C1	0.10	0.12	0.13	0.29	0.29	0.29	0.16	0.28	0.28	0.26	0.35	0.39	0.26	0.27	0.27	0.14	0.26	0.26	0.19	0.28	0.28	0.20	0.27	0.27
C2	0.28	0.33	0.38	0.19	0.22	0.22	0.22	0.22	0.22	0.10	0.11	0.11	0.17	0.29	0.29	0.09	0.27	0.27	0.07	0.09	0.09	0.16	0.22	0.23
C3	0.07	0.07	0.07	0.10	0.15	0.22	0.18	0.28	0.29	0.26	0.36	0.39	0.17	0.29	0.29	0.06	0.15	0.15	0.11	0.20	0.20	0.13	0.21	0.23
C4	0.26	0.30	0.33	0.08	0.25	0.50	0.06	0.15	0.15	0.09	0.12	0.13	0.07	0.10	0.11	0.16	0.27	0.27	0.07	0.25	0.34	0.11	0.21	0.26
C5	0.12	0.18	0.29	0.06	0.08	0.09	0.04	0.12	0.12	0.08	0.08	0.08	0.06	0.10	0.13	0.06	0.15	0.15	0.12	0.25	0.29	0.08	0.14	0.17
	E1-C1.1-C1.4			**E2-C1.1-C1.4**			**E3-C1.1-C1.4**			**E4-C1.1-C1.4**			**E5-C1.1-C1.4**			**E6-C1.1-C1.4**			**E7-C1.1-C1.4**					
C11	0.34	0.48	0.53	0.08	0.16	0.20	0.18	0.33	0.34	0.27	0.37	0.41	0.30	0.52	0.54	0.15	0.29	0.29	0.10	0.16	0.16	0.20	0.33	0.35
C12	0.09	0.20	0.22	0.08	0.16	0.22	0.09	0.15	0.15	0.10	0.12	0.12	0.10	0.17	0.17	0.14	0.28	0.28	0.13	0.16	0.16	0.10	0.18	0.19
C13	0.11	0.11	0.11	0.16	0.46	0.53	0.20	0.31	0.31	0.19	0.45	0.65	0.07	0.14	0.15	0.15	0.24	0.24	0.19	0.35	0.36	0.15	0.29	0.34
C14	0.18	0.25	0.25	0.26	0.26	0.26	0.17	0.28	0.28	0.08	0.08	0.08	0.20	0.23	0.23	0.17	0.26	0.26	0.12	0.40	0.52	0.17	0.25	0.27
	E1-C2.1-C2.2			**E2-C2.1-C2.2**			**E3-C2.1-C2.2**			**E4-C2.1-C2.2**			**E5-C2.1-C2.2**			**E6-C2.1-C2.2**			**E7-C2.1-C2.2**					
C21	0.73	0.73	0.85	0.46	0.46	0.69	0.46	0.46	0.69	0.46	0.46	0.69	0.46	0.46	0.69	0.50	0.67	0.83	0.46	0.46	0.69	0.51	0.53	0.73
C22	0.24	0.24	0.29	0.46	0.46	0.69	0.46	0.46	0.69	0.46	0.46	0.69	0.46	0.46	0.69	0.33	0.33	0.33	0.46	0.46	0.69	0.41	0.41	0.58
	E1-C3.1-C3.4			**E2-C3.1-C3.4**			**E3-C3.1-C3.4**			**E4-C3.1-C3.4**			**E5-C3.1-C3.4**			**E6-C3.1-C3.4**			**E7-C3.1-C3.4**					
C31	0.18	0.18	0.50	0.08	0.08	0.08	0.16	0.32	0.32	0.25	0.31	0.35	0.25	0.41	0.63	0.16	0.32	0.32	0.18	0.18	0.37	0.18	0.26	0.37
C32	0.25	0.25	0.25	0.15	0.43	0.59	0.05	0.17	0.20	0.28	0.31	0.31	0.29	0.29	0.29	0.07	0.15	0.15	0.10	0.33	0.55	0.17	0.28	0.33
C33	0.40	0.40	0.40	0.13	0.18	0.18	0.16	0.34	0.34	0.07	0.07	0.07	0.08	0.14	0.24	0.15	0.30	0.30	0.09	0.09	0.09	0.16	0.22	0.23
C34	0.11	0.11	0.13	0.25	0.35	0.39	0.12	0.26	0.26	0.20	0.32	0.43	0.15	0.15	0.15	0.18	0.32	0.32	0.20	0.28	0.28	0.17	0.25	0.28
	E1-C4.1-C4.2			**E2-C4.1-C4.2**			**E3-C4.1-C4.2**			**E4-C4.1-C4.2**			**E5-C4.1-C4.2**			**E6-C4.1-C4.2**			**E7-C4.1-C4.2**					
C41	0.24	0.24	0.29	0.46	0.46	0.69	0.46	0.46	0.69	0.18	0.20	0.23	0.46	0.46	0.69	0.46	0.46	0.69	0.73	0.73	0.85	0.43	0.43	0.59
C42	0.73	0.73	0.85	0.46	0.46	0.69	0.46	0.46	0.69	0.80	0.80	0.80	0.46	0.46	0.69	0.46	0.46	0.69	0.24	0.24	0.29	0.52	0.52	0.67
	E1-C5.1-C5.3			**E2-C5.1-C5.3**			**E3-C5.1-C5.3**			**E4-C5.1-C5.3**			**E5-C5.1-C5.3**			**E6-C5.1-C5.3**			**E7-C5.1-C5.3**					
C51	0.29	0.44	0.65	0.18	0.18	0.18	0.33	0.33	0.33	0.50	0.57	0.63	0.43	0.57	0.71	0.49	0.49	0.49	0.23	0.23	0.56	0.35	0.40	0.51
C52	0.44	0.44	0.44	0.45	0.54	0.63	0.22	0.33	0.49	0.14	0.14	0.14	0.29	0.29	0.29	0.15	0.15	1.11	0.30	0.30	0.41	0.28	0.31	0.50
C53	0.06	0.11	0.19	0.18	0.27	0.42	0.33	0.33	0.33	0.20	0.28	0.42	0.10	0.14	0.20	0.19	0.19	0.26	0.33	0.38	0.53	0.20	0.24	0.34

Appendix D

Table A3. Linguistic assessments of potential alternatives against the criteria.

		C1.1	C1.2	C1.3	C1.4	C2.1	C2.2	C3.1	C3.2	C3.3	C3.4	C4.1	C4.2	C5.1	C5.2	C5.3
A1	E1	TH	M	M	TH	H	M	H	M	H	M	H	H	TH	H	M
	E2	TF	TH	F	TH	M	TH	TH	TH	TH	TH	TF	TH	TH	TH	TH
	E3	TH	M	M	TH	TH	M	H	TF	TH	TH	F	M	TH	TH	TF
	E4	TH	M	M	TH	TH	M	TH	H	H	TH	H	F	TH	TH	F
	E5	TH	M	F	TH	H	M	TH	H	H	TH	M	TF	TH	H	F
	E6	TH	TH	TH	TH	TH	TH	TH	TH	TH	TH	TH	TH	TH	TH	TH
	E7	TH	H	TH	M	TH	H	TH	M	M	TH	H	H	H	H	M
A2	E1	H	M	F	H	H	M	H	M	H	H	F	TF	M	M	M
	E2	H	H	F	H	M	H	TH	TH	M	M	TF	TF	M	F	TH
	E3	TH	M	F	TH	TH	M	H	TF	H	M	TF	F	TH	M	M
	E4	H	H	F	M	M	H	TH	H	M	TH	H	F	H	H	M
	E5	TH	M	H	TH	H	M	H	H	F	H	TF	M	H	M	F
	E6	F	F	F	F	M	F	M	M	M	M	M	M	F	F	F
	E7	TH	H	TH	TH	TH	H	TH	M	M	TH	H	M	H	H	H
A3	E1	F	M	F	M	H	M	H	M	M	M	F	H	M	M	H
	E2	M	M	F	M	M	M	TH	TH	TH	H	TF	H	M	F	F
	E3	H	M	F	H	TH	M	H	TF	H	TH	TF	F	H	M	H
	E4	TH	TF	M	F	H	TF	TH	H	TH	H	M	F	H	H	TH
	E5	H	M	TF	H	H	M	M	TH	M	H	F	TF	H	M	TH
	E6	TH	TH	TH	TH	TH	TH	TH	TH	TH	TH	TH	TH	TH	TH	TH
	E7	TH	H	TH	TH	TH	H	TH	M	M	TH	H	M	H	H	TH
A4	E1	F	H	M	M	M	H	M	M	TF	F	F	M	M	M	M
	E2	M	M	F	TH	M	M	TH	TF	TF	TF	TF	TF	M	F	H
	E3	H	H	F	M	TH	H	TH	TF	H	M	TF	F	M	M	M
	E4	TH	F	M	M	H	F	TH	H	M	M	M	F	M	M	H
	E5	H	M	M	TH	H	M	M	H	M	H	F	TF	M	F	H
	E6	TH	F	F	F	M	F	M	M	M	M	M	M	F	F	F
	E7	TH	H	TH	TH	TH	H	TH	M	M	TH	H	M	H	H	TH
A5	E1	F	H	M	M	M	H	M	M	F	F	F	M	M	M	M
	E2	M	M	F	TH	M	M	TH	TF	TF	TF	TF	TF	M	F	H
	E3	H	H	F	F	TH	H	TH	TF	H	M	TF	F	F	M	M
	E4	TH	TH	F	M	H	TH	TH	H	F	M	M	F	M	M	F
	E5	H	F	TF	M	H	F	F	F	M	H	TF	F	M	H	TF
	E6	TH	F	F	F	M	F	M	M	M	M	M	M	F	F	F
	E7	TH	TH	TH	M	TH	TH	TH	M	M	M	H	M	M	M	M
A6	E1	TH	F	TF	TH	H	F	H	M	H	F	F	F	H	F	F
	E2	M	H	F	TH	TF	H	TH	TF	TF	TF	TF	TF	TH	M	F
	E3	F	F	F	F	TH	F	TH	TF	H	M	TF	F	F	F	TF
	E4	H	M	F	F	H	M	TH	H	M	F	M	F	M	M	M
	E5	H	TF	TF	TH	M	TF	H	F	M	H	TF	F	M	H	TF
	E6	F	F	F	F	F	F	F	F	F	F	F	F	F	F	F
	E7	H	TH	TH	M	TH	TH	TH	M	M	M	H	H	M	TH	H
A7	E1	TF	F	M	TF	TH	F	H	M	H	F	F	F	TF	F	M
	E2	H	H	F	TF	M	H	TH	TH	H	M	TF	TF	TH	TH	M
	E3	TH	F	F	TH	TH	F	TH	TF	TH	TH	TF	F	TH	TH	TH
	E4	TF	H	F	TF	TH	H	TH	H	H	F	M	F	TF	TF	H
	E5	F	F	TF	TF	M	F	M	F	F	H	TF	F	H	TF	TF
	E6	TH	TH	TH	TH	TH	TH	TH	TH	TH	TH	TH	TH	TH	TH	TH
	E7	TF	H	TH	TH	TH	H	TH	M	M	TH	H	H	TH	TH	M

Appendix E

Table A4. Preference function of alternatives over criteria.

		C1.1	C1.2	C1.3	C1.4	C2.1	C2.2	C3.1	C3.2	C3.3	C3.4	C4.1	C4.2	C5.1	C5.2	C5.3	$\pi(a,b)$		
A1	A2	1	1	1	1	1	0	1	1	1	1	1	1	1	1	0	0.45	0.90	1.26
	A3	1	0	1	1	0	1	1	1	0	1	1	1	1	1	0	0.40	0.78	1.12
	A4	1	0	1	1	1	1	1	1	1	1	1	1	1	1	1	0.51	0.98	1.39
	A5	1	0	1	1	1	1	1	1	0	1	1	1	1	1	1	0.49	0.93	1.34
	A6	1	1	1	1	1	1	1	1	1	1	1	1	1	1	1	0.53	1.03	1.44
	A7	1	1	1	1	0	1	1	1	1	1	1	1	1	1	0	0.44	0.88	1.22
A2	A1	0	0	0	0	0	0	0	0	0	0	0	0	0	0	1	0.02	0.03	0.06
	A3	1	0	0	0	0	1	0	1	0	0	0	0	0	0	1	0.14	0.27	0.36
	A4	1	0	0	0	0	1	1	1	1	1	0	0	1	1	1	0.26	0.52	0.73
	A5	1	0	0	0	0	1	1	1	0	1	1	0	0	1	1	0.26	0.51	0.75
	A6	1	1	1	0	1	1	1	1	1	1	1	0	1	0	1	0.42	0.81	1.11
	A7	1	1	1	1	0	1	0	0	1	1	0	0	1	0	1	0.28	0.56	0.70
A3	A1	0	1	0	1	0	1	0	0	1	1	0	0	0	0	0	0.16	0.31	0.37
	A2	0	1	0	1	1	0	0	0	1	1	0	1	1	1	0	0.29	0.54	0.75
	A4	1	0	0	1	1	0	1	0	1	1	0	1	1	1	1	0.34	0.66	0.94
	A5	1	0	0	1	1	0	1	0	1	1	1	0	1	1	1	0.33	0.65	0.92
	A6	1	1	1	1	1	1	1	1	1	1	1	0	1	1	1	0.47	0.92	1.27
	A7	1	1	1	1	0	1	0	0	1	1	0	0	1	1	0	0.28	0.57	0.73
A4	A1	0	1	1	0	0	1	0	0	1	0	0	0	0	0	0	0.14	0.26	0.33
	A2	0	1	1	1	0	0	0	0	0	0	0	0	0	0	0	0.08	0.19	0.22
	A3	0	0	1	0	0	0	0	1	0	0	0	0	0	0	0	0.05	0.14	0.17
	A5	1	0	1	0	0	0	1	0	0	0	1	0	0	0	0	0.14	0.31	0.43
	A6	0	1	1	0	1	1	0	1	1	1	1	0	0	0	0	0.31	0.58	0.79
	A7	1	1	1	1	0	1	0	0	0	1	0	0	1	0	0	0.24	0.48	0.59
A5	A1	0	1	0	0	0	0	0	0	0	0	0	0	0	0	0	0.02	0.05	0.05
	A2	0	1	0	1	0	0	0	0	1	0	0	1	0	0	0	0.13	0.27	0.35
	A3	0	1	1	1	1	1	1	1	1	0	1	1	1	1	0	0.45	0.85	1.23
	A4	0	1	0	1	0	1	0	1	1	1	0	1	1	1	0	0.29	0.57	0.79
	A6	0	1	1	1	1	1	0	1	1	1	0	0	1	0	0	0.32	0.61	0.79
	A7	1	1	1	1	0	1	0	0	1	1	0	0	1	0	0	0.26	0.53	0.65
A6	A1	0	0	0	0	0	0	0	0	0	0	0	0	0	0	0	0.00	0.00	0.00
	A2	0	0	0	1	0	0	0	0	0	0	0	1	0	0	0	0.09	0.17	0.25
	A3	0	0	0	0	0	0	0	0	0	0	0	0	0	0	0	0.00	0.00	0.00
	A4	1	0	0	0	0	0	0	0	0	0	0	1	1	1	1	0.16	0.32	0.49
	A6	1	0	0	0	0	0	1	0	0	0	1	0	0	1	1	0.15	0.31	0.47
	A7	1	0	1	1	0	1	0	0	0	1	0	0	1	0	0	0.22	0.43	0.54
A7	A1	0	0	0	0	0	0	0	0	0	0	0	0	0	0	1	0.02	0.03	0.06
	A2	0	0	0	0	1	0	1	0	0	0	1	1	0	1	0	0.23	0.41	0.66
	A3	0	0	0	0	0	0	1	1	0	0	1	1	0	0	0	0.15	0.31	0.49
	A4	0	0	0	0	1	0	1	1	1	0	1	1	0	1	1	0.29	0.55	0.85
	A5	0	0	0	0	1	0	1	1	0	0	1	1	0	1	1	0.27	0.50	0.80
	A7	0	1	0	0	1	0	1	1	0	0	1	1	0	1	1	0.29	0.55	0.85

References

1. Dablanc, L. Goods transport in large European cities: Difficult to organize, difficult to modernize. *Transp. Res. Part A Policy Pract.* **2007**, *41*, 280–285. [CrossRef]
2. Aloui, A.; Hamani, N.; Derrouiche, R.; Delahoche, L. Systematic literature review on collaborative sustainable transportation: Overview, analysis and perspectives. *Transp. Res. Interdiscip. Perspect.* **2021**, *9*, 100291. [CrossRef]
3. Aloui, A.; Hamani, N.; Derrouiche, R.; Delahoche, L. Assessing the benefits of horizontal collaboration using an integrated planning model for two-echelon energy efficiency-oriented logistics networks design. *Int. J. Syst. Sci. Oper. Logist.* **2021**, 1–22. [CrossRef]
4. Mrabti, N.; Hamani, N. A Sustainable Collaborative Approach to the Distribution Network Design Problem with CO2 Emissions Allocation. *Int. J. Shipp. Transp. Logist.* **2021**, Forthcoming.

5. Abdelhai, L.; Malhéné, N.; Gonzalez-Feliu, J. Logistique Urbaine Durable: Le CDU, Un Point de Convergence Entre Les Différents Acteurs. In Proceedings of the 1ère Conférence Internationale sur Les Systèmes Industriels et Logistiques (SIL 2014), Mar-rackech, Morocco, 23–24 October 2014. (In French)
6. Silva, O.C.T.; Leite, J.C. Logistics Platform (LP) approach in the stakeholders view. *Int. J. Adv. Eng. Res. Sci.* **2019**, *6*, 436–453. [CrossRef]
7. Agrebi, M. Méthodes D'aide à la Décision Multi-attribut et Multi-acteur Pour Résoudre le Problème de Sélection dans un Envi-ronnement Certain/Incertain: Cas de la Localisation des Centres de Distribution. Ph.D. Thesis, Université de Valenciennes et du Hai-naut-Cambresis, Valenciennes, France, 2018. (In French)
8. Agrebi, M.; Abed, M. Decision-making from multiple uncertain experts: Case of distribution center location selection. *Soft Comput.* **2021**, *25*, 4525–4544. [CrossRef]
9. Kumar, A.; Anbanandam, R. Location selection of multimodal freight terminal under STEEP sustainability. *Res. Transp. Bus. Manag.* **2019**, *33*, 100434. [CrossRef]
10. Yazdani, M.; Chatterjee, P.; Pamucar, D.; Chakraborty, S. Development of an integrated decision making model for location selection of logistics centers in the Spanish autonomous communities. *Expert Syst. Appl.* **2020**, *148*, 113208. [CrossRef]
11. Cheng, Y.-H.; Zhou, S.-Y. Research of distribution center site selection based on fuzzy analytic hierarchy process. In *Proceedings of the 22nd International Conference on Industrial Engineering and Engineering Management 2015*; Atlantis Press: Paris, France, 2016; pp. 335–342.
12. Pamucar, D.S.; Tarle, S.P.; Parezanovic, T. New hybrid multi-criteria decision-making DEMATEL-MAIRCA model: Sustainable selection of a location for the development of multimodal logistics centre. *Econ. Res. Ekon. Istraž.* **2018**, *31*, 1641–1665. [CrossRef]
13. Muravev, D.; Hu, H.; Zhou, H.; Pamucar, D. Location optimization of CR express international logistics centers. *Symmetry* **2020**, *12*, 143. [CrossRef]
14. Ulutaş, A.; Karakuş, C.B.; Topal, A. Location selection for logistics center with fuzzy SWARA and CoCoSo methods. *J. Intell. Fuzzy Syst.* **2020**, *38*, 4693–4709. [CrossRef]
15. He, Y.; Wang, X.; Lin, Y.; Zhou, F.; Zhou, L. Sustainable decision making for joint distribution center location choice. *Transp. Res. Part D Transp. Environ.* **2017**, *55*, 202–216. [CrossRef]
16. Kazançoğlu, Y.; Özbiltekin, M.; Özkan-Özen, Y.D. Sustainability benchmarking for logistics center location decision. *Manag. Environ. Qual. Int. J.* **2019**, *31*, 1239–1260. [CrossRef]
17. Karaşan, A.; Kahraman, C. A novel intuitionistic fuzzy DEMATEL-ANP-TOPSIS integrated methodology for freight village location selection. *J. Intell. Fuzzy Syst.* **2019**, *36*, 1335–1352. [CrossRef]
18. Nardo, M.; Saisana, M.; Saltelli, A.; Tarantola, T. *Tools for Composite Indicators Building*; European Commission: Ispra, Italy, 2005; Volume 15.
19. Saisana, M.; Tarantola, S. *State-of-the-Art Report on Current Methodologies and Practices for Composite Indicator Development*; European Commission, Joint Research Centre: Ispra, Italy, 2002; Volume 214.
20. Munda, G. Multiple criteria decision analysis and sustainable development. In *Multiple Criteria Decision Analysis: State of the Art Surveys*; Springer: New York, NY, USA, 2005; pp. 953–986.
21. Garcia-Bernabeu, A.; Hilario-Caballero, A.; Pla-Santamaria, D.; Salas-Molina, F. A process oriented MCDM approach to construct a circular economy composite index. *Sustainability* **2020**, *12*, 618. [CrossRef]
22. Savić, G.; Martić, M. Composite indicators construction by data envelopment analysis: Methodological background. In *Emerging Trends in the Development and Application of Composite Indicators*; IGI Global: Hershey, PA, USA, 2017; pp. 98–126, ISBN 978-1-5225-0714-7.
23. Gan, X.; Fernandez, I.C.; Guo, J.; Wilson, M.; Zhao, Y.; Zhou, B.; Wu, J. When to use what: Methods for weighting and aggregating sustainability indicators. *Ecol. Indic.* **2017**, *81*, 491–502. [CrossRef]
24. Ruiz, F.; Cabello, J. MRP-PCI: A multiple reference point based partially compensatory composite indicator for sustainability assessment. *Sustainability* **2021**, *13*, 1261. [CrossRef]
25. Triantaphyllou, E. Multi-criteria decision making methods. In *Multi-Criteria Decision Making Methods: A Comparative Study*; Springer: New York, NY, USA, 2000; pp. 5–21, ISBN 978-0-7923-6607-2.
26. Tešić, M.; Hermans, E.; Lipovac, K.; Pešić, D. Identifying the most significant indicators of the total road safety performance index. *Accid. Anal. Prev.* **2018**, *113*, 263–278. [CrossRef]
27. Sharifi, A. A typology of smart city assessment tools and indicator sets. *Sustain. Cities Soc.* **2020**, *53*, 101936. [CrossRef]
28. El Gibari, S.; Gómez, T.; Ruiz, F. Building composite indicators using multicriteria methods: A review. *J. Bus. Econ.* **2018**, *89*, 1–24. [CrossRef]
29. Blancas, F.J.; González, M.; Lozano-Oyola, M.; Pérez, F. The assessment of sustainable tourism: Application to Spanish coastal destinations. *Ecol. Indic.* **2010**, *10*, 484–492. [CrossRef]
30. Saaty, R.M. The analytic hierarchy process—What it is and how it is used. *Math. Model.* **1987**, *9*, 161–176. [CrossRef]
31. Rezaei, J. Best-worst multi-criteria decision-making method. *Omega* **2015**, *53*, 49–57. [CrossRef]
32. Pamučar, D.; Stević, Ž.; Sremac, S. A new model for determining weight coefficients of criteria in MCDM models: Full consistency method (FUCOM). *Symmetry* **2018**, *10*, 393. [CrossRef]
33. Fazlollahtabar, H.; Smailbašić, A.; Stević, Ž. FUCOM method in group decision-making: Selection of forklift in a warehouse. *Decis. Mak. Appl. Manag. Eng.* **2019**, *2*, 49–65. [CrossRef]

34. Pamučar, D.; Lukovac, V.; Bozanic, D.; Komazec, N. Multi-criteria FUCOM-MAIRCA model for the evaluation of level crossings: Case study in the Republic of Serbia. *Oper. Res. Eng. Sci. Theory Appl.* **2019**, *1*, 108–129. [CrossRef]
35. Badi, I.; Abdulshahed, A. Ranking the Libyan airlines by using Full Consistency Method (FUCOM) and Analytical Hierarchy Process (AHP). *Oper. Res. Eng. Sci. Theory Appl.* **2019**, *2*, 1–14. [CrossRef]
36. Matić, B.; Jovanović, S.; Das, D.K.; Zavadskas, E.K.; Stević, Ž.; Sremac, S.; Marinković, M. A new hybrid MCDM model: Sustainable supplier selection in a construction company. *Symmetry* **2019**, *11*, 353. [CrossRef]
37. Stević, Ž.; Brković, N. A novel integrated FUCOM-MARCOS model for evaluation of human resources in a transport company. *Logistics* **2020**, *4*, 4. [CrossRef]
38. Durmić, E. The evaluation of the criteria for sustainable supplier selection by using the FUCOM Method. *Oper. Res. Eng. Sci. Theory Appl.* **2019**, *2*, 91–107. [CrossRef]
39. Puška, A.; Pamucar, D.; Stojanović, I.; Cavallaro, F.; Kaklauskas, A.; Mardani, A. Examination of the Sustainable Rural Tourism Potential of the Brčko District of Bosnia and Herzegovina Using a Fuzzy Approach Based on Group Decision Making. *Sustainability* **2021**, *13*, 583. [CrossRef]
40. Noureddine, M.A.; Ristić, M. Route planning for hazardous materials transportation: Multicriteria decision making approach. *Decis. Mak. Appl. Manag. Eng.* **2019**, *2*, 66–85. [CrossRef]
41. Cao, Q.; Esangbedo, C.O.; Bai, S. Grey SWARA-FUCOM weighting method for contractor selection MCDM problem: A case study of floating solar panel energy system installation. *Energies* **2019**, *12*, 2481. [CrossRef]
42. Bozanic, D.; Tešić, D.; Milić, A. Multicriteria decision making model with Z-numbers based on FUCOM and MABAC model. *Decis. Mak. Appl. Manag. Eng.* **2020**, *3*, 19–36. [CrossRef]
43. Blagojević, A.; Kasalica, S.; Stević, Ž.; Tričković, G.; Pavelkić, V. Evaluation of safety degree at railway crossings in order to achieve sustainable traffic management: A novel integrated fuzzy MCDM model. *Sustainability* **2021**, *13*, 832. [CrossRef]
44. Stojčić, M.; Zavadskas, E.K.; Pamučar, D.; Stević, Ž.; Mardani, A. Application of MCDM methods in sustainability engineering: A literature review 2008–2018. *Symmetry* **2019**, *11*, 350. [CrossRef]
45. Buldeo Rai, H.; van Lier, T.; Macharis, C. Towards data-based mobility policies in flemish cities: Creating an inclusive sustainability index. In Proceedings of the BIVEC/GIBET Transport Research Days, Eindhoven, The Netherlands, 28 May 2015.
46. Gudmundsson, H.; Hall, R.P.; Marsden, G.; Zietsman, J. *Sustainable Transportation: Indicators, Frameworks, and Performance Management*; Springer Texts in Business and Economics; Springer: Berlin/Heidelberg, Germany, 2016; ISBN 978-3-662-46923-1.
47. Asadzadeh, A.; Kötter, T.; Salehi, P.; Birkmann, J. Operationalizing a concept: The systematic review of composite indicator building for measuring community disaster resilience. *Int. J. Disaster Risk Reduct.* **2017**, *25*, 147–162. [CrossRef]
48. Cabello, J.M.; Ruiz, F.; Pérez-Gladish, B. An alternative aggregation process for composite indexes: An application to the Heritage Foundation Economic Freedom Index. *Soc. Indic. Res.* **2021**, *153*, 443–467. [CrossRef]
49. Attardi, R.; Cerreta, M.; Sannicandro, V.; Torre, C.M. Non-compensatory composite indicators for the evaluation of urban planning policy: The Land-Use Policy Efficiency Index (LUPEI). *Eur. J. Oper. Res.* **2018**, *264*, 491–507. [CrossRef]
50. Munda, G.; Nardo, M. Noncompensatory/nonlinear composite indicators for ranking countries: A defensible setting. *Appl. Econ.* **2009**, *41*, 1513–1523. [CrossRef]
51. Mazziotta, M.; Pareto, A. Composite indices construction: The performance interval approach. *Soc. Indic. Res.* **2020**, 1–11. [CrossRef]
52. Munda, G.; Nardo, M.; Saisana, M.; Srebotnjak, T. Measuring uncertainties in composite indicators of sustainability. *Int. J. Environ. Technol. Manag.* **2009**, *11*, 7–26. [CrossRef]
53. De Mare, G.; Granata, M.F.; Nesticò, A. Weak and strong compensation for the prioritization of public investments: Multidimensional analysis for pools. *Sustainability* **2015**, *7*, 16022–16038. [CrossRef]
54. Ziemba, P. Towards strong sustainability management—A generalized PROSA method. *Sustainability* **2019**, *11*, 1555. [CrossRef]
55. Gigović, L.; Pamučar, D.; Bajić, Z.; Milićević, M. The combination of expert judgment and GIS-MAIRCA analysis for the selection of sites for ammunition depots. *Sustainability* **2016**, *8*, 372. [CrossRef]
56. Pamucar, D.; Vasin, L.; Lukovac, V. Selection of railway level crossings for investing in security equipment using hybrid DEMATEL-MARICA model. In Proceedings of the XVI International Scientific-expert Conference on Railway (Railcon '14), Niš, Serbia, 9 October 2014.
57. Brans, J.P.; Vincke, P. Note—A preference ranking organisation method. *Manag. Sci.* **1985**, *31*, 647–656. [CrossRef]
58. Ziemba, P.; Wątróbski, J.; Zioło, M.; Karczmarczyk, A. Using the PROSA method in offshore wind farm location problems. *Energies* **2017**, *10*, 1755. [CrossRef]
59. Mardani, A.; Nilashi, M.; Zavadskas, E.K.; Awang, S.R.; Zare, H.; Jamal, N.M. Decision making methods based on fuzzy aggregation operators: Three decades review from 1986 to 2017. *Int. J. Inf. Technol. Decis. Mak.* **2018**, *17*, 391–466. [CrossRef]
60. van Laarhoven, P.; Pedrycz, W. A fuzzy extension of Saaty's priority theory. *Fuzzy Sets Syst.* **1983**, *11*, 229–241. [CrossRef]
61. Pelissari, R.; Oliveira, M.C.; Abackerli, A.J.; Ben-Amor, S.; Assumpção, M.R.P. Techniques to model uncertain input data of multi-criteria decision-making problems: A literature review. *Int. Trans. Oper. Res.* **2021**, *28*, 523–559. [CrossRef]
62. Awasthi, A.; Chauhan, S.; Goyal, S. A multi-criteria decision making approach for location planning for urban distribution centers under uncertainty. *Math. Comput. Model.* **2011**, *53*, 98–109. [CrossRef]
63. Rao, C.; Goh, M.; Zhao, Y.; Zheng, J. Location selection of city logistics centers under sustainability. *Transp. Res. Part D Transp. Environ.* **2015**, *36*, 29–44. [CrossRef]

64. Mrabti, N.; Hamani, N.; Delahoche, L. The pooling of sustainable freight transport. *J. Oper. Res. Soc.* **2020**, 1–16. [CrossRef]
65. Regmi, M.B.; Hanaoka, S. Location analysis of logistics centres in Laos. *Int. J. Logist. Res. Appl.* **2013**, *16*, 227–242. [CrossRef]
66. Hanaoka, S.; Regmi, M.B. Promoting intermodal freight transport through the development of dry ports in Asia: An environmental perspective. *IATSS Res.* **2011**, *35*, 16–23. [CrossRef]
67. Rai, H.B.; Van Lier, T.; Meers, D.; Macharis, C. An indicator approach to sustainable urban freight transport. *J. Urban. Int. Res. Placemak. Urban Sustain.* **2017**, *11*, 81–102. [CrossRef]
68. Roso, V.; Brnjac, N.; Abramovic, B. Inland intermodal terminals location criteria evaluation: The case of Croatia. *Transportation* **2015**, *54*, 496. [CrossRef]
69. Tong, L.; Pu, Z.; Chen, K.; Yi, J. Sustainable maintenance supplier performance evaluation based on an extend fuzzy PROMETHEE II approach in petrochemical industry. *J. Clean. Prod.* **2020**, *273*, 122771. [CrossRef]
70. Pamučar, D.; Mihajlović, M.; Obradović, R.; Atanasković, P. Novel approach to group multi-criteria decision making based on interval rough numbers: Hybrid DEMATEL-ANP-MAIRCA model. *Expert Syst. Appl.* **2017**, *88*, 58–80. [CrossRef]
71. Boral, S.; Howard, I.; Chaturvedi, S.K.; McKee, K.; Naikan, V. An integrated approach for fuzzy failure modes and effects analysis using fuzzy AHP and fuzzy MAIRCA. *Eng. Fail. Anal.* **2020**, *108*, 104195. [CrossRef]
72. Liu, Z.; Li, L.; Zhao, X.; Sha, L.; Wang, D.; Wang, X.; Liu, P. Selecting the optimal green agricultural products supplier: A novel approach based on GBWM and PROMETHEE II. *Sustainability* **2020**, *12*, 6703. [CrossRef]
73. Segura, M.; Maroto, C. A multiple criteria supplier segmentation using outranking and value function methods. *Expert Syst. Appl.* **2017**, *69*, 87–100. [CrossRef]
74. Brans, J.; Vincke, P.; Mareschal, B. How to select and how to rank projects: The Promethee method. *Eur. J. Oper. Res.* **1986**, *24*, 228–238. [CrossRef]
75. Guo, S.; Zhao, H. Fuzzy best-worst multi-criteria decision-making method and its applications. *Knowl. Based Syst.* **2017**, *121*, 23–31. [CrossRef]
76. Digalwar, A.K.; Date, P.A. Development of fuzzy PROMETHEE algorithm for the evaluation of Indian world-class manufacturing organisations. *Int. J. Serv. Oper. Manag.* **2016**, *24*, 308. [CrossRef]
77. Mousavi, S.M.; Jolai, F.; Tavakkoli-Moghaddam, R. A fuzzy stochastic multi-attribute group decision-making approach for selection problems. *Group Decis. Negot.* **2013**, *22*, 207–233. [CrossRef]

sustainability

Article

Analytic Hierarchy Process-Based Airport Ground Handling Equipment Purchase Decision Model

Yu-Jwo Tao [1], Hsuan-Shih Lee [1] and Chang-Shu Tu [2,*]

[1] Department of Shipping and Transportation Management, National Taiwan Ocean University, No. 2, Pei-Ning Road, Keelung City 202-301, Taiwan; ta794260@gmail.com (Y.-J.T.); hslee@email.ntou.edu.tw (H.-S.L.)
[2] Department of Information Management, Chang Gung University, No .259, WenHwa 1st Road, KweiShan, Taoyuan City 33302, Taiwan
* Correspondence: long.tree@msa.hinet.net; Tel.: +886-927351833

Abstract: The Airport ground handling services (AGHS) equipment supplier provider selection requires a safety guarantee in terms of the daily operations AGHS provider. AGHS providers seek to avoid aircraft damage and airline delays and ensure the provision of reliable and high-quality services. The primary objective of this paper was to develop purchasing decision model of the analytic hierarchy process (AHP), AHP-fuzzy linear programming (FLP), and AHP-Taguchi loss function (TLF) multi-choice goal programming (MCGP) purchase decision models to help the AGHS purchasing managers in selecting the best AGHS equipment supplier provider. The constructed models were assessed, and results obtained for the AHP-FLP and AHP-TLF-MCGP models were compared. We conducted a real-world example of supplier selection by an AGHS company by using the proposed models. The proposed model provides useful information and has practical value for AGHS providers.

Keywords: equipment purchase decision; analytic hierarchy process; Taguchi loss function; fuzzy linear programming; multi-choice goal programming

check for updates

Citation: Tao, Y.-J.; Lee, H.-S.; Tu, C.-S. Analytic Hierarchy Process-Based Airport Ground Handling Equipment Purchase Decision Model. *Sustainability* **2021**, *13*, 2540. https://doi.org/10.3390/su13052540

Academic Editors: Dragan Pamucar, Željko Stević and Marc A. Rosen

Received: 30 December 2020
Accepted: 22 February 2021
Published: 26 February 2021

Publisher's Note: MDPI stays neutral with regard to jurisdictional claims in published maps and institutional affiliations.

Copyright: © 2021 by the authors. Licensee MDPI, Basel, Switzerland. This article is an open access article distributed under the terms and conditions of the Creative Commons Attribution (CC BY) license (https://creativecommons.org/licenses/by/4.0/).

1. Introduction

The Airport ground handling service (AGHS) companies are responsible for varied ground handling service activities. Ground handling operations can be categorized as either terminal or airside operations. In this study, we focused on decision models for identifying the best suppliers of equipment for airside operations. Ramp handling, a major task performed by AGHS companies [1], involves aircraft loading and unloading operations as well as passenger, crew, baggage, and freight (e.g., mail) transport between aircraft and terminal buildings [2]. Ramp handling classified as a logistics service, and it be offered by a third-party service provider ground handler service company (i.e., the AGHS company), an airline company (self-handling), or a ramp handling company within an airport [3,4].

AGHS duties are difficult for ground handlers whose work safety relies on technologically advanced equipment. Furthermore, AGHS companies experience anxieties with the ground handler service performance and quality of equipment. In addition, purchasing managers at AGHS companies select equipment suppliers with favorable reputations, and they have two reasons for doing so. First, the use of high-quality AGHS equipment can increase the safety of ground handling operations. Second, such supplier performs quality assurance, indicating that their equipment is of a high quality, and this can inspire confidence in AGHS providers and the airlines that pay for such services. AGHS equipment supplier selection (AGHSESS) problem has a marked effect on AGHS quality, and it is influenced by several factors. Such factors include a supplier's manufacturing performance, the supplier's industry reputation, and the supplier's produce quality and price [5]. In general, multiple-criteria decision-making (MCDM) methods assistance decision-makers (DMs) make estimations related to a list of options. Depending on the purchasing conditions, certain criteria have different levels of importance, thus necessitating criteria weighting [6,7]. Optimal AGHSESS is an MCDM problem [8]. AGHSESS problem participate a vital role in the decision-making process of an AGHS provider; however, optimal AGHSESS is a time wasting and difficult procedure that requires relevant experience and an understanding of the industry, and this task can be difficult for AGHS purchasing managers [9]. This study motivation has proposed an effective decision model for AGHSESS in order to enable DMs make decisions in scenarios that involve the consideration of numerous factors and the assessment of a substantial body of information and records [10]. The subsequently section presents a review of the related value of literature.

2. Literature Review

Supplier selection (SC) is extensively discussed in the literature. For example, Monczka et al. [11] suggested the use of factor analysis for assessing the value of suppliers. Vonderembse and Tracey [12] conducted a study involving 268 purchasing managers to determine their supplier selection criteria when suppliers constantly endeavored to improve their products. Suppliers can obtain information on buyer requirements, traditions, and decision-making patterns, and this can help them alter and better apply their resources. Such information is advantageous to suppliers [12], because it enables them to clearly understand customer needs. Bhutta and Huq [13] highlighted two main approaches that managers can use to construct decision criteria while considering supplier choices; one of the approaches involves considering the total cost of ownership, and the other is based on a hierarchical procedure. Sarkis and Talluri [14] used an analytic network process (ANP) decision model to perform a strategic supplier selection.

Although the AGHSESS is an important consideration among AGHS providers, publications on this topic are limited. Some studies have provided useful information pertaining to equipment supplier evaluations. For example, Sevkli et al. [5] applied a mixed technique (e.g., AHP-fuzzy linear programming (FLP) model to resolve the vendor choice problems. When the conventional AHP is used, AGHSESS is based on only price and reputation. Specifically, a buyer company must implement optimal supplier selection to increase their efficiency in response to uncertainty and resource constraints. Accordingly, the AHP-FLP technique is more useful than the conventional AHP techniques [7]. Goztepe and Kahraman [15] considered the military DM processes for battlefield operations and operations planning. Ordoobadi [16] combined the AHP and Taguchi loss function (TLF) to rank potential suppliers for outsourcing purposes. Furthermore, Liao and Kao [17] integrated AHP, TLF, and multi-choice goal programming (MCGP) models to resolve provider choice problems. Magdalena [18] combined the TLF and FLP to identify the optimal supplier in a given situation. By combing the TLF and AHP, Ordoobadi [19] provided an approach for ranking technology choices for achievement intentions. However, few investigations of AGHSESS and related techniques in practice have been published. To block this gap in the past literature, this paper applied an AHP-FLP and AHP-TLF-MGCP model in the context of a Taiwanese AGHS provider. Our main goal was to help the purchasing managers of such companies' use straightforward techniques to identify the most suitable equipment suppliers. For the sake of reducing the AHP and AHP-FLP model approaches drawbacks and reach precise results, we are using the AHP-TLF-MCGP model to validate of the AGHSESS problem. Our AHP-TLF-MGCP model can provide a valuable reference that allows DMs to set various desire levels for related equipment suppliers and buyers.

The rest of this manuscript is organized as follows: Section 2 summarizes the proposed methods; Section 3 details the proposed model for AGHSESS problems. In Section 4, presents findings related how to the use of the AHP-FLP and AHP-TLF-MGCP approaches to solve AGHSESS problems. We also explain the practical value of this approach in a real-world case. Lastly, conclusions and implications are presented in Section 5.

3. Methods

In this methods section, we explain how the AHP, AHP-FLP, and AHP-TLF-MGCP purchase decision models can be used to solve the aforementioned AGHSESS problem.

3.1. AHP Technique

The AHP technique is a judgment-based technique established by Saaty [20] for dealing with composite, and MDCM problems. The AHP is founded on three concepts: The formation of the model, a relative decision based on certain criteria, and a combination of the precedence. The AHP has been used to solve diverse MCDM problems [21–24].

Saaty [25] stated that in many real-world cases, the pair-wise decisions of DMs have several levels of ambiguity. Generally, the executive team tasked with AGHSESS is confident about the ranking of the assessment elements but unsure of whether the mathematical assessments related to such decisions are accurate. The traditional AHP-based approach to overcoming this issue is to introduce a distinct linguistic assessment of decisions. In preference to straight allocating numerical values to the comparison ratios, the AGHS equipment executive team selects a suitable linguistic expression that corresponds most to the decision options.

3.2. TLF Method

The TLF method for assessing loss as an outcome of a product does not meet any regulatory terms [26]. Loss calculation is conducted to quantitatively evaluate the loss of quality caused by inconsistencies between the specified quality and actual quality of a product.

In general, three styles of loss function are used to calculate Taguchi loss [27–29]. First, a two-sided loss function is used; in this function, a supposed value is the goal and deviation from either side of the goal is permitted provided that it remains within the speciation limits. In this study, quality loss functions were used to quantify the effect of AGHSESS tasks.

3.3. Fuzzy FLP Method

Bellman and Zadeh [30] proposed a fuzzy programming model for DM in the context of a fuzzy situation. Their method was used by Zimmermann [31] to solve FLP problems [10].

We used an AHP-weighted FLP model (hereafter referred to as AHP-FLP model) in our study to solve AGHSESS problems of practical importance to an AGHS provider in Taiwan. This work represents an attempt to overcome real-world obstacles encountered in the use of an AHP-FLP technique.

3.4. MCGP Approach

Scholars have devised many modified goal programming methods using MCGP. To recover the utility of goal programming (GP) methods, Chang [32] developed a model for resolve the multi-objective decision-making (MODM) problems using multi-choice aspiration levels (MCALs). Chang's proposal to solve MODM problems using MCALs differs considerably from a fuzzy goal programming approach in that their model incorporates membership functions (MFs) to address MODM problems with imprecise goal aspiration levels. The following equations address this issue based on a typical MCGP problem:

$$\text{Minimize} \sum_{i=1}^{n} \left[(d_i^+ + d_i^-) + (e_i^+ + e_i^-) \right] \tag{1}$$

Subject to

$$f_i(X)b_i - d_i^+ + d_i^- = b_i y_i \; i = 1, 2, \ldots, n, \tag{2}$$

$$y_i - e_i^+ + e_i^- = g_{i,\min} \; i = 1, 2, \ldots, n, \tag{3}$$

$$g_{i,\min} \le y_i \le g_{i,\max} \; i = 1, 2, \ldots, n, \tag{4}$$

$$d_i^+, d_i^-, e_i^+, e_i^- \geq 0,\ i = 1, 2, \ldots, n \quad (5)$$

$X \in F$ where F is an achievable set and X has no symbol constraints.

Refer to the case regarding the managerial implications of constraints in [32].

4. Proposed Model

The proposed model is on the foundation of AHP-FLP methodology for solving AGHSESS problems. The model procedure is described as follows.

4.1. Format of the AHP-FLP Purchase Decision Model Process Based on AHP-FLP

The model for the AGHSESS problem and the AHP-FLP model involves three essential phases: (1) Identifying the criteria chosen in the purchase decision model, (2) conducting AHP calculations, (3) and making optimal decisions using the AHP-FLP approach this will help the AGHS provider. The entire AHP process is detailed in [5].

In this proposed model section, the general multi-purpose fuzzy model for AGHSESS is current in the tracking method [5,7,33,34]. The fuzzy multi-objective formulation for the aforementioned selection problems is derived as follows:

Obtain a vector X, where X is $[x_1, x_2, x_3, \ldots x_n]$ that exploits the AGHSESS goal utility function z_k with a number of m criteria [5]:

$$\text{Max } \tilde{z}_k = \sum_{i=1}^{n} (c_{ki} \times x_i) \geq \ \sim \ z_k^0\ k = 1, 2, 3, \ \ldots, \ n \quad (6)$$

The following tracking constraints are applied:

$$\sum_{i=1}^{n} a_{ri} \times x_i \leq b_r \quad (7)$$

where c_{ki}, a_{ri}, and b_r are crisp values.

In this proposed model, ~ signifies the fuzzy situation. The symbol $\geq$ ~ denotes the fuzzified (i.e., unclear or fuzzy) edition of $\geq$ and indicate adequately equipped or better. Z_k^0 is the rank that the DM desires to achieve.

All goal purpose values, $\tilde{z}_k$, are adjusted linearly from $z_k^{\min}$ to $z_k^{\max}$. Such values may be treated similar to fuzzy numerals in relation to the linear membership function (LMF) as presented in Figure 1.

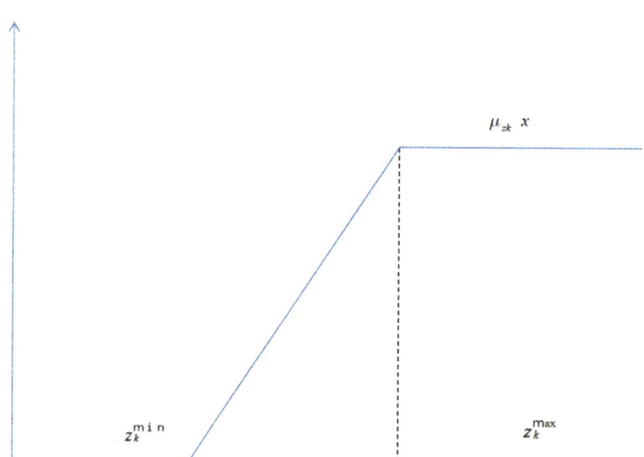

Figure 1. Maximizing objective function as fuzzy number.

As described in the aforementioned LMF, maximization aspirations ($\widetilde{z}_k$) are given as follows:

$$\mu_{z1}(x) = \begin{cases} 1 & \text{for} z_k \geq z_k^{\max} \\ \left(z_k^{\max} - z_k(x)\right) \ / \left(z_k^{\max} - z_k^{\min}\right) \text{ for } & z_k^{\min} < z_k < z_k^{\max} \\ 0 & \text{for} z_k \leq z_k^{\min} \end{cases} \quad (8)$$

The decision model presented in Equations (6) and (7) be present constructed using the weighted additive model, which is broadly used to address vector optimization problems; the essential idea is to use a single efficacy purpose to precise the general preferences of the DM to elucidate the significance of certain criteria [5,35]. In this methodology, all fuzzy MFs must be derived in terms of their relative weights, and subsequently, the derivation results must be added to obtain a linear weighted function.

$$\max \sum_{k=1}^{m} (w_k \times \lambda_k) \quad (9)$$

subject to:

$$\lambda_k \leq \mu_{zk}(x) \quad (10)$$

$$\lambda_k \in [0,1] \text{ and } k = 1,\ 2,\ 3,\ \ldots\ n \quad (11)$$

$$\sum_{k=1}^{m} w_k = 1, w_k \geq 0 \quad (12)$$

$$x_i \geq 0,\ i = 1,\ 2,\ 3,\ \ldots\ m \quad (13)$$

where w_k and $\mu_{zk}(x)$ denote the resolution of the MF and the weighted coefficients that have a crucial relationship with the fuzzy objectives and membership functions.

In the third phased of the proposed approach, the AHP-FLP decision model helps the AGHS provider make optimal purchase decisions. The operating procedures of this AHP-FLP decision model are as follows:

Step 1: An AGHSESS criterion is selected for determining the hierarchical construction for the choice of the finest supplier.
Step 2: The team of experts and researchers conduct weight computation for the criteria at various hierarchical levels to determine the general achievements for each AGHS equipment supplier by conducting pairwise comparisons of the major decision criteria.
Step 3: According to the criteria identified for related equipment selection, the AGHSESS purchase decision model is constructed.
Step 4: Define the lower bound ($z_k^{\min}$) and upper bound ($z_k^{\max}$) multi-objective purchase decision problem is similar to a single-objective linear programming model.
Step 5: $z_k^{\min}$ and $z_k^{\max}$ assessments are used to obtain the LMF for the criterion in Equation (8).
Step 6: On the basis of the weighted additive model, we create the corresponding crispy typical of the fuzzy optimization problem using Equations (6)–(13).
Step 7: We identify the best result vector X, which represents the expert decision on the unique purchase decision problem.
Step 8: We compare the AHP and AHP-FLP models.
Step 9: On the basis of the results of the AHP-FLP model, we construct the AHP-TLF-MCGP model to solve the AGHSESS problem according to Equations (1)–(5). The loss function, weighted Taguchi values, and normalized values derived in the calculation are summarized in Tables 1–12 (further details on the procedure are provided in [19]).
Step 10: The results obtained using the AHP-FLP model and AHP-TLF-MCGP models are compared.

Table 1. Equipment supplier decision criteria.

Equipment Supplier Decision Criteria	Quality Guarantee Definition
Quality management	system (ISO/TS 16 Quality guarantee system(ISO/TS 16949/QS-9000/ISO 14001) policy and domestic quality inspections.
Production capacity and maintenance	Producing ability contain high-quality utilize of statistical process control (SPC), lean manufacturing and a "kanban" system. Supplier novelty abilities comprise hardware, software (CAD/CAE/CAM), information, works and skill. The fix and preservation examine sustains customer agreement.
Product warranty	Suppliers trail assurances and include an assessment procedure to define what forces enhancements in assurance expenses and buyer agreement
Provide technical transfer	The scientific compatibility of the overhaul, the substantial or the parts that are affording to the retail corporation is vital.
Good cooperative relationship and reputation	A durable and flouring buyer/supplier association needs shared reliance and consideration The provider has a fine financial situation in the manufacturing
Reasonable parts price	The supplier provides reasonable parts prices.

Table 2. Criteria matrix.

	Quality	Maintenance	Warranty	Technical	Reputation	Price
Quality	1	2	3	1/2	4	1/3
Maintenance	1/2	1	1/2	1/4	2	1/7
Warranty	1/3	2	1	1/3	2	1/6
Technical	2	4	3	1	6	1/2
Reputation	1/4	1/2	1/2	1/6	1	1/9
Price	3	7	6	2	9	1

Table 3. Pair-wise comparision template criteria.

Supplier Criteria	Weights (w)	λ_{max},CI,RI	CR
SC1 (Quality)	0.151	λ_{max} = 6.521	
SC2 (Maintenance)	0.062		
SC3 (Warranty)	0.079		
SC4 (Technical)	0.241	CI = 0.104 RI = 1.24	0.084
SC5 (Reputation)	0.039		
SC6 (Price)	0.428		

Table 4. Adjusted template [37].

	Quality	Maintenance	Warranty	Technical	Reputation	Price	(Weights Row Average)
Quality	0.141	0.121	0.214	0.120	0.167	0.148	0.152
Maintenance	0.071	0.061	0.036	0.060	0.083	0.063	0.062
Warranty	0.047	0.121	0.071	0.060	0.083	0.074	0.079
Technical	0.282	0.242	0.214	0.240	0.250	0.222	0.241
Reputation	0.035	0.030	0.036	0.040	0.042	0.049	0.039
Price	0.424	0.424	0.429	0.480	0.375	0.444	0.428

Table 5. Pair-wise evaluate assessment of suppliers with regard to all evaluation criteria. [38].

DMU	Supplier1	Supplier2	Supplier3	AHP Weights
Quality				
Supplier1	1	3	5	0.633
Supplier2	1/3	1	3	0.260
Supplier3	1/5	1/3	1	0.106
Consistency ratio				0.033
Maintenance				
Supplier1	1	1/3	1/9	0.077
Supplier2	3	1	1/3	0.231
Supplier3	1/9	3	1	0.692
Consistency ratio				0.000
Warranty				
Supplier1	1	1/5	1/9	0.064
Supplier2	5	1	1/3	0.267
Supplier3	9	3	1	0.669
Consistency ratio				0.025
Technical				
Supplier1	1	1/9	1/7	0.057
Supplier2	9	1	3	0.649
Supplier3	7	1/3	1	0.295
Consistency ratio				0.070
Reputation				
Supplier1	1	1/5	1/4	0.096
Supplier2	5	1	3	0.619
Supplier3	4	1/3	1	0.284
Consistency ratio				0.0923
Price				
Supplier1	1	3	5	0.633
Supplier2	1/3	1	3	0.260
Supplier3	1/5	1/3	1	0.106
Consistency ratio				0.0419

Table 6. Overall score calculation.

	Quality	Maintenance	Warranty	Technical	Reputation	Price	Score
Supplier A1	0.096	+0.005	+0.005	+0.014	+0.004	+0.271	=0.395*
Supplier A2	0.039	+0.014	+0.021	+0.156	+0.024	+0.111	=0.365
Supplier A3	0.016	+0.043	+0.053	+0.071	+0.011	+0.045	=0.239
Rrow Average	0.151**	0.062	0.079	0.241	0.039	0.428	

Notes: 1. 0.395* = 0.633 × 0.151 = 0.096 + 0.077 × 0.062 = 0.005 + 0.064 × 0.079 = 0.005 + 0.057 × 0.241 = 0.014 + 0.096 × 0.039 = 0.004 + 0.633 × 0.428 = 0.271. 2. 0.151** = 0.096 + 0.039 + 0.016.

Table 7. Input data for airport ground handling service equipment supplier selection.

	Quality	Maintenance	Warranty	Technical	Reputation	Price
Supplier1 (x_1)	0.096	0.005	0.005	0.014	0.004	0.271
Supplier2 (x_2)	0.039	0.014	0.021	0.156	0.024	0.111
Supplier3 (x_3)	0.016	0.043	0.053	0.071	0.011	0.045
Row Averages	0.151	0.062	0.079	0.241	0.039	0.428

Table 8. Dataset used for representing the utilities relationships.

Lower (z_k^{min}) and Upper (z_k^{max}) Bounds	z_k^{min} (μ=0)	z_k^{max} (μ=1)
z_1—Quality	0.016	0.096
z_2—Maintenance	0.005	0.043
z_3—Warranty	0.005	0.053
z_4—Technical	0.014	0.156
z_5—Reputation	0.004	0.024
z_6—Price	0.045	0.271

Table 9. Decision-maker's (DM's) perception of supplier performance related to six criteria.

	Quality	Maintenance	Warranty	Technical	Reputation	Price
Supplier1	90	65	90	65	70	92
Supplier2	85	70	94	75	75	90
Supplier3	92	72	96	70	80	94

Table 10. Taguchi parameters for the six criteria.

	Target Value	Range	Specifiction Limit for the Deviation	Loss Coefficient	
Criteria	(%)	(%)	(%)	(k)	Taguchi Loss Function
Quality	100	100~85	15	2500	$L(X) = 2500(X - T)^2$
Maintenance	100	100~70	70	400	$L(X) = 400(X - T)^2$
Warranty	100	100~90	10	10,000	$L(X) = 10,000(X - T)^2$
Technical	100	100~85	15	625	$L(X) = 625(X - T)^2$
Reputation	100	100~80	20	1111	$L(X) = 1111.11(X - T)^2$
Price	100	100~90	10	10,000	$L(X) = 10,000(X - T)^2$

Table 11. Individual and weighted loss scores for the six criteria.

	Quality		Maintenance		Warranty		Technical		Reputation		Price		Weighted Score	Normalized
Supplier	Weight	Loss	Weight	Loss	Weight	Loss	Weight	Loss	Weight	Loss	Weight	Loss		
Supplier1	0.151	25	0.062	49	0.079	100	0.241	76.56	0.039	99.99	0.428	64	64.46	0.381
Supplier2	0.151	56.25	0.062	36	0.079	36	0.241	39.06	0.039	69.44	0.428	100	68.49	0.405
Supplier3	0.151	16	0.062	31.36	0.079	16	0.241	56.25	0.039	44.44	0.428	36	36.32	0.215

Table 12. Analytic hierarchy process (AHP)-Taguchi loss function (TLF)-multi-choice goal programming (MCGP) model solution programming.

AHP-TLF-MCGP Model Solution Programming	AHP-TLF-MCGP Model Goal
Min z=	
$(w_1(0.151(dp_1 + dn_1 + ep_1 + en_1))$	Satisfy quality goal
$+(0.062(dp_2 + dn_2 + ep_2 + en_2))$	Satisfy maintenance goal
$+(0.079(dp_3 + dn_3 + ep_3 + en_3))$	Satisfy warranty goal
$+(0.241(dp_4 + dn_4 + ep_4 + en_4))$	Satisfy technical goal
$+(0.039(dp_5 + dn_5 + ep_5 + en_5))$	Satisfy reputation goal
$+(0.428(dp_6 + dn_6 + ep_6 + en_6))$	Satisfy price goal
$+((w_2(dp_7 + dn_7 + ep_7 + en_7)))$	Satisfy loss function goal
s.t	
$w_1 + w_2 = 1; w_1 = 0.8; w_2 = 0.2$	
$(0.096\, x_1 + 0.039\, x_2 + 0.016\, x_3)\, b_1$ - dp_1+ $dn_1 = y_1 b_1$	For quality goal, the less the better
$y_1 - ep_1 + en_1 = 0.096$	For $\lvert y_1 - g_{1,min}\rvert$
$y_1 <= 0.096$	For bound of the y_1
$0.016 <= y_1$	
$(0.005x_1 + 0.014x_2 + 0.043x_3)\, b_2 - dp_2 + dn_2 = y_2 b_2$	For maintenance goal, the less the better

Table 12. *Cont.*

AHP-TLF-MCGP Model Solution Programming	AHP-TLF-MCGP Model Goal
$y_2 - \text{ep}_2 + \text{en}_2 = 0.043$	For $\lvert y_2 - g_{2,\min}\rvert$
$y_2 <= 0.043$	For bound of the y_2
$0.005 <= y_2$	
$(0.005x_1 + 0.021x_2 + 0.053x_3)\, b_3 - \text{dp}_3 + \text{dn}_3 = y_3b_3$	For warranty goal, the less the better
$y_3 - \text{ep}_3 + \text{en}_3 = 0.053$	For $\lvert y_3 - g_{3,\min}\rvert$
$y_3 <= 0.053$	For bound of the y_3
$0.005 <= y_3$	
$(0.014x_1 + 0.156x_2 + 0.071x_3)\, b_4 - \text{dp}_4 + \text{dn}_4 = y_4b_4$	For technical goal, the less the better
$y_4 - \text{ep}_4 + \text{en}_4 = 0.156$	For $\lvert y_4 - g_{4,\min}\rvert$
$y_4 <= 0.156$	For bound of the y_4
$0.014 <= y_4$	
$(0.004x_1 + 0.024x_2 + 0.011x_3)\, b_5 - \text{dp}_5 + \text{dn}_5 = y_5b_5$	For reputation goal, the less the better
$y_5 - \text{ep}_5 + \text{en}_5 = 0.024$	For $\lvert y_5 - g_{5,\min}\rvert$
$y_5 <= 0.024$	For bound of the y_5
$0.004 <= y_5$	
$(0.271x_1 + 0.111x_2 + 0.045x_3)\, b_6 - \text{dp}_6 + \text{dn}_6 = y_6b_6$	For price goal, the less the better
$y_6 - \text{ep}_6 + \text{en}_6 = 0.271$	For $\lvert y_6 - g_{6,\min}\rvert$
$y_6 <= 0.271$	For bound of the y_6
$0.045 <= y_6$	
$(0.381x_1 + 0.405x_2 + 0.215x_3)\, b_7 - \text{dp}_7 + \text{dn}_7 = y_7b_7$	For loss function goal, the less the better
$y_7 - \text{ep}_7 + \text{en}_7 = 0.215$	
$b_1 = b_2 + b_3 + b_4 + b_5 + b_6 + b_7$	To ensure the quality goal and the others, zero should be achieved.
$b_2 + b_3 + b_4 + b_5 + b_6 + b_7 = 1$	Added auxiliary constraints can force the quality goal and either of the other goals to be achieved.
$x_1 + x_2 + x_3 = 1$	Select a supplier
$x_i >= 0,\ i = 1, 2, 3$	
$d_i^+, d_i^-, e_i^+, e_i^- \geq 0,\ i = 1, 2, \ldots 7$	

The general steps of the AHP-FLP and AHP-TLF-MCGP models are summarized in Figure 2.

4.2. Real-World Application of the Proposed Model

We conducted a case study on Taoyuan International Airport Services Co., Ltd. (TIAS) Taoyuan city, Taiwan, constituting the AGHS. In 2018, TIAS provided ground handling services for 81,789 flights, serviced over 27.2 million passengers, and handled over 1.82 million tons of cargo. TIAS has powered (vehicle quantity: 749) and nonpowered (vehicle quantity: 5584) airport ground handling equipment. The intend of this study was to enable accurate estimations based on all possible decisions and help DMs satisfy company purchasing requirements. DMs often have difficultly choosing the most suitable supplier when various alternatives are available.

To ensure that our model had real-word utility, a decision expert team was structured; it comprised of two vice presidents with responsibility for AGHS company and four mechanical supply and repairs managers employed by TIAS. Ultimately, a foundation for such decisions was provided for applying the aforementioned steps, ensuring optimal decision-based outcomes [36].

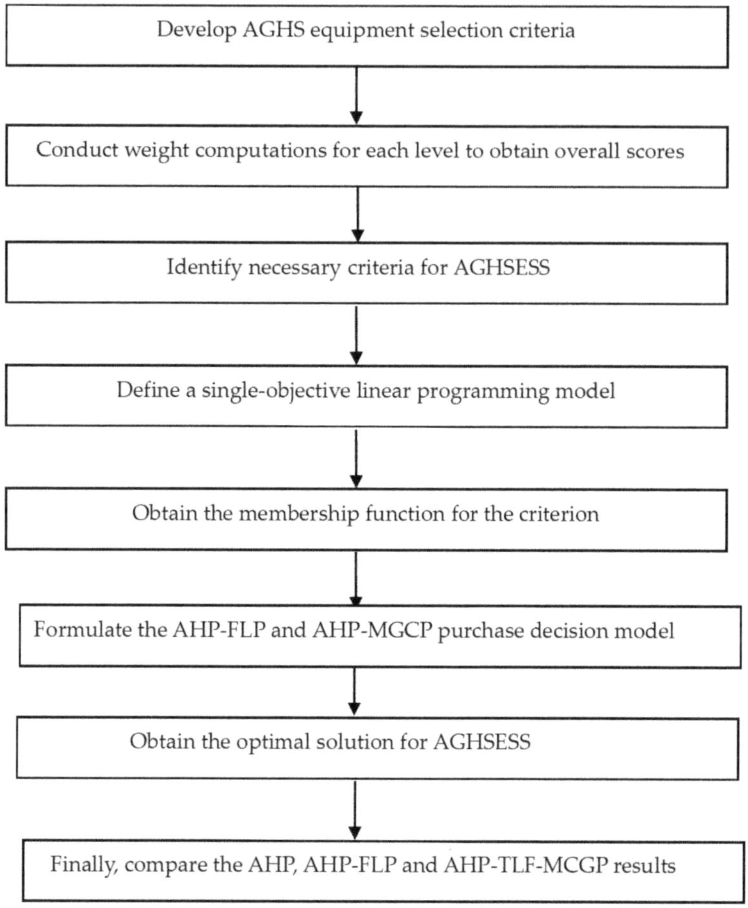

Figure 2. Steps of the proposed method.

4.3. Identification of Necessary Decision Criteria for Equipment Selection

The main decision criteria should relate to the choice of AGHS equipment suppliers, which are ultimately decided by expert team. The experience and qualification of the expert team were harnessed to decide on the vital criteria for AGHSESS. Details of the equipment supplier decision criteria are provided in Table 1 [14].

According to the aforementioned criteria, the AGHS equipment suppliers that were below enlargement or in procedure were examined, and a decision-making team chose six major supplier criteria that matched the company's needs. The six criteria are outlined as follows: Quality management (SC1), good production capacity and maintenance (SC2), product warranty (SC3), technical transfer provision (SC4), good cooperative relationship and reputation (SC5), and reasonable prices for parts (SC6). These criteria were used in our assessment and a decision hierarchy was subsequently constructed. The decision hierarchy based on the chosen AGHS equipment suppliers and their criteria is presented in Figure 3.

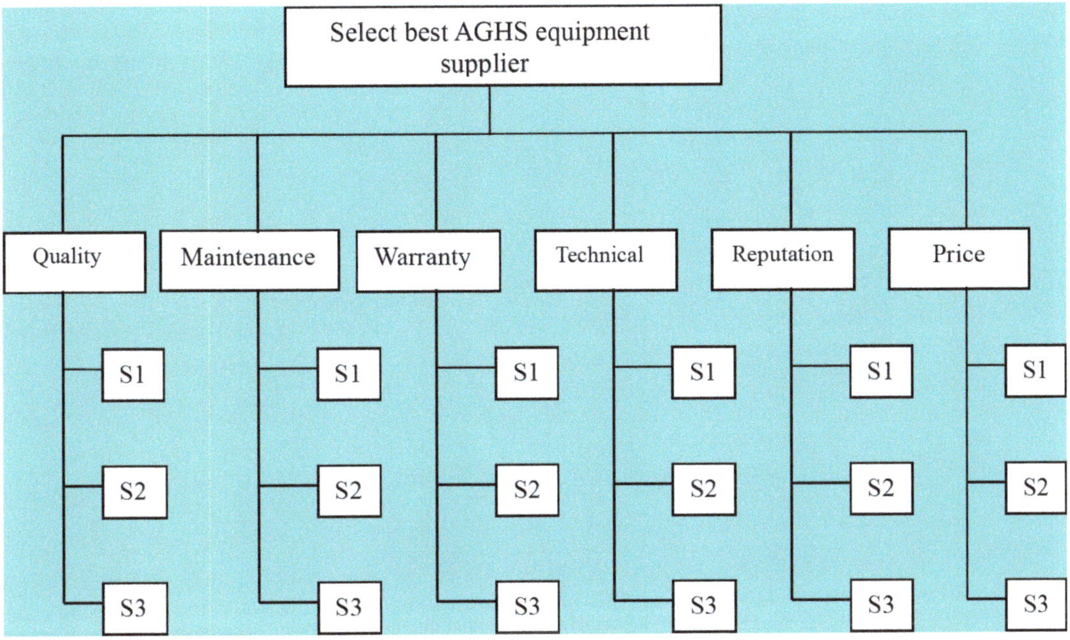

Figure 3. Decision hierarchy.

4.4. *Calculation of Criteria Weights*

Decision hierarchy criteria weights for use to calculate were computed using the AHP. Accordingly, the expert team developed a pairwise comparison matrix using the scale provided in Table 2. The geometric indices related to assessments were determined to establish a pair-wise comparison template, and broad agreement was reached on its content (Table 3). The results of the calculations based in Tables 4 and 5 [24].

Through the AHP, SC1 (quality management), SC4 (product warranty), and SC6 (providing technical transfer) were determined as the three vital criteria in the equipment DM process. A pairwise comparison template of the reliability coefficient of the existed revealed designate 0.084 < 0.1. Therefore, the weights were determined to be reliable, and they were used in the following decision-making procedures [24].

4.4.1. Constructing a Linear Programming Model for Real-World Application

The supplier DM process was founded on the assessments of our decision expert team. This result was that all suppliers could be analyzed by referring to the six aforementioned criteria and pairwise comparison could be made based on the key criteria. Tables 5–7 presents the evaluation criteria for each equipment provider, which was derived with reference to the key supplier decision criteria.

Details of the linear programming model used for AGHSESS are presented in [5]. A linear membership function is used for fuzzifying (i.e., to render unclear or vague) the purpose utilities with the resource constraints indicated in the aforementioned selection problem. The dataset used to set standards for the lower (z_k^{min}) and upper (z_k^{max}) bounds of purpose utilities is presented in Table 8.

4.4.2. Fuzzy Multi-Objective Decision Model

The relationship between the six criteria and related weights are presented herein to maximize the presentation of the providers associated with all key AGHS decision factors. Accordingly, we can derive quality criteria to determine the utility of Z_1.

The utility of z_1 (quality), which is shown in Figure 4, can be calculated according through Equation (8):

$$\mu_{z1}(x) = \begin{cases} 1 & \text{for} \quad z_1 \geq 0.096 \\ (0.096 - z_1(x)) \ /(0.096 - 0.016) & \text{for} \quad 0.016 < z_1(x) < 0.096 \\ 0 & \text{for} \quad z_1 \leq 0.016 \end{cases}$$

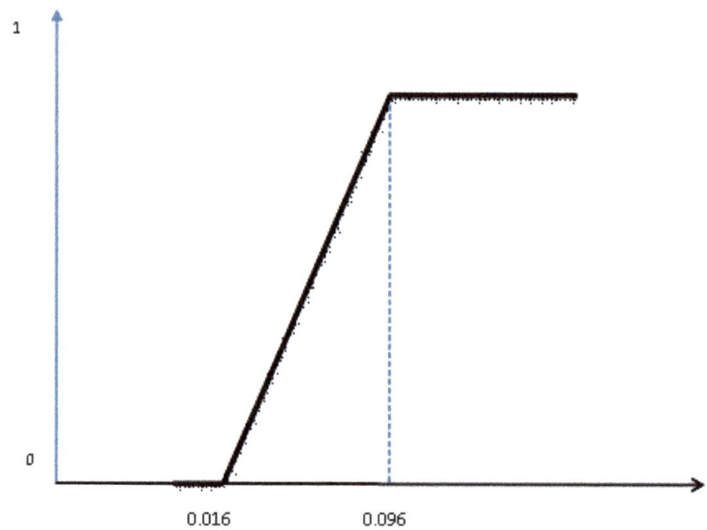

Figure 4. Membership function of quality.

On the basis of this condition, the following fuzzy linear model can be derived: Obtain a vector X, where X = [x_1, x_2, x_3] toward assure:

$$\begin{aligned} \max \widetilde{z}_1 &= 0.096x_1 + 0.039x_2 + 0.016x_3 \sim z_1^0 \\ \max \widetilde{z}_2 &= 0.005x_1 + 0.014x_2 + 0.043x_3 \sim z_2^0 \\ \max \widetilde{z}_3 &= 0.005x_1 + 0.021x_2 + 0.053x_3 \sim z_3^0 \\ \max \widetilde{z}_4 &= 0.014x_1 + 0.156x_2 + 0.071x_3 \sim z_4^0 \\ \max \widetilde{z}_5 &= 0.004x_1 + 0.024x_2 + 0.011x_3 \sim z_5^0 \\ \max \widetilde{z}_6 &= 0.271x_1 + 0.111x_2 + 0.045x_3 \sim z_6^0 \end{aligned}$$

subject to:

$$\begin{aligned} x_1 + x_2 + x_3 &= 1 \\ x_i \geq 0,\ i &= 1, 2, 3. \end{aligned}$$

4.4.3. Formulation of the AHP-FLP Purchase Decision Model

In this case, the weights (w_k) related to the kth objective are obtained pair-wise comparison of the key decision criteria determined using AHP are presented in Table 7, such as "line middling". Tables 5–7 are an indication that the total weights are equivalent to 1 (details of the approach are provided in [14]).

After the formulation of the AHP-FLP decision model and resolution of Equations (6)–(13), the single-objective model, which relates to the aforementioned fuzzy linear purchasing decision model, can be described as follows:

$$\text{Max } 0.151\lambda_1 + 0.062\lambda_2 + 0.079\lambda_3 + 0.241\lambda_4 + 0.039\lambda_5 + 0.428\lambda_6$$

subject to:

$$\lambda_1 \leq \frac{0.096-(0.096x1+0.039x2+0.016x3)}{0.096-0.016}$$
$$\lambda_2 \leq \frac{0.043-(0.005x1+0.014x2+0.043x3)}{0.043-0.005}$$
$$\lambda_3 \leq \frac{0.053-(0.005x1+0.021x2+0.053x3)}{0.053-0.005}$$
$$\lambda_4 \leq \frac{0.1564-(0.014x1+0.156x2+0.071x3)}{0.156-0.014}$$
$$\lambda_5 \leq \frac{0.024-(0.004x1+0.024x2+0.011x3)}{0.024-0.004}$$
$$\lambda_6 \leq \frac{0.271-(0.271x1+0.111x2+0.045x3)}{0.271-0.045}$$
$$\lambda_i \geq 0, \lambda_i \leq 1,\ i = 1,\ 2,\ 3,\ 4,\ 5,\ 6.$$

4.4.4. Solving the AHP-FLP Purchase Decision Problem

After the Lingo [39] was used to resolve this AGHS equipment provider problem, the best result was obtained as follows:

$x_1 = 0$, $x_2 = 0$, $x_3 = 1$, indicating that equipment provider A3 was the optimal choice according to the DMs preferences.

Objective (Z_k) and membership ($\mu_{zk}(x)$ or λ_k) functions for quality assessment can be expressed as follows:

$$z_1 = 0.016,\ z_2 = 0.043,\ z_3 = 0.053,\ z_4 = 0.071,\ z_5 = 0{,}011,\ z_6 = 0.045$$
$$\mu_{z1}(x) = \lambda_1 = 0,\ \mu_{z2}(x) = \lambda_2 = 1,\ \mu_{z3}(x) = \lambda_3 = 1,\ \mu_{z4}(x) = \lambda_4 = 0.401,$$
$$\mu_{z5}(x) = \lambda_5 = 0.350,\ \mu_{z6}(x) = \lambda_6 = 0.$$

The quality assessments showed that the values for Z_3, (product warranty), Z_4, (technical transfer), and Z_6, (reasonable prices for parts) were higher than those for Z_1, (quality management), Z_2, (production capacity and maintenance), and Z_5, (good cooperative relationship and reputation). The success of the model depends on the DM's preferences and the related decision criteria Equipment supplier A3 was the optimal choice in this context.

4.4.5. Comparison of AHP-TLF-FLP and AHP-TLF-MCGP Model Results

For the sake of reducing the AHP and AHP-FLP model approaches' drawbacks and reaching precise results, in this solution, corresponding to DM's preferences (w1 = 0.8, w2 = 0.2), Table 12 presents AHP-TLF-MCGP model solution programming to solve AGHSESS problem.

After, Lingo was used to resolve the aforementioned AGHSESS decision problem, and the optimal resolution derived from the AHP-TLF-MCGP model is outlined as follows: $x_1 = 0$, $x_2 = 0$, $x_3 = 1$, $y_1 = 0.016$, $y_2 = 0$, $y_3 = 0.005$, $y_4 = 0.014$, $y_5 = 0.004$, $y_6 = 0.045$, $y_7 = 0.215$. The finding that supplier A3 was the most appropriate one in the AHP-TLF-MCGP approach confirmed the views of the AGHSESS team.

The AHP, TLF, FLP, and MCGP approaches outperformed the AHP approach in solving the AGHSESS problem when supplier selection criteria were imposed. When AHP model was used without constraints, Supplier A1 was identified as the optimal one. This contradicts the AHP-FLP and AHP-TLF-MCGP models results in that Supplier A3 was determined to be the best option. In AGHSESS problems, the AHP-TLF-MCGP model can effectively handle vague and imprecise input data and criteria with varying relative weights. The AHP-TLF-MCGP model was implemented by integrating the AHP and TLF methods with MCGP to solve AGHSESS problems. This decision model can assist the DMs determine the suitable stock to order from each AGHS supplier provider, and it permits purchasing managers to easily select the optimal supplier in terms of criteria such as quality management, production capacity, product maintenance, product warranty, technical transfer provision, good cooperative relationship, and reputation, and reasonable prices for parts. With a few modifications, the model can also be in the habit of overcome other DM problems, such as those encountered by an overseas AGHS provider or others aviation industry firms.

5. Conclusions and Implications

5.1. Conclusions

The results obtained from the AHP and AHP-FLP decision model are compared in Table 13. The total scores for all providers are provided in this table; such scores were obtained using the aforementioned purchase decision model. As indicated, when the AHP model was used, equipment supplier A1 was recognized as the best choice when no constraints were imposed. When the AHP-FLP model was used and constraints were imposed, equipment supplier A3 was determined to be the best choice. This finding confirmed the postulations of the supplier selection team, giving credence to the argument that the AHP-FLP model would outperform the AHP model in this context.

Table 13. Comparison of the AHP, AHP-fuzzy linear programming (FLP), and AHP-TLF-MCGP model purchase decision model results.

	AHP	AHP-FLP	AHP-TLF-MCGP
AGHS equipment supplier A1	0.385*	0.000	0.000
AGHS equipment supplier A2	0.370	0.000	0.000
AGHS equipment supplier A3	0.097	1.000	1.000

Note: 0.385* = 0.574 × 0.151 + 0.077 × 0.062 + 0.064 × 0.079 + 0.057 × 0.241 + 0.096 × 0.039 + 0.633 × 0.428.

AGHSESS is a pertinent concern and affects AGHS providers in terms of staff members' daily duties and work safety. In our study, we compared AHP and AHP-FLP decision models to evaluate the selection of AGHS equipment suppliers. We used triangular fuzzy numbers to conduct linguistic assessments in order to assess the individual decisions of experts; subsequently, we derived the AHP-FLP and AHP-TLF-MCGP decision models for making group assessments. To provide a real-world example, we conduct a case study on an AGHS supplier in Taiwan. The AHP-FLP and AHP-TLF-MCGP decision model outperformed the AHP technique in terms of AGHSESS under conditions of limited selection criteria. Of the six AGHS equipment suppliers for the AGHS company considered in our case study, supplier A1 was determined to be the optimal supplier when the AHP model was used without constraints imposed. After the AHP-FLP and AHP-TLF-MCGP model was used, supplier A3 was determined to be the most suitable choice. In this case study, this model could deal with inaccurately entered information and criteria with varying relative weights. The results through the AHP-FLP and AHP-TLF-MCGP model corresponded to the real-world AGHSESS decisions of purchasing managers.

5.2. Management Implications

In this study, AHP and AHP-FLP models were compared in terms of how they helped personal deal with supplier selection problems. The integrated model can enable the aviation industry purchasing managers to observe all their options clearly under considerations of quality management, production capacity, product maintenance, product warranty, technical transfer provision, good cooperative relationship and reputation, and reasonable prices for parts. Two advantages are gained from using the AHP-TLF-MCGP model to solve AGHSESS problems: First, it permits DMs to set multiple desire levels related to supplier provider criteria, and second, it facilitates problem solving. Furthermore, the AHP-TLF-MCGP model is possibly beneficial on behalf of solving different MCDM problems, for instance those involving vague data.

5.3. Limitations

We combined the TLF and MCGP models that were integrated with the AHP model to capture the AGHSESS problem of fuzziness in human judgment and account for resource constraints. To reduce the AHP and AHP-FLP model approaches' disadvantages and reach precise outcomes, we compared them with the AHP-TLF-MCGP decision model to verify the AGHSESS problem. Formerly, if DMs use a new AHP technique and a combination of other goal programming methodologies, there can be a dissimilar result in hesitant situations.

5.4. Future Research Directions

Individual constraints in this model are a concern when the multiattribute weighting method is used [40]. The weights chosen to be used in the AHP technique may be regarded as biased. The data envelopment analysis (DEA) technique partly entails an optimization process conducted using linear programming; therefore, it can be used to determined weights. Several authors such as Pitchipoo, [37,41] have indicated that weights may be determined using DEA. These weights can be derived independently and thus involve less bias compared with those derive through other techniques. Accordingly, future research can be conducted on the use of this technique for deriving weights in our proposed model.

Author Contributions: Conceptualization, Y.-J.T.; formal analysis, C.-S.T. and Y.-J.T.; writing; planning all works in the study and supervision, H.-S.L.—original draft. All authors have read and agreed to the published version of the manuscript.

Funding: This research received no external funding.

Institutional Review Board Statement: Not applicable.

Informed Consent Statement: Not applicable.

Data Availability Statement: Not applicable.

Conflicts of Interest: The authors declare no conflict of interest.

References

1. Shen, C.W.; Peng, Y.T.; Tu, C.S. Multi-criteria decision-making techniques for solving the airport ground handling service equipment vendor selection problem. *Sustainability* **2019**, *11*, 3466. [CrossRef]
2. Inan, T. Performance measurement system (PSM) and balanced scorecard (BSC) strategies used in the process related with ramp handling services of ground handling management. *J. Aviat.* **2018**, *2*, 1–9.
3. Fuhr, J.; Beckers, T. Vertical governance between airlines and airport: A transaction cost analysis. *Rev. Netw. Econ.* **2006**, *5*, 386–412. [CrossRef]
4. Schmidberger, S.; Bals, L.; Hartmann, E.; Jahns, C. Ground handling services at European hub airports: Development of a performance measurement system for benchmarking. *Int. J. Prod. Econ.* **2009**, *117*, 104–116. [CrossRef]
5. Sevkli, M.; Koh, S.C.L.; Zaim, S.; Demirbag, M.; Tatoglu, E. Hybrid analytical hierarchy process model for supplier selection. *Ind. Manag. Data Syst.* **2008**, *108*, 122–142. [CrossRef]
6. Dulmin, R.; Mininno, V. Supplier selection using a multi-criteria decision aid method. *J. Purch. Suppl. Manag.* **2003**, *9*, 177–187. [CrossRef]
7. Amid, A.; Ghodsypour, S.H.; O'Brien, C. Fuzzy multiobjective linear model for the supplier selection in a supply chain. *Int. J. Prod. Econ.* **2006**, *104*, 394–407. [CrossRef]
8. Wang, T.C.; Chang, T.H. Application of TOPSIS in evaluating initial training aircraft under a fuzzy environment. *Expert Syst. Appl.* **2007**, *33*, 870–880. [CrossRef]
9. Ayag, Z. An analytic-hierarchy-process based simulation model for implementation and analysis of computer-aided systems. *Int. J. Prod. Res.* **2002**, *40*, 3053–3073. [CrossRef]
10. Ayağ, Z.; Özdemir, R.G. A fuzzy AHP approach to evaluating machine tool alternatives. *J. Intell. Manuf.* **2006**, *17*, 179–190. [CrossRef]
11. Monczka, R.M.; Nichols, E.L., Jr.; Callahan, T.J. Value of supplier information in the decision process. *Int. J. Purch. Mater. Manag.* **1992**, *28*, 20–30.
12. Vonderembse, M.A.; Tracey, M. The impact of supplier selection criteria and supplier involvement on manufacturing performance. *J. Supply Chain Manag.* **1999**, *35*, 33–39. [CrossRef]
13. Bhutta, K.S.; Huq, F. Supplier selection problem: A comparison of the total cost of ownership and analytic hierarchy process approaches. *Supply Chain Manag. Int. J.* **2002**, *7*, 126–135. [CrossRef]
14. Sarkis, J.; Talluri, S. A model for strategic supplier selection. *J. Supply Chain Manag.* **2002**, *38*, 18–28. [CrossRef]
15. Goztepe, K.; Kahraman, C. A new approach to military decision making process: Suggestions from MCDM point of view. In Proceedings of the International Conference on Military and Security Studies, İstanbul, Turkey, 10–11 March 2015; pp. 118–122.
16. Ordoobadi, S.M. Application of AHP and Taguchi loss functions in supply chain. *Ind. Manag. Data Syst.* **2010**, *110*, 1251–1269. [CrossRef]

17. Liao, C.N.; Kao, H.P. Supplier selection model using Taguchi loss function, analytical hierarchy process and multi-choice goal programming. *Comput. Ind. Eng.* **2010**, *58*, 571–577. [CrossRef]
18. Magdalena, R. Supplier selection for food industry: A combination of taguchi loss function and fuzzy analytical hierarchy process. *Asian J. Technol. Manag.* **2012**, *5*, 13–22.
19. Ordoobadi, S.M. Application of AHP and Taguchi loss functions in evaluation of advanced manufacturing technologies. *Int. J. Adv. Manuf. Tech.* **2013**, *67*, 2593–2605. [CrossRef]
20. Saaty, T.L. *The Analytic Hierarchy Process: Planning, Priority Setting, Resource Allocation*; RWS Publications: Pittsburgh, PA, USA, 1988.
21. Chan, F.T.S.; Kumar, N.; Tiwari, M.K.; Lau, H.C.W.; Choy, K.L. Global supplier selection: A fuzzy-AHP approach. *Int. J. Prod. Res.* **2008**, *46*, 3825–3857. [CrossRef]
22. Dağdeviren, M.; Yüksel, I. Developing a fuzzy analytic hierarchy process (AHP) model for behavior-based safety management. *Inform. Sci.* **2008**, *178*, 1717–1733. [CrossRef]
23. Kulak, O.; Durmusoglu, B.; Kahraman, C. Fuzzy multi-attribute equipment selection based on information axiom. *J. Mater. Proces. Technol.* **2005**, *169*, 337–345. [CrossRef]
24. Dağdeviren, M.; Yavuz, S.; Kılınç, N. Weapon selection using the AHP and TOPSIS methods under fuzzy environment. *Expert Syst. Appl.* **2008**, *36*, 8143–8151. [CrossRef]
25. Saaty, T.L. Decision making with the analytic hierarchy process. *Int. J. Serv. Sci.* **2008**, *1*, 83–98. [CrossRef]
26. Taguchi, G.; Phadke, M.S. Quality Engineering through Design Optimization. In *Quality Control, Robust Design, and the Taguchi Method*; Dehnad, K., Ed.; Springer: Boston, MA, USA, 1989.
27. Taguchi, G.; Elsayed, E.; Hsiang, T. *Quality Engineering in Production System*; McGraw-Hill: New York, NY, USA, 1989.
28. Ealey, L.A. *Quality by Design: Taguchi Methods and US Industry*; ASI Press: Bloomington, IN, USA, 1988.
29. Besterfield, D.H.; Besterfield-Michna, C.; Besterfield, G.H.; Besterfield-Sacre, M. *Total Quality Management*, 3rd ed.; Prentice-Hall Inc.: Upper Saddle River, NJ, USA, 2003.
30. Bellman, R.G.; Zadeh, L. A Decision making in a fuzzy environment. *Manag. Sci.* **1970**, *30*, 141–164. [CrossRef]
31. Zimmermann, H.J. Fuzzy programming and linear programming with several objective functions. *Fuzzy Sets Syst.* **1978**, *1*, 45–55. [CrossRef]
32. Chang, C.T. Revised multi-choice goal programming. *Appl. Math. Model.* **2008**, *32*, 2587–2595. [CrossRef]
33. Tiwari, R.N.; Dharmahr, S.; Rao, J.R. Fuzzy goal programming an additive model. *Fuzzy Sets Syst.* **1987**, *24*, 27–34. [CrossRef]
34. Haleh, H.; Hamidi, A. A fuzzy MCDM model for allocating orders to suppliers in a supply chain under uncertainty over a multi-period time horizon. *Exp. Syst. Appl.* **2011**, *38*, 9076–9083. [CrossRef]
35. Lai, Y.J.; Hawang, C.L. *Fuzzy Multiple Objective Decision Making, Methods and Applications*; Springer: Berlin/Heidelberg, Germany, 1994.
36. Mortezaa, Z.; Rezab, F.M.; Seddiqc, M.M.; Shararehd, P.; Jamale, G. Selection of the optimal tourism site using the ANP and fuzzy TOPSIS in the framework of Integrated Coastal Zone Management: A case of Qeshm Island. *Ocean Coast. Manag.* **2016**, *130*, 179–187. [CrossRef]
37. Pitchipoo, P.; Venkumar, S.; Rajakarunakaran, K.M. Modelling and development of decision model for supplier selection in process industry. *Int. J. Comput. Aided Eng. Tech.* **2011**, *3*, 504–516. [CrossRef]
38. Fashoto, S.G.; Akinnuwesi, B.; Owolabi, O.; Adelekan, D. Decision support model for supplier selection in healthcare service delivery using analytical hierarchy process and artificial neural network. *Afr. J. Bus. Manag.* **2016**, *10*, 209–232.
39. Schrage, L. *LINGO Release 8.0*; LINGO System Inc.: Chicago, IL, USA, 2002.
40. Wang, J.J.; Yang, D.L. Using a hybrid multi-criteria decision aid method for information systems outsourcing. *Comput. Oper. Res.* **2007**, *34*, 3691–3700. [CrossRef]
41. Lee, Z.-Y.; Chu, M.T.; Wang, Y.T.; Chen, K.J. Industry performance appraisal using improved MCDM for next generation of Taiwan. *Sustainability* **2020**, *12*, 5290. [CrossRef]

sustainability

MDPI

Article

Application of Exact and Multi-Heuristic Approaches to a Sustainable Closed Loop Supply Chain Network Design

Abdul Salam Khan [1], Qazi Salman Khalid [2,*], Khawar Naeem [2], Rafiq Ahmad [3], Razaullah Khan [4], Waqas Saleem [5] and Catalin Iulian Pruncu [6,7,*]

1 NUST Business School (NBS), National University of Sciences and Technology (NUST), Islamabad 44100, Pakistan; abdul.salam@nbs.nust.edu.pk
2 Department of Industrial Engineering, University of Engineering and Technology, Peshawar 25100, Pakistan; khawar@uetpeshawar.edu.pk
3 Department of Mechanical Engineering, University of Alberta, Edmonton, AB T6G 1H9, Canada; rafiq.ahmad@ualberta.ca
4 Department of Mechanical Engineering Technology, University of Technology, Nowshera 24100, Pakistan; razaullah@uotnowshera.edu.pk
5 Department of Mechanical and Manufacturing Engineering, Institute of Technology, F91 YW50 Sligo, Ireland; waqas95@yahoo.com
6 Design, Manufacturing & Engineering Management, University of Strathclyde, Glasgow G1 1XJ, UK
7 Department of Mechanical Engineering, Imperial College London, London SW7 2AZ, UK
* Correspondence: qazisalman@uetpeshawar.edu.pk (Q.S.K.); catalin.pruncu@strath.ac.uk (C.I.P.)

check for updates

Citation: Khan, A.S.; Khalid, Q.S.; Naeem, K.; Ahmad, R.; Khan, R.; Saleem, W.; Pruncu, C.I. Application of Exact and Multi-Heuristic Approaches to a Sustainable Closed Loop Supply Chain Network Design. *Sustainability* **2021**, *13*, 2433. https://doi.org/10.3390/su13052433

Academic Editor: Jasmine Liu

Received: 24 January 2021
Accepted: 20 February 2021
Published: 24 February 2021

Publisher's Note: MDPI stays neutral with regard to jurisdictional claims in published maps and institutional affiliations.

Copyright: © 2021 by the authors. Licensee MDPI, Basel, Switzerland. This article is an open access article distributed under the terms and conditions of the Creative Commons Attribution (CC BY) license (https://creativecommons.org/licenses/by/4.0/).

Abstract: Closed-loop supply chains (CLSC) are gaining popularity due to their efficiency in addressing economic, environmental, and social concerns. An important point to ponder in the distribution of CLSC is that imperfect refrigeration and bad road conditions may result in product non-conformance during the transit and thus such products are to be returned to the supply node. This may hinder the level of customer satisfaction. This paper presents a sustainable closed-loop supply chain framework coupled with cross-docking subject to product non-conformance. A cost model is proposed to investigate the economic and environmental aspects of such systems. The transportation cost is analyzed in terms of total carbon emissions. A set of metaheuristics are administered to solve the model and a novel lower bound is proposed to relax the complexity of the proposed model. The results of different size problems are compared with the branch and bound approach and the proposed lower bound. The results indicate that the proposed research framework, mathematical model, and heuristic schemes can aid the decision-makers in a closed-loop supply chain context.

Keywords: sustainable logistics scheduling; closed loop supply chain; cross-docking; tabu search; simulated annealing

1. Introduction

Modern manufacturing systems are faced with enormous challenges in a resource constrained environment. These challenges are in the form of supply chain issues, business growth, and responsiveness towards customers. These issues can be addressed through an efficient design of production systems, cost-efficient manufacturing, improved quality of products, and green logistics management. During the last few years, green logistics have attained prime attention and have become an indispensable part of the overall supply chain because it provides a mechanism for end-to-end transportation with social and ecological benefits, i.e., sustainable approaches. The literature shows that the cost related to logistics contributes towards the total sales value by more than 35%.

The performance of organizations involved in closed loop networks and green practices can be affected and fostered by the culture at the heart of an organization and the leadership role offered by the top management. Suryaningtyas et al. [1] argued that a strategic management focus, resilience within an organization and the leadership role play an active role in boosting the overall performance in the wake of disruptive challenges. Further, as CLSC networks are designed to improve sustainable developments, several factors can be identified at the level of an organization that can help the management in achieving the sustainability landmarks. Several management level decisions were examined by Drobyazko et al. [2] such as, financial, psychological, social, and economic factors that can offer an enabling environment to implement sustainable practices at an organization.

In modern times, the notion of sustainability encompasses approaches adopted in manufacturing, transportation, and vehicle distribution. The emissions during manufacturing are related to and can be restricted to the boundary of the manufacturing system; however, more sustainable approaches need to be adopted to cater to the emissions during transportation and vehicle distribution as they are directly related to the well-being of society. In this regard, cost models can be proposed to calculate the impact of emissions caused by vehicles during transportation.

Presently, greenhouse gas emissions (GHG), especially carbon dioxide emissions, have attracted growing concern worldwide. In this perspective, regulatory authorities are trying to enforce strict legislations for a low-carbon economy with clean products [3]. Considering this, closed loop supply chain and dispatching strategies can be tailored to optimize and minimize fuel consumption which in return reduces the carbon emissions. Some excellent research studies have been reported which demonstrated the statistics and relationship among reducing travel mileage, fuel consumption, and saving in GHG emissions [4]. From the studies [5,6]), it is evident that carbon emissions are linearly proportional to fuel consumption. This performance parameter greatly depends on the vehicle condition, type, gross weight, and dispatch time (travel speed). A salient feature of logistics and inventory design is cross-docking which has proved to be effective in controlling the distribution costs and maintaining an utmost level of customer satisfaction [7]. An important problem from the sustainability viewpoint is closed loop supply chain with cross-docking. It represents the most discussed class of combinatorial problems.

Though different contributions have been offered towards the analysis of such problems, the analysis based on defective products and product return by using a lower bound is missing in the concerned literature. The novelty of this study can be defined in terms of framework, methodology, and the analysis of emissions. In the framework, a closed loop supply chain (CLSC) network is designed keeping in view an imperfect product delivery. Traditionally, CLSC networks have been studied where products are delivered in a perfect state of quality. However, the quality-loss of product during transit may occur due to imperfect refrigeration, road conditions, traffic bottlenecks and different quality conditions of multiple products. This quality-loss may result in defective products that are to be returned. In terms of methodology, an initial heuristic is embedded with the simulated annealing to enhance its computational efficiency. The benchmark experiments inform that the initial heuristic embedded with simulated annealing performs well. Further, a novel lower-bound is proposed to relax the computational complexity of the model. It is defined in terms of cost. Lastly, as emissions are central in a sustainable supply chain framework, a comparison between emissions in terms of distance and DEFRA based results is presented. The Department for Environment, Food and Rural Affairs (DEFRA) takes the relative CO_2 per kilometer performance of the European Environment Agency (EEA) CO_2 monitoring database source as the payload capacity information.

This study considers a closed loop supply chain with cross-docking (CLSCCD). A cost-minimization model is proposed to assess the effectiveness of CLSCCD. The presented study considers a defect-based system where forward and returned goods are analyzed by considering the non-conformance of products, penalty, and product return. The cost components used in the model are related to production, transportation, handling, penalty, and the cost of product return. The reduced dispatching cost has been analyzed in terms of fuel savings and reduction in carbon dioxide for a greener environment. A mixed integer non-linear programming (MINLP) model is presented, and multi-heuristic approaches are used for model implementation. Moreover, the presented study is unique in the aspect that it offers a novel lower bound approach for CLSCCD.

The remaining study is organized as follows. Section 2 provides literature related to the CLSCCD, Section 3 outlines the problem description, Section 4 contains the detailed model, Section 5 provides the heuristic approaches, lower bound, parameter tuning, and benchmark experiments. Section 6 provides the results and Section 7 offers the conclusion and future research avenues related to CLSCCD.

2. Literature Review

CLSC frameworks have been thoroughly analyzed in the relevant literature and different models have been proposed to advance this field of research. This section presents the review of literature according to the focus of different objective functions, solution approaches adapted to address the problems, environmental concerns in CLSC, the emergence of green practices in CLSC design and a review of solution approaches.

2.1. CLSC Network, Environmental Concerns, and Green Practices

Several authors have identified key factors that affects the performance of CLSC networks. Choi et al. [8] studied the delivery schedule by employing an Ant Colony Optimization algorithm and predicted a comprehensive strategy to reduce carbon emissions. Likewise, Suzuki [4] proposed that sustainable scheduling with an expert decision-making system for speed and load can significantly reduce the carbon emissions. Qian and Eglese [9] also highlighted that dispatching time (speed) depends on several factors and a road network. They developed a column generation-based TS algorithm and tested real traffic data. Miao et al. [10] developed an adaptive real-time optimization strategy and concluded that vehicle dynamic parameters have a great impact on fuel economy.

Elhedhli and Merrick [11] studied the supply chain problem considering the environmental costs resulting from CO_2 emissions alongside fixed and variable factors. Pishvaee et al. [12] presented a mixed-integer programming (MIP) model for developing a reverse logistics network. They solved the non-polynomial hard (NP-hard) problem by employing a Simulated Annealing (SA) algorithm with neighborhood structures. Kannan et al. [13] studied the model proposed by Pishvaee et al. [12] by incorporating the environmental impacts due to carbon emissions. Faccio et al. [14] proposed a linear programming model to minimize the total cost in a supply chain. Their work evaluated the economic sustainability of the model by performing parametric analysis. Mota et al. [15] argued that while dealing with the sustainable supply chain problems, economic, environmental, and social dimensions are not considered realistically. Fonseca et al. [16] proposed a bi-objective model considering the total costs and environmental impacts with two-phase stochastic programming.

One of the central decisions in a CLSC network is to identify the relevant transportation routes which can help optimize the overall performance of the network. A transportation problem in CLSC can be modeled according to truck capacity, routes, and docks. The truck capacity involves decisions regarding the number and size of trucks to maximize the responsiveness of the supply chain. Routing considers different available options for the product delivery. Docking is the deployment of inter-modal facilities for product storage and consolidation purposes. A decision is made whether a cross dock is needed between the supply and demand node; depending on the capacity, number, and location of cross-docks to minimize the transportation time, and product handling costs. For instance, Vate and Zhang [17] studied a transportation network to optimize the delivery mechanism. Their work investigated the quantity and placement of cross-docks for an efficient transportation network. Noteworthy contributions can be found in the beverage and food industry where different types of routing, such as routing with time window, pickup and deployment, and distance-based routing have been used [18].

An earlier contribution towards the relevant problem type was offered by [19]. It used a mathematical model for minimizing the transportation and operation costs. Musa et al. [20] studied a vehicle routing problem where direct delivery to the customer as well as delivery through cross-docking were analyzed using MILP and ant colony optimization (ACO). In Khan [21], MILP model was proposed to analyze a multi-node transportation network. The model proved to be robust as significant cost reduction was demonstrated.

The CLSC networks are important from the viewpoint of a sustainable distribution design. They have attracted scholarly discussions in the areas of supply chain and cross-docking to address the situations where several destination points are visited by a group of vehicles for demand satisfaction. The demands are satisfied in accordance with the objective functions (cost, time, and responsiveness, etc.) and end-of-life products are retrieved back to the origin. Govindan et al. [22] highlighted the importance of closed-loop supply chain network design due to the emergence of environmental concerns and social issues. It is evident that besides focusing on the production of green products, hazardous emissions of supply chain activities during the transportation operations in both forward and reverse logistics needs to be addressed for enhanced green closed-loop supply chain management (e.g., Rahmani and Yavari [23]). In this perspective, numerous studies have been reported grounding on the mathematical modeling and heuristic optimization of CLSC problems with social-economic objectives (e.g., Kaya and Urek [24]). In this study, the proposed model considers diverse objectives that manipulate the design of closed-loop integration in forwarding/reverse supply.

In continuation to the preceding discussion, the term green closed-loop supply chain has gained prime attention by many researchers due to its undeniable advantages. For example, Talaei et al. [25] worked on a Fuzzy Mixed Integer Programming (FIMP) model for a green closed-loop supply chain and proposed important findings. Likewise, Ramezani et al. [26] developed a stochastic multi-objective model for designing a forward/reverse supply chain network in an uncertain environment of demand and the return rate. Sabulan et al. [27] developed a two-objective model for a closed-loop supply chain by applying Taguchi design experiments and evaluated the effect of different functions on the overall cost. Some researchers have sought the importance of supply chain integration in other domains, like construction [28], tire industry, auto parts manufacturing by including different dimensions of sustainability [29]. Mostafa et al. [30] developed a sustainable closed-loop supply chain network design with discount supposition. They applied a mixed-integer nonlinear programming model to formulate a multi-objective sustainable closed-loop supply chain network design in the transportation costs for a real industrial example in the glass industry.

2.2. Solution Approaches Used in CLSC Network Design

The design of CLSC networks considers multiple decisions that enhance its computational needs and makes it a complex class of problems. It involves multiple transportation options, dispatching facilities, customer locations, and cross-docks which make it computationally hard [31]. To solve such complex problems, different meta-heuristic approaches can be found in the literature [32]. Song and Chen [33] used an NP-hard approach for minimizing the total time in a two-stage based cross dock. Production planning and scheduling were performed by presenting an analogy between cross-docking and a manufacturing setup. In another study, Jayaraman and Ross [34] provided a simulated annealing approach to identify the placement of cross-docks and distribution of demand. In another study, a detailed cross-dock optimization model was proposed which incorporated both cross-dock placement and the identification of improved network paths [35].

Gholizadeh and Fazlollahtabar [36] studied a robust optimization model of a green closed loop supply chain in an uncertain environment. The authors used a modified form of genetic algorithm (GA) that comprised of seven different parts. The modified version of GA was compared with an exact solution approach and the classical GA. The results suggested that the former approach has good convergence ability. Gaur et al. [37] studied the effect of disruption on the performance of the supply chain. A non-linear model was implemented on an automotive case study by using different scenarios. An outer approximation approach was used for implementing the model. Turki et al. [38] studied four different decision aspects in a CLSC network, i.e., the capacity of manufacturing stocks, purchasing warehouse and vehicle and optimal level of returned products. A genetic algorithm was used for implementing the model and for analysing various aspects of decision making.

Cheraghalipour et al. [39] studied the CLSC network for citrus distribution and proposed a model for minimization of cost and enhancing customer responsiveness. A multi-objective Kushtal algorithm was proposed which proved effective when compared to other evolutionary solution approaches. Yavari and Geraeli [40] optimized the objectives of cost and environmental pollutants in a green CLSC network. A heuristic called YAG was implemented to solve the practical size problem instances. The results indicated that YAG had a good solution accuracy and it found solutions for more than one-third of the problem instances. Rad and Nahavandi [41] studied a multi-objective, multi-echelon, and multi-period CLSC model for multiple products. The model was solved by using an exact solution approach in CPLEX. Taleizadeh et al. [42] studied the pricing, quality, and effort in two types of decentralized CLSC networks. A game theoretic approach was used for analysing various aspects of the network design.

To summarize, different exact and evolutionary (meta-heuristic) approaches have been used to solve the CLSC problems. Compared to the earlier approaches, the novelty of present solution approach is two-fold. Firstly, a combination of Tabu search and simulated annealing (SA) approaches are used in this study. SA is used in a hybrid form by embedding an initial heuristic with it. The initial heuristic offers the current solution to obtain a derived solution. Secondly, to reduce the complexity of the presented model, a novel lower-bound (LB) is defined in terms of cost by splitting the model into forward and reverse supply chains.

Table 1 presents the summary of relevant literature according to the choice of objective functions and the application of different solution approaches. It can be observed that different meta-heuristics have been applied to solve the relevant problems; however, there is a dearth of application of multiple and hybrid meta-heuristics. This study uses multiple heuristics in the form of ant colony optimization (ACO) and Tabu search. The ACO is used in a hybrid form by embedding an initial heuristic with it to improve its computational efficiency. In terms of objective functions, the cost is the most opted choice of the objective function. The current study considers novel cost factors due to emissions and product failure. To summarize, there are some imperative differences and contributions in the proposed model from the published studies, for instance, the mathematical model explores closed-loop configuration for determining optimum cost supported by the meta-heuristic algorithm and lower bound approaches.

This paper fills the literature gap by using meta-heuristic approaches to solve a closed-loop Supply Chain Network Design (SCND) considering lower bound and some other novel aspects in this research domain. It addresses a sustainable closed-loop supply chain network along with cross-docking operations. The proposed robust model provides a comprehensive guideline to assist stakeholders in the closed-loop supply chain to make integrated decisions regarding the forward and reverse flows based on realistic assumptions. Although numerous contributions have been offered towards the analysis of cost in a closed loop supply chain, the novelty of current study rests in the following points:

Table 1. Summary of CLSC Literature.

Authors	**Solution Approaches**						**Objectives**						
	NSGA-II	**MILP**	**ACO**	**MOPSO**	**Simulation**	**AMOSA**	**Cost Minimization**	**Truck Selection and Uncertain Demand Environment**	**Analysis Capacitated Problem**	**Risk Management**	**Mechanism Strategic Transport**	**Routing**	**Quantity Discount**
Ozdemir et al. [43]	×	✓	×	×	×	×	×	×	✓	×	×	×	×
Khan [21]	×	✓	×	×	×	×	✓	×	×	×	×	×	×
Hong et al. [44]	×	×	✓	×	×	×	✓	×	×	×	×	×	×
Babaveisi et al. [45]	✓	×	×	✓	×	×	✓	×	×	×	×	×	×
Zhang et al. [46]	✓	×	×	×	×	×	×	✓	×	×	×	×	×
Behzadi et al. [47]	×	✓	×	×	×	×	×	×	×	✓	×	×	×
Bhattacharya et al. [48]	×	✓	×	×	×	×	×	×	×	×	✓	×	×
Cheng et al. [49]	✓	×	×	×	×	×	×	×	✓	×	×	×	×
De Jong et al. [50]	×	×	×	×	✓	×	×	×	×	×	✓	×	×
Yildizbaşi et al. [51]	×	✓	×	×	×	×	×	×	✓	×	×	×	×
Ekren et al. [52]	×	×	×	×	✓	×	×	×	✓	×	×	×	×
Fakhrzad et al. [53]	✓	×	×	×	×	×	✓	×	×	×	×	×	×
Kawamura et al. [54]	×	✓	×	×	×	×	✓	×	×	×	×	×	×
Lu et al. [52,55]	×	✓	×	×	×	×	×	×	✓	×	×	×	×
Mejjaouli et al. [56]	×	✓	×	×	×	×	✓	×	×	×	×	✓	×
Panicker et al. [57]	×	×	✓	×	×	×	×	×	✓	×	×	✓	×
Rais et al. [58]	×	✓	×	×	×	×	×	×	×	×	✓	✓	×
Rad et al. [41]	×	✓	×	×	×	×	×	×	×	×	✓	×	✓
Singh et al. [59]	×	✓	×	×	×	×	×	×	×	×	×	✓	×
Liu et al. [60]	✓	×	×	×	×	✓	✓	×	×	×	×	✓	×
Validi et al. [61]	✓	✓	×	×	×	×	×	×	×	×	×	✓	×
Zheng et al. [62]	✓	✓	×	×	×	×	×	✓	×	✓	×	×	×

- The analysis of non-conformance of products and product return routes has been considered. A mathematical model is developed by integrating different cost components involved in the supply chain of a cross-docking problem. The optimized routing cost has been discussed and analyzed from the perspective of fuel-saving which in turn reduces GHG emissions. Further, a comparison between emissions in terms of distance and DEFRA based results is presented.
- A two-method based approach is presented to solve the simple and complex problems related to CLSCCD. The simple problem instances are solved using the branch-and-bound approach while the complex problem instances are solved using two heuristics namely, Tabu Search and Simulated Annealing (SA). An initial heuristic is embedded with SA to enhance its computational efficiency. The solutions of different approaches are examined using benchmark data.
- A lower bound approach has been proposed. The meta-heuristics are evaluated against the branch-and-bound method (CPLEX) and the proposed lower bound (LB) approach using small and large problem instances, respectively. A comparison is carried out between the current findings and the established results in literature.
- A sensitivity analysis between various aspects of a CLSC in the form of vehicle capacity, conformance of products, number of trucks and customer satisfaction is presented which will help the managers in considering real-time decisions.

3. Problem Description

The schematic description of the CLSC problem with cross-docking is explained with the help of Figure 1. It considers the transportation of goods from node i (transportation node) to node j (delivery node) through node k (cross-dock node) by using distribution trucks. Due to perishability and breakage, some of the products are no longer required by customers and the rejected lot is sent back to the origin (node i). Two flows, i.e., forward and return flows are considered. The forward flow is a two-echelon network in which goods are initially sent to the cross-docking point (k). The items are sorted/consolidated at this point and are shipped to the delivery node (j). The return flow bypasses the cross-docking point, and it ships the non-optimal goods back to node i (this flow is described by the dotted line).

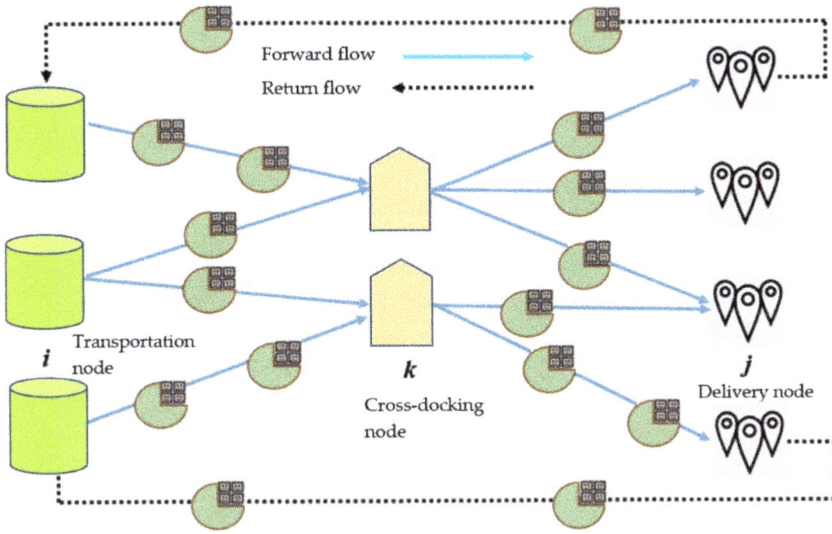

Figure 1. Schematic of the CLSCCD network design.

The facilities have a maximum capacity limit which is to be respected. Also, these locations are placed at a defined distance which is used in the calculation of transportation cost. A fleet of available trucks with loading capacity (LC) is used for delivering the products according to the level of demand. The goal is to accomplish product delivery by minimizing travelling, holding, penalty, return, and manufacturing costs.

4. Mathematical Model

$$minimize\ TC \tag{1}$$

$$TC = TR + DC + RC + HC \tag{2}$$

$$TR = \sum_{k \in K} \sum_{i \in I} NX_{ik} \times QP_{ik} \times cm + \sum_{k \in K} \sum_{i \in I} NX_{ik} \times NW_{ik} \times dx_{ik} + cd \tag{3}$$

$$DC = \sum_{k \in K} \sum_{j \in J} MX_{kj} \times FP_{kj} \times cm + \sum_{k \in K} \sum_{j \in J} MX_{kj} \times MW_{kj} \times dy_{kj} + cd \tag{4}$$

$$RC = \sum_{i \in I} \sum_{j \in J} RX_{ji} \times PR_{ji} \times (cm + cp + cr) + \sum_{i \in I} \sum_{j \in J} RX_{ji} \times TW_{ji} \times dz_{ji} + cd \tag{5}$$

$$HC = ch(\sum_{k\in K}\sum_{i\in I} NX_{ik} \times QP_{ik} - \sum_{k\in K}\sum_{j\in J} MX_{kj} \times FP_{kj}) \quad (6)$$

$$\sum_{k\in K}\sum_{i\in I} NW_{ik} \times LC \geq \sum_{k\in K}\sum_{i\in I} QP_{ik} \quad (7)$$

$$\sum_{k\in K}\sum_{j\in J} MW_{kj} \times LC \geq \sum_{k\in K}\sum_{j\in J} FP_{kj} \quad (8)$$

$$\sum_{i\in I}\sum_{j\in J} TW_{ji} \times LC \geq \sum_{i\in I}\sum_{j\in J} PR_{ji} \quad (9)$$

$$\sum_{i\in I}\sum_{k\in K} NW_{ik} > \sum_{k\in K}\sum_{j\in J} MW_{kj} \quad (10)$$

$$\sum_{i\in I}\sum_{k\in K} QP_{ik} \leq mx \; \forall i \in I \quad (11)$$

$$\sum_{i\in I}\sum_{k\in K} QP_{ik} > \sum_{j\in J} D_j \quad (12)$$

$$\sum_{j\in J}\sum_{k\in K} FP_{kj} \geq \sum_{j\in J} D_j \quad (13)$$

$$\sum_{i\in I}\sum_{k\in K} NW_{ik} \leq W \quad (14)$$

$$\sum_{i\in I}\sum_{j\in J} PR_{ji} = (\mathbf{1} - \alpha) \sum_{k\in K}\sum_{j\in J} FP_{kj} \quad (15)$$

$$QP_{ik},\ FP_{kj},\ PR_{ji},\ NW_{ik},\ MW_{kj},\ TW_{ji} \geq \mathbf{0} \; \forall i \in I, \forall j \in J, \forall k \in K \quad (16)$$

$$NX_{ik},\ MX_{kj},\ RX_{ji} \in \{\mathbf{0},\ \mathbf{1}\} \quad \forall i \in I,\ \forall j \in J,\ \forall k \in K \quad (17)$$

Equation (1) contains the objective function of minimizing the total cost (TC) value which comprises transportation (TR), delivery (DC), return (RC), and holding costs (HC) (Equation (2)). The relationships of respective cost components are provided in Equations (3)–(6). Equations (7) and (8) calculates the number of vehicles for delivering products to cross dock and customer locations, respectively. Equation (9) calculates the number of vehicles required for product return. Since a portion of the transported goods is stored at the cross dock, Equation (10) ensures that the number of vehicles used towards the cross-dock point is greater than the number of vehicles used for delivering products to the customer locations. Equation (11) ensures that the goods transported from node i should respect the capacity of the node. Equations (12) and (13) ensure that the goods transported towards node k and j should at least comply with the demand of customer j. The highest number of goods are transported between i and j, hence it requires more vehicles. Equation (14) ensures that the number of vehicles used between i and k needs to respect the available fleet of vehicles. Finally, Equations (16) and (17) are the domain constraints of non-negativity and binary variables, respectively.

The presented model is a non-linear model (MINLP) as Equations (3)–(6) contain the products of binary and continuous variables. For the nonlinear product XY of variables X and Y, Equations (18)–(20) provides the linear relationships where Z is a big number and A is an auxiliary variable. Equations (21)–(52) provide the linearized relationships for non-linear products in the model.

$$A \leq Z{\cdot}Y \quad (18)$$

$$A \leq X \quad (19)$$

$$A \geq X - Z(1 - Y) \tag{20}$$

$$NQP = NX_{ik} \times QP_{ik} \tag{21}$$

$$NQP \leq Z \cdot QP_{ik} \tag{22}$$

$$NQP \leq NX_{ik} \tag{23}$$

$$NQP \geq NX_{ik} - Z(1 - QP_{ik}) \tag{24}$$

$$NXW = NX_{ik} \times NW_{ik} \tag{25}$$

$$NXW \leq Z \cdot NW_{ik} \tag{26}$$

$$NXW \leq NX_{ik} \tag{27}$$

$$NXW \geq NX_{ik} - Z(1 - NW_{ik}) \tag{28}$$

$$MKP = MX_{kj} \times FP_{kj} \tag{29}$$

$$MKP \leq Z \cdot FP_{kj} \tag{30}$$

$$MKP \leq MX_{kj} \tag{31}$$

$$MKP \geq MX_{kj} - Z\left(1 - FP_{kj}\right) \tag{32}$$

$$MXW = MX_{kj} \times MW_{kj} \tag{33}$$

$$MXW \leq Z \cdot MW_{kj} \tag{34}$$

$$MXW \leq MX_{kj} \tag{35}$$

$$MXW \geq MX_{kj} - Z\left(1 - MW_{kj}\right) \tag{36}$$

$$RPR = RX_{ji} \times PR_{ji} \tag{37}$$

$$RPR \leq Z \cdot PR_{ji} \tag{38}$$

$$RPR \leq RX_{ji} \tag{39}$$

$$RPR \geq RX_{ji} - Z(1 - PR_{ji}) \tag{40}$$

$$RXT = RX_{ji} \times TW_{ji} \tag{41}$$

$$RXT \leq Z \cdot TW_{ji}. \tag{42}$$

$$RXT \leq RX_{ji} \tag{43}$$

$$RXT \geq RX_{ji} - Z(1 - TW_{ji}) \tag{44}$$

$$NQP = NX_{ik} \times QP_{ik} \tag{45}$$

$$NQP \leq Z \cdot QP_{ik} \tag{46}$$

$$NQP \leq NK_{ik} \tag{47}$$

$$NQP \geq NK_{ik} - Z(1 - QP_{ik}) \tag{48}$$

$$MXP = MX_{kj} \times FP_{kj} \tag{49}$$

$$MXP \leq Z \cdot FP_{kj} \tag{50}$$

$$MXP \leq MX_{kj} \tag{51}$$

$$MXP \geq MX_{kj} - Z\left(1 - FP_{kj}\right) \tag{52}$$

Ang with the above listed constraints, following assumptions are considered for modeling the problem.

- The capacity of trucks for transshipment, delivery, and returned products is assumed to be the same.
- The loading and un-loading efforts (such as time and cost) are assumed to be zero.
- The number of trucks in use should be within the range of the available truck fleet.
- The unit distance cost is same for all routes. This applies the assumption that the fuel efficiency of all trucks is approximately in the same range, which causes the same environmental impacts per unit distance travelled.
- The nodes i, k, and j are to be covered once. Similarly, the connection j and i (return flow) is to be covered once as well.
- The demands at delivery points are met in the same period and hence there are no opportunity and backorder costs.
- The returned lot is assumed to be of irreversible nature, and it is considered as waste.
- The rate of conforming products delivered is the same across the network.

5. Solution Approaches

5.1. Tabu Search

Tabu search (TS) is an Artificial Intelligence (AI) based technique that has been frequently applied to the supply chain related problems. For instance, it has been used to solve a heterogeneous fleet routing problem [63], consistent vehicle routing problem [64], and cost minimization problem [65]. It offers the advantage of escaping local solutions by using different neighborhood solutions [66].

The application of TS can be divided into five (5) connected steps. These steps are the identification of an initial feasible solution, formation of neighbor structure, acquiring a tabu list, and the basis for aspiration. The search algorithm provides a feasible solution and moves towards another solution up until the stopping criterion is met. The generated list contains all feasible solutions which have been acquired in recent iterations and the process continues to obtain more targeted solutions. The detailed procedure for implementing TS is presented in Table 2. It is divided into generating initial solution and Tabu Search process.

Table 2. Application procedure of Tabu Search.

a. Initial Solution
1a. Select a truck at the transportation node i.
2a. Compile a list of flows originating from i to k.
3a. Calculate the number of vehicles ($\frac{D_{pt}}{LC}$) which can cover the list created in step 2a.
4a. Identify the combination of delivery nodes.
5a. Select one delivery node j from step 4a and add it to the transportation route.
6a. Consider the demand of added delivery node as part of the truck load.
7a. Repeat 2a and 3a between cross-dock k and delivery j.
8a. Repeat 6a, after subtracting quantity of returned products (route j-i).
b. TS process
1b. Find two routes which have more overall cost between their respective nodes (i-k), compared to other nodes.
2b. Exchange the routes in 1b with other two randomly selected routes.
3b. If cost between nodes of the selected routes is improved, consider the attained solution as an optimal solution, otherwise re-iterate step 1b.
4b. Repeat 1b-3b for routes connecting k-j and j-i.
5b. Identify the best solution by repeating all steps.
6b. Generate the tabu list using solutions found in 5b.
7b. If the limited number of tabu search is reached, stop the procedure, otherwise repeat the tabu search procedure using all previous steps.

5.2. Simulated Annealing

Simulated Annealing (SA) is yet another powerful meta-heuristic tool that has been used due to its robustness and ability to avoid local optima. It is a simple and easily adaptable technique. The logic behind SA is the metallurgical process of cooling called annealing. It has been proposed by Metropolis et al. [67] and later extended by Lai et al. [68].

SA has been implemented in literature for solving the cross-docking problems, such as the application to truck routing and location routing problems [69]. Other applications are based on fixed-charge capacitated network designs. Yoo et al. [70] used SA along with a two-phase heuristic approach for analyzing the network design. Similarly, Yun et al. [71] used hybrid SA for analyzing the network of inland container transportation.

The SA algorithm used in this study proposes an initial heuristic to provide the current solution. A neighborhood search is conducted on a pre-defined current solution to obtain a derived solution. If the derived solution provides better results, it becomes an eligible candidate for the new solution. The current solution is obtained by using the steps explained in the Table 3. The proposed heuristic uses 3 search techniques to provide an optimal combination of routes. These techniques include insertion, reversion, and swap [72]. Insertion selects two nodes at random and identifies a node with a relatively smaller position. This node is then positioned before the other node in the modified solution. Reversion identifies two nodes, and the intermediate nodes are reversed in the modified solution. Swap [73] selects two nodes at random and simply exchanges their original places. The pseudocode of SA is provided in Table 4.

Table 3. Steps for Implementing Simulated annealing.

Steps for Implementing Simulated Annealing
Step 1: Selection of a truck for transportation
Step 2: Addition of cross-docking to the transportation route.
Step 3: Random identification of combination of delivery nodes.
Step 4: Selection of one delivery point from the combination of delivery nodes and adding it to the transportation route.
Step 5: Consider the demand of added delivery node as part of the truck capacity.
Step 6: Assessment of the remaining capacity of truck. If there is left capacity, repeat the last two steps.
Step 7: Repeat step 5, by subtracting the quantity of returned products.
Stage 8: Re-iteration of all prescribed steps, up until the demand of each delivery node is fulfilled.

Table 4. Pseudocode of Simulated Annealing (SA) algorithm.

Algorithm: Pseudocode of SA
1: Define initial and final temp. and cooling rate
2: Generate a starting solution
3: Identification of optimal solution
4: Setting curr_Temp=new_Temp
5: **While** Temp >> Final Temp
6: Insertion, Reversion and Swap
7: Cost of New_Sol < Cost of Initial_Sol
8: New_Sol=Curr_Sol
9: **Else**
10: Δ= Cost New_Sol-Cost of Initial_Sol
11: P=exp. -Δ/T
12: Generation of random real number
13: **Iff**. Real number <1
14: New_Sol=Curr_Sol
15: **If** Cost of Initial_Sol < Cost of best_Sol
16: best Sol=Curr_Sol
17: T=Cooling rate*T
18: **End**

5.3. Lower Bound Procedure

It is difficult and time-consuming to obtain a global optimal solution for large problems. Thus, a lower bound (LB) is normally introduced to examine the solution efficiencies of meta-heuristics. Though the literature on cross-docking and supply chain contains multiple

LB approaches (e.g., in [74]), most of them are related to the calculation of time and they come under the discussion of vehicle routing problem with cross docking (VRPCD) with time windows (VRPCDTW). A lower bound of such problems is developed without considering the consolidation constraints. On the other hand, since the objective of the current study is to optimize the total cost, and LB is proposed using the parameters of cost. The objective function given in (2) combines the costs related to forward and return flows which are separately considered in the LB (i.e., forward and return flows are examined in isolation and then an overall value of LB is obtained). The LB of forward flow is the summation of two-echelons (i.e., from transportation nodes to cross-docks and from cross-docks to delivery nodes), as given in (53):

$$lb_1 = \frac{D_p}{LC} \sum_{j \in J} \sum_{k \in K} \sum_{i \in I} \min\left(dx_{ik} + dy_{jk}\right) \times c_d \tag{53}$$

It considers the product of the number of trucks needed to transport the goods, minimum distance between i, k and j, k and the cost of travelling. Similarly, the LB for return calculates the return, penalty, and travelling costs of non-conforming goods back to node i, as given in (54):

$$lb_2 = (1 - \propto)(c_r + c_p) D_p \sum_{j \in J} \sum_{i \in I} \min(dz_{ji}) \times c_d \tag{54}$$

The total value of the lower bound is calculated as $LB = lb_1 + lb_2$. Based on the LB values, a measure of percent relative deviation (PRD) is calculated to compare the solution efficiencies of meta-heuristics with that of the lower bound. Its relationship is given as:

$$PRD = \frac{(M_{sol} - LB)}{LB} \times 100 \tag{55}$$

where M_{sol} refers to the solution obtained using meta-heuristic. To gain confidence in the results, several iterations ni (equals to 10 in the present analysis) is carried out and an average value of PRD is obtained as:

$$< PRD > = \sum_{ni=1}^{10} \frac{(M_{sol} - LB)}{LB} \times 100.$$

5.4. *Benchmark Experiments and Parameterssetting*

In this section, the computational power of the TS and SA is evaluated in comparison to the established literature by using benchmark data. In this regard, data from [66,75] is used by applying the conversion of parameters as follows.

n = number of nodes.
m = available vehicles.
Q = capacity of vehicles.
p_i = loading capacity at node i;
d_i = un-loading quantity at node j;
c_{ij} = transportation cost from node i to node j considering same environmental impacts

The values of parameters are provided in Table 5. There are three problems (P1–P3) with variable data points related to the number and capacity of trucks, etc., The notations used in the current analysis can be compared to the notations of benchmark data, that include, number of trucks, T = m; capacity of the truck, LC = Q; demand of the product, $D_{pt} = d_i$ and unit cost, cd = c_{ij}. Also, a single period is considered. The respective data related to cost and capacity in Table 5 was generated asymmetrically by using a uniform distribution.

Table 5. Benchmark data from Liao et al. [66].

Parameter Values	P1	P2	P3
N	10	30	50
M	10	20	30
Q	70	150	150
p_i	U(5,50)	U(5,20)	U(5,30)
d_i	U(5,50)	U(5,20)	U(5,30)
c_{ij}	U(48,560)	U(48,480)	U(48,560)

Before implementing the meta-heuristics, it is important to tune their input parameters as the parameters of meta-heuristics are highly sensitive towards changes. A poorly calibrated set of parameters can undermine the power of meta-heuristics to achieve optimal results [76]. The parameters of SA and TS are tuned/calibrated using Taguchi design of experiments (DOE). In SA, population size (*popsize*), maximum number of iterations (*max.It*), initial temperature (*T1*), and final temperature (*T2*) are the list of parameters. TS uses the set of parameters defined by population size (*popsize*), maximum iterations (*max. It*), tabu length (*TL*), and pre-determined number (*n*).

Taguchi's DOE uses an orthogonal array to analyze the set of factors. This set is divided into the sub-sets of controllable/signal and noise factors. A signal to noise ratio (S/N) is used to examine the level of variation. The goal is to reduce noise and a smaller-the-better type of S/N is used as given (56):

$$\frac{S}{N} = -10\log\left(\frac{\sum_{i=1}^{n} Y_i^2}{n}\right) \tag{56}$$

where Y_i is the recorded response in the ith experiment and n denotes the number of orthogonal arrays. To conduct the experiment, 3 levels (i.e., low, mid, and high) were defined for each factor, and then L_9 orthogonal design was conducted in Minitab V 19.0. The range of input parameters was defined based on a hypothetical pilot study. The levels of input parameters/factors are provided in Table 6.

Table 6. Levels of SA and TS parameters.

Meta-Heuristics	Factors	Levels		
		Low	Mid	High
SA				
	popsize	50	70	100
	max.It	500	700	1000
	T1	250	350	450
	T2	60	70	80
TS				
	popsize	50	70	100
	max.It	500	700	1000
	n	5	7	10
	TL	4	6	8

The optimal values of input parameters were selected by using the S/N ratios plots as given in Table 7. The meta-heuristics, along with their tuned parameters were implemented on the benchmark data.

Table 7. Optimal levels of input parameters.

Meta-Heuristics	Factors	Optimal Values
SA	popsize	100
	max.It	1000
	T1	350
	T2	60
TS	popsize	70
	max.It	1000
	N	5
	TL	6

The results (computation time) of ten problem instances of benchmark experiments, in comparison to the results of established literature are provided in Table 8. Each problem instance was executed ten times to gain statistical confidence. The results show the computation time of TS (current study) is better than the results obtained in Lee et al. [74]; however, Liao et al. [65] demonstrated even improved results. It is due to the use of an initial solution that helps in fixing the number of trucks. Also, SA outperformed the TS (current) results, due to an initial heuristic based on neighborhood and diverse search techniques.

Table 8. Computation time comparison between benchmark experiments and current study.

Inst	Lee et al. [75] TS			Liao et al. [66] TS			Current TS			Current SA		
	P1	P2	P3	P1	P2	P3	P1	P2	P3	P1	P2	P3
1	1.52	3.00	5.69	0.22	0.37	0.49	1.14	2.18	4.34	0.86	1.56	3.76
2	1.74	3.55	5.73	0.23	0.16	0.64	1.21	2.63	4.68	0.92	1.74	3.85
3	2.37	4.32	5.80	0.19	0.43	0.30	1.46	3.50	4.21	1.04	2.58	3.42
4	1.60	2.09	5.87	0.27	0.23	0.44	1.30	1.68	5.02	1.13	1.34	4.34
5	2.28	2.26	7.28	0.21	0.39	0.61	1.86	1.97	6.17	1.47	1.18	4.88
6	1.82	2.10	5.38	0.03	0.22	0.56	1.42	1.85	4.84	1.24	1.45	4.32
7	2.80	2.61	7.65	0.01	0.11	0.55	2.24	2.23	6.35	1.58	1.86	5.54
8	1.85	3.00	9.88	0.04	0.16	0.42	1.17	2.54	7.92	0.87	2.03	6.28
9	2.04	3.10	5.54	0.25	0.16	0.38	1.46	2.26	4.37	0.95	2.48	3.84
10	1.82	2.38	5.77	0.36	0.20	0.52	1.19	1.87	4.60	1.12	1.53	3.97

6. Results

The three methods (CPLEX, tabu search, and simulated annealing) were applied to multiple problem instances. These were implemented using Dell core i5, 8th Gen system with 2.8 GHz processing power and 4 GB RAM. The CPLEX solver is a more suitable package for simple problem sizes, but as the problem complexity increases, its computation power is affected. This is presented in Table 9 where 10 problem instances have been tested and the number of trucks (W), objective function value (OBV), and computation time (CPU) are reported for comparison.

The left column describes the problem instances, and the problem size increases in the order I_10 > I_1. The results show that, for small problem instances (I_1 to I_4), CPLEX provides improved results compared to TS and SA. It also takes proportionately less time in solving the problem. However, as the problem complexity increases (I_5 onwards), CPLEX does not prove to be a viable option, due to the combinatorial nature of the problem. For such instances, SA outperforms the computational efficiency of TS and solver. Not only does it take less time in solving the problem, but it also provides a more realistic value of OBV.

A comparison between TS and SA is also conducted for complex problem instances and respective results are provided in Figure 2. Each problem instance was executed 10 times. For smaller instances (PS < 50), there are not any large-scale computation differences between TS and SA. However, as the problem size crosses a certain limit (PS > 70), there is an evident shift between the curves. SA proves to be more computationally robust compared to TS, as much as, when the problem size is 150, SA takes almost half the time taken by TS in solving the problem. These findings agree with the earlier reported results (Table 9).

Table 9. Computation and OBV comparison of 3 methods.

Instance	CPLEX Solver			Tabu Search			Simulated Annealing		
	W	OBV	CPU	W	OBV	CPU	W	OBV	CPU
I_1	5	864	33	5	875	41	5	868	36
I_2	7	1296	45	7	1324	50	7	1304	47
I_3	10	1970	90	10	2034	98	10	1981	92
I_4	9	1836	82	9	1975	90	9	1855	84
I_5	–	–	–	11	2304	121	11	2234	112
I_6	–	–	–	12	2778	144	12	2674	129
I_7	–	–	–	14	2849	171	13	2639	162
I_8	–	–	–	15	3190	224	14	3058	214
I_9	–	–	–	16	3306	278	16	3138	256
I_10	–	–	–	19	3549	320	18	3276	304

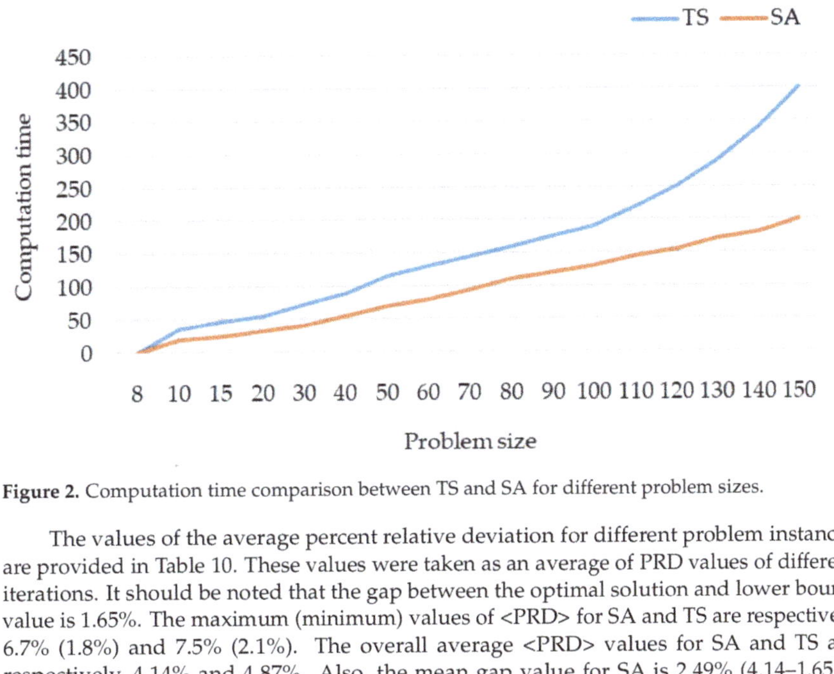

Figure 2. Computation time comparison between TS and SA for different problem sizes.

The values of the average percent relative deviation for different problem instances are provided in Table 10. These values were taken as an average of PRD values of different iterations. It should be noted that the gap between the optimal solution and lower bound value is 1.65%. The maximum (minimum) values of <PRD> for SA and TS are respectively 6.7% (1.8%) and 7.5% (2.1%). The overall average <PRD> values for SA and TS are respectively, 4.14% and 4.87%. Also, the mean gap value for SA is 2.49% (4.14–1.65%) while it is equal to 3.22% (4.87–1.65%) for TS. It can be concluded that, in the presence of LB, SA outperforms TS in terms of maximum, minimum, and average <PRD> values as well as in terms of mean gap value.

A non-conformance rate is used in the proposed model to consider the quantity of returning goods. Figure 3 contains the results of 3 cost factors for different non-conforming rates. The horizontal axis provides different rates of non-conformance (0.1–0.95) and the vertical axis contains the cost values associated with these rates. The horizontal lines represent the cost of docking (DC) and transportation routes (TR) for a forward flow. It can be observed that these routes are not affected by changes in non-conformance. However, the blue line which provides the cost of returned goods for different non-conforming rates is increasing non-linearly. It is evident that, with an increase in the rate of non-conformance, the cost of return route also increases. For instance, when the rate of non-conformance increases from 0.1 to 0.95, the associated cost increases by more than 300%.

Table 10. Results of problem instances of SA and TS in comparison to LB.

Algorithm	Problem Set	<PRD> %	Algorithm	Problem Set	<PRD> %
SA	1	1.8	TS	1	2.1
	2	2.4		2	2.7
	3	2.9		3	3.5
	4	3.5		4	4.0
	5	3.9		5	4.7
	6	4.4		6	5.3
	7	4.7		7	5.8
	8	5.3		8	6.3
	9	5.8		9	6.8
	10	6.7		10	7.5

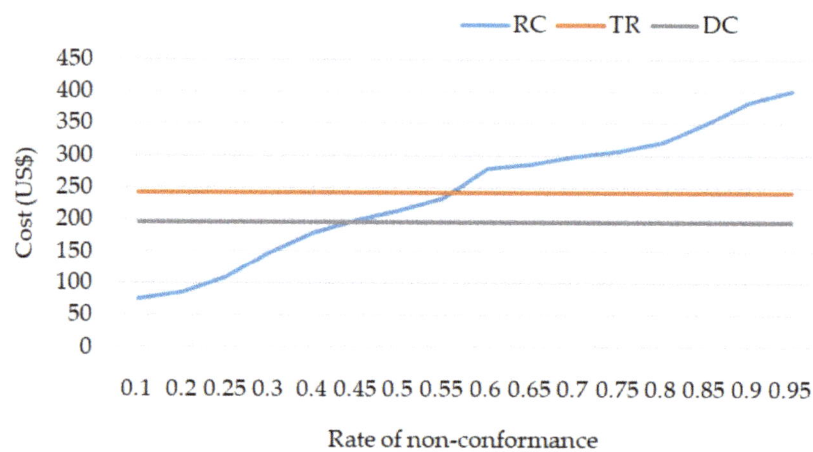

Figure 3. The cost indices of RC, TR, and DC for different rates of non-conformance.

Several other test runs were performed using tabu search. A comparison was made between the findings of the current study and benchmark analysis. The findings of these comparisons are provided in Figures 4–6. The horizontal axis of each figure provides 10 randomly selected instances of the NP-hard problem while the vertical axis provides the solutions. Each instance of the problem was executed 10 times to gain statistical confidence.

Several observations can be made from these illustrations. First, by increasing the number of iterations, the deviation in the results of different instances is minimized. In other words, the curves become flatter as the number of iterations increases. Secondly, the results obtained using our approach are somewhat better than those obtained in Lee et al. [74] but are inadequate than the results reported by Liao et al. [65]. The current study results can be taken as a trade-off measure of the established results in literature.

A comparison is also offered to examine the robustness of SA solution approach. A comparison of obtained results is presented with the benchmark findings [34] by considering two problem sizes. These problem sizes are obtained by multiplying the number of transportations, cross-dock, delivery nodes, and the fleet of vehicles. The problem sizes of 5 × 10 × 30 × 2 and 5 × 15 × 30 × 3 is used. Five annealing cooling rates have been considered within each problem size, i.e., 0.20, 0.45, 0.65, 0.85, and 0.95.

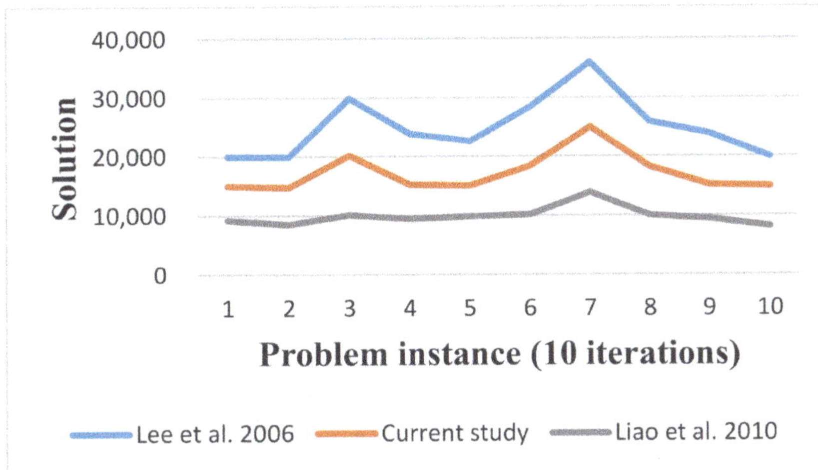

Figure 4. The comparison of obtained results with published data (10 iterations and 10 instances).

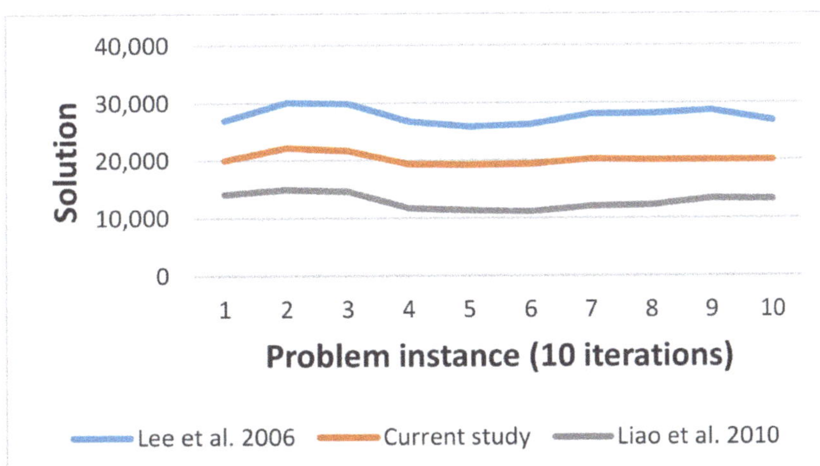

Figure 5. The comparison of obtained results with published data (30 iterations and 10 problem instances).

The findings are provided in Table 11. The results obtained in [34] for small cooling rates are computationally more efficient than our findings, however, for advanced cooling rates (0.85 and 0.95), the current results are attained in relatively less computation time. In other words, the obtained results are more robust when there is a drop in annealing temperature. For assessment of GHG emissions, an estimate can be established considering the distance-based formulation from DEFRA (2019) [76]:

$$\text{GHG emissions} = \text{total distance in km} \times \text{fuel per km} \times \text{fuel conversion factor}$$

DEFRA takes the relative CO_2/km performance of the European Environment Agency (EEA) CO_2 monitoring database source as the payload capacity information. For the present study, a fuel conversion factor of 2.687 kg/l for diesel is taken from the DEFRA report (2019) [76]. For the presented model, the meta-heuristic simulations predicted an average fuel cost of US$250 between node i and j under the assumption that all trucks have the same engine capacity other dynamic performance parameters. By assuming a fixed price of diesel (1 L = $US1.4) and taking an average distance of 12 km per liter for standard 3.5-ton light vehicle goods, the estimated environmental emissions are:

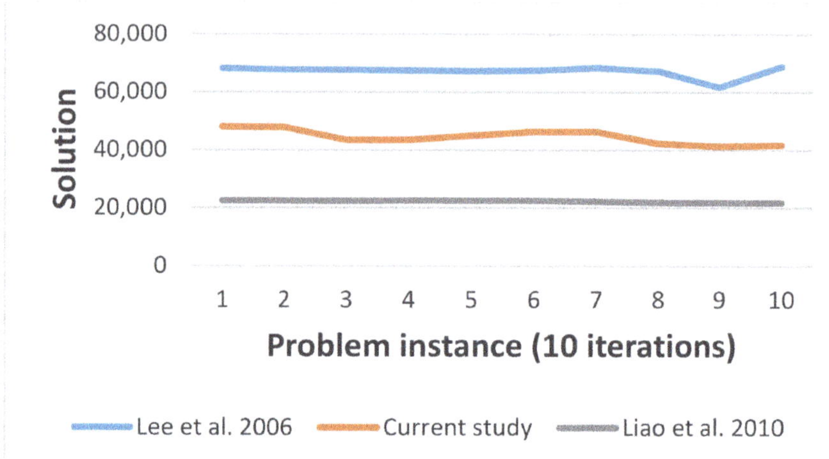

Figure 6. The comparison of obtained results with published data (50 iterations and 10 problem instances).

Table 11. Comparison of SA based obtained results with Jayaraman et al. [34].

Size	Cooling Rate	Jayaraman et al. [34]	Current Findings
	0.20	0.016	0.018
	0.45	0.026	0.029
5 × 10 × 30 × 2	0.65	0.048	0.049
	0.85	0.132	0.123
	0.95	0.376	0.354
	0.20	0.022	0.025
	0.45	0.030	0.037
5 × 15 × 30 × 3	0.65	0.066	0.069
	0.85	0.158	0.153
	0.95	0.498	0.487

In Table 12, a comparison of the predicted GHG emission per km and the DEFRA established emissions show a good corroboration. The percent difference between the predicted and actual established total emissions is found to be 5.55, which indicates the conformance of the mathematical model and meta-heuristic studies. Overall, the proposed methodology presents a good guideline for determining an optimal solution for sustainable vehicle routing problems considering the economic and environmental impacts.

Table 12. Estimated guideline for GHG emissions.

Total Distance (km)	Fuel per km	Fuel Conversion Factor [76]		
250 × 1.4	1.4/12	kg CO_2e	kg CH_4	kg N_2O
2160	0.11667	2.68697	0.00030	0.03425
Model based predicted GHG Emissions per km				
	kg CO_2e	kg CH_4	kg N_2O	Total (kg)
	0.31348	0.00004	0.00400	0.31751
DEFRA, Table 19 [75], GHG emissions per km for 3.5-t Goods Transportation Vans				
	0.26400	0.00001	0.00185	0.26200
% difference (Simulation predicted values—DEFRA estimated values)				
	4.95	0.00	0.21	5.55

6.1. Sensitivity Analysis

The presented model is based on CLSC network design in the presence of product non-conformance subject to capacity and loading constraints. This sub-section offers an insight into the effect of different model parameters on the sensitivity of various decision aspects of a CLSC network. Figure 7a–d presents the results of sensitivity analysis of various performance indices of the network. Figure 7a shows the impact of an increase in the level of conformance (α) on the number of delivered/returned products. When $\alpha = 0$, it means that all transported products are failed during the transit. Such failure can be attributed to reasons such as imperfect refrigeration and road conditions etc. Thus, all products are returned, and none are delivered. As the level of conformance increases, the quantity of conforming products delivered increases and the returned products level diminishes. A trade-off is beneficial for managers to balance the quantity of delivered and returned products.

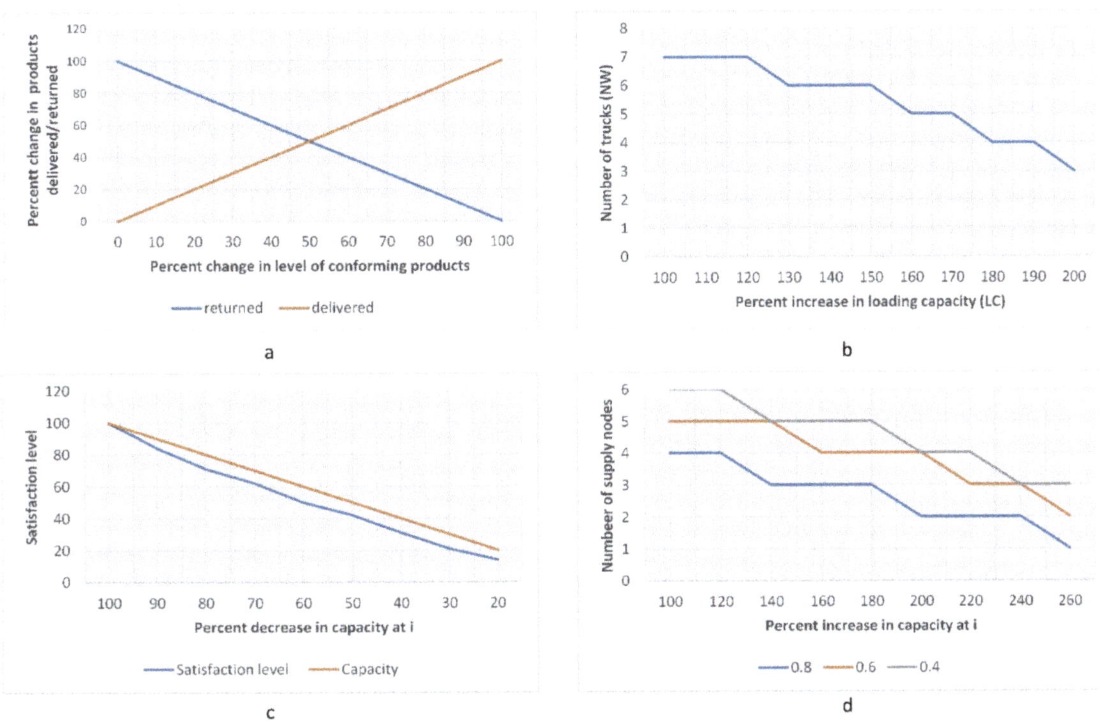

Figure 7. (**a**) relationship between change in the level of conformance and quantity delivered/returned, (**b**) relationship between the loading capacity and the number of trucks, (**c**) relationship between the capacity at supply node and customer satisfaction and (**d**) relationship between the capacity at supply node, level of conformance and number of supply nodes.

A homogeneous fleet of trucks was used in the CLSC network design. It is imperative to examine how the model reacts towards changes in the capacity of trucks. In this regard, Figure 7b presents a relationship between a percent increase in the truck capacity on the number of trucks used. The 100% LC refers to the initial truck capacity. It can be observed that a straightforward relationship between these two aspects does not exist, rather a reduction in the number of trucks occurs after a certain level of increase in the loading capacity of trucks. Under the current circumstances, it is not feasible to increase the loading capacity of trucks from 100%, 130%, 160% and 180% to 120%, 150%, 170% and 190%, respectively as it will not help in reducing the number of trucks. Thus, although managers may invest in bigger trucks in certain cases, such decisions may not payback and are non-viable. A viable option will be to increase the loading capacity from 120%, 150%, 170% and 190% to 130%, 160%, 180% and 200%, respectively. In these cases, the number of vehicles can be reduced while meeting the required level of demand.

The satisfaction of customers is an integral part of an efficient supply chain network design. It can be measured in terms of high-quality delivery, and responsiveness, etc. It can also be measured in terms of the level of products delivered to customers. To demonstrate this aspect, constraints (7) and (12) were relaxed so that the demand and truck capacity constraints are omitted. Following this, initially, the number of available supply nodes was restricted to 4, i.e., $I = 4$. A comparison was drawn between the capacity levels at node i and the customer satisfaction i.e., the number of products delivered. Figure 7c provides the relationship between the decrease in capacity at node i and the percent change in customer satisfaction. It is worth noticing that the data point of 100% on either axis refers to the baseline data when the constraints are not relaxed. It can be observed that as the capacity level at the nodes decreases, customer satisfaction decreases at a faster rate. It is because that portion of the transported products are non-conforming and are returned. Thus, not all the units at supply node i are delivered and hence a relatively sharp decline in customer satisfaction is observed. Thus, the level of conformance and non-conformance affects customer satisfaction. To further investigate this aspect, the model was re-run by using three different values i.e., 0.8, 0.6, and 0.4. The relevant results are provided in Figure 7d. The main difference between the analysis in Figure 7c and d is that the latter is based on demand fulfilment constraints. The results indicate that a reduction in the number of supply nodes is observed as the capacity at node i increases. Further, a higher number of supply nodes are needed when the rate of conformance decreases. This is logical because a lower conformance means that a higher quantity is returned, and thus more product units are to be sent to meet the level of demand at the destination. Hence, a higher number of supply nodes will be needed to accommodate such higher product quantities in the presence of capacity constraints.

6.2. Managerial Implications

The CLSC network designs have offered an opportunity for practitioners to reduce the carbon footprints and retrieve products back from the customers. This study offers the following guidelines for managers and practitioners working in the relevant industry. In the presence of higher non-conformance, the return route cost increases. Thus, either the conformance of products is to be ensured by using high quality products through adequate refrigeration or the return route is to be properly designed. One way to do this is by opening a refurbishing plant close to the customer locations so that the return distance can be reduced.

Managers are always interested in reducing the overall cost of the network. It can be reduced by using a limited number of trucks by expanding the loading capacity of trucks. However, the current findings inform that not all expansion decisions are viable, as in some cases, minor expansion in capacity does not reduce the number of trucks. This will offer an opportunity for managers in deciding when and to what extent an investment decision in larger trucks should be made. Lastly, a relationship was established between the capacity at the supply node and customer satisfaction. In some cases, higher capacity levels may be needed to ensure higher customer satisfaction.

7. Conclusions

This study analyzed a CLSCCD problem for optimizing the objective of cost which comprised of costs related to production, transportation, handling, penalty, and the cost of the product return. Two flow types, i.e., forward and return flows were considered as part of the analysis by considering the non-conformance of products. A mathematical model and two heuristic approaches were adapted to analyze the problem. The branch-and-bound method was used for small problem instances whereas meta-heuristics were used for higher-order problem sizes. The analysis was presented by using parameter tuning, benchmark experiments, and a lower bound approach. A comparison between Tabu search and Simulated Annealing proved that the latter was a computationally efficient and robust measurement approach due to an initial heuristic embedded with it for refining the solutions.

A higher rate of conformance of product delivery can enhance customer satisfaction. It can be ensured by employing perfect refrigeration conditions and monitoring the road conditions. Similarly, attention needs to be paid to the capacity expansion of trucks and its impact on the number of trucks used. Not all expansion decisions are viable in reducing the number of trucks. However, it is pertinent to reduce the number of trucks in transit to reduce emissions.

Since an indirect measure was used to analyze the product quality, therefore, there is a dearth of using a direct measure to capture the quality of products during transit. Future research can define a dedicated objective function that defines the quality of in-transit products subject to imperfect refrigeration. This study made assumptions regarding the traveling distance, the capacity of trucks, fuel efficiency, and environmental impacts. It was assumed that the unit distance cost is the same for all routes, loading capacity is same for all trucks as well as for all nodes. Future research can relax these assumptions by considering variable distance cost and variable capacity of a truck for different routes. The presented model was deterministic with respect to the behavior of different aspects of the analysis. It will be interesting to formulate a robust variant of this model by considering stochastic aspects. Such practice will help in comparing the deterministic and stochastic behavior of various input parameters to consider a more informed decision. The initial heuristic embedded SA outperformed the Tabu search approach. Future research may embed an initial heuristic with Tabu search to understand how well it performs subject to modifications. Similarly, the discussion can be extended by comparing the results of meta-heuristics with other methods, such as genetic algorithms and the Petri-net approach.

The proposed modeling and simulation approach can be used for strategic and operational decisions in the supply chain, for instance, designing the sustainable network configuration, resource management, and optimal routing to reduce the GHG emissions. The mathematical model and meta-heuristic algorithms can be extended by incorporating other complicated objectives simultaneously, such as vehicle fuel efficiency parameters, traffic condition of a road network, and availability of alternate vehicles.

Author Contributions: Conceptualization, A.S.K. and K.N.; methodology, A.S.K.; software, R.A.; validation, Q.S.K., R.K. and W.S.; formal analysis, A.S.K. and W.S.; investigation, C.I.P.; resources, R.K.; data curation, K.N.; writing—original draft preparation, A.S.K.; writing—review and editing, A.S.K. and R.K.; visualization, R.A.; supervision, C.I.P.; project administration, W.S.; funding acquisition, C.I.P. All authors have read and agreed to the published version of the manuscript.

Funding: This study received no external funding.

Institutional Review Board Statement: Not applicable.

Informed Consent Statement: Not applicable.

Data Availability Statement: Not applicable.

Conflicts of Interest: The authors declare no conflict of interest.

Nomenclature

Indices	
i	*set of supply nodes* $i = \{1, 2, \ldots I\}$
j	*set of customer locations* $j = \{1, 2, \ldots J\}$
k	*set of cross docking nodes* $k = \{1, 2, \ldots K\}$
Parameters	
W	*number of available homogeneous trucks*
dx_{ik}	*distance between supply and cross docking nodes*
dy_{kj}	*distance between cross docking and delivery nodes*
dz_{ji}	*distance between delivery and supply nodes*
LC	*loading capacity of a truck*
α	*rate of conforming products delivered*
cd	*unit cost of distance travelled (considering fuel cost only with fixed environmental impacts)*
ch	*unit handling cost at cross docking point*
cp	*penelty cost pre unit defective product*
cr	*unit cost of returned product*
cm	*per unit manufacturing cost*
D_j	*demand of product at customer location j*
mx	*max. capacity of product at node i*
Decision variables	
NX_{ik}	*1, if goods are transhipped from node i to crossdock k, otherwise 0*
MX_{kj}	*1, if goods are delivered to customer node j from crossdock k, otherwise 0*
RX_{ji}	*1, if goods are returned from node j to node i, otherwise 0*
QP_{ik}	*number of transhipped products between i and cross dock k*
FP_{kj}	*number of products delivered to customers*
PR_{ji}	*number of returned products*
NW_{ik}	*number of track used between i and k*
MW_{kj}	*number of track used between k and j*
TW_{ji}	*number of track used between j and i*
TR	*total cost of transhippment (with fixed environmental impacts)*
DC	*total cost of delivery (with fixed environmental impacts)*
RC	*total cost of return (with fixed environmental impacts)*
HC	*total cost of holding (with fixed environmental impacts)*
TC	*total cost of logistic network (with fixed environmental impacts)*

References

1. Suryaningtyas, D.; Sudiro, A.; Eka, T.A.; Dodi, I.W. Organizational resilience and organizational performance: Examining the mediating roles of resilient leadership and organizational culture. *Acad. Strateg. Manag. J.* **2019**, *18*, 1–7.
2. Drobyazko, S.; Okulich-Kazarin, V.; Rogovyi, A.; Goltvenko, O.; Marova, S. Factors of influence on the sustainable development in the strategy management of corporations. *Acad. Strateg. Manag. J.* **2019**, *18*, 1–5.
3. Zhang, L.; Wang, J.; You, J. Consumer environmental awareness and channel coordination with two substitutable products. *Eur. J. Oper. Res.* **2015**, *241*, 63–73. [CrossRef]
4. Suzuki, Y. A new truck-routing approach for reducing fuel consumption and pollutants emission. *Transp. Res. Part D Transp. Environ.* **2011**, *16*, 73–77. [CrossRef]
5. Bektaş, T.; Laporte, G. The Pollution-Routing Problem. *Transp. Res. Part B Methodol.* **2011**, *45*, 1232–1250. [CrossRef]
6. Van Fan, Y.; Perry, S.; Klemeš, J.J.; Lee, C.T. A review on air emissions assessment: Transportation. *J. Clean. Prod.* **2018**, *194*, 673–684. [CrossRef]
7. Stalk, G.; Evans, P.; Shulman, L.E. Competing on capabilities. In *Understanding Business: Processes*; Routledge: New York, NY, USA, 2001; pp. 42–49.
8. Choi, J.; Xuelei, J.; Jeong, W. Optimizing the Construction Job Site Vehicle Scheduling Problem. *Sustainability* **2018**, *10*, 1381. [CrossRef]
9. Qian, J.; Eglese, R. Fuel emissions optimization in vehicle routing problems with time-varying speeds. *Eur. J. Oper. Res.* **2016**, *248*, 840–848. [CrossRef]
10. Miao, C.; Liu, H.; Zhu, G.G.; Chen, H. Connectivity-based optimization of vehicle route and speed for improved fuel economy. *Transp. Res. Part C Emerg. Technol.* **2018**, *91*, 353–368. [CrossRef]

11. Elhedhli, S.; Merrick, R. Green supply chain network design to reduce carbon emissions. *Transp. Res. Part D Transp. Environ.* **2012**, *17*, 370–379. [CrossRef]
12. Pishvaee, M.S.; Rabbani, M.; Torabi, S.A. A robust optimization approach to closed-loop supply chain network design under uncertainty. *Appl. Math. Model.* **2011**, *35*, 637–649. [CrossRef]
13. Kannan, D.; Diabat, A.; Alrefaei, M.; Govindan, K.; Yong, G. A carbon footprint based reverse logistics network design model. *Resour. Conserv. Recycl.* **2012**, *67*, 75–79. [CrossRef]
14. Faccio, M.; Persona, A.; Sgarbossa, F.; Zanin, G. Sustainable SC through the complete reprocessing of end-of-life products by manufacturers: A traditional versus social responsibility company perspective. *Eur. J. Oper. Res.* **2014**, *233*, 359–373. [CrossRef]
15. Mota, B.; Gomes, M.I.; Carvalho, A.; Barbosa-Povoa, A.P. Towards supply chain sustainability: Economic, environmental and social design and planning. *J. Clean. Prod.* **2015**, *105*, 14–27. [CrossRef]
16. Fonseca, M.C.; García-Sánchez, Á.; Ortega-Mier, M.; Saldanha-da-Gama, F. A stochastic bi-objective location model for strategic reverse logistics. *Top* **2010**, *18*, 158–184. [CrossRef]
17. Vate, H.D.R.J.V.; Zhang, M. Network Design for Load-Driven Cross-Docking Systems. GIT Technical Report. 1999. Available online: http://www2.isye.gatech.edu/~{}jvandeva/Classes/6203/Load_driven8-4-02.doc (accessed on 2 June 2020).
18. De Oliveira, H.C.B.; Vasconcelos, G.C. A hybrid search method for the vehicle routing problem with time windows. *Ann. Oper. Res.* **2010**, *180*, 125–144.
19. Wen, M.; Larsen, J.; Clausen, J.; Cordeau, J.-F.; Laporte, G. Vehicle routing with cross-docking. *J. Oper. Res. Soc.* **2009**, *60*, 1708–1718. [CrossRef]
20. Musa, R.; Arnaout, J.-P.; Jung, H. Ant colony optimization algorithm to solve for the transportation problem of cross-docking network. *Comput. Ind. Eng.* **2010**, *59*, 85–92. [CrossRef]
21. Khan, A.S. Mixed Integer Linear Programming Optimization of Multi-Node Transportation Transshipment Network. *J. Appl. Emerg. Sci.* **2018**, *8*, 70–76.
22. Govindan, K.; Jafarian, A.; Khodaverdi, R.; Devika, K. Two-echelon multiple-vehicle location–routing problem with time windows for optimization of sustainable supply chain network of perishable food. *Int. J. Prod. Econ.* **2014**, *152*, 9–28. [CrossRef]
23. Rahmani, K.; Yavari, M. Pricing policies for a dual-channel green supply chain under demand disruptions. *Comput. Ind. Eng.* **2019**, *127*, 493–510. [CrossRef]
24. Kaya, O.; Urek, B. A mixed integer nonlinear programming model and heuristic solutions for location, inventory and pricing decisions in a closed loop supply chain. *Comput. Oper. Res.* **2016**, *65*, 93–103. [CrossRef]
25. Talaei, M.; Moghaddam, B.F.; Pishvaee, M.S.; Bozorgi-Amiri, A.; Gholamnejad, S. A robust fuzzy optimization model for carbon-efficient closed-loop supply chain network design problem: A numerical illustration in electronics industry. *J. Clean. Prod.* **2016**, *113*, 662–673. [CrossRef]
26. Ramezani, M.; Kimiagari, A.M.; Karimi, B.; Hejazi, T.H. Closed-loop supply chain network design under a fuzzy environment. *Knowledge-Based Syst.* **2014**, *59*, 108–120. [CrossRef]
27. Subulan, K.; Taşan, A.S.; Baykasoğlu, A. Designing an environmentally conscious tire closed-loop supply chain network with multiple recovery options using interactive fuzzy goal programming. *Appl. Math. Model.* **2015**, *39*, 2661–2702. [CrossRef]
28. Kamal, A.; Azfar, R.W.; Salah, B.; Saleem, W.; Abas, M.; Khan, R.; Pruncu, C.I. Quantitative Analysis of Sustainable Use of Construction Materials for Supply Chain Integration and Construction Industry Performance through Structural Equation Modeling (SEM). *Sustainability* **2021**, *13*, 522. [CrossRef]
29. Sahebjamnia, N.; Fathollahi-Fard, A.M.; Hajiaghaei-Keshteli, M. Sustainable tire closed-loop supply chain network design: Hybrid metaheuristic algorithms for large-scale networks. *J. Clean. Prod.* **2018**, *196*, 273–296. [CrossRef]
30. Hajiaghaei-Keshteli, M.; Fard, A.M.F. Sustainable closed-loop supply chain network design with discount supposition. *Neural Comput. Appl.* **2019**, *31*, 5343–5377. [CrossRef]
31. Fonseca, G.B.; Nogueira, T.H.; Ravetti, M.G. A hybrid Lagrangian metaheuristic for the cross-docking flow shop scheduling problem. *Eur. J. Oper. Res.* **2019**, *275*, 139–154. [CrossRef]
32. Heidari, F.; Zegordi, S.H.; Tavakkoli-Moghaddam, R. Modeling truck scheduling problem at a cross-dock facility through a bi-objective bi-level optimization approach. *J. Intell. Manuf.* **2018**, *29*, 1155–1170. [CrossRef]
33. Song, K.; Chen, F. Scheduling Cross Docking Logistics Optimization Problem with Multiple Inbound Vehicles and One Outbound Vehicle. In Proceedings of the 2007 IEEE International Conference on Automation and Logistics, Jinan, China, 18–21 August 2007; pp. 3089–3094.
34. Jayaraman, V.; Ross, A. A simulated annealing methodology to distribution network design and management. *Eur. J. Oper. Res.* **2003**, *144*, 629–645. [CrossRef]
35. Gumuş, M.; Bookbinder, J.H. Cross-docking and its implications in location-distribution systems. *J. Bus. Logist.* **2004**, *25*, 199–228. [CrossRef]
36. Gholizadeh, H.; Fazlollahtabar, H. Robust optimization and modified genetic algorithm for a closed loop green supply chain under uncertainty: Case study in melting industry. *Comput. Ind. Eng.* **2020**, *147*, 106653. [CrossRef]
37. Gaur, J.; Amini, M.; Rao, A.K. The impact of supply chain disruption on the closed-loop supply chain configuration profit: A study of sourcing policies. *Int. J. Prod. Res.* **2019**, *58*, 5380–5400. [CrossRef]
38. Turki, S.; Didukh, S.; Sauvey, C.; Rezg, N. Optimization and analysis of a manufacturing–remanufacturing–transport–warehousing system within a closed-loop supply chain. *Sustainability* **2017**, *9*, 561. [CrossRef]

39. Cheraghalipour, A.; Paydar, M.M.; Hajiaghaei-Keshteli, M. A bi-objective optimization for citrus closed-loop supply chain using Pareto-based algorithms. *Appl. Soft Comput.* **2018**, *69*, 33–59. [CrossRef]
40. Yavari, M.; Geraeli, M. Heuristic method for robust optimization model for green closed-loop supply chain network design of perishable goods. *J. Clean. Prod.* **2019**, *226*, 282–305. [CrossRef]
41. Rad, R.S.; Nahavandi, N. A novel multi-objective optimization model for integrated problem of green closed loop supply chain network design and quantity discount. *J. Clean. Prod.* **2018**, *196*, 1549–1565. [CrossRef]
42. Taleizadeh, A.A.; Moshtagh, M.S.; Moon, I. Pricing, product quality, and collection optimization in a decentralized closed-loop supply chain with different channel structures: Game theoretical approach. *J. Clean. Prod.* **2018**, *189*, 406–431. [CrossRef]
43. Özdemir, D.; Yücesan, E.; Herer, Y.T. Multi-location transshipment problem with capacitated transportation. *Eur. J. Oper. Res.* **2006**, *175*, 602–621. [CrossRef]
44. Hong, J.; Diabat, A.; Panicker, V.V.; Rajagopalan, S. A two-stage supply chain problem with fixed costs: An ant colony optimization approach. *Int. J. Prod. Econ.* **2018**, *204*, 214–226. [CrossRef]
45. Babaveisi, V.; Paydar, M.M.; Safaei, A.S. Optimizing a multi-product closed-loop supply chain using NSGA-II, MOSA, and MOPSO meta-heuristic algorithms. *J. Ind. Eng. Int.* **2018**, *14*, 305–326. [CrossRef]
46. Zhang, M.; Hu, Q. An effective evolutionary algorithm for the biobjective full truckload transportation service procurement problem. *Comput. Ind. Eng.* **2019**, *127*, 1012–1023. [CrossRef]
47. Behzadi, G.; O'Sullivan, M.; Olsen, T.; Zhang, A. Allocation flexibility for agribusiness supply chains under market demand disruption. *Int. J. Prod. Res.* **2017**, *56*, 3524–3546. [CrossRef]
48. Bhattacharya, A.; Kumar, S.A.; Tiwari, M.K.; Talluri, S. An intermodal freight transport system for optimal supply chain logistics. *Transp. Res. Part C Emerg. Technol.* **2014**, *38*, 73–84. [CrossRef]
49. Cheng, C.; Qi, M.; Zhang, Y.; Rousseau, L.-M. A two-stage robust approach for the reliable logistics network design problem. *Transp. Res. Part B Methodol.* **2018**, *111*, 185–202. [CrossRef]
50. De Jong, G.; Ben-Akiva, M. A micro-simulation model of shipment size and transport chain choice. *Transp. Res. Part B Methodol.* **2007**, *41*, 950–965. [CrossRef]
51. Yildizbaşi, A.; Çalik, A.; Paksoy, T.; Farahani, R.Z.; Weber, G.-W. Multi-level optimization of an automotive closed-loop supply chain network with interactive fuzzy programming approaches. *Technol. Econ. Dev. Econ.* **2018**, *24*, 1004–1028. [CrossRef]
52. Ekren, B.Y.; Heragu, S.S. Simulation based optimization of multi-location transshipment problem with capacitated transportation. In Proceedings of the 2008 Winter Simulation Conference, Miami, FL, USA, 7–10 December 2008; pp. 2632–2638.
53. Fakhrzad, M.B.; Alidoosti, Z. A realistic perish ability inventory management for location-inventory-routing problem based on Genetic Algorithm. *J. Ind. Eng. Manag. Stud.* **2018**, *5*, 106–121.
54. Kawamura, M.S.; Ronconi, D.P.; Yoshizaki, H. Optimizing transportation and storage of final products in the sugar and ethanol industry: A case study. *Int. Trans. Oper. Res.* **2006**, *13*, 425–439. [CrossRef]
55. Lu, H.; Li, L.; Zhao, X.; Cook, D. A model of integrated regional logistics hub in supply chain. *Enterp. Inf. Syst.* **2018**, *12*, 1308–1335. [CrossRef]
56. Mejjaouli, S.; Babiceanu, R.F. Cold supply chain logistics: System optimization for real-time rerouting transportation solutions. *Comput. Ind.* **2018**, *95*, 68–80. [CrossRef]
57. Panicker, V.V.; Vanga, R.; Sridharan, R. Ant colony optimisation algorithm for distribution-allocation problem in a two-stage supply chain with a fixed transportation charge. *Int. J. Prod. Res.* **2013**, *51*, 698–717. [CrossRef]
58. Rais, A.; Alvelos, F.; Carvalho, M. New mixed integer-programming model for the pickup-and-delivery problem with transshipment. *Eur. J. Oper. Res.* **2014**, *235*, 530–539. [CrossRef]
59. Singh, M.; Aggarwal, D.; Kumar, V. Benders Decomposition for Distribution Networks with Cross-docking Centre. *Pertanika J. Sci. Tech.* **2018**, *26*, 1081–1093.
60. Liu, M.; Liu, R.; Zhu, Z.; Chu, C.; Man, X. A Bi-Objective Green Closed Loop Supply Chain Design Problem with Uncertain Demand. *Sustainability* **2018**, *10*, 967. [CrossRef]
61. Validi, S.; Bhattacharya, A.; Byrne, P.J. Sustainable distribution system design: A two-phase DoE-guided meta-heuristic solution approach for a three-echelon bi-objective AHP-integrated location-routing model. *Ann. Oper. Res.* **2020**, *290*, 191–222. [CrossRef]
62. Zheng, Y.-J.; Ling, H.-F. Emergency transportation planning in disaster relief supply chain management: A cooperative fuzzy optimization approach. *Soft Comput.* **2012**, *17*, 1301–1314. [CrossRef]
63. Li, X.; Leung, S.C.; Tian, P. A multistart adaptive memory-based tabu search algorithm for the heterogeneous fixed fleet open vehicle routing problem. *Expert Syst. Appl.* **2012**, *39*, 365–374. [CrossRef]
64. Tarantilis, C.D.; Stavropoulou, F.; Repoussis, P.P. A template-based tabu search algorithm for the consistent vehicle routing problem. *Expert Syst. Appl.* **2012**, *39*, 4233–4239. [CrossRef]
65. Liao, C.-J.; Lin, Y.; Shih, S.C. Vehicle routing with cross-docking in the supply chain. *Expert Syst. Appl.* **2010**, *37*, 6868–6873. [CrossRef]
66. Barbarosoglu, G.; Ozgur, D. A tabu search algorithm for the vehicle routing problem. *Comput. Oper. Res.* **1999**, *26*, 255–270. [CrossRef]
67. Metropolis, N.; Rosenbluth, A.W.; Rosenbluth, M.N.; Teller, A.H.; Teller, E. Equation of state calculations by fast computing machines. *J. Chem. Phys.* **1953**, *21*, 1087–1092. [CrossRef]

68. Lai, K.; Chan, J.W. Developing a simulated annealing algorithm for the cutting stock problem. *Comput. Ind. Eng.* **1997**, *32*, 115–127. [CrossRef]
69. Vincent, F.Y.; Lin, S.W. Multi-start simulated annealing heuristic for the location routing problem with simultaneous pickup and delivery. *Appl. Soft Comput.* **2014**, *24*, 284–290.
70. Yoo, W.S.; Kim, J.G.; Kim, C.M. A combined heuristic approach to the nonlinear fixed charge capacitated network design problem. *Int. J. Ind. Eng.* **2016**, *23*, 83–93.
71. Yun, W.Y.; Wang, W.F.; Ha, B.H. A hybrid SA algorithm for inland container transportation. *Int. J. Ind. Eng.* **2013**, *20*, 12–23.
72. Birim, Ş. Vehicle Routing Problem with Cross Docking: A Simulated Annealing Approach. *Procedia Soc. Behav. Sci.* **2016**, *235*, 149–158. [CrossRef]
73. Baniamerian, A.; Bashiri, M.; Zabihi, F. Two phase genetic algorithm for vehicle routing and scheduling problem with cross-docking and time windows considering customer satisfaction. *J. Ind. Eng. Int.* **2018**, *14*, 15–30. [CrossRef]
74. Lee, Y.H.; Jung, J.W.; Lee, K.M. Vehicle routing scheduling for cross-docking in the supply chain. *Comput. Ind. Eng.* **2006**, *51*, 247–256. [CrossRef]
75. Mousavi, S.M.; Sadeghi, J.; Niaki, S.T.A.; Alikar, N.; Bahreininejad, A.; Metselaar, H.S.C. Two parameter-tuned meta-heuristics for a discounted inventory control problem in a fuzzy environment. *Inf. Sci.* **2014**, *276*, 42–62. [CrossRef]
76. 2018 Government GHG Conversion Factors for Company Reporting, Methodology Paper for Emission Factors: Final Report. Available online: https://www.gov.uk/government/collections/government-conversion-factors-forcompany-reporting (accessed on 2 July 2018).

sustainability

MDPI

Article

An Analytic Hierarchy Process Approach for Prioritisation of Strategic Objectives of Sustainable Development

Zsuzsanna Katalin Szabo [1,*], Zsombor Szádoczki [2,3], Sándor Bozóki [2,3], Gabriela C. Stănciulescu [4] and Dalma Szabo [5]

1 Faculty of Economics and Law, "George Emil Palade" University of Medicine, Pharmacy, Science and Technology of Târgu-Mures, 540566 Tîrgu Mureș, Romania
2 Research Group of Operations Research and Decision Systems, Research Laboratory on Engineering and Management Intelligence, Institute for Computer Science and Control (SZTAKI), Eötvös Loránd Research Network (ELKH), 1111 Budapest, Hungary; szadoczki.zsombor@sztaki.hu (Z.S.); bozoki.sandor@sztaki.hu (S.B.)
3 Department of Operations Research and Actuarial Sciences, Corvinus University of Budapest, 1093 Budapest, Hungary
4 Faculty of Business and Tourism, Bucharest University of Economic Studies, 010374 Bucharest, Romania; cecilia.stanciulescu@gmail.com
5 Freelance Marketer, 90408 Nürnberg, Germany; k.dalma.szabo@gmail.com
* Correspondence: szabo.zs.katalin@gmail.com

Abstract: Sustainability is one of the world's fundamental objectives, and a wide variety of information types, parameters, and uncertainties need to be appraised and managed to assess it. In the present paper, Multi-Criteria Decision Analysis (MCDA) is used to prioritise the criteria of sustainable development based on regularly published indicators. In line with most approaches in the literature, the main criteria are Economy, Society and Environment. Complex criteria are decomposed into subcriteria until the performance with respect to them can be measured directly. Weights of importance are calculated by the Analytic Hierarchy Process (AHP), in decision support system PriEsT. The model is flexible to both the modification of criteria and re-weighting, and the PriEsT file is supplemented to the paper. Moreover, the results can also be applied in decisions on resource allocation. The proposed methodology has the potential of resulting in a new composite index to measure, compare or rank countries and regions regarding sustainable development or one of its subcriteria, as well as to track, year by year, the improvements or the impact of the policies introduced.

Keywords: sustainable development; indicators; multi-criteria decision making; analytic hierarchy process; PriEsT

check for updates

Citation: Szabo, Z.K.; Szádoczki, Z.; Bozóki, S.; Stănciulescu, G.C.; Szabo, D. An Analytic Hierarchy Process Approach for Prioritisation of Strategic Objectives of Sustainable Development. *Sustainability* **2021**, *13*, 2254. https://doi.org/10.3390/su13042254

Academic Editors: Dragan Pamucar and Željko Stević
Received: 30 December 2020
Accepted: 10 February 2021
Published: 19 February 2021

Publisher's Note: MDPI stays neutral with regard to jurisdictional claims in published maps and institutional affiliations.

Copyright: © 2021 by the authors. Licensee MDPI, Basel, Switzerland. This article is an open access article distributed under the terms and conditions of the Creative Commons Attribution (CC BY) license (https://creativecommons.org/licenses/by/4.0/).

1. Introduction

Sustainability is essential for a straightforward reason: the quality of life and the Earth's ecosystems' prosperity cannot be maintained if it is not addressed. Therefore, sustainability is one of the world's fundamental objectives, driving the vision and establishing the policies and treaties for sustainable development worldwide. Governments must act responsibly and embrace the ethical aspects of evolution, and policies must address the tension between the pursuit of endless economic growth and sustainable development.

1.1. Historical Overview

The first document that tackled the subject is the Club Rome's Report published in 1972, which had a powerful impact on the concept of economic development. The report [1] revealed the results of a research group obtained using a mathematical model built on the theory of dynamic systems known as World3. It presented the conflict between the unlimited growth and consumption and the limited resources of the Earth and therefore drew the attention of researchers, politicians, and governmental institutions. Consequently, high political level meetings were organised worldwide, agreements and treaties were signed, several research papers were published, and research institutions and universities hosted an increased number of scientific conferences and events. In 2000, the declaration on promoting sustainability science [2] was adopted.

The Living Planet Report [3] highlights the continuous decline of the living planet index/indicators and claims, "World leaders must take urgent action to protect and restore nature as the foundation for a healthy society and thriving economy" [3] (p. 5).

In 2015, the post-2015 development agenda was adopted by more than 150 world leaders who participated in the UN Sustainable Development Summit organised at the UN headquarters in New York. The outcome of the Summit was entitled "Transforming our world: the 2030 Agenda for Sustainable Development" [4], and it came officially into force on 1 January 2016. The Agenda became a support for the international community to promote prosperity and well-being for the next period and achieve sustainable development by 2030. It comprises 17 SD Goals and 169 targets to be achieved until 2030 and must be applied by all countries worldwide.

In 2019, the first quadrennial report "The Future is Now: Science for Achieving Sustainable Development" [5] was published bearing the vital component of the follow-up and review process of the 2030 Agenda, as Member States decided in the Ministerial Declaration of the High-Level Political Forum in 2016 on sustainable development [6].

At the Paris Climate Conference (COP21) in December 2015, the first universal agreement on global climate change [7] was adopted. It entered into force on 4 November 2016.

The Sustainable Development Goals (SDGs) and the Paris Agreement present a new global consensus on sustainable development. "A life of dignity for all within the planet's limits and reconciling economic efficiency, social inclusion and environmental responsibility is at the essence of sustainable development" [8].

In almost 50 years after the publication of the Club Rome's Report, many definitions of sustainable development were articulated. According to Faber et al., the "definitions of sustainability show conceptual developments". The authors of [9] identified about 50 definitions and circumscriptions of sustainability and consequently created a conceptual framework. Nevertheless, it must not be forgotten that Carsons' book, "Silent Spring", published in 1962 [10], started the discussions on sustainability before using the term "sustainability". Coomers' sustainable society "is one that lives within the self-perpetuating limits of its environment ... is not a non-growth society. It is, rather, a society that recognises the limits of growth ... a society that looks for alternative ways of growing" [11]. According to Faber [9], "Coomer attributes explicit limits to a sustainable society that is aimed at, his approach is labelled absolute".

Different definitions of sustainable development are presented in the book of Rumen Gechev [12]; a comparative analysis of the emerging definitions was published by Ciegis [13]; and the evolution of the concept of sustainable development, and EU strategies were presented in a paper published in 2011 [14]. Moreover, a study on the history and chronological overview of the concept's meaning was published in 2018 [15] and a literature review on sustainable development was published in 2019 [16]. Although formulated differently, all these definitions have some common ideas, and they refer to some extent to surviving and assuring needs in the long run. It is well known that sustainable development focuses on ensuring economic, human, and environmental well-being. Different approximations define the notion of sustainability in a weak sense and a strong one, respectively.

According to Coomer [11], "the sustainable society recognises that there is one primary environment—the physical environment—within which all other environments function. All other environments—political, social, economic, to name three major ones—exist within and act upon the primary environment". It can be considered that sustainable development intends to harmonize the realization of economic growth and environmental concerns simultaneously.

In 1987, the "Brundtland Report" [17] was published by the World Commission on Environment and Development (WCED), also known as "Our Common Future" (WCED, 1987). G.H. Brundtland, a year after the Chernobyl disaster, formulated a definition of sustainable development, one that has become widely known. "Sustainable development is a development that meets the needs of the present without compromising the ability of future generations to meet their own needs" [17] (p. 41). The Report pleads for reconciliation between economy and the environment. "Strategy for sustainable development is the need to integrate economic and ecological considerations in decision making . . . economic and ecological concerns are not necessarily in opposition" [17] (p. 55).

1.2. *The Multi-Criteria Nature of Sustainable Development*

From the analysed literature, the authors highlight Barbier's definition from 1987 [18], which assessed sustainable development as "simultaneous maximization" of the objectives of the biological system (genetic diversity, biological productivity, flexibility), the objectives of the economic system (satisfying basic needs, equity increase, growth in goods and services) and the objectives of the social system (cultural diversity, institutional durability, social equity). Munda [19] (p. 954) noted that Barbier's definition "correctly points out that sustainable development is a multi-dimensional concept".

The need for weighting is highlighted by Prescott-Allen [20]. According to Munda, "sustainability management and planning is essentially a conflict analysis", a "tool for conflict management" [19] (p. 956), and the multi-criteria evaluation, multi-criteria decision method (MCDM) is a suitable tool due to its flexibility and ease to enable dialogue between stakeholders, analysts, and scientists [19] (p. 957). Thus, when analysing sustainable development, the relations between the economic, social, and environmental dimensions should be taken into consideration and treated as an integrative whole. In this respect, according to Ciegis et al. [13], "a hierarchical framework with coherent sustainability logic is needed".

The Analytic Hierarchy Process (AHP) [21] is undoubtedly among the most often applied multi-criteria decision methods, permitting to model a complex problem in a hierarchical structure showing the relationships between the goal, criteria-objectives, sub-criteria and alternatives, giving in this way a general overview of the problem for decision makers. Moreover, it is probably the easiest, among multi-criteria decision models (e.g., PROMETHEE, ELECTRE, TOPSIS [22]), to communicate to both experts and non-experts, since the hierarchical system of criteria follows the structure of many other well-known analogies (categorisation, organisation), and the weighting by pairwise comparisons is likewise reasonable. Therefore, a significant advantage of using AHP is that the end-users (decision-makers) do not necessarily need to go into the mathematical details of the calculations.

According to Prescott-Allen, "No country is sustainable or even close to sustainability" [20] (p. 107), thus, tools, methods are needed to assist and achieve sustainable development. Therefore, creating a sustainable development framework using publicly available databases presents an efficient way to monitor progress on national, regional, and international levels, because it enables identifying the barriers towards progress quickly and addressing them through dedicated policies.

1.3. *The Outline of the Paper*

This paper aims to give a method and methodology to prioritise the criteria and determine the strategic objectives for sustainable development. Hence, the paper intends to identify a suitable method to rank the criteria with the highest overall impact on sustainable development, thus providing an action plan of priorities.

The present study focuses on the following research questions:

- How can policymakers identify the criteria with the highest overall impact on the efficient implementation of sustainable development?
- How can sustainable development be inserted into policy and resource allocation?

- What is the potential of a weighted system of criteria in the assessment concerning sustainable development?

This paper presents a general, complex, five-level hierarchical criteria system to analyse the economic, social and environmental pillars' (subsystems') impact on sustainable development. The authors focus on applying the AHP on strategic level decision-making, and on the prioritisation and ranking of criteria with the highest overall impact to implement sustainable development successfully.

The effectiveness of sustainable development policies must be measured in their multi-objective and multi-level complexity, considering all pillars. Therefore, the authors use a Multi-Criteria Decision-Making (MCDM) approach, applying the AHP to explore different scenarios, evaluate the outcomes of prioritisation and select the strategic objectives of sustainable development. Furthermore, the framework permits to analyse all hypothetical scenarios' effects on all pillars (subsystems). The method serves as a useful instrument for policymakers giving a solid, scientific base for the decision-making and optimal resource allocation. For decision-making the authors use the Priority Estimation Tool, PriEsT [23]. The strategic objectives from the authors' point of view are considered to be the criteria with the highest overall impact on the efficient implementation of sustainable development.

The article is organised as follows: Section 2, specifically Section 2.1, presents a historical review of the sustainability indicators having a significant role in evaluating and determining whether a region or a country is heading in a sustainable direction, while Section 2.2 presents a critical review of the literature to prove that the multi-criteria decision methodology, especially AHP, is a suitable tool for the sustainability policy problems. Section 3 describes the AHP method, the PriEsT software and data used. Section 4 presents the authors' results, the multi-criteria framework to analyse, organise and prioritise the criteria, and concludes and raises the question of evaluation concerning each criterion and the scenarios, the numerical results being obtained with PriEsT. Section 5 concludes with an emphasis on the need for sophisticated assessment methodologies concerning the criteria.

2. Literature Review

2.1. Sustainable Development Indicators

The significant role of indicators to help countries make the right decisions regarding sustainable development was recognized in 1992 at the UN Conference, having the topic "Environment and Development" organized in Rio de Janeiro. The conference resulted in an action plan entitled "Agenda 21" [24]. The Commission on Sustainable Development (CSD) created the first two sets of indicators between 1994 and 2001. The third, revised set of sustainable development indicators to measure progress in achieving sustainable development, including the MDGs, was published in 2007 [25].

In 2002, the OECD published a comparative analysis of the sustainable development indicators elaborated by different International Institutions [26] (p. 51). These indicators are generally enrolled in social, environmental, economic and institutional dimensions.

In 2004, a study on the development and usage of sustainable development indicators (SDIs) in the EU Member States was issued [27]. In 2006, the European Council adopted the renewed EU sustainable development strategy, which contoured how sustainable development indicators will monitor the implementation. In 2007, a study covering the sets of SDIs used by countries to control the National Reform Programmes and SD strategies was published [28]. In 2005 the joint UNECE/OECD/Eurostat Working Group on Statistics for SD (WGSSD) was founded with the aim to develop a conceptual framework for measuring sustainable development. The first Report was published in 2009 [29]. As the basis for measuring sustainability, it had capital at its centre—economic, natural, human and social capital [29] (p. 2). The next framework with recommended indicators appeared in 2013. The suggested SDIs are used for international comparisons and are defined on three dimensions: human well-being ("here and now"), capital ("later"), transboundary impacts ("elsewhere") [30]. Some characteristics of the analysed frameworks are presented in Appendix A.

Sustainable development is generally built upon the three-pillar concept. In the literature, other structures can be identified as well, for example the reference framework for sustainable cities (RFSC) [31], which relies on five dimensions (spatial, governance, social, economic (green growth, circular economy, innovation, cooperation, sustainable production, and consumption), and environmental) and 30 objectives for the European vision of tomorrow's cities. According to Prescott-Allen [20], "sustainability" is "the good life"; thus, he considered it as a combination of human and ecosystem wellbeing and published a dual wellbeing assessment framework in the book [20], which contains indicators and the Barometer of Sustainability.

The SDIs sets are the result of "rigorous consultation inside and outside government to ensure that different perspectives on sustainable development are taken into account" [29] (p. 29).

According to Pupphachai and Zuidema [32], sustainable indicators "have the potential to support adaptive processes of learning-by-doing on the path towards sustainable development". It can be done through active management, which requires indicators to be easily accessible and understandable. Moreover, they state that "it is advisable to develop well-categorized online databases that are supported by search engines, explanatory reports".

The role, the importance of SDIs, the characteristics of "good" and "effective" indicators is presented in the paper published by Garrett [33]. According to them "the creation of good sustainability indicators can greatly aid policy and management decision making".

At present, the 2030 Agenda for Sustainable Development with 17 SDGs is the universal standard for development [34].

The evaluation procedure of the progress toward sustainability has to follow scientific methods, which require indicators. Measurement methods, the elaboration of a set of multi-dimensional indicators are needed to evaluate and determine whether a region or a country is heading toward a sustainable direction. The results depend on the quality of information available, the choice of indicators and the relative importance of used indicators—if they have the same significance or weighted—mathematical aggregation—pairwise comparison and ranking in a complete pre-order [35].

Agenda 21 stipulates the necessity for all the countries to be concerned with elaborating and identifying indicators for sustainable development that may assure a solid foundation for decisions at all levels for a particular country's general and socio-economic evolution. It has been suggested that such indicators be set at a regional, national, and global level.

Indicators are used to obtain a simplified, consistent, reasonable, and multi-dimensional view of sustainability and sustainable development. Therefore, through the quantitative aggregation of many indicators, indexes are defined, which are crucial to monitor sustainability progress. The identification of the collection of the most suitable indicators and indexes is the most critical, decisive procedure for decision making. "The quality of decisions, the result of the application of the exact method to support the decision will always depend on the quality of the input data" [36]. Creating an appropriate conceptual framework with the corresponding indicators is mainly dependent on the purpose to be served.

Faber et al. [9] elaborated a conceptual framework to examine the lists of indicators for sustainability, "the practical application of sustainability". The authors analysed the indicator lists chronologically (from 1995 to 2003) and identified the developments. "Both entity and construct-oriented indicator lists developed over time from an absolute to a relative approach towards sustainability. Whereas the construct-oriented indicator lists hold on to a static approach, the entity-oriented lists adopted a dynamic approach regarding sustainability".

How to get closer to sustainable economic development? There are huge differences between the socio-economic and national level within the UN countries, and there are inequalities at the regional and local levels. Thus, to implement sustainable development goals, global policies cannot be applied; particularities at the national, regional, and local levels must be considered. Therefore, to reach the SDGs, strategies need to be formulated. "Biodiversity conservation is more than an ethical commitment for humanity: it is a non-negotiable and strategic investment to preserve our health, wealth and security" [3].

A thorough analysis, a real evaluation of the situation, and the needs became necessary to identify adequate policies with the maximal overall SD impact. This procedure asks for scientific-based treatment, and therefore, MCDM is acknowledged as an efficient tool.

According to Bonissone et al. [37], MCDM "is the conjunction of three components: search, preference trade-offs, and interactive visualisation". Policymaking, strategic planning and general problems that ask for multi-criteria decisions are unstructured. According to Moshkovich et al., using decision-making methods to not well formulated problems "should be to help the decision-maker to structure the problem (form a set of alternatives and elaborate a set of relevant criteria) and work out a consistent policy for evaluating, comparing multiple criteria alternatives" [38] (p. 607).

2.2. Multi-Criteria Decision Analysis for Sustainable Development

Due to the rapidly evolving world, decision-making is more complex than ever. Therefore, to assess sustainability, various information types, parameters, and uncertainties need to be appraised and managed.

The introduction of exact decision-making methods is essential since, in its absence, managers and policymakers cannot make rational decisions [36]. The scientific literature analyses the efficiency of intuition-based decision-making [39] and recognises that a data-based decision is required in the case of complex problems. In instances where "stakes are extremely high, human perceptions and judgments are involved and whose solutions have long term repercussions" are enrolled in the strategic level decision-making category and demand a scientific-based approach for solutions [40].

According to Munda [19,41] (p. 1240) Multi-Criteria Decision Analysis is an adequate approach toward sustainability issues, being an efficient tool "to implement a multi/interdisciplinary approach". Munda used in this respect theoretical arguments, practical experience, and examples of "good practice".

A critical literature review revealed that multi-criteria evaluation proved its usefulness in many different management and sustainability policy problems.

The Multi-Criteria Decision Making (MCDM) methods [22], and especially the Analytic Hierarchy Process (AHP), have been used in connection with sustainable development in many different fields, for instance, in the mining industry [42–44] in agriculture [45], in electricity generation [46], in flooring systems [47], in strategic planning, and management of companies [48], in the reuse of the industrial heritage [49], in environmental problems [50,51], in smart city [52], in energy management [53,54] and in well-being [55].

A multi-criteria approach was used by Nijkamp and Vindigni [56] to study integrated assessment procedures for evaluating the effectiveness of agri-environmental policy strategies. Akgün et al. [57] published a multi-criteria approach of regional sustainable resource policy. A systematic review [58] of the literature on using multi-criteria analysis in a sustainable development context was published in 2019.

Dos Santos et al. (2019) [59] focuses on the applications of the AHP in sustainable development and finds manufacturing, urban/public, business, construction, and energy the most frequent areas of applications. Cinelli et al. [60] in the paper published in 2014 analysed the performance of AHP method in respect to "ten crucial criteria that sustainability assessments tools should satisfy: scientific soundness (use of qualitative and quantitative data, life cycle perspective, weights typology, threshold values, compensation degree, uncertainty treatment/sensitivity analysis, robustness); feasibility (software support and feasibility, ease of use) and utility (learning dimension)". Recently applications of the AHP method in the selection of production project based on sustainable development [21] and in sustainable rural development [61] were published.

The survey of Vaidyaa and Kumar [62] presents an analysis of 150 scientific paper and group the applications by theme (selection, evaluation, benefit-cost analysis, allocations, planning and development, priority and ranking, decision making); by specific applications (in forecasting, medicine, and related fields); and applications combined with other methodology (AHP applied with Quality Function, Deployment (QFD)). The areas of applications of the analysed research papers are personal, social, manufacturing, political, engineering, education, industry, and government.

AHP is undoubtedly among the most often applied multi-criteria decision models. A short survey of surveys on applications and case studies can be found in the following papers: [21,63–65]. Forman and Gass [66] published a study on the history, development, and an overview of AHP applications in different areas, such as choice, ranking, prioritisation/evaluation, resource allocation, benchmarking, quality management, public policy, health care, strategic planning. Strategic level decision-making problems in business, defence, and governance are presented in the book of Navneet and Kanwal [40].

To strengthen the credibility and authenticity, and real-life connection of the pairwise comparison, interdisciplinary groups of specialists should be formed, and citizens should be involved in the decision-making process. The versatile nature of AHP is examined in some cases, considering that the method enables researchers to arrange different alternatives based on the decisions' requirements.

Professional computer application software is required to apply AHP in complex situations. The decision support system Expert Choice [67] was applied in most of the literature reviewed above. It is, however, to be noted that, the authors of this paper applied PriEsT [23].

3. Materials and Methods

3.1. Analytic Hierarchy Process (AHP)

The methodology of the Analytic Hierarchy Process (AHP) [68] builds on (i) structuring and decomposing the goal; (ii) applying pairwise comparisons to quantify the priorities. Two of the first reported applications of the hierarchical structure of criteria, also known as the tree of criteria, are the multi-attribute decision model PATTERN (Planning Assistance Through Technical Evaluation of Relevance Numbers) Alderson [69], and Reverse Factor Analysis [70]. Pairwise comparisons in preference modelling can be traced back to Llull, a 13th-century polymath [71].

Pairwise comparison on a ratio scale shows how many times a criterion (alternative) is more important (better) than the other one. An $n \times n$ pairwise comparison matrix $A = [a_{ij}], i, j = 1, \ldots, n$ consists of the numerical answers (which are often converted from verbal responses) to these questions, where reciprocity $a_{ij} = 1/a_{ji}$ is assumed. A pairwise comparison matrix is called consistent if cardinal transitivity holds, i.e., $a_{ij} \cdot a_{jk} = a_{ik}$ for all i, j, k. Otherwise, the matrix is inconsistent. The decision maker is not at all supposed to provide consistent pairwise comparison matrices, and this flexibility is one of the reasons why AHP has been often applied in a wide range of decision problems.

Once a pairwise comparison matrix is filled in, the goal is to find a positive weight vector $w = (w_1, w_2, \ldots, w_n)$, so that the ratios w_i/w_j are close to the estimations a_{ij} given by the decision-maker in some specified sense. Saaty [68] proposed the eigenvector method for the calculation of priorities from the pairwise comparison matrix. The weight vector w is calculated from the eigenvalue-eigenvector equation:

$$A \cdot w = \lambda_{max} \cdot w, \quad (1)$$

where λ_{max} denotes the maximal eigenvalue of A.

Since the eigenvector can be computed for pairwise comparison matrices with arbitrarily high inconsistency, several inconsistency indices have been proposed (i) to detect if the matrix is not acceptable for further calculations, and (ii) to localize the primary sources of inconsistency and involve the decision-maker to revise those critical matrix elements [72]. Saaty [68] introduced the following inconsistency index:

$$CR = (\lambda_{max} - n)/(\Lambda - n), \quad (2)$$

where Λ denotes the average of randomly generated pairwise comparison matrices' maximal eigenvalues. Saaty suggested the threshold $CR \leq 0.1$ (ten percent rule) of acceptable inconsistency.

The hierarchical structure makes the system of criteria structured and transparent and enables the decision-makers to handle global and local weights of importance. A criterion is called a leaf criterion if it has no subcriteria, otherwise it is called parent criterion. Global weights (abbreviated with G) of all the leaf criteria sum up to 1, and they show their shares within the whole list of leaf criteria. Local weights (abbreviated with L) show the relative importance of criteria directly below a parent (not leaf) criteria; in other words, the relative weights of a criterion's subcriteria. Both the global and the local weights of the goal (the root of the tree) sum up to 100%, and both the global and local weights of the main criteria coincide. An illustrative example is given in Appendix B.

The Priority Estimation Tool (PriEsT) [23] was used to obtain numerical results for decision-making. The choice is supported by the facts that (i) it has no size limitation of the tree, (ii) the eigenvector method can be chosen for the calculation of weights from a pairwise comparison matrix, (iii) it is portable, runs without installation, (iv) it is free, (v) finally it is indeed easy to download and use without registration (even on a mobile). The PriEsT file and authors' short manual on running the PriEsT application are attached to this paper (Supplementary Materials, available under Supplementary Materials); thus, the reader has full access, including to modify the criteria, re-weight and last but not least, add other alternatives.

3.2. *Data*

To build up a hierarchical criteria system of sustainable development, publicly available reports and databases were used, such as:

- European Innovation Scoreboard (EIS) [73], which gives a comparative assessment of EU member states' innovation performance and ranks countries according to four performance groups.
- The Global Competitiveness Index (GCI) [74], introduced by Sala-i-Martin (2004), which assigns the economies to development stages, the countries rank is published annually by the World Economic Forum (WEF), using different indicators grouped in 12 pillars. The 2018 edition introduces the new Global Competitiveness Index 4.0., which sheds light on an emerging set of drivers of productivity and long-term growth in the era of the Fourth Industrial Revolution.
- The Global Information Technology Report 2013 is a project within the framework of WEF and INSEAD, which publishes the Network Readiness Index (NRI) [75,76]. NRI 2020 is the second edition of NRI renewed model using 60 variables grouped in four pillars: technology, people, governance, impact.
- The Global Innovation Index (GII) [77] published by INSEAD and WIPO and measures the economy's innovation performance since 2007.
- The Sustainable Development Report (SDR) [34], formerly the SDG Index & Dashboards, includes 115 indicators in 2020. It assesses worldwide the countries' achievements on the Sustainable Development Goals performance. The latest SD report was elaborated considering Covid-19, the pandemic having huge implications on progress towards SDGs.
- The eco-innovation scoreboard (Eco-IS) [78] published since 2010 by European Commission (EC), measures and compares the eco-innovation performance of EU Member States and facilitates the transition process of member states to the green economy. Applying 16 indicators grouped by five dimensions: eco-innovation input, eco-innovation activities, eco-innovation output, resource efficiency outcomes and socio-economic outcomes "aims to promote a holistic view on economic, environmental and social performance".
- The Integrated Coastal Zone Management (ICZM) was defined by the EU and is planned to promote and monitor the sustainable management of coastal zones. The

set of 54 of indicators "measures the general state of the coast and the general trend towards, or away from, sustainability" [79].

- The EC defined a set of indicators grouped into three categories: sustainable resource management, societal behaviour and business operations. They measure "performance in several areas that directly or indirectly contribute to the Circular Economy development" [80].
- The Global Entrepreneurship Monitor (GEM) is published since 1999 and "carries out survey-based research on entrepreneurship and entrepreneurship ecosystems around the world". The GEM presents data on necessity (entrepreneurs may be pushed to start their business because they have no other work options and need a source of income)—and opportunity (entrepreneurs who recognise market opportunities)—driven entrepreneurship [81].
- Eurostat, Waste Management Indicators [82].
- World Bank (WB) [83].
- The Global Indicator Framework for the SDGs and targets of the 2030 Agenda for SD [84].
- Our world in data, which publishes every four years data on the ozone layer [85].
- The Environmental Performance Index (EPI) using 32 performance indicators ranks countries on environmental health and ecosystem vitality and sustainability issue performance [86].
- Quality of Air Index (IQAir) [87].
- European Environment Agency (EEA) currently manages 122 indicators. EEA presents data on the quality and quantity of Europe's water resources [88].
- Biodiversity Indicators Dashboard published by the NatureServe organization [89].
- The World Energy Balance report published by the International Energy Agency (IEA) and founded in 1974 [90].
- The Population Stability Index [91].
- The Prosperity Index, published by the Legatum Institute from the UK, developed "as a practical tool to help identify what specific action needs to be taken to contribute to strengthening the pathways from poverty to prosperity globally". It is built upon 12 pillars, on 65 actionable policy areas and 294 indicators [92].
- Environment at a Glance Indicators (OECD) [93].
- The 2020 global Social Progress Index launched in 2014 by Social Progress Imperative. The index uses 50 indicators grouped into three broad dimensions of social progress [94].
- The Living Planet Report 2020 published by Worldwide Fund (WWF) [3].

In several indices listed above the indicators and their weights are revised regularly. However, the main criteria and their weights are more or less unchanged. Although the evaluation methodologies concerning criteria are beyond this paper's scope, the authors note that the rules of assessment seem to be more static.

4. Results

In this section, the AHP is applied for the prioritisation to select the strategic objectives of sustainable development. At first, the procedure of choosing relevant indicators and their structuring into a hierarchy is described. Then, to quantify the importance, to calculate the local and global weights and to determine the leaf criteria of the highest global weights, the construction of pairwise comparison matrices is presented. Moreover, a discussion on the obtained results is also included.

Thus, the section is organised as follows:

Section 4.1:

- ○ identifying possible indicators
- ○ structuring them into a tree

Section 4.2:

- ○ filling in pairwise comparison matrices for every parent node
- ○ calculating global weights
- ○ determining the leaf criteria of the highest global weights

Section 4.3:

- ○ discussion

4.1. The Proposed Hierarchical System of Criteria of Sustainable Development

The authors propose a hierarchical criteria system (summarised in Figure 1 with a further detailed presentation of the selected criteria in Tables 1–5 from Section 4.2) of sustainable development. The chosen criteria of the multi-criteria framework are indicators selected from relevant literature. An integrative literature review was employed in this regard and sets of indicators used to measure sustainable development were consulted and analysed. Criteria, being too complex for direct evaluation, were divided into specific subcriteria, and this decomposition stopped when a suitable criterion was reached for measurement. The indicators were chosen to be feasible and fully representative to cover the most important aspects of sustainable development. It was a vital standpoint to be consistently defined, measured across countries, and regularly published by different institutions in publicly available databases. The authors intended to obtain a continuous monitoring and measuring possibility of achievements on the implementation of sustainable economic development. The proposed hierarchy is sufficiently general and flexible: criteria can be eliminated or added, and subtrees (e.g., that of Economy) themselves allow reasonable analysis.

The framework is built upon three main criteria: the economic, social, and environmental sustainability pillars.

In the authors' framework, the Economic pillar consists of three subcriteria: Enablers, Firm Activities, and Output. They accelerate and track the progress to achieve targets of Sustainable Development Goals (SDGs) such as SDG8 (Decent work & economic growth), SDG9 (Industry, Innovation & Infrastructure), SDG10 (Reduced inequalities (GINI Index [95]), and SDG17 (Partnerships for the Goals).

For each subcriterion, sub-subcriteria were formulated which are detailed in the Tables 1–3. The selected indicators from the reviewed sources monitor and measure the progress towards a sustainable, innovation-driven economy, a developed knowledge-based economy, and information society that can manage change [96], use advanced technologies, and is equipped for the future of production. According to the World Economic Forum (WEF), the readiness for "the future of production" has a significant role in ensuring sustainable economic development. "Rapidly emerging technologies—such as the Internet of Things, artificial intelligence, wearables, robotics, and additive manufacturing—are spurring the development of new production techniques, business models, and value chains that will fundamentally transform global production. Both the speed and scope of change add a layer of complexity to the already challenging task of developing and implementing industrial strategies that promote productivity and inclusive growth" [97].

The Society pillar is built upon six subcriteria, and 20 sub-subcriteria. When constructing the framework for this pillar, the indicators were selected to monitor the progress in reaching the SDG1, SDG2, SDG3, SDG4, SDG5, SDG6, SDG9, SDG11, SDG15, SDG16 targets as well.

The Environment pillar comprises six subcriteria, and 16 sub-subcriteria corresponding to the hierarchy. When constructing the framework at this pillar, the indicators were also selected to monitor the progress in reaching the targets: SDG6, SDG7, SDG11, SDG12, SDG13, SDG14, SDG15.

The subcriteria till the third level are displayed in Figure 1, while at the fourth and fifth levels (if applicable) are expressed in detail in Tables 1–5. The chosen "indicators" are defined in the listed sources.

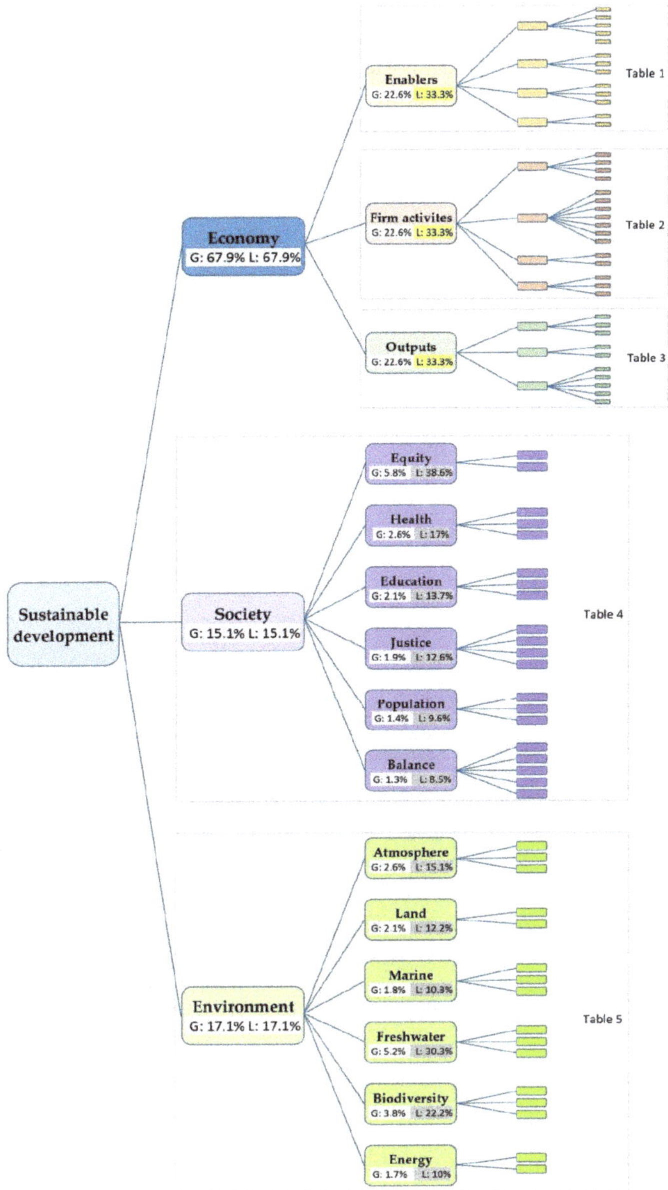

Figure 1. The proposed hierarchical system of criteria. Framed parts are given in detail in Tables 1–5.

4.2. *The Quantification of the Priorities*

The criteria with the highest overall impact on sustainable development in the hierarchical system of criteria is identified through the following steps: at first the pairwise comparison matrices for every parent node are constructed, then the local and global weights are calculated and finally the leaf criteria of the highest global weights is determined.

The construction of pairwise comparison matrices reflects the authors' point of view, aiming to demonstrate that AHP can be an implementable instrument for policymakers to prioritise and identify the strategic objectives, the criteria with highest overall impact.

Global weights (abbreviated by G in the tables) quantify the overall importance within the whole system of criteria. The global weight of a parent criterion is distributed among its subcriteria. Global weights of all leaf criteria sum up to 100% as before. Local weights (abbreviated by L in the tables) quantify the relative importance within the subcriteria of the same parent criterion.

Global and local weights of criteria are calculated by PriEsT [23] from 30 (=number of parent criteria) pairwise comparison matrices and are given in Tables 1–5. Weight vectors are calculated by the eigenvector method [19].

The procedure of weights' calculation is presented through two parent criteria and their subcriteria. In the case of the three main criteria, weights are calculated from the pairwise comparison matrix of size 3 × 3 below, and local and global weights coincide (see also in Figure 2a,b):

- Economy: 67.9%
- Society: 15.1%
- Environment: 17.1%

	Economy	Society	Environment		eigenvector, local & global weights
Economy	1	6	3		67.9%
Society	1/6	1	1.18	⇨	15.1%
Environment	1/3	1/1.18	1		17.1%

(a)

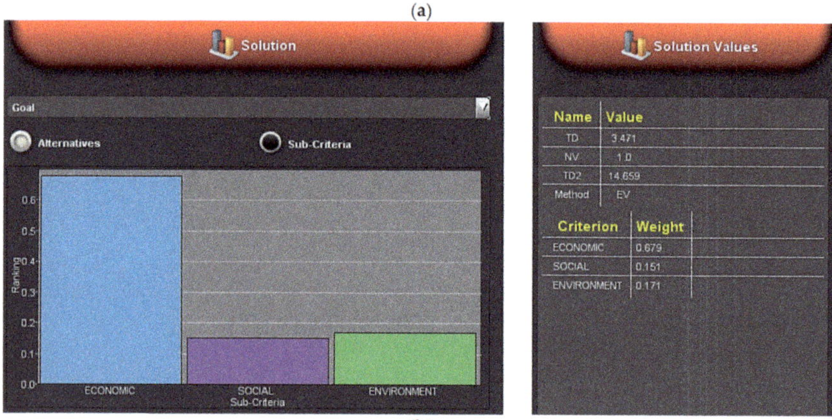

(b)

	Equity	Health	Education	Justice	Population	Balance		eigen-vector (local weights)	multiplied by 0.151, the global weight of Society	global weights
Equity	1	4.04	1.91	13.87	1.14	1.15		38.6%		5.8%
Health	0.248	1	2.64	4.25	1.99	1.14		17%		2.6%
Justice	0.524	0.379	1	3.25	1.97	1.89	⇨	13.7%		2.1%
Education	0.072	0.235	0.308	1	6	3		12.6%	⇨	1.9%
Population	0.877	0.503	0.508	0.167	1	2.42		9.6%		1.4%
Balance	0.87	.0877	0.529	0.333	0.413	1		8.5%		1.3%

(c)

Figure 2. Global (=local) weights of the three main criteria calculated by PriEsT. (**a**) Global (=local) weights of the three main criteria; (**b**) Three main criteria calculation results in PriEsT. (**c**) Subcriterion global weight calculation.

Global weight of any subcriterion is calculated from the product of its local weight and the global weight of its parent criterion. For example, the global weight of Equity is equal to its local weight (38.6%), calculated from the pairwise comparison matrix of size 6 × 6 in Figure 2c, multiplied by the global weight of its parent criterion Society (15.1%), resulting in 5.8%.

It is to be emphasized that the pairwise comparisons, available in the PriEsT file supplemented, (available under Supplementary Materials), were made by the authors, who are not experts in all specific areas. Thus, their preferences were undoubtedly influenced by the studies reviewed.

Table 1 presents the sub and sub-subcriteria at the fourth and fifth levels of criterion Enablers in concordance with Figure 1. Moreover, it contains the local (L) and global (G) weights calculated from pairwise comparison matrices by PriEsT.

Table 1. The subcriteria and sub-subcriteria of Enablers.

3rd Level Criteria	Subcriteria	Sub-Subcriteria	Source
		New doctorate graduates G: 0.1% L: 2.2%	EIS [73]
		Population aged 25-34 with completed tertiary education G: 0.8% L: 13.2%	EIS [73] GII [77] NRI [75,76]
	Human resources G: 5.7% L: 25%	Youth aged 20-24 with at least upper secondary education G: 0.4% L: 7%	EIS [73]
		Number of students enrolled at mathematics, sciences G: 0.1% L: 2.2%	GII [77]
		Duration of studies G: 4.3% L: 75.4%	EIS [73] GII [77] GCI [74] NRI [75,76]
		International scientific co-publications G: 4% L: 69.8%	EIS [73] GCI [74] SDR [33] EC [78]
Economy-Enablers G: 22.6% L: 33.3%	Open, excellent, attractive research systems G: 5.7% L: 25%	Top 10% most cited scientific publications G: 1% L: 16.7%	EIS [73] Scopus WOS
		Foreign doctorate students G: 0.8% L: 13.5%	EIS [73] Eurostat [82]
		R&D expenditure in the public sector G: 2% L: 34.7%	EIS [73] Eurostat [82]
	Finance and support G: 5.7% L: 25%	Financing of SMEs G: 1.8% L: 31.3%	GCI [74]
		Venture capital G: 1.9% L: 34.1%	EIS [73] GII [77] GCI [74]
	Business environment G: 5.7% L: 25%	Time to start businesses G: 3.9% L: 68.8%	GCI [74]
		Ease of resolving insolvency G: 1.8% L: 31.2%	GCI [74]

Some remarks to Table 1.

Subcriterion "human resources" listed in Table 1 is essential in the 21st-century and has an influential role in social and economic inclusion. Hence, it must be of high quality and therefore measured by its preparedness and qualification level. The sub-subcriterion "duration of studies" can be replaced with indicators from other listed sources such as "school life expectancy", "lifelong learning participation", and the "skillset of graduates".

An innovative and innovation-friendly environment demands an "open, excellent, attractive research system". In this respect, the number of international scientific co-publications has a significant role, revealing the level of international recognition. Moreover, here, the number of "eco-innovation related academic publications" as a percentage of the total number of international scientific co-publications can be considered as well. Furthermore, instead of the "venture capital" indicator, selected as sub-subcriterion of "finance and support" from the listed source, the indicators "venture capital expenditure", "deals", and "availability" can also be used.

Table 2 presents the sub and sub-subcriteria at the fourth and fifth levels of criterion Firm Activities in concordance with Figure 1. Moreover, it contains the local (L) and global (G) weights calculated from pairwise comparison matrices by PriEsT. The selected indicators evaluate the innovative activities in companies and investments in R&D.

Table 2. The subcriteria and sub-subcriteria of Firm Activities.

3rd Level Criteria	Subcriteria	Sub-Subcriteria	Source
Economy-Firm Activities G: 22.6% L: 33.3%	Firm investments G: 4.6% L: 20.5%	R&D expenditures in the business sector G: 0.9% L: 19.8%	EIS [73] NRI [75,76] EC [78]
		Non-R&D innovation expenditure G: 1.3% L: 27.7%	EIS [73]
		The intensity of research collaboration between universities and enterprises/industry G: 1% L: 22%	GII [77]
		Advanced technologies procurement by the government G: 1.4% L: 30.5%	NRI [75,76] EC [78]
	Linkages & entrepreneurship G: 9.9% L: 43.6%	SMEs in-house innovation G: 1.5% L: 14.9%	EIS [73] EC [78]
		Technological readiness G: 1% L: 9.6%	GCI [74] NRI [75,76]
		Availability of modern technologies G: 1.5% L: 14.8%	NRI [75,76]
		Technology absorption G: 0.6% L: 5.9%	GCI [74] NRI [75,76]
		State of cluster development G: 1.8% L: 18.4%	GII [77]
		Innovative SMEs collaborating with others G: 1.6% L: 15.7%	EIS [73]
		Public-private co-publications G: 2% L: 20.7%	EIS [73]
	Intellectual assets G: 2% L: 8.8%	PCT patent applications G: 1.8% L: 87.6%	EIS [73] NRI [75,76]
		PCT patent applications in climate changes and societal challenges G: 0.2% L: 12.4%	EIS [73] GCI [74] EC [78]
	Market efficiency G: 6.1% L: 27.1%	Goods market efficiency G: 0.8% L: 12.6%	GCI [74] GII [77]
		Labour market efficiency G: 1.1% L: 18.5%	GCI [74] GII [77] EC [78]
		Financial market development G: 4.2% L: 69%	GCI [74] GII [77]

Regarding Table 2 and the listed sources from Section 3.2, the following observations could be deducted: in the case of the indicator "R&D expenditures in the business sector", the total value of "green early-stage investments" could be considered as well, since it provides information on the level of implementation of sustainable policies. Moreover, instead of the indicator "advanced technologies procurement by the government", the indicator "governments environmental and energy R&D appropriations and outlays" could be utilised. When it comes to the indicator "SMEs in-house innovation" indicator, it is beneficial to verify the indicator "implementing resource efficiency actions among SMEs" from the mentioned source as well.

Regarding the selected "PCT patent applications in climate changes and societal challenges per 1 million inhabitants" indicator, it is valuable to consider "the number of eco-innovation related patents" whose value is given by the European Eco-innovation Action Plan. Concerning the "labour market efficiency" indicator, it can be useful to verify "the employment in environmental protection and resource management activities" indicator too.

Table 3 presents the sub and sub-subcriteria at the fourth and fifth levels of criterion Outputs in concordance with Figure 1. Moreover, it contains the local (L) and global (G) weights calculated from pairwise comparison matrices by PriEsT. The selected indicators measure the immediate output of (eco)innovation activities.

Table 3. The subcriteria and sub-subcriteria of Outputs.

3rd Level Criteria	Subcriteria	Sub-Subcriteria	Source
Economy-Outputs G: 22.6% L: 33.3%	Innovators G: 3.8% L: 17%	Innovative activities, product or process innovation G: 1.6% L: 41.5%	EIS [73]
		Ratio of opportunity-driven and necessity-driven entrepreneurship G: 0.3% L: 7.5%	GEM [81] GCI [74] EIS [73]
		Online creativity, intangible assets G: 1.9% L: 51%	GII [77]
	Economic effects G: 5.8% L: 25.6%	Employment in knowledge-intensive activities G: 5.1% L: 87.4%	EIS [73]
		License and patent revenues from abroad G: 0.7% L: 12.6%	EIS [73]
	Business sophistication G: 13% L: 57.4%	Local supplier quantity and quality G: 1.8% L: 14.2%	GCI [74] GII [77]
		Production process sophistication G: 1.9% L: 14.9%	GCI [74] GII [77]
		Nature of competitive advantage G: 1.9% L: 14.3%	GCI [74] GII [77]
		Willingness to delegate authority G: 0.8% L: 5.8%	GCI [74] GII [77]
		Attitudes toward entrepreneurial risk G: 6.6% L: 50.8%	GCI [74] GII [77]

Remarks on Table 3.

In case of sub-subcriterion "ratio of opportunity-driven vs. necessity-driven entrepreneurship", the authors reckon on considering the value of the following indicators from the listed source as well: "the growth of innovative companies" and "opportunity-driven entrepreneurship".

Table 4 presents the sub and sub-subcriteria at the third and fourth levels of main criterion Society in concordance with Figure 1. Moreover, it contains the local (L) and global (G) weights calculated from pairwise comparison matrices by PriEsT.

Table 4. The subcriteria, sub-subcriteria of Society.

2nd Level Criteria	Subcriteria	Sub-Subcriteria	Source
		Poverty G: 4.4% L: 75.2%	SDR [33]
	Equity G: 5.8% L: 38.6%	Gender equality G: 1.4% L: 24.8%	SDR [33] NRI [75,76]
		Waste and recycling management G: 0.3% L: 11.7%	Eurostat [82]
	Health G: 2.6% L: 17%	Access to safe drinking water G: 0.9% L: 36.4%	SDR [33]
		Health services, access to basic sanitation G: 1.4% L: 52%	SDR [33] NRI [75,76] GCI [74]
		Mean years of schooling G: 0.6% L: 26.7%	SDR [33]
	Education G: 2.1% L: 13.7%	Illiteracy and functional illiteracy G: 0.2% L: 10.9%	SDR [33] NRI [75,76]
		Expenditure on education, % GDP G: 1.3% L: 62.4%	SDR [33]
		Criminality, organized crime G: 0.5% L: 26.7%	GCI [74]
Society G: 15.1% L: 15.1%	Justice and security institutions G: 1.9% L: 12.6%	Judicial independence G: 0.6% L: 29.8%	GCI [74]
		Efficiency of legal framework G: 0.5% L: 25.3%	GCI [74]
		Reliability of police services G: 0.3% L: 18.2%	SDR [33] GCI [74]
		Digital skills of the population G: 0.3% L: 19%	GCI [74]
	Population G: 1.4% L: 9.6%	Population stability, zero hunger G: 0.4% L: 28.9%	PSI [91]
		Accessibility-access to information, internet users G: 0.7% L: 52.2%	NRI [75,76]
		Good governance G: 0.6% L: 42.6%	GII [77] NRI [75,76] LegatumInst [92]
		Labour market efficiency G: 0.3% L: 25.6%	NRI [75,76]
	A safe and balanced society G: 1.3% L: 8.5%	Population growth rate trend G: 0.1% L: 5.6%	WB [83] UN [84] EIS [73]
		Average income G: 0.2% L: 15.4%	NRI [75,76]
		National debt G: 0.1% L: 10.8%	GCI [74]

Remarks on Table 4.

The selected sub-subcriterion "health services, access to basic sanitation" in the listed source might appear as "good health and well-being".

The subcriterion "education" has a significant role, one of the enablers' developments and a prerequisite for an inclusive society. The sub-subcriterion "illiteracy and functional illiteracy" reflects the "quality of education" and refers to the "analphabetism", "entry rate in primary school" and "digital analphabetism" as well.

The subcriterion "justice and security institutions" monitors the progress on reaching the target SDG16 (Peace, Justice and Strong Institutions).

In the case of the indicator "population growth rate trend", selected for the framework, the values of "ratio of land consumption rate to population growth rate", and "population density" could also be used from the listed sources, which provide useful information to monitor the course on achieving the targets SDG11 and SDG15 (life on land).

Table 5 presents the sub and sub-subcriteria at the third and fourth levels of main criterion Environment in concordance with Figure 1. Moreover, it contains the local (L) and global (G) weights calculated from pairwise comparison matrices by PriEsT.

Table 5. The subcriteria and sub-subcriteria of Environment.

2nd Level Criteria	Subcriteria	Sub-Subcriteria	Source
Environment G: 17.1% L:17.1%	Atmosphere G: 2.6% L: 15.1%	Ozone-depleting emissions G: 0.4% L: 16.5%	OWD [85]
		Air quality G: 1.1% L: 41.9%	EPI [86] IQAir [87] SDR [33] OECD [93]
		Greenhouse gas emissions G: 1.1% L: 41.6%	SDR [33] EC [78]
	Land G: 2.1% L: 12.2%	Land quality G: 0.7% L: 33.3%	GCI [74]
		Land occupied by waste G: 1.4% L: 66.7%	SDR [33] EC [78,80] EPI [86]
	Marine G: 1.8% L: 10.3%	Coastal zone G: 0.9% L: 48.6%	EC [79]
		Fisheries yield G: 0.5% L: 26.3%	SDR [33] EPI [86]
	Freshwater G: 5.2% L: 30.3%	Water quantity and quality G: 0.5% L: 25.1%	SDR [33]
		River quality G: 2.2% L: 42.7%	EEA [88]
		Quantity G: 1.6% L: 31.3%	GCI [74] EC [78] EPI [86]
		Expenditure on water treatment and distribution G: 1.4% L: 26%	GCI [74] EPI [86]
	Biodiversity G: 3.8% L: 22.2%	Degree of urbanization G: 1% L: 27.5%	NRI [75,76]
		Biodiversity conservation G: 2.2% L: 56.7%	Naturserve [89] SDR [33] EPI [86]
		Ecological footprint G: 0.6% L: 15.8%	GCI [74] GII [77]
	Affordable energy and clean energy G: 1.7% L: 10%	Usage of renewable energy (proportion) G: 0.9% L: 50%	GCI [74] IEA [90]
		Energy from waste G: 0.4% L: 25%	EC [78,80]
		Transport energy use G: 0.4% L: 25%	GCI [74]

Remarks on Table 5. The subcriterion "atmosphere" could be used to track progress on the SDG13 (Climate Action), identified in the listed sources as "climate action"; the subcriterion "land" can be used to monitor the progress to reach the target of SDG15 (life on land); while the subcriterion "affordable energy and clean energy" measures the target of SDG7 (Affordable and Clean Energy).

The sub-subcriterion "air quality" is measured in the listed source also by indicators "nitrogen oxides and sulfur oxide emissions", "production-based SO_2 emissions, (kg/capita)", "production-based nitrogen emissions (kg/capita)", "responsible consumption and production" and "pollution emission". These indicators can be used to monitor the progress to reach the target of SDG12 (Responsible Consumption and Production).

Instead of the sub-subcriterion "greenhouse gas emissions", the indicators "climate action" and "Green House Gas emissions intensity" can also be considered.

The chosen indicator "land quality" is substitutable with "the quality of land administration" indicator measured by the World Economic Forum (WEF) and can be used to monitor the progress to reach the target of SDG12.

The selected indicator "coastal zone" can be replaced with the indicator "sustainable development at coastal zone", and the sub-subcriterion "fisheries yield with the indicators "life below water" and "fisheries". All these can be used to monitor the course to achieve the SDG14 (Life below water) target.

The indicator "water quantity and quality" can be substituted by the "ocean health index: clean waters score", while sub-subcriteriaon "river quality" by the indicator "oxygen-consuming substances in European rivers".

Regarding the subcriterion "freshwater", the indicators "quantity", "exposure to unsafe drinking water", "water productivity (GDP/total freshwater abstraction)" and "sanitation and drinking water" can also be considered. These can be used to monitor the progress to reach the SDG6 (Clean water and Sanitation) target. Instead of the indicator "expenditure on water treatment and distribution", the "reliability of water supply" and "water resources" indicators could also be used.

The sub-subcriterion "degree of urbanization" can be used to monitor the course to achieve the SDG11 (Sustainable cities and communities). The indicator "biodiversity conservation" meets the SDG15 (Life on land, biodiversity & habitat) target.

To monitor the sub-subcriterion "ecological footprint", the indicators "environment-related treaties in force" and "ecological sustainability" could also be considered.

"Usage of renewable energy (proportion)" is compatible with other indicators from the listed sources such as "renewable energy regulation" and "energy efficiency regulation". "Energy from waste" can be changed with "societal behaviours" from the listed source. Quantitative data on "transport energy use" are found in the mentioned sources under the indicator "transport infrastructure".

Therefore, based on the calculations presented before and results listed in Tables 1–5, the four most important leaf criteria (taking into consideration their global weights) are considered strategic objectives. They are:

- Attitudes toward entrepreneurial risk (within Economy—Outputs—Business sophistication) with global weight 6.6%
- Employment in knowledge-intensive activities (within Economy—Outputs—Economic effects) with global weight 5.1%
- Poverty (within Society—Equity) with global weight 4.4%
- Duration of studies (within Economy—Enablers—Human resources) with global weight 4.3%.

4.3. Discussion

The AHP method was used to analyse the economic, social and environmental subsystems' impact on sustainable development. As mentioned earlier, all the 30 pairwise comparison matrices (= number of parent nodes in the tree of criteria) were filled in by the authors.

As it was also highlighted before, global weights do not directly depend on how deep or wide the tree of criteria is; they are determined by their own local and their parents' global weights.

The primary goal was to present that weighting by pairwise comparisons is suitable even in a complex hierarchy in Figure 1 rather than to emphasize the weights themselves.

Nevertheless, it can still be observed that in this illustrative system, the global weights of the 75 leaf criteria are between 0.1% and 6.6%. This means that none of the leaf criteria have smaller [larger] global weight than the average global weight, 100%/75 = 1.33%, multiplied by 1/13.3 [66], respectively.

Re-weighting of criteria would naturally lead to different global weights and possibly other strategic objectives.

Since the hierarchical system of criteria follows the structure of several approaches listed in Section 3.2, and the weighting can be performed, as presented, by the Analytic Hierarchy Process, it has been demonstrated that such prioritisation does not necessarily require advanced knowledge of multi-criteria decision modelling from the decision/policymakers.

Although our approach primarily refers to the macroeconomic level (countries or regions), the proposed hierarchy can be a good starting point at the microeconomic level (for large organizations like international companies) as well.

Strategic objectives may rise in governmental and corporate decisions, and this approach can support not only ranking or selecting the most important leaf criteria but also distributing resources.

5. Conclusions

The main result of this paper is to give a method and a methodology on how the strategic objectives of sustainable development can be prioritised. A sufficiently general, flexible, yet implementable decomposition of sustainable development is proposed. A tree structure of the criteria was built, and priorities were calculated with pairwise comparisons. This multi-criteria framework is provided to policymakers to analyse, organise and prioritise essential sustainability indicators/criteria in order to identify the strategic objectives of sustainable development, which is indeed needed in such a complex problem.

The authors demonstrated that the multi-criteria approach of the AHP is a useful tool to find those criteria that have the highest overall impact (weight) on the main goal. Using this method, any of the considered criteria can become a strategic objective, depending on the weights of the highest priority for a successful implementation. Thus, a priority action plan can be formulated for an efficient implementation of sustainable development, where numerical priorities also support the decisions on the distribution of the resources.

In the framework presented in this paper, only those indicators were collected for which there is available data. Their values are published and updated regularly, thus can be used to plan rigorous interventions. Nevertheless, attention is needed, because the criteria are often reprioritised, sustainability is a moving target in case of most indices (e.g., [74,75,77,86]).

Having quantitative data, one can monitor the efficiency of decisions, the level of implementation and the progress made to achieve sustainable development. Using specific, measurable indicators as criteria, or decomposing the main goal to measurable criteria and prioritisation gives better implementable, and achievable strategic objectives.

It is also demonstrated that the weights of criteria can be estimated and calculated from pairwise comparisons even in the case of a deep (5 levels) and wide (75 leaf criteria) tree. However, a limitation of the results regarding a possible future application is that the model certainly requires re-weighting, by experts. Nevertheless, their subjectivity remains a limitation in the pairwise comparisons.

The proposed system of criteria can be applied in various fields of sustainable development, as it can be reduced according to the goals and extended if specific criteria, e.g., environment, are more in focus.

Given any composite index of sustainable development, if a country is looking for the possible ways of improvement in order to be ranked higher, then it should not only focus on the criteria with the highest global priority, but also on its own potential of performing better with respect to the criteria. A significant improvement with respect to a not extremely important criterion might be overall more beneficial than a slight improvement with respect to a very important criterion.

Another factor is the interaction of criteria: it is practically impossible to restrict improvements to a single leaf criterion. Any intervention inevitably impacts not only on the targeted goal, but also on seemingly unrelated goals.

Future research includes the structured ways of evaluating the alternatives (strategies, policies, countries, regions, etc.) with respect to each criterion in the criterion tree's leaf nodes. In contrast to the selection of criteria, evaluation relies more on the data available, consequently requiring a customized approach. A review of the best practices would inherently support the design of the rules of evaluation. In this respect the performance of criteria can be measured on each objective (data is published), and a composite index can be calculated, thus countries and regions (as in [98]) can be compared if and in what extent they reduced the distance to sustainable development.

Supplementary Materials: The PriEsT file is available online at https://www.mdpi.com/2071-1050/13/4/2254/s1.

Author Contributions: Conceptualization, Z.K.S., Z.S., S.B., G.C.S. and D.S.; methodology, Z.K.S., Z.S., S.B., G.C.S. and D.S.; software, Z.K.S., Z.S. and S.B.; validation, Z.K.S., Z.S., S.B., G.C.S. and D.S.; formal analysis, Z.K.S., Z.S., S.B., G.C.S. and D.S.; investigation, Z.K.S., Z.S., S.B., G.C.S. and D.S.; resources, Z.K.S., Z.S., S.B., G.C.S. and D.S.; data curation, Z.K.S., Z.S. and S.B.; writing—original draft preparation, Z.K.S., Z.S., S.B., G.C.S. and D.S.; writing—review and editing, Z.K.S., Z.S., S.B., G.C.S. and D.S.; visualization, Z.K.S., Z.S. and S.B.; supervision, Z.K.S., Z.S., S.B., G.C.S. and D.S.; project administration, Z.K.S.; funding acquisition, Z.K.S. and G.C.S. All authors have read and agreed to the published version of the manuscript.

Funding: This research received no external funding.

Institutional Review Board Statement: Not applicable.

Informed Consent Statement: Not applicable.

Data Availability Statement: All data are available in the Supplementary Materials.

Acknowledgments: The research of Z. Szádoczki and S. Bozóki was supported by the Hungarian National Research, Development and Innovation Office (NKFIH) under Grant NKFIA ED_18-2-2018-0006.

Conflicts of Interest: The authors declare no conflict of interest.

Appendix A

Table 1. Frameworks to Assess Sustainable Development.

Frameworks for Assessing SD	
UN CSD indicators Agenda 21 [24] 1994–2001	Quadruple: 132 indicators (23 economic, 55 environment, 39 social, and 15 institutions)
OECD SDIs [26] 2002	Quadruple: economic (2 theme, 8 indicators), environment (5 theme, 16 indicators), social (6 theme, 17 indicators), institutions (2 theme, 6 indicators)
CSD SDIs UN 2007 [25]	Quadruple: 15 theme (2 economic, 5 environment, 6 social, 2 institutions)
EU-15 SDIs EC [27] 2004	Triple: Comparative study across EU15 Member states, (p. 66–69)
CSD SDIs EC [28] 2007	Triple: Comparative analysis across Europe (25 countries) on indicators used by the National Reform Programme (NRP) and Sustainable Development Strategy (SDS) (pp. 157–158)
UNECE/OECD/ Eurostat SDIs 2009 [29]	Triple: Comparative analysis across EU, where 11 common themes and 27 common indicators were identified
UNECE/OECD/ Eurostat SDIs [30] 2013	Triple: 39 theme, 3 conceptual dimensions: human wellbeing (HWB), capital and transboundary impact (p. 75)

Appendix B

AHP Method, an Illustrative Example

An illustrative example is given in Figure A1d, the leaf criteria being framed in orange. Although the main criterion 3 (also being a leaf node) is positioned at a higher level in the tree compared to the subcriterion 1.1, the latter has larger global weight.

Priorities are calculated from pairwise comparison matrices in Figure A1a–c, the corresponding weights being highlighted with the same colours.

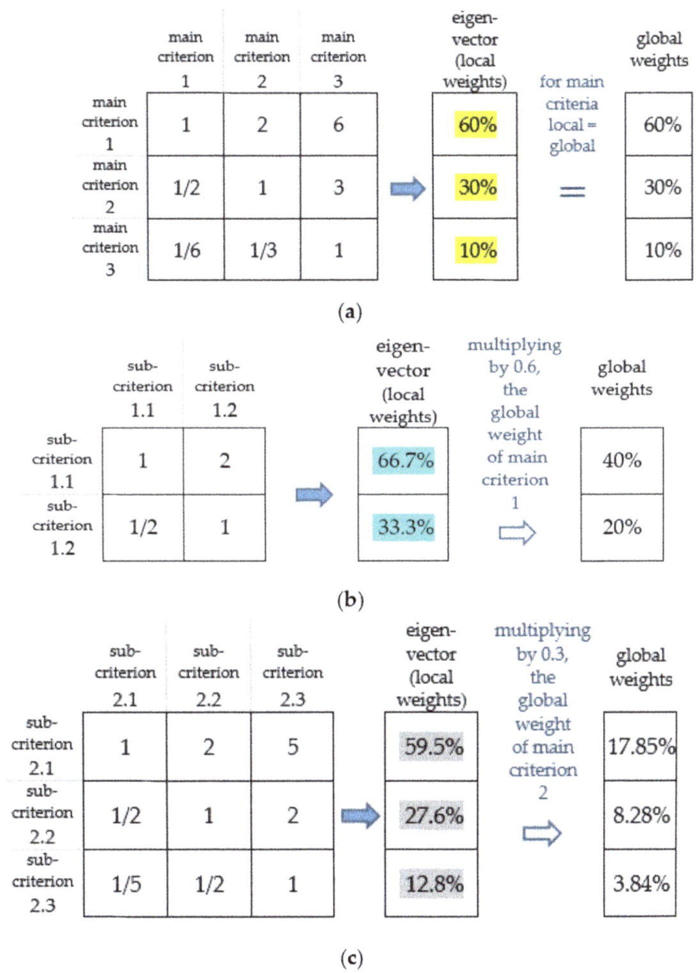

Figure A1. *Cont.*

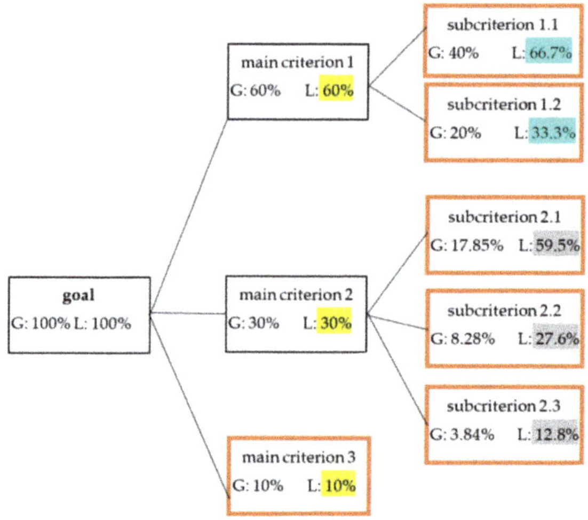

(d)

Figure A1. Global and local weights from pairwise comparison matrices' eigenvectors. (**a**) Global and local weights of three main criterions; (**b**) Subcriterion global and local weights of main criterion 1; (**c**) Subcriterion global and local weights of main criterion 2; (**d**) Global and local weights from pairwise comparison matrices' eigenvectors.

References

1. Meadows, H.D.; Meadows, L.D.; Randers, J.; Behrens, W.W. *The Limits to Growth*; Potomac Associates, Universe Books: New York, NY, USA, 1972; Available online: http://www.donellameadows.org/wp-content/userfiles/Limits-to-Growth-digital-scan-version.pdf (accessed on 12 August 2020).
2. Kates, W.R.; Clark, C.W.; Corell, J.R.; Hall, M.; Jaeger, C.C.; Lowe, I.; McCarthy, J.J.; Schellnhuber, H.J.; Bolin, B.; Huntley, B.; et al. *Sustainability Science*; Research and Assessment Systems for Sustainability Program Discussion Paper 2000-33. Environment and Natural Resources Program, Belfer Center for Science and International Affairs; Kennedy School of Government, Harvard University: Cambridge, MA, USA, 2000; Available online: http://rwkates.org/pdfs/a2001.01.pdf (accessed on 9 November 2020).
3. WWF. Living Planet Report 2020—Bending the Curve of Biodiversity Loss. 2020. Available online: https://www.zsl.org/sites/default/files/LPR%202020%20Full%20report.pdf (accessed on 9 November 2020).
4. UN. Transforming Our World: The 2030 Agenda for Sustainable Development. New York: UN General Assembly. 2015. Available online: https://www.un.org/ga/search/view_doc.asp?symbol=A/69/L.85&Lang=E (accessed on 15 October 2020).
5. UN. The Future is Now—Science for Achieving Sustainable Development. Department of Economic and Social Affairs. New York: Independent Group of Scientists appointed by the Secretary-General, Global Sustainable Development Report 2019. 2019. Available online: https://sustainabledevelopment.un.org/content/documents/24797GSDR_report_2019.pdf (accessed on 17 November 2020).
6. UN. United Nations Headquarters, New York. Zero-Draft, Ministerial Declaration of the 2016 High-Level Political Forum on Sustainable Development Convened under the Auspices of the Economic and Social Council on the Theme "Ensuring than No One is Left Behind". 2016. Available online: https://sustainabledevelopment.un.org/content/documents/10521HLPF%202016%20outcome%20-%20zero%20draft%2020%20June.pdf (accessed on 28 November 2020).
7. UN. Climate Change. The Paris Agreement. 2015. Available online: https://unfccc.int/process-and-meetings/the-paris-agreement/the-paris-agreement (accessed on 6 November 2020).
8. EC. 29 January 2020. Available online: https://ec.europa.eu/environment/sustainable-development/index_en.htm (accessed on 12 August 2020).
9. Faber, N.; Jorna, R.; Engelen, J.M. The sustainability of "sustainability"—A study into the conceptual foundations of the notion of "sustainability". *J. Environ. Assess. Policy Manag.* **2005**, *7*, 1–33. [CrossRef]
10. Carson, R. *Silent Spring*; Fawcett Publications Inc.: Greenwich, CT, USA, 1962; Available online: http://library.uniteddiversity.coop/More_Books_and_Reports/Silent_Spring-Rachel_Carson-1962.pdf (accessed on 9 November 2020).
11. Coomer, C.J. *Quest for a Sustainable Society*; Pergamon Press Inc.: Oxford, UK, 1979.
12. Rumen, G. *Sustainable Development, Economic Aspects*; University of Indianapolis Press: Bloomington, IN, USA, 2005.

13. Ciegis, R.; Ramanauskiene, J.; Martinkus, B. The Concept of Sustainable Development and its Use for Sustainability Scenarios. *Inz. Ekon. Eng. Econ.* **2009**, *62*, 28–37. Available online: https://inzeko.ktu.lt/index.php/EE/article/view/11609 (accessed on 23 October 2020).
14. Szabo, K.Z. Analysis of Research on Sustainable Development the Goals of Sustainable Development, Practical and Theoretical Framework in EU and Romania. Juridical Current, 253–262. 2011. Available online: http://revcurentjur.ro/old/arhiva/attachments_201104/recjurid114_22F.pdf (accessed on 21 October 2020).
15. Klarin, T. The Concept of Sustainable Development: From its Beginning to the Contemporary Issues. *Zagreb Int. Rev. Econ. Bus.* **2018**, *21*, 67–94. [CrossRef]
16. Mensah, J. *Sustainable Development: Meaning, History, Principles, Pillars, and Implications for Human Action: Literature Review*; Cogent Social Sciences; Taylor&Francis: Abingdon-on-Thames, UK, 2019; pp. 1–21. [CrossRef]
17. WCED. Our Common Future. Report of the World Commission on Environment and Development, 374. 4 August 1987. Available online: https://digitallibrary.un.org/record/139811#record-files-collapse-header (accessed on 12 October 2020).
18. Barbier, B.E. The Concept of Sustainable Economic Development. *Environ. Conserv.* **1987**, *14*, 101–110. [CrossRef]
19. Munda, G. Multiple Criteria Decision Analysis and Sustainable development. In *Multiple Criteria Decision Analysis. State of the Art Surveys*; International Series in Operations Research & Management Science; Figueria, J., Greco, S., Ehrgott, M., Eds.; Springer: New York, NY, USA, 2005; Chapter 23; Volume 78, pp. 953–983. [CrossRef]
20. Prescott-Allen, R. The Wellbeiing of Nations. A Country-by-Country Index of Quality of Life and the Environment. Island Press: Washington, DC, USA, 2001.
21. Saaty, T.L.; Vargas, L.G. *Models, Methods, Concepts & Applications of the Analytic Hierarchy Process*; Springer: New York, NY, USA, 2012.
22. Greco, S.; Ehrgott, M.; Figueira, J.R. (Eds.) *Multiple Criteria Decision Analysis: State of the Art Surveys*, 2nd ed.; International Series in Operations Research and Management Science; Springer: New York, NY, USA, 2016; Volume 233.
23. Siraj, S.; Mikhailov, L.; Keane, J.A. PriEsT: An interactive decision support tool to estimate priorities from pairwise comparison judgments. *Int. Trans. Oper. Res.* **2015**, *22*, 217–235. [CrossRef]
24. UN. Agenda 21. In Proceedings of the United Nations Sustainable Development, Conference on Environment & Development, Rio de Janerio, Brazil, 3–14 June 1992; Available online: https://sustainabledevelopment.un.org/content/documents/Agenda21.pdf (accessed on 4 December 2020).
25. UN. *Indicators of Sustainable Development: Guidelines and Methodologies*, 3rd ed.; United Nations Publication: New York, NY, USA, 2007; Available online: http://www.sustentabilidad.uai.edu.ar/pdf/sde/guidelines.pdf (accessed on 12 November 2020).
26. Hass, J.L.; Brunvoll, F.; Hoie, H. *Overview of Sustainable Development Indicators Used by National and International Agencies*; OECD Publishing: Paris, France, 2002; Volume 2. [CrossRef]
27. EC. *EU Member State Experiences with Sustainable Development Indicators*; Eurostat Working Paper; Office for Official Publications of the European Communities: Luxemburg, 2004; Available online: https://ec.europa.eu/eurostat/documents/3888793/5829493/KS-AU-04-001-EN.PDF.pdf/830b2fe8-8f24-457c-9937-fd2e0f29faa5?t=1414779058000 (accessed on 27 November 2020).
28. EC. *Analysis of National Sets of Indicators Used in the National Reform Programmes and Sustainable Development Strategies*; Eurostat Methodologies and Working Papers; Office for Official Publications of the European Communities: Luxemburg, 2007; Available online: https://ec.europa.eu/eurostat/documents/3888793/5844237/KS-RA-07-028-EN.PDF/d10749ca-fa4d-471c-b9c6-3c9dfc7d92bf?version=1.0 (accessed on 4 December 2020).
29. UNECE; OECD; Eurostat. Measuring Sustainable Develeopment. Geneva. 2009. Available online: https://unece.org/fileadmin/DAM/stats/publications/Measuring_sustainable_development.pdf (accessed on 15 December 2020).
30. UNECE; Eurostat; OECD. Framework and Suggested Indicators to Measure Sustainable Development. 27 May 2013. Available online: https://communityindicators.net/resources/framework-and-suggested-indicators-to-measure-sustainable-development/54 (accessed on 25 November 2020).
31. ICLEI. The Reference Framework for Sustainable Cities 5 Dimensions and 30 Objectives for a European Vision of Tomorrow's Cities. Sustainable Cities Platform. Europe: Sustainable Cities Platform. 2020. Available online: http://rfsc.eu/wp-content/uploads/2016/03/30-objectives-1.pdf (accessed on 28 November 2020).
32. Pupphachai, U.; Zuidema, C. Sustainability indicators: A tool to generate learning and adaptation in sustainable urban development. *Ecol. Indic.* **2017**, *72*, 784–793. [CrossRef]
33. Garrett, R.; Latawiec, E.A. What Are Sustainability Indicators For? In *Sustainability Indicators*; Latawiec, A.E., Ed.; De Gruyter: Berlin, Germany, 2015. [CrossRef]
34. Sachs, J.; Schmidt-Traub, G.; Kroll, C.; Lafortune, G.; Fuller, G.; Woelm, F. *SDGR. The Sustainable Development Goals and COVID-19*; Sustainable Development Report; Cambridge University Press: Cambridge, UK, 2020; Available online: https://s3.amazonaws.com/sustainabledevelopment.report/2020/2020_sustainable_development_report.pdf (accessed on 14 November 2020).
35. Munda, G. "Measuring Sustainability": A Multi-Criterion Framework. *Environ. Dev. Sustain.* **2005**, *7*, 117–134. [CrossRef]
36. Jurík, L.; Horňáková, N.; Šantavá, E.; Cagáňová, D.; Sablik, J. Application of AHP method for project selection in the context of sustainable development. *Wirel. Netw.* **2020**. [CrossRef]
37. Bonissone, P.P.; Subbu, R.; LizziJ. Multi Criteria Decision Making (MCDM): A Framework for Research and Applications. *IEEE Comput. Intel. Mag.* **2009**, *58*, 48–61. [CrossRef]

38. Moshkovich, H.; Mechitov, A.O.D. Verbal Decision Analysis. In *International Series in Operations Research & Management Science: Multiple Criteria Decision Analysis. State of the Art Surveys*; Greco, S.M., Ehrgott, J.M., Figueira, R., Eds.; Springer: New York, NY, USA, 2016; Volume 1. [CrossRef]
39. Liebowitz, J. *Bursting the Big Data Bubble the Case for Intuition-Based Decision Making*; Auerbach Publications: Boca Raton, FL, USA, 2019.
40. Navneet, B.; Kanwal, R. *Strategic Decision Making Applying the Analytic Hierarchy Process*; Springer: Berlin, Germany, 2004; Available online: https://archive.org/details/springer_10.1007-b97668/page/n3/mode/2up (accessed on 24 November 2020).
41. Munda, G. Multiple Criteria Decision Analysis and Sustainable Development. In *International Series in Operations Research & Management Science: Multiple Criteria Decision Analysis. State of the Art Surveys*, 2nd ed.; Greco, S.M., Ehrgott, J.M., Figueira, R., Price, C.C., Eds.; Springer: New York, NY, USA, 2016; Volume 233, pp. 1235–1267. [CrossRef]
42. Shen, L.; Muduli, K.; Barve, A. Developing a sustainable development framework in the context of mining industries: AHP approach. *Resour. Policy* **2015**, *46*, 15–26. [CrossRef]
43. Salvado, M.F.; Azevedo, S.G.; Matias, J.C.O.; Ferreira, L.M. Proposal of a Sustainability Index for the Automotive Industry. *Sustainability* **2015**, *7*, 2113–2144. [CrossRef]
44. Patalas-Maliszewska, J.; Łosyk, H. An Approach to Assessing Sustainability in the Development of a Manufacturing Company. *Sustainability* **2020**, *12*, 8787. [CrossRef]
45. Rezaei-Moghaddam, K.; Karami, E. A multiple criteria evaluation of sustainable agricultural development models using AHP. *Environ. Dev. Sustain.* **2008**, *10*, 407–426. [CrossRef]
46. Ahmad, S.; Mat Tahar, R. Selection of renewable energy sources for sustainable development of electricity generation system using analytic hierarchy process: A case of Malaysia. *Renew. Energy* **2014**, *63*, 458–466. [CrossRef]
47. Reza, B.; Sadiq, R.; Hewage, K. Sustainability assessment of flooring systems in the city of Tehran: An AHP-based life cycle analysis. *Constr. Build. Mater.* **2011**, *25*, 2053–2066. [CrossRef]
48. Calabrese, A.; Costa, R.; Levialdi, N.; Menichini, T. Integrating sustainability into strategic decision-making: A fuzzy AHP method for the selection of relevant sustainability issues. *Technol. Forecast. Soc. Change* **2019**, *139*, 155–168. [CrossRef]
49. Claver, J.; García-Domínguez, A.; Sebastián, M. Multicriteria Decision Tool for Sustainable Reuse of Industrial Heritage into Its Urban and Social Environment. Case Studies. *Sustainability* **2020**, *12*, 7430. [CrossRef]
50. Gómez-Limón, J.A.; Arriaza, M.; Guerrero-Baena, M.D. Building a Composite Indicator to Measure Environmental Sustainability Using Alternative Weighting Methods. *Sustainability* **2020**, *12*, 4398. [CrossRef]
51. Ramos-Quintana, F.; Tovar-Sánchez, E.; Saldarriaga-Noreña, H.; Sotelo-Nava, H.; Sánchez-Hernández, J.P.; Castrejón-Godínez, M.-L. A CBR–AHP Hybrid Method to Support the Decision-Making Process in the Selection of Environmental Management Actions. *Sustainability* **2019**, *11*, 5649. [CrossRef]
52. Myeong, S.; Jung, Y.; Lee, E. A Study on Determinant Factors in Smart City Development: An Analytic Hierarchy Process Analysis. *Sustainability* **2018**, *10*, 2606. [CrossRef]
53. Cheng, W.; Mo, D.; Tian, Y.; Xu, W.; Xie, K. Research on the Composite Index of the Modern Chinese Energy System. *Sustainability* **2018**, *11*, 150. [CrossRef]
54. Nakthong, V.; Kubaha, K. Development of a Sustainability Index for an Energy Management System in Thailand. *Sustainability* **2019**, *11*, 4587. [CrossRef]
55. Hienuki, S.; Noguchi, K.; Shibutani, T.; Saigo, T.; Miyake, A. The Balance of Individual and Infrastructure Values in Decisions Regarding Advanced Science and Technology. *Sustainability* **2019**, *11*, 3385. [CrossRef]
56. Nijkamp, P.; Vindigni, G. Impact assessment of qualitative policy scenarios. *Manag. Environ. Qual. Int. J.* **2003**, *14*, 108–133. [CrossRef]
57. Akgün, A.A.; Van Leeuwen, E.; Nijkamp, P. A multi-actor multi-criteria scenario analysis of regional sustainable resource policy. *Ecol. Econ.* **2012**, *78*, 19–28. [CrossRef]
58. Kandakoglu, A.; Frini, A.; Ben Amor, S. Multicriteria decision making for sustainable development: A systematic review. *J. Multi Criteria Decis. Anal.* **2019**, *26*, 202–251. [CrossRef]
59. Dos Santos, P.H.; Neves, S.M.; Sant'Anna, D.O.; De Oliveira, C.H.; Carvalho, H.D. The analytic hierarchy process supporting decision making for sustainable development: An overview of applications. *J. Clean. Prod.* **2019**, *212*, 119–138. [CrossRef]
60. Cinelli, M.; Coles, S.R.; Kirwan, K. Analysis of the potentials of multi criteria decision analysis methods to conduct sustainability assessment. *Ecol. Indic.* **2014**, *46*, 138–148. [CrossRef]
61. Baffoe, G. Exploring the utility of Analytic Hierarchy Process (AHP) in ranking livelihood activities for effective and sustainable rural development interventions in developing countries. *Eval. Prog. Plan.* **2019**, *72*, 197–204. [CrossRef] [PubMed]
62. Vaidya, O.S.; Kumar, S. Analytic hierarchy process: An overview of applications. *Eur. J. Oper. Res.* **2006**, *169*, 1–29. [CrossRef]
63. Vargas, L.G. An overview of the analytic hierarchy process and its applications. *Eur. J. Oper. Res.* **1990**, *48*, 2–8. [CrossRef]
64. Ho, W.; Xu, X.; Dey, P.K. Multi-criteria decision making approaches for supplier evaluation and selection: A literature review. *Eur. J. Oper. Res.* **2010**, *202*, 16–24. [CrossRef]
65. Subramanian, N.; Ramanathan, R. A review of applications of Analytic Hierarchy Process in operations management. *Int. J. Prod. Econ.* **2012**, *138*, 215–241. [CrossRef]
66. Forman, E.H.; Gass, S.I. The analytical hierarchy process—An exposition. *Oper. Res.* **2001**, *49*, 469–487. [CrossRef]
67. Forman, E.H.; Saaty, T.L.; Selly, M.A.; Waldron, R. *Expert Choice*; Decision Support Software Inc.: McLean, VA, USA, 1983.
68. Saaty, T.L. A Scaling Method for Priorities in Hierarchical Structures. *J. Math. Psychol.* **1977**, *15*, 234–281. [CrossRef]

69. Alderson, R.C.; Sproull, W.C. Requirement Analysis, Need Forecasting, and Technology Planning Using the Honeywell PATTERN Technique. *Technol. Forecast. Soc. Change* **1972**, *3*, 255–265. [CrossRef]
70. Martin, W.T.; Sharp, J.M. Reverse factor analysis: A modification of relevance tree techniques. *Technol. Forecast. Soc. Change* **1973**, *4*, 355–373. [CrossRef]
71. Hägele, G.; Pukelsheim, F. Llull's writings on electoral systems. *Studia Lul.* **2001**, *41*, 3–38.
72. Brunelli, M. A survey of inconsistency indices for pairwise comparisons. *Int. J. Gen. Syst.* **2018**, *47*, 751–771. [CrossRef]
73. Hollanders, H. European Innovation Scoreboard Methodology Report. EC. 2020. Available online: https://ec.europa.eu/commission/presscorner/detail/en/QANDA_20_1150 (accessed on 25 November 2020).
74. Schwab, K. The Global Competitiveness Report. Geneva. Available online: www.weforum.org/gcr (accessed on 18 December 2020).
75. Dutta, S.; Lanvin, B. (Eds.) *NRI. Network Readiness Index: Towards a Future-Ready Society*; Portulans Institute: Washington DC, USA, 2019; Available online: https://networkreadinessindex.org/wp-content/uploads/2020/03/The-Network-Readiness-Index-2019-New-version-March-2020.pdf (accessed on 20 October 2020).
76. Dutta, S.; Lanvin, B. Network Readiness Index, Accelerating Digital Transformation in a post-COVID Global Economy. Portulans Institute. 2020. Available online: https://networkreadinessindex.org/wp-content/uploads/2020/10/NRI-2020-Final-Report-October2020.pdf (accessed on 11 December 2020).
77. Soumitra, D.; Lanvin, B.; Wunsch-Vincent, S. (Eds.) Global Innovation Index 2020. In *Who Will Finance Innovation?* Cornell University: Ithaca, NY, USA; INSEAD: Fontainebleau, France; WIPO: Geneve, Switzerland, 2020; Available online: https://www.wipo.int/edocs/pubdocs/en/wipo_pub_gii_2020.pdf (accessed on 18 October 2020).
78. EU; EC. Environment. Eco-innovation Action Plan. 2020. Available online: https://ec.europa.eu/environment/ecoap/indicators/index_en (accessed on 27 November 2020).
79. EC. Environment, Our Oceans, Seas and Coasts. 2020. Available online: https://ec.europa.eu/environment/iczm/expert_group.htm (accessed on 3 November 2020).
80. EC. Environment. Eco-Innovation Action Plan. Circular Economy Indicators. 2020. Available online: https://ec.europa.eu/environment/ecoap/indicators/sustainable-resource-management_en (accessed on 26 November 2020).
81. Bosma, N.; Hill, S.; Ionescu-Somers, A.; Kelley, D.; Levie, J.; Tarnawa, A. Global Entrepreneurship Monitor. London: Global Entrepreneurship Research Association, London Business School. 2019–2020. Available online: https://www.gemconsortium.org/report/gem-2019-2020-global-report (accessed on 21 December 2020).
82. Eurostat. Waste Management Indicators. 2020. Available online: https://ec.europa.eu/eurostat/statistics:https://ec.europa.eu/eurostat/statistics-explained/index.php/Waste_management_indicators (accessed on 18 December 2020).
83. WB. Population Growth Rate. The World Bank Group. 2020. Available online: https://data.worldbank.org/indicator/SP.POP.GROW? (accessed on 17 October 2020).
84. UN. Global Indicator Framework for the Sustainable Development Goals and Targets of the 2030 Agenda for Sustainable Development. 2020. Available online: https://unstats.un.org/sdgs/indicators/Global%20Indicator%20Framework%20after%202020%20review_Eng.pdf (accessed on 19 October 2020).
85. Ritchie, H.; Roser, M. Ozone Layer. Our worldindata. 2018. Available online: https://ourworldindata.org/ozone-layer (accessed on 5 November 2020).
86. Wendling, Z.A.; Emerson, J.W.; de Sherbinin, A. Environmental Performance Index. In *Global Metrics for the Environmnet: Ranking Country Performance on Sustainability Issues*; Yale Center for Environmental Law&Policy: Mountain View, CA, USA; New Haven, CT, USA, 2020; Available online: https://epi.yale.edu/downloads/epi2020report20210112.pdf (accessed on 4 February 2021).
87. IQAir. Quality of Air Index. 2020. Available online: https://www.iqair.com/world-air-quality-ranking (accessed on 26 October 2020).
88. EEA. European Environment Agency. 2020. Available online: https://www.eea.europa.eu/data-and-maps/indicators/oxygen-consuming-substances-in-rivers/oxygen-consuming-substances-in-rivers-9 (accessed on 26 November 2020).
89. Naturserve. Biodiversity Indicators Dashboard. 2020. Available online: https://www.natureserve.org/conservation-tools/biodiversity-indicators-dashboard (accessed on 27 November 2020).
90. Paris: International Energy Agency, IEA. World Energy Balances: Overview. 2020. Available online: https://www.iea.org/reports/world-energy-balances-overview (accessed on 29 November 2020).
91. Deepanshu, B. Population Stability Index and Characteristic Analysis. 2015. Available online: https://www.listendata.com/2015/05/population-stability-index.html (accessed on 28 November 2020).
92. London, UK: The Legatum Institute Foundation, LegatumInst. Prosperity Index. 2019. Available online: https://prosperitysite.s3-accelerate.amazonaws.com/8115/8635/0367/The_Legatum_Prosperity_Index_2019.pdf (accessed on 5 October 2020).
93. OECD. *Environment at a Glance Indicators*; OECD Publishing: Paris, France, 2020. [CrossRef]
94. SPI. Social Progress Index. Washington, DC: The Social Progress Imperative. 2020. Available online: https://www.socialprogress.org/ (accessed on 12 October 2020).
95. Statista. GINI Index. 2020. Available online: https://www.statista.com/forecasts/1171540/gini-index-by-country (accessed on 23 November 2020).
96. KPMG. Change Readiness Index. 2019. Available online: https://home.kpmg/xx/en/home/insights/2019/06/2019-change-readiness-index.html (accessed on 23 November 2020).

97. WEF; Kearney, A.T. Readiness for the Future of Production Report. Geneva: World Economic Forum's System Initiative on Shaping the Future of Production. 2018. Available online: http://www3.weforum.org/docs/FOP_Readiness_Report_2018.pdf (accessed on 23 November 2020).
98. Zuo, X.; Hua, H.; Dong, Z.; Hao, C. Environmental performance index at the provincial level for China 2006–2011. *Ecol. Indic.* **2017**, *75*, 48–56. [CrossRef]

sustainability

MDPI

Article

A Hesitant Fuzzy Combined Compromise Solution Framework-Based on Discrimination Measure for Ranking Sustainable Third-Party Reverse Logistic Providers

Arunodaya Raj Mishra [1], Pratibha Rani [2], Raghunathan Krishankumar [3], Edmundas Kazimieras Zavadskas [4,*], Fausto Cavallaro [5] and Kattur S. Ravichandran [3,6]

1 Department of Mathematics, Government College, Satna, Jaitwara 485221, Madhya Pradesh, India; arunodaya87@outlook.com
2 Department of Mathematics, National Institute of Technology, Warangal 506004, Telangana, India; pratibha138@gmail.com
3 Department of Computing, Sastra University, Thanjavur 613401, Tamil Nadu, India; krishankumar@sastra.ac.in (R.K.); raviks@sastra.edu (K.S.R.)
4 Institute of Sustainable Construction, Vilnius Gediminas Technical University, Sauletekio Ave. 11, LT-10223 Vilnius, Lithuania
5 Department of Economics, University of Molise, Via De Sanctis, 86100 Campobasso, Italy; cavallaro@unimol.it
6 Administrative Area, Rajiv Gandhi National Institute of Youth Development, Sriperumbudur 602105, Tamil Nadu, India
* Correspondence: edmundas.zavadskas@vgtu.lt

check for updates

Citation: Mishra, A.R.; Rani, P.; Krishankumar, R.; Zavadskas, E.K.; Cavallaro, F.; Ravichandran, K.S. A Hesitant Fuzzy Combined Compromise Solution Framework-Based on Discrimination Measure for Ranking Sustainable Third-Party Reverse Logistic Providers. *Sustainability* **2021**, *13*, 2064. https://doi.org/10.3390/su13042064

Academic Editor: Andrea Appolloni

Received: 12 January 2021
Accepted: 9 February 2021
Published: 14 February 2021

Publisher's Note: MDPI stays neutral with regard to jurisdictional claims in published maps and institutional affiliations.

Copyright: © 2021 by the authors. Licensee MDPI, Basel, Switzerland. This article is an open access article distributed under the terms and conditions of the Creative Commons Attribution (CC BY) license (https://creativecommons.org/licenses/by/4.0/).

Abstract: Customers' pressure, social responsibility, and government regulations have motivated the enterprises to consider the reverse logistics (RL) in their operations. Recently, companies frequently outsource their RL practices to third-party reverse logistics providers (3PRLPs) to concentrate on their primary concern and diminish costs. However, to select the suitable 3PRLP candidate requires a multi-criteria decision making (MCDM) process involving uncertainty owing to the presence of many associated aspects. In order to choose the most appropriate sustainable 3PRLP (S3PRLP), we introduce a hybrid approach based on the classical Combined Compromise Solution (CoCoSo) method and propose a discrimination measure within the context of hesitant fuzzy sets (HFSs). This approach offers a new process based on the discrimination measure for evaluating the criteria weights. The efficiency and practicability of the present approach are numerically demonstrated by solving an illustrative case study of S3PRLPs selection under a hesitant fuzzy environment. Moreover, sensitivity and comparative studies are presented to highlight the robustness and strength of the introduced methodology. The result of this work concludes that the introduced methodology can recommend a more feasible performance when facing with determinate and inconsistent knowledge and qualitative data.

Keywords: hesitant fuzzy sets; discrimination measure; multi-criteria decision-making; combined compromise solution

1. Introduction

Over the last two decades, supply chain management (SCM) has been popularized as a significant part of every company, no matter whether the size of the company is large or small. SCM encompasses the planning and management of all supply chain activities engaged in sourcing and procurement, conversion, and all logistics management activities with the objective of maximizing customer value, creating net value, and achieving a sustainable competitive advantage [1]. Logistics, one of the core components of effective SCM, is not only delivering the merchandise but also proffering the opportunity for stock to be returned to the suppliers through a feedback loop. Along with the requirement or prospective for the recycling of redundant stock, emergent global environmental concerns and increasing revenues from effective product returns strategies have encouraged some industrialists and researchers to originate a novel concept termed as "reverse logistics (RL)". It stands for all the processes or activities associated with the reuse of products and materials. RL is the process of scheduling, implementing, and controlling the well-organized, cost-effective flow of raw materials, in-process inventory, finished merchandise, and related information for returned products in reverse flow of the classical supply chain in order to recover value and get the competitive advantage [2]. Effective RL is believed to result in direct benefits, consisting of enhanced customer satisfaction, decreased resource investment levels, increased protection level of the environment with recycling and proper disposal, and reductions in storage and distribution costs [3,4].

The pressure of increasing costs of enterprises, commercial competition, globalization of business activities, legislative pressure, and customers are compelling the companies to pay increasing attention to their RL activities. The companies have three different modes to handle their RL operations, which are (i) handle them in-house; (ii) manage them through a reliable logistics subsidiary or group company, either by setting up or buying an associated logistics firm [1], and (iii) outsource this function to an exterior provider [5]. Nonetheless, several enterprises struggle for implementing RL in house because they do not have adequate capacity to manage this function. Moreover, there are massive risks involved in setting up or buying a logistics firm, and doing so may have a negative consequence on the company's core business, particularly its economic and operational features. Indeed, outsourcing the company's RL operations may be important to facilitate a company for pursuing financial efficiency, operational investments, and sustainable long-term development. One of the essential decisions that must be made with RL activities is the selection of appropriate third party reverse logistics provider (3PRLP). The suitable 3PRLP candidate offers benefits in terms of operational and resource efficiency, cost reduction, improvement in performance, organizational competitiveness, and long-term development of the company [2,4]. Thus, the 3PRLP selection decision can be a strategic issue faced by supply chain executives in retaining organizational strategic competitive advantage. The process of choosing a proper sustainable 3PRLP (S3PRLP) can be treated as a multi-criteria decision making (MCDM) due to the presence of multiple qualitative and quantitative criteria.

Uncertainty common occurs in the S3PRLP selection process. The fuzzy set theory [6] has been successfully implemented to solve the uncertain 3PRLP selection problem. Sometimes, the decision-making experts (DMEs) might assess the Belongingness Degree (BD) of an object to a set of various distinct degrees in numerous real-life circumstances because of their individual attention, time restrictions, and deficiency of information. For example, if a group of DMEs is required to offer the BD of a particular opinion to an adult age cluster, the first DME wishes 0.65, another 0.70, and the last one does not recommend the BD due to the time restrictions and deficiency of knowledge/information/data [7]. To handle this issue, the doctrine of HFSs was given by Torra and Narukawa [8], which offers the BD to comprise various distinct assessment degrees. As the extension of FSs, HFSs have gained much interest from the researchers in dealing with the ambiguity that occurred in daily life problems. It is represented by a BD and signified by a set of possible degrees. Recently, it is revealing that the HFSs have powerfully associated to the existing concepts, namely, Intuitionistic Fuzzy Sets (IFSs) [9], Type-2 Fuzzy Sets (T2FSs) [10], and Fuzzy Multi Sets (FMSs) [11]. As stated by Torra [7], the prime concern to invent HFSs is that when describing the BD of an object, the complexity of generating the BD is not a margin of error (previously seen in IFSs) or specific possibility distribution (previously seen in T2FSs) on the possibility degrees, but a possible degrees' set. Owing to the association with mentioned extended FSs, HFSs are distinguished as IFSs when it is a non-void closed interval; in specific concerns, HFSs can describe FMSs, even if the laws for FMSs do not implement properly to HFSs. Since its appearance, a great number of studies have been conducted within the HFSs context. However, there is no study regarding the proposed hesitant fuzzy-based decision-making methodology for selecting a desirable S3PRLP alternative over a set of multiple conflicting criteria.

In the process of selecting an ideal S3PRLP alternative, it is expected to deal with the compromise of performance values of partner candidates over incompatible or diverse influencing factors. In several situations, a comprehensive study on the fundamental properties of compromise solutions can assist the MCDM procedure. Presently, lots of MCDM techniques have been introduced to investigate the compromise solutions, such as the VlseKriterijumska Optimizacija I Kompromisno Resenje (VIKOR), Technique for Order of Preference by Similarity to Ideal Solution (TOPSIS), COmbinative Distance-based Assessment (CODAS), EDAS (Evaluation Based on Distance from Average Solution), and Complex Proportional Assessment (COPRAS). Nonetheless, the preference ordering of the candidates attained by these techniques may fluctuate extensively owing to the variation of weight distribution of the criteria. In other words, the consistency and stability of the outcomes obtained by these approaches are restricted [12,13]. To conquer this issue, Yazdani et al. [14] introduced the idea of Combined Compromise Solution (CoCoSo) approach, which mingles the combined compromise decision-making algorithm with some aggregation strategies to get a versatile compromise solution. The compromise solution acquired by the CoCoSo approach is consistent with that attained by other MCDM approaches, but it is not simply influenced by the amendment of weight distribution of criteria. This means that the CoCoSo approach has high reliability and stability regarding the ranking of alternatives.

In a real-life S3PRLP selection process, there are various cases wherein people feel doubtful or hesitant for one or another reason, which makes it complex for them to reach agreement on a final judgment. In other words, it is often hard for a panel of experts that comprises numerous DMEs to acquire suitable and realistic outcomes against a set of provider candidates over multiple qualitative and quantitative criteria. Moreover, in reality, it is difficult to evaluate the 3PRLP candidates and choose a desirable one(s) because of the existence of multiple conflicting factors, lack of information, and time limitation. Consequently, there is a need to establish an appropriate framework for handling the 3PRLP selection problem with uncertainty from sustainable viewpoints. In order to overcome the above-discussed shortcomings, this study focuses under a hesitant fuzzy sets (HFSs) context, as the HFS theory is one of the remarkable tools for handling the hesitant information that arises in practical applications. For ranking of the alternatives, the classical CoCoSo method has high stability and reliability regarding the preference ordering of the alternatives. Thus, in this work, we develop an integrated hesitant fuzzy-based CoCoSo methodology that can efficiently tackle the intrinsic uncertainty and the hesitancy in DME's opinions and provide a compromise solution for the S3PRLP selection problem. Since the weight determination of criteria has a momentous impact on the decision outcomes, thus, this paper proposes a new hesitant fuzzy discrimination measure-based formula to calculate the criteria weights. The proposed method in this study offers an easy calculation process with accurate and consistent results for assessing S3PRLPs. In this perspective, the major contributions are listed as follows:

1. A novel integrated HF-CoCoSo methodology is introduced for solving MCDM problems with HFSs.
2. An HF-discrimination measure based framework is employed for evaluating the criteria weights.
3. The proposed method is implemented to choose an optimal S3PRLP for an automobile manufacturing company within the HFS context.
4. At last, comparison with extant approaches and sensitivity assessment are studied to confirm the reliability and practicability of the outcomes.

The present manuscript is systematized as follows: Section 2 confers the comprehensive reviews associated to the present study. Section 3 discusses the fundamentals of the HFSs. Section 4 pioneers an innovative HF-CoCoSo technique for solving MCDM problems under an HFSs environment. Section 5 discusses a decision-making example related to the S3PRLPs selection problem with uncertainty. In addition, comparative and sensitivity analyses are conferred at the end of this section. Lastly, Section 6 suggests the conclusions and scope for future research.

2. Preliminaries

The current section presents comprehensive literatures related to this study.

2.1. Hesitant Fuzzy Set (HFS)

As the extension of FSs, HFSs [7,8] have gained much interest from the researchers in dealing with the ambiguity occurred in daily life problems. It is represented by a BD and signified by a set of possible degrees. Next, a great number of studies have been considered and applied under the HFSs context. For instance, Xia and Xu [15] and Xia et al. [16] presented a set of Aggregation Operators (AOs) on HFSs. Xu and Xia [17,18] pioneered the distance measure on HFSs and discussed the relationship between different information measures. He et al. [19] and Sun [20] presented the HF-power geometric and normalized geometric Bonferroni mean operators and used them to handle MCDM problems under HFS settings. Liao et al. [21] introduced the measures of correlation on HFSs. Li et al. [22] and Hu et al. [23] discussed several hesitant fuzzy information measures. Cuiping et al. [24] constructed several novel prioritized AOs. Yu [25] firstly offered the notion of HF-Heronian mean operators. Lv et al. [26] firstly suggested the conception of a feature vector and then studied the hesitant fuzzy information measures. Wang [27] used the combination of similarity measures and applied the synthetical procedure for HFSs. In addition, they used their formula for clustering analysis within the HFSs context.

Currently, HFS has extensively been employed in decision-making applications owing to its usefulness in articulating the ambiguous information. Afterward, numerous techniques have been presented to elucidate the MCDM problems under the HFSs environment. Lan et al. [28] suggested an innovative MCDM model with a new priority degree formula under HFSs environment. Mousavi et al. [29] introduced the HF-Elimination and Choice Expressing the REality (HF-ELECTRE) technique to assess the renewable energy source assessment problem. Wu et al. [30] employed a modified HF-VIKOR-based model for prioritizing engineering characteristics of an electric vehicle. It is worthwhile to say that various theoretical and practical reasoning can be simplified with the help of a uniformly typical HFS, which was articulated by Alcantud and Torra [31]. Cheng [32] recommended a novel autocratic approach using group recommendations for HFSs to deal with the green hotel assessment problems. Liu et al. [33] initiated the hesitant fuzzy cognitive maps and showed how to use it to explore the risk factors that arise in an electric power system. With the implementation of the HF-Complex Proportional Assessment (HF-COPRAS) method, Mishra et al. [34] assessed and presented the ranking order of the service quality alternatives for vehicle insurance companies. In a further study, Mishra et al. [35] designed an innovative framework within the HFSs environment to evaluate and prioritize green supplier alternatives. Farhadinia and Herrera-Viedma [36] re-defined the existing definition of extended HFS by employing the Cartesian product of HFS. Wang et al. [37] extended the group emergency decision-making (GEDM) method using hesitant fuzzy numbers. Mokhtia et al. [38] adapted the Ridge, least absolute shrinkage and selection operator (LASSO), and Elastic Net regression techniques for the task of choosing a feature. Çolak and Kaya [39] proposed a hybrid technique to solve the energy storage technology selection problem within the HFS setting. Mardani et al. [40] proposed a combined structure based on different approaches with HFS to assess the digital health intervention during the COVID-19 pandemic. Wu et al. [41] designed a trust-based social network group decision-making methodology with HF preference relations to the Water–Energy–Food nexus estimation. Narayanamoorthy et al. [42] gave a latest ranking scheme, named as HF-Multi-Objective Optimization on the basis of Simple Ratio Analysis (HF-MOOSRA), for evaluating and choosing an ideal bio-medical waste disposal treatment method. Rani et al. [43] assessed and prioritized the sustainable suppliers by employing the HF-Step-wise Weight Assessment Ratio Analysis-COPRAS methodology. Demirel et al. [44] proposed an innovative Choquet integral-based hesitant fuzzy decision-making method to prevent soil erosion. To deal with the hydrogen underground storage site selection decision-making problem under an interval type-2 HFS domain, Deveci [45] introduced a multi-criteria decision support system based on interval type-2 HFS. In the literature, several other models have been discussed within the HFSs setting [46,47].

2.2. Combined Compromise Solution (CoCoSo) Method

Existing studies offer various MCDM approaches for dealing with the practical MCDM issues. However, the preference results obtained by the extant decision-making approaches [48,49] have low reliability and stability due to change of weight distributions of criteria, which may be irrational for DEs to opt an ideal option. To surmount this limitation, Yazdani et al. [14] initiated a novel CoCoSo method that offers effective and simple calculation steps to deal with the uncertain information in a logical way. It integrates the two most prominent MCDM models, which are simple additive weighting (SAW) and exponentially weighted product (EWP). In recent times, copious scholarly articles have focused their interest on the development and application of the CoCoSo method. For instance, Yazdani et al. [50] introduced the decision making trial and evaluation laboratory (DEMATEL) and best worst method (BWM)-based gray extension of the CoCoSo approach to select the ideal supplier in construction management. In a further study, Wen et al. [13] originated a collective methodology by combining the CoCoSo model and hesitant linguistic fuzzy set and then applied it to a decision-making problem. Peng and Smarandache [51] developed a novel neutrosophic soft CoCoSo method for China's rare earth industry security assessment. Liao et al. [52] introduced an innovative Pythagorean fuzzy CoCoSo method for the selection of cold chain logistics delivery center. Alrasheedi et al. [53] suggested a novel IVIF-CoCoSo technique for assessing the green growth indicators to accomplish sustainable development in the manufacturing region. Rani and Mishra [54] gave a single-valued neutrosophic CoCoSo method for evaluating sustainable waste electrical and electronics equipment recycling partners. In a recent study, Liu et al. [55] evaluated and selected an appropriate medical waste treat technology by using a Pythagorean fuzzy CoCoSo approach.

From the aforementioned discussion, we can see that an integrated method based on the CoCoSo approach with HFSs can additionally improve the practicality of the CoCoSo approach with uncertain decision-making settings.

2.3. Sustainable Third-Party Reverse Logistics Provider (S3PRLP) Assessment

The S3PRLP assessment problem is a complex MCDM problem due to the participation of numerous decisive factors. To choose the optimal S3PRLP option and enhance the precision and consistency of the outcomes, lots of new MCDM procedures have been considered and utilized to several fields [56–58]. FSs and their generalizations have expansively been employed to manage the uncertain and vague information that occurs in realistic MCDM processes. Efendigil et al. [59] designed a two-way method by integrating artificial neural networks and fuzzy logic for the evaluation of an ideal 3PRLP alternative. Govindan [60] designed a fuzzy-based procedure for assessing the 3PRLP selection problem. Govindan and Murugesan [61] used the fuzzy extent assessment approach for choosing the desirable 3PRLP for a battery manufacturing industry. Corresponding to AHP and TOPSIS approaches, Senthil et al. [62] suggested a combined model for evaluating an ideal reverse logistics contractor. Uygun et al. [63] planned and selected an outsourcing provider for a telecommunications business by employing DEMATEL and ANP methods for FSs. Mavi et al. [64] presented the SWARA method for weighting the assessment criteria of 3PRLP in the plastics industry and further ranked the S3PRLP alternatives through the MOORA model within FSs. Tavana et al. [65] suggested a combined method with the integration of ANP and gray superiority and inferiority methods on IFSs for the assessment of the 3PRLPs selection process. Li et al. [66] used a combined cumulative prospect doctrine with hybrid information MCDM methodology for the evaluation of 3PRLPs from sustainability perspectives. Zarbakhshnia et al. [67] weighted the assessment criteria through the fuzzy-SWARA method and ranked the S3PRLPs by employing the COPRAS method under a fuzzy environment. Zhang and Su [68] introduced a dominance degree-based heterogeneous linguistic model to assess the best S3PRLP for a car manufacture industry. On the basis of fuzzy AHP and gray MOORA, Zarbakhshnia et al. [69] originated a collective MCDM method for assessing the S3PRLP alternatives. Chen et al. [1] suggested a new multi-perspective multi-attribute decision-making support system for the evaluation and prioritization of 3PRLPs under an HF linguistic term set environment. Mishra et al. [70] gave a combined framework with CRITIC and EDAS methods for Fermatean fuzzy sets (FFSs) to deal the S3PRLP selection problem in which the attributes and DEs' weights are completely unknown.

3. Preliminaries

The present section is dedicated to describing the fundamental definitions and operations of HFSs. The concept of HFSs was pioneered by Torra [7] and Torra and Narukawa [8], to handle the problems where the BD of an element to a given set includes some diverse values.

Definition 1 [7]. *An HFS on the fixed universal set Ω is often expressed by the following mathematical form:*

$$S = \{\langle v, h_S(v)\rangle : v \in \Omega\}, \tag{1}$$

where $h_S(v)$, called as a hesitant fuzzy element (HFE), is a set of some values in $[0,1]$, denoting the possible BD of an element $v \in \Omega$ to the set S. Based on the set of all HFEs of S, the HFS S can be defined as $S = \left\{\left\langle v, \underset{k\in h_S(v)}{\cup}\{k\}\right\rangle : v \in \Omega\right\}$.

Example 1. *If $\Omega = \{v_1, v_2, v_3\}$ is the fixed universal set, $h(v_1) = \{0.2, 0.4, 0.5\}$, $h(v_2) = \{0.3, 0.4\}$ and $h(v_3) = \{0.3, 0.2, 0.5, 0.6\}$ are the HFEs of $v_i (i = 1,2,3,4)$ to a set S, respectively. Then, the set S can be considered as an HFS and mathematically represented as $S = \{\langle v_1, \{0.2, 0.4, 0.5\}\rangle, \langle v_2, \{0.3, 0.4\}\rangle, \langle v_3, \{0.3, 0.2, 0.5, 0.6\}\rangle\}$.*

Definition 2 [7,15]. *Let $h, h_1, h_2 \in HFEs(Y)$. Then, the operations on HFEs are discussed as follows:*

(a) $h^c = \underset{\kappa\in h}{\cup}\{1-\kappa\};$

(b) $h_1 \cup h_2 = \underset{\kappa_1\in h_1, \kappa_2\in h_2}{\cup} \max\{\kappa_1, \kappa_2\};$

(c) $h_1 \cap h_2 = \underset{\kappa_1\in h_1, \kappa_2\in h_2}{\cup} \min\{\kappa_1, \kappa_2\};$

(d) $\lambda h = \underset{\kappa\in h}{\cup}\left\{1-(1-\kappa)^\lambda\right\}, \lambda > 0;$

(e) $h^\lambda = \underset{\kappa\in h}{\cup}\left\{\kappa^\lambda\right\}, \lambda > 0;$

(f) $h_1 \oplus h_2 = \underset{\kappa_1\in h_1, \kappa_2\in h_2}{\cup}\{\kappa_1 + \kappa_2 - \kappa_1\kappa_2\};$

(g) $h_1 \otimes h_2 = \underset{\kappa_1\in h_1, \kappa_2\in h_2}{\cup}\{\kappa_1\kappa_2\}.$

Definition 3 [21]. *For a given HFE $h(v)$, the score and variance functions are given by the following expressions (2) and (3):*

$$\wp(h(v)) = \frac{1}{l(h(v))}\sum_{\kappa\in h(v)} \kappa, \tag{2}$$

$$\Im(h(v)) = \frac{1}{l(h(v))}\sqrt{\sum_{\kappa_i,\kappa_j\in h} (\kappa_i - \kappa_j)^2}, \tag{3}$$

where $l(h(v))$ is the number of elements in $h(v)$.

Inspired by the idea of score and variance functions, a comparative procedure is obtained as follows:

(i) If $\wp(h_1) > \wp(h_2)$, then $h_1 > h_2$,
(ii) If $\wp(h_1) = \wp(h_2)$, then
 a) If $\Im(h_1) > \Im(h_2)$, then $h_1 < h_2$,
 b) If $\Im(h_1) = \Im(h_2)$, then $h_1 = h_2$.

Definition 4 [15,21]. *Consider a set of HFEs $E = \{h_1, h_2, \ldots, h_n\}$, then, the HF weighted average (HFWA) and HF weighted geometric (HFWG) operators are as follows:*

$$HFWA(h_1, h_2, \ldots, h_n) = \overset{n}{\underset{j=1}{\oplus}} \phi_j h_j = \underset{\kappa_1 \in h_1, \kappa_2 \in h_2, \ldots, \kappa_n \in h_n}{\cup} \left\{ 1 - \prod_{j=1}^{n} (1-\kappa_j)^{\phi_j} \right\}, \quad (4)$$

$$HFWG(h_1, h_2, \ldots, h_n) = \overset{n}{\underset{j=1}{\otimes}} (h_j)^{\phi_j} = \underset{\kappa_1 \in h_1, \kappa_2 \in h_2, \ldots, \kappa_n \in h_n}{\cup} \left\{ \prod_{j=1}^{n} (\kappa_j)^{\phi_j} \right\}. \quad (5)$$

The measure of the discrimination of two HFSs is defined axiomatically with the following axioms similar to the case of FSs and IFSs [71–73].

- It is a nonnegative and symmetric mapping of the two compared sets.
- It becomes zero when the two sets coincide.
- It decreases when the two subsets become "more similar" in some sense.

Definition 5 [74]. *For all $R, S, T \in HFSs(\Omega)$, a function $D : HFS(\Omega) \times HFS(\Omega) \to \mathbb{R}$ is said to be an HF—Discrimination measure if it satisfies*

(J_1). $D(R,S) = D(S,R)$;
(J_2). $D(R,R) = 0$;
(J_3). $\max\{D(R \cup T, S \cup T), D(R \cup T, S \cap T)\} \leq D(R,S)$.

Definition 6. *Consider $R, S \in HFSs(\Omega)$. In the following, Xu and Xia [17] presented a HF-distance measure:*

$$d(R,S) = \frac{1}{n} \sum_{i=1}^{n} \left[\frac{1}{l(h(v_i))} \sum_{j=1}^{l_y} \left| \hbar_R^{\sigma(j)}(v_i) - \hbar_S^{\sigma(j)}(v_i) \right| \right]. \quad (6)$$

4. New Discrimination Measure for HFSs

The divergence measure is one of the important research contents in FS theory and has a broad application background. For HFSs, Xu and Xia [18] firstly pioneered the concept of divergence measure. Mishra et al. [34,35] proposed some entropy and divergence measures for HFSs and applied them to a service quality assessment and green supplier selection problem, respectively. Kobza et al. [70] studied a new axiomatic definition of HF divergence measure and proposed some divergence measures under an HFS context. Recently, Mishra et al. [75] studied an HF divergence measure-based Additive Ratio Assessment approach for a COVID-19 medication problem. To overcome the drawbacks of existing measures, this section introduces a new discrimination measure for HFSs.

In the following, a parametric symmetric discrimination measure between HFSs R and S with $\gamma > 0$ $(\gamma \neq 2)$ is proposed:

$$D(R,S) = \frac{1}{n}\sum_{i=1}^{n}\left(\frac{1}{l(h(v_i))\left(2^{(1-\gamma/2)}-1\right)}\sum_{j=1}^{l(h(v_i))}\left[\left(\frac{\left(h_R^{\sigma(j)}(v_i)\right)^2+\left(h_S^{\sigma(j)}(v_i)\right)^2}{2}\right)^{\gamma/2}-\frac{\left(h_R^{\sigma(j)}(v_i)\right)^\gamma+\left(h_S^{\sigma(j)}(v_i)\right)^\gamma}{2}\right]\right). \quad (7)$$

Theorem 1. *Let $R, S \in HFSs(\Omega)$, then measure $D(R,S); \gamma > 0 (\gamma \neq 2)$, as presented by (7), which is a discrimination measure for HFSs.*

Proof. To prove this theorem, the function $D(R,S); \gamma > 0 (\gamma \neq 2)$ has to satisfy all the postulates of Definition 5. □

(D_1). For any two real numbers $\alpha, \beta \in \mathbb{R}$, the following inequalities assure:

$$\left(\frac{\alpha^2+\beta^2}{2}\right)^{\gamma/2} \geq \frac{\alpha^\gamma+\beta^\gamma}{2}, \text{for}\gamma < 2\text{and}\left(\frac{\alpha^2+\beta^2}{2}\right)^{\gamma/2} \leq \frac{\alpha^\gamma+\beta^\gamma}{2}, \text{for}\gamma > 2. \quad (8)$$

In addition, we have

$$\frac{1}{n\left(2^{(1-\gamma/2)}-1\right)} > 0, \text{for}\gamma < 2\text{and}\frac{1}{n\left(2^{(1-\gamma/2)}-1\right)} < 0, \text{for}\gamma > 2. \quad (9)$$

Let $R, S \in HFSs(\Omega)$, we know that $h_R^{\sigma(j)}, h_S^{\sigma(j)} : \Omega \to [0,1]$, which implies that

$$\left(\frac{\left(h_R^{\sigma(j)}(v_i)\right)^2+\left(h_S^{\sigma(j)}(v_i)\right)^2}{2}\right)^{\gamma/2} \geq \frac{\left(h_R^{\sigma(j)}(v_i)\right)^\gamma+\left(h_S^{\sigma(j)}(v_i)\right)^\gamma}{2}, \text{for}\gamma < 2 \quad (10)$$

and

$$\left(\frac{\left(h_R^{\sigma(j)}(v_i)\right)^2+\left(h_S^{\sigma(j)}(v_i)\right)^2}{2}\right)^{\gamma/2} \geq \frac{\left(h_R^{\sigma(j)}(v_i)\right)^\gamma+\left(h_S^{\sigma(j)}(v_i)\right)^\gamma}{2}, \text{for}\gamma > 2. \quad (11)$$

Therefore, for each $v_i \in \Omega$, we combine Equations (9)–(11), and therefore, we have

$$\frac{1}{n}\sum_{i=1}^{n}\left(\frac{1}{l(h(v_i))\left(2^{(1-\gamma/2)}-1\right)}\sum_{j=1}^{l(h(v_i))}\left[\left(\frac{\left(h_R^{\sigma(j)}(v_i)\right)^2+\left(h_S^{\sigma(j)}(v_i)\right)^2}{2}\right)^{\gamma/2}-\frac{\left(h_R^{\sigma(j)}(v_i)\right)^\gamma+\left(h_S^{\sigma(j)}(v_i)\right)^\gamma}{2}\right]\right) \geq 0. \quad (12)$$

From Equation (7), we obtain $D(R,S) \geq 0$.

(D_2). It is obvious from Definition 5.

(D_3). Suppose that $R = S$, therefore, $h_R^{\sigma(j)}(v_i) = h_S^{\sigma(j)}(v_i)$, for each $i = 1,2,\ldots,n, j = 1,2,\ldots,l(h(v_i))$. Then, Equation (7) becomes

$$D(R,S) = \frac{1}{n}\sum_{i=1}^{n}\left(\frac{1}{l(h(v_i))\left(2^{(1-\gamma/2)}-1\right)}\sum_{j=1}^{l(h(v_i))}\left[\left(\frac{\left(h_R^{\sigma(j)}(v_i)\right)^2+\left(h_S^{\sigma(j)}(v_i)\right)^2}{2}\right)^{\gamma/2}-\frac{\left(h_R^{\sigma(j)}(v_i)\right)^\gamma+\left(h_S^{\sigma(j)}(v_i)\right)^\gamma}{2}\right]\right) = 0.$$

Conversely, assume $D(R,S) = 0$,

$$\Leftrightarrow \frac{1}{n}\sum_{i=1}^{n}\left(\frac{1}{l(h(v_i))(2^{(1-\gamma/2)}-1)}\sum_{j=1}^{l(h(v_i))}\left[\left(\frac{\left(h_R^{\sigma(j)}(v_i)\right)^2+\left(h_S^{\sigma(j)}(v_i)\right)^2}{2}\right)^{\gamma/2}-\frac{\left(h_R^{\sigma(j)}(v_i)\right)^\gamma+\left(h_S^{\sigma(j)}(v_i)\right)^\gamma}{2}\right]\right)=0,$$

$$\Leftrightarrow \left(\frac{\left(h_R^{\sigma(j)}(v_i)\right)^2+\left(h_S^{\sigma(j)}(v_i)\right)^2}{2}\right)^{\gamma/2}-\frac{\left(h_R^{\sigma(j)}(v_i)\right)^\gamma+\left(h_S^{\sigma(j)}(v_i)\right)^\gamma}{2}=0.$$

This is possible if and only if $h_R^{\sigma(j)}(v_i)=h_S^{\sigma(j)}(v_i), i=1,2,\ldots,n, j=1,2,\ldots,l(h(v_i))$. This proves that $R=S$.

Theorem 2. *Let $R,S,T\in HFSs(\Omega)$. Then, the discrimination measure $D(R,S);\gamma>0(\gamma\neq 2)$, given by (6), holds the following postulates:*

(P1). $D(R,S)\leq D(R,T)$ and $D(S,T)\leq D(R,T)$, for $R\subseteq S\subseteq T$;

Proof. Since $R\subseteq S\subseteq T$, therefore, $h_R^{\sigma(j)}(v_i)\leq h_S^{\sigma(j)}(v_i)\leq h_T^{\sigma(j)}(v_i)$, and thus, from Equation (7), we obtain. □

$$D(R,S)=\frac{1}{n}\sum_{i=1}^{n}\left(\frac{1}{l(h(v_i))(2^{(1-\gamma/2)}-1)}\sum_{j=1}^{l(h(v_i))}\left[\left(\frac{\left(h_R^{\sigma(j)}(v_i)\right)^2+\left(h_S^{\sigma(j)}(v_i)\right)^2}{2}\right)^{\gamma/2}-\frac{\left(h_R^{\sigma(j)}(v_i)\right)^\gamma+\left(h_S^{\sigma(j)}(v_i)\right)^\gamma}{2}\right]\right)$$

$$\leq \frac{1}{n}\sum_{i=1}^{n}\left(\frac{1}{l(h(v_i))(2^{(1-\gamma/2)}-1)}\sum_{j=1}^{l(h(v_i))}\left[\left(\frac{\left(h_R^{\sigma(j)}(v_i)\right)^2+\left(h_T^{\sigma(j)}(v_i)\right)^2}{2}\right)^{\gamma/2}-\frac{\left(h_R^{\sigma(j)}(v_i)\right)^\gamma+\left(h_T^{\sigma(j)}(v_i)\right)^\gamma}{2}\right]\right)=D(R,T).$$

Thus, $D(R,S)\leq D(R,T),\forall T\in HFS(\Omega)$. In the same way, we can prove $D(S,T)\leq D(R,T),\forall T\in HFS(\Omega)$.

(P2). $D(R\cup S,R\cap S)=D(R,S)$;

(P3). $D(R\cup S,T)\leq D(R,T)+D(S,T),\forall T\in HFS(\Omega)$;

(P4). $D(R\cap S,T)\leq D(R,T)+D(S,T),\forall T\in HFS(\Omega)$;

Proof. To verify (P2)–(P4), we partition the universe of discourse Ω into two disjoint sets Ω_1 and Ω_2, where $\Omega_1=\{v_i|v_i\in\Omega,R(v_i)\subseteq S(v_i)\}$ and $\Omega_2=\{v_i|v_i\in\Omega,S(v_i)\subseteq R(v_i)\}$. □

$$D(R\cup S,R\cap S)$$

$$=\frac{1}{n}\sum_{i=1}^{n}\left(\frac{1}{l(h(v_i))(2^{(1-\gamma/2)}-1)}\sum_{j=1}^{l(h(v_i))}\left[\left(\frac{\left(h_{R\cup S}^{\sigma(j)}(v_i)\right)^2+\left(h_{R\cap S}^{\sigma(j)}(v_i)\right)^2}{2}\right)^{\gamma/2}-\frac{\left(h_{R\cup S}^{\sigma(j)}(v_i)\right)^\gamma+\left(h_{R\cap S}^{\sigma(j)}(v_i)\right)^\gamma}{2}\right]\right)$$

$$= \frac{1}{n}\sum_{v_i\in\Omega_1}\left(\frac{1}{l(h(v_i))\left(2^{(1-\gamma/2)}-1\right)}\sum_{j=1}^{l(h(v_i))}\left[\left(\frac{\left(h_S^{\sigma(j)}(v_i)\right)^2+\left(h_R^{\sigma(j)}(v_i)\right)^2}{2}\right)^{\gamma/2}-\frac{\left(h_S^{\sigma(j)}(v_i)\right)^\gamma+\left(h_R^{\sigma(j)}(v_i)\right)^\gamma}{2}\right]\right)$$

$$+\frac{1}{n}\sum_{v_i\in\Omega_2}\left(\frac{1}{l(h(v_i))\left(2^{(1-\gamma/2)}-1\right)}\sum_{j=1}^{l(h(v_i))}\left[\left(\frac{\left(h_R^{\sigma(j)}(v_i)\right)^2+\left(h_S^{\sigma(j)}(v_i)\right)^2}{2}\right)^{\gamma/2}-\frac{\left(h_R^{\sigma(j)}(v_i)\right)^\gamma+\left(h_S^{\sigma(j)}(v_i)\right)^\gamma}{2}\right]\right).$$

This implies that $D(R\cup S, R\cap S) = D(R,S)$. Hence, (P2) is proved.
Again, from Equation (7), for each $T \in HFS(\Omega)$, we get

$$D(R\cup S, T)$$

$$= \frac{1}{n}\sum_{i=1}^{n}\left(\frac{1}{l(h(v_i))\left(2^{(1-\gamma/2)}-1\right)}\sum_{j=1}^{l(h(v_i))}\left[\left(\frac{\left(h_{R\cup S}^{\sigma(j)}(v_i)\right)^2+\left(h_T^{\sigma(j)}(v_i)\right)^2}{2}\right)^{\gamma/2}-\frac{\left(h_{R\cup S}^{\sigma(j)}(v_i)\right)^\gamma+\left(h_T^{\sigma(j)}(v_i)\right)^\gamma}{2}\right]\right)$$

$$= \frac{1}{n}\sum_{v_i\in\Omega_1}\left(\frac{1}{l(h(v_i))\left(2^{(1-\gamma/2)}-1\right)}\sum_{j=1}^{l(h(v_i))}\left[\left(\frac{\left(h_S^{\sigma(j)}(v_i)\right)^2+\left(h_T^{\sigma(j)}(v_i)\right)^2}{2}\right)^{\gamma/2}-\frac{\left(h_S^{\sigma(j)}(v_i)\right)^\gamma+\left(h_T^{\sigma(j)}(v_i)\right)^\gamma}{2}\right]\right)$$

$$+\frac{1}{n}\sum_{y_i\in Y_2}\left(\frac{1}{l(h(v_i))\left(2^{(1-\gamma/2)}-1\right)}\sum_{j=1}^{l(h(v_i))}\left[\left(\frac{\left(h_R^{\sigma(j)}(v_i)\right)^2+\left(h_T^{\sigma(j)}(v_i)\right)^2}{2}\right)^{\gamma/2}-\frac{\left(h_R^{\sigma(j)}(v_i)\right)^\gamma+\left(h_T^{\sigma(j)}(v_i)\right)^\gamma}{2}\right]\right).$$

It implies that $D(R\cup S, T) \le D(R,T) + D(S,T), \forall T \in HFS(\Omega)$. Hence, the property (P3) is proved.
Similarly, we can prove the property (P4).
(P5). $D(R\cap T, S\cap T) \le D(R,S), \forall T \in HFS(\Omega)$;
(P6). $D(R\cup T, S\cup T) \le D(R,S), \forall T \in HFS(\Omega)$.

Proof. To prove (P5) and (P6), the universal set Ω is partitioned into the following eight subsets:

$$\Omega = \{v_i\in\Omega | R(v_i)\le S(v_i) = T(v_i)\}\cup\{v_i\in\Omega|R(v_i) = T(v_i)\le S(v_i)\}$$
$$\cup\{v_i\in\Omega|R(v_i)\le S(v_i) < T(v_i)\}\cup\{v_i\in\Omega|R(v_i)\le T(v_i) < S(v_i)\}$$
$$\cup\{v_i\in\Omega|S(v_i) < R(v_i)\le T(v_i)\}\cup\{v_i\in\Omega|S(v_i)\le T(v_i) < R(v_i)\}$$
$$\cup\{v_i\in\Omega|T(v_i) < R(v_i)\le S(v_i)\}\cup\{v_i\in\Omega|T(v_i) < S(v_i) < R(v_i)\},$$

which are denoted by $\Delta_1, \Delta_2, \ldots, \Delta_8$. According to Mishra et al. [34,35], for every $\Delta_j; j = 1(1)8$,

$$\left(\frac{\left(h_R^{\sigma(j)}(v_i)\right)^2+\left(h_S^{\sigma(j)}(v_i)\right)^2}{2}\right)^{\gamma/2} \ge \frac{\left(h_R^{\sigma(j)}(v_i)\right)^\gamma+\left(h_S^{\sigma(j)}(v_i)\right)^\gamma}{2}, \text{for}\gamma < 2 \quad (13)$$

$$\left(\frac{\left(h_R^{\sigma(j)}(v_i)\right)^2 + \left(h_S^{\sigma(j)}(v_i)\right)^2}{2}\right)^{\gamma/2} \geq \frac{\left(h_R^{\sigma(j)}(v_i)\right)^\gamma + \left(h_S^{\sigma(j)}(v_i)\right)^\gamma}{2}, \text{for} \gamma > 2. \quad (14)$$

□

Therefore, from (P1), we obtain
$D(R \cap T, S \cap T) \leq D(R, S)$ and $D(R \cup T, S \cup T) \leq D(R, S), \forall T \in HFS(\Omega)$.
Since measure $D(R, S)$ holds all the essential postulates of Definition 5, hence, $D(R, S)$ is a valid discrimination measure for HFSs.

5. Hesitant Fuzzy Combined Compromise Solution (HF-CoCoSo) Approach

In the current section, we offer an extended decision-making model, named as the hesitant fuzzy CoCoSo (HF-CoCoSo) approach, for dealing with the MCDM problems under an HF environment. For this purpose, we combine the CoCoSo approach [76–78] with the notions and operations of the HFNs, score function, and proposed discrimination measure within an HFS setting. The calculation procedure of the developed approach is presented as follows (see Figure 1):

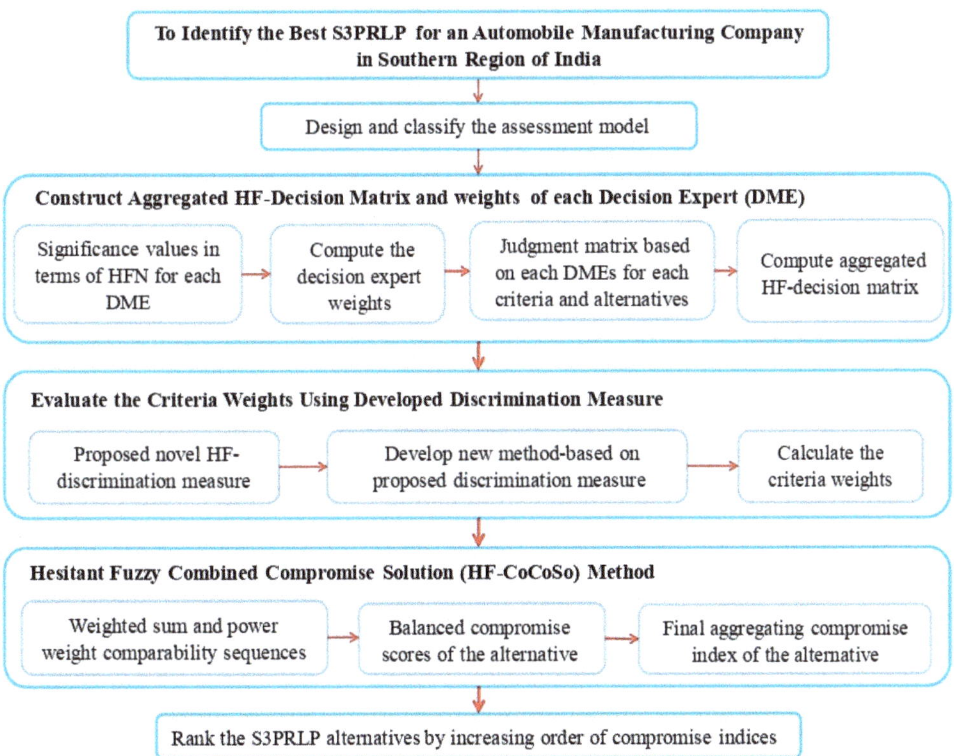

Figure 1. Schematic diagram of proposed hesitant fuzzy Combined Compromise Solution (HF-CoCoSo) method.

Step 1: Problem description.

Consider a team of ℓ DMEs $\{P_1, P_2, \ldots, P_\ell\}$ to determine the best choice among a set of 'm' alternatives $\{M_1, M_2, \ldots, M_m\}$ over a criteria set $\{H_1, H_2, \ldots, H_n\}$. Create a decision matrix $Z = (z_{ij})$ based on linguistic terms for assessing the alternatives over the criteria, where $i = 1, 2, \ldots, m, j = 1, 2, \ldots, n$. Convert the linguistic matrix into an HF-decision matrix (HF-DM) $\Pi = (x_{ij}), i = 1, 2, \ldots, m, j = 1, 2, \ldots, n$.

Step 2: Determine crisp DMEs' weights.

For the evaluation of kth DME, let $h_k(v) = \{\kappa : \kappa \in h_k(v)\}$ be the hesitant fuzzy number (HFN). Then, the weight of the kth DME is obtained by using the expression

$$\lambda_k = \frac{\sum_{\kappa_a \in h} \kappa_a}{\sum\limits_{k=1}^{\ell} \left(\sum_{\kappa_a \in h} \kappa_a\right)}, k = 1, 2, \ldots, \ell, a = 1, 2, \ldots, l(h(v)). \tag{15}$$

Clearly, $\lambda_k \geq 0$ and $\sum\limits_{k=1}^{\ell} \lambda_k = 1$.

Step 3: Create the aggregated HF-DM.

Here, we combine all the individuals' matrices $\Pi_k = \left(x_{ij}^k\right); k = 1, 2, \ldots, l(h(v))$ and create the aggregated HF-DM $M = (y_{ij}),\ i = 1, 2, \ldots, m,\ j = 1, 2, \ldots, n$, where $M = \bigoplus\limits_{k=1}^{\ell} \lambda_k M_k$, and

$$y_{ij} = \bigcup_{\kappa_1 \in y_{ij}^k, \kappa_2 \in h_{ij}^k, \ldots, \kappa_a \in h_{ij}^k} \left\{1 - \prod_{k=1}^{\ell} \left(1 - y_{ij}^k\right)^{\lambda_k}\right\}. \tag{16}$$

Step 4: Determine the criteria weights.

In the following, we present a combined criteria weight-determining method based on objective and subjective weighting methods.

Step 4.1: To estimate the objective criteria weights, we execute the following process by employing the proposed HF-discrimination measure (7):

$$\varpi_j = \frac{\sum\limits_{i=1}^{m} \left(\frac{1}{m-1} \sum\limits_{k=1}^{m} D\left(y_{ij}, y_{kj}\right)\right)}{\sum\limits_{j=1}^{n} \sum\limits_{i=1}^{m} \left(\frac{1}{m-1} \sum\limits_{k=1}^{m} D\left(y_{ij}, y_{kj}\right)\right)}, j = 1, 2, \ldots, n. \tag{17}$$

Step 4.2: Calculate the subjective criteria weights.

To compute the subjective criteria weight, first, we compute the importance degree (κ_j) of each criterion H_j. For this, create the individual importance degree matrix $\left(\eta_j^k\right)$ for kth expert by using the process

$$\eta_j^k = \begin{bmatrix} \eta_1^k \\ \eta_2^k \\ \vdots \\ \eta_n^k \end{bmatrix}_{1 \times n}, \tag{18}$$

where η_j^k is the importance degree of criterion H_j given by kth expert, $j = 1, 2, \ldots, n$, $k = 1, 2, \ldots, \ell$. To estimate the subjective weight, we get

$$\vartheta_j = HFWA_\lambda\left(\eta_j^1, \eta_j^2, \ldots, \eta_j^\ell\right). \tag{19}$$

Let $\eta_j^k = \left(h_{ij}^k\right)$ be the decision importance, where $h_{ij}^k = \left\{\kappa_k : \kappa_k \in h_{ij}^k\right\}, k = 1, 2, \ldots, \ell$ is an HFN. Then, we find the overall importance degree as follows:

$$\vartheta_j = (h_{ij})^* = \bigcup_{\kappa_1 \in h_{i1}^k, \kappa_2 \in h_{i2}^k, \ldots, \kappa_n \in h_{in}^k} \left\{1 - \prod_{k=1}^{\ell} (1 - \kappa_k)^{\lambda_k}\right\}. \tag{20}$$

Here, $\kappa_j = (h_{ij})^*$ is an HFN.

Step 4.3: Evaluate the overall importance degree (η_j) of criteria H_j.

Next, we calculate the score value $s(\vartheta_j)$ by employing Equation (2). Thus, we calculate the importance degree as follows:

$$\eta_j = \frac{s(\vartheta_j)}{\sum_{j=1}^{n} s(\vartheta_j)}; j = 1, 2, \ldots, n. \tag{21}$$

Step 4.4: Determine the final weight of criteria.

With the use of using Equations (16)–(20), the formula for the computation of combined weight is

$$w_j = \tau \varpi_j + (1-\tau)\eta_j, \tag{22}$$

where ϖ_j represents the objective weight obtained by the proposed discrimination measure of evaluation of the jth criterion, η_j represents the subjective weight of the jth criterion expressed by DEs, $\tau(0 \le \tau \le 1)$ represents the adjustment coefficient, and its value can be chosen along with the actual situation of the group decision and $0 \le w_j \le 1$.

Step 5: Construct the normalized aggregated HF-DM.

To create the normalized aggregated HF-DM $\mathbb{R} = (\varsigma_{ij})_{m \times n}$, we have utilized the following expression:

$$\varsigma_{ij} = \begin{cases} y_{ij}, & \text{for benefit criterion} \\ (y_{ij})^c = 1 - y_{ij}, & \text{for cost criterion.} \end{cases} \tag{23}$$

Step 6: Determine the weighted sum and power weight comparability sequences.

In the following, we present the weighted sum measure (WSM) (or sum of the weighted comparability sequence) $\Lambda_i^{(1)}$ for each candidate:

$$\Lambda_i^{(1)} = \overset{n}{\underset{j=1}{\oplus}} w_j \varsigma_{ij}. \tag{24}$$

Next, we present the weighted product measure (WPM) (or amount of the power weight of comparability sequence) $\Lambda_i^{(2)}$ for each candidate, which is given as

$$\Lambda_i^{(2)} = \overset{n}{\underset{j=1}{\otimes}} w_j \varsigma_{ij}. \tag{25}$$

Step 7: Computation of relative weights or balanced compromise scores.

Next, three appraisal score strategies are presented to formulate the relative weights of the alternative, which are derived as follows:

$$C_i^{(1)} = \frac{s\left(\Lambda_i^{(1)}\right) + s\left(\Lambda_i^{(2)}\right)}{\sum_{i=1}^{m}\left(s\left(\Lambda_i^{(1)}\right) + s\left(\Lambda_i^{(2)}\right)\right)}, \tag{26}$$

$$C_i^{(2)} = \frac{s\left(\Lambda_i^{(1)}\right)}{\min_i s\left(\Lambda_i^{(1)}\right)} + \frac{s\left(\Lambda_i^{(2)}\right)}{\min_i s\left(\Lambda_i^{(2)}\right)}, \tag{27}$$

$$C_i^{(3)} = \frac{\gamma s\left(\Lambda_i^{(1)}\right) + (1-\gamma) s\left(\Lambda_i^{(2)}\right)}{\gamma \max_i s\left(\Lambda_i^{(1)}\right) + (1-\gamma) \max_i s\left(\Lambda_i^{(2)}\right)}, \tag{28}$$

where $\gamma \in [0,1]$ is compromise decision mechanism coefficient. Generally, we take $\gamma = 0.5$. Here, Equation (26) articulates the arithmetic mean of sums of WSM and WPM scores, whilst Equation (27) articulates a sum of relative scores of WSM and WPM compared to the best. Equation (28) expresses the balanced compromise of WSM and WPM models scores.

Step 8: Estimate the final aggregating compromise index.

The final aggregating compromise index C_i is computed as

$$C_i = \left(C_i^{(1)} C_i^{(2)} C_i^{(3)}\right)^{\frac{1}{3}} + \frac{1}{3}\left(C_i^{(1)} + C_i^{(2)} + C_i^{(3)}\right). \quad (29)$$

Lastly, prioritize the candidates according as the increasing values of aggregating compromise index C_i.

Step 9: End.

6. Case Study of S3PRLP Selection

To show the application of the introduced MCDM framework, an empirical case study of 3PRLP selection is carried out for an automobile manufacturing company (XYZ). This company is located in the southern region of India that endorses economic and social development. Its manufacturing chain is long and has a great level of relevance to other businesses. This automobile company is effectual for promoting employment and motivating consumption, and it acts as a momentous part in the social and economic growth of the country. Simultaneously, it entails large-scale production escorted by a huge flow of raw materials, parts, finished products, and scrap materials. Reverse logistics is vastly associated to scrap waste and flawed products related to the manufacturing process. As a result of the speedy growth of the automotive businesses, the general public has paid great interest to the automobile reverse logistics. The management of recollected defective and scrapped cars is strongly associated with the individual security and ecological protection, and consequently, study on the development of automobile reverse logistics is a significant process to fulfill the customers' demands, enhance the corporate competitiveness, conserve the natural resources, and protect the atmosphere.

The company XYZ currently has a well-built marketing system and a trustworthy reverse logistics network. The company lacks material resources, manpower, experience, and capital in logistics practices, particularly reverse logistics practices, and this has motivated the company to outsource its reverse logistics business to a proficient 3PRLP for reducing costs and enhancing its core competitiveness. The company executives are facing some difficulties in choosing the 3PRLP and in managing successful reverse logistics practices in a green and sustainable supply chain during the network. The improper reverse logistics practices may result in environmental pollution, global warming, etc. In order to choose a suitable 3PRLP option, the company XYZ posted a tender notice to invite 3PRLPs to participate in the reverse logistics services. Accordingly, lots of potential alternatives are contributed in the tender. To assess the candidate 3PRLPs appropriately, the company formed a team of three professionals (P_1, P_2, P_3). After establishing the expert team, six candidate providers (M_1, M_2, M_3, M_4, M_5, M_6) are identified for a further evaluation process. These alternatives are evaluated based on the following 13 criteria: Green warehousing (H_1), Pollution control cost (H_2), Green product and eco-design cost (H_3), RL cost (H_4), Green R&D and innovation (H_5), Air emissions (H_6), Environmental management system (H_7), Flexibility (H_8), Quality (H_9), Financial risk (H_{10}), Health and safety practices (H_{11}), Social responsibility (H_{12}), and Employment Practices (H_{13}). In this example, the criteria S_2, S_3, S_4, S_6, and S_{10} are of cost type, and the rest all are of the benefit type. Next, the implementation procedures of the proposed HF-CoCoSo model for assessing the S3PRLPs are expressed as follows.

Tables 1 and 2 present the linguistic terms (LTs) and their consequent HFEs for the rating of relative significance of criteria and 3PRLP candidate assessment to the given decision-making problem, respectively. As a result of the qualitative [34] and quantitative [79] nature of evaluation factors, imprecise information, and lack of time, it is easy for the DMEs to articulate their opinions based on linguistic values (LVs) [33].

Table 1. Linguistic ratings for the importance of criteria and experts.

LVs	HFEs	DMEs Risk Preferences		
		Pessimist	Moderate	Optimist
Very high (VH)	[0.85, 1.00]	0.85	0.925	1.00
High (H)	[0.70, 0.85]	0.70	0.775	0.85
Medium (M)	[0.55, 0.70]	0.55	0.625	0.70
Low (L)	[0.40, 0.55]	0.40	0.475	0.55
Very low (VL)	[0.25, 0.40]	0.25	0.325	0.40

Table 2. Linguistic variables for the importance of criteria and alternatives.

LVs	HFEs	DMEs Risk Preferences		
		Pessimist	Moderate	Optimist
Extremely preferable (EP)	[0.90, 1.00]	0.90	0.95	1.00
Strong preferable (SP)	[0.75, 0.90]	0.75	0.825	0.90
Preferable (P)	[0.60, 0.75]	0.60	0.675	0.75
Moderately preferable (MP)	[0.50, 0.60]	0.50	0.55	0.60
Moderate (M)	[0.40, 0.50]	0.40	0.45	0.50
Moderately undesirable (MU)	[0.30, 0.40]	0.30	0.35	0.40
Undesirable (U)	[0.20, 0.30]	0.20	0.25	0.30
Strong undesirable (SU)	[0.10, 0.20]	0.10	0.15	0.20
Extremely undesirable (EU)	[0.00, 0.10]	0.00	0.05	0.10

Suppose the DMEs' weights are given by LVs, which as {H, VH, M}. Consequently, with the use of Table 2, Equations (6) and (15), the crisp weights of the DMEs are obtained as $\{\lambda_1 = 0.3333, \lambda_2 = 0.2689, \lambda_3 = 0.3978\}$. The linguistic decision matrix is presented in Table 3.

Table 3. Linguistic decision matrix given by different experts for sustainable third-party reverse logistics provider (S3PRLP) evaluation.

Criteria	DMEs	M_1	M_2	M_3	M_4	M_5	M_6
	P_1	MP	M	MU	MP	M	MP
H_1	P_2	M	MU	P	MU	P	SU
	P_3	P	MP	M	P	MU	MU
	P_1	MP	MU	MP	MP	P	U
H_2	P_2	SP	MP	P	M	MU	MP
	P_3	P	U	MU	P	P	U
	P_1	U	M	M	P	P	M
H_3	P_2	MP	SP	MU	M	MU	P
	P_3	M	P	M	P	M	MU
	P_1	M	MU	U	M	P	M
H_4	P_2	M	U	P	MU	U	P
	P_3	MP	MU	U	U	MP	M
	P_1	M	M	M	MP	MP	M
H_5	P_2	SP	M	MU	M	P	MP
	P_3	M	SU	M	SU	MU	M
	P_1	MP	P	M	MP	SP	SP
H_6	P_2	M	MP	MU	U	MU	U
	P_3	P	M	SP	P	U	MP
	P_1	MU	P	MP	M	SU	P
H_7	P_2	MP	M	SP	SU	MP	M
	P_3	P	SP	P	M	M	SU
	P_1	SP	EP	MP	P	M	M
H_8	P_2	M	P	P	MP	P	P
	P_3	P	MP	P	M	SP	P

Table 3. *Cont.*

Criteria	DMEs	M_1	M_2	M_3	M_4	M_5	M_6
	P_1	MP	EP	SP	M	MP	SP
H_9	P_2	SP	MP	MP	MU	M	M
	P_3	P	MU	U	M	M	M
	P_1	MU	MP	P	MU	P	SP
H_{10}	P_2	MP	M	M	MP	U	M
	P_3	M	P	MU	P	MU	MU
	P_1	MP	P	M	SP	MU	P
H_{11}	P_2	P	MP	MP	MP	P	MP
	P_3	SP	P	MU	M	MU	SP
	P_1	U	P	M	P	P	M
H_{12}	P_2	SP	SP	SP	M	MP	U
	P_3	M	MU	P	MP	M	MP
	P_1	MU	SP	MU	P	MP	MP
H_{13}	P_2	SP	MP	P	MU	P	MU
	P_3	P	M	SP	SU	M	M

With the use of Equation (16), we combine the individuals' opinions given in Table 3 and then create an aggregated HF-DM in Table 4.

Table 4. Aggregated HF-decision matrix (HF-DM) for 3PRLP candidates over criteria.

	M_1	M_2	M_3	M_4	M_5	M_6
H_1	0.584	0.461	0.497	0.499	0.476	0.401
H_2	0.680	0.424	0.543	0.564	0.557	0.396
H_3	0.472	0.641	0.461	0.581	0.484	0.476
H_4	0.527	0.372	0.470	0.384	0.524	0.524
H_5	0.581	0.368	0.455	0.425	0.539	0.523
H_6	0.548	0.571	0.579	0.527	0.526	0.575
H_7	0.527	0.652	0.665	0.423	0.439	0.472
H_8	0.660	0.740	0.663	0.555	0.652	0.581
H_9	0.665	0.699	0.581	0.424	0.505	0.579
H_{10}	0.444	0.603	0.507	0.539	0.449	0.552
H_{11}	0.686	0.621	0.488	0.619	0.466	0.675
H_{12}	0.559	0.609	0.653	0.561	0.539	0.470
H_{13}	0.651	0.606	0.632	0.416	0.555	0.477

In order to determine the significance degrees of considered criteria, firstly, we have calculated the objective weights, which are given as ϖ_j = (0.0479, 0.1329, 0.0630, 0.0808, 0.0918, 0.0082, 0.1315, 0.0370, 0.1137, 0.0493, 0.0973, 0.0438, and 0.1027).

Based on the above-discussed objective weights (ϖ_j) and the subjective weights given by Table 5, the combined weights of the criteria are estimated as below:

Table 5. Linguistic values (LVs) for criteria performance.

Criteria	LVs Given by DMEs			HFNs Given by DMEs			$s(\kappa_j)$	η_j
	P_1	P_2	P_3	P_1	P_2	P_3		
H_1	MU	MP	P	0.35	0.55	0.675	0.575	0.0769
H_2	MP	M	P	0.50	0.45	0.75	0.625	0.0835
H_3	U	MP	M	0.25	0.55	0.50	0.472	0.0631
H_4	MP	M	P	0.50	0.45	0.675	0.584	0.0781
H_5	M	MP	M	0.40	0.60	0.50	0.528	0.0706
H_6	MU	P	P	0.35	0.60	0.675	0.592	0.0791
H_7	U	P	SP	0.30	0.60	0.75	0.623	0.0833
H_8	SP	M	SP	0.75	0.40	0.825	0.734	0.0981

Table 5. *Cont.*

Criteria	LVs Given by DMEs			HFNs Given by DMEs			$s(\kappa_j)$	η_j
	P_1	P_2	P_3	P_1	P_2	P_3		
H_9	P	MP	P	0.60	0.55	0.675	0.639	0.0854
H_{10}	U	P	MP	0.30	0.60	0.55	0.524	0.0700
H_{11}	MP	M	SP	0.50	0.45	0.75	0.625	0.0835
H_{12}	MU	MP	M	0.40	0.55	0.50	0.509	0.0680
H_{13}	MP	M	MU	0.50	0.45	0.35	0.452	0.0604

w_j = (0.0624, 0.1082, 0.0630, 0.0794, 0.0812, 0.0437, 0.1074, 0.0675, 0.0996, 0.0597, 0.0904, 0.0559, 0.0815).

As the criteria H_2, H_3, H_4, H_6, and H_{10} are of non-benefit, and the rest all are of the benefit-type criteria; therefore, by using Table 4 and Equation (23), the normalized aggregated HF-DM is depicted in Table 6.

Table 6. Normalized aggregated HF-DM for S3PRLP selection.

	M_1	M_2	M_3	M_4	M_5	M_6
H_1	0.584	0.461	0.497	0.499	0.476	0.401
H_2	0.320	0.576	0.457	0.436	0.443	0.604
H_3	0.528	0.359	0.539	0.419	0.516	0.524
H_4	0.473	0.628	0.530	0.616	0.476	0.476
H_5	0.581	0.368	0.455	0.425	0.539	0.523
H_6	0.452	0.429	0.421	0.473	0.474	0.425
H_7	0.527	0.652	0.665	0.423	0.439	0.472
H_8	0.660	0.740	0.663	0.555	0.652	0.581
H_9	0.665	0.699	0.581	0.424	0.505	0.579
H_{10}	0.556	0.397	0.493	0.461	0.551	0.448
H_{11}	0.686	0.621	0.488	0.619	0.466	0.675
H_{12}	0.559	0.609	0.653	0.561	0.539	0.470
H_{13}	0.651	0.606	0.632	0.416	0.555	0.477

On the basis of Equations (24)–(29), the overall computational steps of the HF-CoCoSo approach has been determined and presented in Table 7. Along with the values of increasing values of aggregating compromise index C_i, the prioritization order of the S3PRLP candidates is $M_2 \succ M_1 \succ M_3 \succ M_6 \succ M_5 \succ M_4$ and thus, the option M_2 is the suitable, sustainable 3PRLP option. Here, Figure 2 shows the variation of compromise degrees with different measures of the proposed HF-CoCoSo method.

Table 7. Overall aggregation and combined compromise ranking results.

Option	$\Lambda_i^{(1)}$	$\Lambda_i^{(2)}$	$C_i^{(1)}$	$C_i^{(2)}$	$C_i^{(3)}$	C_i	Ranking
M_1	0.5676	0.5437	0.1747	2.3004	0.9818	1.8857	2
M_2	0.5811	0.5508	0.1779	2.3428	1.0000	1.9206	1
M_3	0.5551	0.5414	0.1724	2.2701	0.9688	1.8608	3
M_4	0.4888	0.4773	0.1519	2.0000	0.8535	1.6394	6
M_5	0.5079	0.5015	0.1587	2.0899	0.8918	1.7130	5
M_6	0.5292	0.5172	0.1645	2.1663	0.9245	1.7758	4

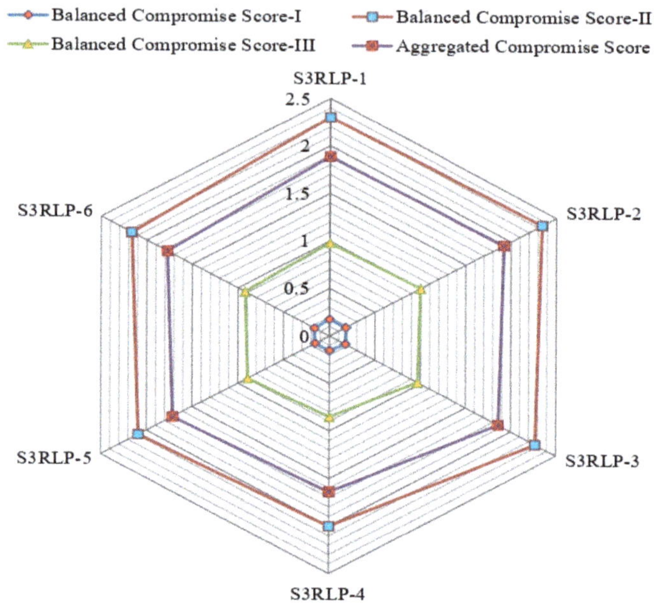

Figure 2. Variation of compromise degrees with different measures of the proposed HF-CoCoSo method.

6.1. Comparative Study

In the current section, we compare the introduced HF-CoCoSo method with some of the existing methods, namely HF-TOPSIS developed by Xu and Zhang [80] and HF-COPRAS presented by Mishra et al. [34].

6.1.1. HF-TOPSIS Method

The structure of the HF-TOPSIS method is as follows:

Step 1–4: Same as HF-CoCoSo methodology.

Step 5: Determine the HF-positive ideal solution (HF-PIS) and HF-negative ideal solution (HF-NIS).

The HF-PIS and HF-NIS are computed by the following expressions (30) and (31):

$$\xi^+ = \begin{cases} \max\limits_{i=1,2,\ldots,m} y_{ij} \text{ for benefit criterion } H_b \\ \min\limits_{i=1,2,\ldots,m} y_{ij} \text{ for cost criterion } H_n \end{cases} \tag{30}$$

$$\xi^- = \begin{cases} \min\limits_{i=1,2,\ldots,m} y_{ij} \text{ for benefit criterion } H_b \\ \max\limits_{i=1,2,\ldots,m} y_{ij} \text{ for cost criterion } H_n, \end{cases} \tag{31}$$

Step 6: Calculation of discrimination measures from HF-PIS and HF-NIS

By means of Equation (7), calculate the weighted discrimination measure $D(y_{ij}, \xi^+)$ among the options $M_i (i = 1, 2, \ldots, m)$ and the HF-PIS ξ^+ and the Discrimination measure $D(y_{ij}, \xi^-)$ among the options $M_i (i = 1, 2, \ldots, m)$ and the HF-AIS ξ^-.

Step 7: Estimation off relative closeness coefficient (CF).

With the use of Equation (30), the relative CC of each candidate is calculated, which is given as:

$$CF(M_i) = \frac{D(y_{ij}, \xi^-)}{D(y_{ij}, \xi^-) + D(y_{ij}, \xi^-)}, i = 1, 2, \ldots, m. \tag{32}$$

Step 8: Selection of optimal alternative

Opt the maximum value, symbolized by $CF(M_k)$, among the values $CF(M_i), i = 1, 2, \ldots, m$. Thus, M_k is the optimal choice.

Step 9: End.

From Table 4 and Equations (30) and (31), HF-PIS and HF-NIS are appraised. Next, the entire procedural steps of HF-TOPSIS [80] framework are described in Table 8.

Table 8. Overall outcomes of HF-Technique for Order of Preference by Similarity to Ideal Solution (HF-TOPSIS) framework.

Option	$D(y_{ij},\xi^{+})$	$D(y_{ij},\xi^{-})$	$CF(M_i)$	Ranking
M_1	0.132	0.133	0.502	3
M_2	0.093	0.145	0.610	1
M_3	0.119	0.104	0.465	4
M_4	0.153	0.083	0.352	6
M_5	0.142	0.090	0.387	5
M_6	0.101	0.137	0.577	2

Thus, the final ordering of the S3PRLP candidates is $M_2 \succ M_6 \succ M_1 \succ M_3 \succ M_5 \succ M_4$. Therefore, the most suitable alternative is M_2.

6.1.2. HF-COPRAS Method

The structure of the HF-COPRAS method is as follows:

Step 1–4: Same as HF-CoCoSo framework.

Step 5: Sum the values of criteria for benefit and cost.

Let $\Delta = \{1, 2, \ldots, k\}$ be a collection of criteria, the greater values of which are finer. Then, determine the following index for each candidate:

$$\delta_i^{(1)} = \bigoplus_{j=1}^{k} w_j y_{ij}, i = 1, 2, \ldots m. \tag{33}$$

Let $\nabla = \{k+1, k+2, \ldots, n\}$ be a set of criteria, the lesser values of which are finer. Then, estimate the following index for each candidate:

$$\delta_i^{(2)} = \bigoplus_{j=k+1}^{n} w_j y_{ij}, i = 1, 2, \ldots, m. \tag{34}$$

Here, k, n, and w_j denote the number of benefit criteria, total number of criteria, and weight of the jth criterion.

Step 6: Determine the relative weight.

The relative weight of ith candidate is computed by

$$\lambda_i = s\left(\delta_i^{(1)}\right) + \frac{\min_i s\left(\delta_i^{(2)}\right) \sum_{i=1}^{m} s\left(\delta_i^{(2)}\right)}{s\left(\delta_i^{(2)}\right) \sum_{i=1}^{m} \frac{\min_i s\left(\delta_i^{(2)}\right)}{s\left(\delta_i^{(2)}\right)}}, i = 1, 2, \ldots, m. \tag{35}$$

where $s\left(\delta_i^{(1)}\right)$ and $s\left(\delta_i^{(2)}\right)$ are the score values of $\delta_i^{(1)}$ and $\delta_i^{(2)}$, respectively.

Step 7: Compute the priority order.

The greater relative weight λ_i of the alternative, the higher the priority of the alternative. In the case of $\lambda_{\max}$, the satisfaction degree is maximum.

Step 8: Evaluate the utility degree.

The utility degree of each alternative is

$$\eta_i = \frac{\lambda_i}{\lambda_{\max}} \times 100\%, i = 1, 2, \ldots, m. \tag{36}$$

Step 9: End.

With the use of Equations (33)–(36), the overall procedures of the HF-COPRAS approach are calculated and presented in Table 9.

Table 9. The overall computational results of the HF-Complex Proportional Assessment (HF-COPRAS) method.

Option	$\delta_i^{(1)}$	$\delta_i^{(2)}$	λ_i	η_i	Ranking
M_1	0.464	0.254	0.670	96.63	2
M_2	0.460	0.224	0.693	100.00	1
M_3	0.435	0.224	0.667	96.30	3
M_4	0.352	0.230	0.578	83.45	6
M_5	0.377	0.227	0.607	87.59	5
M_6	0.391	0.213	0.635	91.70	4

Thus, the ranking of the S3PRLP candidates is $M_2 \succ M_1 \succ M_3 \succ M_6 \succ M_5 \succ M_4$. The ranking reflects that the alternative M_2 is the best alternative. Figure 3 presents the comparison of score values of S3PRLP alternatives with various methods.

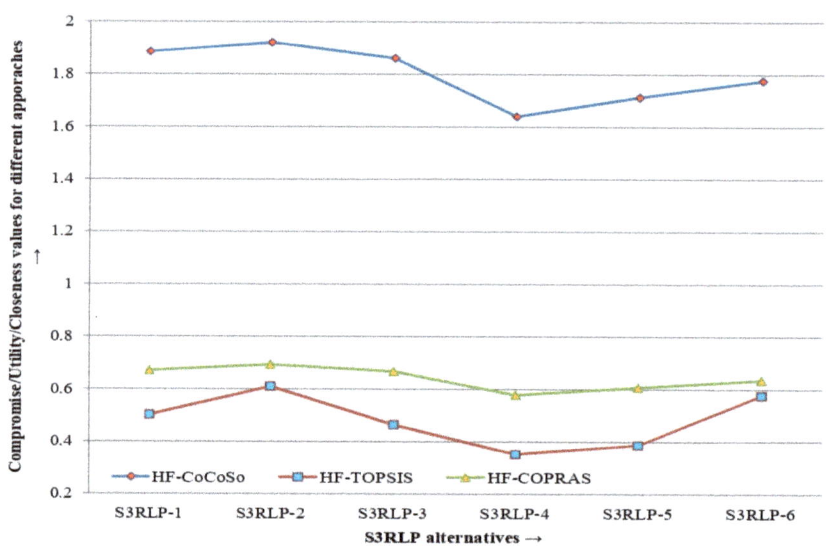

Figure 3. Comparison of score values of S3PRLP options with various methods.

The outcomes of the developed method are given as follows:

- The proposed approach utilizes a comparability sequence, and then, the weights are aggregated through two manners. One of them follows the usual multiplication rule, and the second one narrates the weighted power of the distance from the comparability sequence. To validate the preference order, we have defined three different measures (aggregation strategy) for a given alternative. At the ultimate, a cumulative equation reports a ranking. There is not any algorithm among MCDM tools supporting this kind of aggregation. Each strategy would offer a ranking score, which would be further improved by a complete ranking index. This procedure is based on a combination of compromise attitudes.
- The CoCoSo methodology has the potential to entail the DMEs' estimations into decision-making processes, whilst the COPRAS technique is incapable of doing so.

- The hesitancy degree is considered independently significant in the complete implementation process, and preference ordering of the candidates is obtained using the transaction degrees of all three parameters.
- Assessment of criteria weights is one of the main challenges in the MCDM process. In the developed method, the objective weights, which are determined by new discrimination measure-based formula and the subjective weights, which are expressed by DMEs, are combined, and the aggregated weights are employed in the extended CoCoSo framework.

7. Conclusions

Increasing environmental awareness and the rapidly growing utilization of natural resources have motivated businesses to focus on sustainable SCM. As a result of the restraints related to their technology, own capital, or knowledge, some businesses outsource their RL activities. However, the selection of the correct S3PRLP alternative among a set of alternatives over multiple qualitative and quantitative criteria is an important and uncertain MCDM problem. The objective of the study is to recommend an MCDM methodology for evaluating the S3PRLP assessment problem with HFNs information. To do this, firstly, an innovative HF-CoCoSo approach is introduced by combining the traditional CoCoSo method, HFSs' operators, and discrimination measure under HFSs settings. The weights of the criteria have been assessed by combining a proposed discrimination measure-based objective weighting method and a subjective method given by experts. Then, an illustrative case study of S3PRLP evaluation and selection problem has been discussed within HFS settings, which expresses the applicability and viability of the current method. The effectiveness of the introduced approach has been justified by comparison with extant methods. Finally, sensitivity assessment has also been discussed to show the stability of the proposed approach.

The shortcomings of this study are (i) the number of DMEs engaged in the present case study was small; (ii) the interrelationships between the criteria were not taken into deliberation, and (iii) the BDs of a certain element are not necessarily real numbers; this may bound the application range of the developed technique to some extent. Thus, further study will need to be carried out to enhance the precision of the decision-making selection problem by considering these factors. We can extend our study by incorporating objective and subjective criteria weights under different fuzzy environment. In addition, we will suggest some methods such as GLDS, DNMA, or MULTIMOORA to assess the S3PRLP selection problem. In addition, we will apply the developed methodology to other applications, such as sustainable biomass crop selection, renewable energy technology selection, and others.

Author Contributions: The paper has been read by all authors, and they agree for its submission to the journal. A.R.M., P.R., and R.K. conceptualized the idea for logistic provider selection. Later, K.S.R., E.K.Z., and F.C. gave suggestions on the choice of preference structure, methods, and the criteria for evaluation of logistics providers. A.R.M. and P.R. coded the idea by framing the mathematical base that was fine tuned and improvised by R.K., K.S.R. and E.K.Z. checked the result and F.C. provided valuable suggestions on the presentation of figures and tables. A.R.M., P.R., and R.K. made the initial draft of the manuscript that was fine tuned by K.S.R. and F.C. Language was edited by E.K.Z. and F.C. Symbol clarity and equation readability was checked by K.S.R. All authors have read and agreed to the published version of the manuscript.

Funding: This research received no external funding.

Conflicts of Interest: The authors declare no conflict of interest.

References

1. Chen, Z.S.; Zhang, X.; Govindan, K.; Wang, X.J.; Chin, K.S. Third-party reverse logistics provider selection: A computational semantic analysis-based multi-perspective multi-attribute decision-making approach. *Expert Syst. Appl.* **2020**. [CrossRef]
2. Govindan, K.; Kadzinski, M.; Ehling, R.; Miebs, G. Selection of a sustainable third-party reverse logistics provider based on the robustness analysis of an outranking graph kernel conducted with ELECTRE-I and SMAA. *Omega* **2019**, *85*, 1–15. [CrossRef]

3. Peretti, U.; Tatham, P.; Wu, Y.; Sgarbossa, F. Reverse logistics in humanitarian operations: Challenges and opportunities. *J. Humanit. Logist. Supply Chain Manag.* **2015**, *5*, 253–274. [CrossRef]
4. Zarbakhshnia, N.; Soleimani, H.; Goh, M.; Razavi, S.S. A novel multi-objective model for green forward and reverse logistics network design. *J. Clean. Prod.* **2019**, *208*, 1304–1316. [CrossRef]
5. Razzaque, M.A.; Sheng, C.C. Outsourcing of logistics functions: A literature survey. *Int. J. Phys. Distrib. Logist. Manag.* **1998**, *28*, 89–107. [CrossRef]
6. Zadeh, L.A. Fuzzy sets. *Inf. Control* **1965**, *8*, 338–353. [CrossRef]
7. Torra, V. Hesitant fuzzy sets. *Int. J. Intell. Syst.* **2010**, *25*, 529–539. [CrossRef]
8. Torra, V.; Narukawa, Y. On hesitant fuzzy sets and decision. In Proceedings of the 18th IEEE International Conference on Fuzzy Systems, Jeju Island, Korea, 20–24 August 2009.
9. Atanassov, K.T. Intuitionistic fuzzy sets. *Fuzzy Sets Syst.* **1986**, *20*, 87–96. [CrossRef]
10. Zadeh, L.A. The concept of a linguistic variable and its application to approximate reasoning Part-I. *Inf. Sci.* **1975**, *8*, 199–249. [CrossRef]
11. Yager, R.R. On the theory of bags. *Int. J. Gen. Syst.* **1986**, *13*, 23–37. [CrossRef]
12. Peng, X.; Krishankumar, R.; Ravichandran, K.S. A novel interval-valued fuzzy soft decision-making method based on CoCoSo and CRITIC for intelligent healthcare management evaluation. *Soft Comput.* **2021**. [CrossRef]
13. Wen, Z.; Liao, H.; Zavadskas, E.K.; Al-Barakati, A. Selection third-party logistics service providers in supply chain finance by a hesitant fuzzy linguistic combined compromise solution method. *Econ. Res. Ekon. Istraživanja* **2019**, *32*, 4033–4058. [CrossRef]
14. Yazdani, M.; Zarate, P.; Zavadskas, E.K.; Turskis, Z. A combined compromise solution (CoCoSo) method for multi-criteria decision-making problems. *Manag. Decis.* **2019**, *57*, 2501–2519. [CrossRef]
15. Xia, M.; Xu, Z.S. Hesitant fuzzy information aggregation in decision making. *Int. J. Approx. Reason.* **2011**, *52*, 395–407. [CrossRef]
16. Xia, M.; Xu, Z.S.; Chen, N. Some hesitant fuzzy aggregation operators with their application in group decision making. *Group Decis. Negot.* **2013**, *22*, 259–279. [CrossRef]
17. Xu, Z.S.; Xia, M. Distance and similarity measures for hesitant fuzzy sets. *Inf. Sci.* **2011**, *181*, 2128–2138. [CrossRef]
18. Xu, Z.S.; Xia, M. Hesitant fuzzy entropy and cross entropy and their use in multi attribute decision-making. *Int. J. Intell. Syst.* **2012**, *27*, 799–822. [CrossRef]
19. He, Y.; He, Z.; Wang, G.; Chen, H. Hesitant fuzzy power bonferroni means and their application to multiple attribute decision making. *IEEE Trans. Fuzzy Syst.* **2015**, *23*, 1655–1668. [CrossRef]
20. Sun, M. Normalized geometric bonferroni operators of hesitant fuzzy sets and their application in multiple attribute decision making. *J. Inf. Comput. Sci.* **2013**, *10*, 2815–2822. [CrossRef]
21. Liao, H.; Xu, Z.S.; Zeng, X.J. Novel correlation coefficients between hesitant fuzzy sets and their application in decision making. *Knowl. Based Syst.* **2015**, *82*, 115–127. [CrossRef]
22. Li, D.Q.; Zeng, W.Y.; Li, J.H. New distance and similarity measures on hesitant fuzzy sets and their applications in multiple criteria decision making. *Eng. Appl. Artif. Intell.* **2015**, *40*, 11–16. [CrossRef]
23. Hu, J.; Zhang, X.; Chen, X.; Liu, Y. Hesitant fuzzy information measures and their applications in multi-criteria decision making. *Int. J. Syst. Sci.* **2016**, *47*, 62–76. [CrossRef]
24. Cuiping, W.; Zhiliang, R.; Xijin, T.; Jiangang, F. Information aggregation operators based on hesitant fuzzy sets and prioritization relationship. *J. Intell. Fuzzy Syst.* **2016**, *31*, 765–774.
25. Yu, D. Hesitant fuzzy multi-criteria decision making methods based on Heronian mean. *Technol. Econ. Dev. Econ.* **2015**, *23*, 296–315. [CrossRef]
26. Lv, J.H.; Guo, S.C.; Guo, F.F. Study on Hesitant Fuzzy Information Measures and Their Clustering Application. *Comput. Intell. Neurosci.* **2019**, *2019*, 5370763. [CrossRef]
27. Wang, F. Aggregation similarity measure based on hesitant fuzzy closeness degree and its application to clustering analysis. *J. Syst. Sci. Inf.* **2019**, *7*, 70–89. [CrossRef]
28. Lan, J.; Jin, R.; Zheng, Z.; Hu, M. Priority degrees for hesitant fuzzy sets: Application to multiple attribute decision making. *Oper. Res. Perspect.* **2017**, *4*, 67–73. [CrossRef]
29. Mousavi, M.; Gitinavard, H.; Mousavi, S.M. A soft computing based modified ELECTRE model for renewable energy policy selection with unknown information. *Renew. Sustain. Energy Rev.* **2017**, *68*, 774–787. [CrossRef]
30. Wu, S.M.; Liu, H.C.; Wang, L.E. Hesitant fuzzy integrated MCDM approach for quality function deployment: A case study in electric vehicle. *Int. J. Prod. Res.* **2017**, *55*, 4436–4449. [CrossRef]
31. Alcantud, J.C.R.; Torra, V. Decomposition theorems and extension principles for hesitant fuzzy sets. *Inf. Fusion* **2018**, *41*, 48–56. [CrossRef]
32. Cheng, S.H. Autocratic decision making using group recommendations based on hesitant fuzzy sets for green hotels selection and bidders selection. *Inf. Sci.* **2018**, *467*, 604–617. [CrossRef]
33. Liu, X.D.; Wang, Z.W.; Zhang, S.T.; Liu, J.S. A novel approach to fuzzy cognitive map based on hesitant fuzzy sets for modeling risk impact on electric power system. *Int. J. Comput. Intell. Syst.* **2019**, *12*, 842–854. [CrossRef]
34. Mishra, A.R.; Rani, P.; Pardasani, K.R. Multiple-criteria decision-making for service quality selection based on Shapley COPRAS method under hesitant fuzzy sets. *Granul. Comput.* **2019**, *4*, 435–449. [CrossRef]

35. Mishra, A.R.; Rani, P.; Pardasani, K.R.; Mardani, A. A novel hesitant fuzzy WASPAS method for assessment of green supplier problem based on exponential information measures. *J. Clean. Prod.* **2019**, *238*. [CrossRef]
36. Farhadinia, B.; Herrera-Viedma, E. Multiple criteria group decision making method based on extended hesitant fuzzy sets with unknown weight information. *Appl. Soft Comput.* **2019**. [CrossRef]
37. Wang, Z.; Nie, H.; Zhao, H. An extended GEDM method with heterogeneous reference points of decision makers and a new hesitant fuzzy distance formula. *Comput. Ind. Eng.* **2020**. [CrossRef]
38. Mokhtia, M.; Eftekhari, M.; Saberi-Movahed, F. Feature selection based on regularization of sparsity based regression models by hesitant fuzzy correlation. *Appl. Soft Comput.* **2020**. [CrossRef]
39. Çolak, M.; Kaya, İ. Multi-criteria evaluation of energy storage technologies based on hesitant fuzzy information: A case study for Turkey. *J. Energy Storage* **2020**. [CrossRef]
40. Mardani, A.; Saraji, M.K.; Mishra, A.R.; Rani, P. A novel extended approach under hesitant fuzzy sets to design a framework for assessing the key challenges of digital health interventions adoption during the COVID-19 outbreak. *Appl. Soft Comput.* **2020**. [CrossRef]
41. Wu, N.; Xu, Y.; Liu, X.; Wang, H.; Herrera-Viedma, E. Water-Energy-Food nexus evaluation with a social network group decision making approach based on hesitant fuzzy preference relations. *Appl. Soft Comput.* **2020**. [CrossRef]
42. Narayanamoorthy, S.; Annapoorani, V.; Kang, D.; Baleanu, D.; Jeon, J.; Kureethara, J.V.; Ramya, L. A novel assessment of bio-medical waste disposal methods using integrating weighting approach and hesitant fuzzy MOOSRA. *J. Clean. Prod.* **2020**. [CrossRef]
43. Rani, P.; Mishra, A.R.; Krishankumar, R.; Mardani, A.; Cavallaro, F.; Ravichandran, K.S.; Balasubramanian, K. Hesitant Fuzzy SWARA-Complex Proportional Assessment Approach for Sustainable Supplier Selection (HF-SWARA-COPRAS). *Symmetry* **2020**, *12*, 1152. [CrossRef]
44. Demirel, T.; Öner, S.C.; Tüzün, S.; Deveci, M.; Öner, M.; Demirel, N.C. Choquet integral-based hesitant fuzzy decision-making to prevent soil erosion. *Geoderma* **2018**, *313*, 276–289. [CrossRef]
45. Deveci, M. Site selection for hydrogen underground storage using interval type-2 hesitant fuzzy sets. *Int. J. Hydrog. Energy* **2018**, *43*, 9353–9368. [CrossRef]
46. Li, C.; Zhao, H.; Xu, Z.S. Hesitant fuzzy psychological distance measure. *Int. J. Mach. Learn. Cybern.* **2020**. [CrossRef]
47. Ding, Q.; Wang, Y.M.; Goh, M. An extended TODIM approach for group emergency decision making based on bidirectional projection with hesitant triangular fuzzy sets. *Comput. Ind. Eng.* **2020**. [CrossRef]
48. Tus, A.; Adali, E.A. The new combination with CRITIC and WASPAS methods for the time and attendance software selection problem. *Opsearch* **2019**, *56*, 528–538. [CrossRef]
49. Mishra, A.R.; Rani, P.; Pardasani, K.R.; Mardani, A.; Stević, Z.; Pamučar, D. A novel entropy and Discrimination measures with multi-criteria service quality assessment using interval-valued intuitionistic fuzzy TODIM method. *Soft Comput.* **2020**, *24*, 11641–11661. [CrossRef]
50. Yazdani, M.; Wen, Z.; Liao, H.; Banaitis, A.; Turskis, Z. A grey combined compromise solution (CoCoSo-G) method for supplier selection in construction management. *J. Civ. Eng. Manag.* **2019**, *25*, 858–874. [CrossRef]
51. Peng, X.; Smarandache, F. A decision-making framework for China's rare earth industry security evaluation by neutrosophic soft CoCoSo method. *J. Intell. Fuzzy Syst.* **2020**. [CrossRef]
52. Liao, H.; Qin, R.; Wu, D.; Yazdani, M.; Zavadskas, E.K. Pythagorean fuzzy combined compromise solution method integrating the cumulative prospect theory and combined weights for cold chain logistics distribution center selection. *Int. J. Intell. Syst.* **2020**, *35*, 2009–2031. [CrossRef]
53. Alrasheedi, M.; Mardani, A.; Mishra, A.R.; Streimikiene, D.; Liao, H.; Alnefaie, A.H. Evaluating the green growth indicators to achieve sustainable development: A novel extended interval-valued intuitionistic fuzzy-combined compromise solution approach. *Sustain. Dev.* **2020**. [CrossRef]
54. Rani, P.; Mishra, A.R. Novel single-valued neutrosophic combined compromise solution approach for sustainable waste electrical and electronics equipment recycling partner selection. *IEEE Trans. Eng. Manag.* **2020**. [CrossRef]
55. Liu, P.; Rani, P.; Mishra, A.R. A novel Pythagorean fuzzy combined compromise solution framework for the assessment of medical waste treatment technology. *J. Clean. Prod.* **2021**. [CrossRef]
56. Li, W.; Wu, H.; Jin, M.; Lai, M. Two-stage remanufacturing decision makings considering product life cycle and consumer perception. *J. Clean. Prod.* **2017**, *161*, 581–590. [CrossRef]
57. Prajapati, H.; Kant, R.; Shankar, R. Bequeath life to death: State-of-art review on reverse logistics. *J. Clean. Prod.* **2019**, *11*, 503–520. [CrossRef]
58. Yu, H.; Solvang, W.D. Incorporating flexible capacity in the planning of a multiproduct multi-echelon sustainable reverse logistics network under uncertainty. *J. Clean. Prod.* **2018**, *198*, 285–303. [CrossRef]
59. Efendigil, T.; Onut, S.; Kongar, E. A holistic approach for selecting a third-party reverse logistics provider in the presence of vagueness. *Comput. Ind. Eng.* **2008**, *54*, 269–287. [CrossRef]
60. Govindan, K. Fuzzy approach for the selection of third party reverse logistics Provider. *Asia Pac. J. Mark. Logist.* **2009**, *21*, 397–416.
61. Govindan, K.; Murugesan, P. Selection of third-party reverse logistics provider using fuzzy extent analysis. *Benchmarking Int. J.* **2011**, *18*, 149–167. [CrossRef]

62. Senthil, S.; Srirangacharyulu, B.; Ramesh, A. A robust hybrid multi-criteria decision making methodology for contractor evaluation and selection in third-party reverse logistics. *Expert Syst. Appl.* **2014**, *41*, 50–58. [CrossRef]
63. Uygun, Ö.; Kaçamak, H.; Kahraman, Ü.A. An integrated DEMATEL and Fuzzy ANP techniques for evaluation and selection of outsourcing provider for a telecommunication company. *Comput. Ind. Eng.* **2015**, *86*, 137–146. [CrossRef]
64. Mavi, R.K.; Goh, M.; Zarbakhshnia, N. Sustainable third party reverse logistic provider selection with fuzzy SWARA and fuzzy MOORA in plastic industry. *Int. J. Adv. Manuf. Technol.* **2017**, *91*, 2401–2418. [CrossRef]
65. Tavana, M.; Zareinejad, M.; Santos-Arteaga, F.J. An intuitionistic fuzzy-grey superiority and inferiority ranking method for third-party reverse logistics provider selection. *Int. J. Syst. Sci. Oper. Logist.* **2018**, *5*, 175–194. [CrossRef]
66. Li, Y.L.; Ying, C.S.; Chin, K.S.; Yang, H.T.; Xu, J. Third-party reverse logistics provider selection approach based on hybrid-information MCDM and cumulative prospect theory. *J. Clean. Prod.* **2018**, *195*, 573–584. [CrossRef]
67. Zarbakhshnia, N.; Soleimani, H.; Ghaderi, H. Sustainable third-party reverse logistics provider evaluation and selection using fuzzy SWARA and developed fuzzy COPRAS in the presence of risk criteria. *Appl. Soft Comput.* **2018**, *65*, 307–319. [CrossRef]
68. Zhang, X.; Su, T. The Dominance Degree-Based Heterogeneous Linguistic Decision-Making Technique for Sustainable 3PRLP Selection. *Complexity* **2020**, 6102036. [CrossRef]
69. Zarbakhshnia, N.; Wu, Y.; Govindan, K.; Soleimani, H. A novel hybrid multiple attribute decision-making approach for outsourcing sustainable reverse logistics. *J. Clean. Prod.* **2020**. [CrossRef]
70. Mishra, A.R.; Rani, P.; Pandey, K. Fermatean fuzzy CRITIC-EDAS approach for the selection of sustainable third-party reverse logistics providers using improved generalized score function. *J. Ambient. Intell. Humaniz. Comput.* **2021**. [CrossRef] [PubMed]
71. Rani, P.; Mishra, A.R.; Mardani, A.; Cavallaro, F.; Alrasheedi, M.; Alrashidi, A. A novel approach to extended fuzzy TOPSIS based on new divergence measures for renewable energy sources selection. *J. Clean. Prod.* **2020**, *257*, 120352. [CrossRef]
72. Kumari, R.; Mishra, A.R. Multi-criteria COPRAS method based on parametric measures for intuitionistic fuzzy sets: Application of green supplier selection. *Iran. J. Sci. Technol. Trans. Electr. Eng.* **2020**, *44*, 1645–1662. [CrossRef]
73. Mishra, A.R.; Mardani, A.; Rani, P.; Zavadskas, E.K. A novel EDAS approach on intuitionistic fuzzy set for assessment of health-care waste disposal technology using new parametric divergence measures. *J. Clean. Prod.* **2020**, *272*, 122807. [CrossRef]
74. Kobza, V.; Janis, V.; Montes, S. Divergence measures on hesitant fuzzy sets. *J. Intell. Fuzzy Syst.* **2017**, *33*, 1589–1601. [CrossRef]
75. Mishra, A.R.; Rani, P.; Krishankumar, R.; Ravichandran, K.S.; Kar, S. An extended fuzzy decision-making framework using hesitant fuzzy sets for the drug selection to treat the mild symptoms of Coronavirus Disease 2019 (COVID-19). *Appl. Soft Comput.* **2021**. [CrossRef]
76. Mi, X.; Liao, H. Renewable energy investments by a combined compromise solution method with stochastic information. *J. Clean. Prod.* **2020**, *123351*. [CrossRef]
77. Luo, Y.; Zhang, X.; Qin, Y.; Yang, Z.; Liang, Y. Tourism Attraction Selection with Sentiment Analysis of Online Reviews Based on Probabilistic Linguistic Term Sets and the IDOCRIW-COCOSO Model. *Int. J. Fuzzy Syst.* **2020**. [CrossRef]
78. Peng, X.; Luo, Z. Decision-making model for China's market bubble warning: The CoCoSo with picture fuzzy information. *Artif. Intell. Rev.* **2021**. [CrossRef]
79. Rani, P.; Mishra, A.R. Single-valued neutrosophic SWARA-VIKOR framework for performance assessment of eco-industrial thermal power plants. *Icses Trans. Neural Fuzzy Comput.* **2020**, *3*, 1–9.
80. Xu, Z.S.; Zhang, X. Hesitant fuzzy multi-attribute decision making based on TOPSIS with incomplete weight information. *Knowl. Based Syst.* **2013**, *52*, 53–64. [CrossRef]

sustainability

MDPI

Article

A New Sustainable Warehouse Management Approach for Workforce and Activities Scheduling

Vlado Popović [1,*], Milorad Kilibarda [1], Milan Andrejić [1], Borut Jereb [2] and Dejan Dragan [2]

[1] Faculty of Transport and Traffic Engineering, University of Belgrade, Vojvode Stepe 305, 11000 Belgrade, Serbia; m.kilibarda@sf.bg.ac.rs (M.K.); m.andrejic@sf.bg.ac.rs (M.A.)
[2] Faculty of Logistics, University of Maribor, Mariborska Cesta 7, 3000 Celje, Slovenia; borut.jereb@um.si (B.J.); dejan.dragan@um.si (D.D.)
* Correspondence: vlado.popovic@yahoo.com; Tel.: +381-11-30-91-304

Abstract: Sustainable engineering is very important for logistics systems. Nowadays, sustainable warehouse management is a key factor in market success. Workforce fluctuation and inverting the number of customers' demands make a lot of problems in distribution warehouses. This study addresses a sustainable approach for the workforce scheduling problem recognized in a real distribution warehouse. The problem arises from the high variability of demand for workers over one workday, which causes workforce surplus in some periods of the workday and shortages in others. Engineering managers of the distribution warehouse already use different full-time and part-time shifts, and schedule workers on different activities, but they still have significant workforce surpluses or shortages in some periods. This study proposes the scheduling of activities' execution together with workers to face that variability and decrease the cost of the workforce. This idea comes from the fact that some activities in a distribution warehouse can be done in a specific time period after the need for them occurs. In this way, the variability of demand for workers can be decreased, and a lower workforce cost may be ensured. Based on this idea, the entire problem is modeled as integer linear programming. The real example of the problem is solved, and the proposed model is tested on randomly generated instances of the problem in Python by means of the PuLP linear programming package. The results indicate different positive effects in the manner of sustainable warehouse management: lower workforce costs, time savings, better utilization of all types of resources and equipment, increased employee satisfaction, and so on. For even 61% of instances of the introduced problem, the obtained cost of the workforce is lower by more than 20% if activities' executions are scheduled together with employees.

Keywords: sustainable warehouse management; workforce; optimization; scheduling; activities; warehouse

check for updates

Citation: Popović, V.; Kilibarda, M.; Andrejić, M.; Jereb, B.; Dragan, D. A New Sustainable Warehouse Management Approach for Workforce and Activities Scheduling. *Sustainability* **2021**, *13*, 2021. https://doi.org/10.3390/su13042021

Academic Editor: Dragan Pamucar
Received: 28 December 2020
Accepted: 8 February 2021
Published: 13 February 2021

Publisher's Note: MDPI stays neutral with regard to jurisdictional claims in published maps and institutional affiliations.

Copyright: © 2021 by the authors. Licensee MDPI, Basel, Switzerland. This article is an open access article distributed under the terms and conditions of the Creative Commons Attribution (CC BY) license (https://creativecommons.org/licenses/by/4.0/).

1. Introduction

The warehouse is one of the most important parts of companies with distribution functions, on whose sustainability the entire company's sustainability largely depends. It has a particularly important impact on economic and social sustainability as a part of the company that most absorbs the consequences of changes in demand. This has been best demonstrated nowadays where there are large differences in demand. Recently, more and more attention has also been paid to the ecological sustainability of warehouses. Warehouses are the most responsible for the increase in greenhouse gas emissions in the supply chain [1], primarily due to the increase in energy consumption for lighting, heating, cooling, air conditioning, and the handling of goods [2]. In this regard, companies are being invested in to provide renewable energy sources, green technologies, recycling systems, and so on. In addition, the number of works on the topic of green warehousing has increased significantly in recent years [1].

Nevertheless, economic, social, and environmental sustainability are equally important for achieving overall warehouse sustainability [2]. Tan et al. [3] describe in detail their importance, and point out that any change in goals in one type affects the achievement of the goals of the other two types of sustainability. Therefore, it is very important to maintain a balance between the factors of all three types of sustainability [4]. In this regard, engineering has the task to go a step further than seeking sustainable resources and offer systems in which sustainable resources are used in a sustainable way [5]. Operations research should get more attention in this process [6].

The sustainability of the warehouse largely depends on the utilization of the warehouse's resource (i.e., on their schedule in space and time). If the warehouse is observed only between four walls, then its main resources are space, equipment, and workforce. The schedule in the use of space and stationary equipment is more a tactical or strategic decision and generally does not require daily optimization. On the other hand, workers and the mobile part of the equipment are scheduled every day. Without the efficient use of these resources, there is no sustainable order picking. Since order picking is an extremely intensive process that consumes a lot of energy, capital, and human resources, its impact on overall warehouse sustainability is extremely high. Andriansyah et al. [7] point out that an efficient use of resources is a prerequisite for sustainable order picking and thus for sustainable warehousing.

Numerous papers deal with scheduling, routing, and other methods that optimize the use of mobile resources in the warehouse. The aim is to minimize their costs, energy consumption, and greenhouse emission while performing activities. Burinskiene et al. [8] scheduled forklift driving and picking vehicle routes to reduce travel time and cost in replenishment and order picking. Boenzi et al. [9] optimized forklift operation in terms of both energy consumption and environmental impacts. Pashkevich et al. [10] examined which factors cause an increase in forklift energy consumption. Lorenc and Lehrer [11] scheduled products in warehouses of modern companies to achieve more efficient use of resources in the order-picking process. On the other hand, studies such as [12–14] and others have scheduled workers. In general, workers have been scheduled on tasks, jobs, activities, days, and so on, following a work execution plan. Papers dedicated to the workforce in the warehouse and workforce scheduling on activities are represented in the literature review in more detail.

However, to achieve optimal utilization of mobile resources, it is not enough just to optimally schedule them according to a work execution plan; the work execution must also be optimally scheduled when scheduling mobile resources. With this approach, this paper goes a step ahead compared to the existing literature, which deals only with the effects of scheduling of mobile resources on work. In addition, this paper gives preference to the workforce over the mobile equipment in the scheduling process in the warehouse and defines all types of works in the warehouse as activities. In a similar sense as in this paper, Carli et al. [15] scheduled activities and forklifts in a warehouse to optimize the charging batteries. Their aim was to charge forklift batteries less often during peak hours. In contrast, this paper takes into account the economic and social aspects when scheduling activities. It exploits the possibility of postponing the execution of some activities and the use of workers in different types of activities to lower workforce costs. It provides a more even use of workers over the workday and a break on time, and eliminates overcrowding in the warehouse to improve working conditions. All of that increases economic and social sustainability and creates a potential for investment in environmental sustainability solutions.

The workforce scheduling and planning is hence very important but also a complicated task for companies with distribution warehouses (DW). Like in many other service systems, the complexity comes from the variability of workload during a planning horizon. As a result, a different number of workers is needed in different periods of the planning horizon. However, on a daily basis, this problem might be resolved in a DW better than in other service systems. During the analysis of activities in the real DW, it was noticed that many activities could be postponed completely or partially. It means that the execution of these activities can be scheduled in a specific time period after the need for their execution occurs.

In that way, the variability of demand for workers may be decreased, and a lower workforce cost may be ensured. The main aim of this study was to incorporate this specificity in the scheduling of workers on activities in DW and provide a solution with a lower cost of the workforce. At the same time, the solution should contribute to a more balanced workload for workers during the day, respecting the period for a break. In this way, the solution should help increase the efficiency and sustainability of the warehouse.

The idea for this study came from cooperating with engineering managers of the real DW to decrease the cost of the workforce. After analyzing activities in their warehouse, and how they schedule workers, the workforce scheduling problem in DW (DW-WSP) was defined. Based on the defined DW-WSP, the idea of applying an activities execution scheduling (AES) for decreasing the cost of the workforce in DW was developed. This idea and the DW-WSP are presented in the rest of this section, together with the main contributions of the paper and implications for engineering managers. The position of the study among the existing literature and related studies are discussed in the next section. The approach to solving the introduced problem, which allows the scheduling of workers and execution of activities at the same time, is presented next. This approach explains the concept for solving the problem and a mathematical model of the problem, named a workforce and activities execution scheduling model (the WAES model). The WAES model is tested for the input values obtained from the considered DW, and for the 100 randomly generated instances of the problem. Each instance of the problem is solved with the WAES model and the model that excludes the AES (the WS model). The results of the testing, final discussion, and future research conclude the paper.

1.1. The DW-WSP

The DW-WSP is introduced as follows. The workforce has to be determined by shifts and by worker profiles, along with scheduling workers' activities over the workday. Jobs of the same type perform an activity that lasts the entire workday and requires a different number of workers in different intervals of a workday. The required number of workers for executing an activity in an interval is equal to the sum of workers needed for performing all jobs belonging to that activity that occurred in that interval. For each activity, there is a deadline for finishing. The needed number of workers by intervals and by activities is calculated based on the history of jobs' occurrence. The exceptions are activities whose workload depends on the workload of one or more other activities. For such an activity, the needed number of workers in an interval is calculated based on the number of workers that are scheduled on the activities it depends on. In DW, there are full-time and part-time shifts. Each full-time worker needs to take their break within a prescribed time window, which depends on the shift. Different worker profiles are employed, which can execute different activities following their qualifications. Based on the overall definition, this problem can be characterized as a relatively extended version of the shift-scheduling problem, in which a worker can perform more than one activity in the same shift. Ultimately, the conclusion can be drawn that the defined problem belongs to the group of multi-activity shift scheduling problems (MASSPs).

1.2. The AES

The scheduling of activities' execution, together with workforce scheduling, is the main idea for decreasing the cost of the workforce in DW. The AES in this paper refers to the scheduling of some activities' execution in a time period after the need for their execution occurs. This scheduling is possible due to the opportunities that some activities provide in terms of their completion. The execution of these activities can be completely or partially postponed. The following example explains the AES from the workforce scheduling perspective. Let it be prescribed that each truck checked in for the unloading has to be unloaded in three hours. The truck can be unloaded completely in the first, the second, or the third hour after arrival. The unloading can also be scheduled partly in two or three of these hours. Accordingly, the requirements for workers of unloading activity can be fulfilled in different ways after they occur.

Further, assume it is measured that four workers can unload the truck in one hour. If the workday is divided into one-hour intervals, then a truck checked in for unloading creates the job that requires four workers to work for one hour. As the unloading has to be finished in three hours, all four workers can be scheduled to unload the truck in the first, in the second, or the third hour after the truck is checked in. However, the truck can also be unloaded by one worker over all three hours and by another worker during one of these three hours, and so on. Therefore, there are many combinations to schedule the execution of activities (i.e., to fulfill requirements for workers of activities that have a deadline for finishing). The idea of using the AES for decreasing the cost of the workforce in DW is represented by the example in Figure 1.

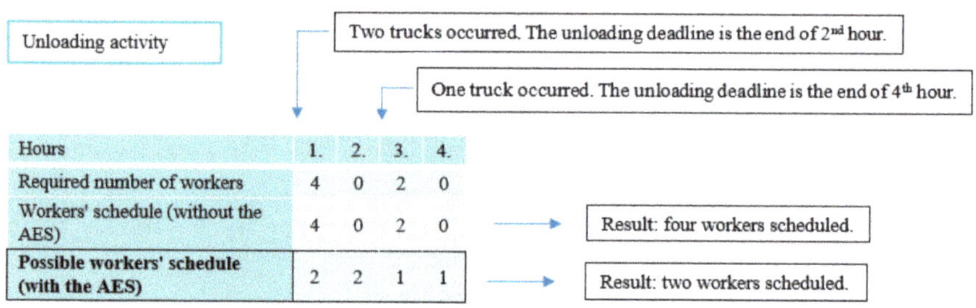

Figure 1. The impact of the activities execution scheduling (AES) on the workforce scheduling solution.

1.3. Main Contributions

It is believed that this study might contribute to knowledge and practice by offering novelties that were not identified in other research. Careful investigation of the literature showed that many studies are dealing with MASSPs, but trying to develop a better technique to solve a familar MASSP. In this work, the MASSP model is adjusted for application in DW and increasing the sustainability thereof. Despite the overall MASSP complexity, the entire problem is extended with the AES to decrease the cost of the workforce in DW. Therefore, it is believed that the study has two potential contributions:

- The first is practical, and refers to developing a model for workforce scheduling in DW, which the literature lacks, and which can decrease the cost of the workforce in DW and increase its sustainability.
- The second is methodological, and refers to the introduction of the AES to an MASSP as a novelty that might provide a better solution of MASSPs in general.

2. Literature Review

This study can be positioned among the studies that consider workforce scheduling problems, activity scheduling problems, and warehouse sustainability.

Regarding workforce scheduling problems, this study can be positioned among studies that consider shift scheduling problems. According to Ernst et al. [16], "shift scheduling deals with the problem of selecting from a potentially large pool of candidates; what shifts are to be worked together with an assignment of the number of employees to each shift to meet demand." The position of the break within selected shifts is also considered unless it is fixed. Usually, the problem is considered in a single-day planning horizon divided into a certain number of equal intervals. If the number of available workers is specified, then the objective is usually to find a schedule that minimizes understaffing and overstaffing. If the number of available workers is not provided, the aim is to find a schedule that provides the minimal cost of the workforce. Some of the other objectives in use are minimizing a gap between assignments and employees' preferences, balancing a workload between employees, minimizing workforce size, and maximizing employees' satisfaction.

This study can be classed among the studies that consider MASSPs. MASSPs are a relatively extended type of shift scheduling problem. The multi-activity context in shift scheduling problems extends these problems to determining when and which activity each worker performs. Demand for workers is represented by activities and by intervals of a workday. Hence, MASSPs generally imply solving two problems. One problem is to select working shifts and, subsequently, assign workers to them. The second one is to assign workers to activities by intervals of shifts. If activities require different qualified workers, then for each selected shift, the number of every worker profile needs to be determined.

Finally, this study can be classed among the studies that consider warehouse sustainability. These studies are related to the ideas of green warehousing, sustainable warehouse management, development of operational research tools for solving sustainable engineering problems in the warehouse, and so on.

2.1. Literature Review of Workforce Scheduling in Warehouses

Despite the significant number of studies addressing the workforce scheduling problem, a small number of studies have investigated that problem in a warehouse. Out of more than 700 papers about workforce scheduling published between 1954 and 2003, included and described in the review paper by Ernst et al. [17], not one mentions the warehouse. In the second review paper by Van der Bergh et al. [18], which includes papers from 2003 to 2012, the warehouse is mentioned in four papers. After 2012, to the best of the authors' knowledge, only two papers mention the warehouse. All of these papers are briefly described below.

De Causmaecker et al. [19] classified real-world personnel planning problems by investigating a small yet representative sample of Belgian companies. In a warehouse company, a call center, a fast-food restaurant, and an employment agency, they encountered personnel planning based on fluctuating demands. As the required number of personnel fluctuated, they described the problem in these companies as a fluctuation-centered personnel planning problem. Authors have pointed out that these companies maintain a historical record for demand forecasting, whereas the warehouse company has at least three workers qualified for each task and hires temporary employees.

Günther and Nissen [20,21] scheduled workers on workstations (jobs) in one German logistics provider, where most of the jobs are related to the warehouse. Both papers have solved the same problem, but used different techniques to do so, with both aimed at improving the existing schedule of workers during the day. Accordingly, the objectives of the studies were obtained by allowing workers to perform more than one job during the day. The objective function was the minimizing of over-satisfying and under-satisfying of demands for workers, as well as changing jobs. Bard and Wan [22] presented a model for determining the optimal size of the permanent workforce, where workstation groups specify the demand for workers for up to 24 h a day, seven days a week. The model was tested using the data provided by the U.S. Postal Service (USPS) mail processing and distribution center (P&DC) in Dallas.

Ladier et al. [23] defined three phases of making weekly and daily schedules for workers. In the first phase, the number of temporary workers of every profile and the number of hours by task for each worker (temporary or permanent employee) are determined for each day of the one-week horizon. In the second phase, for each employee, the shift and the percentage of time spent on the assigned task are calculated for each day and each hour. In the third phase, an assignment of tasks to employees is made for each interval. The solution in each phase is obtained by minimizing the deviation of the solution from the given demands, conditions, rules, and so on. Ladier and Alpan [24] solved the cross-dock truck scheduling problem and the employee rostering problem in an integrated manner. Two iterative procedures were proposed for solving the problem, encompassing employees-first and trucks-first.

2.2. Literature Review of Multi-Activity Shift Scheduling

An in-depth review of literature in the workforce planning and scheduling sphere revealed that the multi-activity context had been rarely noted. Moreover, the term "activity" is quite newly introduced. Referring to the study conducted by Ernst et al. [17], it is uncovered that none of the papers that have been reviewed described jobs as activities. Besides, Van den Bergh et al. [18] conducted a review study in which the scheduling of workers on activities was only identified in cited studies such as [25–27]. However, before these, it is also worth mentioning the papers in which jobs are activities following this paper's definition, but they were not described in that way.

Ritzman et al. [28] and Loucks and Jacobs [29] are among the first papers that introduce the multi-activity context. Ritzman et al. [28] assigned every worker to a particular work, every hour, within a post office. Similarly, Loucks and Jacobs [29] allocated different types of work (called tasks) to restaurant workers, where each task required workers to have adequate skills. Bard and Wan [22] scheduled workers on workstations in a way that the demand for workers of every workstation was defined in half-hour intervals, every day, five days a week. The scheduling of workers on workstations, which corresponded to the activities, was conducted in a study by Günther and Nissen [21]. The results of the study indicate that on each workstation, in each interval of the day, a certain number of workers are needed. In a study conducted by Ladier et al. [24], tasks in a warehouse were considered in two ways, which is to say, they defined their workload precisely (hour-by-hour) and for a whole time slot. Tasks defined for the whole time slot could be realized at any time. Given the definition of activity from this paper, the first group of tasks corresponds to activities, and the second one to tasks.

It can be asserted that Demassey et al. [12,25] and Cote et al. [30] first introduced the term "activities" in reference to workforce problems. However, the problems in their papers are described as the employee timetabling problem. Besides, these studies also introduced the constraints that are subsequently used in almost every paper concerning MASSP and are presented below (ended with Lequy et al. [31]). Accordingly, the constraints can be regulated as allowed and forbidden assignments of activities to shift intervals; a minimum and a maximum number of consecutive periods assigned to an activity or break for an employee; allowed and forbidden activity changes; and a required number of break periods after changing activities or consecutive working periods; and so on. Followings studies could have solved a quite similar MASSP, but using different solving techniques.

Demassey et al. [12,25] determined valid schedules to which an employee can be assigned (valid shifts) by using constraint programming considering two sequences. Firstly, they have fulfilled intervals of valid shifts with activities and a break/lunch according to various constraints. Secondly, they have deployed the column generation approach to cover demand. They identified the required number of workers to work in specific shifts. Additionally, Demassey et al. [12] solved this problem when the set of valid shifts is determined for each worker separately. In studies conducted by Cote et al. [30,32], they determined, for each worker, which activity to perform in which interval, without using a set of feasible shifts. For each employee, a shift was formed by using various constraints (i.e., an implicit way was used to form shifts).

Cote et al. [32] and the subsequent papers described a solved problem as MASSP, or by a slightly modified name. Cote et al. [26] continued to use the implicit way for defining shifts. However, this time, the decision variable denoted the number of workers assigned to an activity in an interval. Cote et al. [13] selected a working shift for each worker from his/her set of feasibility shifts by using a column generation approach. Each valid shift for an employee was fulfilled with a break, a lunch, or an activity, according to defined constraints. The pricing sub-problem was modeled by using a context-free grammar that allowed extracting the set of feasible shifts for each employee.

Lequy et al. [14] assumed that shifts (with breaks) are already assigned to workers. Accordingly, they have assigned activities to shifts (i.e., to the periods of shifts). An activity was assigned to a period of a shift, considering only if there was an adequate number of qualified workers to perform the activity in that shift. Ultimately, they have proposed three integer programming formulations of the problem, and subsequently, they have developed three heuristics based on branch and bound or column generation methods. This problem was extended in Lequy et al. [31] by assigning tasks to shifts besides activities. Tasks were assigned to shifts in the same way as activities; however, they obeyed different rules. Similarly, Boyer et al. [33] extended the model of Cote et al. [13] by including tasks besides activities in the problem.

Dahmen and Rekik [34] solved an MASSP over a multi-day planning horizon, where workdays were known for each worker. The study aimed at constructing admissible shifts followed by assigning them to workers in a way that minimizes under-staffing and over-staffing costs over their workdays. They have proposed a tabu-search-based constructive algorithm and a mixed integer linear programming (MILP) model of the shift scheduling problem. Restrepo et al. [27] solved another real case of MASSP. The problem was concerned with finding the schedule for security staff and cashiers over 100 parking lots in the city of Bogotá. They modeled parking lots as work activities and the staff movements between parking lots as changes between work activities. The solution was articulated as the number of workers in selected shifts and the assignment of activities to time slots of those shifts.

2.3. Literature Review of Sustainable Warehouse Studies

Tan et al. [3] define a sustainable warehouse as a warehouse where all economic, environmental, and social inputs are integrated, balanced, and managed. In their work, they developed a model for testing warehouse sustainability. The model also provides decision support that leads to sustainable warehousing. Bank and Murphy [35] emphasize the importance of warehouse sustainability for the sustainability of the entire supply chain. They introduced warehouse sustainability metrics, measurements, and guidelines. Similarly, Torabizadeh et al. [36] provided a list of 33 performance indicators that can be used to assess the sustainability of a sustainable warehouse management system. Bartolini et al. [1] pointed out that warehouses make a major contribution to increasing greenhouse gas in supply chains. For that reason, they provided an overview of the literature dealing with green warehousing and warehouse sustainability. They identified 38 papers related to this topic. Carli et al. [15] scheduled mobile electrical equipment in the warehouse, taking into account the cost of electricity and the cost of penalties related to the makespan. In this way, they ensured a reduction in the impact of mobile electrical equipment on the economic and environmental sustainability of the warehouse. Freis et al. [37] proposed a methodology to determine the energy demand of logistics centers and evaluate the impacts of design options to create energy-efficient and CO_2-neutral systems. They concluded that the energy demand and CO_2 emissions of logistics centers can be reduced if the energy interrelations between the sub-systems intra-logistics, building technology, and building skin are taken into account when improving the base elements of these sub-systems. They also provided energy-saving measures, depending on the degree of automation of the logistics center. Kamarulzaman et al. [38] made a list of green initiatives that can be applied when performing activities in the warehouse and evaluated Malaysian food-based manufacturers according to the application of these initiatives. Amjed et al. [39] defined a sustainable warehouse model that shows elements and constructs that impact sustainability. Elements were grouped into nine constructs, and the dimensions of sustainability that impacted each element were defined. They named the constructs as warehouse facility design, warehouse layout, inventory management, mechanical handling equipment, warehouse staff, warehouse operations, onsite facilities, and warehouse management system. Malinowska et al. [40] defined criteria for warehouse assessment in regards to sustainable functioning, and created a roadmap to gain a better level of sustainable warehouse. The road map for each criterion contains activities to achieve better results of the evaluation criteria and advantages of their application. Durmić [41] determined the importance of various factors and dimensions of sustainability in the selection of a sustainable supplier in BIH. Based on expert assessments, it was obtained that economic factors and sustainability are the most important when selecting a supplier. Đalić et al. [42] presented a model for evaluation and supplier selection based on seven environmental criteria: environmental image, recycling, pollution control, environmental management system, environmentally friendly products, resource consumption, and green competencies.

3. The Sustainable Warehouse Management Approach for Workforce and Activities Scheduling

3.1. Problem Statement

The problem has its roots in selecting working shifts, as well as in determining the number of workers for each profile in each selected shift. Accordingly, the aim is to figure out a solution so that the cost of the workforce is minimized, and every constraint and demand for workers is satisfied over a workday. Depending on activity, demand for workers has to be fulfilled within the prescribed time window or by the end of the prescribed interval of a workday. Since the solution to this problem is robustly tied to the scheduling of workers on activities, then obtaining such a schedule is also part of the problem. Besides, the prospective schedule has to specify a break for each worker within a prescribed time window. Therefore, the overall solution is expected to address the following issues:

- How many workers of each profile are needed to work in each selected shift?
- Which activity is performed by each of the workers in each interval of their shift, and when do they take a rest?

3.2. Conceptual Framework for Solving the Problem

Based on the introduced issues, an entire conceptual framework for solving the problem is developed. It implies solving the WAES model first and then making the schedule for each worker. The solution of WAES provides the number of workers employed in each shift by profile, and the number of workers assigned to each activity in each interval by profile. However, the solution does not specify personally for each worker which activity he performs in which interval and when he has a break. For that need, an algorithm is developed in Python. Based on the obtained solution, the algorithm for each worker assigns an activity or a break in each interval of his shift. The entire conceptual framework for solving the problem is presented in Figure 2.

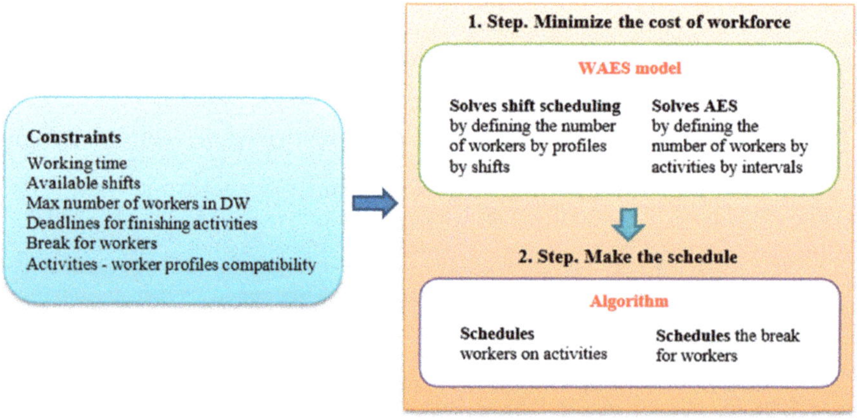

Figure 2. The conceptual framework for solving the distribution warehouses workforce scheduling problem (DW-WSP).

3.3. Assumptions, Limitations, and Notation

The problem is considered over one workday, which is divided into n intervals of equal length. Each interval is represented by an ordinal number, where n is also the ordinal number of the last interval in the workday. Intervals of a workday are contained in set N (i.e., $N = \{1, 2, \ldots, n\}$). The workday can have full-time and part-time shifts. All possible full-time shifts are contained in set M_1, and all possible part-time shifts are contained in set M_2. Together, they form the set of all possible shifts M. The working environment of the warehouse can include independent and dependent activities. Independent activities are included in set B, whereas dependent activities are incorporated into set C. All considered activities are included in set A.

For activity $a \in B$ and interval $k \in N$, a required number of workers by intervals of the workday is assumed to be known. This is represented as demand for workers (d_{ai}). On the other hand, for activity $a \in C$ and interval $k \in N$, the demand for workers (d_{ai}) is equal to the sum of the percentages of the number of workers assigned to each activity from set G_a in interval k. Set G_a contains activities on which activity $a \in C$ depends. Therefore, for activity $a \in C$ and activity $a' \in G_a$, there is a percent of dependence $(s_{a'})$. Since demand for workers of dependent activities can be a non-integer number, it has to be fulfilled by an integer number that is equal to or greater than demand.

Each activity has to be executed within a defined time window. The time window for fulfilling a demand for workers on an activity is set according to the time window available for that activity's execution. There are two types of time windows. For activity $a \in A_1$, demand for workers needs to be fulfilled within the exact number of intervals (v_a), starting from the one in which it occurs. For example, if the policy of a distribution company is to unload each truck within two intervals after checking in, then each demand for workers of the unloading activity needs to be fulfilled within two intervals.

For activity $a \in A_1$, demand for workers generated before interval r_a needs to be fulfilled by the end of that interval. Requests for performing these activities are received until the interval r_a or an interval before it starts. For example, if a distribution company operates until 22:00, and all orders received by 16:00 are distributed in the morning of the subsequent workday, then all orders received by 16:00 have to be prepared by 22:00. Accordingly, each demand for workers of an order-picking activity needs to be fulfilled by 22:00.

Different worker profiles are employed in the warehouse. All considered worker profiles are represented by set T. For each activity, it is defined which of the workers' profiles are allowed to perform it. Set T_a denotes the set of worker profiles who are allowed to perform an activity a. Similarly, set H_t denotes activities which a worker of the profile $t \in T$ is allowed to perform. A worker of profile t costs c_t. The maximal number of workers who can work at the same tim (i.e., the maximal number of active warehouse workers) is q. In each full-time shift, there is a break of p intervals long, which needs to be taken by each worker employed in a full-time shift within the time window made of f intervals. For activity $j \in M_1$, intervals available for the break are given in set O_j.

Demand for workers on independent activities over a workday is considered predictable based on a previous archived record about the workload. However, workload prediction and its conversion into demand for workers are not considered in this paper. The model is also limited to application in distribution warehouses.

Mathematical notations used in the model are presented in Table 1, separately for model inputs and outputs. The model inputs are presented in the context of time, activities, deadlines for finishing activities, and workers.

Table 1. Table of notation.

Inputs	
Time	
N	Set of intervals of a workday
M	Set of shifts
M_1	Set of full-time shifts
M_2	Set of part-time shifts
n	Ordinal number of the last interval in a workday
b_{kj}	A parameter that has a value of 1 if an interval $k \in N$ is part of a shift $j \in M$; otherwise, it is 0.
Activities	
A	Set of all activities
B	Set of independent activities
C	Set of dependent activities
G_a	Set of activities that activity $a \in C$ depends on
d_{ai}	Demand for workers of activity a in interval i
$s_{a'}$	Percent of the dependence of activity $a \in C$ on activity $a' \in G_a/100$
Deadlines	
A_1	Set of activities with a deadline represented by the time window
v_a	Duration of the time window for finishing activity $a \in A_1$ (in intervals)
A_2	Set of activities with a deadline represented by number of interval
r_a	The interval by which demand for workers of activity $a \in A_2$ needs to be fulfilled
Workers	
T	Set of all worker profiles
T_a	Set of worker profiles who are allowed to perform activity a
H_t	Set of activities which a worker of profile $t \in T$ is allowed to perform
c_t	Cost of worker of profile t
q	Maximal number of workers who can work at the same time
p	Duration of the break (in intervals)
f	Duration of time window available for breaks (in intervals)
O_j	Intervals available for the break in shift $j \in M_1$
w	Percent of workers employed in part-time shifts/100
Outputs	
y_{tj}	Number of workers of a profile $t \in T$ assigned to shift $j \in M$
y_{tija}	Number of workers of profile $t \in T$ assigned to interval $i \in N$ of shift $j \in M$ on activity $a \in A$
x_{aijk}	Number of workers provided in interval $i \in N$ of shift $j \in M$ to fulfill the demand for workers of activity $a \in A$ from interval $k \in N$

3.4. *Optimization Model*

Before introducing the optimization model, the working mechanism of the proposed approach for solving DW-WSP is presented to indicate the position of the model in the process of reaching a solution (Figure 3).

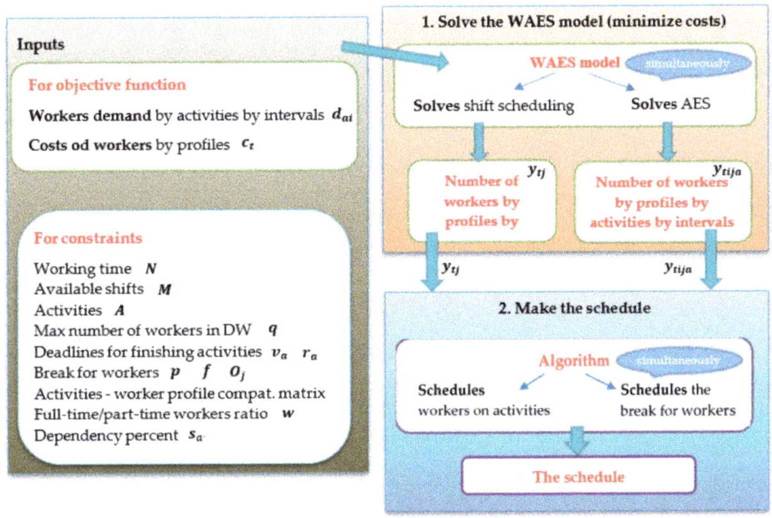

Figure 3. The working mechanism for solving the DW-WSP.

Following the conceptual framework, the working mechanism for solving the DW-WSP, and the introduced notation, the WAES model of the problem is developed in the form of a linear program. The WAES model has the following form:

$$min: \sum_{j\in M}\sum_{t\in T} c_t * y_{tj} \tag{1}$$

Subject to:

$$\sum_{j\in M}\sum_{k=i}^{until} x_{aijk} * b_{kj} = d_{ai},\ \ \forall i \in N,\ \forall a \in B \cap A_1;\ until = \min(i + v_a,\ n) \tag{2}$$

$$\sum_{j\in M}\sum_{k=i}^{until} x_{aijk} * b_{kj} = d_{ai},\ \ \forall i \in N,\ \forall a \in B \cap A_2; until = r_a \tag{3}$$

$$\sum_{j\in M}\sum_{k'=k}^{until} x_{akjk'} * b_{kj} \geq \sum_{a'\in G_a} s_{a'} * \sum_{j\in M}\sum_{i\in N} x_{aijk},\ \forall k \in N,\ \forall a \in C \cap A_1;\ until = \min(k + v_a,\ n) \tag{4}$$

$$\sum_{j\in M}\sum_{k'=k}^{until} x_{akjk'} * b_{kj} \geq \sum_{a'\in G_a} s_{a'} * \sum_{j\in M}\sum_{i\in N} x_{aijk},\ \forall k \in N,\ \forall a \in C \cap A_2;\ until = r_a \tag{5}$$

$$\sum_{i=1}^{k} x_{aijk} \leq \sum_{t\in T_a} y_{tija},\ \ \forall k \in N; \forall j \in M,\ \forall a \in A \tag{6}$$

$$\sum_{a\in H_t} y_{tija} \leq y_{tj},\ \ \forall i \in N,\ \forall j \in M,\ \forall t \in T \tag{7}$$

$$\sum_{t\in T}\sum_{j\in M}\sum_{a\in A} y_{tija} \leq q,\ \ \forall i \in N \tag{8}$$

$$\sum_{\forall i\in O_j}\sum_{a\in H_t} y_{tija} \leq (f - p) * y_{tj},\ \ \forall t \in T, \forall j \in M_1 \tag{9}$$

$$\sum_{j\in M}\sum_{t\in T} y_{tj} \leq w * \sum_{j\in M_2}\sum_{t\in T} y_{tj} \tag{10}$$

$$integer: \ x_{aijk}, y_{tija}, y_{tj}, \ \ \forall i \in N, \ \forall j \in M, \ \forall a \in A, \ \forall t \in T, \ \forall k \in N \quad (11)$$

The *objective function* (1) aims at minimizing the total cost of the workforce. *Constraints* (2)–(5) ensure that demand for workers on all activities is fulfilled within a prescribed time. *Constraint* (6) regulates the overall fulfillment of the demand for workers on an activity in an interval not to be larger than the number of workers of all profiles assigned to that activity in that interval. *Constraint* (7) regulates, for each worker profile, the number of workers by intervals of a shift not to be larger than the number of workers assigned to that shift. *Constraint* (8) ensures that the number of workers in the warehouse does not exceed the maximum allowed number in any interval of the workday. *Constraint* (9) ensures that every worker has a break of the defined duration in the defined period. *Constraint* (10) prevents a situation in which workers are coming and leaving all the time. *Constraint* (11) defines the nature of the variables.

4. Numerical Experiments

The potential of the WAES model to provide workforce scheduling solutions with a lower workforce cost is firstly tested for the input values provided by engineering managers of the real DW and then on randomly generated instances.

4.1. The Real DW-WSP

The DW operates from 8:00 to 22:00 and employs full-time workers and part-time workers. Full-time shifts consist of 8 h of work and 1 h of break. Part-time shifts are 4-h shifts without a break. Let the minimal time interval of the day be 1 h. Accordingly, there are 14 intervals in one workday and six potential full-time shifts (i.e., $n = 14$, $p = 1$, and $M_1 = \{1, \ldots, 6\}$). The number of potential part-time shifts is nine, and they are denoted by numbers following the number of the last full-time shift (i.e., $M_2 = \{7, \ldots, 15\}$). The number of workers employed in part-time shifts is limited to 30% of the total number of employed workers (i.e., $w = 0.3$).

There are ten different activities in the DW and nine profiles of workers that can be employed. The compatibility matrix (Table 2) specifies for each worker profile the activities which they can perform. The costs of worker profiles were not available, but the ratio between the cost of each worker profile and the lowest cost among them were available. These ratios are used as costs of worker profiles in the model. The last column of the compatibility matrix presents these costs. The costs are valid for the engagement in full-time shifts in one workday. For part-time shifts, the costs of workers' engagement are half of the presented costs. For example, the cost of full-time engagement of one worker of profile 1 is 1.28 for one workday.

Table 2. Compatibility matrix.

Worker Profile	Activities										Cost
	1	2	3	4	5	6	7	8	9	10	
1	*	*	*	*							1.28
2	*	*	*	*							1.4
3	*	*	*	*							1
4	*	*	*	*							1.28
5	*	*	*	*	*				*	*	1.6
6	*					*					1.6
7	*						*	*		*	1.4
8	*	*	*	*					*		1.36
9	*	*	*	*			*	*	*	*	1.4

Legend **Worker profiles** 1- Order picker 2- Controller 3- Auxiliary worker 4- Packer	5- Supervisor 6- Controller at reception 7- Forklift driver 8- WMS receiver 9- WMS forklift driver	**Activities** 1- Order picking 2- Control 3- Auxiliary jobs 4- Preparing goods for loading 5- Supervision	6- Control at reception 7- Loading vehicles 8- Unloading vehicles 9- WMS goods receiving 10-Putting away and extracting of goods

* Worker is qualified for a certain activity

For the sake of discussion and ease of presentation, the problem is restricted to activities 1, 6, 7, and 8, and worker profiles 1, 6, and 7. Activities 6, 7, and 8 belong to set A_1, whereas activity 1 belongs to set A_2. Activity 6 depends on activity 8 with a dependence of 50% (i.e., $G_6 \in \{8\}$ and $s_8 = 0.5$). The other three activities are independent. Demand for workers of independent activities was obtained from the engineering managers of the DW and is presented in Table 3. From the compatibility matrix, it follows that $T_1 \in \{1,6,7\}$, $T_6 = \{6\}$, $T_7 = \{1\}$, and $T_8 = \{1\}$, while $c_1 = 1.28$, $c_6 = 1.6$, and $c_7 = 1.4$. In addition, it follows that $H_1 \in \{1\}$, $H_6 \in \{1,6\}$, and $H_7 \in \{1,7,8\}$. Each full-time worker needs to have the fourth, fifth, or sixth interval of his shift free (for example, $O_2 = \{5, 6, 7\}$). Accordingly, it follows that $p = 1$ and $f = 3$. The maximum number of active workers in the warehouse is set to be 30 (i.e., $q = 30$).

Table 3. The solution of the real DW-WSP.

Interval		Shifts																		Workers Demand		
		1					2								8		11					
No.	h.	e7	e7	e7	e7	e7	e1	e1	e1	e1	e1	e1	e1	e1	e7	e7	e6	e6	e6	a7	a1	a8
1	8	a7	a7	a7	a7	a7	/	/	/	/	/	/	/	/	/	/	/	/	/	7	11	0
2	9	a7	a7	a7	a7	a7	a1	a1	a1	a1	a1	a1	a1	a1	a7	a7	/	/	/	8	10	0
3	10	a7	a7	a7	a7	a7	a1	a1	a1	a1	a1	a1	a1	a1	a7	a7	/	/	/	5	7	0
4	11	a7	B	B	a7	a7	a1	a1	a1	a1	a1	a1	a1	a1	a7	a7	/	/	/	4	8	0
5	12	a8	a8	a8	a8	B	B	B	B	B	B	B	B	B	a8	a8	a6	a6	a6	0	7	6
6	13	B	a8	a8	B	a8	a1	a1	a1	a1	a1	a1	X	X	a8	a8	a6	a6	a6	0	6	5
7	14	a1	a1	a8	a8	a8	a1	a1	a1	a1	a1	a1	a1	X	/	/	a1	a6	a6	0	9	3
8	15	a1	a1	a1	a8	a8	a1	a1	a1	a1	a1	a1	a1	a1	/	/	a1	a1	a6	0	10	2
9	16	a1	a1	a1	a1	a1	a1	a1	a1	a1	a1	a1	a1	a1	/	/	a1	a1	a1	0	9	0
10	17	/	/	/	/	/	a1	a1	a1	a1	a1	a1	a1	a1	/	/	/	/	/	0	0	0
11	18	/	/	/	/	/	/	/	/	/	/	/	/	/	/	/	/	/	/	0	0	0
12	19	/	/	/	/	/	/	/	/	/	/	/	/	/	/	/	/	/	/	0	0	0
13	20	/	/	/	/	/	/	/	/	/	/	/	/	/	/	/	/	/	/	0	0	0
14	21	/	/	/	/	/	/	/	/	/	/	/	/	/	/	/	/	/	/	0	0	0

Legend

Worker profiles
e1—Order picker
e6—Controller at receptione7—Forklift driver

Activities
a1—Order picking
a6—Control at reception
a7—Loading vehicles
a8—Unloading vehicles

Other
B—Break
X—Without work
/—Interval is not part of the shift

The policy of the distribution company defines that trucks for loading can wait up to two intervals to be loaded. Trucks for unloading need to be unloaded within one interval, starting from the interval in which they checked in for unloading. During the unloading, shipments are controlled at the reception without delay. The distribution company dispatches all orders received before 16:00 in the morning of the following workday. As trucks do not arrive strictly at the beginning of an interval, a truck checked in for unloading or loading during an interval is considered as checked in at the beginning of the subsequent interval. Therefore, the execution of these four activities needs to be finished according to the following parameters: $v_6 = 0, v_7 = 1, v_8 = 0$, and $r_1 = 14$. The exceptions are trucks checked in for unloading in the next to last interval of the workday. They need to be processed until the end of the last interval. Trucks that arrive during the last interval are not processed.

Currently, engineering managers of the DW do not consider the AES and strive to execute each activity when the need for it occurs. To examine the cost of the DW-WSP solution in such circumstances, the problem is solved with the WS model of the problem that excludes the possibility of the AES. This model is obtained by setting all deadline parameters to zero ($v_6 = 0$, $v_7 = 0$, $v_8 = 0$ and $r_1 = 0$) in the WAES model. In that case, the obtained solution of DW-WSP has a cost of 22.95. If the WAES model is used—that is, if deadline parameters have real values ($v_6 = 0$, $v_7 = 1$, $v_8 = 0$, and $r_1 = 14$) and the AES is possible—a solution with the cost of 21.04 is obtained. In both cases, the problem is solved in Python by means of the PuLP linear programming package. The integrality gap is set to be up to 3%. The workforce schedule that follows from the last obtained solution is demonstrated in Table 3. For making the schedule from the solution of the model, an algorithm in Python was developed.

The solution implies a schedule with three full-time shifts and two part-time shifts. Full-time shifts are 1 and 2. Shift 1 starts at 8 o'clock and shift 2 at 9 o'clock. Part-time shifts are 8 and 11. Shift 8 starts at 9:00 and shift 11 at 12:00. In shift 1, there are five workers of profile 7, in shift 2 there are eight workers of profile 1, and so on. If one of the workers is observed individually, it can be seen what activity he is engaged in at each interval of his shift and when he has a break. For example, the first worker in shift 1 works on activity 7 in the first four intervals; in the fifth interval, he works on activity 8; in the seventh interval, he has a rest; and then until the end of the workday, he works on activity 1. If the execution of an activity is observed, it can be seen which workers execute the activity in each interval. For example, activity 7 in the first interval requires seven workers of profile 7. According to the schedule, five workers perform this activity in the first interval. The need for two workers is transferred to the next interval because this activity allows a delay in execution of one hour. The solution also contains three intervals in which some workers do not work due to lack of work. This is the sixth interval for the seventh and the eighth worker in shift 2, and the seventh interval for the eighth worker in the same shift. With the application of the WAES model, the number of such intervals is reduced and the costs of the solutions are decreased. The main reason for this is the application of the AES within the WAES model.

4.2. *Randomly Generated Instances of the DW-WSP*

The idea of using the AES to decrease the cost of the workforce has been shown to be very helpful in the solved example. To generalize this observation, 100 instances of the same problem were solved as in the real example. Instances of the problem differ in the number of required workers by intervals for activities. The required numbers of workers are randomly generated for each interval of the workday and each independent activity. Randomly generated values are set to be between 0 (included) and 10 (included). All other features and input values are the same for each instance of the problem and correspond to those in the solved example. Results of the testing show that for each instance of the problem, a lower cost is obtained if the WAES model is used for solving the problem. For 61% of instances of the problem, the obtained cost of the workforce is lower by more than 20% compared to the cost obtained by using the WS model (Figure 4).

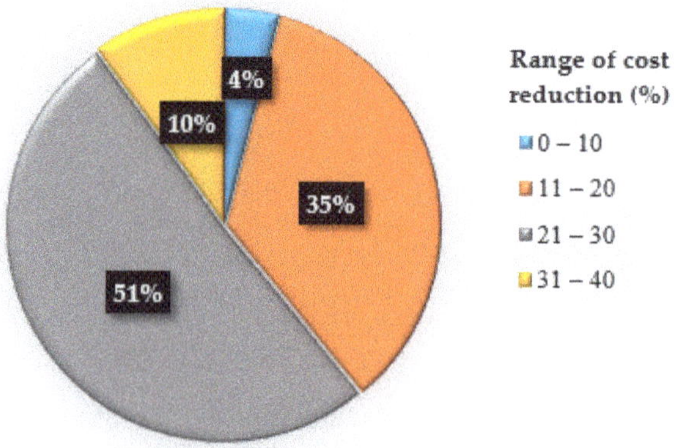

Figure 4. Cost reduction with the workforce activities execution scheduling (WAES) model in the percent of instances.

If we consider efficiency as the ratio of realized costs to costs that are theoretically sufficient to perform a given workload, then changes in the efficiency of the workforce can also be calculated to examine the effects of solving the problem as the WAES model (i.e., performing workforce scheduling together with the AES). For an activity, the theoretical lowest costs can be calculated by multiplying the hours required to perform the activity with the cost per hour of the cheapest profile of a worker who can perform the activity. By using theoretical costs of activities, the overall theoretical costs are obtained. For the real example of DW-WSP, the overall theoretical costs are 18.24. The efficiency of the workforce in the case of using the WS model for solving the problem is 79%, whereas in the case of using the WAES model, it is 87%. Therefore the efficiency is increased by 9%. The increases in efficiency for the randomly generated instances of the same problem are presented in the next figure (Figure 5).

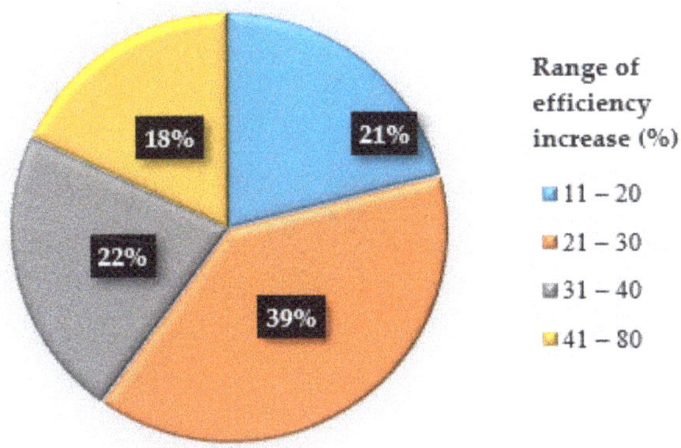

Figure 5. Efficiency increase with the WAES model in the percent of instances.

5. Discussion

The solutions of the example and the randomly generated instances indicate the importance of taking into account the possibility of AES in workforce scheduling in DW. It has proved to be good assistance for minimizing the variability of demand for workers and for better utilization of workers. The number of intervals without work for workers is decreased, and the use of workers in activities that are primarily intended for them is increased. The consequence is decreasing the cost of the workforce in DW if the WAES model is used for the DW-WSP.

Taking into account that the execution of some activities in DW can be completely or partly postponed, the model allows engineering managers to schedule workers in a better way and optimize the cost of the workforce. In workforce planning and scheduling models, jobs are usually considered as tasks that last only a certain period and require the same number of workers during that time. Interrupting a task that is being performed or replacing an assigned worker by another is not considered, or it is penalized. In other words, demand for workers has to be fulfilled at the time when it occurs. In this model, the fulfillment of some activities' demand for workers is scheduled following the available time for fulfillment. In addition, a worker scheduled on an activity can be replaced by another during the same shift. These features of some activities in DW enable more flexible workforce scheduling and provide an additional opportunity for optimizing the cost of the workforce.

At the same time, the model allows engineering managers to shape the final solution by setting certain parameters of the model. With these parameters, they can set the period allowed for a break during a shift, the break duration, the maximum allowed number of active workers, deadlines for finishing activities, the percent of part-time workers, the duration of shifts, and so on. The model can also be used for determining the number of part-time workers by days. If a workweek includes seven workdays, then the model can be used for obtaining inputs for a day-off scheduling problem as well. A model for solving such a problem, that also includes different worker profiles, can be found in Billionnet et al. [43].

Previously, not many studies have been devoted to raising warehouse sustainability through scheduling the execution of activities over time, especially not with limitations that take into account working conditions. Many studies dealing with sustainable warehousing are dedicated to building models for assessing warehouse sustainability. Other studies deal with energy savings in the process of handling goods. This paper offers a model that enables the raising of economic sustainability in a specific way, taking into account the social aspects of sustainability as well. A special feature of the paper is that it takes into account the needs of workers in optimizing work processes. It not only improves social sustainability, but indirectly enables the achievement of economic benefits as well. Workers are more productive, fewer costs are spent on training new workers, expenses due to workers going to competing companies are lowered, expenses caused by absences from work are decreased, and so on. According to Andriansyah et al. [7], sustainable development implies an efficient use of both natural and human resources. Human capital is a company's greatest asset [44].

6. Conclusions

This study provides engineering managers with a sustainable engineering model that integrates workforce scheduling of workers on activities and of the execution of activities. By integrating the scheduling of the workforce and activities, greater opportunities are achieved for optimizing workforce utilization and raising the sustainability of warehouses to a higher degree. Frequent changes in demand have forced companies with distribution functions to attach more importance to the sustainability of their warehouses today. The success of these companies largely depends on the sustainability of their warehouses. Given that insufficient use of resources leads to losses and the unsustainability of the warehouse, it is very important to plan and schedule activities in the warehouse and of its resources.

This article provides a solution that allows for the reduction of costs in the warehouse and an increase in its efficiency and sustainability. Optimal scheduling of activities in the warehouse over time reduces the necessary resources and costs and opens space for processing even larger quantities of goods with the same or fewer resources. By optimal scheduling, the stress on the warehouse system in certain periods is reduced, and thus the possibility of errors in operation, damage of goods and equipment, delays in delivery, and so on. All of this provides opportunities for raising economic sustainability. Optimal scheduling of activities and resources in the warehouse also reduces the stress on external systems such as the power grid or roads in certain periods, thus increasing environmental sustainability. In addition, the workload of workers is scheduled more evenly over time, as well as between workers. Each worker has a break in the planned period, and no more workers operate at the same time than allowed for the given space. In this way, it is possible to increase the productivity of workers, reduce their mistakes, turnover, and injuries, and increase the overall social sustainability of the warehouse.

This paper contributes primarily to raising the economic and social sustainability of the warehouse. From that point of view, it could be said that this is a limitation of this paper. However, if we take into account approaches for raising environmental sustainability, then it can be said that this paper complements these approaches. The environmental sustainability within the warehouse can be raised by reducing energy consumption, primarily for cooling, heating, air conditioning, and handling of goods. Some studies [37,45,46] have dealt with how to accomplish this. Moreover, environmental sustainability in the warehouse can be raised by reducing waste production and water consumption, and through the recycling and reuse of packaging, pallets, paper, and so on. Other studies [39,47,48] have dealt with these issues. By providing a solution for increasing social and economic sustainability, this paper provides a sustainable basis for investing in solutions that improve the environmental sustainability of the warehouse.

Having all of the above in mind, the new sustainable warehouse management approach for workforce and activities scheduling should be a useful tool for increasing the sustainability of the warehouse. In addition, the achievements of this study represent a good base for further research, which should make this model even more useful in practice. The cost of the workforce can be decreased even more if some of the parameters in the model also become the subject of optimization. In future models, it would be useful to pay more attention to other aspects of sustainability in DW using similar approaches, but also using hybrid approaches that combine several different methods. Future research may also be aimed at predicting the warehouse's workload and the needed number of workers. One of the most important tasks should be to define the factors influencing the prediction of workload and the needed number of workers in addition to historical data on the workload. The current situation, government measures, and announcements regarding the COVID-19 crisis should be recognized in this prediction.

Author Contributions: Conceptualization, V.P., M.K., M.A., D.D., and B.J.; methodology, V.P., M.K., M.A., D.D., and B.J.; software, V.P., M.K., M.A., D.D., and B.J.; validation V.P., M.K., M.A., D.D., and B.J.; formal analysis, V.P., M.K., M.A., D.D., and B.J.; investigation, V.P., M.K., M.A., D.D., and B.J.; resources, V.P., M.K., M.A., D.D., and B.J.; data curation, V.P., M.K., M.A., D.D., and B.J.; writing—original draft preparation, V.P., M.K., M.A., D.D., and B.J.; writing—review and editing, V.P., M.K., M.A., D.D., and B.J. All authors have read and agreed to the published version of the manuscript.

Funding: This research received no external funding.

Institutional Review Board Statement: Not applicable.

Informed Consent Statement: Not applicable.

Data Availability Statement: Not applicable.

Conflicts of Interest: The authors declare no conflict of interest.

References

1. Bartolini, M.; Bottani, E.; Grosse, E.H. Green warehousing: Systematic literature review and bibliometric analysis. *J. Clean. Prod.* **2019**, *226*, 242–258. [CrossRef]
2. Minashkina, D.; Happonen, A. Decarbonizing warehousing activities through digitalization and automatization with WMS integration for sustainability supporting operations. In Proceedings of the E3S Web of Conferences, Melbourne, Australia, 18–20 December 2020; p. 158.
3. Tan, K.S.; Ahmed, M.D.; Sundaram, D. Sustainable Warehouse Management. In Proceedings of the International Workshop on Enterprises & Organizational Modeling and Simulation, Amsterdam, The Netherlands, 8–9 June 2009; pp. 1–15.
4. Indrawati, S.; Miranda, S.; Pratama, A.B. Model of Warehouse Performance Measurement Based on Sustainable Warehouse Design. In Proceedings of the 4th International Conference on Science and Technology (ICST), Yogyakarta, Indonesia, 7–8 August 2018.
5. Rosen, M.A. Engineering Sustainability: A Technical Approach to Sustainability. *Sustainability* **2012**, *4*, 2270. [CrossRef]
6. Dekker, R.; Bloemhof, J.; Mallidis, I. Operations Research for Green Logistics—An Overview of Aspects, Issues, Contributions and Challenges. *Eur. J. Oper. Res.* **2012**, *219*, 671–679. [CrossRef]
7. Andriansyah, R.; Etman, L.F.P.; Rooda, J.E. On sustainable operation of warehouse order picking systems. In Proceedings of the XIV Summer School 'Francesco Turco, Porto Giardino, Italy, 15–19 September 2009.
8. Burinskiene, A.; Lorenc, A.; Lerher, T. A simulation study for the sustainability and reduction of waste in warehouse logistics. *Int. J. Simul. Model.* **2018**, *3*, 485–497. [CrossRef]
9. Boenzi, F.; Digiesi, S.; Facchini, F.; Mossa, G.; Mummolo, G. Sustainable Warehouse Logistics: A NIP Model for non-road vehicles and storage configuration selection. In Proceedings of the XX Summer School "Francesco Turco-Operational, Excellence, Experiences", Napoli, Italy, 16–18 September 2015.
10. Pashkevich, N.; Haftor, D.; Karlsson, M.; Chowdhury, S. Sustainability through the Digitalization of Industrial Machines: Complementary Factors of Fue Consumption and Productivity for Forklifts with Sensors. *Sustainability* **2019**, *11*, 6708. [CrossRef]
11. Lorenc, A.; Lerher, T. PickupSimulo—Prototype of Intelligent Software to Support Warehouse Managers Decisions for Product Allocation Problem. *Appl. Sci.* **2020**, *10*, 8683. [CrossRef]
12. Demassey, S.; Pesant, G.; Rousseau, L.M. Constraint programming based column generation for employee timetabling. In Proceedings of the 2nd International Conference on Integration of AI and OR Techniques in Constraint Programming for Combinatorial Optimization Problems, Prague, Czech Republic, 31 May–1 June 2005.

13. Côté, M.C.; Gendron, B.; Rousseau, L.M. Grammar-based column generation for personalized multi-activity shift scheduling. *Manage. Sci.* **2012**, *57*, 151–163. [CrossRef]
14. Lequy, Q.; Bouchard, M.; Desaulniers, G.; Soumis, F.; Tachefine, B. Assigning multiple activities to work shifts. *J. Sched.* **2010**, *15*, 239–251. [CrossRef]
15. Carli, R.; Dotoli, M.; Digiesi, S.; Facchini, F.; Mossa, G. Sustainable Scheduling of Material Handling Activities in Labor-Intensive Warehouses: A Decision and Control Model. *Sustainability* **2020**, *12*, 3111. [CrossRef]
16. Ernst, A.T.; Jiang, H.; Krishnamoorthy, M.; Sier, D. Staff scheduling and rostering: A review of applications, methods and models. *Eur. J. Oper. Res.* **2004**, *153*, 3–27. [CrossRef]
17. Ernst, A.T.; Jiang, H.; Krishnamoorthy, M.; Owens, B.; Sier, D. An annotated bibliography of personnel scheduling and rostering. *Ann. Oper. Res* **2004**, *127*, 121–144. [CrossRef]
18. Van den Bergh, J.; Beliën, J.; De Bruecker, P.; Demeulemeester, E.; De Boeck, L. Personnel scheduling: A literature review. *Eur. J. Oper. Res.* **2013**, *226*, 367–385. [CrossRef]
19. De Causmaecker, P.; Demeester, P.; Vanden, B.G.; Verbeke, B. Analysis of real-world personnel scheduling problems. In Proceedings of the 5th International Conference on the Practice and Theory of Automated Timetabling, Pittsburgh, PA, USA, 18–20 August 2004.
20. Günther, M.; Nissen, V. A comparison of three heuristics on a practical case of sub-daily staff scheduling. In Proceedings of the 8th International Conference on the Practice and Theory of Automated Timetabling, Belfast, UK, 10–13 August 2010.
21. Günther, M.; Nissen, V. Sub-daily staff scheduling for a logistics service provider. *KI Künstliche Intell.* **2010**, *24*, 105–113. [CrossRef]
22. Bard, J.F.; Wan, L. Workforce design with movement restrictions between workstation groups. *Manuf. Serv. Oper. Manag.* **2008**, *40*, 24–42. [CrossRef]
23. Ladier, A.L.; Alpan, G.; Penz, B. Joint employee weekly timetabling and daily rostering: A decision-support tool for a logistics platform. *Eur. J. Oper. Res.* **2014**, *234*, 278–291. [CrossRef]
24. Ladier, A.L.; Alpan, G. Integrating truck scheduling and employee rostering in a cross-docking platform—An iterative approach. In Proceedings of the 6th International Conference on Industrial Engineering and Systems Management, Seville, Spain, 21–23 October 2015.
25. Demassey, S.; Pesant, G.; Rousseau, L.M. A cost-regular based hybrid column generation approach. *Constraints* **2006**, *11*, 315–333. [CrossRef]
26. Côté, M.C.; Gendron, B.; Rousseaum, L.M. Grammar-based integer programming models for multi-activity shift scheduling. *Electron. Notes Discret. Math* **2010**, *36*, 727–734. [CrossRef]
27. Restrepo, M.I.; Lozano, L.; Medaglia, A.L. Constrained network-based column generation for the multi-activity shift scheduling problem. *Int. J. Prod. Econ.* **2012**, *140*, 466–472. [CrossRef]
28. Ritzman, L.; Krajewski, L.J.; Showalter, M.J. The disaggregation of aggregate manpower plans. *Manag. Sci.* **1976**, *22*, 1204–1214. [CrossRef]
29. Loucks, J.; Jacobs, F.R. Tour scheduling and task assignment of a heterogeneous workforce: A heuristic approach. *Decision Sci.* **1991**, *22*, 719–739. [CrossRef]
30. Côté, M.C.; Gendron, B.; Rousseau, L.M. Modeling the regular constraint with integer programming. In Proceedings of the 4th International Conference on Integration of AI and OR Techniques in Constraint Programming for Combinatorial Optimization Problems, Brussels, Belgium, 23–26 May 2007.
31. Lequy, Q.; Desaulniers, G.; Solomon, M.M. A two-stage heuristic for multi-activity and task assignment to work shifts. *Comput. Ind. Eng.* **2012**, *63*, 831–841. [CrossRef]
32. Côté, M.C.; Gendron, B.; Quimper, C.; Rousseau, L.M. Formal languages for integer programming modeling of shift scheduling problems. *Constraints* **2009**, *16*, 54–76. [CrossRef]
33. Boyer, V.; Gendron, B.; Rousseau, L.M. A branch-and-price algorithm for the multi-activity multi-task shift scheduling problem. *J. Sched.* **2014**, *17*, 185–197. [CrossRef]
34. Dahmen, S.; Rekik, M. Solving multi-activity multi-day shift scheduling problems with a hybrid heuristic. *J. Sched.* **2014**, *18*, 207–223. [CrossRef]
35. Bank, R.; Murphy, R. Warehousing Sustainability Standards Development. In Proceedings of the IFIP International Conference on Advances in Production Management Systems, State College, PA, USA, 9–12 September 2013.
36. Torabizadeh, M.; Yusof, N.M.; Ma'aram, A.; Shaharoun, A.M. Identifying sustainable warehouse management system indicators and proposing new weighting method. *J. Clean. Prod.* **2019**, *248*, 119190. [CrossRef]
37. Freis, J.; Vohlida, P.; Guthner, V.A. Low-Carbon Warehousing: Examining Impacts of Building and Intra-Logistics Design Options on Energy Demand and the CO_2 Emissions of Logistics Centers. *Sustainability* **2016**, *8*, 448. [CrossRef]
38. Kamarulzaman, N.K.; Hussin, H.; Mahir Abdullah, A.M.; Rahman, A.A. Green warehousing initiatives towards environmental sustainability: Adoption and performance in the malaysian food-based industry. In Proceedings of the ICAM, Jember, Indonesia, 25–26 June 2012.
39. Amjed, T.W.; Harrison, N.R. A model for sustainable warehousing: From theory to best practices. In Proceedings of the EventInternational Decision Sciences Institute Conference and Asia Pacific Decision Sciences Institute Conference, Bali, Indonesia, 9–13 July 2013.

40. Malinowska, M.; Rzeczycki, A.; Sowa, M. Roadmap to sustainable warehouse. In Proceedings of the SHS Web of Conferences InfoGlob, Gdańsk, Nynäshamn, Poland, 18–20 September 2018.
41. Durmić, E. The evaluation of the criteria for sustainable supplier selection by using the fucom method. *ORESTA* **2019**, *2*, 91–107. [CrossRef]
42. Đalić, I.; Stević, Ž.; Karamasa, C.; Puška, A. A Novel Integrated Fuzzy PIPRECIA—Interval Rough Saw Model: Green Supplier Selection. *DMAME* **2020**, *3*, 126–145. [CrossRef]
43. Billionnet, A. Integer programming to schedule a hierarchical workforce with variable demands. *Eur. J. Oper. Res.* **1999**, *114*, 105–114. [CrossRef]
44. Merriman, K.K. *Valuation of Human Capital—Quantifying the Importance of an Assembled Workforce*; Palgrave Macmillan, Manning School of Business, University of Massachusetts Lowell, USA Springer Nature: New York, NY, USA, 2017.
45. Ene, S.; Küçükoglu, I.; Aksoy, A.; Oztürk, N. A genetic algorithm for minimizing energy consumption in warehouses. *Energy* **2016**, *114*, 973–980. [CrossRef]
46. Lorenc, A.; Lerher, T. Effectiveness of product storage policy according to classification criteria and warehouse size. *FME Trans.* **2019**, *47*, 142–150. [CrossRef]
47. Luu, M. Developing the implementation of green warehousing at IKEA Finland. Bachelor's Thesis, Haaga Helia University of Applied Sciences, Helsinki, Finland, 2016.
48. Coelho, P.M.; Corona, B.; Klooster, R.; Worrell, E. Sustainability of reusable packaging—Current situation and trends. *Resour. Conserv. Recycl. X* **2020**, *6*, 100037.

sustainability

MDPI

Article

Mathematical Analysis of Criteria for Maintenance of Technical Systems in the Function of Achieving Sustainability

Goran Otić [1], Oliver Momčilović [2], Ljiljana Radovanović [3,*], Goran Jovanov [4], Dragica Radosav [5] and Jasmina Pekez [3]

1 Military Medical Academy, Crnotravska 17, 11000 Belgrade, Serbia; goran.otic@gmail.com
2 Department of Engineering Management, Faculty of Information Technology and Engineering, University "Union-Nikola Tesla", Belgrade, Jurija Gagarina 149a, 11070 New Belgrade, Serbia; oliver.momcilovic@fiti.edu.rs
3 Department of Mechanical Engineering, Technical Faculty "Mihajlo Pupin", University of Novi Sad, Đure Đakovića bb, 23000 Zrenjanin, Serbia; jasmina.pekez@tfzr.rs
4 Department of Forensics, University of Criminal Investigation and Police Studies, Cara Dušana 196, 11080 Belgrade, Serbia; goran.jovanov@kpu.edu.rs
5 Department of Information Technology, Technical Faculty "Mihajlo Pupin", University of Novi Sad, Đure Đakovića bb, 23000 Zrenjanin, Serbia; dragica.radosav@tfzr.rs
* Correspondence: ljiljana.radovanovic@tfzr.rs

Abstract: Achieving sustainable development requires strategic efforts involving the entire organization. Maintenance efforts also play an important role. Company management needs to understand and develop an appropriate strategy to achieve sustainable development by applying maintenance performance measurements. The aim of this paper is to present possible ways of analyzing and ranking the impact of certain criteria with respect to achieving sustainability. The paper uses the method of Structural Equation Modeling—SEM in order to determine the most influential variable on the sustainability of maintenance of technical systems. Based on the set theoretical system model, for all its variables in the model, statements were made that describe them, on which 136 respondents gave their views (from 1 to 5, Likert scale) in the territory of the Republic of Serbia. An intuitive F-DEMATEL method was also used to prioritize variables. A team of 10 experts in the field of maintenance of technical systems was compared the criteria A—Application of technical diagnostics, B—Management of maintenance resources, C—Maintenance process planning, and the dependent variable D—Sustainability of maintenance of technical systems. According to experts, the importance of the criteria coincides with the results obtained by a survey with 136 respondents.

Keywords: criteria; maintenance; sustainability; SEM; F-DEMATEL

check for updates

Citation: Otić, G.; Momčilović, O.; Radovanović, L.; Jovanov, G.; Radosav, D.; Pekez, J. Mathematical Analysis of Criteria for Maintenance of Technical Systems in the Function of Achieving Sustainability. *Sustainability* **2021**, *13*, 1680. https://doi.org/10.3390/su13041680

Academic Editor: António Abreu
Received: 31 December 2020
Accepted: 30 January 2021
Published: 4 February 2021

Publisher's Note: MDPI stays neutral with regard to jurisdictional claims in published maps and institutional affiliations.

Copyright: © 2021 by the authors. Licensee MDPI, Basel, Switzerland. This article is an open access article distributed under the terms and conditions of the Creative Commons Attribution (CC BY) license (https://creativecommons.org/licenses/by/4.0/).

1. Introduction

Maintenance is the basic logistics of industrial systems, both in terms of their working capacity and in terms of meeting the requirements of environmental protection and sustainable development of the maintenance as a whole. It is a multidisciplinary set of indirect (preparatory) and direct (executive) activities to predict, prevent and eliminate failures of machinery and equipment in order to achieve the optimal degree of system effectiveness [1].

Sustainable Maintenance (SM) should contribute to the minimization of environmental and social impacts of a system, the reduction of life cycle costs and enhancement of equipment durability and socioeconomic well-being [2].

Some papers suggest a preliminary framework to integrate sustainability issues into the maintenance performance measurement in automotive companies [3,4] and in maintenance dashboards [5] or rank attributes/indicators that contribute to a Sustainable Maintenance performance evaluation [6–8]. In research [9], the sustainability improvements achieved, relative to the company's initial situation after implementing a lean and green manufacturing system, instead propose the new OEEE (Overall Environmental Equipment Effectiveness) indicator to evaluate the environmental impact of the asset life cycle. A common general classification of sustainable performance indicators for maintenance is not found using the Scoping Literature Review. It will be necessary the integration of such indicators in maintenance policies to achieve a Sustainable Maintenance management [10].

This paper assesses the relative importance of each of the best criteria for the sustainable management of the maintenance of national heritage buildings in Malaysia. The results show that "training and expertise of maintenance staff" are the most important criteria that respondents consider to be key in supporting sustainable best practice [11]. Sustainable maintenance has introduced a new category into the two previously existing ones (economic and environmental). The third category is social issues and the search for a balance between three aspects: financial, environmental, and social. Consequently, natural maintenance is included in the realization of a sustainable production approach. From practical point of view, it requires changes in approach to maintenance represented by managers and changes in actions performed within maintenance area [12].

Mathematically, if sustainable maintenance practice is denoted by an acronym (SMP), then the connecting equation expressing the SMP is given in SMP = F(M, D, P, C, F, M, E) where M is the machine records keeping, D is diagnostic technique, P stands for prognostic technique, and C is the machine condition monitoring technology. These four factors are inherent and internal factors associated with data keeping. F, M, and E stand for machines functionality, manufacturability, and environmental impact, respectively. These three factors are external factors imposed on the machines due to demand pressure and rivalry competitions that may arise in the course of production, which will in one way or another affect the machinery functionalities [13].

The main goal of research [14] is to define a sustainable approach for the maintenance of asphalt pavement construction. Predictive maintenance is inevitable for sustainable smart manufacturing in I4.0 in research [15].

2. Objectives of the Maintenance Process

Maintenance is a process that enables the management of technical condition and reliability throughout the entire life cycle of the system. Furthermore, maintenance goals, among other things, go into the sphere of business economics, so they are expressed in the form of rationalization and are in principle measurable. The main goals to be achieved in the maintenance process which are also in the function of sustainability are [16] the following:

- Providing (maximizing) the required level of reliability of technical systems in the process of exploitation (reliability is the probability of operation without failure during time period, with the greatest possible availability of the system and with as few failures as possible);
- Minimization of total maintenance costs (direct and indirect);
- Prevention and limitation of obsolescence of technical systems;
- Joint participation (with all functions in the company) in production and financial management (integration into the CIM production system, i.e., computer-integrated production);
- Achieving better product quality;
- Increasing labor productivity in production;
- Increasing the level of motivation for work (faulty technical systems cause bad conditions and relationships, and even accidents, both in the company and its environment);
- Keeping all available resources in the company;
- Knowledge management;
- Delivery of products to customers on time, etc.

3. Formulating a New Vision of Maintenance

The main purpose of the maintenance system is to ensure the efficient operation of the company with the shortest possible downtime. Creating a vision of one of the strategically capable maintenance systems is the basic driving force of the company. The goal of such a conceptualized maintenance is to incorporate this fact into all company processes. This means that maintenance employees, regardless of their expertise, must understand that their primary role is to contribute to the company profitability.

The basic functions of the organizational maintenance system are [17] the following:

- Diagnosing the causes of technical system failure;
- Elimination of the resulting failure condition.

However, basic functions cannot exist on their own. Continuous functional maintenance means that the system must contain the following set of functions:

- A set of functions to ensure continuity;
- A set of operational service functions;
- A set of operational management functions.

The basic aspect of achieving the functioning and survival of the organizational maintenance system is to provide a financial function. A financial function is needed to provide the necessary resources in the form of equipment, technical and other materials. However, in addition to the necessary financial resources, a functional maintenance process must also include materials, equipment, and technical apparatus in order to be able to perform basic functions. All these items belong to the function of providing material and technical resources.

The fourth necessary function to achieve continuity of maintenance is the required human resource. Well-trained and professional staff should be behind the quality functioning of maintenance. Professional, technical staff of various profiles with accompanying administrative staff make the maintenance system as it is. Hiring staff, predicting the required number of employees, middle technical staff to perform the function F0 with the accompanying administrative staff for the implementation of smooth maintenance work is a function of providing human resources. The financing of salaries of employees is within the scope of the financial function and is paid on the basis of the required gross funds for the payment of salaries at the company level.

The following set of functions that plays an important role in achieving maintenance continuity is a set of operational service functions [17]:

- Information provision function;
- Function for monitoring economic and financial trends;
- Function of ensuring the correctness of technical support;
- Legal problem solving function;
- Function of ensuring the correctness of functioning.

4. Identifying Maintenance Criteria for Achieving Sustainability

Based on the review of domestic literature [16,18–24], foreign literature [25–28], research in our country [29–32], the following aspects or elements of maintenance for which is important to establish an adequate management system are listed [17]:

- Maintenance staff;
- Maintenance organization;
- Maintenance planning;
- Quality control in maintenance;
- Application of information technologies in the maintenance process;
- Spare parts, equipment, and maintenance materials;
- Existing technical systems, protection, supervision, and diagnostic equipment;
- Technical diagnostics;
- Development of maintenance technology;
- Application of modern maintenance methodologies;

- Decision making in the maintenance process;
- Maintenance costs.

The impact of each of the mentioned maintenance criteria on sustainability will not be considered in this paper. Namely, this is only a proposal of certain criteria that are used as a framework for defining influential or relevant criteria for sustainability for a particular company in a particular industry.

5. Materials and Methods

5.1. Theoretical Settings, Structural Equation Modeling—SEM

During the 1970s, a lot of work was done on the development of mathematical-statistical causal analysis that would enable testing and proving causal hypotheses in the field of social sciences. For these purposes, Structural Equation Modeling (hereinafter SEM) is often present [33–35]. In current research in various fields of psychology, SEM is an integral part of research methodology [36–38]. Modeling with structural equations presents a series of hypotheses about how the variables in the analysis are generated and how they relate to each other [39,40]. SEM includes analysis of latent variables and pathway analysis to test hypothetical models and detect the relationship of manifest and latent variables [34]. Modeling by structural equations in this paper is used to study the correlation and regression analyzes of variables of the system model set. Correlation analysis is used to researches and quantifies the connection between the observed phenomena, i.e., variables. The correlation coefficient shows the degree of statistical correlation of phenomena, i.e., variables. In this research, the coefficient of multiple linear correlation is used, which means that the degree of existence of a linear relationship of one in relation to two or more variables is studied. This degree is calculated by (Formula (1)):

$$r = \frac{\sigma_{xy}}{\sigma_x \cdot \sigma_y} \tag{1}$$

wherein

σ_x and σ_y—standard deviations of variables X and Y;
σ_{xy}—covariance (arithmetic mean of the product of the deviation of variables from their arithmetic means).

Regression analysis determines the analytical relationship between the phenomenon, i.e., the variables. If the model expresses a linear relationship between dependent variables and two or more variables, it is a multiple linear regression model. The simplest form of dependence is given (Formula (2)):

$$y = ax_1 + ax_2 + ax_3 + \cdots ax_k + b \tag{2}$$

wherein

a—regression coefficient—which shows how much the dependent variable changes on average, if the independent variable changes by one;
b—is a constant and shows the value of the dependent variable in the case when the independent variable equals zero.

5.2. Theoretical Assumptions of the F-DEMATEL Method

Proposed intuitive F-DEMATEL method for determining the priority of variables (hereinafter criteria) in a typical decision-making problem, which consists of 10 steps by [41–43]:

Step 1. In order to achieve the goals of decision-making, it is necessary to establish a team of experts. This team must include experts with good experience and knowledge in the appropriate field of decision making.

Step 2. Assessing alternatives would be insignificant without considering a set of criteria. An effective set of criteria will provide a better assessment. Therefore, it is necessary to identify criteria for evaluating the problem, using a fuzzy matrix of direct assessments. Experts should determine the relationships between the criteria by giving their views, based on the Intuitistic Fuzzy Linguistic Scale of five comparisons (Table 1), as follows: 0 (no impact), 1 (small impact), 2 (medium impact), 3 (high impact), and 4 (very high impact) (Table 1). Based on this scale, experts give their linguistic assessments for the interrelationships between the defined criteria. Trapezoidal fuzzy numbers are used for evaluation.

Table 1. Intuistic Fuzzy linguistic scale.

Linguistic Phrase	The Result of the Impact	Intuitive Trapezoidal Fuzzy Number	Expected Crips Value
no influence (NI)	0	(0; 0; 0; 0)	0
low influence (LI)	1	(0; 0.1; 0.2; 0.3)	0.15
medium influence (MI)	2	(0.3; 0.4; 0.5; 0.6)	0.45
high influence (HI)	3	(0.7; 0.8; 0.9; 1)	0.85
very high influence (VHI)	4	(1; 1; 1; 1)	1

Step 3. Since trapezoidal fuzzy numbers are not suitable for matrix operations, it is necessary to convert or dephase trapezoidal fuzzy numbers into crips numbers. The fuzzy intuitionistic direct relative matrix (M) is given (in Formula (3)).

$$M = \frac{1}{4}(a_1 + a_2 + a_3 + a_4) \quad (3)$$

Step 4: Normalized matrix of the direct relation (N) derived by using the (Formulas (4) and (5)) [44]:

$$L = Min\left[\frac{1}{Max \sum_{j=1}^{n} a_{ij}}, \frac{1}{Max \sum_{i=1}^{n} a_{ij}}\right] \quad (4)$$

$$N = L \times M \quad (5)$$

Step 5. Construction of a matrix of total relations. Standardized direct relative matrix (N) is transformed into the matrix of total relations (T) based on (Formula (6)):

$$T = N \times (I - N)^{-1} \quad (6)$$

Step 6. Obtaining causal parameters is done using the following formulas, where D stands for the sum of columns, and R the sum of rows (Formulas (7) and (8)):

$$D = \left[\sum_{j=1}^{n} a_{ij}\right]_{1 \times n} \quad (7)$$

$$R = \left[\sum_{i=1}^{n} a_{ij}\right]_{n \times 1} \quad (8)$$

Step 7—Obtaining the importance of the criterion "can be calculated by taking into account the values of (D + R) and (DR)" by [45,46], and is given (in Formula (9)):

$$\omega_i = \left\{(D + R)^2 + (D - R)^2\right\}^{\frac{1}{2}} \quad (9)$$

Step 8—The importance of any criterion can be obtained by normalization shown (in Formula (10)):

$$W_i = \frac{\omega_i}{\sum_{i=1}^{n} \omega_i} \quad (10)$$

where $i = 1, 2, \ldots, n$.

Step 9. The adjustment of the threshold value (α) is calculated from the matrix of the total relations (T) using Formula (11) and aims to eliminate some criteria in the matrix:

where N is the arithmetic mean of the quantities from the matrix of total relations (T)

$$\alpha = \frac{\sum_{i=1}^{n}\sum_{j=1}^{n}[a_{ij}]}{N} \quad (11)$$

Step 10. Drawing a diagram of causality and effect consists of (D + R) and (D − R)) values.

6. Results

In addition to providing material resources for maintenance, the company also has the problem of providing the necessary expertise and knowledge of employees involved in maintenance. Technological advance and consequently the need to upgrade equipment require different company management strategies. Demands that existing equipment be used and exploited for as long as possible and the desire to own the best available equipment are opposed. It is no secret that manufacturers of equipment and machinery generate a large part of their income through the sale of spare parts and maintenance services. Especially, it is not easy to replace expensive and valuable equipment. Products and machines are not standardized, i.e., there is a large number of manufacturers of different products. This diversity complicates the problem of equipment and machine maintenance. A particularly important problem is the age of the equipment where the parent manufacturers stopped producing spare parts. From all of the above, many users of equipment and machines decide to rent equipment where the rental price is also the price of proper operation. The company then focuses on its core process, leaving maintenance to a kind of outsourcing. Thanks to significantly advanced IT technologies and means of transport, special services and maintenance services can react quickly in case of need.

In accordance with the aforementioned, authors decided to single out the three criteria that will be analyzed from the aspect of impact on sustainability, and they are the following:

- Application of technical diagnostics;
- Maintenance resource management;
- Maintenance process planning.

Figure 1 shows the theoretical system model, and it consists of the following independent variables: A—Application of technical diagnostics (hereinafter variable A), B—Management of maintenance resources (hereinafter variable B), C—Maintenance process planning (hereinafter variable C), and the dependent variable D—Sustainability of maintenance of technical systems (hereinafter variable D).

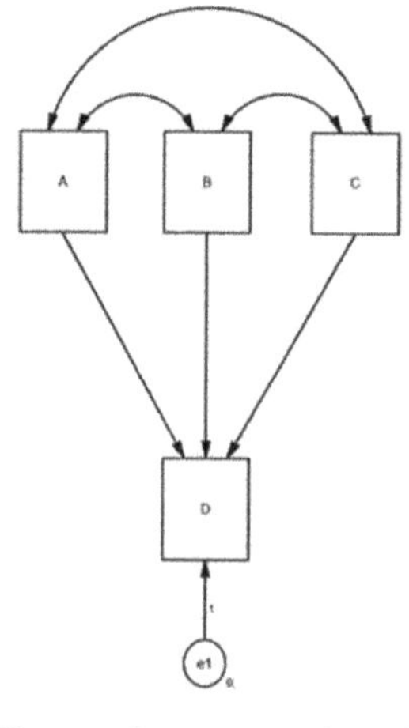

Figure 1. System research model.

Based on the set theoretical system model, for all its variables in the model, statements were made that describe them and on which 136 respondents gave their views (from 1 to 5, Likert scale) in the period from 1 August 2020 to 1 November 2020, on the territory of the Republic of Serbia, as follows:

- Variable A—Application of technical diagnostics consists of the following statements:
  - A_1—Validation affects the maintenance of technical systems;
  - A_2—Functional checking affects the maintenance of technical systems;
  - A_3—Security checking affects the maintenance of technical systems;
  - A_4—Monitoring of technical systems affects the maintenance of technical systems.
- Variable B—Management of maintenance resources consists of the following statements:
  - B_1—Human resource productivity has an impact on the maintenance of technical systems;
  - B_2—Management of spare parts for maintenance systems of technical systems;
  - B_3—Proper functioning of equipment has an impact on the maintenance of technical systems;
  - B_4—Assessment of the service life of equipment used to maintain technical systems;
  - B_5—Operating conditions with equipment load used to maintain technical systems;
  - B_6—Equipment renting (outsourcing) is used to maintain technical systems;
  - B_7—The application of new technologies is used to maintain technical systems;
  - B_8—Supply system is used for maintenance of technical systems.
- Variable C—Maintenance process planning consists of the following statements:
  - C_1—Maintenance cost planning affects the maintenance of technical systems;
  - C_2—Resource planning (materials, equipment, spare parts, labor) affects the maintenance of technical systems;
  - C_3—Synchronization of maintenance plans and the production plan affects the maintenance of technical systems.
- Variable D—Sustainability of maintenance of technical systems consists of the following statements:
  - D_1—Proper maintenance of technical systems contributes the greater environmental safety;
  - D_2—Permanent monitoring of maintenance performance has an impact on the sustainability of maintenance of technical systems;
  - D_3—Reliability of technical systems has an impact on the sustainability of maintenance.

The main null hypothesis of the research is H_0: Levels A—Application of technical diagnostics, B—Maintenance resources management, and C—Maintenance process planning significantly affect level D—Sustainability of maintenance of technical systems.

The values of descriptive statistics according to the work experience of the respondents are given in Table 2. The highest number of respondents was with 21–30 years of work experience, 60 or 41.11%, and the lowest number of respondents was with less than 10 years of work experience, 12 or 8.82%, out of the total number of 136 respondents.

Table 2. Descriptive statistics according to work experience of the respondents.

Work Experience of the Respondents	<10	11–20	21–30	>30	All
Frequency	12	28	60	36	136
Percentage	8.82	20.58	41.11	26.47	100.0

The values of descriptive statistics according to the educational background of the respondents are given in Table 3. More respondents had a high school diploma, 120 or 88.23%, and fewer had a college or university degree, 16 or 11.76%, out of the total number of 136 respondents.

Table 3. Descriptive statistics according to work experience and educational background of the respondents.

Educational Background	High School	College or Faculty	All
Frequency	120	16	136
Percentage	88.23	11.76	100.00

Cross-sections according to work experience and educational background of the respondents are given in Table 4 as follows: most of them, 52 or 38.23%, are the respondents with the work experience of 21–30 years and with a high school degree, and the least number of respondents, 2 or 16.67%, out of 136 respondents, have work experience of less than 10 years and a college or faculty degree. We can say that the most respondents, 52 of them, or 86.67%, out of a total of 60 respondents, are in the group with work experience of 21–30 years, or 43.33%, out of a total of 120 respondents with a high school degree. We can say that at least 2 respondents, or 1.47%, out of 136 respondents, have less than 10 years of work experience or a college or faculty degree. Also, we can say that 2 respondents, or 16.67%, out of 12 respondents, have work experience less than 10 years or a college or faculty degree, or 12.5%, out of 16 respondents have a college or faculty degree.

Table 4. Crossed descriptive quantities according to educational background of the respondents.

Work Experience of the Respondents/Educational Background of the Respondents	<10			11–20			21–30			>30			All		
	N	Column %	Row %	N	Column %	Row %	N	Column %	Row %	N	Column %	Row %	N	Column %	Row %
High School	10	83.33	8.33	25	89.29	20.83	52	86.67	43.33	33	91.67	27.5	120	88.24	100
College or faculty	2	16.67	12.5	3	10.71	18.75	8	13.33	50	3	8.33	18.75	16	11.76	100
All	12	100	8.82	28	100	20.59	60	100	44.12	36	100	26.47	136	100	100

The values of descriptive statistics for the group of statements for variable A are given in Table 5. Statement A_2—has the highest mean score of 3.9852941, and statement A_1—the lowest mean score of 3.7058824. The mean score of variable A is 3.8694853.

Table 5. Descriptive statistics for all variables related to A.

A	A_1	A_2	A_3	A_4	All Variable A
Mean	3.7058824	3.9852941	3.9117647	3.8750000	3.8694853
Std Dev	1.0264476	0.9426935	1.0918594	1.1510864	0.9874054
Std Err Mean	0.0880172	0.0808353	0.0936262	0.0987048	0.0846693
Upper 95% Mean	3.8799532	4.1451615	4.0969285	4.0702078	4.0369351
Lower 95% Mean	3.5318115	3.8254268	3.7266010	3.6797922	3.7020354
N	136	136	136	136	136

The values of descriptive statistics for the group of statements for variable B are given in Table 6. Statement B_4—has the highest mean score of 4.0220588, and statement B_8—the lowest mean score of 3.5955882. The average score of variable B is 3.8566176.

Table 6. Descriptive statistics for all variables related to B.

B	B_1	B_2	B_3	B_4
Mean	3.9338235	3.6985294	3.9117647	4.0220588
Std Dev	1.0968116	1.2130149	0.8816700	0.9772745
Std Err Mean	0.0940508	0.1040152	0.0756026	0.0838006
Upper 95% Mean	4.1198271	3.9042394	4.0612833	4.1877906
Lower 95% Mean	3.7478200	3.4928194	3.7622461	3.856327
N	136	136	136	136

Table 6. *Cont.*

B	B_5	B_6	B_7	B_8
Mean	4.0000000	3.9044118	3.7867647	3.5955882
Std Dev	0.9737290	0.9728056	0.9615427	1.0064877
Std Err Mean	0.0834966	0.0834174	0.0824516	0.0863056
Upper 95% Mean	4.1651305	4.0693857	3.9498286	3.7662742
Lower 95% Mean	3.8348695	3.7394378	3.6237008	3.4249023
N	136	136	136	136
B		**All variables B**		
Mean		3.8566176		
Std Dev		0.6407333		
Std Err Mean		0.0549424		
Upper 95% Mean		3.9652769		
Lower 95% Mean		3.7479584		
N		136		

The values of descriptive statistics for the group of statements for variable C are given in Table 7. Statement C_1—has the highest mean score of 3.9191176, and statement C_3—the lowest mean score of 3.7720588. The mean score of variable C is 3.8357843.

Table 7. Descriptive statistics for all variables related to C.

C	C_1	C_2	C_3	All Variable C
Mean	3.9191176	3.8161765	3.7720588	3.8357843
Std Dev	0.9510629	1.0968116	1.1480541	1.0126739
Std Err Mean	0.0815530	0.0940508	0.0984448	0.0868361
Upper 95% Mean	4.0804043	4.0021800	3.9667524	4.0075193
Lower 95% Mean	3.7578310	3.6301729	3.5773652	3.6640493
N	136	136	136	136

The values of descriptive statistics for the group of statements for the variable D are given in Table 8. Statement D_1 has the highest mean score of 4.0000000, and statement D_2 has the lowest mean score of 3.7794118. The mean score of variable D is 3.8921569.

Table 8. Descriptive statistics for all variables related to D.

D	D_1	D_2	D_3	All Variable D
Mean	4.0000000	3.7794118	3.8970588	3.8921569
Std Dev	0.9888265	1.0797200	1.1371140	0.7199478
Std Err Mean	0.0847912	0.0925852	0.0975067	0.0617350
Upper 95% Mean	4.1676908	3.9625168	4.0898971	4.0142497
Lower 95% Mean	3.8323092	3.5963067	3.7042205	3.7700640
N	136	136	136	136

6.1. Results of Multiple Correlation and Regression Analysis (SEM)

The basic standard evaluation of the system model was performed (Appendix A Figure A1, figure on the left). The coefficient of determination is 0.868535, which means that the dependent variable D—Sustainability of maintenance of technical systems—can be explained by other independent variables with 86.85% of the variability. The correlation of the variables is strong. The values of the correlation coefficients are also given, where we can see that the largest correlation between the independent variables A and B is 0.7243 and it is of medium strength. The smallest correlation is between the independent variables A and C. It is to −0.0117, and it is negative and insignificant. The independent variable A, which is 0.5289, has the largest impact on the dependent variable D—Sustainability of maintenance of technical systems and then then the variable B which is 0.3818. The least impact has the independent variable C which is 0.2622.

The assessment of statistical significance is given in Table 9, and it is [$F(3132) = 290.6897$, $p < 0.0001$].

Table 9. ANOVA.

Source	DF	Sum of Squares	Mean Square	F Ratio
Model	3	60.774739	20.2582	290.6897
Error	132	9.199117	0.0697	**Prob > F**
C. Total	135	69.973856		<0.0001

Based on the data from (Table 9), the main null hypothesis H_0 can be confirmed: Levels A—Application of technical diagnostics, B—Management of maintenance resources, and C—Maintenance process planning, significantly affect the level D—Sustainability of maintenance of technical systems.

Non-standard contribution values for the set system model are given in (Appendix A Figure A1, figure on the right). The highest mean score is for the independent variable A and is 3.8695, and the lowest for the independent variable C is 3.8358. The largest value for the variance is the size of the independent variable C 1.0180, and the smallest variance is for the dependent variable D—Sustainability of maintenance of technical systems—and it is 0.0676. The largest value of covariance is between the independent variables A and B, and it is 0.4549, and the smallest is between the independent variables A and C, and it is −0.0116. The size of the intercept is 0.0303. The highest non-standardized value has the variable B, and it is 0.429054, followed by the variable A with the value of 0.3856187. The lowest value has the variable C 0.1864136.

Based on the data from (Appendix A Figure A1, figure on the right), a multiple regression equation (Formulas (12) and (13) to seven decimal places) can be formed:

$$y = 0.0302714 + 0.3856187 \cdot x_1 + 0.429054 \cdot x_2 + 0.1864136 \cdot x_3 \tag{12}$$

or

$$\begin{gathered} D(Sustainability\ of\ maintenance\ of\ technical\ systems) = \\ = 0.0302714 + 0.3856187 \cdot A(Application\ of\ technical\ diagnostics) + \\ + 0.429054 \cdot B(Management\ of\ maintenance\ resources) + \\ + 0.1864136 \cdot C(Maintenance\ process\ planning) \end{gathered} \tag{13}$$

The Diagram of the multiple regression equation is given in (Appendix B Figure A2), based on which we can give a sample of the research and set the system model to predict the variable D—Sustainability of maintenance of technical systems.

6.2. Results of F-DEMATEL Method

A team of 10 experts in the field of maintenance of technical systems conducted the survey. They were provided by the authors of the paper. Experts compared the criteria A—Application of technical diagnostics, B—Management of maintenance resources, and C—Maintenance process planning.

This part of the research was done after the responses of 136 the respondents had been analyzed in the period from 11 November 2020 to 1 December 2020. The following is a calculation for the F-DEMATEL method. Linguistic values of the influence of the above mentioned 10 experts for the variables (criteria) A, B, and C are given in Table 10.

Table 10. Opinions of 10 experts on the variables set—linguistic values.

	A	B	C
A	NI	HI, MI, HI, VHI, VHI, MI, HI, HI, VHI, VHI	MI, HI, LI, HI, MI, HI, HI, MI, VHI, LI
B	HI, VHI, HI, VHI, HI, VHI, MI, VHI, HI, HI	NI	MI, HI, LI, MI, HI, HI, LI, HI, MI, MI
C	LI, MI, HI, HI, VHI, HI, LI, HI, HI, MI	MI, LI, VHI, HI, MI, MI, LI, MI, MI, HI	NI

Legend: NI—no influence; LI—low influence; MI—medium influence; HI—high influence; VHI—very high influence.

The results of the influence of 10 experts for the variables A, B, and C are given in Table 11.

Table 11. Results of the influence of the opinion of 10 experts on the variables set.

	A	B	C
A	0	3,2,3,4,4,2,3,3,4,4	2,3,1,3,2,3,3,2,4,1
B	3,4,3,4,3,4,2,4,3,3	0	2,3,1,2,3,3,1,3,2,2
C	1,2,3,3,4,3,1,3,3,2	2,1,4,3,2,2,1,2,2,3	0

The average opinions of 10 experts for the variables A, B, and C are given in Table 12. They are derived as the mean value of the response for the variable expressed by the crips number.

Table 12. Average opinions of 10 experts on the variables set.

	A	B	C	Σ
A	0.0000	0.8300	0.6050	1.4350
B	0.8700	0.0000	0.5500	1.4200
C	0.6450	0.5250	0.0000	1.1700
Σ	1.5150	1.3550	1.1550	4.0250

The values (N) of the normalized initial matrix of influence of 10 experts for variables A, B, and C are given in Table 13.

Table 13. Normalized initial matrix of influence of 10 experts on set variables.

	A	B	C
A	0.000000000	0.578397213	0.421602787
B	0.606271777	0.000000000	0.383275261
C	0.449477352	0.365853659	0.000000000

The procedure for determining the matrix of total relations for the variables A, B, and C is given in (Appendix C, formulas (A1)–(A7)). The total relations for the variables A, B, and C are given in Table 14.

Table 14. Matrix of the total relations for the variables A, B, and C.

	A	B	C	D_i	R_j	$D_i + R_j$	$(Di + Rj)_{rank}$	$D_i - R_j$	$(D_i - R_j)_{rank}$	ω_i	W_i	$W_{i\ rank}$
A	5.80	5.79	5.09	16.68	17.26	33.94	1	−0.58	1	33.95	35.59%	1
B	6.16	5.41	5.05	16.62	16.15	32.77	2	0.47	2	32.77	34.36%	2
C	5.31	4.95	4.13	14.39	14.27	28.67	3	0.12	3	28.67	30.05%	3
					Mean	31.79	**Mean**	0.00	Σ	95.39	100.00%	

The threshold value is derived from the arithmetic mean from the matrix of total relations (T) and used via formula (11), and it is $\alpha = 5.30$. A causal diagram of the performance and significance of the variables set (criteria) for A, B, and C is given in (Figure 2) according to (Vafadarnikjoo et al. 2015).

Based on the obtained values that represent the importance and influence of the examined variables, i.e., criteria, and their normed rank values, we can see that the most important criterion for the number of experts set is the criterion A—Application of technical diagnostics 33.95 or (35.59%, (1)), then the criterion B—Management of maintenance resources 32.77 or (34.36%, (2)), and finally the criterion C—Maintenance process planning 28.67 or (30.05%, (3)).

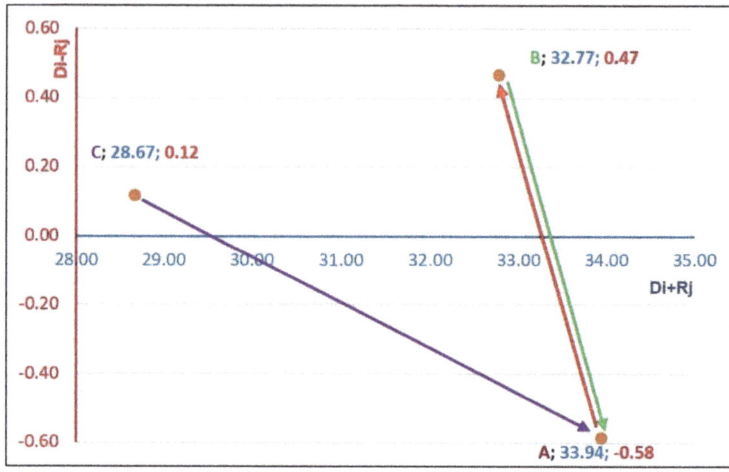

Figure 2. Causal diagram of the performance and significance of the variables set.

7. Conclusions

According to 136 respondents, the independent variable A—Application of technical diagnostics—has the greatest impact on the Sustainability of technical system maintenance, followed by the variable B—Management of maintenance resources. The independent variable C—Maintenance process planning—has the smallest impact. The main null hypothesis, H0, has been confirmed, namely, that Levels A—Application of technical diagnostics, B—Management of maintenance resources, and C—Maintenance process planning and the dependent variable significantly affect the level D—Sustainability of technical system maintenance. The F-DEMATEL method has proven to be suitable for solving problems of ranking the impact of certain criteria in the function of achieving sustainability with group decision-making in a gentle environment. This method is excellent and important for decision makers in the areas of technical systems maintenance because it can be used to investigate any complex technical decision problem. The results obtained based on the opinion of 10 experts on the criteria set are shown in the Causal Diagram. The importance of the criterion rank is shown on the ($D_i + R_j$) axis. It determines the success factor, which is ranked according to the following importance: the criterion A—Application of technical diagnostics > B—Management of maintenance resources > C—Maintenance process planning. The criterion A has the greatest impact on the sustainability of maintenance of technical systems. According to experts, the importance of the criteria coincides with the results obtained by a survey with 136 respondents.

This study showed that the criteria: A—Application of technical diagnostics, B—Management of maintenance resources, and C—Maintenance process planning are ranked by importance based on the highest ($D_i + R_j$) values of 33.94, 32.77, and 28.67. Criteria B Management of maintenance resources and C—Maintenance process planning are in the group of causes based on their positive ($D_i - R_j$) values of 0.47 and 0.12. Criterion A—Application of technical diagnostics is in the group of effects, considering its negative ($D_i - R_j$) value of -0.58. From (Figure 2) we can see that criterion B—Management of maintenance resources—is the most critical because it directly affects criterion A—Application of technical diagnostics, followed by criterion C—Maintenance process planning, which also affects criterion A—Application of technical diagnostics. Criterion B—Management of maintenance resources—has the most effect on criterion A—Application of technical diagnostics and these criteria directly interact with each other.

Author Contributions: Conceptualization, G.O., O.M., and L.R.; methodology, O.M.; validation, G.O., O.M., and G.J.; formal analysis, O.M. and L.R.; investigation, G.O., O.M., and L.R.; resources,

G.O, D.R., and J.P.; data curation, G.J., D.R., and J.P.; writing—original draft preparation, G.O., O.M., and L.R.; writing—review and editing, O.M., L.R., D.R., and J.P.; visualization, G.J.; supervision, G.J., D.R., and J.P. All authors have read and agreed to the published version of the manuscript.

Funding: This research received no external funding.

Acknowledgments: The authors would like to thank the Ministry of Education, Science and Technological Development of the Republic of Serbia for financial support.

Conflicts of Interest: The authors declare no conflict of interest.

Appendix A. Standard (Figure on the Left) and Non-Standard (Figure on the Right) Contribution Sizes of the System Model

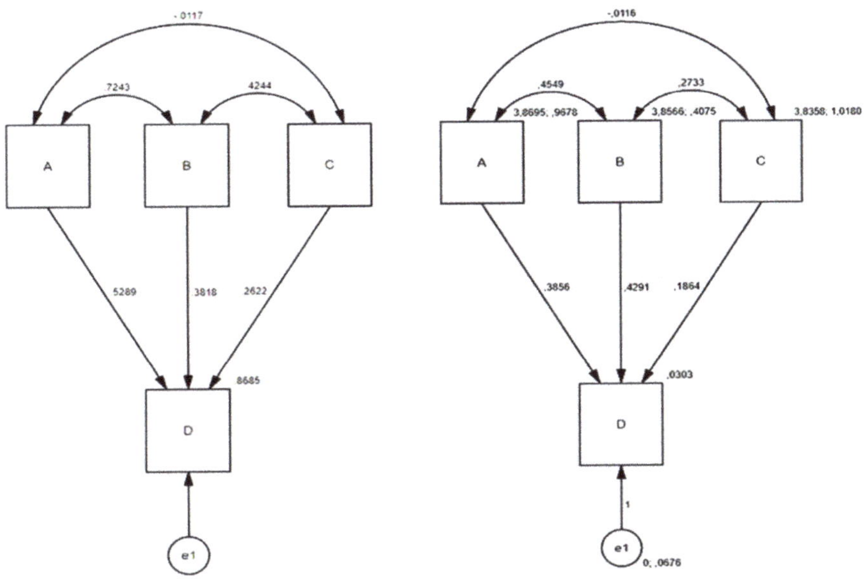

Figure A1. Standard (figure on the left) and non-standard (figure on the right) contribution sizes of the System Model.

Appendix B. Diagram of Multiple Regression Equation

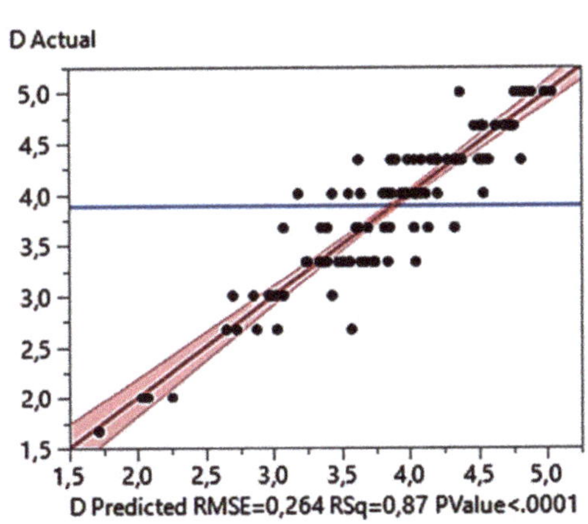

Figure A2. Diagram of multiple regression equation.

Appendix C. Procedure for Determining the Matrix of Total Relations for Variables A—Application of Technical Diagnostics, B—Management of Maintenance Resources, C—Maintenance Process Planning

The procedure for determining the matrix of total relations for the variables A—Application of technical diagnostics, B—Management of maintenance resources, C—Maintenance process planning is shown using the following (formulas (A1)–(A7)).

$$T = N \times (I-N)^{-1} \tag{A1}$$

wherein:

$$(I)\text{ unit matrix } \begin{pmatrix} 1 & 0 & 0 \\ 0 & 1 & 0 \\ 0 & 0 & 1 \end{pmatrix} \tag{A2}$$

$$(N)\text{ standardized matrix } \begin{pmatrix} 0 & 0.578397213 & 0.421602787 \\ 0.606271777 & 0 & 0.383275261 \\ 0.449477352 & 0.365853659 & 0 \end{pmatrix} \tag{A3}$$

$$(I-N) \begin{pmatrix} 1 & 0 & 0 \\ 0 & 1 & 0 \\ 0 & 0 & 1 \end{pmatrix} - \begin{pmatrix} 0 & 0.578397213 & 0.421602787 \\ 0.606271777 & 0 & 0.383275261 \\ 0.449477352 & 0.365853659 & 0 \end{pmatrix} = \begin{pmatrix} 1 & -0.578 & -0.422 \\ -0.606 & 1 & -0.383 \\ -0.449 & -0.366 & 1 \end{pmatrix} \tag{A4}$$

$$(I-N)^{-1} \left(\begin{pmatrix} 1 & 0 & 0 \\ 0 & 1 & 0 \\ 0 & 0 & 1 \end{pmatrix} - \begin{pmatrix} 0 & 0.578397213 & 0.421602787 \\ 0.606271777 & 0 & 0.383275261 \\ 0.449477352 & 0.365853659 & 0 \end{pmatrix} \right)^{(-1)} = \begin{pmatrix} 6.80 & 5.79 & 5.09 \\ 6.16 & 6.41 & 5.05 \\ 5.31 & 4.95 & 5.13 \end{pmatrix} \tag{A5}$$

$$T = N \times (I-N)^{-1} \text{ matrix of total relations} \tag{A6}$$

$$\begin{pmatrix} 0 & 0.578397213 & 0.421602787 \\ 0.606271777 & 0 & 0.383275261 \\ 0.449477352 & 0.365853659 & 0 \end{pmatrix} * \left(\begin{pmatrix} 1 & 0 & 0 \\ 0 & 1 & 0 \\ 0 & 0 & 1 \end{pmatrix} - \begin{pmatrix} 0 & 0.578397213 & 0.421602787 \\ 0.606271777 & 0 & 0.383275261 \\ 0.449477352 & 0.365853659 & 0 \end{pmatrix} \right)^{(-1)} = \begin{pmatrix} 5.80 & 5.79 & 5.09 \\ 6.16 & 5.41 & 5.05 \\ 5.31 & 4.95 & 4.13 \end{pmatrix} \tag{A7}$$

References

1. Bulatović, M. Održavanje u funkciji održivog razvoja. In Proceedings of the 2th Conference Maintenance 2012, Zenica, Bosnia and Herzegovina, 10–13 June 2012. (In Montenegrin).
2. Afrinaldi, F.; Tasman, A.M.; Zhang, H.C.; Hasan, A. Minimizing economic and environmental impacts through an optimal preventive replacement schedule: Model and application. *J. Clean. Prod.* **2017**, *143*, 882–893. [CrossRef]
3. Sari, E.; Shaharoun, A.B.M.; Ma'aram, A.B. Preliminary Framework of Sustainable Maintenance Performance Measurement Systems for Automotive Companies. In *Advanced Materials Research*; Trans Tech Publications: Stafa-Zurich, Switzerland, 2014; Volume 845, pp. 590–595.
4. Sari, E.; Shaharoun, A.M.; Ma'aram, A.; Yazid, A.M. Sustainable maintenance performance measures: A pilot survey in Malaysian automotive companies. *Procedia CIRP* **2015**, *26*, 443–448. [CrossRef]
5. Sénéchal, O. Research directions for integrating the triple bottom line in maintenance dashboards. *J. Clean. Prod.* **2017**, *142*, 331–342. [CrossRef]
6. Amrina, E.; Aridharma, D. Sustainable maintenance performance evaluation model for cement industry. In Proceedings of the International Conference on Industrial Engineering and Engineering Management (IEEM), Bali, Indonesia, 4–7 December 2016; pp. 350–354. [CrossRef]
7. Heravi, G.; Fathi, M.; Faeghi, S. Evaluation of sustainability indicators of industrial buildings focused on petrochemical projects. *J. Clean. Prod.* **2015**, *109*, 92–107. [CrossRef]
8. Pires, S.P.; Sénéchal, O.; Loures, E.F.R.; Jimenez, J.F. An approach to the prioritization of sustainable maintenance drivers in the TBL framework. *IFAC-PapersOnLine* **2016**, *49*, 150–155. [CrossRef]
9. Domingo, R.; Aguado, S. Overall environmental equipment effectiveness as a metric of a lean and green manufacturing system. *Sustainability* **2015**, *7*, 9031–9047. [CrossRef]
10. Franciosi, C.; Iung, B.; Miranda, S.; Riemm, S. Maintenance for Sustainability in the Industry 4.0 context: A Scoping Literature Review. *IFAC-PapersOnLine* **2018**, *51*, 903–908. [CrossRef]

11. Sodangi, M.; Khamdi, M.F.; Idrus, A.; Hammad, D.B.; Umar, A.A. Best Practice Criteria for Sustainable Maintenance Management of Heritage Buildings in Malaysia. *Procedia Eng.* **2014**, *77*, 11–19. [CrossRef]
12. Jasiulewicz-Kaczmarek, M. The role and contribution of maintenance in sustainable manufacturing. In Proceedings of the FAC Conference on Manufacturing Modelling, Management, and Control, Saint Petersburg State University and Saint Petersburg National Research University of Information Technologies, Mechanics, and Optics, Saint Petersburg, Russia, 19–21 June 2013.
13. Kanisuru, A.M. Sustainable Maintenance Practices and Skills for Competitive Production System. Chapter 3. Available online: https://www.intechopen.com/books/skills-development-for-sustainable-manufacturing/sustainable-maintenance-practices-and-skills-for-competitive-production-system (accessed on 27 December 2020).
14. Kilić Pamuković, J.; Rogulj, R.; Dumanić, D.; Jajac, N. A Sustainable Approach for the Maintenance of Asphalt Pavement Construction. *Sustainability* **2021**, *13*, 109. [CrossRef]
15. Çınar, Z.M.; Nuhu, A.A.; Zeeshan, Q.; Korhan, O.; Asmael, M.; Safaei, B. Machine Learning in Predictive Maintenance towards Sustainable Smart Manufacturing in Industry 4.0. *Sustainability* **2020**, *12*, 8211. [CrossRef]
16. Adamović, Ž. *Tehnologija Održavanja*; University of Novi Sad, Technical Faculty "Mihajlo Pupin": Zrenjanin, Serbia, 2004. (In Serbian)
17. Radovanović, L. Metodologije Održavanja sa Aspekta Njihove Primene u Industriji. Ph.D. Thesis, University of Novi Sad, Technical Faculty "Mihajlo Pupin", Zrenjanin, Serbia, 2012. (In Serbian).
18. Todorović, J. *Inženjerstvo Održavanja Tehničkih Sistema*; Jugoslovensko Društvo za Motore i Vozila: Beograd, Serbia, 1993. (In Serbian)
19. Todorović, J. *Osnovi Teorije Održavanja*; Mašinski Fakultet Beograd: Beograd, Serbia, 1984. (In Serbian)
20. Todorović, J. Upravljanje održavanjem na bazi rizika. *J. Appl. Eng. Sci.* **2003**, *1*, 23–32. (In Serbian)
21. Stanivuković, D.; Kecojević, S. *Održavanje-Prilaz Projektovanju i Upravljanju*; IIS—Istraživački i Tehnološki Centar; FTN-Institut za Industrijske Sisteme: Novi Sad, Serbia, 1995. (In Serbian)
22. Stanivuković, D.; Zelenović, D.; Kecojević, S. *Održavanje-IIS Prilaz*; IIS—Istraživački i Tehnološki Centar; FTN-Institut za Industrijske Sisteme: Novi Sad, Serbia, 1997. (In Serbian)
23. Vasić, B. *Upravljanje Održavanjem*; OMO: Beograd, Serbia, 1997. (In Serbian)
24. Jeremić, B.; Todorović, T.; Mačudžić, I. *Inženjering Održavanja-Skripta*; Mašinski Fakultet Kragujevac: Kragujevac, Serbia, 2008. (In Serbian)
25. Wireman, T. *World Class Maintenance Management*; Industrial Pres, Inc.: New York, NY, USA, 1990.
26. Wireman, T. *Total Productive Maintenance: An American Approach*; Industrial Pres, Inc.: New York, NY, USA, 1991.
27. Wireman, T. *Developing Performance Indicators for Managing Maintenance*; Industrial Pres, Inc.: New York, NY, USA, 2005.
28. Liyanage, J.P.; Kumar, U. Towards a value-based view on operations and maintenance performance management. *J. Qual. Maint. Eng.* **2003**, *9*, 333–350. [CrossRef]
29. Milosavljević, P. Modeliranje Procesa Održavanja Tehničkih Sistema na Konceptu Totalnog Produktivnog Održavanja. Ph.D. Thesis, University of Niš, Mechanical Faculty, Niš, Serbia, 2005. (In Serbian).
30. Šćepanović, S. Efikasnost Modela Održavanja Termoelektrana. Ph.D. Thesis, University of Novi Sad, Technical Faculty "Mihajlo Pupin", Zrenjanin, Serbia, 2006. (In Serbian).
31. Nikolić, D. Optimizacija Postupaka Tehničke Dijagnostike Hidroelektrana. Ph.D. Thesis, University of Novi Sad, Technical Faculty "Mihajlo Pupin", Zrenjanin, Serbia, 2007. (In Serbian).
32. Kutin, M. Optimizacija Primene Dijagnostičkih Tehnika i Njihov Uticaj na Pouzdanost Tehničkih Sistema. Ph.D. Thesis, University of Novi Sad, Technical Faculty "Mihajlo Pupin", Zrenjanin, Serbia, 2010. (In Serbian).
33. Anderson, J.; Gerbing, D. Structural equation modeling in practice: A review and recommended two-step approach. *Psychol. Bull.* **1988**, *103*, 411–423. [CrossRef]
34. DiLalla, L. Structural equation modeling: Uses and issues. In *Handbook of Applied Multivariate Statistics and Mathematical Modeling*; Academic Press: Cambridge, MA, USA, 2000; pp. 439–464.
35. Fajgelj, S.; Kuzmanović, B.; Đukanović, B. *Priručnik za Socijalna Istraživanja*; CID: Podgorica, Montenegro, 2014. (In Montenegrin)
36. Fassinger, R. Use of structural equation modeling in counseling psychology research. *J. Couns. Psychol.* **1987**, *34*, 425. [CrossRef]
37. Long, B.; Kahn, S.; Schutz, R. Causal model of stress and coping: Women in management. *J. Couns. Psychol.* **1992**, *39*, 227–239. [CrossRef]
38. Smederevac, S.; Mitrović, D.; Čolović, P.; Nikolašević, Ž. Validation of the measure of revised reinforcement sensitivity theory constructs. *J. Individ. Differ.* **2014**, *35*, 12–21. [CrossRef]
39. Iacobucci, D. Structural equations modeling: Fit indices, sample size, and advanced topics. *J. Consum. Psychol.* **2010**, *20*, 90–98. [CrossRef]
40. Hu, L.; Bentler, M. Cutoff criteria for fit indexes in covariance structure analysis: Conventional criteria versus new alternatives. *Struct. Equ. Model. A Multidiscip. J.* **1999**, *6*, 1–55. [CrossRef]
41. Govindan, K.; Khodaverdi, R.; Vafadarnikjoo, A. A grey DEMATEL approach to develop third-party logistics provider selection criteria. *Ind. Manag. Data Syst.* **2016**, *116*, 690–722. [CrossRef]
42. Vafadarnikjoo, A.; Mobin, M.; Salmon, C.; Javadian, N. An integrated gray-fuzzy cause-effect approach to determine the most significant categories of project risks. In Proceedings of the Industrial and Systems Engineering Research Conference (ISERC), Nashville, TN, USA, 30 May–2 June 2015.
43. Zadeh, L.A. Fuzzy sets. *Inf. Control* **1965**, *8*, 338–353. [CrossRef]

44. Lin, R.J. Using fuzzy DEMATEL to evaluate the green supply chain management practices. *J. Clean. Prod.* **2013**, *40*, 32–39. [CrossRef]
45. Baykasoglu, A.; Kaplanoğlu, V.; Durmusoglu, Z.; Şahin, C. Integrating fuzzy DEMATEL and fuzzy hierarchical TOPSIS methods for truck selection. *Expert Syst. Appl.* **2013**, *40*, 899–907. [CrossRef]
46. Dalalah, D.; Hayajneh, M.; Batieha, F. A Fuzzy Multi-Criteria Decision-Making Model for Supplier Selection. *Expert Syst. Appl.* **2011**, *38*, 8384–8391. [CrossRef]

sustainability

MDPI

Article

Evaluation of Safety Degree at Railway Crossings in Order to Achieve Sustainable Traffic Management: A Novel Integrated Fuzzy MCDM Model

Aleksandar Blagojević [1], Sandra Kasalica [1], Željko Stević [2,*], Goran Tričković [1] and Vesna Pavelkić [1]

1 Academy of Technical and Artistic Professional Studies Belgrade, College of Railway Engineering, Zdravka Čelara 14, 11000 Belgrade, Serbia; aleksandar.blagojevic@vzs.edu.rs (A.B.); sandra.kasalica@vzs.edu.rs (S.K.); goran.trickovic@vzs.edu.rs (G.T.); vesna.pavelkic@vzs.edu.rs (V.P.)

2 Faculty of Transport and Traffic Engineering, University of East Sarajevo, Vojvode Mišića 52, 74000 Doboj, Bosnia and Herzegovina

* Correspondence: zeljkostevic88@yahoo.com or zeljko.stevic@sf.ues.rs.ba

Abstract: Sustainable traffic system management under conditions of uncertainty and inappropriate road infrastructure is a responsible and complex task. In Bosnia and Herzegovina (BiH), there is a large number of level crossings which represent potentially risky places in traffic. The current state of level crossings in BiH is a problem of the greatest interest for the railway and a generator of accidents. Accordingly, it is necessary to identify the places that are currently a priority for the adoption of measures and traffic control in order to achieve sustainability of the whole system. In this paper, the Šamac–Doboj railway section and passive level crossings have been considered. Fifteen different criteria were formed and divided into three main groups: safety criteria, road exploitation characteristics, and railway exploitation characteristics. A novel integrated fuzzy FUCOM (full consistency method)—fuzzy PIPRECIA (pivot pairwise relative criteria importance assessment) model was formed to determine the significance of the criteria. When calculating the weight values of the main criteria, the fuzzy Heronian mean operator was used for their averaging. The evaluation of level crossings was performed using fuzzy MARCOS (measurement of alternatives and ranking according to compromise solution). An original integrated fuzzy FUCOM–Fuzzy PIPRECIA–Fuzzy MARCOS model was created as the main contribution of the paper. The results showed that level crossings 42 + 690 (LC4) and LC8 (82 + 291) are the safest considering all 15 criteria. The verification of the results was performed through four phases of sensitivity analysis: resizing of an initial fuzzy matrix, comparative analysis with other fuzzy approaches, simulations of criterion weight values, and calculation of Spearman's correlation coefficient (SCC). Finally, measures for the sustainable performance of the railway system were proposed.

Keywords: sustainable traffic; level crossing; fuzzy FUCOM method; fuzzy MARCOS method

check for updates

Citation: Blagojević, A.; Kasalica, S.; Stević, Ž.; Tričković, G.; Pavelkić, V. Evaluation of Safety Degree at Railway Crossings in Order to Achieve Sustainable Traffic Management: A Novel Integrated Fuzzy MCDM Model. *Sustainability* **2021**, *13*, 832. https://doi.org/10.3390/su13020832

Received: 25 December 2020
Accepted: 11 January 2021
Published: 15 January 2021

Publisher's Note: MDPI stays neutral with regard to jurisdictional claims in published maps and institutional affiliations.

Copyright: © 2021 by the authors. Licensee MDPI, Basel, Switzerland. This article is an open access article distributed under the terms and conditions of the Creative Commons Attribution (CC BY) license (https://creativecommons.org/licenses/by/4.0/).

1. Introduction

At the very beginning of its development, the railway encountered traffic safety problems. This problem still exists today. Railway safety is a remarkably complex and diversely comprehensive concept. It is one of the most important, if not the most important criteria for the sustainable performance of the railway as a complex transportation system. Without traffic safety, there is no high quality traffic system. Safer traffic is certainly a prerequisite for satisfying a society as a whole. However, traffic, with its dynamism, can also cause undesirable consequences by taking human lives and creating huge material damage. Roads that intersect with railways at the same level, i.e., level crossings, represent risk points, i.e., places where undesirable consequences occur and can occur for both road and railway traffic.

Level crossings should be accepted as a necessary evil and all available measures should be taken in order to reduce the negative consequences of their presence. Problems related to the existence of level crossings are numerous. At level crossings, the continuity of road vehicles is disturbed, and in some cases of railway vehicles too, which results in increased energy consumption, higher environmental pollution, lost time of passengers and staff, increased time of engaging mobile means, and reduced traffic capacity. Negative effects also include the costs of maintaining the level crossing as well as the costs related to certain investment projects. There are obviously many problems related to level crossings, but the main issue is sustainable traffic safety management.

Due to the above, and with the aim of sustainable traffic safety management, certain measures are being taken to reduce this danger to a minimum. The best solutions are when the intersection of roads is in uneven mode (via overpasses and underpasses). However, the fact is that in Bosnia and Herzegovina (BiH), there are a large number of road-railway crossings that are unlikely to be at two levels because their reconstruction requires very large investments, but often the crossings cannot be avoided for technical, historical, and other reasons, so the application of modern technical means is required for their security. The technical means used to secure level crossings are in fact a "warning" to road users that they are approaching a crossing over railway tracks, and these are not means to prevent irresponsible road users from crossing over railway tracks at a time when it should be crossed by a railway vehicle.

If we consider the importance of traffic and the degree of its safety in all spheres of human activity, then it is understandable why there is so much interest of all traffic participants, traffic organizations, and society as a whole in solving safety problems at railway crossings. As a first step, in a function of sustainable railway safety management, it is necessary to solve the issue of level crossings. In this paper, the degree of traffic safety at railway crossings has been evaluated using the example of the Šamac–Doboj railway section.

These are only some of the problems that manifest themselves in the field of traffic safety since they usually arise under the influence of a number of criteria that are not correlated. Considering the above, and the importance and relevance of the field of research, the following goals of the paper stand out. The first goal is the need to form a model for evaluating safety at level crossings and the possibility of forming a set of measures to improve the situation. Further, the second goal involves the development of a novel integrated fuzzy FUCOM (full consistency method; based on a fuzzy Heronian mean operator)–fuzzy PIPRECIA (pivot pairwise relative criteria importance assessment)–fuzzy MARCOS (measurement of alternatives and ranking according to compromise solution) model as a contribution to multi-criteria decision-making processes. The third goal of this paper is to create a novel integrated model for determining the significance of the criteria. All the above mentioned aims describe the originality of the developed integrated fuzzy model, its importance, and achieved contribution of the paper. The advantage of the developed integrated fuzzy FUCOM–fuzzy PIPRECIA–fuzzy MARCOS model is the fact that the created model eliminated the possibility of inadequate ranking compared to similar approaches as we stated further through the paper.

The rest of the paper is divided into five other sections. In Section 2, a review of the literature related to the field of application is performed. Section 3 presents materials and methods with a research flow diagram first presented explaining the overall methodology in detail and describing the parameters of the fuzzy multi-criteria decision-making (MCDM) model. The results of the research are presented in Section 4 with a brief overview of the most important calculations. Section 5 includes a sensitivity analysis and discussion of the results obtained. Finally, in Section 6, which refers to the concluding remarks, the most important contributions of the paper are summarized and an overview of future research and a set of improvement measures are given.

2. Literature Review

Many accidents and incidents, especially those with a fatal consequence, occur at the intersections of railways and roads. In order to reduce the number of accidents at level crossings and maintain the safety level at these places, numerous studies have been conducted. Ci Lang et al. [1] point out that level crossings are one of the most critical issues for the sustainable management of traffic safety and that road accidents account for a third of all railway accidents in Europe. Based on statistics on road accidents, the authors propose Bayesian networks for risk analysis at level crossings in order to establish a framework for improving safety at level crossings. Blagojević et al. [2] developed a novel integrated entropy-fuzzy PIPRECIA-DEA (data envelopment analysis) model for determining the state of railway safety in BiH under some conditions of uncertainty. The obtained results have been verified through a sensitivity analysis, which implies a change in the impact of five most significant criteria and a comparison with two MCDM methods. Obradović et al. [3] investigated level crossings in the territory of the Republic of Serbia, where they included 245 level crossings. The research was conducted in order to determine the current situation and identify deficiencies in the railway network. Based on the identified deficiencies, they proposed adequate design and management measures to improve traffic safety at level crossings. Kasalica et al. [4] applied statistical models to estimate the frequency of accidents, the severity of the consequences of accidents and the empirical risk at level crossings in order to identify places of high risk on the railway network. The authors identified variables that were significantly related to the number and consequences of accidents. The proposed models of frequency and consequences of accidents were used to assess the reduction of accidents at level crossings by applying appropriate technical measures to raise safety levels. In addition, Kasalica et al. [5] investigated the direct behavior of road traffic participants at passively secured railway crossings. The research was conducted on 61 road vehicle drivers. The results have shown that drivers who have limited visibility cannot estimate the speed of the approaching train well and, as a result, make riskier decisions and cause accidents at level crossings. Djordjević et al. [6] have developed a new approach for assessing safety at railway level crossings based on the non-radial DEA model. The developed non-radial DEA model is used to assess the railway efficiency of European countries in terms of sustainable safety at level crossings considering desirable and undesirable criteria. The results of the sensitivity analysis of the developed model indicated certain weaknesses related to the number of criteria, as well as the accuracy of the input and output criteria. Pamučar et al. [7] developed a group multi-criteria FUCOM–MAIRCA (full consistency method–multi-attributive ideal-real comparative analysis) model for selecting a level crossing which requires investing in its equipment in order to increase traffic safety. The FUCOM-MAIRCA model was tested in a case study that included the assessment of ten level crossings within the railway infrastructure in the Republic of Serbia. The developed approach enables bridging the gap that currently exists in the methodology for assessing the level crossings that require investing in their equipment for sustainable traffic safety. Salmon et al. [8] used the basic principles of the Systems Theoretic Accident Model and Processes (STAMP) control structure method which were then added to the Event Analysis of Systemic Teamwork (EAST) framework and used to examine the sustainable management of level crossing safety. Task, social, and information networks for a life cycle of the railway crossing were developed along with an additional control network showing safety controls and their interrelations. The analysis of the networks points to a need to strengthen activities and controls around proactive risk management and performance monitoring, tightening the links between organizations responsible for safety management and increasing the flexibility of design standards. In order to improve safety at level crossings, Pedro and Márquez [9] propose the use of advanced information technology, i.e., intelligent failure monitoring and reliability on level crossing safety devices. In addition, Djordjević et al. [10] propose an information subsystem concept of level crossings within the existing information system of the Serbian Railways infrastructure. The main goal of the paper is to create conditions for more efficient management of level crossings in order to improve traffic safety. The concept of the information subsystem included solutions for monitoring the technical data of accidents at level crossings. Bester et al. [11] defined a methodology based on stationary, homogeneous, and ergodic Markov processes to assess reliability and safety of devices at level crossings. The authors performed a hazard analysis and risk assessment at level crossings for the purpose of sustainable traffic safety using the tolerable hazard rate (THR) method and calculated the mean time between failures of device using the Windchill Quality Solutions software. Lutovac and Lutovac [12] proposed a solution to eliminate the shortcomings of existing diagnostic systems and event loggers at level crossings. They developed software for local and remote reading and processing of data using a computer. It enables a more efficient system of maintenance and registration of events at level crossings in order to improve safety. Rudin-Brown et al. [13] investigated drivers' behavior at level crossings. Twenty-five drivers aged between 20 and 50 participated in a driving simulator study that compared the efficiency and drivers' subjective perception of two traffic control devices at an active level crossing: flashing lights with boom barriers and standard traffic lights. Due to its common usage in most Australian states, a stop sign-controlled level crossing served as a passive referent. Although crossing violations were less likely at the level crossings controlled by active devices than at those controlled by stop signs, both types of active control were associated with a similar number of violations. Furthermore, the majority (72%) of drivers stated that they preferred flashing lights to traffic lights. Collectively, the results show that the installation of traffic lights at level crossings could not offer safety advantages over those already provided by flashing lights with boom barriers. Additionally, Lenné et al. [14] compared drivers' behavior at two railway level crossings with active controls, flashing red lights and traffic signals, to behavior at the current standard passive control of level crossings, a stop sign. Participants drove the advanced MUARC driving simulator for 30 min. During the simulated drive, participants had three level crossing scenarios. Each scenario consisted of one of three types of level crossing control. The results of the research have shown that traffic signals at level crossings do not provide any safety benefits. Evans [15] investigates fatal accidents with fatalities at level crossings in Great Britain over the 64-year period 1946–2009. The number of fatal accidents and fatalities per year fell by about 65% in the first half of that period, but since then they have remained more or less constant at about 12 fatalities per year. The paper classifies level crossings into three types: railway-controlled, automatic, and passive. The safety performance of the three types of crossings was very different. Railway-controlled crossings are the best-performing type of crossing, with a declining number of fatal accidents. Automatic crossings have a higher accident rate per crossing than railway controlled or passive crossings, and the accident rates have not been reduced. Passive crossings are by far the most numerous, but many have low usage by road users. Their fatal accident rate remained remarkably constant over the period at about 0.9 fatal accidents per 1000 crossings per year. Hu So-Ren et al. [16] developed an interesting generalized logit model in order to identify the key parameters responsible for different levels of consequences of level crossing accidents in Taiwan. The model has shown that the technical measures used in Taiwan to raise safety at level crossings, such as separators and obstacle detectors, do not affect the prevention of accidents with more severe consequences. Park et al. [17] developed multi-stage accident prediction models using a linear regression technique (Poisson regression), as well as a recursive partitioning (RPART) non-parametric method for sustainable traffic safety. The authors classified level crossings into groups according to similar physical and traffic characteristics and used separate accident prediction models for each group. Saccomano et al. [18] apply a risk-based model to identify level crossings that pose a high risk to sustainable traffic safety. The model consists of two prediction components: predicting the frequency of accidents and predicting the consequences of accidents at level crossings. Austin and Carson [19] developed an accident prediction model using negative binomial regression. The instrumental variables technique was used in the model in order to correct the influence of variables on each other. However, the effects of several factors are not logical. For example, it has been observed that the presence of "stop" signals and traffic lights affects the increase in the predicted number of accidents, so it is contradictory to expect effects if these measures are applied to raise the level of safety.

3. Materials and Methods

The overall research flow and applied methodology can be presented through four phases composed of a total of 15 steps (Figure 1).

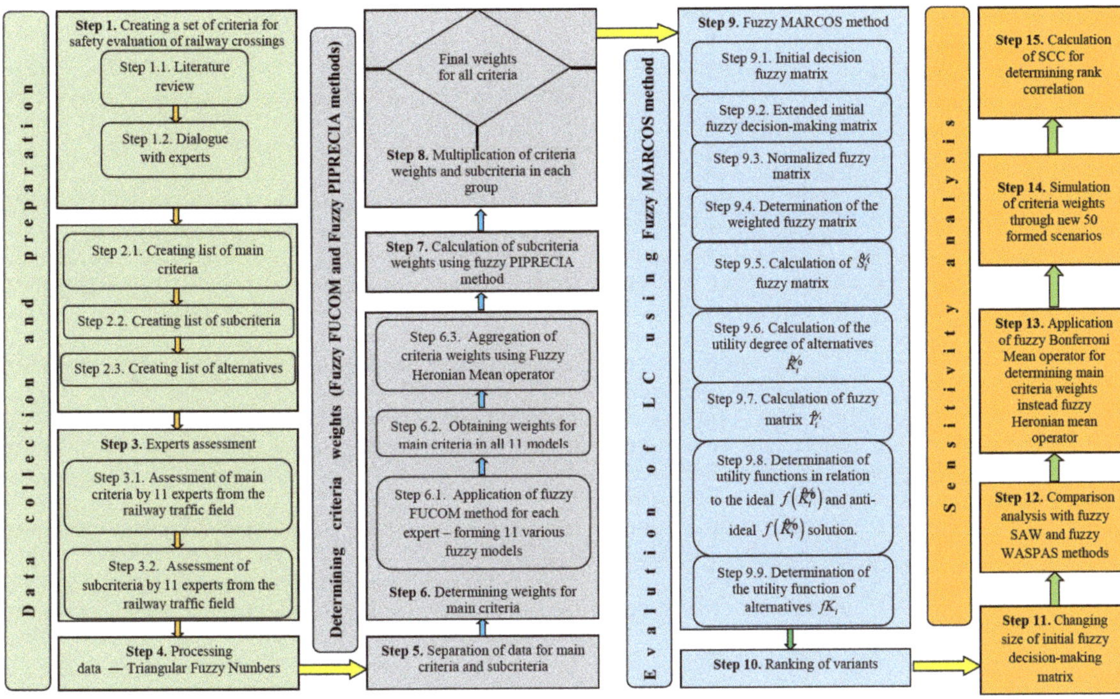

Figure 1. The research flow with the proposed methodology.

Figure 1 presents a diagram of research and a proposed integrated fuzzy MCDM model for the evaluation of level crossings on the Šamac–Doboj (Republic of Srpska–RS) railway section. The first phase is data collection and processing and consists of four steps. In the first step, based on a review of the literature of similar studies and discussion with experts in this field, a list of evaluation criteria was created. The second step involved creating a hierarchical structure by defining three main criteria: safety C1, road exploitation characteristics C2, and railway exploitation characteristics C3 at the first level of the hierarchical structure. At the second level, five sub-criteria were defined within each of the main groups of criteria. Safety criteria refer to the types of adverse events at level crossings:

C11—the number of serious accidents, it represents any collision or derailment of trains resulting in the death of at least one person, serious injuries to five or more persons. In addition, it refers to extensive damage to rolling stock (damage of at least EUR 2 million), infrastructure, or the environment, as well as other similar accidents that have an obvious impact on railway safety or on sustainable safety management.

C12—the number of accidents, which represents an unwanted sudden event or a specific series of such events that have serious consequences (collisions, derailments, accidents at level crossings, accidents involving persons caused by railway vehicles in motion, fires ...) [2].

C13—the number of incidents, which includes any event that is not classified as an accident or serious accident, but which is related to the traffic of trains or shunting rolling stock, and thus affects traffic safety.

C14—the number of persons killed, which means that all persons died immediately or within 30 days as a result of the adverse event.

C15—the number of injured persons who were hospitalized for more than 24 h after an accident, serious accident, or incident, other than suicide attempt.

Criteria related to road exploitation characteristics include a group of the following five sub-criteria:

C21—category of the road, which is defined by legal acts and depends on the parameters of the road itself. Three categories of roads appear in this paper: regional, local, and uncategorized.

C22—frequency of road traffic, which means the number of road vehicles on a daily basis that cross the level crossing.

C23—speed of movement of road vehicles, which is expressed in km/h and implies the speed of movement at which the road vehicle approaches the level crossing.

C24—visibility triangle, it represents a space where there are no visible obstacles, i.e., drivers of road vehicles are enabled unobstructed visibility of the railway on both sides. The size of safety triangle, i.e., the visibility of the railway track for road vehicle drivers depends on the road category itself.

C25—the slope of the road, it represents the geometric characteristic of the road and is expressed as a percentage.

When it comes to sub-criteria belonging to a group of railway exploitation characteristics as the third main criterion, the following are singled out:

C31—train speed, expressed in km/h, and it means the speed of the train when approaching and passing over the level crossing.

C32—frequency of railway traffic, it means the number of trains circulating on a daily basis on certain sections of the railway.

C33—the angle of intersection of the railway and the road, it means the angle at which the two types of traffic intersect and is related to the criterion referring to the triangle of visibility.

C34—method of security, when it comes to level crossings that are passively secured, the following methods of security are possible: visibility triangle, signal sign-Andreja's cross, and stop traffic sign, i.e., some of the combinations of the mentioned methods of security.

C35—the width of the level crossing is expressed in meters and mainly depends on the number of tracks at the crossing. In urban zones and in the station area, there are road crossings where a road crosses several tracks. All this increases the probability of accidents at the level crossing.

As the last activity of the second step in the first phase, a certain number of level crossings were selected to be evaluated on the basis of the previously explained criteria. In this paper, the Šamac-Doboj railway section was considered, as a section that stands out with a considerable number of adverse events at level crossings. Figure 2 shows a graphical representation of all level crossings on a given section considering only those where some kind of accident occurred. The total number of level crossings on this section is 53, out of which 48 are passively and 5 are actively secured. In Figure 2, actively secured crossings are marked in red, while level crossings where some kind of accident occurred are marked in black (Figure 3) and there are a total of eight of them which are further considered in the MCDM model. Other (40) level crossings at which there were no accidents in the observed period are marked in orange.

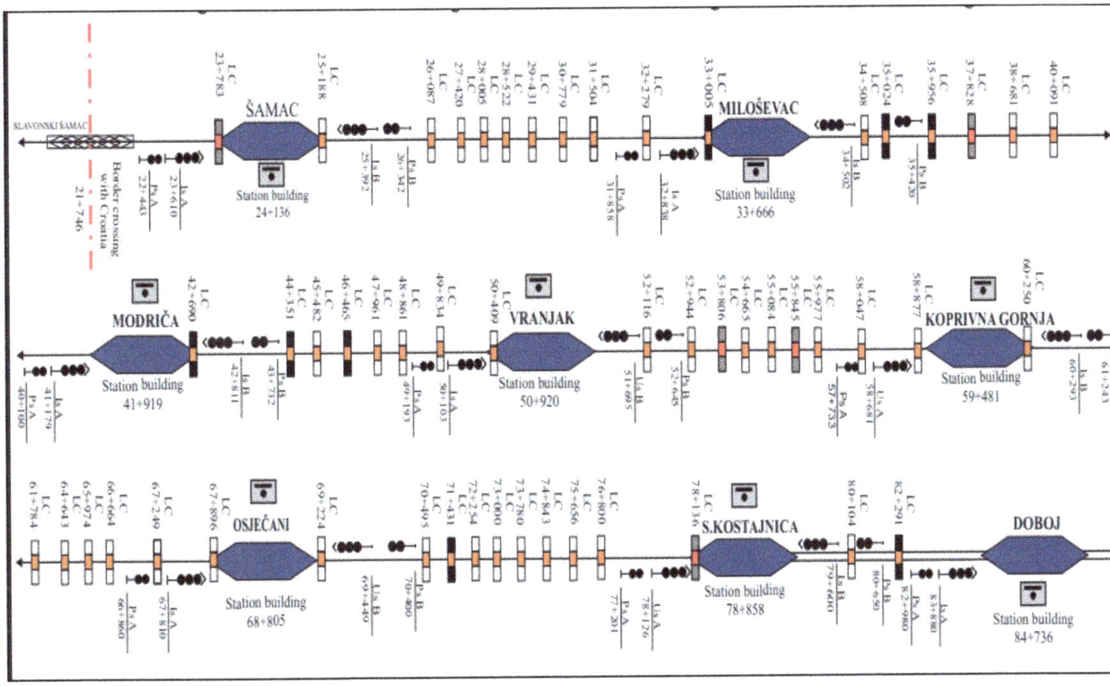

Figure 2. Graphical representation of the Šamac-Doboj section with the structure of all level crossings.

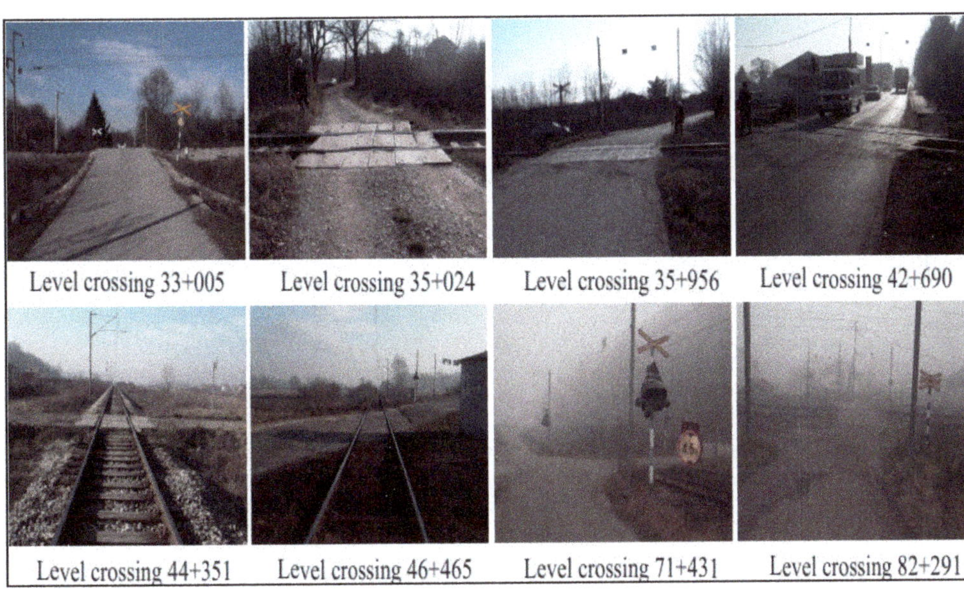

Figure 3. Overview of level crossings considered in the multi-criteria decision-making (MCDM) model.

Today, there are a total of 491 level crossings on the railway network in BiH, with a total length of 1048 km, out of which 57 are actively secured and 434 are passively secured level crossings. The current state of level crossings in BiH is a problem of the greatest interest for the railway and is a generator of accidents. There are 296 level crossings in the

Republic of Srpska, out of which 13 are actively secured and 283 are passively secured level crossings. Figure 3 shows the level crossings that have been considered and evaluated in this study.

Eight most dangerous level crossings were analyzed, i.e., the crossings where accidents occurred. Field surveys were carried out, checking traffic safety, photographing and recording the crossing of the road over the railway with a video camera, and counting traffic participants. The basic characteristics of the considered level crossings are described in detail.

The level crossing at railway kilometer 33 + 005, by its location, belongs to the municipality of Modriča. The crossing is on an open railway and crosses one track. The functional significance of the road is local. The crossing is on a curve and the angle of intersection between the road and track is of 60 degrees. The slope of the road in relation to the crossing is 11%. The width of the road in the crossing zone is 4 m. The permitted speed over the crossing for road and railway vehicles is 50 km/h. The average annual daily traffic on the road is 78 vehicles per day, and for railway vehicles it is 4 trains per day. The crossing is secured only by the visibility triangle and there is not a sufficient visibility zone for safe traffic. There was one accident at this crossing in 2019.

The level crossing at railway kilometer 35 + 024, by its location, belongs to the municipality of Modriča. The crossing is on an open railway and crosses one track. The functional significance of the road is local. The crossing is on a curve and the angle of the intersection between the road and track is of 60 degrees. The slope of the road in relation to the crossing is 8.5%. The width of the road in the crossing zone is 4 m. The permitted speed over the crossing for road and railway vehicles is 50 km/h. The average annual daily traffic on the road is 105 vehicles per day, and for railway vehicles it is 4 trains per day. The crossing is secured with traffic signs and the visibility triangle. There is not a sufficient visibility zone for safe traffic at the crossing. There was one accident at this crossing in 2019.

The level crossing at railway kilometer 42 + 690, by its location, belongs to the municipality of Modriča. The crossing is on an open railway and crosses one track. The functional significance of the road is regional. The crossing is in a straight direction and the angle of the intersection between the road and track is of 90 degrees. The slope of the road in relation to the crossing is 0%. The width of the road in the crossing zone is 6 m. The permitted speed over the crossing for road vehicles is 60 km/h and for railway vehicles it is 50 km/h. The average annual daily traffic on the road is 562 vehicles per day and for railway vehicles it is 4 trains per day. The crossing is secured with traffic signs and the visibility triangle. There is not a sufficient visibility zone for safe traffic at the crossing. There were two incidents at this crossing in 2019.

The level crossing at railway kilometer 44 + 351, by its location, belongs to the municipality of Modriča. The crossing is on an open railway and crosses one track. The functional significance of the road is uncategorized. The crossing is in a straight direction and the angle of the intersection between the road and track is of 90 degrees. The slope of the road in relation to the crossing is 5.5%. The width of the road in the crossing zone is 3 m. The permitted speed over the crossing for road vehicles is 30 km/h and for railway vehicles it is 50 km/h. The average annual daily traffic on the road is 10 vehicles per day and for railway vehicles it is 4 trains per day. The crossing is secured only by the visibility triangle. There is not a sufficient visibility zone for safe traffic at the crossing. In 2019, there were three serious accidents with two people killed at this crossing.

The level crossing at railway kilometer 46 + 465, by its location, belongs to the municipality of Modriča. The crossing is on an open railway and crosses one track. The functional significance of the road is regional. The crossing is located in a straight direction and the angle of the intersection between the road and track is of 80 degrees. The slope of the road in relation to the crossing is 9.5%. The width of the road in the crossing zone is 6 m. The permitted speed over the crossing for road vehicles is 60 km/h and for railway vehicles it is 50 km/h. The average annual daily traffic on the road is 860 vehicles per day, and for railway vehicles it is 4 trains per day. The crossing is secured only by the visibility triangle.

There is a sufficient visibility zone at the crossing for safe traffic. There was one incident at this crossing in 2019.

The level crossing at railway kilometer 71 + 431, by its location, belongs to the municipality of Doboj. The crossing is on an open railway and crosses one track. The functional significance of the road is local. The crossing is in a straight direction and the angle of the intersection between the road and track is of 90 degrees. The slope of the road in relation to the crossing is 0%. The width of the road in the crossing zone is 4 m. The permitted speed over the crossing for road and railway vehicles is 50 km/h. The average annual daily traffic on the road is 28 vehicles per day and for railway vehicles it is 4 trains per day. The crossing is secured with traffic signs and the visibility triangle. There is not a sufficient visibility zone for safe traffic at the crossing. In 2019, there was one serious accident with three people killed at this crossing.

The level crossing at railway kilometer 82 + 291, by its location, belongs to the municipality of Doboj. The crossing is on an open railway and crosses over two tracks. The functional significance of the road is local. The crossing is in a straight direction and the angle of the intersection between the road and track is of 100 degrees. The slope of the road in relation to the crossing is 0%. The width of the road in the crossing zone is 3 m. The permitted speed over the crossing for road and railway vehicles is 50 km/h. The average annual daily traffic on the road is 14 vehicles per day and for railway vehicles it is 4 trains per day. The crossing is secured only by the visibility triangle. There is not a sufficient visibility zone for safe traffic at the crossing. There was one incident at this crossing in 2019.

The analysis of serious accidents, accidents, and incidents, with killed and injured persons at level crossings in BiH in the last five years is shown in Figure 4. It can be seen that there are variations in the number of accidents in the last five years and that in 2019 there was the largest number of injured persons, while the number of incidents decreased compared to the previous two years.

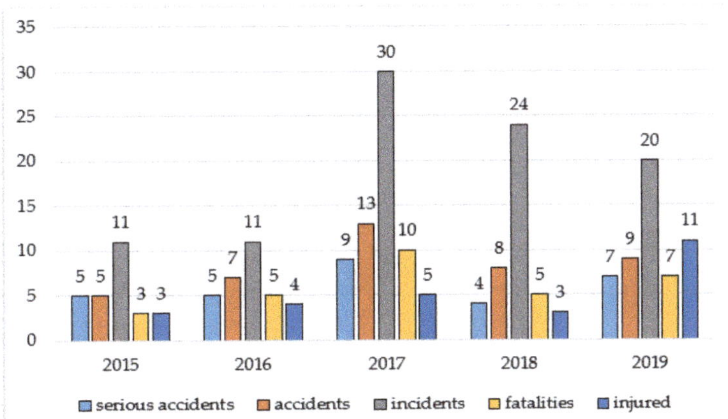

Figure 4. The total number of serious accidents, accidents, incidents, fatalities, and injuries in the last five years.

In the third step, the evaluation of criteria was performed by 11 decision-makers in a group decision-making process. In particular, the main group of criteria and all sub-criteria within the main groups were evaluated, which is a prerequisite for the application of this methodology. In the fourth step, data processing, i.e., fuzzy triangular numbers, was performed. The next phase involved the application of the fuzzy FUCOM method based on the fuzzy Heronian mean operator to determine the values of the main criteria. Then, fuzzy PIPRECIA was applied to determine the values of the sub-criteria.

The fuzzy FUCOM method [20,21] is an extension of the original FUCOM method [22–24] with fuzzy triangular numbers (TFN) and consists of the following steps:

Step 1. Ranking of criteria using expert judgment.

Step 2. Determining the vector of comparative significance of the evaluation criteria using TFN.

Step 3. Defining the constraints of a nonlinear optimization model. The values of the weighting coefficients should satisfy two conditions, namely:

1. Condition 1. The ratio of the weight coefficients is equal to the comparative significance between the observed criteria.
2. Condition 2. The final values of the weighted coefficients should satisfy the condition of mathematical transitivity.

Step 4. Defining a model for determining the final values of the weighting coefficients of the evaluation criteria.

Step 5. Solving the model and obtaining the final weight of the criteria/sub-criteria

Since it is a group decision-making, the fuzzy form of the Heronian mean operator [25] was used to average the weights of the main criteria.

$$\widetilde{w}_j = \left(w_j^l, w_j^s, w_j^u\right) = \begin{cases} w_j^l = \left(\frac{2}{e(e+1)} \sum\limits_{i=1}^{n} \sum\limits_{j=i}^{n} \zeta_i^{lp} \zeta_j^{lq}\right)^{\frac{1}{p+q}} \\ w_j^s = \left(\frac{2}{e(e+1)} \sum\limits_{i=1}^{n} \sum\limits_{j=i}^{n} \zeta_i^{sp} \zeta_j^{sq}\right)^{\frac{1}{p+q}} \\ w_j^u = \left(\frac{2}{e(e+1)} \sum\limits_{i=1}^{n} \sum\limits_{j=i}^{n} \zeta_i^{up} \zeta_j^{uq}\right)^{\frac{1}{p+q}} \end{cases}$$

The fuzzy PIPRECIA method [26–28] consists of the steps given below.

Step 1. Forming a set of criteria and sorting the criteria according to marks from the first to the last. Step 2. Each decision-maker individually evaluates pre-sorted criteria by starting from the second criterion. Step 3. Determining the coefficient $\overline{k_j}$. Step 4. Determining the fuzzy weight $\overline{q_j}$. Step 5. Determining the relative weight of the criterion $\overline{w_j}$. In the following steps, the inverse methodology of fuzzy PIPRECIA method needs to be applied. Step 6. Performing the assessment, but this time starting from a penultimate criterion. Step 7. Determining the coefficient $\overline{k_j}'$. Step 8. Determining the fuzzy weight $\overline{q_j}'$. Step 9. Determining the relative weight of the criterion $\overline{w_j}'$. Step 10. In order to determine the final weights of criteria, it is first necessary to perform the defuzzification of the fuzzy values $\overline{w_j}$ and $\overline{w_j}'$. Step 11. Checking the results obtained by applying Spearman and Pearson correlation coefficients.

Primarily, 11 fuzzy models were formed to determine the weights of the main criteria using the fuzzy FUCOM method. The obtained values were aggregated using the fuzzy Heronian mean operator to obtain the final weights of the main criteria. After determining the weights of all sub-criteria using the fuzzy PIPRECIA method, the final weights of all criteria were determined by multiplying the values of the main criteria by the values of the sub-criteria. The next phase involved the application of the fuzzy MARCOS method in order to determine safety degree at the level crossings of the Šamac–Doboj section. The procedure for solving the MCDM model using the fuzzy MARCOS method [29] is as follows (Figure 5). In the first step, the initial fuzzy matrix is formed. After that, an extended fuzzy decision matrix, which implies the formation of an anti-ideal and ideal solution depending on the type of criteria, is created. The next step involves the normalization of the initial fuzzy matrix, which depends on the type of criteria, followed by weighting with the weight values of the criteria obtained using the fuzzy FUCOM–fuzzy PIPRECIA model based on averaging the main criteria with the fuzzy Heronian Mean operator. In the fifth step, the fuzzy values are summed by rows and a matrix of 1 × n is obtained, and it is used for further calculation of the function degrees of the alternatives. Then, the following matrix is calculated, which is the sum of the elements of the utility degree matrix. The maximum fuzzy triangle number is chosen and its defuzzification is performed in order to calculate the utility functions in relation to the ideal and anti-ideal solution.

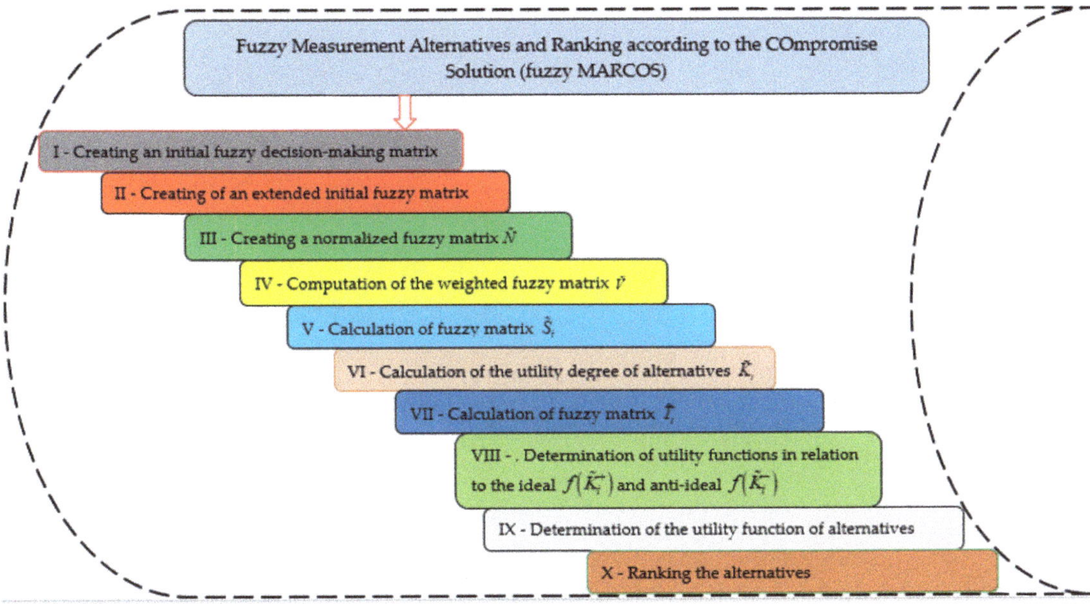

Figure 5. Steps of the fuzzy MARCOS (measurement of alternatives and ranking according to compromise solution) method.

After applying the overall previously described methodology, in the tenth step, the variants were ranked and the safety degree of level crossings was determined on the basis of the values obtained. The fourth phase involved the verification of the obtained results through four phases of sensitivity analysis. In the eleventh step, a change in the results caused by the influence of a change (decrease) in the size of the initial fuzzy matrix was determined. The twelfth step involved a comparative analysis of the original fuzzy FUCOM–Fuzzy PIPRECIA–fuzzy MARCOS model with two fuzzy approaches, namely: simple additive weighting (SAW) and weighted aggregated sum product assessment (WASPAS), and, in step 13, a change of an aggregator to average the values of the main criteria: Fuzzy MARCOS based on the fuzzy Bonferroni mean operator (Fuzzy MARCOS–Fuzzy BMO). Then, a significant sensitivity analysis was performed through the next step, which involves the formation of 30 scenarios in which new fuzzy values of all criteria are simulated, and in the last 15th step, the Spearman's correlation coefficient (SCC) was calculated in order to determine the correlation of ranks.

4. Results

4.1. Determining the Criteria Weights

In this section of the paper, criterion weights are shown. First, using the fuzzy FUCOM method, the weights of the main criteria (C1, C2, and C3) were calculated for all 11 experts who participated in the group decision-making. After that, using the fuzzy power Heronian mean operator, they were averaged in order to obtain the final values of the main criteria, as shown below.

$$FPHO^{p=1,q=1}\{(0.37,0.49,0.61),(0.38,0.48,0.58),(0.27,0.28,0.28),...,(0.31.0,31,0.31)\} =$$

$$\begin{cases} w^{l}_{C_1} = \left(\frac{2}{11(11+1)}\sum\limits_{i=1}^{n}\sum\limits_{j=i}^{n}\xi_i^{lp}\xi_j^{lq}\right)^{\frac{1}{1+1}} = (0.015(0.37^1\cdot0.37^1+0.37^1\cdot0.38^1+0.37^1\cdot0.27^1+...+0.31^1\cdot0.31^1))^{\frac{1}{1+1}} = 0.324 \\ w^{m}_{C_1} = \left(\frac{2}{11(11+1)}\sum\limits_{i=1}^{n}\sum\limits_{j=i}^{n}\xi_i^{lp}\xi_j^{lq}\right)^{\frac{1}{1+1}} = (0.015(0.49^1\cdot0.49^1+0.49^1\cdot0.48^1+0.49^1\cdot0.28^1+...+0.31^1\cdot0.31^1))^{\frac{1}{1+1}} = 0.421 \\ w^{u}_{C_1} = \left(\frac{2}{11(11+1)}\sum\limits_{i=1}^{n}\sum\limits_{j=i}^{n}\xi_i^{lp}\xi_j^{lq}\right)^{\frac{1}{1+1}} = (0.015(0.61^1\cdot0.61^1+0.61^1\cdot0.58^1+0.61^1\cdot0.28^1+...+0.31^1\cdot0.31^1))^{\frac{1}{1+1}} = 0.501 \end{cases}$$

After calculating the weights of the main criteria using the fuzzy FUCOM method based on the fuzzy Heronian Mean operator, the calculation of the weights of the sub-criteria within each group was started. For this purpose, the fuzzy PIPRECIA method was applied, the results of which are shown in Table 1. First, the aggregation of estimates was performed by the expert team using the average mean as highlighted in the steps of this method. As an example of the results, the sub-criteria of the safety group are presented (Table 1).

Table 1. Calculation and results obtained by applying the fuzzy-PIPRECIA (pivot pairwise relative criteria importance assessment) method for determining the criteria weights of traffic safety group.

	$\overline{s_j}$	$\overline{k_j}$	$\overline{q_j}$	$\overline{w_j}$	**DF**
C11		(1,1,1)	(1,1,1)	(0.19,0.24,0.29)	0.240
C12	(0.71,0.82,0.98)	(1.02,1.18,1.29)	(0.77,0.85,0.98)	(0.15,0.2,0.29)	0.208
C13	(0.64,0.76,0.9)	(1.1,1.24,1.36)	(0.57,0.68,0.9)	(0.11,0.16,0.26)	0.171
C14	(1.14,1.29,1.35)	(0.65,0.71,0.86)	(0.67,0.96,1.38)	(0.13,0.23,0.4)	0.242
C15	(0.44,0.54,0.67)	(1.33,1.46,1.56)	(0.43,0.66,1.04)	(0.08,0.16,0.3)	0.170
SUM			(3.44,4.15,5.3)		
	$\overline{s'_j}$	$\overline{k'_j}$	$\overline{q'_j}$	$\overline{w'_j}$	DF
C11	(0.9,1,1.1)	(0.9,1,1.1)	(0.76,1.21,1.67)	(0.11,0.21,0.36)	0.216
C12	(1.01,1.12,1.18)	(0.82,0.88,0.99)	(0.84,1.21,1.5)	(0.12,0.21,0.32)	0.211
C13	(0.56,0.69,0.77)	(1.23,1.31,1.44)	(0.83,1.06,1.22)	(0.12,0.18,0.26)	0.185
C14	(1.16,1.28,1.33)	(0.67,0.72,0.84)	(1.19,1.39,1.5)	(0.17,0.24,0.32)	0.241
C15		(1,1,1)	(1,1,1)	(0.15,0.17,0.22)	0.174
SUM			(4.62,5.87,6.88)		

Where $\overline{s_j}$ is a group decision matrix obtained by aggregating the estimates of all experts, starting from the second criterion according to the standard procedure of applying the fuzzy PIPRECIA method, $\overline{k_j}$ is a coefficient obtained when $\overline{s_j}$ is subtracted from number two, except for $\overline{s_1}$. $\overline{q_j}$ is a fuzzy weight, $\overline{w_j}$ is a relative weight of the criterion and DF is a defuzzified value.

The same methodology was applied for the other two groups of criteria: C2–road exploitation characteristics, C3–railway exploitation characteristics, and the final values of all criteria were obtained, which are shown in Table 2.

The most significant criteria belonged to each of the main groups, which means that the most significant criterion was C34–the method of securing the level crossing, followed by C14–the number of killed at the level crossing and C24–the visibility triangle. The obtained results were further included in the calculation with the fuzzy MARCOS method when evaluating the variants.

Table 2. Final values of all criteria after applying the fuzzy FUCOM (full consistency method)–fuzzy Heronian mean operator–fuzzy PIPRECIA model.

	Local Values	Global Values	Rank
C11	(0.15,0.223,0.326)	(0.048,0.094,0.163)	4
C12	(0.134,0.205,0.305)	(0.043,0.086,0.153)	7
C13	(0.114,0.173,0.263)	(0.037,0.073,0.132)	8
C14	(0.15,0.234,0.363)	(0.048,0.098,0.182)	2
C15	(0.113,0.165,0.259)	(0.037,0.069,0.13)	9
C21	(0.082,0.126,0.205)	(0.015,0.031,0.069)	15
C22	(0.125,0.21,0.341)	(0.023,0.052,0.115)	11
C23	(0.11,0.184,0.309)	(0.02,0.045,0.104)	13
C24	(0.168,0.298,0.518)	(0.031,0.073,0.174)	3
C25	(0.113,0.182,0.326)	(0.021,0.045,0.11)	12
C31	(0.082,0.126,0.205)	(0.022,0.05,0.116)	10
C32	(0.125,0.21,0.341)	(0.028,0.068,0.16)	6
C33	(0.11,0.184,0.309)	(0.026,0.067,0.161)	5
C34	(0.168,0.298,0.518)	(0.037,0.101,0.249)	1
C35	(0.113,0.182,0.326)	(0.021,0.043,0.097)	14

4.2. Evaluation of Level Crossings Using the Fuzzy MARCOS Method

After determining the final weight values of the criteria, conditions were created for determining the safety degree of level crossings on the Šamac–Doboj railway section in order to achieve sustainable railway traffic management. As stated and explained above, a total of eight level crossings were evaluated using the fuzzy MARCOS method. In the first step, the initial fuzzy matrix was formed consisting of eight level crossings, which were evaluated on the basis of a total of 15 criteria arranged in three basic groups. After that, an extended fuzzy decision matrix was created, implying the formation of an anti-ideal and ideal solution depending on the type of criteria. Criteria C11–C15, C22, C23, C25, C31, and C32 belong to the group of criteria with a minimum value as desirable, so the ideal solution of the extended fuzzy decision matrix included the lowest value for these criteria. The other five criteria belong to the group of criteria with a maximum value as desirable, so the highest value for each criterion entered the ideal solution. Regarding the formation of the anti-ideal solution, the situation was reversed, i.e., the criterion with a maximum value as desirable had the lowest value and the criterion with a minimum value as desirable had the highest value. The next step involves the normalization of the initial fuzzy matrix, which depends on the type of criteria, followed by weighting with the weight values of the criteria obtained using the fuzzy FUCOM–fuzzy PIPRECIA model based on averaging the main criteria with the fuzzy Heronian mean operator. In the fifth step, the fuzzy values were summed by rows and a matrix of 1×8 was obtained, and it was used for further calculation of the function degrees of alternatives. Then, the following matrix was calculated, representing the sum of the elements of utility degree matrices. The maximum fuzzy triangle number was chosen and its defuzzification was performed in order to calculate the utility functions in relation to the ideal and anti-ideal solution. The results of the described methodology are shown in Table 3.

The results given in Table 3 show that the level crossing 42 + 690 (LC4) represents the crossing with the highest level of safety considering all 15 criteria. At this level crossing, two incidents occurred, which is the only adverse event when it comes to the criteria belonging to the safety group. When it comes to road exploitation characteristics, the LC belongs to the regional category of roads with a fairly high frequency of road traffic, speed of 60 km/h, limited visibility triangle, and horizontal ground (without slope). Regarding the railway exploitation characteristics, it is important to point out that for this passive level crossing, the angle of intersection is at a right angle of 90 degrees, the largest width of the crossing, and the way of securing which includes a combination of visibility triangle, signal sign, and stop sign.

Table 3. Results obtained by applying the integrated fuzzy FUCOM–fuzzy PIPRECIA–fuzzy MARCOS model.

	$f(\tilde{K}_i^-)$	$f(\tilde{K}_i^+)$	K-	K+	fK-	fK+	Ki	Rank
LC1	(0.02,0.13,0.86)	(0.05,0.37,2.43)	3.609	1.286	0.236	0.662	1.029	6
LC2	(0.02,0.14,0.86)	(0.05,0.39,2.43)	3.659	1.303	0.239	0.671	1.061	5
LC3	(0.02,0.13,0.93)	(0.04,0.36,2.62)	3.733	1.329	0.243	0.684	1.108	4
LC4	(0.03,0.15,0.94)	(0.05,0.43,2.65)	4.006	1.427	0.261	0.734	1.298	1
LC5	(0.02,0.12,0.88)	(0.04,0.34,2.49)	3.531	1.257	0.230	0.647	0.980	8
LC6	(0.03,0.15,0.91)	(0.05,0.42,2.57)	3.907	1.391	0.255	0.716	1.227	3
LC7	(0.02,0.13,0.85)	(0.04,0.37,2.41)	3.573	1.272	0.233	0.655	1.006	7
LC8	(0.03,0.15,0.95)	(0.05,0.41,2.67)	3.969	1.413	0.259	0.727	1.271	2

The second safest level crossing is LC8 (82 + 291), which has the following characteristics. In terms of all the safety-related parameters, one incident occurred, without participation in other classification groups of adverse events. Further, it belongs to the category of local roads with a very low frequency of road traffic, with a speed of 50 km/h, a limited visibility triangle, and a horizontal ground. When it comes to railway exploitation characteristics, the angle of intersection is 100°, the way of securing is the visibility triangle, and the width of the crossing is three meters.

The level crossing that is the riskiest in terms of sustainable traffic management is LC5 (44 + 351), which had three serious accidents and two fatalities, followed by LC7 (71 + 431) with one serious accident with three people killed.

5. Sensitivity Analysis and Discussion

A sensitivity analysis was performed in several phases. The first phase involved changing the size of ranking of the initial matrix, forming a total of seven additional scenarios. In each of the scenarios, the last-ranked alternative was eliminated, and the calculation was performed again with the size of the matrix $n-1$ comparing to the initial solution. Figure 6 shows the results by comparing to the initial ranks and values with each scenario formed.

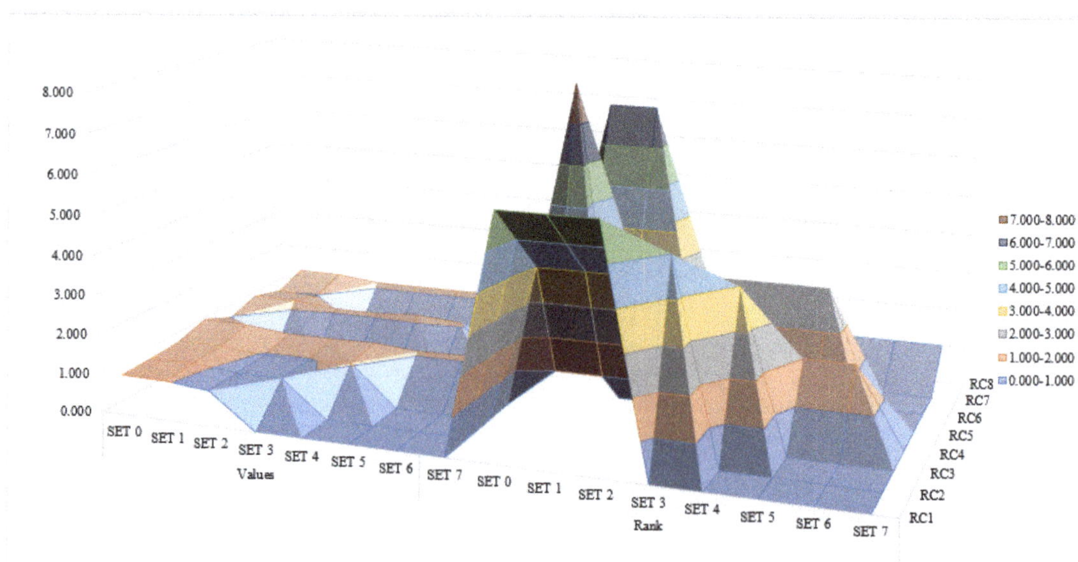

Figure 6. Results of sensitivity analysis when changing the size of the initial matrix.

Figure 6 shows the values (left) and ranks (right) through the formed scenarios in terms of the size of the fuzzy rank of the initial decision matrix. It can be observed that the dynamic conditions of the fuzzy matrix affect the results. Specifically, in the first four scenarios when LC5, LC7, LC1, and LC2 were eliminated, respectively for each scenario separately, there was no change of ranks compared to the initial rank. When it comes to scenarios S5–S7, one change occurred when the first and second best alternatives changed their positions, which was caused by eliminating the third alterative solution LC3.

In the second part of the sensitivity analysis, a comparative analysis of the original fuzzy FUCOM–fuzzy PIPRECIA–fuzzy MARCOS model was performed with three fuzzy approaches, namely: fuzzy SAW [30], fuzzy WASPAS [30], and fuzzy MARCOS based on fuzzy BMO operator [31,32] (fuzzy MARCOS–fuzzy BMO).

Figure 7 shows the results of a comparative analysis with three fuzzy approaches. Considering the obtained results, it can be concluded that there were certain changes in the ranks, which is a consequence of a different normalization approach in applying other approaches. Furthermore, one of the causes of changes in the rankings is reflected in a very small difference in the values of some alternative solutions obtained in the original model. When it comes to a comparative analysis with the fuzzy SAW method, the only change of rank occurred when LC2 and LC3 change their positions, LC2 is in the fourth position and vice versa regarding LC3. When it comes to the comparison with the fuzzy WASPAS method, there were slightly larger changes that are reflected in the change of ranks of LC1, LC2 and LC3 that change their ranks by one position. Observing the new changes, it can be concluded that the ranks still remain highly correlated. When the original fuzzy model is compared to a fuzzy model in which the fuzzy Bonferroni mean operator is used instead of the fuzzy Heronian mean operator, the initial ranks are retained. This means that changing the averaging operator in the main group of criteria does not play a role in the final result. In order to check rankings similarity, we calculated the WS coefficient (Figure 8) which was developed in the paper [33]. Advantage of WS coefficient in fact that positions at the top of the ranking have a more significant impact on the similarity than those further away, which is right in the decision-making domain.

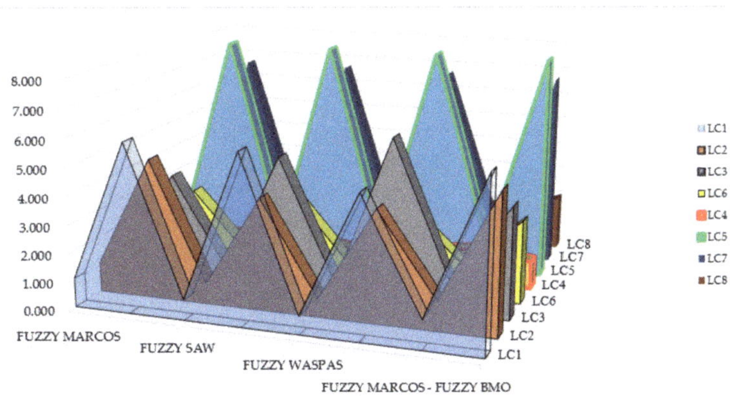

Figure 7. Comparative analysis with other fuzzy MCDM approaches.

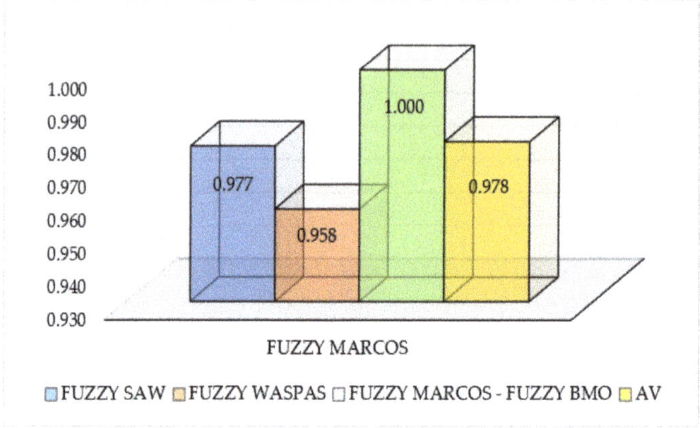

Figure 8. Calculated WS coefficient of rankings similarity in comparative analysis.

As can be seen in Figure 8, previous discussion is proven, where WS coefficient has values 0.977 (Fuzzy SAW), 0.958 (Fuzzy WASPAS), and 1.000 (Fuzzy MARCOS–Fuzzy BMO) which show extremely high correlation in ranking alternatives.

The third part of the sensitivity analysis involved the formation of 30 new scenarios with the reduction of the three most significant criteria and the verification of the impact on the final results. Thus, in the first 10 scenarios, the criterion C34 was reduced in the range of 5–95%, in S11–S20, the criterion C14 was reduced in the same range, and the criterion C24 was also reduced in the range of 5–95% in S21–S30.

The results obtained through the formed scenarios (Figure 9) show the variation caused by a large percentage decrease in the values of three most significant criteria. When it comes to the safest level crossing LC4 in scenarios S3–S10 when the most significant criterion was reduced by 25–95% respectively for each scenario by 10%, it loses the first position. Then, the second-placed LC8 in the original model takes the first position. In scenarios S9 and S10, LC4 is in third place, while LC8 is still in the first place. In other scenarios, there is no change in the rank of the safest level crossing, which means that the change in the value of criteria C14 and C24 does not play any role in these alternative solutions. The third-placed LC6 changes its position in four scenarios, in S9 and S10 when it is in the second place caused by the reduction of criterion C14 by 85 and 95%; then, in scenarios S19 and S20 when it is in the fourth and fifth place, respectively. In these scenarios, criterion C14 is reduced by 85% and 95%, respectively. Alternatives LC1, LC2, and LC3 do not change their positions compared to the initial ranking in the first 11, 12, and 14 scenarios, which means that the change of criterion C34 does not affect their positions. Changes in these alternatives occur when the value of criterion C14 is reduced.

Since, in the newly formed scenarios, there was a change in the results comparing to the initial solution, i.e., the model proved to be sensitive to the effects of the criteria, the SCC was calculated to determine the correlation of the initial solution with all scenarios.

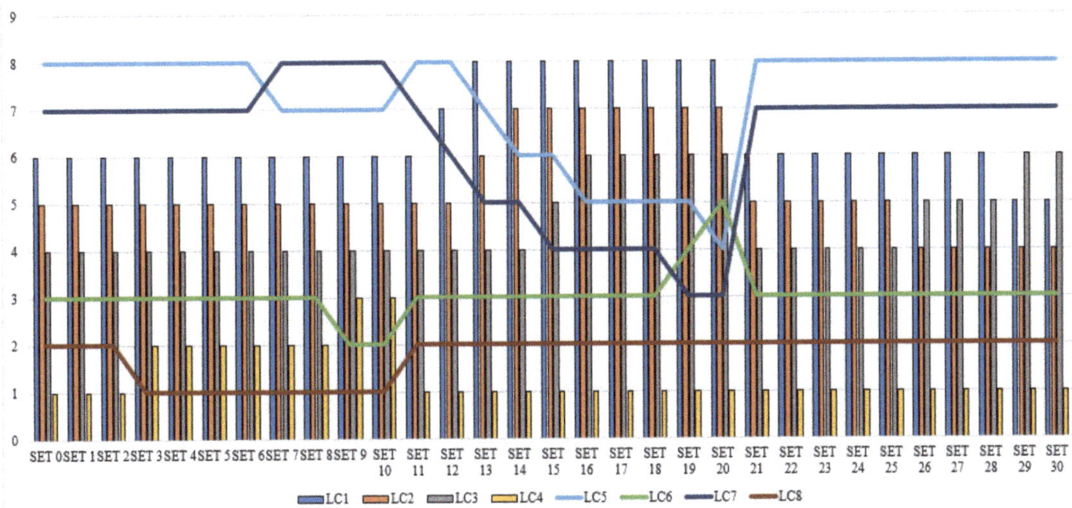

Figure 9. Results of sensitivity analysis in simulation of weight values of criteria through 30 scenarios.

Figure 10 shows a statistical correlation—SCC [34] and WS [33,35] of the initial rank results in relation to 30 scenarios in which the weight values of the criteria have been changed. A total correlation was observed in a total of eight scenarios out of 30 (SCC = 1.00). In the same number of scenarios, the SCC was 0.976, which means that there is a change in the ranks of the alternative solutions. In two scenarios (S7 and S8), SCC = 0.952, which means that there were two changes in ranks, while the value of the correlation coefficient was 0.929 in three scenarios. In other scenarios, the value of SCC was significantly lower and ranges 0.429–0.905, which means that there was a low correlation individually in some scenarios. In general, if all correlation coefficients are taken into account and the average SCC is calculated, it can be seen that the ranks were still highly correlated (SCC = 0.890). WS coefficient showed different values than SCC depending on rank changes because positions at the top of the ranking have a more significant impact on the similarity than those further away. For example, if we consider the third scenario where the two best alternatives changing their places, WS is less than SCC for 0.089 (SCC = 0.976, WS = 0.887). On the contrary, if we consider scenario 20 which has more changes in ranks, but not for best alternatives, WS has much higher value than SCC (SCC = 0.429, WS = 0.889). In general, if all correlation coefficients are taken into account and the average WS is calculated, it can be seen that the ranks are very highly correlated (WS = 0.943).

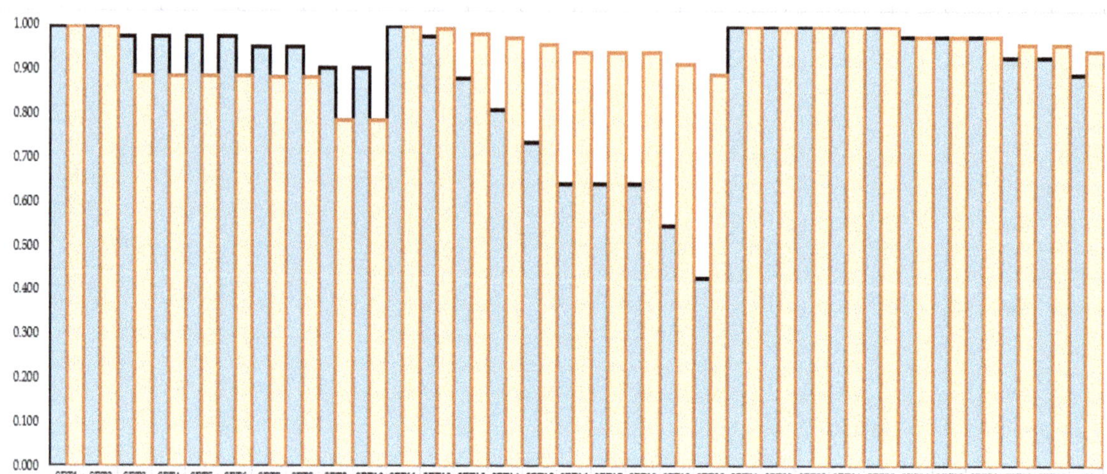

	SET1	SET2	SET3	SET4	SET5	SET6	SET7	SET8	SET9	SET10	SET11	SET12	SET13	SET14	SET15	SET16
SCC	1	1	0.976	0.976	0.976	0.976	0.952	0.952	0.905	0.905	1	0.976	0.881	0.810	0.738	0.643
WS	1	1	0.887	0.887	0.887	0.887	0.885	0.885	0.789	0.789	1	0.996	0.983	0.974	0.957	0.941

	SET17	SET18	SET19	SET20	SET21	SET22	SET23	SET24	SET25	SET26	SET27	SET28	SET29	SET30	AV
SCC	0.643	0.643	0.548	0.429	1	1	1	1	1	0.976	0.976	0.976	0.929	0.929	0.890
WS	0.941	0.941	0.915	0.889	1	1	1	1	1	0.977	0.977	0.977	0.958	0.958	0.943

SCC WS

Figure 10. Spearman's correlation coefficient (SCC) and WS for obtained ranks through simulated values of criterion weights in 30 scenarios.

6. Conclusions

In this research, an analysis of safety degree at some level crossings on the Šamac-Doboj railway section was conducted. The total number of crossings on the whole section was observed, as well as individual crossings with a passive way of security where some kind of accident occurred. A multi-criteria fuzzy model consisting of 15 criteria and 8 alternatives was formed. For the purpose of their evaluation, an original integrated fuzzy FUCOM–fuzzy PIPRECIA–fuzzy MARCOS model was applied. In addition, a fuzzy Heronian mean operator was used to average the weights of the main criteria. The results obtained in the research were verified through an extensive sensitivity analysis. This research can serve as a benchmark for other sections of the railway in BiH. Regarding the area of research and the results obtained, the following measures are presented in order to increase the sustainability of the overall transportation system.

There are many level crossings in Bosnia and Herzegovina where unwanted consequences for both road and railway traffic occur and can occur. Level crossings are not in accordance with modern standards and requirements in terms of slope, surface quality, and providing visibility triangles. The sustainable level of traffic safety is significantly reduced due to such state of level crossings. In BiH, it is necessary to take measures to improve the traffic safety degree at level crossings, as follows:

1. reducing level crossings by merging two or more level crossings into one,
2. providing necessary visibility from roads to tracks,
3. technical security of level crossings,
4. abolition of level crossings, i.e., their replacement with overpasses or underpasses,
5. lighting of level crossings,
6. application of modern technology, i.e., technical and technological improvements on the crossing infrastructure, such as the installation of various types of sensors (audio, video, radar, and lasers) for timely detection of potentially dangerous situations,
7. introduction of a video surveillance system at level crossings. Video surveillance can regularly monitor the functioning of traffic at crossings, and it can react quickly and efficiently in the event of any incident. In addition, drivers of road vehicles will

behave more responsibly when they know that they will be sanctioned if they do not comply with legal regulations, and

8. creating a culture of safety. All users (both road and rail, and especially road users) must be aware of dangers at level crossings.

The imperative is that some of above mentioned measures for improvement of the traffic safety degree at level crossings be implemented in short possible time. Future research that represents the continuation of this study refers to the formation of such or similar models for determining the safety level of other railway lines in Bosnia and Herzegovina. It is necessary to create unique model for all railway crossings on the complete railway line. Some other approaches for risk analysis can be applied.

Author Contributions: Conceptualization, A.B. and S.K.; methodology, A.B. and Ž.S.; validation, S.K. and V.P.; formal analysis, G.T.; investigation, A.B. and G.T.; writing—original draft preparation, Ž.S. and A.B.; writing—review and editing, S.K. and V.P. All authors have read and agreed to the published version of the manuscript.

Funding: This research received no external funding.

Institutional Review Board Statement: Not applicable.

Informed Consent Statement: Not applicable.

Data Availability Statement: Restrictions apply to the availability of these data. Data was obtained from the Railway Regulatory Board Bosnia and Herzegovina and are available [from the authors] with the permission of the Railway Regulatory Board Bosnia and Herzegovina

Conflicts of Interest: The authors declare no conflict of interest.

References

1. Liang, C.; Ghazel, M.; Cazier, O.; Bouillaute, L. Advanced model-based risk reasoning on automatic railway level crossings. *Saf. Sci.* **2020**, *124*, 104592. [CrossRef]
2. Blagojević, A.; Stević, Ž.; Marinković, D.; Kasalica, S.; Rajilić, S. A novel entropy-fuzzy PIPRECIA-DEA model for safety evaluation of railway traffic. *Symmetry* **2020**, *12*, 1479. [CrossRef]
3. Obradović, M.; Jevremović, S.; Trpković, A.; Milosavljević, M. Traffic-spatial analysis of road-rail crossings on state roads in the Republic of Serbia. *J. Road Traffic Eng.* **2020**, *66*, 35–40.
4. Kasalica, S.; Obradović, M.; Blagojević, A.; Jeremić, D.; Vuković, M. Models for ranking railway crossings for safety improvement. *Oper. Res. Eng. Sci. Theory Appl.* **2020**, *3*, 85–100. [CrossRef]
5. Kasalica, S.; Vukadinović, R.; Lučanin, V. Study of drivers behaviour at a passive railway crossing. *Promet Traffic Transp.* **2012**, *24*, 193–201. [CrossRef]
6. Djordjević, B.; Krmac, E.; Mlinarić, T.J. Non-radial DEA model: A new approach to evaluation of safety at railway level crossings. *Saf. Sci.* **2018**, *103*, 234–246. [CrossRef]
7. Pamučar, D.; Lukovac, V.; Božanić, D.; Komazec, N. Multi-criteria FUCOM-MAIRCA model for the evaluation of level crossings: Case study in the Republic of Serbia. *Oper. Res. Eng. Sci. Theory Appl.* **2018**, *1*, 108–129. [CrossRef]
8. Salmon, P.M.; Read, G.J.M.; Walker, G.H.; Goode, N.; Grant, E.; Dallat, C.; Carden, T.; Naweed, A.; Stantond, N. STAMP goes EAST: Integrating systems ergonomics methods for the analysis of railway level crossing safety management. *Saf. Sci.* **2018**, *110*, 31–46. [CrossRef]
9. Márquez, F.P.G.; Pedregal, D.J.; Roberts, C. New methods for the condition monitoring of level crossings. *Int. J. Syst. Sci.* **2015**, *46*, 878–884. [CrossRef]
10. Đorđević, D.; Atanasković, P.; Gopčević, Š.; Mikić, T. Concept of level crossing Information subsystem. In Proceedings of the Railcon 2008—Scientific-expert conference on railways, Niš, Serbia, 9–10 October 2008; pp. 277–280.
11. Bester, L.; Toruń, A. Modeling of reliability and safety at level crossing including in Polish railway conditions. In Proceedings of the International Conference on Transport Systems Telematics, Katowice/Kraków/Ustroń, Poland, 22–25 October 2014.
12. Lutovac, T.; Lutovac, D. Development of a diagnostic system for computer controlled level crossing systems. *Info M* **2013**, *11*, 11–17.
13. Rudin-Brown, C.M.; Lenné, M.G.; Edquist, J.; Navarro, J. Effectiveness of traffic light vs. boom barrier controls at road–rail level crossings: A simulator study. *Accid. Anal. Prev.* **2012**, *45*, 187–194. [CrossRef] [PubMed]
14. Lenné, M.G.; Rudin-Brown, C.M.; Navarro, J.; Edquist, J.; Trotter, M.; Tomasevic, N. Driver behaviour at rail level crossings: Responses to flashing lights, traffic signals and stop signs in simulated rural driving. *Appl. Ergon.* **2011**, *42*, 548–554. [CrossRef] [PubMed]

15. Evans, A.W. Fatal accidents at railway level crossings in Great Britain 1946–2009. *Accid. Anal. Prev.* **2011**, *43*, 1837–1845. [CrossRef] [PubMed]
16. Hu, S.-R.; Li, C.-S.; Lee, C.-K. Investigation of key factors for accident severity at railroad grade crossings by using a logit model. *Saf. Sci.* **2010**, *48*, 186–194. [CrossRef] [PubMed]
17. Park, Y.-J.P.; Saccomanno, F.F. Collision frequency analysis using tree- based stratification. *Transp. Res. Rec. J. Transp. Res. Board* **2005**, *1908*, 121–129. [CrossRef]
18. Saccomano, F.; Fu, L.; Miranda-Moreno, M.L. Risk-based model for identifying highway-rail grade crossing blackspots. *Transp. Res. Rec. J. Transp. Res. Board* **2004**, *1862*, 127–135. [CrossRef]
19. Austin, R.; Carson, J. An alternative accident prediction model for highway-rail interfaces. *Accid. Anal. Prev.* **2002**, *34*, 31–42. [CrossRef]
20. Pamucar, D.; Ecer, F. Prioritizing the weights of the evaluation criteria under fuzziness: The fuzzy full consistency method—FUCOM-F. *Facta Univ. Ser. Mech. Eng.* **2020**, *18*, 419–437.
21. Simić, J.M.; Stević, Ž.; Zavadskas, E.; Bogdanović, V.; Subotić, M.; Mardani, A. A novel CRITIC-Fuzzy FUCOM-DEA-Fuzzy MARCOS model for safety evaluation of road sections based on geometric parameters of road. *Symmetry* **2020**, *12*, 2006. [CrossRef]
22. Pamučar, D.; Stević, Ž.; Sremac, S. A new model for determining weight coefficients of criteria in MCDM models: Full consistency method (FUCOM). *Symmetry* **2018**, *10*, 393. [CrossRef]
23. Božanić, D.; Tešić, D.; Kočić, J. Multi-criteria FUCOM–Fuzzy MABAC model for the selection of location for construction of single-span Bailey bridge. *Decis. Mak. Appl. Manag. Eng.* **2019**, *2*, 132–146. [CrossRef]
24. Nenadić, D. Ranking dangerous sections of the road using MCDM model. *Decis. Making: Appl. Manag. Eng.* **2019**, *2*, 115–131. [CrossRef]
25. Beliakov, G.; Pradera, A.; Calvo, T. *Aggregation Functions: A Guide for Practitioners*; Springer: Heidelberg, Germany, 2007; Volume 221.
26. Stević, Ž.; Stjepanović, Ž.; Božičković, Z.; Das, D.K.; Stanujkić, D. Assessment of conditions for implementing information technology in a warehouse system: A novel fuzzy piprecia method. *Symmetry* **2018**, *10*, 586. [CrossRef]
27. Đalić, I.; Stević, Ž.; Karamasa, C.; Puška, A. A novel integrated fuzzy PIPRECIA–interval rough SAW model: Green supplier selection. *Decis. Mak. Appl. Manag. Eng.* **2020**, *3*, 126–145. [CrossRef]
28. Memiş, S.; Demir, E.; Karamaşa, Ç.; Korucuk, S. Prioritization of road transportation risks: An application in Giresun province. *Oper. Res. Eng. Sci. Theory Appl.* **2020**, *3*, 111–126. [CrossRef]
29. Stanković, M.; Stević, Ž.; Das, D.K.; Subotić, M.; Pamučar, D. A new fuzzy MARCOS method for road traffic risk analysis. *Mathematics* **2020**, *8*, 457. [CrossRef]
30. Petrović, G.; Mihajlović, J.; Ćojbašić, Ž.; Madić, M.; Marinković, D. Comparison of three fuzzy MCDM methods for solving the supplier selection problem. *Facta Univ. Ser. Mech. Eng.* **2019**, *17*, 455–469. [CrossRef]
31. Pamucar, D.; Deveci, M.; Canıtez, F.; Božanić, D.I. A fuzzy full consistency Method-Dombi-Bonferroni model for prioritizing transportation demand management measures. *Appl. Soft Comput.* **2020**, *87*, 105952. [CrossRef]
32. Pamucar, D. Normalized weighted Geometric Dombi Bonferoni Mean Operator with interval grey numbers: Application in multicriteria decision making. *Rep. Mech. Eng.* **2020**, *1*, 44–52. [CrossRef]
33. Sałabun, W.; Urbaniak, K. A new coefficient of rankings similarity in decision-making problems. In Proceedings of the International Conference on Computational Science, Amsterdam, The Netherlands, 3–5 June 2020; pp. 632–645.
34. Subotić, M.; Stević, B.; Ristić, B.; Simić, S. The selection of a location for potential roundabout construction–a case study of Doboj. *Oper. Res. Eng. Sci. Theory Appl.* **2020**, *3*, 41–56. [CrossRef]
35. Shekhovtsov, A.; Kołodziejczyk, J.; Sałabun, W. Fuzzy model identification using monolithic and structured approaches in decision problems with partially incomplete data. *Symmetry* **2020**, *12*, 1541. [CrossRef]

sustainability

MDPI

Article

Research on Speeds at Roundabouts for the Needs of Sustainable Traffic Management

Slavko Davidović [1], Vuk Bogdanović [2], Nemanja Garunović [2], Zoran Papić [2] and Dragan Pamučar [3,*]

1 Banja Luka City Administration, Department of Traffic and Roads, 78000 Banja Luka, Bosnia and Herzegovina; davidovic.slavko@hotmail.com
2 Faculty of Technical Sciences, University of Novi Sad, 21000 Novi Sad, Serbia; vuk@uns.ac.rs (V.B.); garunovic@uns.ac.rs (N.G.); njele@uns.ac.rs (Z.P.)
3 Department of Logistics, Military Academy, University of Defence in Belgrade, 11000 Belgrade, Serbia
* Correspondence: dragan.pamucar@va.mod.gov.rs

Abstract: Knowledge of the characteristics of speed at roundabouts is very important in design procedures, simulation models, and determining the influence of these roundabouts on traffic conditions in a street network. Sustainable management in the street network means the influence of all its parts on traffic conditions and travel time. In order to reliably determine roundabouts parameters in the procedures of planning and the choice of sustainable method of traffic management, it is very important to know the values of the traffic flow parameters, particularly the speeds at the entry and exit leg, as well as in the circulatory zone. This article analyses the speed characteristics in traffic flow at urban roundabouts with different geometrical characteristics in the city of Banja Luka. The applied method for traffic data collecting in this research was the method of video recording processing, which excludes any influence on driver behavior. Furthermore, statistical analysis was conducted, which established the correlation between the achieved speeds and geometrical characteristics of the intersection. Due to roundabout characteristics, the research was focused on the access, that is, the entry leg, the segment of the circulatory zone and the exit leg. The research results showed there is a significant dependence between geometrical characteristics of certain elements of the roundabout on speeds. In the further course of the research, it was proved that the variation of speeds at the segments of roundabouts significantly affects the size of time losses and emission of pollutants, i.e., parameters based on which it is possible to objectively assess the possibility of sustainable implementation of the planned solution of roundabouts of similar geometry within the street network in cities similar to Banja Luka.

Keywords: urban roundabout; speed; sustainable management; geometrical characteristics

check for updates

Citation: Davidović, S.; Bogdanović, V.; Garunović, N.; Papić, Z.; Pamučar, D. Research on Speeds at Roundabouts for the Needs of Sustainable Traffic Management. *Sustainability* **2021**, *13*, 399. https://doi.org/10.3390/su13010399

Received: 24 November 2020
Accepted: 28 December 2020
Published: 4 January 2021

Publisher's Note: MDPI stays neutral with regard to jurisdictional claims in published maps and institutional affiliations.

Copyright: © 2021 by the authors. Licensee MDPI, Basel, Switzerland. This article is an open access article distributed under the terms and conditions of the Creative Commons Attribution (CC BY) license (https://creativecommons.org/licenses/by/4.0/).

1. Introduction

The speed is a parameter of traffic flow which defines functionality and the quality of traffic conditions on the part of a road or street network. Since the first scientific and expert articles, for the parts of road and street network with the conditions of continuous traffic flows, functional connection between geometrical and constructive characteristics of the road and traffic flow speed has been determined. On all road types, through different dependence and correlations the influence of the width and the number of traffic lanes, lateral clearances, curvature characteristics and grade on the speed are expressed [1]. Due to the functional dependence of speed with other traffic flow parameters, primarily with flow and density, it is well known that traffic flow speed affects roads capacity; thus, the quality of traffic conditions, expressed through level of service.

The influence of geometrical characteristics and environment on speed and traffic conditions was also proved at signalized intersections, although they are ruled by the conditions of interrupted traffic flow. When saturated flow volume is being determined, the grade is considered, as well as the number and the width of traffic lanes, but also the type of maneuver (straight, left or right) because in case of turning, trajectory radius directly affects speed. Apart from that, other features are also considered, such as parking, public transportation characteristics, disturbances by pedestrians and cyclists, etc. [2].

When traffic conditions are determined at all types of non-signalized intersections, usually Harders [3] or Siegloch [4] models are applied, based on the probability of accepting critical gap occurring in the priority flow by the drivers who perform minor maneuver. This method is also applied for the analysis of traffic conditions at roundabouts where the right of way is given to the traffic flow in the circulatory zone [1]. These models do not consider intersection geometrical characteristics, that is, their influence on traffic flow parameters, capacity and traffic conditions. Since the end of the previous century, traffic conditions at roundabouts where vehicles in the circulatory zones have priority, the aforementioned models have also been applied for their determination. In the street network, there are roundabouts with significantly different geometrical characteristics, which quite certainly can affect the way of movement and vehicle speed [5].

Since the beginning of this century, roundabouts have been applied with more frequency, even to the arterials of the primary street network. With the aim of having a sustainable traffic management in the street network, that is, the influence of certain segments on the time and emission of pollutants, it is necessary to know characteristic values of traffic flow parameters [6–8]. Roundabouts, due to their specific traffic mode, can significantly affect traffic conditions, not only on approaches to the roundabout, but also in the surrounding street network, particularly in the conditions of a saturated traffic flow when even small disturbances at any segment can cause congestions and the effects of shock waves. For this reason, when traffic is planned, as well as in the procedures of optimal management, it is particularly important to know the values of speeds at different segments of roundabouts, so as to approach near the results of analyses and simulations to real traffic conditions. Quality analysis of traffic conditions in the street network and the choice of optimal, that is, sustainable method of traffic management demands, requires the knowledge of the characteristics of the traffic flow and the geometry of roundabouts on traffic flow speed [9]. Accordingly, the main aim of this paper is to determine whether there are functional correlations between geometrical characteristics of roundabouts and speed in a traffic flow.

Methodological approach to the problem of the influence of the geometry of roundabouts on the traffic flow speed, as well as the regression model for the calculation of the speed of the flow in circulation, can be applied to all typical segments of roundabouts of different geometries, similar to those that have already been the subject of research and analyses. Thus, considering the average delays generated at roundabouts, and which are determined through standard models, it is possible to determine the influence of different types of roundabouts on the emission of pollutants in a more precise way.

In addition to the application of the results achieved in this paper in the field of Traffic Management and Transport Planning, the results can also be applied in the field of Traffic Safety. Namely, these results can be used to determine the impact of the roundabout on the level of service as well as the selection of an optimal method of traffic regulation and the segment of the street network, which is extremely important for traffic safety management. Furthermore, speed projections by roundabout segments can be used to assess risk and define concrete measures to increase the safety of road users [10].

This article consists of six sections. Introduction provides the basic observations and initial guidelines related to the topic the article deals with. The second section deals with the analysis of the previous research studies relevant to this subject matter. The third section gives the description of the analyzed locations, as well as the methodology of vehicle speed data collecting and processing. The fourth section represents a section with the display of descriptive parameters of the analyzed samples, as well as the results of the statistical processing with a discussion. The fifth section analyses the influence of the research results on traffic flow parameters and the emission of pollutants, which are used as a basis for the assessment of roundabout application on the parts of the street network in the procedures of planning and operational analysis. The last section contains the main conclusions and guidelines for further research.

2. Previous Research Studies

The first concept of circular traffic flow at intersections appeared in the United States of America in the early 1930s. The new method of regulation meant signposts putting into the intersection center, alternatively called "Dummy Cop", around which there was circulatory traffic. At roundabouts, at the beginning of their application, there were no concrete rules for defining vehicle circulation. With the first applied principle, the rule of the right, it was defined that the vehicles which enter the intersection have priority over those already being in the intersection. This regulation method resulted in congestion of intersections with increased traffic volume [11]. The concept of contemporary roundabouts was developed in 1960 in Great Britain when smaller roundabouts were proposed, with a greater vehicle lane deflection. In 1966, a rule was adopted that circulatory vehicles in roundabouts have the priority; that is, the vehicles at the intersection access are obliged to give way to the vehicles in the circulatory zone [12].

Previous research studies imply that there is a significant influence of local traffic conditions on the value of traffic flow parameters, thus the value of capacity of all types of priority intersections, which include circular priority intersections. [13]. Research studies [14] conducted just before writing the manual HCM 2016 (Highway Capacity Manual 6) established the differences in capacities all over the USA. They imply that possible differences occur due to local cultural behavior of drivers or different geometrical characteristics of roundabouts. Geometrical characteristics of roundabouts mean the number of traffic lanes, the width of the entry leg, the values of the inscribed circle diameter etc. Therefore, there is a possibility of expanding and correcting a thorough analysis conducted in a certain country for analyzing it in other countries.

Analyzing the available literature [15–32], it was determined that many research studies emphasize speed as one of the most significant parameters in roundabouts geometrical shaping, as well as the key parameter in micro-simulating traffic models. In order to evaluate the influence of the roundabout geometry, certain simulation models enable the distributions of speed, that is, defining the deceleration zone when approaching the roundabout, circulatory speed and acceleration speed to the common values at the links [15]. In many research studies, it has been shown that there is a connection between geometrical characteristics of the roundabout, speed and capacity, traffic flow parameters and traffic conditions [16,17]. The combination of the influence of geometry, traffic flow parameters, and driver behavior make the traffic conditions at a roundabout very complex with significant variations of the speed and other parameters of traffic flow. According to the research study from 2011, it is necessary to use the parameter of roundabout diameter for the calculation of roundabouts capacity [18], whereby bigger roundabouts use the exponential coefficients of the model, which directly regresses.

Chen, Persaud and Lyon [19] established the relationship between some geometrical characteristics of roundabouts and average speed which was defined as the average value of entrance speed, the speed within the roundabout and the exit speed. Statistical analysis included 15 different geometrical parameters, and by means of linear regression they highlighted three main parameters for the definition of average speed: Diameter of the inscribed circle, entrance width and the width of lanes within the roundabout. For a well-controlled going through a roundabout, geometrical elements are applied for vehicles' speed reducing at the access to the intersection, it also limits the speed in the roundabout and enables reaching a normal speed after leaving the roundabout [20].

Prediction of speed at roundabouts is of essential importance in the design process due to different design elements defining and dimensioning. After the study carried out on roundabouts in Australia, Great Britain and the USA [21], it was concluded that entrance radius and minimal radius of the central circular island are two basic parameters that define frameworks for roundabouts design. The results of this study showed that the higher value of entrance radius, the lower entrance speed into the roundabout, and the correlation between the increase of minimal value of the central circular island, which is dependent on the increase of design speed at the approach to the roundabout.

The research conducted on roundabouts in the Municipality of Cuneo, located in the Northwest of Italy, studied the influence of the main geometrical parameters of the roundabout on vehicles' speed. This research included the analysis of vehicle speed on the approach to the intersection, in the very roundabout flow and at the exit segment. The research results showed that geometrical elements of the roundabout mostly affect the vehicles' speed in the roundabout, as well as that this speed can be dependent on the inner diameter of the roundabout, roundabout roadway width and the width of inflow roadway lane [22]. Another method for determining the speed of vehicles going through the roundabout is based on defining the vehicles' path [23]. With this approach, vehicle speed at roundabouts is correlated to vehicle path radius, super elevation and side friction factor. In other articles, the influence of intersection geometry on speed was confirmed, and a significant difference in speed on the approach within the circulatory zone and the exit leg was proven [24].

Taking into consideration the models based on acceptable follow-up gaps, many research studies proved a correlation among geometrical characteristics of roundabouts, speed and capacity—thus, traffic conditions as well [25,26].

Speed is one of the basic parameters defined in simulation models to describe real traffic conditions at roundabouts [27]. In order to evaluate the influence of intersection geometry, certain simulation models enable speed distribution, that is, define the zone of speed reduction when approaching the roundabout, the circulatory speed and the acceleration to the common values at the links [15]. For this reason, it is very important to know speed distribution depending on the geometrical characteristics of intersections.

The research conducted at four roundabouts from the group of mini and small urban roundabouts on the territory of the City of Zagreb [28], was carried out with the aim of determining the connection of the design speed and vehicles trajectory, that is, comparison of the measured speeds in the field and the adopted design speeds for the corresponding trajectory. The results have shown certain deviations, which the authors explained with a specific position of the intersections in the street network, elements of the geometrical design of the intersections, the structure of traffic flow. Also, in the same town, for the needs of defining the model of trajectories of the vehicles within the roundabout, a research study was conducted [29] that determined the connection among the radius, deflection angle and the radius of central island.

In the town of Livorno [30], geometrical characteristics at roundabouts were analyzed and those intersections were built more than fifty years ago, with the outer diameter of more than 100 m, with the aim of the reconstruction from the aspect of the geometrical parameters correction. Thus, through simulation software packages, a significantly better level of service was achieved.

Certain research studies have shown that the probability of a traffic accident occurrence is proportional to the speed squaring [31]. Consequently, the research on flow speed is also justified, in this case at roundabouts. Other authors, using the data on traffic accidents and vehicle trajectories processed through software packages AIMSUN and VISSIM, conducted the comparative analysis and formed the model for predicting collisions, that is, the assessment of roundabouts safety in Slovenia [32].

Research on the impact of road geometry and traffic flow parameters on traffic safety has shown that there is a significant dependence [33]. However, in the Southeastern Europe (SEE) region, surveys about roundabouts are still rare and are mainly focused on determining the influence of drivers' behavior on traffic flow parameters [34]. This paper is a result of the first research conducted in Bosnia and Herzegovina (BiH) with the aim of considering the influence of geometry on the speed of traffic flow at different segments of roundabouts with the possibility of application in the field of Traffic Management and Transport Planning.

3. Methodology

3.1. Key Roundabout Dimensions

In order to describe a roundabout as an object on road and street network, it is necessary to previously define basic dimensions, that is, key geometrical characteristics. In the process of design and detail analyses of roundabouts, it is necessary to define numerous geometrical parameters, whereby the following elements are defined as the key ones [35]:

- Inscribed circle diameter (D1), basic parameter used to define the size of a roundabout. It is measured between the outer edges of the circulatory roadway.
- Central island circle diameter (D2), measured between the internal edges of the circulatory roadway.
- Circulatory roadway width (W_cr), defines the roadway width for vehicle circulation around the central island. It is measured as the width between the outer edge of this roadway and the central island. It does not include the width of any mountable apron, which is defined to be part of the central island. For speed and traffic conditions, the number of traffic lanes in circulatory zone (N_cr), also matters, as well as width of the lane on circulatory roadway (W_lnc). Accordingly, radius of center line of circulatory lane (R_lnc) is also a parameter that can influence movement speed in the circulatory zone.
- Approach width, the width of the roadway used by approaching traffic upstream of any changes in width associated with the roundabout. This parameter depends on the number of lanes on the approach (N_lne) and width of the lane on the approach (W_lne).
- Departure width, width of the roadway used by departing traffic downstream of any changes in width associated with the roundabout. This parameter depends on the number of lanes on the departure (N_lnx) and width of the lane on departure (W_lnx).
- Entry width (W_en), defines the width of the entry where it meets the inscribed circle. It is measured perpendicularly from the right edge of the entry to the intersection point of the left edge line and the inscribed circle.
- Exit width (W_ex), defines the width of the exit where it meets the inscribed circle. It is measured perpendicularly from the right edge of the exit to the intersection point of the left edge line and the inscribed circle.
- Entry radius (R_en) the minimum radius of curvature of the outside curb at the entry.
- Exit radius (R_ex) the minimum radius of curvature of the outside curb at the exit.

In the following figure (Figure 1), there are labeled roundabout key elements.

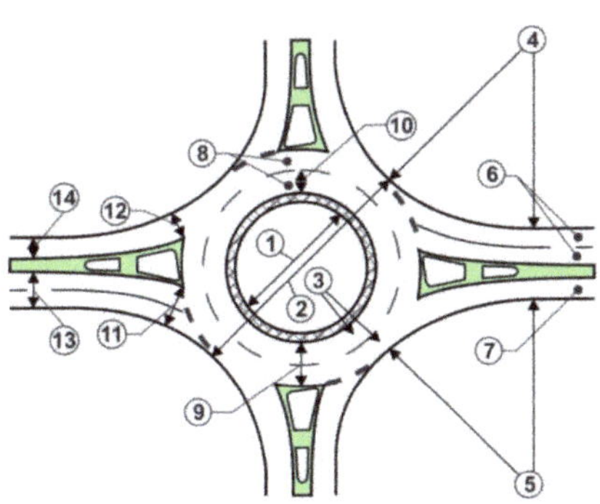

1. Central island circle diameter
2. Inscribed circle diameter
3. Radius of center line of circulatory lane
4. Entry radius
5. Exit radius
6. Number of lane on approach
7. Number of lane on departure
8. Number of lane on circulatory roadway
9. Circulatory roadway width
10. Width of the lane on circulatory roadway
11. Entry width
12. Exit width
13. Approach width
14. Departure width

Figure 1. Drawing of key roundabout dimensions.

The research analyzed the influence of all aforementioned geometrical elements on the speed at the entry leg, circulatory zone and bottleneck exit leg.

3.2. Description of the Research Location

Data collecting was conducted by recording real traffic flows at four urban roundabouts. The recording was carried out by means of a drone "DJI Mavic 2 Pro", which was set at a certain height above the analyzed roundabouts and it recorded traffic. The analyzed intersection is in Banja Luka (Bosnia and Herzegovina), a town with around 180,000 inhabitants. The intersections used for data collecting are in the urban area of the town, and their type belongs to mid-sized roundabouts and large town roundabouts. In the City of Banja Luka, there are a total of 15 roundabouts, out of which 9 are urban roundabouts. A sample of 4 out of 9 urban roundabouts covered all types, taking into account the characteristic geometrical parameters to be analyzed. The researched intersections are at the edge of the central zone of Banja Luke, with AADT of around 20,000 veh/day. Traffic for trucks is prohibited on these roundabouts. In addition, during the observed period, no occurrence of buses was recorded at the intersections or the number of buses was negligible. Given such a structure of traffic flow, only the speeds of passenger cars were analyzed in the paper, in other words, the analysis was performed on the basis of a homogeneous traffic flow composed of passenger cars.

The recording was conducted in July and August 2019. During the recording, it was sunny with the temperature of 25–35 °C, without fog, rain, gales or any other unfavorable climate conditions, the roadway was dry and without any damages, and the traffic without the situations which would influence unhindered and safe traffic, that is, there were no factors which would cause negative effects on drivers' behavior and vehicles' movement. Figures 2 and 3 show the position of the analyzed intersections in the street network of the town of Banja Luka, as well as the observed roundabouts appearance. Table 1 gives the data on the exact position of the intersections, and Table 2 the basic geometrical characteristics of the observed intersections, that is, approaches.

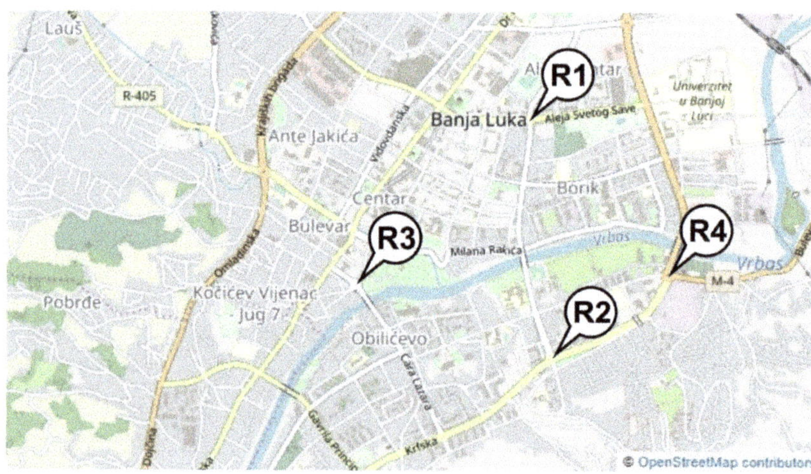

Figure 2. Positions of the analyzed intersections in the street network of the town of Banja Luka.

Figure 3. Ortho-photo image of the analyzed intersections: (**a**) R1: Gundulićeva–Aleja svetog Save; (**b**) R2: Majke Jugovića–Bul. Desanke Maksimović; (**c**) R3: Patre, Cara Lazara–Teodora Kolokotronisa, Isaije Mitrovića; (**d**) R4: Bul. Desanke Maksimović–Bul. Vojvode Stepe Stepanovića.

Table 1. Data on the position of the analyzed intersections.

Position of Intersection in WGS-84 Coordinate System	R1	R2	R3	R4
Latitude	44.773963	44.762172	44.765897	44.766366
Longitude	17.199593	17.201245	17.187834	17.209049

Table 2. Geometrical characteristics of the analyzed intersections.

Geometrical Characteristics	R1	R2	R3	R4
D1—Inscribed circle diameter (m)	33.60	32.00	43.00	57.20
D2—Central island circle diameter (m)	22.00	16.00	31.00	34.80
W_cr—Circulatory roadway width (m)	5.80	8.00	6.00	9.40
N_cr—Number of lane on circulatory roadway	1	2	1	2
W_lnc—Width of lane on the circulatory roadway (m)	5.80	4.00	6.00	4.70
W_en—Entry width (m)	5.00	7.50	5.10	7.50
N_lne—Number of lane on approach	1	2	1	2
W_lne—Width of lane on approach (m)	4.00	3.50	4.50	3.75
W_ex—Exit width (m)	5.20	8.40	5.00	9.00
N_lnx—Number of lane on departure	1	2	1	2
W_lnx—Width of lane on departure (m)	4.00	3.50	4.50	3.80
R_en—Entry radius (m)	15.00	12.00	13.00	23.00
R_ex—Exit radius (m)	15.00	16.00	40.00	26.00

3.3. Data Collection Method

For determining the vehicles' speed, the method of video recording processing and analysis was used. Video recording processing proved to be an extremely efficient method for traffic flow parameters measuring, because its use minimizes the influence of the research on the drivers. The applied method enabled the analysis in real traffic flow, in which the speed of traffic flow depends exclusively on the degree of interaction between vehicles and the geometric characteristics of the intersection. Data collected and processed in this way represent an efficient and practical basis in many research studies of traffic flow parameters [36,37]. The collected video material of the frequency 24 fps was processed with the software package Kinovea (Figure 4), which enables referential lines marking, the analysis of the objects movements in the recording, and the recording survey by sequences in the accuracy of approximately 0.042 s. The measured speed represents the mean spatial speed measured between two referential lines at the measuring area.

Figure 4. Display of video recording processing.

During the period of collecting data on vehicle speed, it was adopted that a vehicle goes through three typical segments: entry leg–A road part immediately before the circular traffic flow; circulating–A road part in the lane within the circular traffic flow; and the exit leg–A road part immediately after the circular traffic flow. Accordingly, the speed of vehicles' going through the roundabout was divided in the following way.

- Vehicles' speed at the inflow leg (approach)to the roundabout, entry speed (S_en),
- Vehicles' speed in the roundabout, circulatory speed (S_cr),
- Vehicles' speed at the outflow leg of the roundabout, exiting speed (S_el).

During samples forming, the recording was made only for the speeds of the vehicles whose movement was not influenced by the interaction with other participants in traffic flow.

Obstructions that could have appeared as the consequence of these interactions are:

- Vehicles stopping or deceleration, which occurs as a consequence of the change in the movement of the preceding vehicle;
- Vehicles stopping or deceleration on the entry leg to the intersection with the aim of giving way to a pedestrian or a vehicle which is within the roundabout;
- Circulating vehicles stopping or deceleration with the aim of speed adjustment to the speed of the vehicle that is leaving the roundabout flow or which is changing lanes within the roundabout with two traffic lanes;
- Vehicles stopping or deceleration at the exit leg of the intersection with the aim of giving way to a pedestrian;
- Vehicles stopping or deceleration, which occurs as the consequence of unforeseen actions in the intersection zone (improper pedestrians going across the road outside the pedestrian crossing, improper stopping and parking on the roadway, etc.)

If the vehicle was hindered in any of the aforementioned ways, its speed, in that case, is not taken as relevant. In this way, the only collected and considered speeds were the speeds of the vehicles whose movement was influenced only by geometrical characteristics of the intersection.

Methodology steps in the process of data collecting for the needs of the subject analysis have been shown by means of the algorithm (Figure 5). The given algorithm followed by the drawing showing the zones where speed measuring was conducted.

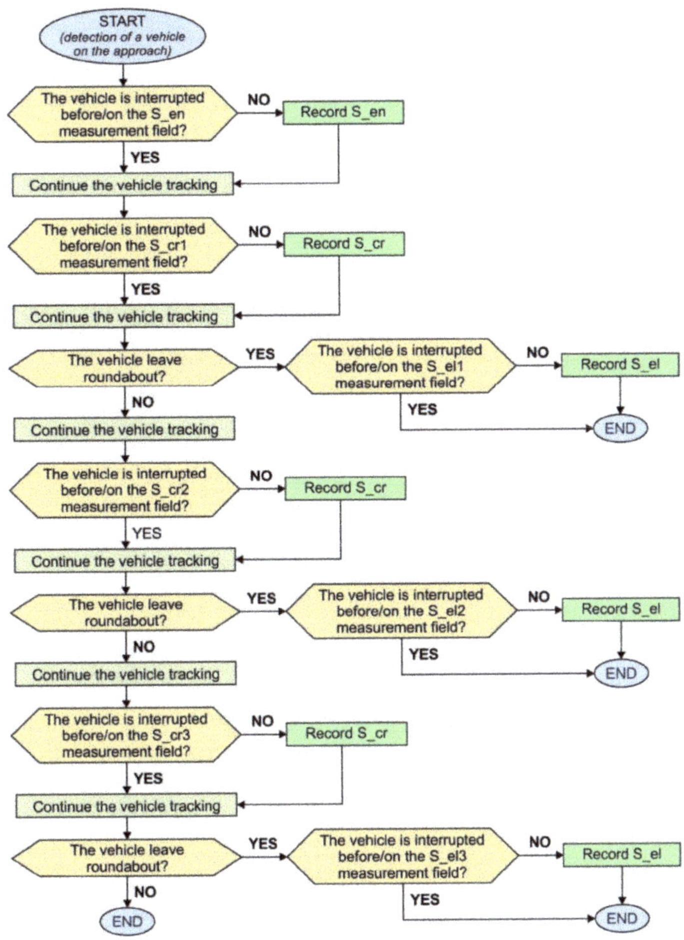

Figure 5. Algorithm of data collection.

As it can be seen in Figure 6, the average measuring points were positioned at all inflow and outflow bottlenecks, and in the circulatory zone at all segments between neighboring approaches. In accordance with the cross-sectional layout where the measuring was conducted for the unhindered vehicles at the inflow and outflow bottleneck, one-speed measuring was conducted at the cross-section. However, in the circulatory zone, speed measuring was conducted at all cross-sections, for the vehicles which turn right at one cross-section (e.g., S_cr1), for the vehicles that go straight through the intersection at two cross-sections (e.g., S_cr1 and S_cr2) and for the vehicles that turn left at three cross-sections (e.g., S_cr1, S_cr2 and S_cr3).

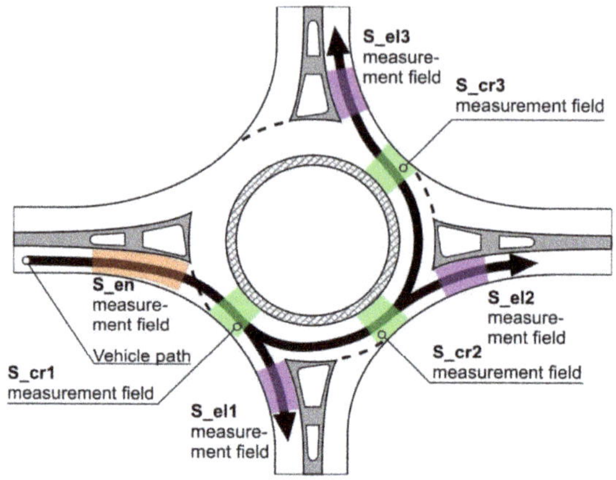

Figure 6. Zones of typical vehicles' speeds measuring at the roundabout.

4. Research Results and Discussion

This section shows the research results which means descriptive statistical parameters of the collected samples and the results of statistical analyses of the interconnection of the speed and geometrical characteristics of the intersections. In order to describe the characteristics of the observed sample, the collected data are presented through the following descriptive:

- **N**—Total number of observations;
- **Mean**—The sum of all the observations divided by the number of observations (arithmetic average);
- **SEMean**—Standard error of the mean;
- **StDev**—Standard deviation of data;
- **Min**—Lowest value in observed data set;
- **Q1**—First quartile of observed data set, 25% of the data are less than or equal to this value;
- **Median**—The middle of the range of data, 50% of the data are less than or equal to this value;
- **Q3**—Third quartile of observed data set, 75% of the data are less than or equal to this value;
- **Max**—Highest value in observation data set.

All statistical analyses have been conducted by means of Minitab® statistical software.

4.1. *Vehicles Speed at the Entry Leg*

For the needs of analysis which was conducted within this paper, the surveyed speeds at the entry legs were only for the vehicles which, arriving at the intersection, were not hindered by other circulating vehicles. The analysis was carried out with the sample of 537 vehicles with the entrance speeds of the following characteristics (Table 3):

Table 3. Results of the measured vehicles' speeds on leg.

Variable [1]/Roundabout	Descriptive Statistic Parameters								
	N	Mean	SEMean	StDev	Min	Q1	Median	Q3	Max
S_en/R1	104	23.03	0.35	3.59	15.31	20.96	22.94	24.92	32.05
S_en/R2	171	28.67	0.39	5.16	16.02	24.99	28.79	32.14	46.09
S_en/R3	103	25.29	0.38	3.89	18.06	22.50	24.77	26.87	39.13
S_en/R4	159	29.26	0.44	5.61	12.49	25.20	28.80	32.63	52.17

[1] Variable abbreviations: S_en—vehicle speed on entry leg (km/h)); R1, R2, R3 and R4—identification mark of analyzed roundabout.

What was further surveyed was how the entrance speed (S_en) depends on the following geometrical parameters of the roundabout: inscribed circle diameter (D1), central island circle diameter (D2), entry radius (R_en), entry width (W_en), number of lane on leg (N_lne) and width of lane on leg (W_lne). The correlation of the achieved S_en depending on the aforementioned intersection geometrical characteristics was determined by means of the Pearson correlation coefficient (Table 4).

Table 4. Pearson correlations of S_en and factors D1, D2, R_en, W_en, N_lne, W_lne.

Factor	D1	D2	R_en	W_en	N_lne	W_lne
Pearson correlation	0.191	0.048	0.172	0.437	0.434	−0.303
p-Value	0.000	0.269	0.000	0.000	0.000	0.000

On the basis of the conducted analysis, it can be stated that, taking into consideration all the aforementioned parameters of geometrical characteristics of roundabout, speed S_en is mostly influenced by entry width (W_en) and the number of lanes (N_lne) (Figure 7). According to the determined coefficients, it can be said that between speed S_en and these factors there is a real and significant linear correlation [38]. The increase of these two factors directly influences the increase of average entrance speed of vehicles.

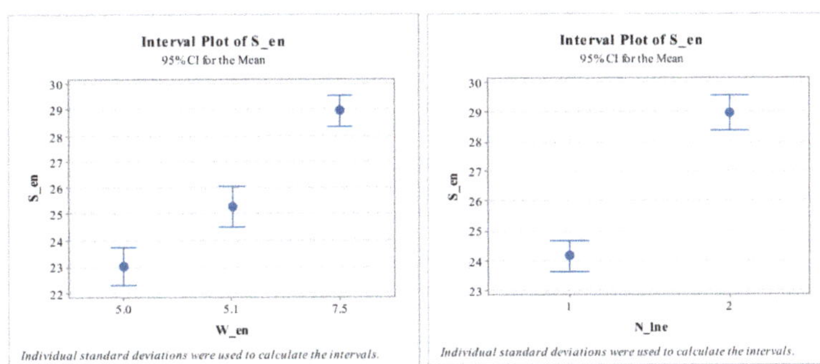

Figure 7. Interval plot of S_en: (**a**) depending on the factor W_en; (**b**) depending on the factor N_lne.

4.2. Vehicles Speed in the Roundabout

The speed of vehicles in the roundabout (S_cr) represents the average speed in the measurement zone in the lane designated for vehicles circulating. The analysis was conducted on the sample of the speed of 1095 vehicles that moved unhindered in circulatory lanes (Table 5).

Table 5. Results of the measured vehicles' velocities on circulatory roadway.

Variable [1]/Roundabout	Descriptive Statistic Parameters								
	N	Mean	SEMean	StDev	Min	Q1	Median	Q3	Max
S_cr/R1	235	20.49	0.19	2.95	9.35	18.65	20.22	22.25	29.33
S_cr/R2/inter. lane	142	19.28	0.31	3.67	9.82	16.86	18.91	21.45	34.87
S_cr/R2/exter. lane	180	21.36	0.27	3.68	7.85	18.98	21.29	23.68	30.90
S_cr/R3	249	22.40	0.20	3.15	15.15	20.22	22.09	24.05	33.86
S_cr/R4/inter. lane	163	25.15	0.28	3.55	16.33	22.50	25.23	27.38	37.44
S_cr/R4/exter. lane	126	26.15	0.38	4.23	5.64	23.14	26.26	28.92	35.01

[1] Variable abbreviations: S_cr—vehicle speed on circulatory roadway (km/h); R1, R2, R3 and R4—identification mark of analyzed roundabout.

The speed of the vehicles in the circulatory lane was analyzed depending on the following geometrical characteristics of the intersection: inscribed circle diameter (D1), central island circle diameter (D2), width of the circulatory roadway (W_cr), number of lanes on circulatory roadway (N_lnc) and width of the lane on circulatory roadway (W_lnc). Apart from the aforementioned geometrical parameters, what was also surveyed was speed dependence in relation to the radius of center line of circulatory lane (R_lnc), which practically represents the radius of the path of the circulating vehicle. Table 6 shows the values of Pearson correlation coefficient between the speed S_cr and the aforementioned geometrical characteristics of the intersection.

Table 6. Pearson correlations of S_cr and factors D1, D2, R_lnc, W_cr, N_lnc, w_lnc.

Factor	D1	D2	R_lnc	W_cr	N_lnc	W_lnc
Pearson correlation	0.518	0.467	0.527	0.331	0.169	0.000
p-Value	0.000	0.000	0.000	0.000	0.000	0.996

According to the established correlation coefficients, it can be said that between speed S_cr and the factor D1, D2 and R_lnc, there is a real and significant linear connection, while with the other factors that connection is smaller. For the needs of further analysis factor R_lnc and D1 are distinguished as the two with the highest correlation coefficient. The relationship of these two parameters and the value of the average speed in circulating for concrete values of the chosen factors have been shown in Figure 8.

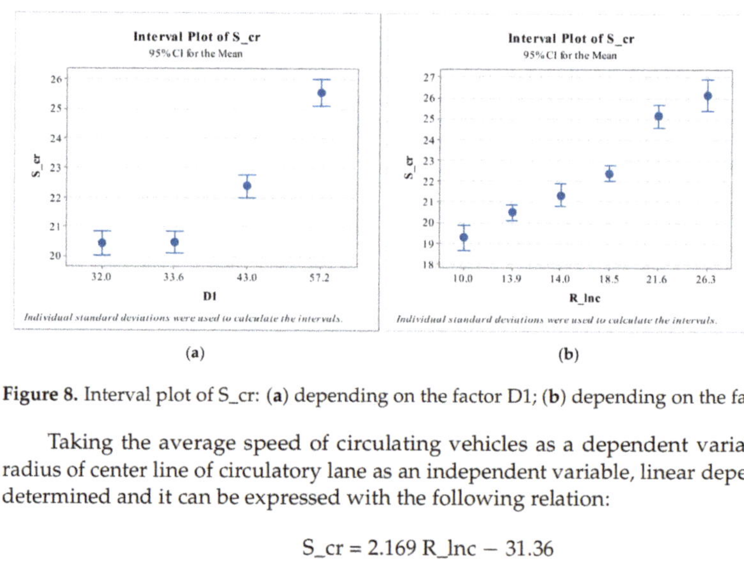

(**a**) (**b**)

Figure 8. Interval plot of S_cr: (**a**) depending on the factor D1; (**b**) depending on the factor R_lnc.

Taking the average speed of circulating vehicles as a dependent variable and the radius of center line of circulatory lane as an independent variable, linear dependence was determined and it can be expressed with the following relation:

$$S\_cr = 2.169\ R\_lnc - 31.36$$

where: S_cr—Mean speed value of the circulating vehicle (km/h), R_lnc—Radius of center line of circulatory lane (m).

This relation has been determined by means of simple regression on the basis of the analysis conducted in statistical software Minitab®. Figure 9 shows the line which describes the dependence of the aforementioned variables, with 96.1% confidence interval (CI) and prediction interval (PI). Coefficient of determination R2 which in this case is equal to 96.1% indicates that the model from the equation (1) has a good fit with observed speed values. A visual inspection of the plot (Figure 8) reveals that the data are randomly spread about the regression line, implying no systematic lack-of-fit. The R2-adjusted for this regression is 95.2%.

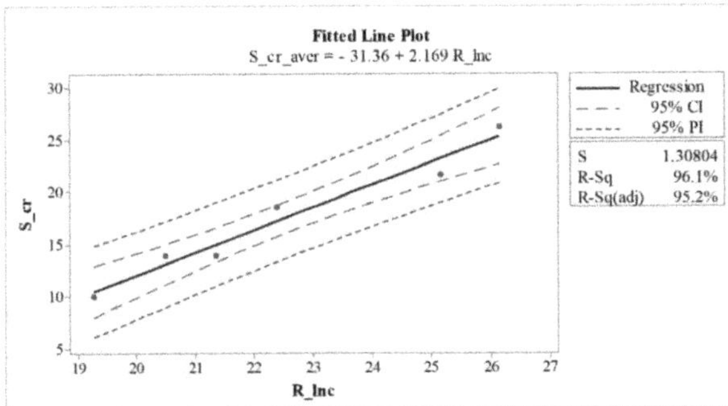

Figure 9. Fitted line plot of average S_cr depending on the factors R_lnc.

After the analysis of the interconnection of the average circulatory speed and the factor with which it achieves the highest correlation, the multiple regression analysis was conducted, which includes other factors given in Table 6. Further analysis defined the multiple regression subset models that produce the highest R2 values from full set of the specified predictor variables from Table 6. Table 7 shows five top-rated models. Generally observing, the introduction of more factors, in this case, does not give significant improvement of the very regressive model considering the fact that the increase of R2 is extremely small, and the value of R2-adjusted is decreasing.

Table 7. Best subset multiple regression models.

Number of Predictors	Multiple Regression Equation	S	R^2	R^2-Adjusted
2	S_cr_aver = −32.41 + 2.378 R_lnc − 0.468 W_cr	1.377	96.8	94.6
2	S_cr_aver = −33.8 + 2.366 R_lnc − 0.046 D1	1.493	96.2	93.7
3	S_cr_aver = −40.6 + 2.94 R_lnc − 0.354 D1 + 0.270 D2	1.515	97.4	93.5
3	S_cr_aver = −40.0 + 3.05 R_lnc − 0.133 D1 − 0.694 W_cr	1.532	97.3	93.4
4	S_cr_aver = −41.2 + 2.72 R_lnc − 0.73 D1 + 0.74 D2 + 1.2 W_cr	1.127	97.4	87.2

4.3. Vehicles Speed at Exit Leg

The speed of vehicles on the exit leg (S_el) represents the average speed in the measurement zone immediately after the traffic lane designated for circulating. The analysis was conducted on the sample of 773 velocities of vehicles that left the circulatory lane unhindered. The observed sample has the following general characteristics (Table 8):

Table 8. Results of the measured vehicles' speeds on departure.

Variable [1]/Roundabout	Descriptive Statistic Parameters								
	N	Mean	SEMean	StDev	Min	Q1	Median	Q3	Max
S_el/R1	180	27.66	0.47	6.31	4.33	24.12	27.45	30.77	65.16
S_el/R2	193	29.62	0.43	6.02	15.55	25.47	29.75	33.72	44.55
S_el/R3	224	27.15	0.38	5.72	3.09	23.58	26.18	29.92	49.09
S_el/R4	176	30.24	0.34	4.55	14.75	27.34	30.25	32.80	42.60

[1] Variable abbreviations: S_cr—vehicle speed on exiting leg (km/h); R1, R2, R3 and R4—identification mark of analyzed roundabout.

The speed of vehicles at the exit was analyzed depending on the following geometric characteristics of the intersection: inscribed circle diameter (D1), central island circle diameter (D2), exit radius (R_ex), exit width (W_ex), number of lane on exit leg (N_lnx) and width of lane on exit leg (W_lnx). Pearson correlation coefficient between the speed S_el and the aforementioned geometrical characteristics of the intersection has the following values (Table 9).

Table 9. Pearson correlations of S_el and factors D1, D2, N_lnx, w_lnx.

Factor	D1	D2	R_ex	W_ex	N_lnx	W_lnx
Pearson correlation	0.080	−0.011	−0.099	0.221	0.217	−0.192
p-Value	0.026	0.753	0.006	0.000	0.000	0.000

On the basis of the determined p-values and correlation coefficient it can be said that the correlation between the speed S_el and factor D1 and D2 does not exist while the biggest, but a slight correlation is with factors W_ex and N_lnx (Figure 10). On the basis of the stated facts, it can be concluded that the speed of the exit vehicles is not correlated to a large degree with the surveyed geometrical characteristics of the intersection.

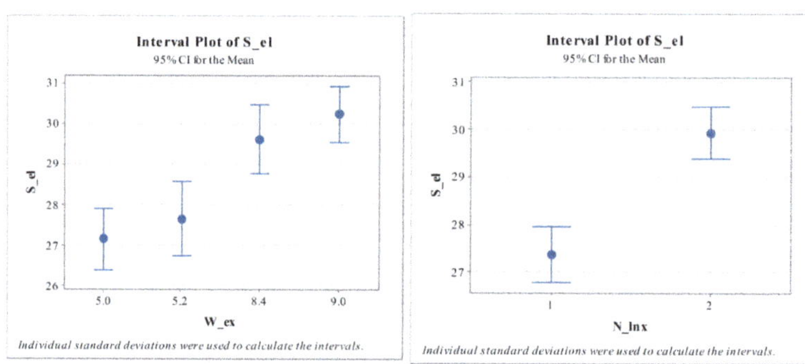

(**a**) (**b**)

Figure 10. Interval plot of S_el: (**a**) depending on the factor W_ex; (**b**) depending on the factor N_lnx.

The analysis of the research results showed that certain factors have different effects on the speeds in the roundabout zone. Also, in order to examine the impact of as many geometric parameters as possible on the flow velocity, the subject research included intersections with significantly different geometric characteristics. If we consider the individual impact of factors, say D1 (Inscribed circle diameter) and D2 (Central island circle diameter), we could conclude that:

- If there are two intersections that are identical in all characteristics, except for parameter D1, there may be a difference in speeds S_en and S_cr, (p-value = 0.000), whilst this parameter will not have a significant impact on S_el (p-value = 0.026).
- If there are two intersections that are identical in all characteristics, except for parameter D2, there may be a difference in speed S_cr, (p-value = 0.000), whilst in the case of S_en and S_el it would have no effect (p-value = 0.269 and 0.753 respectively).

If a certain parameter affects the flow velocity in only one segment of the roundabout, it is clear that it will also affect the average speed observed at the level of the entire intersection. The significance of the influence of a certain element on the average speed observed at the level of the whole intersection would depend on the influence on a specific speed in the specific area of the intersection (S_en, S_cr or S_el). In addition, it must be taken into account that the data set formed in this way implies the possibility of a combined influence of different parameters, so the individual influence of a certain geometric factor must be considered separately. A specific and detailed examination of the influence of an individual factor in real conditions would involve analysis of at least 2 roundabouts, which differ only in 1 parameter. This type of analysis could be the subject of further research.

5. Application of the Research Results in the Procedure of Analysis of the Parameters of the Sustainable Functioning of Roundabouts

For the evaluation of sustainability of the proposed solutions of roundabouts in the procedures of planning analysis, and for operational analysis of traffic conditions at roundabouts, parameters which should be considered are total delays generated at the roundabout, fuel consumption, as well as pollutants emission. All models and applicative software packages for the analysis of traffic conditions (SINCRO, SIDRA, VISSIM, etc.) allow the possibility of defining basic access speeds and the speeds in the circulatory zone for unhindered vehicles. These speeds depend, among other things, on geometrical characteristics of the roundabout [12,13,15,18,19,39]. In the situations when the value of flow demand reaches the value of the access capacity, that is, when saturated flows and high-density flows are formed on approaches, the interaction degree between vehicles most affects the speed. For this reason, in the procedures of sustainability evaluation, through Traffic Impact Assessment Analysis (TIA) and Traffic Environmental Impact Analysis (TEA), the influence of basic vehicular speeds is neglected. Namely, operators most usually adopt the values of access speeds and the speeds in the circulatory zone, given as default. In order to consider the influence of the established values of basic speeds on the approaches to roundabouts and in the circulatory zone on the sustainable functioning of roundabouts, the analysis within the program "Trafficware Synchro" was conducted, by two scenarios:

- **Scenario SC1**- Analysis of traffic conditions by default values of access speeds on approaches to roundabouts and in the circulatory zone.
- **Scenario SC2**- Analysis of traffic conditions by the values of access speeds on approaches to roundabouts and in the circulatory zone, which have been established in the research (Figure 11).

Figure 11. The sequence of SC2 "Trafficware Synchro" analysis of R1 in 3D view.

Emission factors, as well as the parameters which depend on driver behavior (e.g., critical gap and follow-up time), adopted by basic values, they were defined in the program, since they were not the subject of the analyses. The results of the conducted analyses, by defined scenarios for all four intersections, are shown in Table 10.

The displayed data in the previous table refer to one analyzed approach at each roundabout. Analysis of geometrical characteristics of other accesses at the observed roundabouts does not imply significant differences. Therefore, it can be expected that there is a similar influence of the change of average speed of the flow on the parameters of traffic conditions on the approaches with similar characteristics as the analyzed ones.

The analyses results imply that there are significant differences between delays and the established emission of pollutants by two considered scenarios in the situation when traffic conditions on accesses to the roundabout are nearly like traffic conditions in the saturated flow R1 and R3. In SC2, that is, at lower speeds on the accesses and in the circulatory zone, all parameters used as a basis for the evaluation of the sustainability of traffic conditions at roundabouts are significantly more unfavorable. At lower traffic load, that is, in traffic conditions close to free flow, favorable traffic conditions are obtained by SC2. This result is the consequence of a fact that at lower speeds there is a lower emission of pollutants for the vehicles which go through the roundabout without delays, and the influence of the emission which is the result of vehicles stopping on accesses is significantly lower compared to traffic conditions of higher density flow.

Table 10. Level of service parameters and pollutant emission at intersections.

Measured Factors	R1		R2		R3		R4	
on Analyzed Approach	SC1	SC2	SC1	SC2	SC1	SC2	SC1	SC2
Volume (pc/h)	611		371		718		306	
V/C Ratio	0.98		0.21		1.27		0.39	
Denied Delay (hr)	40.9	71.6	0	0	83.7	111	0	0
Denied Del/Veh (s)	225.6	393.4	0.2	0.3	418.8	552.8	0.2	0.2
Total Delay (hr)	3.8	3.5	0.5	0.8	3.1	3.0	0.2	0.2
Total Del/Veh (s)	22.3	22.6	4.2	6.8	18.9	19.8	2.6	2.3
Stop Delay (hr)	4.2	4.3	0.3	0.8	3.4	3.5	0.1	0.1
Stop Del/Veh (s)	24.6	27.1	2.1	6.3	21	23.3	1.2	1.6
Travel Time (hr)	45.1	75.9	0.9	1.4	87.1	114.5	0.4	0.4
Avg Speed (kph)	4	4	16	10	4	4	16	15
Fuel Used (l)	40	66.7	1.7	1.8	76.4	100.1	1	0.8
Fuel Eff. (kpl)	0.4	0.2	7.9	7.8	0.2	0.1	6.1	7.1
CO Emissions (g)	556	884	87	44	997	1290	55	34
NOx Emissions (g)	9	9	13	6	8	7	9	6

The following graphs (Figure 12) show the differences in parameters values for the evaluation of traffic condition for the defined scenarios.

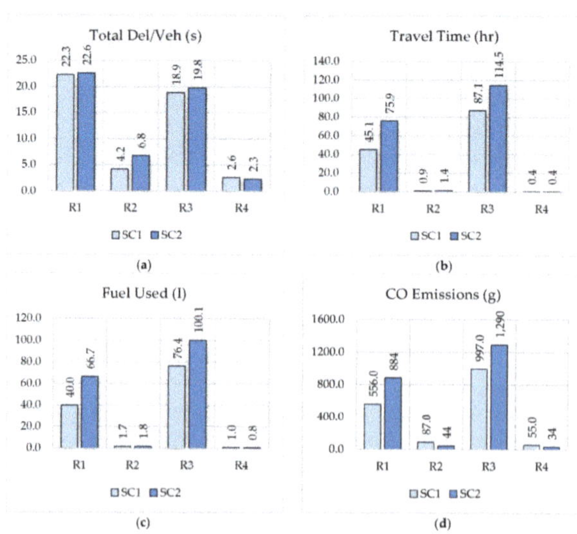

Figure 12. Basic differences of service parameters and pollutant emission: (**a**) Total Del/Veh (s); (**b**) Travel Time (hr); (**c**) Fuel Used (L); (**d**) R4: CO Emissions (g).

6. Conclusions

The research was conducted in a real traffic flow, at four roundabouts with significantly different geometrical characteristics, by means of the method of video recording processing, which excludes the possibility of the influence of the conducted research on driver behavior. The applied research methodology was oriented to determination of speeds characteristics at typical segments of roundabouts, and not to speed variations of particular vehicles when going through the roundabout. During sample forming the chosen speeds were only those of the vehicles whose movement on the approach, in circulatory flow and the exit leg from the intersection was not influenced by the movement of other vehicles in the flow, pedestrian movement in the intersection zone. This approach enabled the determination of the basic speed of traffic flow at typical segments of the roundabout. In this way, the impact of geometry on the traffic flow speed at roundabout segments was determined.

Research results analysis proved that between the speed at the exit leg and entry width (W_en) and the number of lanes (N_lne), there is a real and significant correlation. The increase of these two factors directly affects the increase of the average vehicle speed at the entry leg.

Therewith, the results analysis showed that in the circulatory zone there is a real and significant correlation between the speed and inscribed circle diameter (D1), central island circle diameter (D2), and radius of center line of circulatory lane (R_lnc), while with other analyzed factors the correlation is significantly smaller or there is no correlation. Taking into consideration the high degree of the correlation between the radius of center line of circulatory lane (R_lnc) and the speed of circulatory vehicles, a regression equation was formed which with a high degree of determination describes the connection between these two parameters.

The analysis of the speeds was limited to defining the dependence of speeds at the defined segments and geometrical characteristics of the intersection. However, this research did not determine the significant connection between geometrical characteristics of the intersection and the speed at the exit leg, which should be analyzed in further research. In addition, in further research studies there should be analyses on evaluation of the correlation between speeds in different sections, with the radius inserted in the "less effort trajectory".

In a special section, there is the analysis of the influence of the research results on traffic flow parameters and pollutants emission at roundabouts, used as a basis for sustainability evaluation within operative or planning Traffic Impact Assessment Analysis and Traffic Environmental Impact Analysis. The analysis was conducted in the program "Trafficware Synchro" by two scenarios depending of the traffic flow speed on the roundabouts. The results of the analyses showed that there is a significant influence of the speed values on the traffic flow parameters used for defining the level of service, and the significant influence of pollutants emission.

With these research results, which imply certain dependence of speeds and geometrical parameters of the roundabout, it is possible to expect that traffic flow speed will change in similar patterns at other roundabouts with similar geometrical characteristics. Within this paper, a regression model was defined, which describes, with an acceptable degree of determination, the dependence of the circulatory flow speed and certain geometrical roundabout parameters. Thus, the developed model can be used for a more precise determination of the influence of this segment of roundabouts with the geometry of similar roundabouts which were the subject of research, on the emission of all types of pollutants along street arterials. However, the characteristics of flow speeds at the exit and entry segment of the roundabout were adopted as empirically determined, and the further research will be directed to defining a model for determining the characteristics of traffic flow speed at these segments.

The research results shown in this paper can be used in the procedure of the analysis of the influence of roundabouts on the level of service of street network segments, that is, determination of the sustainability of the application of roundabouts at the observed segment of the street network in the procedures of planning or operational analysis. Unlike other parts of street network, the influence of roundabouts on the level of service of segments of street network in the existing methodologies has not been completely defined. The result of the analyses conducted within this paper showed that, apart from delays, for defining the influence of roundabouts on the level of service of segments of street network, it is necessary to know the characteristics of speeds on the access, circulatory zone and the exit leg. The application of the real values of speeds on the segments of the roundabout enables a better quality assessment of traffic conditions at the roundabout in the context of sustainability.

Further research, as a continuation of this paper, should be focused on the research of the speed impact and its dependence on the geometric parameters of roundabouts, on traffic safety, taking into account traffic safety factors, vehicle and road. Such research would enable an objective consideration of the impact of roundabouts on traffic conditions and traffic safety in cities that are similar in size to Banja Luka. In addition, it is necessary to determine in further research the impact of the structure of traffic flow on the speed of traffic flow at certain segments of roundabouts. Namely, in accordance with the proven impact of the traffic flow structure on the value of traffic flow parameters on other parts of the road and street network, one can objectively expect a significant mathematical dependence of geometric elements of roundabouts and individual vehicle categories, i.e., traffic flow structure. Also, One of the further research directions is to develop of a hybrid fuzzy methodology which would eliminate shortcomings of the crisp approach [40–42].

Author Contributions: S.D. and V.B. conceived of the presented idea. S.D. and Z.P. carried out the experiment. N.G. and D.P. performed the statistical analysis. V.B. and N.G. performed the traffic flow simulation. All authors provided critical feedback and helped shape the research, analysis and manuscript. All authors have read and agreed to the published version of the manuscript.

Funding: This research received no external funding.

Institutional Review Board Statement: Not applicable.

Informed Consent Statement: Not applicable.

Data Availability Statement: The data presented in this study are available on request from the corresponding author.

Conflicts of Interest: The authors declare no conflict of interest.

References

1. Transportation Research Board of The National Research Council. *Highway Capacity Manual*; Transportation Research Board of The National Research Council: Washington, DC, USA, 2010; Volume 2.
2. Transportation Research Board of The National Research Council. *Highway Capacity Manual*; Transportation Research Board of The National Research Council: Washington, DC, USA, 2010; Volume 3, Chapter 18.
3. Harders, J. *Die Leistungsfähigkeit nicht Signalregelter Städtischer Verkshrsknoten (Capacity of Unsignalized Urban Intersections)*; Bundesminister für Verkehr: Bonn, Germany, 1968.
4. Siegloch, W. *Die Leistungsermittlung an Knotenpunkten ohne Lichtsignalanlagen (Capacity Calculations at Unsignalized Intersections)*; Series Strassenbau und Strassenverkehrstechnik; Bundesminister für Verkehr: Bonn, Germany, 1973; Volume 154.
5. Bassani, M.; Sacchi, E. The effect of road geometrics and traffic regulations on driver—Preferred speeds in northern Italy. An exploratory analyses. *Transp. Res. Part A* **2014**, *25*, 10–26. [CrossRef]
6. Hyden, C.; Varheyli, A. The Effects on Safety, Time Consumption and Environment of Large Scale Use of Roundabouts in an Urban Area: A Case Study. *Accid. Anal. Prev.* **2000**, *32*, 11–23. [CrossRef]
7. Varheyli, A. The Effects of Small Roundabout on Emissions and Fuel Consumption: A Case Study. *Transp. Res. D Trans. Environ.* **2002**, *7*, 65–71. [CrossRef]
8. Coelho, M.C.; Farias, T.L.; Rouphail, N.M. Effect of Roundabout Operations on Pollutant Emissions. *Transp. Res. D Trans. Environ.* **2006**, *11*, 333–343. [CrossRef]
9. Al-Ohmari, B.H.; Ghuzlan, K.A.; Al-Helo, L.B. Modeling through Traffic Speed at Roundabouts along Urban and Suburban Street Arterials. *J. Trans. Res. Forum* **2014**, *53*, 7–19. [CrossRef]
10. Pešić, D.; Antić, B.; Smajlović, E.; Todosijević, B. The impact of the average traffic flow speed on occurrence risk of road crash. *J. Road Traffic Eng.* **2019**, *65*, 29–36. [CrossRef]

11. Daniels, S.; Wets, G. Traffic Safety Effects of Roundabouts: A Review with Emphasis on Bicyclist's Safety. In Proceedings of the 18th ICTCT Workshop, Helsinki, Finland, 27–28 October 2005.
12. Waddell, E. Evolution of Roundabout Technology: A History-Based Literature Review. In Proceedings of the Institute of Transportation Engineers 67th Annual Meeting, Boston, MA, USA, 3–7 August 1997.
13. Gazzarri, A.; Martello, M.T.; Pratelli, A.; Souleyrette, R.R. *Gap Acceptance Parameters for HCM 2010 Roundabout Capacity Model Applications in Italy. Transport Systems and Traffic Engineering*, 1st ed.; WIT Press: Southampton, UK, 2013; pp. 1–6.
14. Rodegerdts, L.A.; Malinge, A.; Marnell, P.S.; Beaird, S.G.; Kittelson, M.J.; Mereszczak, Y.S. *Assessment of Roundabout Capacity Models for the Highway Capacity Manual: Volume 2 of Accelerating Roundabout Implementation in the United States*; Report FHWA-SA-15-070; Federal Highway Administration: Washington, DC, USA, 2015.
15. Vaiana, R.; Gallelli, V.; Iuele, T. Methodological Approach for Evaluation of Roundabout Performances trought Microsimulation. *Appl. Mech. Mater.* **2013**, *1956*, 253–255.
16. Vincenzo, G.; Rosolino, V.; Teresa, I. Comparison between simulated and experimental crossing speed profiles on roundabout with different geometric features. In Proceedings of the EWGT2013—16th Meeting of the EURO Working Group on Transportation, Porto, Portugal, 4–6 September 2013.
17. Rodegerdts, L.; Bansen, J.; Tiesler, C.; Knudsen, J.; Myers, E.; Johnson, M.; O'Brien, A. *National Cooperative Highway Research Program (NCHRP) Report 672 Roundabouts: An Informational Guide*, 2nd ed.; Transportation Research Board of the National Academies: Washington, DC, USA, 2010.
18. Brilon, W. Studies on Roundabouts in Germany: Lessons Learned. In Proceedings of the 3rd International TRB Roundabout Conference, Carmel, IN, USA, 17–20 May 2011.
19. Yongsheng, C.; Bhagwant, P.; Lyon, C. Effect of Speed on Roundabout Safety Performance: Implications for Use of Speed as Surrogate Measure. In Proceedings of the Transportation Research Board 90th Annual Meeting, Washington, DC, USA, 23–27 January 2011.
20. Trueblood, M.; Dale, J. Simulating Roundabouts with VISSIM. In Proceedings of the 2nd Urban Street Symposium: Uptown, Downtown, or Small Town: Designing Urban Streets that Work, Anaheim, CA, USA, 28–30 July 2003.
21. Arndt, O. Speed control at roundabouts. In Proceedings of the 23rd ARRB Conference, Adelaide, Australia, 30 July–1 August 2008.
22. Bassani, M.; Sacchi, E. Ivestigation into speed performance and consistency of urban roundabouts: An Italian case study. In Proceedings of the TRB 3rd International Roundabout Conference, Carmel, IN, USA, 28–20 May 2011.
23. Aashto, A. *Policy on Geometric Design of Highways and Streets*; American Association of State Highway and Transportation Officials: Washington, DC, USA, 2004.
24. Wanga, W.; Yangb, X. Research on Capacity of Roundabouts in Beijing. *Procedia Soc. Behav. Sci.* **2012**, *43*, 157–168. [CrossRef]
25. Robinson, B.; Rodegerdts, L. *Capacity and Performance of Roundabouts: A Summary of Recommendations in the FHWA Roundabout Guide*; Kittelson & Associated, Inc: San Francisco, CA, USA, 2000.
26. Akçelik, R. Some common and differing aspects of alternative models for roundabout capacity and performance estimation. Proceedings of International Roundabout Conference, Transportation Research Board, Carmel, IN, USA, 17–20 May 2011.
27. Nikolic, G.; Pringle, R.; Bragg, K. Evaluation of analytical tools used for the operational analysis of roundabouts. In Proceedings of the Annual Conference of the Transportation Association of Canada, Halifax, NS, Canada, 26–29 September 2010.
28. Pilko, H.; Brčić, D.; Šubić, N. Study of vehicle speed in the design of roundabouts. *Građevinar* **2014**, *4*, 407–416.
29. Šurdonja, S.; Dragčević, V.; Deluka-Tibljaš, A.; Korlaet, Ž. Model of vehicle path radius at roundabout center. *Građevinar* **2010**, *71*, 163–175.
30. Pratelli, A.; Sechi, P.; Souleyrette, R.R. Upgrading traffic circles to modern roundabouts to improve safety and efficiency—Case studies from Italy. *Promet Traffic Transp.* **2018**, *30*, 217–229. [CrossRef]
31. Kulović, M.; Drašković, D.; Petrović, T. Upravljanje brzinama na putevima primjenom ITS-a. In Proceedings of the IX Međunarodni Naučni Skup Saobraćajne Nezgode, Zlatibor, Serbia, 14–16 May 2015.
32. Giuffrè, O.; Granà, A.; Tumminello, M.L.; Giuffrè, T.; Trubia, S.; Sferlazza, A.; Rencelj, M. Evaluation of Roundabout Safety Performance through Surrogate Safety Measures from Microsimulation. *Hindawi J. Adv. Transp.* **2018**, *2018*, 1–14. [CrossRef]
33. Mitrović Simić, J.; Stević, Ž.; Zavadskas, E.K.; Bogdanović, V.; Subotić, M.; Mardani, A. A Novel CRITIC-Fuzzy FUCOM-DEA-Fuzzy MARCOS Model for Safety Evaluation of Road Sections Based on Geometric Parameters of Road. *Symmetry* **2020**, *12*, 2006. [CrossRef]
34. Stanimirović, D.; Bogdanović, V.; Davidović, S.; Zavadskas, E.K.; Stević, Ž. The Influence of the Participation of Non-Resident Drivers on Roundabout Capacity. *Sustainability* **2019**, *11*, 3896. [CrossRef]
35. Robinson, B.W.; Rodegerds, L.; Scarborough, W.; Kittelson, W.; Troutbeck, R.; Brilon, W.; Bondizo, L.; Courage, K.; Kyte, M.; Mason, J.; et al. *Roundabouts: An Informational Guide*; Report No. FHWA-RD-00-067; U.S. Department of Transportation—Federal Highway Administration: Washington, DC, USA, 2000.
36. Bogdanović, V.; Ruškić, N.; Kulović, M.; Han, L. Toward a Capacity Analysis Procedure for Nonstandard Two-Way Stop-Controlled Intersections. *Transp. Res. Rec. J. Transp. Res. Board* **2013**, *2395*, 132–138. [CrossRef]
37. Bogdanović, V.; Papić, Z.; Ruškić, N.; Jeftić, N. Karakteristike brzina na signalisanim raskrižjima. *Suvremeni Promet* **2011**, *31*, 196–200.
38. Petz, B. *Osnovne Statističke Metode za Nematematičare*; Udžbenici Sveučilišta u Zagrebu: Zagreb, Croatia, 1985; Volume 2.

39. Bunevska Talevska, J.; Ristov, M.; Malenkovska Todorova, M. Development of Methodology for the Selection of the Optimal Type of Pedestrian Crossing. *Decis. Mak. Appl. Manag. Eng.* **2019**, *2*, 105–114. [CrossRef]
40. Precup, R.-E.; Preitl, S.; Petriu, E.; Bojan-Dragos, C.-A.; Szedlak-Stinean, A.-I.; Roman, R.-C.; Hedrea, E.-L. Model-Based Fuzzy Control Results for Networked Control Systems. *Rep. Mech. Eng.* **2020**, *1*, 10–25. [CrossRef]
41. Blagojević, A.; Vesković, S.; Kasalica, S.; Gojić, A.; Allamani, A. The application of the fuzzy AHP and DEA for measuring the efficiency of freight transport railway undertakings. *Oper. Res. Eng. Sci. Theor. Appl.* **2020**, *3*, 1–23. [CrossRef]
42. Petrovic, G.; Mihajlovic, J.; Cojbasic, Z.; Madic, M.; Marinkovic, D. Comparison of three fuzzy MCDM methods for solving the supplier selection problem. *FU Mech. Eng.* **2019**, *17*, 455–469. [CrossRef]

sustainability

MDPI

Article

Determining the Competitiveness of Spa-Centers in Order to Achieve Sustainability Using a Fuzzy Multi-Criteria Decision-Making Model

Miroslav Mijajlović [1], Adis Puška [2], Željko Stević [3,*], Dragan Marinković [4], Dragan Doljanica [5], Saša Virijević Jovanović [5], Ilija Stojanović [6] and Jasminka Beširović [2]

[1] Faculty of Mechanical Engineering, University of Niš, A. Medvedeva 14, 18000 Niš, Serbia; miroslav.mijajlovic@masfak.ni.ac.rs

[2] Government of Brčko district B&H, Bulevara mira 1, 76100 Brčko District, Bosnia and Herzegovina; adispuska@yahoo.com (A.P.); jasnahba@yahoo.com (J.B.)

[3] Faculty of Transport and Traffic Engineering, University of East Sarajevo, Vojvode Mišića 52, 74000 Doboj, Bosnia and Herzegovina

[4] Faculty of Mechanical Engineering and Traffic Systems, TU Berlin, Str. d. 17. Juni 135, 10623 Berlin, Germany; dragan.marinkovic@tu-berlin.de

[5] MEF Faculty of Applied Management, Economics and Finance, University Business Academy in Novi Sad, Jevrejska 24, 11000 Belgrade, Serbia; dragan.doljanica@mef.edu.rs (D.D.); sasa.virijevic@mef.edu.rs (S.V.J.)

[6] College of Business Studies, Al Ghurair University, Dubai International Academic City PO Box 37374, UAE; ilija.stojanovic1976@gmail.com

* Correspondence: zeljkostevic88@yahoo.com or zeljko.stevic@sf.ues.rs.ba

Received: 4 September 2020; Accepted: 13 October 2020; Published: 16 October 2020

check for updates

Abstract: Bosnia and Herzegovina (B&H) possess many natural resources that can be exploited for the development of medical tourism. The offer of medical tourism in B&H is focused on spa tourism. B&H has 16 registered spa-centers offering different types of services. This study provides a complete overview of the assessment of the current state of spa-centres using expert decision-making and methods of multi-criteria analysis. An innovative and novel MCDM model based on integration of the FUCOM (full consistency method) and fuzzy MARCOS (Measurement of Alternatives and Ranking according to COmpromise Solution) methods was used. The model consists of 16 alternatives and eight sustainable criteria. The results of this research have shown that the spa-centers of Ilidža near Sarajevo, Fojnica and Vrućica have the best assessments of the current situation and prerequisites for sustainable business. These spa-centers should be a benchmark to other spas providing direction on how to improve their business to be more sustainable and competitive in the market. These esults were confirmed by a sensitivity analysis with two approaches used. The first approach was to compare the results obtained by the fuzzy MARCOS method with other fuzzy methods, and the second approach was to examine the influence of the application of different weights on the final ranking of the spa. The results of this study can serve spa-canter managers to understand the position of their spa-centers in order to exploit advantages they have and eliminate the shortcomings to improve their business.

Keywords: spa-centers; Bosnia and Herzegovina; sustainable tourism; FUCOM method; fuzzy MARCOS method

1. Introduction

Contemporary trends in tourism require specific tourism products that emphasize authenticity, uniqueness and intact resources [1]. Bosnia and Herzegovina (B&H) owns significant resources that are not used adequately and in a sustained way. Especially, B&H possesses many natural, thermal and thermo-mineral springs and peloids, among them are very rare mineral springs, which are known in professional circles around the world [2]. B&H has significant comparative advantages in the field of tourism where competitiveness can be improved. The offer of spas is a type of medical tourism that needs to be improved in order to improve tourism to achieve sustainability in B&H.

The use of thermo-mineral waters in B&H dates to the distant past. Since the ancient times of the Greeks and Romans, the healing properties of geothermal springs have been noted [2]. The spa offer in B&H is being built on these natural resources. The main activity of spa-centers is the health function of treatments, which is the most important and oldest function of spas. However, more attention is paid to the importance of the spa offer for the purpose of sustaining tourism development. The priority of spa development should be built on the health and tourism function available to spa-centers in B&H. Furthermore, the development of spa-centers is conducive with development of the local tourist offer [3].

Although B&H has a rich tradition in medical tourism, in previous studies, a complete overview of the spas on offer in B&H has not been given. This paper seeks to evaluate the entire offer of medical tourism in B&H through the offering of spas. In this way, we will provide an insight into the current state of medical tourism and what B&H currently has in terms of medical tourism. Obtaining this information is necessary for the implementation of future activities to improve medical tourism in B&H by improving the spa offer. In addition, this paper will indicate the possibilities for improving medical tourism in B&H.

The aim of this paper is to develop guidelines for improving the spa offers in B&H in order to improve the competitiveness of spas. Improvements in competitiveness will be achieved by attracting more tourists and patients and young people to spas. In order to improve the spas on offer in B&H, it is necessary to evaluate the current and potential spas on offer. The evaluation of the spas on offer in B&H was performed by using an innovative multi-criteria decision-making model. The goal of this model is not to determine which is the best spa-center in B&H, but to determine the advantages and disadvantages that spas have. The results could form a solid basis to improve the competitiveness of spas in B&H.

This paper aims to address the following questions: (a) What is the current situation of the spa sector in B&H? (b) What are the advantages and disadvantages of spas? (c) What are the fundamentals on which to build a competitive advantage in spa-centres? The expected scientific contribution of this study is to evaluate the current situation and provide guidelines for the development of the spas on offer in B&H. Based on these guidelines, every spa manager can develop a business plan to improve competitiveness of their spa-center on the market. In addition, they will gain insights into the position of their spa-center relative to other spas. This study will assist in developing medical tourism in order to build competitive spa offerings in B&H. By strengthening the competitiveness of medical tourism, it will strengthen the overall tourism in B&H.

In order to achieve the study goals, the evaluation of selected spa-centers was performed by using a fuzzy approach. The fuzzy approach was adapted to human thinking because grading is done by applying linguistic values. Selected spa-centers, through a multi-criteria evaluation model, were evaluated by the expert. This evaluation model was used to evaluate spas in B&H. Additionally, a contribution of this paper is the development of one integrated Full COnsistency Method (FUCOM)-Fuzzy MARCOS (Measurement of Alternatives and Ranking according to COmpromise Solution) model.

2. A Theoretical Framework

The offer of Spas is a representative form of medical tourism [4], where various spa treatments are provided that include alternative therapies such as: homeopathy, osteopathy, acupuncture, yoga, counselling, fitness, aromatherapy, beauty treatments, aesthetic treatments, cosmetic surgery, liposuction, and chiropractic treatments [5]. Spa tourism is a narrower term than health tourism and implies a type of health tourism that is carried out in spa centers in order to treat certain diseases, improve psycho-physical health or relaxation of the body [6]. Spas are used not only by those seeking a cure for diseases such as arthritis, back pain, obesity, trauma, asthma, sterility, and surgical rehabilitation, but also by those seeking relaxation, beauty and longevity treatments [7].

New health concepts and wider activities that emerged from the trends of modern Western societies, have contributed to the growth of spas and the development of the specialization and segmentation of spas [8]. According to the International Spa Association (ISPA), spas are classified into seven main categories [9]:

- Daily spas: include facial and body treatment services where overnight stays are not provided.
- Hotel spas, spa resorts: offer fitness and health services and spa services with the possibility of overnight stay.
- Spa destinations: the main goal of these spas is to direct visitors to an individual healthy lifestyle. This service can be achieved through a comprehensive program that includes spa services, fitness activities, health education, healthy cuisine and programs of special interest to users.
- Medical spas: are health centers in which professional on-site medical workers provide comprehensive medical and health services that integrate all types of therapies and treatments.
- Spa clubs: are facilities that have the purpose of fitness and offer professional spa services that are offered daily. It should be noted that hotels, gyms and fitness clubs are not spas, unless they explicitly offer spa products and services as an added offer.
- Mineral spas: offer natural mineral, thermal or sea water used in hydrotherapy treatments.
- Spas on cruise ships: provide professional spa services, fitness and health components and a selection of spa menus on cruise ships.

Development of the spa is considered a natural response to human desire for treatments in the context of the evolution of consciousness, globalization, and various global crises [10]. The diversity of spas has been influenced by the fact that spas are used not only by people seeking treatment for various diseases, but also by guests who want relaxation, beauty and longevity treatments [7]. On this basis, spa tourism has developed as one of the sectors of health tourism. Spa tourism is one of the oldest types of tourism that has been constantly evolving. Complex and various forms have developed under the influence of political and economic systems, on the one hand, and changes in social options and tourist interests in relation to this type of tourism, on the other [11].

Health tourism is a comprehensive concept for the application of medical and treatment tourism in which the main motivation of tourists is to improve and maintain their health [12]. However, there is no single definition of what spa tourism is [13]. Spa tourism can be defined as a part of health tourism that refers to the provision of specific services that include mineral and thermal waters, but is also used for leisure, because it offers accommodation services [7]. Jahić and Selimović [6] point out that spa tourism is a narrower term than health tourism and implies a type of health tourism that is carried out in spa centers in order to treat certain diseases, improve psycho-physical health or relax the body. As a result, more and more people visit spa centers to improve their health. On this basis, spa tourism is currently one of the fastest growing sub sectors in health tourism [14]. In many Western European countries, spa tourism is an important factor in local and regional development [13].

B&H is known for its sources of mineral water, which has been used in the treatment of various types of diseases [2]. Spa centers in B&H with a rich history have been built near these springs. There are currently 16 registered medical spas in B&H. These spas are the basis for development of medical tourism in B&H. Therefore, it is necessary to actively invest in the modernization of spa facilities in order to be able to offer guests quality spa services. It is necessary to improve medical treatments in spas and expand the range of treatments in order to improve the competitiveness of spas in B&H. In previous studies, only certain spas in B&H were included and no comparison was made between them. In previous scientific publications, medical tourism has not been made the focus of attention and this paper represents a shift at improving this form of tourism. In the last ten years, only a few papers on medical tourism in B&H have been published, and some of them are presented below.

In his paper, Segić [15] provided a more precise definition of spa and recreational health tourism on a scientific and professional basis for the purpose of developing health tourism in the Republic of Srpska through foreign investments. Zelenbabić [16] took Spa Vrućica as an example of how the combination of natural resources and financial investments in the long run will produce positive results for: service providers, users and the local community. In their paper, Operta and Hyseni [17] presented the problems encountered with spa tourism in B&H, as well as the possibilities for improving tourist destinations. Bodiroža and Ćerketa [18] pointed out that the further development of health, spa and climate tourism should be based on modern world achievements, but above all in the construction of modern spa centers.

Spahić and Temimović [19] pointed out that tourism development in B&H should be based on the offer of spas which represents the backbone of the tourism development strategy. Jahić and Selimović [6] focused on the balenology and health tourism in Fojnica, which is based on thermal water, springs resources. In their paper, they provided guidelines on how to improve the competitiveness of the spa in Fojnica by offering modern treatments in the form of expanding health treatments to sports and recreational tourism. It is necessary to look at the current offer of spas in B&H, and they provided recommendations on how to improve and make this branch of tourism more competitive. Milinković et al. [20] based on the example of Spa Vrućica, tried to answer whether this spa-center can meet the modern needs of the tourist market, and thus contribute to economic growth and development of the national and local economy. Puška et al. [2] presented the spa offer in B&H in their paper, but they did not compare individual spas, rather they found some areas in which great opportunities are present for improving the spa offer in B&H.

Based on this literature review, it can be concluded that B&H has a rich range of spas on offer that can be used for the development of health tourism, but it is necessary to improve and enhance this offer of spas. Therefore, it is necessary to look at the current state of the spas on offer in B&H and provide guidelines on how to develop individual spa centers in order to be more competitive in the market, thus contributing to further development of health tourism in B&H. For this purpose, a multi-criteria approach for evaluation of the spa offer in B&H was used.

3. Methodology

The phases and methodology used in this paper consists of the following phases (Figure 1):

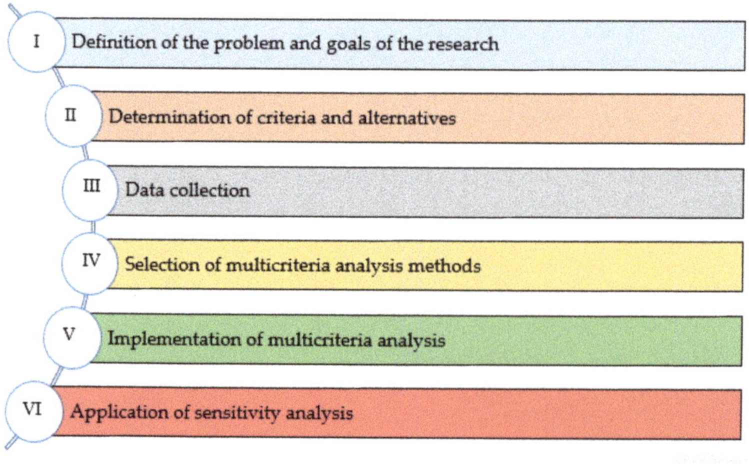

Figure 1. Phases of the research.

The initial phase of any study is to define the problem and goals of the research. The problem of this research is the evaluation of the current spa offer as a part of the health tourism industry in B&H. The goal of this paper is to develop guidelines for improving the spas on offer in B&H in order to improve the competitiveness of spa-centers. In order to evaluate the existing spa offer, it is necessary to define the criteria for evaluation of the current spa offer. When using this approach, a multi-criteria analysis of research alternatives is used. All registered spa-centers in B&H were taken as alternatives. The criteria used in this analysis were taken in collaboration with experts, as follows:

- Accommodation capacity (C1) is the number of rooms and beds that the spa offers
- Food and beverage offer (C2) represent various food and beverage services offered to guests of these spa-centers
- Internet promotion (C3) includes the existence of its own website and presence on specialized websites
- The offer of treatment and therapy (C4) includes the existence of different treatments and the availability of therapies in these spas
- Cosmetic treatments (C5) includes a comprehensive range of cosmetic and beauty and rejuvenation treatments
- Recreation and relaxation (C6) include the offer of the spa in a form of various services aimed at the relaxation of spa users
- Education and events (C7) include the possibility of conducting various seminars, congresses and various entertainment events in spas
- Natural conditions (C8) includes the comprehensiveness of natural resources available to individual spas.

In order to collect the data necessary for evaluation of the current spa offer in B&H, expert evaluation was used. Experts were selected from the pool of researchers who worked on the project: "Development and promotion of health spa tourism in the cross-border area of Bosnia and Herzegovina–Serbia." A total of three experts in the field of health tourism were selected. A two-part questionnaire was sent to these experts. The first part was related to the weight of the criteria of this model, while the second part of the questionnaire was related to the current state of the spa according to the presented criteria. These estimates are presented in the form of linguistic value (Table 1). Fuzzy logic was used to transform these linguistic values. The use of fuzzy logic was done with the fuzzy MARCOS method to rank the spas according to expert evaluation. The values of the criteria weights were determined using the FUCOM model.

Table 1. Linguistic values and affiliation function to fuzzy number [21].

Linguistic Values	Fuzzy Numbers
Very bad (VB)	(0,0,1)
Bed (B)	(0,1,3)
Medium bed (MB)	(1,3,5)
Medium (M)	(3,5,7)
Medium good (MG)	(5,7,9)
Good (G)	(7,9,10)
Very good (VG)	(9,10,10)

After the data were collected from the experts, the methods from multi-criteria analysis were implemented and the research results were obtained. Sensitivity analysis was performed to confirm the results. Sensitivity analysis in this paper had the following tasks: to examine the sensitivity of the rank order to changing the weights of the criteria and to examine the sensitivity of the data on application of different methods of multicriteria analysis. More details on the steps of the above methods of multicriteria analysis are shown below.

3.1. FUCOM (Full Consistency Method) Method

The FUCOM method is a new model for determining the weight of criteria in multicriteria decision making. The FUCOM method uses the comparison of paired criteria and the validation of results by deviating from the maximum consistency, developed by Pamučar et al. [22]. Using this method, the subjectivity in the decision-making process is reduced. This method, in relation to other methods for determining the subjective weights of criteria, has the following main advantages: reduction in the number of pairs for comparison, consistency in comparing criteria and contribution to rational judgment [23,24]. The FUCOM method is implemented using the following steps [25,26]:

Step 1. Ranking of criteria/sub-criteria using expert judgment.

Step 2. Determining the vector of comparative significance of the evaluation criteria.

Step 3. Defining the constraints of a nonlinear optimization model. The values of the weighting coefficients should satisfy two conditions, namely:

- Condition 1. The ratio of the weight coefficients is equal to the comparative significance between the observed, that the condition is fulfilled: $w_k/w_{k+1} = \varphi_{k/(k+1)}$
- Condition 2. The final values of the weighted coefficients should satisfy the condition of mathematical transitivity $\varphi_{k/(k+1)} \times \varphi_{(k+1)/(k+2)=}[yellow]\varphi_{k/(k+2)}$

Step 4. Defining a model for determining the final values of the weighting coefficients of the evaluation criteria.

Step 5. Solving the model and obtaining the final weight of the criteria/sub-criteria $(w_1, w_2, \ldots, w_n)^T$.

3.2. Fuzzy MARCOS (Measurement of Alternatives and Ranking According to Compromise Solution) Method

The MARCOS method was developed by Stević et al. [27] and it represents a new method of multicriteria decision making. The MARCOS method is based on a defined relationship between alternatives and reference values that are presented as ideal and anti-ideal points. Decision making using the MARCOS method is based on the utility function [28]. The utility function looks at an alternative to an ideal and anti-ideal solution. The best alternative is the one that is closest to the ideal and at the same time the furthest from the anti-ideal solution.

The steps for calculating the fuzzy MARCOS method are as follows [29]:

Step 1. Forming an initial fuzzy decision matrix.

Step 2. Expanding the initial fuzzy decision matrix. In this step, the initial matrix is expanded with anti-ideal (AAI) and ideal solution (AI). The anti-ideal solution (AAI) is an alternative with the worst characteristic depending on the type of criteria. The ideal solution (AI) is an alternative with the best characteristic.

The anti-ideal solution (AAI) is obtained by applying the following expression:

$$\widetilde{A}(AI) = \min_i \widetilde{x}_{ij} \; if \; j \in B \; and \; \max_i \widetilde{x}_{ij} \; if \; j \in C \tag{1}$$

while the ideal solution (AI) is obtained using the following expression:

$$\widetilde{A}(ID) = \max_i \widetilde{x}_{ij} \; if \; j \in B \; and \; \min_i \widetilde{x}_{ij} \; if \; j \in C \tag{2}$$

B represents the criteria that need to be maximized, while C represents the criteria that need to be minimized.

Step 3. Normalizing the initial fuzzy decision matrix. Normalization is performed using the following expressions:

$$\widetilde{n}_{ij} = \left(n^l_{ij}, n^m_{ij}, n^u_{ij}\right) = \left(\frac{x^l_{id}}{x^u_{ij}}, \frac{x^l_{id}}{x^m_{ij}}, \frac{x^l_{id}}{x^l_{ij}}\right) \; if \; j \in C \tag{3}$$

$$\widetilde{n}_{ij} = \left(n_{ij}^{l}, n_{ij}^{m}, n_{ij}^{u}\right) = \left(\frac{x_{ij}^{l}}{x_{id}^{u}}, \frac{x_{ij}^{m}}{x_{id}^{u}}, \frac{x_{ij}^{u}}{x_{id}^{u}}\right) if\ j \in B \tag{4}$$

where l is the first fuzzy number, m is the second fuzzy number and u is the third fuzzy number.

Step 4. Weighting the normalized decision matrix. The weighting is done using the following expression:

$$\widetilde{v}_{ij} = \left(v_{ij}^{l}, v_{ij}^{m}, v_{ij}^{u}\right) = \widetilde{n}_{ij} \otimes \widetilde{w}_{j} = \left(n_{ij}^{l} \times w_{j}^{l}, n_{ij}^{m} \times w_{j}^{m}, n_{ij}^{u} \times w_{j}^{u}\right) \tag{5}$$

Step 5. Calculation of the S_i matrix which implies summation of values by rows or alternatives including alternatives for anti-ideal and ideal solution by the following expression:

$$\widetilde{S}_{i} = \sum_{i=1}^{n} \widetilde{v}_{ij} \tag{6}$$

Step 6. Calculation of the degree of usefulness K_i according to the anti-ideal and ideal solution using the following expressions:

$$\widetilde{K}_{i}^{-} = \frac{\widetilde{S}_{i}}{\widetilde{S}_{ai}} = \left(\frac{s_{i}^{l}}{s_{ai}^{u}}, \frac{s_{i}^{m}}{s_{ai}^{m}}, \frac{s_{i}^{u}}{s_{ai}^{l}}\right) \tag{7}$$

$$\widetilde{K}_{i}^{+} = \frac{\widetilde{S}_{i}}{\widetilde{S}_{id}} = \left(\frac{s_{i}^{l}}{s_{id}^{u}}, \frac{s_{i}^{m}}{s_{id}^{m}}, \frac{s_{i}^{u}}{s_{id}^{l}}\right) \tag{8}$$

Step 7. Calculation of the fuzzy matrix $\widetilde{T}_i$. using the following expression:

$$\widetilde{T}_{i} = \widetilde{t}_{i} = \left(t_{i}^{l}, t_{i}^{m}, t_{i}^{u}\right) = \widetilde{K}_{i}^{-} \oplus \widetilde{K}_{i}^{+} = \left(k_{i}^{-l} + k_{i}^{+l}, k_{i}^{-m} + k_{i}^{+m}, k_{i}^{-u} + k_{i}^{+u}\right) \tag{9}$$

Determining the fuzzy number $\widetilde{D}$ using the following expression:

$$\widetilde{D} = \left(d^{l}, d^{m}, d^{u}\right) = \max_{i} \widetilde{t}_{ij} \tag{10}$$

Step 8. De-fuzzification of fuzzy numbers using the following expression:

$$df_{crisp} = \frac{l + 4m + u}{6} \tag{11}$$

Step 9. Determining the utility function $f\left(\widetilde{K}_i\right)$ through the aggregation of utility functions according to the anti-ideal solution (a) and the ideal solution (b).

- Utility function according to the anti-ideal solution

$$f\left(\widetilde{K}_{i}^{+}\right) = \frac{\widetilde{K}_{i}^{-}}{df_{crisp}} = \left(\frac{k_{i}^{-l}}{df_{crisp}}, \frac{k_{i}^{-m}}{df_{crisp}}, \frac{k_{i}^{-u}}{df_{crisp}}\right) \tag{12}$$

- Utility function according to the ideal solution

$$f\left(\widetilde{K}_{i}^{-}\right) = \frac{\widetilde{K}_{i}^{+}}{df_{crisp}} = \left(\frac{k_{i}^{+l}}{df_{crisp}}, \frac{k_{i}^{+m}}{df_{crisp}}, \frac{k_{i}^{+u}}{df_{crisp}}\right) \tag{13}$$

Step 10. Calculation of the final utility function:

$$f(K_{i}) = \frac{K_{i}^{+} + K_{i}^{-}}{1 + \frac{1 - f\left(K_{i}^{+}\right)}{f\left(K_{i}^{+}\right)} + \frac{1 - f\left(K_{i}^{-}\right)}{f\left(K_{i}^{-}\right)}}; \tag{14}$$

Step 11. Ranking alternatives. The best alternative is the one with the highest value, while the worst is the alternative with the lowest value.

4. Results

When assessing the current condition of spas in B&H, it was first necessary to determine the weights of the main criteria and we used the FUCOM method in this study.

4.1. Criteria Weights Obtained Using FUCOM Method

According to the steps from FUCOM method, the experts first had to determine which, in their opinion, is the most important criterion and this criterion was assigned a value of one. The other criteria were assigned values in relation to the importance towards the most important criterion. Experts had at their disposal, values from one to nine, with the possibility to assign decimal values to the criteria. The less important the criterion, its closer its designated value to nine. It should be noted that it was possible that the criteria have the same importance and in such cases they were assigned the same value. Based on these steps, the experts evaluated the criteria (Table 2).

Table 2. Evaluation of criteria by experts.

Expert 1	C4	C8	C2	C6	C5	C7	C1	C3
	1	1	2	2	3	3	4	4
Expert 2	C1	C4	C8	C6	C2	C5	C3	C7
	1	1.1	1.1	1.4	1.6	1.6	2	2.5
Expert 3	C4	C8	C7	C6	C5	C1	C2	C3
	1	1	5	6	7	8	8	8

As can be seen from the value of the criteria, there is no agreement between the experts on are the most important criteria for assessing the current state of spas in B&H. This can be noticed from the example in which expert one and expert three have indicated the criterion C4 (treatment offer and therapy) as the most important criterion, while expert two has indicated the criterion C1 (accommodation capacity). In order to harmonize these opinions, individual values of weights were calculated for the criteria by individual experts, and these values were harmonized using a geometric mean. The procedure for implementing the FUCOM method has been previously explained, and in this part of the paper only the final weights by criteria are presented.

The most important criterion according to the expert is C4 (treatment offer and therapy), followed by criterion C8 (natural conditions), and the least important criterion according to them is criterion C6 (recreation and relaxation) (Table 3).

Table 3. Criteria weight values.

Expert 1	C1	C2	C3	C4	C5	C6	C7	C8
	0.1078	0.1274	0.1078	0.1557	0.1168	0.1274	0.1168	0.1402
Expert 2	C1	C2	C3	C4	C5	C6	C7	C8
	0.1545	0.1312	0.1264	0.1377	0.1312	0.0605	0.1209	0.1377
Expert 3	C1	C2	C3	C4	C5	C6	C7	C8
	0.1000	0.1000	0.1000	0.1889	0.1063	0.1133	0.1214	0.1700
Joint Grade	C1	C2	C3	C4	C5	C6	C7	C8
	0.1199	0.1200	0.1121	0.1612	0.1189	0.0966	0.1210	0.1502

4.2. Ranking Alternatives Using Fuzzy MARCOS Method

After determining a weight of the criteria, the experts had to assess the current state of the spas on offer in B&H. Experts evaluated each spa-center and assigned an appropriate linguistic value (Table 1).

The experts obtained the necessary information for evaluation of these spas by searching the internet, visiting these spas and using similar methods. Based on this knowledge, the experts evaluated these spas and the current state of the spa sector in B&H (Table 4). In order to obtain the results of the assessment, it was necessary to transform the linguistic values into fuzzy numbers using the affiliation functions to fuzzy numbers (Table 1).

Table 4. Linguistic assessment of the current state of spas in B&H by experts.

Expert 1	C1	C2	C3	C4	C5	C6	C7	C8
A1—Spa Dvorovi	G	G	MG	MG	MG	G	MG	G
A2—Spa Fojnica	VG	G	VG	VG	MG	MG	VG	VG
A3—Spa Gata	MG	G	MG	G	G	G	MB	G
A4—Spa Guber	VB	B	VB	VB	VB	VB	VB	VG
A5—Spa Ilidža–Gradačac	MG	MG	G	G	MG	MG	G	G
A6—Spa Akvaterm–Olovo	MG	MG	G	G	G	G	MG	MG
A7—Spa Ilidža–Sanski Most	MG	MG	MG	MG	MB	G	MB	VG
A8—Spa Ilidža near Sarajevo	VG	VG	VG	G	VG	G	VG	VG
A9—Spa Kulaši	G	G	G	MG	MG	G	MB	MG
A10—Spa Laktaši	MG	G	G	G	MG	G	MB	G
A11—Spa Lješljani	MB	G	VB	B	VB	MB	VB	MG
A12—Spa Mlječanica	G	MB	B	MB	B	G	G	G
A13—Spa Slatina	VG	G	MB	VG	MB	MB	MG	G
A14—Spa Vilina Vlas	MG	MG	MG	MG	MB	MG	MG	G
A15—Spa Vrućica	VG	G	G	G	MG	G	VG	VG
A16—Slana Spa	MG	G	MG	G	MG	G	G	MG
Expert 2	**C1**	**C2**	**C3**	**C4**	**C5**	**C6**	**C7**	**C8**
A1—Spa Dvorovi	G	G	G	MG	G	MG	G	MG
A2—Spa Fojnica	VG	G	VG	G	VG	G	VG	G
A3—Spa Gata	MG	MG	MB	MG	G	G	G	G
A4—Spa Guber	VB	B	B	VB	VB	VB	VB	VG
A5—Spa Ilidža–Gradačac	G	G	VG	G	G	G	G	G
A6—Spa Akvaterm–Olovo	MG	MG	MG	G	MG	MG	MG	G
A7—Spa Ilidža–Sanski Most	MG	G	MG	MG	G	G	MB	G
A8—Spa Ilidža near Sarajevo	VG	G	VG	G	VG	G	VG	G
A9—Spa Kulaši	G	G	G	MG	G	G	G	G
A10—Spa Laktaši	G	G	G	MG	G	MG	G	MG
A11—Spa Lješljani	MB	MG	B	B	B	MB	VB	G
A12—Spa Mlječanica	MG	MG	MB	G	G	G	MG	G
A13—Spa Slatina	VG	G	MB	VG	MG	MG	MG	G
A14—Spa Vilina Vlas	MG	G	MG	MG	MG	G	MG	VG
A15—Spa Vrućica	VG	G	G	G	VG	G	G	G
A16—Slana Spa	G	G	MG	G	VG	G	G	MG
Expert 3	**C1**	**C2**	**C3**	**C4**	**C5**	**C6**	**C7**	**C8**
A1—Spa Dvorovi	G	MG	MG	MG	MG	MG	G	MG
A2—Spa Fojnica	VG	G	VG	VG	G	G	VG	VG
A3—Spa Gata	MG	MG	MG	G	G	MG	G	G
A4—Spa Guber	B	B	MB	G	G	G	MG	VG
A5—Spa Ilidža–Gradačac	MG	G	G	G	MG	MG	MG	MG
A6—Spa Akvaterm–Olovo	G	G	G	MG	MG	MG	MG	G
A7—Spa Ilidža–Sanski Most	MG	MG	MG	MG	G	G	MB	MB
A8—Spa Ilidža near Sarajevo	VG	VG	VG	VG	G	G	VG	VG
A9—Spa Kulaši	MG	G	MG	G	MG	G	G	G
A10—Spa Laktaši	G	MG	G	MG	G	G	G	MG
A11—Spa Lješljani	B	MB	B	B	B	MB	VB	G
A12—Spa Mlječanica	MB	G	MB	MG	G	MB	G	MG
A13—Spa Slatina	VG	MG	MB	G	G	MG	G	G
A14—Spa Vilina Vlas	G	G	MG	G	MG	G	MG	G
A15—Spa Vrućica	G	VG	G	VG	G	G	G	G
A16—Slana Spa	MG	G	G	VG	G	G	G	MG

Since certain spas received the values of Very Bad (VB) and Poor (P), it was not possible to use a geometric mean to match these scores, but it an arithmetic mean was applied, and a common fuzzy decision matrix was formed (Table 5). Once this matrix was formed, the next step was to expand this decision matrix by calculating the ideal and anti-ideal solution. Following this, the next step was to normalize the data. Since the score for each of the used criteria should have been as high as possible, these criteria were normalized using expression (3). The next step was to make the normalized decision matrix (expression 5).

Table 5. Initial fuzzy decision matrix.

	C1	C2	C3	C4	C5	C6	C7	C8
A1	4.3 6.3 8.0	6.3 8.3 9.7	5.7 7.7 9.3	5.0 7.0 9.0	5.7 7.7 9.3	5.7 7.7 9.3	6.3 8.3 9.7	5.7 7.7 9.3
A2	9.0 10.0 10.0	7.0 9.0 10.0	9.0 10.0 10.0	8.3 9.7 10.0	7.0 8.7 9.7	6.3 8.3 9.7	9.0 10.0 10.0	8.3 9.7 10.0
A3	5.0 7.0 9.0	5.7 7.7 9.3	3.7 5.7 7.7	5.0 7.0 8.7	3.0 5.0 7.0	3.7 5.7 7.7	2.3 4.3 6.3	4.3 6.3 8.0
A4	0.0 0.3 1.7	0.0 1.0 3.0	0.3 1.3 3.0	1.0 1.7 3.0	1.0 1.7 3.0	1.0 1.7 3.0	1.7 2.3 3.7	9.0 10.0 10.0
A5	5.7 7.7 9.3	6.3 8.3 9.7	7.7 9.3 10.0	7.0 9.0 10.0	5.7 7.7 9.3	5.7 7.7 9.3	5.0 7.0 8.7	6.3 8.3 9.7
A6	5.7 7.7 9.3	5.7 7.7 9.3	6.3 8.3 9.7	6.3 8.3 9.7	4.3 6.3 8.3	5.7 7.7 9.3	5.0 7.0 9.0	3.7 5.7 7.7
A7	5.0 7.0 9.0	5.7 7.7 9.3	5.0 7.0 9.0	5.0 7.0 9.0	2.3 4.3 6.3	4.3 6.3 8.0	1.0 3.0 5.0	5.7 7.3 8.3
A8	9.0 10.0 10.0	8.3 9.7 10.0	9.0 10.0 10.0	7.7 9.3 10.0	8.3 9.7 10.0	7.0 9.0 10.0	9.0 10.0 10.0	8.3 9.7 10.0
A9	6.3 8.3 9.7	7.0 9.0 10.0	6.3 8.3 9.7	5.7 7.7 9.3	5.7 7.7 9.3	7.0 9.0 10.0	5.0 7.0 8.3	6.3 8.3 9.7
A10	6.3 8.3 9.7	6.3 8.3 9.7	7.0 9.0 10.0	4.3 6.3 8.3	5.0 7.0 8.7	5.0 7.0 8.7	2.3 4.3 6.3	4.3 6.3 8.3
A11	0.7 2.3 4.3	3.0 5.0 7.0	0.0 0.7 2.3	0.0 1.0 3.0	0.0 0.7 2.3	1.0 3.0 5.0	0.0 0.0 1.0	6.3 8.3 9.7
A12	4.3 6.3 8.0	3.0 5.0 7.0	0.7 2.3 4.3	4.3 6.3 8.0	2.0 3.7 5.7	3.7 5.7 7.3	5.0 7.0 8.7	6.3 8.3 9.7
A13	9.0 10.0 10.0	6.3 8.3 9.7	1.0 3.0 5.0	8.3 9.7 10.0	3.0 5.0 7.0	3.7 5.7 7.7	5.7 7.7 9.3	7.0 9.0 10.0
A14	5.7 7.7 9.3	6.3 8.3 9.7	5.0 7.0 9.0	4.3 6.3 8.3	3.7 5.7 7.7	5.0 7.0 8.7	5.0 7.0 9.0	7.7 9.3 10.0
A15	8.3 9.7 10.0	7.7 9.3 10.0	7.0 9.0 10.0	7.7 9.3 10.0	7.0 8.7 9.7	7.0 9.0 10.0	7.7 9.3 10.0	7.7 9.3 10.0
A16	5.7 7.7 9.3	7.0 9.0 10.0	5.7 7.7 9.3	7.7 9.3 10.0	7.0 8.7 9.7	7.0 9.0 10.0	7.0 9.0 10.0	5.0 7.0 9.0

What distinguishes the fuzzy MARCOS method from other similar methods is the calculation of the utility and the utility function. To implement this, it was necessary to sum up the values by alternatives for individual fuzzy numbers. The calculation of the degree of usefulness was done using expressions (7) and (8) (Table 6). The next step is calculating the fuzzy matrix $\widetilde{T}_i$ (Table 6) (expression 9). The utility values of the anti-ideal and ideal solution were summed for the alternatives and the maximum values of the individual fuzzy numbers were determined. After this, these maximum values of fuzzy numbers (expression 11) were de-fuzzified and the values were calculated df_{crisp}.

Table 6. Summarizing and calculating the degree of utility and Fuzzy matrix $\widetilde{T}_i$.

	S_i	K_i^-	K_i^+		$\widetilde{T}_i$
Ideal	0.85 0.98 1.00	0.85 1.00 1.17	2.72 6.57 15.44		
A1	0.56 0.76 0.92	0.56 0.77 1.08	1.77 5.08 14.21		2.32 5.86 15.29
A2	0.81 0.95 0.99	0.81 0.97 1.16	2.56 6.36 15.33		3.37 7.32 16.50
A3	0.41 0.61 0.80	0.41 0.63 0.94	1.32 4.13 12.35		1.73 4.76 13.29
A4	0.20 0.27 0.40	0.20 0.28 0.47	0.63 1.83 6.14		0.82 2.11 6.60
A5	0.62 0.82 0.95	0.62 0.84 1.12	1.97 5.49 14.71		2.59 6.33 15.83
A6	0.53 0.73 0.90	0.53 0.75 1.06	1.69 4.91 13.92		2.22 5.66 14.98
A7	0.43 0.63 0.80	0.43 0.64 0.94	1.37 4.21 12.42		1.80 4.85 13.36
A8	0.83 0.97 1.00	0.83 0.99 1.17	2.65 6.50 15.44		3.48 7.49 16.61
A9	0.61 0.81 0.95	0.61 0.83 1.11	1.95 5.46 14.65		2.56 6.30 15.76
A10	0.50 0.70 0.87	0.50 0.72 1.02	1.59 4.71 13.39		2.10 5.43 14.40
A11	0.15 0.27 0.44	0.15 0.28 0.52	0.47 1.84 6.85		0.62 2.12 7.37
A12	0.38 0.57 0.75	0.38 0.59 0.87	1.21 3.85 11.52		1.59 4.43 12.39
A13	0.57 0.75 0.87	0.57 0.77 1.02	1.83 5.05 13.49		2.40 5.82 14.52
A14	0.54 0.73 0.90	0.54 0.75 1.05	1.71 4.92 13.86		2.25 5.67 14.91
A15	0.75 0.92 1.00	0.75 0.94 1.17	2.39 6.20 15.38		3.15 7.15 16.55
A16	0.65 0.84 0.97	0.65 0.86 1.13	2.07 5.65 14.91		2.72 6.51 16.04
Anti-ideal	0.06 0.15 0.31	0.06 0.15 0.37	0.21 1.00 4.86	Max	3.48 7.49 16.61

Calculating the value df_{crisp} was done as follows: $df_{crisp} = (3.48 + 7.49 + 16.61)/6 = 8.34$. This value was used to calculate the utility function according to anti-ideal and ideal solutions. This was followed by calculating the value of the utility function. Calculating expressions $f(\tilde{K}_i^-)$ was done by taking the values of the degree of usefulness according to the anti-ideal solution (K_i^-) and dividing with the value df_{crisp}. The value of the expression $f(\tilde{K}_i^+)$ was calculated by dividing the utility values according to the ideal solution (K_i^+) with the value df_{crisp} (Table 7).

Table 7. Utility functions.

	$f(\tilde{K}_i^-)$	$f(\tilde{K}_i^+)$
A1	0.21 0.61 1.70	0.07 0.09 0.13
A2	0.31 0.76 1.84	0.10 0.12 0.14
A3	0.16 0.49 1.48	0.05 0.08 0.11
A4	0.08 0.22 0.74	0.02 0.03 0.06
A5	0.24 0.66 1.76	0.07 0.10 0.13
A6	0.20 0.59 1.67	0.06 0.09 0.13
A7	0.16 0.50 1.49	0.05 0.08 0.11
A8	0.32 0.78 1.85	0.10 0.12 0.14
A9	0.23 0.66 1.76	0.07 0.10 0.13
A10	0.19 0.57 1.61	0.06 0.09 0.12
A11	0.06 0.22 0.82	0.02 0.03 0.06
A12	0.14 0.46 1.38	0.05 0.07 0.10
A13	0.22 0.61 1.62	0.07 0.09 0.12
A14	0.20 0.59 1.66	0.06 0.09 0.13
A15	0.29 0.74 1.84	0.09 0.11 0.14
A16	0.25 0.68 1.79	0.08 0.10 0.14

After all the necessary parameters were calculated, the fuzzy numbers were de-fuzzified, namely the values of the degree of utility (DSK^- and DSK^+), the functions of the degree of utility ($Df(K^-)$ and $Df(K^+)$) and the value of the final utility function were calculated (K_i) (Table 8).

Table 8. Results of the fuzzy MARCOS method.

	DSK^-	DSK^+	$Df(\tilde{K}_i^-)$	$Df(\tilde{K}_i^+)$	K_i	Rank
A1	0.788	6.051	0.725	0.094	0.6238	7
A2	0.973	7.220	0.866	0.117	0.9391	2
A3	0.644	5.030	0.603	0.077	0.4170	13
A4	0.296	2.347	0.281	0.035	0.0860	16
A5	0.847	6.441	0.772	0.102	0.7182	5
A6	0.763	5.877	0.705	0.091	0.5851	10
A7	0.656	5.106	0.612	0.079	0.4319	12
A8	0.994	7.347	0.881	0.119	0.9779	1
A9	0.842	6.409	0.768	0.101	0.7104	6
A10	0.731	5.640	0.676	0.088	0.5362	11
A11	0.298	2.448	0.293	0.036	0.0904	15
A12	0.600	4.687	0.562	0.072	0.3598	14
A13	0.779	5.921	0.710	0.093	0.6028	8
A14	0.764	5.877	0.705	0.092	0.5861	9
A15	0.949	7.097	0.851	0.114	0.8978	3
A16	0.870	6.596	0.791	0.104	0.7580	4

According to the results obtained by applying the integrated model of the FUCOM method and fuzzy MARCOS, we have found that the best ratings have been assigned to the alternative A8 (Spa Ilidža near Sarajevo), followed by the alternative A2 (Spa Fojnica), while spa A4 (Spa Guber) has rated with the worst position.

4.3. Sensitivity Analysis

Sensitivity analysis was performed in two ways. The first way was focused to compare the results obtained using the fuzzy MARCOS method with the other five fuzzy methods, namely: simple additive weighting (SAW) [30], weighted aggregated sum product assessment (WASPAS) [30], multi-attributive border approximation area comparison (MABAC) [26], additive ratio assessment (ARAS) [30] and technique for order preference by similarity ideal solution (TOPSIS) [31]. Another way was focused to change the weights of the criteria and the impact of these weights on the ranking order of the alternatives [32].

The first way of conducting sensitivity analysis has aimed to show the sensitivity of the ranking order based on the use of different methods of multi-criteria analysis. The results of this method of conducting sensitivity analysis examined the obtained results of the primary method used in the analysis. The results of this sensitivity analysis have shown that the ranking order of the first seven alternatives have been the same by applying any fuzzy method (Figure 2). The correlation between the ranking orders of alternatives using different fuzzy methods was examined using the Spearman correlation coefficient (SCC) [33] (Table 9).

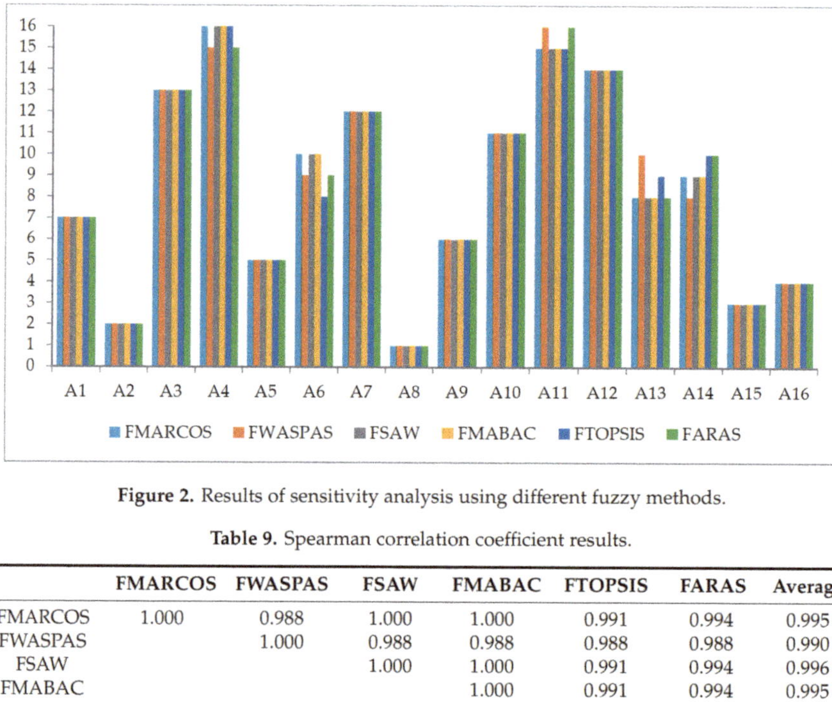

Figure 2. Results of sensitivity analysis using different fuzzy methods.

Table 9. Spearman correlation coefficient results.

	FMARCOS	FWASPAS	FSAW	FMABAC	FTOPSIS	FARAS	Average
FMARCOS	1.000	0.988	1.000	1.000	0.991	0.994	0.995
FWASPAS		1.000	0.988	0.988	0.988	0.988	0.990
FSAW			1.000	1.000	0.991	0.994	0.996
FMABAC				1.000	0.991	0.994	0.995
FTOPSIS					1.000	0.994	0.997
FARAS						1.000	1.000

Another way to conduct sensitivity analysis is to change the weight of the criteria and observe the effects of change on rankings of the alternatives. This sensitivity analysis eliminates the subjective attitude of experts regarding the importance of the weights of individual criteria. The implementation of this method of sensitivity analysis was implemented as follows. First, scenarios were formed with the assumption: one criterion is three times more important than the other criteria and thus eight scenarios are formed. The ninth scenario assumes that all criteria are of equal importance (Table 10).

Table 10. Scenarios in conducting sensitivity analysis.

	C1	C2	C3	C4	C5	C6	C7	C8
Scenario 1	0.300	0.100	0.100	0.100	0.100	0.100	0.100	0.100
Scenario 2	0.100	0.300	0.100	0.100	0.100	0.100	0.100	0.100
Scenario 3	0.100	0.100	0.300	0.100	0.100	0.100	0.100	0.100
Scenario 4	0.100	0.100	0.100	0.300	0.100	0.100	0.100	0.100
Scenario 5	0.100	0.100	0.100	0.100	0.300	0.100	0.100	0.100
Scenario 6	0.100	0.100	0.100	0.100	0.100	0.300	0.100	0.100
Scenario 7	0.100	0.100	0.100	0.100	0.100	0.100	0.300	0.100
Scenario 8	0.100	0.100	0.100	0.100	0.100	0.100	0.100	0.300
Scenario 9	0.125	0.125	0.125	0.125	0.125	0.125	0.125	0.125

Applying the scenario setting, a sensitivity analysis was performed (Figure 3).

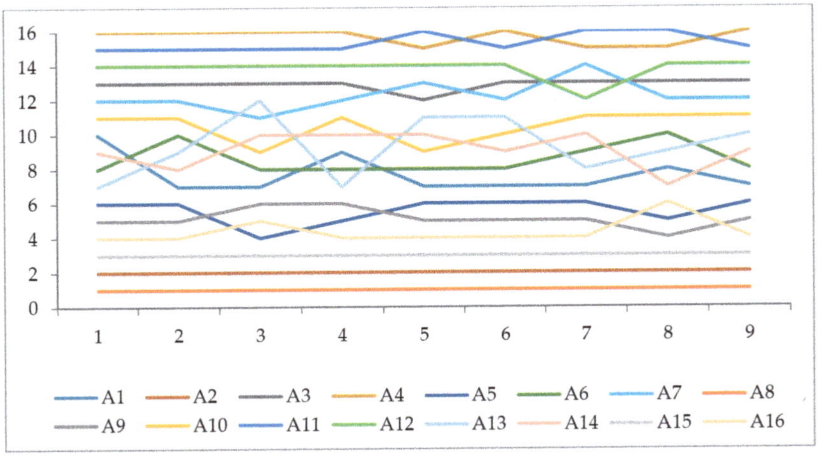

Figure 3. Results of sensitivity analysis for set scenarios.

Managers of all spa-centers can use the study results to understand the current state of their spa compared to the position of other spa-centers in B&H. Based on this insight, they are able to perceive their advantages and disadvantages, take competitive advantages and eliminate the shortcomings.

5. Discussion

Based on the obtained results using FUCOM-Fuzzy MARCOS model, it can be noticed that three spas are especially distant from the others: Ilidža near Sarajevo, Fojnica and Vrućica. This is because these spas have used natural resources and have incorporated these resources within the offer of their services. This is the reason why these spa-centers differ from other spas. The situation is the same at the bottom of the ranking list, where two spas especially deviate from other spas: Guber and Spa Lješljani. These spas have good natural resources, but accommodation capacities are inadequate or not exist, and the supply is weak or non-existent. These are the main reasons why such research results have been obtained.

After conducting the first phase of sensitivity analysis, it can be concluded that only five alternatives underwent a change in ranking order using different methods. The fuzzy WASPAS method has deviated the most in the ranking order of five alternatives, the fuzzy ARAS method in the ranking order of four alternatives and the fuzzy TOPSIS method in ranking order of three alternatives. Other fuzzy methods MARCOS, SAW and MABAC have provided the same ranking order for all alternatives.

The results of the Spearman correlation coefficient have shown that the fuzzy method MARCOS, MABAC and SAW deviates the most with the fuzzy WASPAS method (r = 0.988), followed by the fuzzy TOPSIS method (r = 0.991). One of the characteristics of the fuzzy WASPAS method is not having a perfect connection with other methods because the ranking order obtained by this method deviates from other methods, and this is also present in the fuzzy TOPSIS and ARAS methods. Thus, the results obtained using the fuzzy MABAC method and the ranking of spa-centers in the analysis were confirmed.

The results of the first method of sensitivity analysis have shown that there is a small difference between the used fuzzy methods (average r = 0.996), thus the results of the study were confirmed. Alternatives A6, A13 and A14 have provided the largest deviation in the ranking. We have found that alternatives A6, A13 and A14 have similar characteristics in terms of assessing the current situation; therefore, this difference in ranking orders has occurred by individual fuzzy methods. We found a similar example with alternatives A4 and A11 which have been ranked as the worst alternatives.

The results of the second part of the sensitivity analysis have shown that alternatives A8, A2 and A15 are not sensitive to changes in criterion weights, and these spas maintained the same ranking order. This finding has shown that the assessment of the current condition of these spas is the best in relation to other spas and that they are insensitive to changes in the weight of the criteria. These spas have also shown the best characteristics, and in order to further improve their competitiveness, it is necessary to improve those criteria where they did not receive maximum ratings. As an example, Ilidža Spa near Sarajevo should improve the offer of treatments and therapies (C4) and cosmetic treatments (C5); however this spa should also improve the criterion of natural conditions (C8), but this cannot be improved because one spa cannot change the environment where it is located and all the predispositions offered by that environment.

The second group of spas are spa-centers A5, A9 and A16, in which there were changes in the rankings in relation to individual scenarios. Thus, the spa A16 has been placed in fourth place in seven scenarios, while in scenarios three and eight it has been placed in fourth place. These results have shown that the spa A9 has better natural conditions than the spa A16, while the spa A5 has better internet promotion. In order to improve its competitiveness, the A16 spa should primarily work on internet promotion and be more accessible to all potential and future users as well as current users of the services of this spa. The spa A9 has shown the best ratings in natural conditions and this is the advantage from which this spa can benefit in order to improve competitiveness. Similarly, all evaluated spa-centers can use the results of this study to work on improving their business. It should be noted that this sensitivity analysis has shown that all 16 spas can be grouped into five groups. In order to move to a better group, the spa must invest in its development, and only in that way it can be more competitive and achieve better business results.

The disadvantages of this analysis can be found in a restricted number of experts involved and only certain MCDM methods used in this study. However, the aim of this paper was not to include as many experts as possible, to assess the current state of the spa offer, in order to improve the competitiveness of medical tourism in B&H. Moreover, the involvement of a larger number of experts potentially leads to greater inconsistencies in the opinions and possibilities of obtaining even conflicting opinions of experts. Therefore, the opinions of experts who worked together on the project "Development and promotion of health spa tourism in the cross-border area of Bosnia and Herzegovina–Serbia" were taken. Through the completion of this research, they are informed about the current situation regarding the spa offers in B&H. Therefore, the opinion of these experts, who are deeply involved in this area, is fully competent to provide assessments of the current state of medical tourism in B&H.

The use of different methods showed that the results obtained using the fuzzy MARCOS method are similar to the results obtained by other methods, and there are no significant deviations in the results. In this way, these results were confirmed, which served to provide guidelines for improving the competitiveness of medical tourism in B&H. Based on the findings from this study, it is necessary to implement certain measures in order to improve the spa offers in B&H, and thus improve medical tourism in B&H, especially due to the fact that spas offer the highest volume of services in medical tourism in B&H.

6. Conclusions

In order to improve their business, all business entities must first assess the current situation in relation to the competition to understand their own advantages and disadvantages. This paper provides a complete overview of the current state of all registered spa-centers in B&H. An innovative decision model based on the integration of the FUCOM and fuzzy MARCOS methods was implemented. The results have shown that the spa-centers of Ilidža near Sarajevo, Fojnica and Vrućica have the best assessments of the current situation. These spa-centers should be an example to other spas on how to use their resources to improve their business. The conducted sensitivity analysis confirmed the results of our research and provided guidelines on which criteria need to be improved by individual spas.

In addition to assessment of the current situation, a new method for multi-criteria analysis was used in this paper, namely fuzzy MARCOS. With sensitivity analysis we compared the results of this method with the results given by other fuzzy methods. It has been found that these results did not differ significantly. Thus, it has been proven that the fuzzy MARCOS method can be used in solving problems that require multi-criteria decision making. This paper also has shown that different methods can be integrated into a single model that will make decision-making easier for decision makers.

The limitation of this research is that the analysis was based on expert opinion. This decision-making is subjective due to the potential tendency that experts sometimes aspire to give better grades to some alternatives. Therefore, in future decisions it is necessary to include the opinions of users of these spas. However, this research is the first to give a complete picture of the current state of spas in B&H. Previous studies have analyzed only a partial overview of individual spas; thus, this paper makes a significant contribution to improving the business of spas in B&H. Another shortcoming of this paper are the criteria used. However, these criteria were chosen by experts in order to reduce the volume of used criteria, but keeping relevant numbers of those adequate for getting appropriate results to assess the current state of the spa sector in B&H.

In future research, it is necessary to include more criteria in order to get the best possible grade and to provide additional information to spa-centers' managers. The proposed current situation assessment model has shown significant flexibility and could be used in future research where multi-criteria decision-making needs to be applied. Furthermore, additional studies in the field of the spa-sector of B&H should be implemented in order to get specific guidelines on how to use the natural resources in B&H in strengthening medical tourism with the final aim to improve this branch of the economy.

Author Contributions: Conceptualization, A.P. and J.B.; methodology, A.P. and Ž.S.; validation, D.M. and I.S.; formal analysis, D.M.; investigation, A.P. and J.B.; writing—original draft preparation, A.P., M.M. and D.D.; writing—review and editing, Ž.S. and S.V.J.; supervision, D.D. and M.M. All authors have read and agreed to the published version of the manuscript.

Funding: This research received no external funding.

Conflicts of Interest: The authors declare no conflict of interest.

References

1. Valjarević, A.; Vukoičić, D.; Valjarević, D. Evaluation of the tourist potential and natural attractivity of the Lukovska Spa. *Tour. Manag. Perspect.* **2017**, *22*, 7–16. [CrossRef]
2. Puška, A.; Beganović, A.I.; Šadić, S. Situation and prospects of balneo-climatic treatment in Bosnia and Herzegovina. *Serbian J. Eng. Manag.* **2017**, *2*, 61–69. [CrossRef]
3. Kucukusta, D.; Guillet, B.D. Measuring spa-goers' preferences: A conjoint analysis approach. *Int. J. Hosp. Manag.* **2014**, *41*, 115–124. [CrossRef]
4. Giritlioglu, I.; Jones, E.; Avcikurt, C. Measuring food and beverage service quality in spa hotels. *Int. J. Contemp. Hosp. Manag.* **2014**, *26*, 183–204. [CrossRef]
5. Adongo, C.A.; Amuquandoh, F.E.; Amenumey, E.K. Modelling spa-goers' choices of therapeutic activities. *J. Hosp. Tour. Manag.* **2017**, *31*, 105–113. [CrossRef]

6. Jahić, H.; Selimović, M. Balneological tourism in Fojnica–state and prospects. *Acta Geogr. Bosniae Herzeg.* **2015**, *2*, 93–104.
7. Hall, C.M. Health and medical tourism: A kill or cure for global public health? *Tour. Rev.* **2011**, *66*, 4–15. [CrossRef]
8. Gustavo, N.S. A 21st-Century Approach to Health Tourism Spas: The Case of Portugal. *J. Hosp. Tour. Manag.* **2010**, *17*, 127–135. [CrossRef]
9. Joppe, M. One Country's Transformation to Spa Destination: The Case of Canada. *J. Hosp. Tour. Manag.* **2010**, *17*, 117–126. [CrossRef]
10. Chen, K.-H.; Chang, F.-H.; (Kenny), C.W. Investigating the wellness tourism factors in hot spring hotel customer service. *Int. J. Contemp. Hosp. Manag.* **2013**, *25*, 1092–1114. [CrossRef]
11. Erdeli, G.; Dinca, A.I.; Gheorghilaş, A.; Surugiu, C. Romanian spa tourism: A communist paradigm in a post communist era. *J. Stud. Res. Hum. Geogr.* **2011**, *5*, 41–56. [CrossRef]
12. Csapó, J.; Marton, G. The Role and Importance of Spa and Wellness Tourism in Hungary's Tourism Industry. *Czech J. Tour.* **2017**, *6*, 55–68. [CrossRef]
13. Borović, S.; Marković, I. Utilization and tourism valorisation of geothermal waters in Croatia. *Renew. Sustain. Energy Rev.* **2015**, *44*, 52–63. [CrossRef]
14. Loureiro, S.M.C.; Almeida, M.; Rita, P. The effect of atmospheric cues and involvement on pleasure and relaxation: The spa hotel context. *Int. J. Hosp. Manag.* **2013**, *35*, 35–43. [CrossRef]
15. Segić, S. Razvoj banjsko rekreativno zdravstvenog turizma u Republici Srpskoj kao mogućnost za strane ulagače. *EMC Rev. Čas. za ekon. trž. komun.* **2011**, *1*, 189–200. [CrossRef]
16. Zelenbabić, S. Upotreba prirodnih resursa u kombinaciji sa savremenim oblicima zdravstvenog turizma–ključni faktor uspjeha zdravstveno turističkog centra Banja Vrućica. In *Zbornik Radova sa Stručno-Naučne Rasprave „Turizam u Funkciji Lokalnog Razvoja"*; Fakultet Poslovne Ekonomije: Fojnica, Bosnia and Herzegovina, 2011; pp. 231–244.
17. Operta, M.; Hyseni, S. Thermal mineral waters in Bosnia and Herzegovina as well as the potential for tourism development. *Int. J. Water Res.* **2013**, *1*, 25–29.
18. Bodiroža, M.M.; Ćerketa, L. Zdravstveni, banjsko-klimatski i rekreacioni turizam u Bosni i Hercegovini sa posebnim osvrtom na Banju Slatina kod Banjaluke. *Nauka i Tehnologija* **2014**, *2*, 66–73.
19. Spahić, M.; Temimović, M. Thermomineral waters of Bosnia and Herzegovina in a function of balneological tourism. *Acta Geogr. Bosniae Herzeg.* **2014**, *1*, 65–75.
20. Milinković, Z.; Simović, S.; Ljubojević, A.; Jovanovic, J.; Babić, K.P. Strategic Analysis of the External and Internal Environment of Sport Tourism at the "Banja Vrućica" Spa. *Pol. J. Sport Tour.* **2017**, *24*, 252–258. [CrossRef]
21. Shen, L.; Olfat, L.; Govindan, K.; Khodaverdi, R.; Diabat, A. A fuzzy multi criteria approach for evaluating green supplier's performance in green supply chain with linguistic preferences. *Resour. Conserv. Recycl.* **2013**, *74*, 170–179. [CrossRef]
22. Pamučar, D.; Stević, Ž.; Sremac, S. A New Model for Determining Weight Coefficients of Criteria in MCDM Models: Full Consistency Method (FUCOM). *Symmetry* **2018**, *10*, 393. [CrossRef]
23. Nenadić, D. Ranking dangerous sections of the road using MCDM model. *Decis. Mak. Appl. Manag. Eng.* **2019**, *2*, 115–131. [CrossRef]
24. Stević, Ž.; Brković, N. A Novel Integrated FUCOM-MARCOS Model for Evaluation of Human Resources in a Transport Company. *Logist.* **2020**, *4*, 4. [CrossRef]
25. Badi, I.; Abdulshahed, A. Ranking the Libyan airlines by using Full Consistency Method (FUCOM) and Analytical Hierarchy Process (AHP). *Oper. Res. Eng. Sci. Theory Appl.* **2019**, *2*, 1–14. [CrossRef]
26. Božanić, D.; Tešić, D.; Kočić, J.; Defence, S.A.F.M.O. Multi-criteria FUCOM—Fuzzy MABAC model for the selection of location for construction of single-span bailey bridge. *Decis. Mak. Appl. Manag. Eng.* **2019**, *2*, 132–146. [CrossRef]
27. Stević, Ž.; Pamučar, D.; Puška, A.; Chatterjee, P. Sustainable supplier selection in healthcare industries using a new MCDM method: Measurement of alternatives and ranking according to COmpromise solution (MARCOS). *Comput. Ind. Eng.* **2020**, *140*, 106231. [CrossRef]
28. Puška, A.; Stojanović, I.; Maksimović, A.; Osmanović, N. Project Management Software Evaluation by Using the Measurement of Alternatives and Ranking According to Compromise Solution (MARCOS) Method. *Oper. Res. Eng. Sci. Theory Appl.* **2020**, *3*, 89–101. [CrossRef]

29. Stanković, M.; Stević, Ž.; Das, D.K.; Subotić, M.; Pamučar, D. A New Fuzzy MARCOS Method for Road Traffic Risk Analysis. *Math.* **2020**, *8*, 457. [CrossRef]
30. Petrović, G.; Mihajlović, J.; Ćojbašić, Ž.; Madić, M.; Marinković, D. Comparison Of Three Fuzzy Mcdm Methods For Solving The Supplier Selection Problem. *Facta Univ. Ser. Mech. Eng.* **2019**, *17*, 455–469. [CrossRef]
31. Chatterjee, P.; Stević, Ž. A two-phase fuzzy AHP-fuzzy TOPSIS model for supplier evaluation in manufacturing environment. *Oper. Res. Eng. Sci. Theory Appl.* **2019**, *2*, 72–90. [CrossRef]
32. Puška, A.; Kozarević, S.; Stević, Ž.; Stovrag, J. A New Way of Applying Interval Fuzzy Logic in Group Decision Making For Supplier Selection. *Econ. Comput. Econ. Cybern. Stud. Res.* **2018**, *52*, 217–234. [CrossRef]
33. Zolfani, S.; Yazdani, M.; Pamucar, D.; Zaraté, P. A VIKOR and TOPSIS focused reanalysis of the MADM methods based on logarithmic normalization. *Facta Universitatis* **2020**, *18*, 341–355.

Publisher's Note: MDPI stays neutral with regard to jurisdictional claims in published maps and institutional affiliations.

© 2020 by the authors. Licensee MDPI, Basel, Switzerland. This article is an open access article distributed under the terms and conditions of the Creative Commons Attribution (CC BY) license (http://creativecommons.org/licenses/by/4.0/).

sustainability

MDPI

Article

Will Ending the One-Child Policy and Raising the Retirement Age Enhance the Sustainability of China's Basic Pension System?

Pu Liao [1,*], Hui Su [1] and Dragan Pamučar [2]

[1] China Institute for Actuarial Science/School of Insurance, Central University of Finance and Economics, Beijing 102206, China; suuhuii@163.com

[2] Department of logistics, University of Defence, 11000 Belgrade, Serbia; dpamucar@gmail.com

* Correspondence: liaopu@cufe.edu.cn

Received: 12 September 2020; Accepted: 30 September 2020; Published: 3 October 2020

check for updates

Abstract: The sustainability of China's Basic Pension System (CBPS) has been challenged by the ageing of the population and the decline in economic growth. This article establishes a Markov model for CBPS to examine whether the reforms, including ending the one-child policy and raising retirement the age, will shrink the negative income–expenditure gap. We find that the negative income–expenditure gap will destroy CBPS in the future in the absence of fiscal transfer or reform. Ending the one-child policy will increase the number of contributors and then reduce the gap in the short term but will worsen the gap in the long term. Raising the retirement age will have several positive effects overall while increasing expenditures in certain periods. The contributions of this article are describing CBPS in detail and establishing a precise model to analyze the effectiveness of reforms.

Keywords: China's Basic Pension System (CBPS); Markov model; ending one-child policy; raising retirement age

1. Introduction

After several decades of development and reforms, China has established a three-pillar old-age security program comprising China's Basic Pension System (CBPS), Supplementary Pension System, and Personal Saving. Sponsored by the Government, CBPS plays the most important role because of the history of the planned economy. CBPS contains several sub-systems for different groups of people: an urban employees' basic pension system for workers in urban areas, an urban residents' basic pension system for workers who are not employed, and a rural residents' basic pension system for people in rural areas (now urban residents and rural residents are combined in the urban and rural residents' basic pension system). The urban employees' basic pension system was the benchmark system, which began in 1997 and included all employees in China (see [1]). Therefore, it is the focus of this article.

A mixed scheme was designed for China's urban employees' basic pension system, comprising a public account and a private account, for example [1–3]. The public account holds funds from the pay-as-you-go plan contributed by all employers at a rate of 20% (the ratio was decreased to 16% to reduce the burden of employers in 2019) of employees' wages and benefiting all the retirees. The benefit for public accounts is calculated at the base of an individual's history of contributions (related to the historical wages) and the average wages in retirement, awarding 1% for 1 year's contribution. The private account holds the funded plan contributions by individuals at a rate of 8% of wages and benefiting the individuals after retirement. The benefit for private accounts is calculated by dividing the planned periods from individuals' cumulative value at retirement.

China's urban employees' basic pension system does not work very well because there was no fiscal transfer into the start-up fund at the outset, but the system paid benefits to retirees at that time (see [2]). Consequently, the money in the private account was embezzled. With the aging of the population, a gap appeared and has grown. In the 2000s, the problem was not apparent because of the high economic growth, thus providing ample public revenue and millions of subsidies. In fact, the government awarded large bonuses to retirees by increasing their public account benefits during 2005 to 2015 (the annual average growth rate was above 10% during 2005 to 2015).

Alarm bells rang in the early 2010s. The sustainability of China's urban employees' basic pension system was eroded by the population's aging, particularly the sudden reduction in new labor because of the one-child policy beginning in 1978. In addition, economic growth has continued to decline, thus resulting in a decrease in fiscal subsidies. Therefore, there is an understanding that the income–expenditure gap not is inevitable and reforms are necessary, for example [1,2,4,5].

The government issued several measures quickly. The one-child policy was ended in 2016, and couples are now allowed to have two children (see [6]). In addition, the compulsory retirement age will be increased by 5 years gradually over the course of several years, to 65 and 60 for males and females (see [7]). In addition, parts of stated-owned assets have been transferred to China's urban employees' basic pension system, thereby allowing the assets' return to complement the benefits. Furthermore, a disbursement mechanism allows surplus provinces to subsidize provinces with deficits.

Will these measures enhance CBPS sustainability? The third measure will help because the transfer will result in cash inflow, whereas the fourth measure will not help because it is only a rebalancing among provinces. However, what about ending the one-child policy and raising the retirement age? This is an interesting question because of its details: CBPS income and expenditure are not actuarial-balanced. Therefore, there may be a surplus or a deficit for each individual, and the surplus or the deficit may accumulate if the number of contributors increases. Furthermore, raising the retirement age will shorten the benefit period but increase the benefit level. Thus, whether the expenditure will decrease is unclear.

This article established a Markov model to simulate CBPS cashflows according to detailed rules and then analyzed the impacts of population policies and reforms on the sustainability of CBPS. We found that first, the negative income–expenditure gap will destroy CBPS in the future if there is no outside transfer or reform; second, ending the one-child policy will increase the number of contributors and reduce the gap in the short term but will worsen the gap in the long-term; and third, raising the retirement age will have some positive effects overall while increasing the expenditures in certain periods.

The remainder of the article is organized as follows. The Section 2 reviews the literature and summarizes the innovations of this article. The Section 3 establishes an actuarial model of cash inflow and expenditure, assuming that the contribution is not always continuous and the contributor could take early retirement. The Section 4 calibrates the parameters in the model. The Section 5 tests the model and parameters with the data from 2009 to 2015, forecasts CBPS cashflows from 2018 to 2058, and then discusses how policies and reforms affect the cashflows. The Section 6 describes the conclusions.

2. Literature Review

The prediction of income and expenditure for pension systems has attracted substantial interest. Some official models have been established to forecast the impacts of reform on sustainability. For example, the Office of the Chief Actuary of the US Social Security Administration has built an actuarial model (the OCA Stochastic Model, OSM) to forecast the income and expenditure of the US Social Security Fund in the next 75 years. The Congressional Budget Office also predicts the cashflows with a micro-stochastic model. A special report on the short-term and long-term cashflows of Social Security Funds in the UK has been released by the Government Actuary's Department. Unfortunately, similar modeling is not conducted in China, although many researchers have made attempts to do so, for example [1–3].

The aging of the population has led to many studies of the pension system's sustainability; ref. [8] has proposed the Aaron condition, based on the Samuelson model, in which intergenerational equilibrium for a pay-as-you-go pension system occurs when the growth rates of the population and wages exceed the interest rates. However, the pension system will inevitably have a deficit because of the low fertility rate in an aging society, for example [9–11]. Using an OLG model to calculate the gap in Slovenia, ref. [12] has shown that the gap increases as the population ages, and ref. [13] believes that pension system reforms are essential to solving the aging problem, e.g., delaying retirement and increasing the fertility rate and the contribution rate. Delayed retirement is generally accepted to enhance a pension system's sustainability: see [14–16]. In addition, on the basis of an endogenous fertility model, ref. [17] has suggested that a pension system's sustainability will be enhanced if the fertility rate is increased; ref. [18] have claimed that financial sustainability of unfunded pension systems will be eroded by population ageing and analyzed the impact of immigration flows on the financial sustainability, taking Spain as a test case.

Some studies have focused on the sustainability of CBPS. For an aging population with a pay-as-you-go system, ref. [1] have claimed that CBPS will have a current deficit in 2025 if there are no reforms; ref. [9] believes that the income and expenditure will be balanced if the compulsory retirement age is 65 years; ref. [4] analyzes the sustainability of China's basic pension fund under two macroeconomic scenarios and indicates that China's present basic pension system faces insolvency crisis; ref. [6] has claimed that if more than 54% of qualified couples have a second child, the accumulated deficit of pension insurance fund will not appear before 2090; ref. [19] has suggested that China's pension plan is expected to finish the transition in 2081. Postponing the retirement age will delay the peak time of debt, but the total debt will increase substantially. Ref. [20] has assessed the effects of rates on enterprises' incentive to participate in CBPS. The findings suggest that a lower contribution rate motivates enterprises to participate in CBPS and increases the fund's revenue; ref. [21] explores the financial self-balancing ability of the individual accounts of China's urban enterprise employees' pension plan and indicates that the self-balancing ability of individual accounts will be improved if a stochastic bookkeeping rate is adopted; ref. [5] studies the effect of poverty alleviation policy on the sustainability of China's basic pension funding for urban and rural residents in the short term (2020–2025) and medium term (2025–2050) and suggests that the sustainability of the fund will inevitably face challenges.

Various aspects of CBPS have been studied, but some problems have yet to be solved. No micro-stochastic actuarial model has been developed for CBPS. Although many models have been established for CBPS, they do not consider break-off and early retirement. Moreover, CBPS is not a standard pay-as-you-go system, and it includes many additional bonuses. Existing studies have ignored the particularities of CBPS, thus potentially resulting in ineffectiveness in delaying retirement and increasing the fertility rate.

Our article introduces CBPS, describes a Markov actuarial model built to predict the income and expenditure of CBPS in the future, and discusses the impact of delaying retirement and increasing fertility rate.

3. Model and Assumption

3.1. Population Model

3.1.1. Markov Model for Contributors and Retirees

Figure 1 shows the state space and the transitions for the contributors and the retirees.

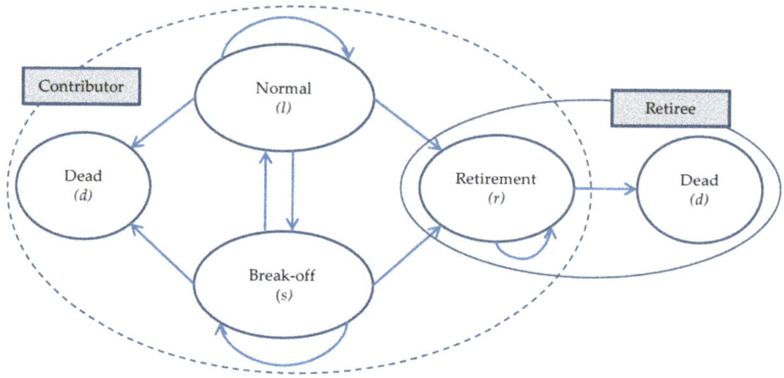

Figure 1. Markov model for contributors and retirees.

All CBPS participants are divided into two groups: contributors and retirees. There are four states for each contributor: "normal" describes the state in which the contributor contributes to CBPS (including the contribution of the employer and employee). "Break-off" describes the state in which the contributor belongs to CBPS but does not contribute to the system, for example, because of unemployment. Break-off is a serious problem for CBPS, and the official data show that the break-off rate is 20%. However, a break-off contributor can re-contribute to CBPS. "Dead" and "retirement" are two absorbing states. Although there are compulsory retirement ages for men and women, many contributors have reason to apply for retirement before the compulsory age. Early retirement is taken by 40% of men and 50% of women. There are two states for each retiree: "retirement" and "dead". "Retirement" means that the retiree is alive and receives pension from CBPS.

According to the state transition diagrams, it is easy to construct the transition probabilities matrix for the contributor, i.e.,

$$P_x(t) = \begin{pmatrix} p_x^{ll}(t) & p_x^{ls}(t) & p_x^{lr}(t) & p_x^{ld}(t) \\ p_x^{sl}(t) & p_x^{ss}(t) & p_x^{sr}(t) & p_x^{sd}(t) \\ 0 & 0 & 1 & 0 \\ 0 & 0 & 0 & 1 \end{pmatrix} \quad (1)$$

where $P_x(t)$ denotes the matrix of an x-year-old contributor at period t and $p_x^{ij}(t)$ denotes the probability of an x-year-old contributor transitioning from state i to state j at period t.

The matrix for the retiree is

$$Q_x(t) = \begin{pmatrix} q_x^{rr}(t) & q_x^{rd}(t) \\ 0 & 1 \end{pmatrix} \quad (2)$$

where $Q_x(t)$ denotes the matrix of an x-year-old retiree at period t and $q_x^{ij}(t)$ denotes the probability of an x-year-old retiree transitioning from state i to state j at period t.

For simplicity, we assume that the Markov models are time homogeneous. Thus, the transition probabilities depend only on age. Therefore,

$$P_x = \begin{pmatrix} p_x^{ll} & p_x^{ls} & p_x^{lr} & p_x^{ld} \\ p_x^{sl} & p_x^{ss} & p_x^{sr} & p_x^{sd} \\ 0 & 0 & 1 & 0 \\ 0 & 0 & 0 & 1 \end{pmatrix}, \; Q_x = \begin{pmatrix} q_x^{rr} & q_x^{rd} \\ 0 & 1 \end{pmatrix}$$

3.1.2. The Number of Participants in the Future

We forecast the number of participants in the future on the basis of the present population structure and the transition probability matrix.

Participants in the future comprise two groups. The first is those who have already participated in CBPS at period 0. Let LA_x^i denote the number of contributors who are x years old and are in state i, and let LR_y denote the number of retirees who are y years old. The second is those who enter CBPS afterward. We assume that all new participants are c years old and contribute to CBPS at the beginning. Let $NLA_c(t)$ denote the number of new participants at period t.

With those assumptions, the expected number of "normal" contributors at period t is:

$$\sum_{x=x_s}^{x_f} LA_x^l \cdot {}_tp_x^{ll} + \sum_{x=x_s}^{x_f} LA_x^s \cdot {}_tp_x^{sl} + \sum_{k=1}^{t} NLA_c(k) \cdot {}_{t-k}p_c^{ll} \tag{3}$$

where x_s and x_f represent the minimum and maximum age of the contributor, respectively.

The expected number of "retirement" retirees at period t is

$$\begin{aligned} &\sum_{y=y_s}^{y=y_f} LR_y \cdot {}_tq_y^{rr} + \sum_{x=x_s}^{x=x_f}\left[LA_x^l \cdot \sum_{t_r=1}^{t} P(R_x^l = t_r)_{t-t_r}q_{x+t_r}^{rr}\right] + \sum_{x=x_s}^{x=x_f}\left[LA_x^s \cdot \sum_{t_r=1}^{t} P(R_x^s = t_r)_{t-t_r}q_{x+t_r}^{rr}\right] \\ &+ \sum_{k=1}^{t}\left[NLA_c(k) \cdot \sum_{t_r=k}^{t} P(R_c(s) = t_r)_{t-t_r}q_{c+t_r-k}^{rr}\right] \end{aligned} \tag{4}$$

where $P\left(R_x^l = t_r\right)$ (or $P(R_x^s = t_r)$) denotes the probability of an x-year-old "normal" (or "break-off") contributor retiring at period t_r, $P(R_c(s) = t_r)$ denotes the probability of new participants at period s retiring at period t_r, and y_s and y_f represent the minimum and maximum ages of retirees, respectively.

3.2. *Inflow Model of CBPS*

CBPS specifies that the contribution base is an individual's average wage in the previous year. However, if a person's wage exceeds 300% of the social average wage, the contribution base is restricted to 300% of the social average wage, and if a person's wage is less than 60% of the social average wage, the contribution base is restricted to 60% of the social average wage. In addition, we assume that there is no difference in wages for people at the same age. The individual contributes λ (8%) of the contribution base to the private account, while the employer contributes η (20%) of the base to the public account. Therefore, the cash inflows for the public account TIN^t and the private account PIN^t at period t are:

$$TIN^t = \sum_{x=x_s}^{x=x_f} (\eta B_{x+t}^t \cdot LA_x^l \cdot {}_tp_x^{ll}) + \sum_{x=x_s}^{x=x_f} (\eta B_{x+t}^t \cdot LA_x^s \cdot {}_tp_x^{sl}) + \sum_{k=1}^{t} (\eta B_{c+t-s}^t NLA_c(k) \cdot {}_{t-k}p_c^{ll}) \tag{5}$$

$$PIN^t = \sum_{x=x_s}^{x=x_f} (\lambda B_{x+t}^t \cdot LA_x^l \cdot {}_tp_x^{ll}) + \sum_{x=x_s}^{x=x_f} (\lambda B_{x+t}^t \cdot LA_x^s \cdot {}_tp_x^{sl}) + \sum_{k=1}^{t} (\lambda B_{c+t-s}^t NLA_c(k) \cdot {}_{t-k}p_c^{ll}) \tag{6}$$

respectively, where B_x^t is the contribution base of an x-year-old contributor at period t.

Either TIN^t or PIN^t contains three parts. The first part of the inflow is contributed by those who are "normal" at the beginning and are "normal" at period t. The second part is contributed by those who are "break-off" at the beginning and are "normal" at period t. The third part is contributed by those who entered in CBPS from period 1 to period t and are "normal" at period t.

3.3. *Expenditure Model of CBPS*

CBPS specifies that the expenditure of the private account includes individual pensions and unfinished amounts in the plan-month, and the expenditure of the public account includes the public pension, individual pension out of the plan-month, and the funeral favor, a bonus covering the funeral expenses of retirees. The structure of expenditures is shown in Figure 2.

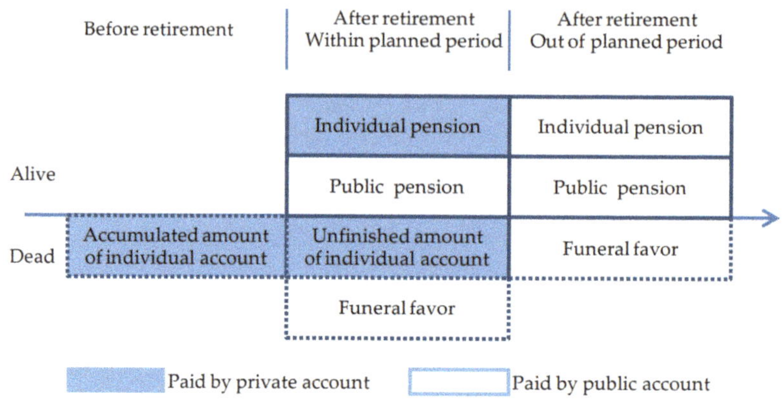

Figure 2. The expenditure structure of China's Basic Pension System (CBPS).

In addition, the retirees at period t come from three groups of people:

- Type I: those retired at period 0;
- Type II: those employed at period 0;
- Type III: those who will join CBPS from period 1 to period t.

3.3.1. Expenditure Models for Type I

For participants in type I, their benefits at period 0 have already been determined. Therefore, we take the living status and the growth rate of public pension into consideration. The expenditures of the private account and public account for type I are as follows.

$$PRC^t = \sum_{y=y_s}^{y_f} \left[{}_tq_y^{rr} \cdot AVP_y \cdot (1 - I_{\{t>G_y\}}) + {}_{t-1}q_y^{rr} q_{y+t-1}^{rd} \cdot AVP_y \cdot \max(G_y - t, 0) \right] \cdot LR_y \quad (7)$$

$$TRC^t = \sum_{y=y_s}^{y_f} \left[{}_tq_y^{rr} \cdot AVB_y \cdot \prod_{s=1}^{t} (1 + \alpha_s) + {}_{t-1}q_y^{rr} q_{y+t-1}^{rd} \cdot D \cdot \prod_{s=1}^{t} (1 + \beta_s) + {}_tq_y^{rr} \cdot AVP_y \cdot I_{\{t>G_y\}} \right] \cdot LR_y \quad (8)$$

where ${}_tq_y^{rr}$ denotes the survival probability of retirees from y years old to $y + t$ years old, AVP_y and AVB_y are the average individual pension and public pension of y-year-old participants at period 0, respectively, $I_{\{\cdot\}}$ is the indicator function, G_y is the average remaining years in the planned period, α_t denotes the growth rate of the public pension at period t, D is the funeral favor, and β_t is the growth rate of the funeral favor at period t.

3.3.2. Expenditure Models for Type II

Among participants in type I, the benefits are partially determined by their contribution history but still depend on their contributions in the future. Therefore, we must forecast their benefits on the basis of assumptions regarding their wages and status. Before that, we will review the rules of CPBS. CPBS specifies that the individual pension equals the accumulated amount of an individual's private account divided by the length of the planned period, which is set by the government, i.e.,

Individual pension = accumulated amount of private account/length of planned period
In addition, the public pension is calculated as follows:
Public pension = (years of contribution/100) × ...
(Average wage of individual contribution + average wage of all employees)/2

With these specifications, the expenditures of the private account and public account for type II are as follows:

$$PLC^t = \sum_{i=l,s}\left\{\sum_{x=x_s}^{x=x_f}\left[LA_x^i \cdot \sum_{t_r=1}^{t} P(R_x^i = t_r)PLC_x^{t,t_r,i} + LA_x^i \cdot P(T_x^i = t)PA_x^{t,i}\right]\right\} \tag{9}$$

$$TLC^t = \sum_{i=l,s}\left\{\sum_{x=x_s}^{x=x_f}\left[LA_x^i \cdot \sum_{t_r=1}^{t} P(R_x^i = t_r)TLC_x^{t,t_r,i}\right]\right\} \tag{10}$$

where $P\left(R_x^i = t_r\right)$ is the probability that an i-status x-year-old individual retires at period t_r, $P\left(T_x^i = t\right)$ is the probability that an i-status x-year-old individual dies at period t, $PLC_x^{t,t_r,i}$ and $TLC_x^{t,t_r,i}$ are the expenditures at period t of the private account and public account for an i-status x-year-old individual who retires at period t_r, and $PA_x^{t,i}$ is the accumulated amount in the private account for an i-status x-year-old individual at period t. $PLC_x^{t,t_r,i}$ is calculated by

$$PLC_x^{t,t_r,i} = {}_{t-t_r}q_{x+t_r}^{rr}PLC_x^{t_r,i}(1 - I_{\{t-t_r \geq g(x+t_r)\}}) + {}_{t-1}q_{x+t_r}^{rr}q_{x+t_r+t-1}^{rd}PAS_x^{t,t_r,i} \tag{11}$$

$$PLC_x^{t_r,i} = \frac{\sum\limits_{k=0}^{t_r}(1+r)^{t_r-k}\lambda B_{x+kk}^{k}p_x^{il} + PAY_x^i(1+r)^{t_r}}{g(x+t_r)} \tag{12}$$

$$PAS_x^{t,t_r,i} = PLC_x^{t_r,i}\max(g(x+t_r) - (t-t_r), 0) \tag{13}$$

where $PLC_x^{t_r,i}$ is the individual pension of an i-status x-year-old individual who retires at period t_r, $g(x+t_r)$ is the length of the planned period for those who retired at age $x+t_r$, PAY_x^i is the accumulated amount at period 0 of an i-status x-year-old individual, and $PAS_x^{t,t_r,i}$ is the unfinished amount at period t for an individual account of an i-status x-year-old individual who retires at period t_r.

$TLC_x^{t,t_r,i}$ is calculated by

$$\begin{aligned} TLC_x{}^{t,t_r,i} = {} & {}_{t-t_r}q_{x+t_r}^{rr} \cdot TLC_x{}^{t_r,i} \cdot \prod_{s=t_r}^{t}(1+\alpha_s) + {}_{t-1}q_{x+t_r}^{rr}q_{x+t_r+t-1}^{rd} \cdot \\ & D \cdot \prod_{s=1}^{t}(1+\beta_s) + {}_{t-t_r}q_{x+t_r}^{rr} \cdot PLC_x^{t_r,i} \cdot I_{\{t-t_r \geq g(x+t_r)\}} \end{aligned} \tag{14}$$

$$\begin{aligned} TLC_x{}^{t_r,i} = {} & \overline{W}^{t_r-1} \times z_x^{equ,i} \times n_x^{equ,i} \times 1\% + \overline{W}^{t_r-1} \times z_x^{real92-98,i} \times n_x^{real92-98,i} \times 1\% \\ & +1\% \times 0.5(\overline{W}^{t_r-1} + \overline{W}^{t_r-1} \times (z_x^{real99-08,i} + EZ_x^{real09-16,i} + EZ_x^{real17-t_r,i}) \\ & \times (n_x^{real99-08,i} + EN_x^{real09-16_r,i} + EN_x^{real17-t_r,i}) \end{aligned} \tag{15}$$

where $TLC_x{}^{t_r,i}$ is the public pension of an i-status x-year-old individual who retires at period t_r and Equation (15) is the calculation of $TLC_x{}^{t_r,i}$.

$PA_x^{t,i}$ is calculated by

$$PA_x^{t,i} = \sum_{k=0}^{t}(1+r)^{t-k}\lambda B_{x+kk}^{k}p_x^{il} \tag{16}$$

3.3.3. Expenditure Models for Type III

The expenditure models for new participants (type III) are similar to those for type II, except that the participants are aged c at period s instead of x at period 0. Therefore, we omit the explanations for these formulas. The expenditures of the private account and public account for type III are as follows.

$$PLNC^t = \sum_{s=1}^{t}\left[NLA_c(s) \cdot \sum_{t_r=s}^{t} P(R_c(s) = t_r)PLC_c^{t,t_r}(s) + NLA_c(s) \cdot P(T_c^d(s) = t)PA_c^t(s)\right] \tag{17}$$

$$TLNC^t = \sum_{s=1}^{t}\left[NLA_c(s)\cdot\sum_{t_r=s}^{t}P(R_c(s)=t_r)TLC_c^{t,t_r}(s)\right] \tag{18}$$

where

$$PLC_c^{t,t_r}(s) = {}_{t-t_r}q_{c+t_r-s}^{rr}PLC_c^{t_r}(s)(1-I_{\{t-t_r\geq g(x+t_r)\}}) + {}_{t-1}q_{c+t_r-s}^{rr}q_{c+t_r-s+t-1}^{rd}PAS_c^t(s) \tag{19}$$

$$PLC_c^{t_r}(s) = \frac{\sum\limits_{k=s}^{t_r}(1+r)^{t-k}\lambda B_{c+kk-s}^{k}p_c^{ll}}{g(c+t_r-s)} \tag{20}$$

$$PAS_c^t(s) = PLC_c^{t_r}(s)\max(g(c+t_r-s)-(t-t_r),0) \tag{21}$$

$$PA_c^t(s) = \sum_{k=s}^{t}(1+r)^{t-k}\lambda B_{c+kt-s}^{k}p_c^{ll} \tag{22}$$

and

$$TLC_c{}^{t,t_r}(s) = {}_{t-t_r}q_{c+t_r-s}^{rr}\cdot TLC_c{}^{t_r}(s)\cdot\prod_{m=t_r}^{t}(1+\alpha_m) \tag{23}$$

$$TLC_c{}^{t_r}(s) = 1\% \times 0.5(\overline{W}^{t_r-1} + \overline{W}^{t_r-1} \times EZ_x^{reals-t_r}) \times EN_x^{real09-t_r} \tag{24}$$

3.4. Balance Model of CBPS

With the population model, inflow model, and expenditure model, the balance models at period t for private account and public account are

$$TB^t = TIN^t - TRC^t - TLC^t - TLNC^t \tag{25}$$

$$TP^t = PIN^t - PRC^t - PLC^t - PLNC^t \tag{26}$$

4. Calibration

4.1. Participants in the Future

The population structure in the future depends on the population structure at period 0 and the mortality and fertility rates. Let L_y^t denote the population of y-year-old people at period t. Then, $L_0^{t+1} = n_t\sum\limits_y L_y^t$ and $L_{y+1}^{t+1} = L_y^t\cdot p_y$, where n_t is the fertility rate of the total population, and p_y is the probability of an individual y years old surviving to $y+1$ years old. Therefore, we must estimate the population structure at period 0 and the mortality and fertility rates.

We calculate the population structure at period 0, $\{L_y^0\}$, according to the tabulation of the 2010 Population Census of China.

We must estimate a national life table because one is not currently available. According to the China Life Insurance Mortality Table (2010–2013) and the actual population and fertility rate issued by the China National Bureau of Statistics, the actual mortality rate is 1.2 times that in Table 1, China Life Insurance Mortality Table (2010–2013). The comparison of the actual population and the estimated population is shown in Table 1.

Table 1. Comparison of the actual population and estimated population in China from 2010 to 2018.

	2010	2011	2012	2013	2014	2015	2016	2017	2018
Actual population	13,409	13,474	13,540	13,607	13,678	13,746	13,827	13,901	13,954
Estimated population	13,328	13,406	13,482	13,560	13,632	13,715	13,788	13,838	13,908
Accuracy rate	99.4%	99.5%	99.6%	99.7%	99.7%	99.8%	99.7%	99.6%	99.7%

Estimating the fertility rate in the future in China is highly difficult. The one-child policy was issued in 1978, and couples were allowed to have only one child. However, the policy was ended in 2016, and a new policy allows couples to have two children. The reform of the one-child policy has attracted great interest. At the beginning of the new policy, many studies claimed that the total fertility rate would increase rapidly in the short term, e.g., [22,23]. However, those expectations did not come true. According to data from China National Bureau of Statistics, the total fertility rate was 1.094, which is the lowest in the history of China. In this study, we sought to examine the impact of the increasing population on the sustainability of CBPS. Therefore, we assume that the fertility rate will increase in the short term and then decline in the long term. According to the data from a population sample survey released by the National Bureau of Statistics, we assume a fertility rate of 1.4 from 2019 to 2022 and 1.25 after 2022.

With these assumptions, the population will increase first and then decrease after the 2038 peak, as shown in Figure 3.

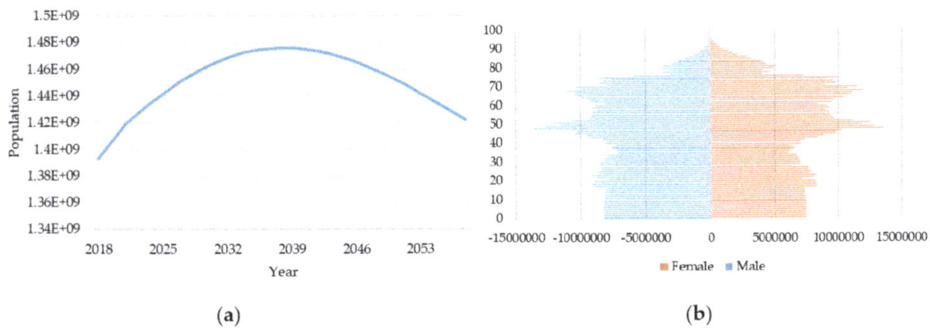

Figure 3. The future population in China. (**a**) The population in China (2018–2058). (**b**) The population structure in China (2038).

As described before, only employees participate in CBPS, and the populations who join in CBPS must be estimated.

Step 1: estimating the structure of populations that have already joined CBPS. We obtained the number of contributors and retirees from the China Labor Statistical Yearbook 2018. From the unpublished social security database of Beijing, we obtained the proportions of participants of all ages. Then, we estimated the number of participants of all ages in China, which are LA_x^i and LR_y at period 0.

Step 2: estimating the numbers of new participants in all periods. We assumed that the number of new participants is the product of the new workforce, urbanization rate, and employment rate. The numbers of the new workforce come from the population in China. According to the China National Population Development Plan (2016–2030), the urbanization rate in 2030 will be 70%. Given that the urbanization rate in 2018 was 58.52%, we assume that the urbanization rate will increase 1% per year from 2019 to 2030 and then remain unchanged after 2030. In addition, the unemployment rate in China was between 4% and 4.3% from 2002, according to the China National Bureau of Statistics. Thus, we assume that the unemployment rate in the future will be 4.14%, and the employment rate will be 95.86%.

Step 3: estimating the transition probability matrix. We collected the unpublished social security database of Beijing, stratified the data by age and status, and then estimated the probabilities (Appendix A).

4.2. *Wages*

Following [24], we assume the following rules for wages:

- $W^t_{x+t} = W^t_x e_{x+t}$. The human capital e_{x+t} determines the difference in wages of individuals who are x years old and $x + t$ years old in the same period.
- $W^t_y = W^{t-1}_y(1+\varepsilon)$. The wages of individuals of the same age increase ε per year.

Therefore, we have $W^t_{x+t} = W^0_x e_{x+t}(1+\varepsilon)^t$, where e_{x+t} and ε will be estimated.

According to the unpublished social security database of Beijing, we calculated the average wages for all ages and then obtained e_{x+t}, which is shown in Figure 4.

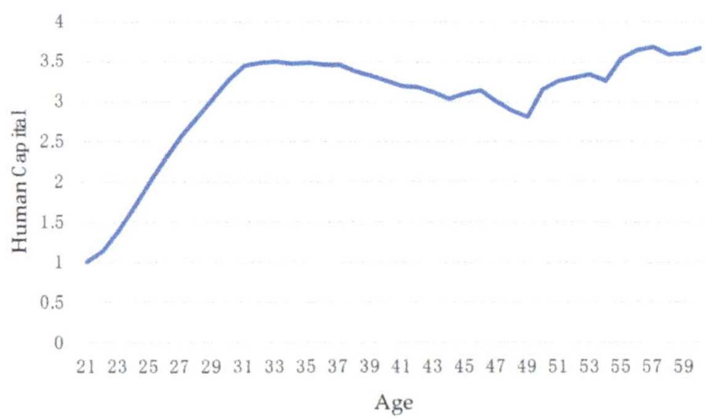

Figure 4. Human capital for all ages (e_x= 1).

The average growth rate of wages ε reached to 13.83% during 2000–2014. However, the subsequent growth rate will not be as high because of the decline of economic growth in China. According to [2], we assume that $\varepsilon = 7.7\%$ for 2018–2020, $\varepsilon = 6.6\%$ for 2021–2025, and $\varepsilon = 5.7\%$ for 2026–2068.

4.3. Parameters in CBPS

We assume that the age of the youngest employees is 21, and the maximum age of retirees is 100, according to the social security database of Beijing. CBPS specifies an individual contribution rate of 8% and an employer contribution rate of 20% (the employer contribution rate has been decreased to 16% to reduce the employer's burden and promote the development of the private sector in China. However, the policy is temporary because of the large CBPS gap. Therefore, we still assume 20% but perform a sensitivity test for the parameter.). The length of the planned period depends on the age at which the participant retires; for example, the length of the planned period for an individual retiring at 60 years is 139 months (approximately 12 years). Appendix B shows the relationship between the length of the planned period and age.

Official documents regulated the interest rate of the private account in 2017, and the interest rate was above 8%. We assume an interest rate of 8.13% after 2017, according the Annual Report of National Council of Social Security Funds. Before 2017, the money in the private account was embezzled. The return of the private account was not recovered and for bookkeeping purposes was considered to have accrued at the lowest interest rate. Therefore, we assume that the interest rate was 2.84% before 2017.

The public pension increases every year after individuals retire. The increases are a bonus to retirees to ensure that the standard of living of retirees is not affected by inflation. However, the growth rate of the public pension was much higher than the inflation rate during 2006–2015, reaching 10% per year. However, the growth rate began to fall in 2016 and was 6.5%, 5.5%, 5%, and 5% in 2016, 2017, 2018, and 2019, respectively. Therefore, we assume a 5% future growth rate.

The funeral favor, another bonus, differs among provinces. For example, the funeral favor is 5000 CNY in Beijing and four times the average wage in Chengdu. According to [25], we assume that the funeral favor is 60% of the national average wage.

All parameter values are shown in Table 2.

Table 2. Parameter values.

Description	Value
Mortality rate	1.2 times the value in Table 1, China Life Insurance Mortality Table (2010–2013)
Fertility rate	The fertility rate is 1.4 from 2019 to 2022 and 1.25 after 2022
Population structure at period 0	The tabulation on the 2010 Population Census of China
Urbanization rate	The urbanization rate in 2018 was 58.52%, which will increase 1% per year from 2019 to 2030 and then remain unchanged.
Unemployment rate	4.14%
Average growth rate of wages	7.7% for 2018–2020, 6.6% for 2021–2025, and 5.7% after 2025
Interest rate	2.84% before 2017 and 8.13% after 2017
Funeral favor	60% of the national average wage
Public pension growth rate	5% after 2018

5. Model Testing and Policy Analysis

5.1. Model Testing

We used the data from 2009 to 2017 to test our model to ensure that the model, parameters, and codes are correct. The comparison of the actual data and simulated data is shown in Table 3.

Table 3. Comparison of actual data and simulated data (2009–2017).

(0.1 Billion CNY)	Actual Income [1]	Simulated Income	Ratio	Actual Expenditure	Simulated Expenditure	Ratio
2009	9534.0	10,924.7	114.59%	8894.4	8965.7	100.80%
2010	11,110.0	12,211.2	109.91%	10,554.9	10,888.1	103.20%
2011	13,956.0	14,199.8	101.75%	12,765.0	13,249.9	103.80%
2012	16,467.0	16,184.8	98.29%	15,561.8	16,049.5	103.10%
2013	18,634.0	18,324.3	98.34%	18,470.4	19,644.0	106.40%
2014	20,434.0	20,844.6	102.01%	21,754.7	23,435.2	107.70%
2015	23,016.0	23,617.4	102.61%	25,812.7	28,030.7	108.60%
2016	26,768.0	27,158.3	101.46%	31,853.8	32,310.4	101.40%
2017	33,403.0	31,453.2	94.16%	38,051.5	37,027.4	97.30%

[1] Data source: China Labor Statistical Yearbook 2018; the actual income does not contain the subsidies.

Table 3 shows that our model fits CBPS well. The error rate of income is 4.62% and that of expenditure is 4.19%. Our model appears to be reliable.

5.2. Forecasting

Taking 2017 as period 0, we predict the income and expenditure of CBPS from 2018 to 2058. The results are shown in Table 4.

Table 4. Income and expenditure forecasting of CBPS (2018–2058).

(0.1 Billion CNY)	Income	Expenditure	Gap in Single Year
2018	35,288.4	42,417.6	−7129.2
2023	59,106.1	80,415.9	−2,1309.8
2028	88,826.7	152,188.6	−6,3361.8
2033	126,820.4	274,725.0	−147,904.7
2038	177,121.9	459,464.7	−282,342.9
2043	246,078.1	729,378.4	−483,300.3
2048	336,225.5	1,076,058.5	−739,833.0
2053	452,095.0	1,534,990.1	−1,082,895.1
2058	617,958.4	2,093,026.2	−1,475,067.9

Table 4 shows that the contributions are less than the benefits every year from 2018, and CBPS is unsustainable if there are no reforms or external transfers. In addition, the gap grows in the future.

As there is a positive accumulated balance at the end of 2017, the accumulated balance will not become negative in 2018. The accumulated balances from 2018 to 2028 are shown in Table 5.

Table 5. Accumulated balance of CBPS (2018–2028).

	Accumulated Balance (0.1 Billion CNY)
2018	40,323.2
2019	34,605.7
2020	26,469.1
2021	14,911.0
2022	−769.2
2023	−22,079.0
2024	−49,565.3
2025	−83,610.6
2026	−125,697.6
2027	−177,370.3
2028	−240,732.2

Table 5 shows that the accumulated balance of CBPS will become negative in 2022, and the deficit is 76.9 billion CNY. Subsequently, the deficits grow, reaching 24,073.2 billion CNY in 2028, which is 27% of the GDP in 2018.

5.3. Policy Analysis

5.3.1. The Decline in Employer Contribution Rate

As described before, the government decreased the employer contribution rate from 20% to 16% in 2019, to ease employer burden and promote economic growth. However, this change is not beneficial for the unsustainable CBPS. There is no doubt that the deficits will become larger than before. The results are shown in Figure 5.

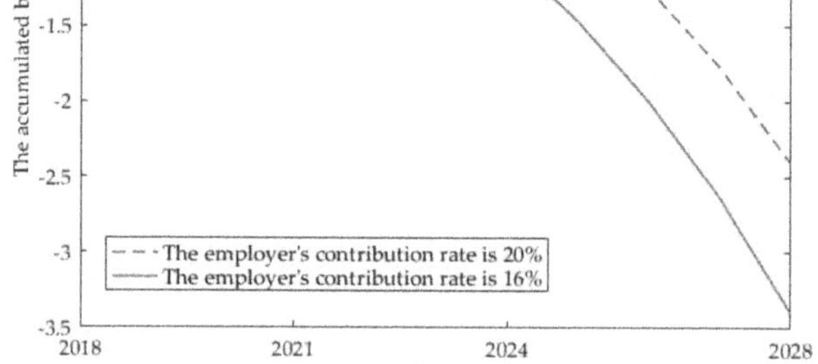

Figure 5. Accumulated balance when the employer contribution rate is decreased to 16%.

As expected, the deficits grow. Owing to the decrease in the employer contribution rate from 20% to 16%, the time at which the accumulated balance will become negative becomes 2021 instead of 2022. In addition, the deficit will be 1192.8 billion CNY, instead of 76.9 billion CNY.

However, the impact of the decline in the employer contribution rate may be substantial. The decline will increase employer profit and then increase the wages and employment rate, which in turn, will increase the income of CBPS. However, this topic is beyond the scope of our article.

5.3.2. The End of One-Child Policy

China is getting older faster than anywhere else worldwide. Ending the one-child policy was one measure taken to alleviate the aging problem. Simultaneously, an increase in fertility rate will also enhance the sustainability of CBPS, according to the experience of other countries. Here, we study the impact of ending the one-child policy on the sustainability of CBPS.

We examine three situations:

- High fertility rate: 2 from 2019 to 2022 and 1.8 after 2022.
- Moderate fertility rate (baseline situation): 1.4 from 2019 to 2022 and 1.25 after 2022.
- Low fertility rate: 1.4 from 2019 to 2022 and 1.1 after 2022.

Changes in fertility will affect the income and expenditure of CBPS for more than 100 years. Therefore, we must extend the prediction period from 2018–2057 to 2018–2107. The results are shown in Figure 6.

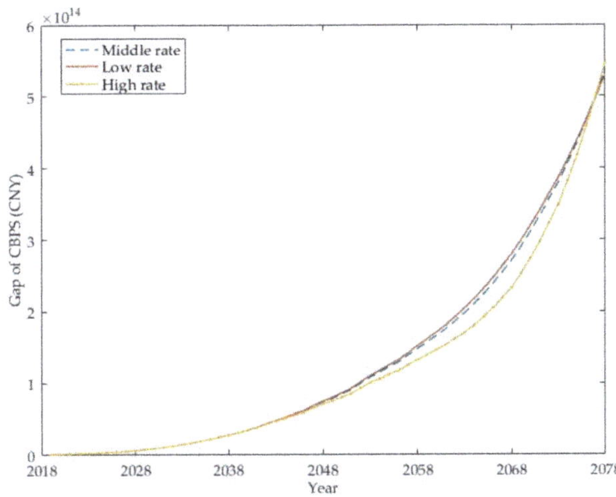

Figure 6. CBPS gap with different fertility rates (2018–2078).

Figure 6 shows that fertility rate have no effects in the short term because the new generation does not participate in CBPS until 2038. After these individuals are employed and contribute to CBPS during 2038–2078, the income will increase, and the gap between expenditure and income will decrease if the fertility rate increases. Therefore, the increase in fertility rate will also enhance the sustainability of CBPS in the short to medium term.

However, Figure 6 shows that the sustainability will substantially decrease when the new generation retires. This conclusion is contrary to those of many other researchers, who have claimed that the increase in fertility rate will enhance the sustainability of CBPS at all times, e.g., [6]. The reason for this discrepancy is that CBPS is not actuarially balanced. In other words, every participant receives more than his/her contributions, as shown in Table 6. Therefore, CBPS has a greater deficit if there are more participants.

Table 6. The actuarial present value of the benefits and contributions of a male participant who enters the system in 2018 (assuming an interest rate of 7%).

Entry Year	Contributions (CNY)	Benefits (CNY)
2018	61,710	79,420

In conclusion, ending the one-child policy may enhance the sustainability of CBPS in the short to medium term but will erode the sustainability of CBPS in the long term.

5.3.3. The Increase in Retirement Age

Raising the retirement age is another way to enhance the sustainability of the pension system. China also issued a delayed retirement policy specifying that the retirement age will increase 1 year every 2 years, and the final compulsory age will be 65 for men and 60 for women. We consider the policy as the experimental group and the baseline situation as the control group. The results are shown in Figure 7.

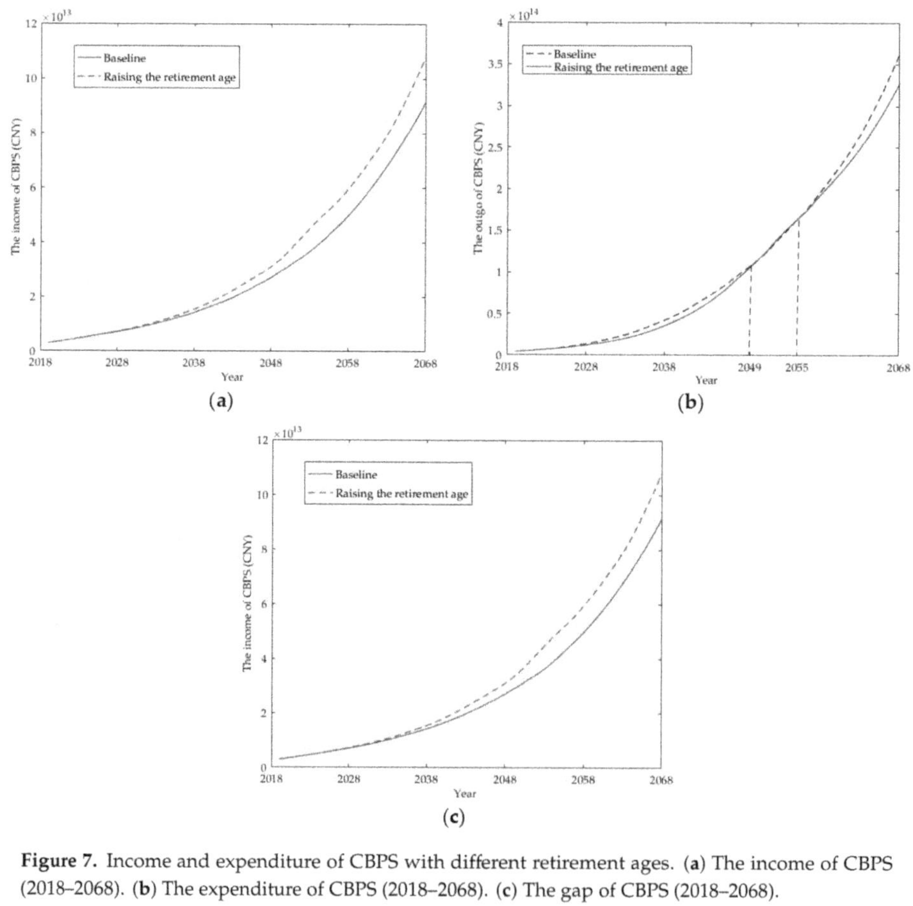

Figure 7. Income and expenditure of CBPS with different retirement ages. (**a**) The income of CBPS (2018–2068). (**b**) The expenditure of CBPS (2018–2068). (**c**) The gap of CBPS (2018–2068).

Figure 7 shows that the income will increase because individuals will contribute longer if the retirement ages are increased. However, the change in expenditure is slightly more than we expected. We expected that the expenditure would decrease if the retirement age increased. Although this is true in most years, the expenditure increases between 2049 and 2055, although the gap decreases in all years because of delayed retirement.

We sought to determine why the expenditure increases between 2049 and 2055. Taking a male participant who enters the system in 2018 as an example, we calculated the actuarial present value of his total contributions and benefits with different retirement ages. The results are shown in Figure 8. Although delayed retirement results in a delay in pension payment for part of the working staff, the amount of pension payment for this part of the insured staff will increase in the future, thus potentially leading to an increase in the total amount of pension funds paid in the future, and in contrast, increasing the pressure on pension funds to pay.

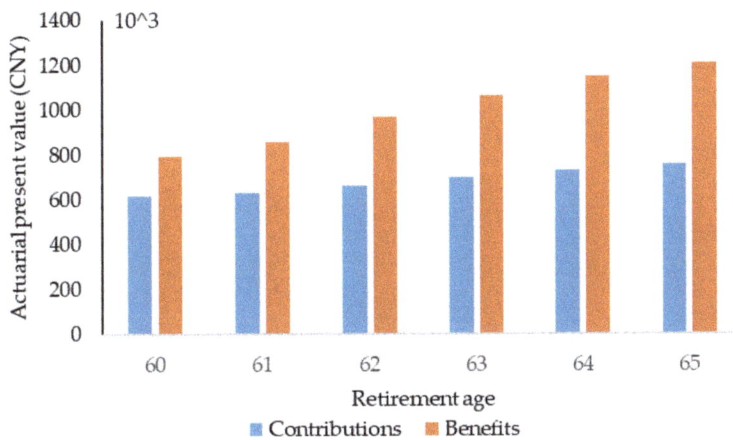

Figure 8. The actuarial present value of the benefits and contributions of a male participant who enters the system in 2018 with different retirement ages (assuming an interest rate of 7%).

5.3.4. Comprehensive Analysis

There are interactions if the above policies are simultaneously affected. Therefore, we comprehensively analyzed the effects of the decline in employer contribution rate, the end of the one-child policy, and the increase in retirement age. The results are shown in Figure 9.

Figure 9 shows that the income will increase if all policies go into effect simultaneously. There will be many more participants because of the ending of the one-child policy. All participants will contribute for longer times because of delayed retirement. Although the income will decrease because of the decline in the employer contribution rate, it will be offset by the increases resulting from the other two policies.

However, the change in expenditure is not consistent. The impacts on expenditure of ending the one-child policy and increasing the retirement age are opposite. Moreover, the decline in the employer contribution rate does not affect the expenditure. Therefore, the expenditure will decrease because of fewer participants, whereas the expenditure will increase when the population exceeds a threshold, which will occur in 2078.

The changes in income and expenditure determine the gap. Our results show that the gap will shrink before 2088 and subsequently increase rapidly.

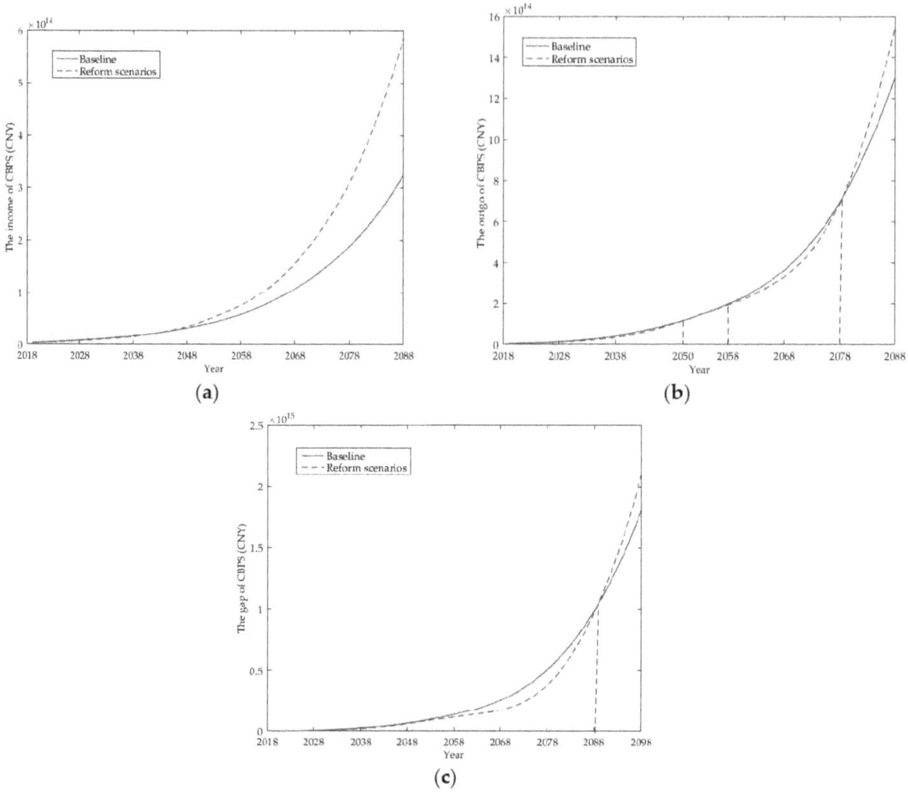

Figure 9. Income and expenditure of CBPS with reforms. (**a**) The income of CBPS (2018–2088). (**b**) The expenditure of CBPS (2018–2088). (**c**) The expenditure of CBPS (2018–2098).

6. Conclusions

This article describes CBPS and discusses the effectiveness of reforms on its sustainability. It provides a description of the pension system in China. In addition, a Markov actuarial model, which had not previously been developed for China, was established and verified for CBPS.

This article found that: (1) The contributions are less than the benefits every year from 2018, and the accumulated balance will be negative from 2022. (2) Ending the one-child policy may enhance the sustainability of CBPS in the short to medium term but will erode the sustainability of CBPS in the long term. (3) Raising the retirement age will shrink the gap of CBPS although the expenditure increases in a certain period. (4) Comprehensively, the gap will shrink before 2088 and subsequently increase rapidly if all of the policies are implemented.

From the above results, we believe that CBPS will not be sustainable in the long term if there are no subsidies or other financial transfers, even with delayed retirement and the increase in fertility. Moreover, the rise in fertility will shrink the gap of CBPS in the medium to short term but will increase the gap in the long term. In addition, raising the retirement age may not help, because the benefits depend on the length of contributions. It is the system itself that is not sustainable. In other words, the system is not actuarially balanced, by giving everyone extra money. Therefore, thorough reform is necessary to make the system self-financed, such as the methods to calculate the benefits of the public and private account.

There are two limitations of the article. Firstly, there are a lot of parameters and assumptions when forecasting the future cashflows of CBPS. Because of the uncertainties in future, the actual parameters and assumptions may be different with the settings in the paper. Secondly, the implementation of some reforms will not only directly affect the sustainability of CBPS but also have indirect impacts by changing macroeconomic variables. It is an interesting topic for future works. In addition, our model can also be applied to projecting the cashflows of other pension systems.

Author Contributions: Conceptualization, resources, and writing—original raft preparation, P.L.; investigation, data curation, H.S.; writing—review and editing, D.P. All authors have read and agreed to the published version of the manuscript.

Funding: This research was funded by China National Social Science Fund, grant number 17CSH018, and Program for Innovation Research in Central University of Finance and Economics.

Conflicts of Interest: The authors declare no conflict of interest.

Appendix A

Table A1. The transition probability matrix of contributors.

Age	ll	ls	lr	ld	sl	ss	sr	sd
21	0.85463	0.14527	0.00000	0.00010	0.75794	0.24043	0.00000	0.00163
22	0.87759	0.12232	0.00000	0.00009	0.78286	0.21565	0.00000	0.00149
23	0.88040	0.11952	0.00000	0.00008	0.77670	0.22153	0.00000	0.00177
24	0.86935	0.13054	0.00000	0.00011	0.73936	0.25817	0.00000	0.00247
25	0.86049	0.13940	0.00000	0.00011	0.69870	0.29815	0.00000	0.00315
26	0.86177	0.13810	0.00000	0.00013	0.64215	0.35340	0.00000	0.00445
27	0.86705	0.13282	0.00000	0.00013	0.59016	0.40370	0.00001	0.00613
28	0.87679	0.12304	0.00000	0.00016	0.55109	0.44093	0.00000	0.00798
29	0.88109	0.11874	0.00001	0.00017	0.52666	0.46534	0.00004	0.00796
30	0.88877	0.11102	0.00000	0.00021	0.50992	0.48211	0.00000	0.00797
31	0.89559	0.10421	0.00002	0.00018	0.49817	0.49361	0.00008	0.00814
32	0.90243	0.09739	0.00002	0.00016	0.49603	0.49378	0.00008	0.01010
33	0.90865	0.09106	0.00005	0.00023	0.50305	0.48558	0.00021	0.01116
34	0.91690	0.08278	0.00012	0.00019	0.49749	0.48979	0.00049	0.01222
35	0.92598	0.07361	0.00011	0.00030	0.50250	0.48356	0.00047	0.01347
36	0.93114	0.06832	0.00022	0.00032	0.49364	0.48995	0.00100	0.01542
37	0.93625	0.06316	0.00025	0.00034	0.49680	0.48432	0.00116	0.01772
38	0.93906	0.06015	0.00037	0.00041	0.49965	0.47797	0.00179	0.02059
39	0.94039	0.05864	0.00050	0.00047	0.49556	0.47992	0.00261	0.02191
40	0.93309	0.06574	0.00068	0.00049	0.49382	0.47941	0.00360	0.02316
41	0.93850	0.06011	0.00082	0.00057	0.50950	0.46298	0.00445	0.02307
42	0.94005	0.05820	0.00113	0.00062	0.50980	0.45816	0.00625	0.02579
43	0.93781	0.06039	0.00120	0.00060	0.50734	0.45419	0.00726	0.03121
44	0.93427	0.06333	0.00157	0.00083	0.48147	0.46866	0.01078	0.03910
45	0.83349	0.07278	0.09273	0.00099	0.27621	0.27516	0.42243	0.02619
46	0.91394	0.06067	0.02438	0.00101	0.40522	0.39219	0.15931	0.04328
47	0.93032	0.06133	0.00724	0.00112	0.45130	0.43657	0.05695	0.05517
48	0.93977	0.05364	0.00540	0.00119	0.45080	0.44098	0.04425	0.06396
49	0.94059	0.05373	0.00394	0.00175	0.43647	0.46080	0.03496	0.06777
50	0.65188	0.05800	0.28844	0.00167	0.09661	0.12411	0.75741	0.02187
51	0.90594	0.06468	0.02752	0.00186	0.31975	0.41671	0.18588	0.07767
52	0.92773	0.06163	0.00879	0.00184	0.34389	0.49763	0.06597	0.09250
53	0.92894	0.06080	0.00789	0.00237	0.32010	0.51606	0.06332	0.10052
54	0.91843	0.07275	0.00667	0.00215	0.30672	0.53597	0.05690	0.10040
55	0.50985	0.05435	0.43390	0.00189	0.05214	0.09147	0.83271	0.02368
56	0.84429	0.05271	0.10077	0.00223	0.19373	0.29432	0.42343	0.08851

Table A1. *Cont.*

Age	ll	ls	lr	ld	sl	ss	sr	sd
57	0.92097	0.04898	0.02684	0.00322	0.26167	0.43847	0.16313	0.13673
58	0.94414	0.03671	0.01598	0.00317	0.25210	0.50030	0.10868	0.13892
59	0.94620	0.04017	0.00969	0.00394	0.24076	0.53257	0.07314	0.15352
60	0.03196	0.04342	0.92153	0.00310	0.00617	0.04715	0.92728	0.01940
61	0.17245	0.18253	0.64166	0.00336	0.03069	0.35778	0.42889	0.18263
62	0.53000	0.14500	0.32000	0.00500	0.03488	0.63081	0.09302	0.24128
63	0.58182	0.11818	0.30000	0.00000	0.03795	0.64286	0.07366	0.24554
64	0.55556	0.13889	0.30556	0.00000	0.02609	0.74348	0.09565	0.13478
65	0.00000	0.00000	1.00000	0.00000	0.00000	0.00000	1.00000	0.00000

"l", "s", "r", "d" denote "Normal", "Break-off", "Retirement" and "Dead", respectively. And "ij" denotes the 1-year transition probability from state i to state j. For example, "ll" denotes the the 1-year transition probability from Normal to state Normal. And the denotations the following table are same.

Table A2. The transition probability matrix of retirees.

Age	rr	rd	Age	rr	rd
41	0.97575	0.02425	71	0.96806	0.03194
42	0.98830	0.01170	72	0.96310	0.03690
43	0.98209	0.01791	73	0.95697	0.04303
44	0.97702	0.02298	74	0.95318	0.04682
45	0.99574	0.00426	75	0.94776	0.05224
46	0.99525	0.00475	76	0.94279	0.05721
47	0.99395	0.00605	77	0.93871	0.06129
48	0.99497	0.00503	78	0.93138	0.06862
49	0.99320	0.00680	79	0.92154	0.07846
50	0.99722	0.00278	80	0.91088	0.08912
51	0.99533	0.00467	81	0.90527	0.09473
52	0.99519	0.00481	82	0.89749	0.10251
53	0.99543	0.00457	83	0.88268	0.11732
54	0.99436	0.00564	84	0.86553	0.13447
55	0.99508	0.00492	85	0.85774	0.14226
56	0.99310	0.00690	86	0.83706	0.16294
57	0.99223	0.00777	87	0.81914	0.18086
58	0.99153	0.00847	88	0.79904	0.20096
59	0.99092	0.00908	89	0.78832	0.21168
60	0.99163	0.00837	90	0.77597	0.22403
61	0.98948	0.01052	91	0.73368	0.26632
62	0.98779	0.01221	92	0.71470	0.28530
63	0.98577	0.01423	93	0.71448	0.28552
64	0.98497	0.01503	94	0.68311	0.31689
65	0.98356	0.01644	95	0.67688	0.32312
66	0.98134	0.01866	96	0.65587	0.34413
67	0.97979	0.02021	97	0.70186	0.29814
68	0.97593	0.02407	98	0.56842	0.43158
69	0.97566	0.02434	99	0.46429	0.53571
70	0.97087	0.02913	100	0.00000	1.00000

"l", "s", "r", "d" denote "Normal", "Break-off", "Retirement" and "Dead", respectively. And "ij" denotes the 1-year transition probability from state i to state j. For example, "ll" denotes the the 1-year transition probability from Normal to state Normal. And the denotations the following table are same.

Appendix B

Table A3. The relationship of the length of planned period and the age.

Retirement Age	The Length of Planned Period (in Months)	Retirement Age	The Length of Planned Period (in Months)
40	233	53	180
41	230	54	175
42	226	55	174
43	223	56	164
44	220	57	158
45	216	58	152
46	212	59	145
47	208	60	139
48	204	61	132
49	199	62	125
50	195	63	117
51	195	64	109
52	185	65	101

References

1. McCarthy, F.D.; Zheng, K. Population aging and Pension Systems: Reform Options for China (May 1996). Available online: https://ssrn.com/abstract=636131 (accessed on 4 September 2020).
2. Yang, Z.; Shi, C. Financial burden on pensions for employees in urban enterprises in China. *Econ. Sci.* **2016**, *2*, 42–52. (In Chinese)
3. Zheng, S.; Liao, P. Actuarial model for urban public pension fund in China. *Syst. Eng. Theory Pract.* **2017**, *31*, 10–17. (In Chinese)
4. Yu, H.; Shen, H. Analysis on the sustainability of China's basic pension fund. In Proceedings of the 2011 2nd International Conference on Artificial Intelligence, Management Science and Electronic Commerce (AIMSEC), Zhengzhou, China, 8–10 August 2011; pp. 6553–6556. [CrossRef]
5. Sun, L.; Su, C.; Xian, X. Assessing the Sustainability of China's Basic Pension Funding for Urban and Rural Residents. *Sustainability* **2020**, *12*, 2833. [CrossRef]
6. Zeng, Y.; Zhang, X.; Liu, L. From "selective two-child policy" to universal two-child policy: Will the payment crisis of China's Pension System be solved? *China Financ. Econ. Rev.* **2017**, *5*, 1–17. [CrossRef]
7. Fan, R. The Path to Select Chinese Progressive Delay Retirement Age Policy. *Stud. Asian Soc. Sci.* **2015**, *2*, 19–22. [CrossRef]
8. Aaron, H. The social insurance paradox. *Can. J. Econ. Political Sci.* **1966**, *33*, 371–374. [CrossRef]
9. Corbo, V. Policy challenges of population aging and pension systems in Latin America. *Proc. Econ. Policy Symp. Jackson* **2004**, *Aug*, 257–280.
10. Sin, Y. *China: Pension Liabilities and Reform Options for Old Age Insurance*; The World Bank Working Paper; The World Bank: Washington, DC, USA, 2005.
11. Grech, A.G. Assessing the sustainability of pension reforms in Europe. *J. Int. Comp. Soc. Policy* **2013**, *29*, 143–162. [CrossRef]
12. Verbič, M.; Majcen, B.; Nieuwkoop, R.V. Sustainability of the Slovenian Pension System: An analysis with an overlapping generations general equilibrium model. *East. Eur. Econ.* **2006**, *44*, 60–81.
13. Barr, N.; Diamond, P. The economics of pensions. *Oxf. Rev. Econ. Policy* **2006**, *22*, 15–39. [CrossRef]
14. Samwick, A.A. New evidence on pensions, social security, and the timing of retirement. *J. Public Econ.* **1998**, *70*, 207–236. [CrossRef]
15. Breyer, F.; Hupfeld, S. On the fairness of early retirement provisions. *Ger. Econ. Rev.* **2010**, *11*, 60–77. [CrossRef]
16. Fehr, H.; Kallweit, M.; Kindermann, F. Pension reform with variable retirement age: A simulation analysis for Germany. *J. Pension Econ. Financ.* **2012**, *11*, 389–417. [CrossRef]

17. Conde-Ruiz, J.I.; Giménez, E.L.; Pérez-Nievas, M. Endogenous fertility-pension programs and intergenerational welfare. In *Modeling and Control of Economic Systems 2001*; Neck, R., Ed.; Elsevier Science: Amsterdam, The Netherlands, 2003; pp. 157–161.
18. Serrano, F.; Eguía, B.; Ferreiro, J. Public pensions' sustainability and population ageing: Is immigration the solution? *Int. Labour Rev.* **2011**, *150*, 63–79. [CrossRef]
19. Zhao, Y.; Bai, M. Stressing testing the transition debt of pension plan in China. *J. Quant. Tech. Econ.* **2018**, *4*, 80–96. (In Chinese)
20. Han, J.; Meng, Y. Decreased contribution rates increase public pension fund revenue: Evidence from China. *J. Aging Soc. Policy* **2019**, 1–18. [CrossRef]
21. Chen, X.; Yang, Z. Stochastically Assessing the Financial Sustainability of Individual Accounts in the Urban Enterprise Employees' Pension Plan in China. *Sustainability* **2019**, *11*, 3568. [CrossRef]
22. Zhai, Z.; Zhang, X.; Jin, Y. Demographic consequences of an immediate transition to a universal two-child policy. *Popul. Res.* **2014**, *38*, 3–17. (In Chinese)
23. Qi, M.; Dai, M.; Zheng, Y. Discuss the impact and tendency of the universal two-child policy on China's birth rate fluctuation. *China Popul. Resour. Environ.* **2016**, *9*, 1–10. (In Chinese)
24. Kulish, M.; Kent, C.; Smith, K. Aging, retirement, and savings: A general equilibrium analysis. *BE J. Macroecon.* **2010**, *10*, 1–32. [CrossRef]
25. Liao, P. Does China's social pension system enhance the participants' welfare? *J. World Econ.* **2016**, *11*, 147–171. (In Chinese)

© 2020 by the authors. Licensee MDPI, Basel, Switzerland. This article is an open access article distributed under the terms and conditions of the Creative Commons Attribution (CC BY) license (http://creativecommons.org/licenses/by/4.0/).

sustainability

MDPI

Article

Sustainability of the Motivation Policy Model for Employees in State Administration

Milan Kovačević, Srđan Blagojević * and Bojan Kuzmanović *

Military Academy, University of Defense, 11000 Belgrade, Serbia; kacaiaca@ptt.rs

* Correspondence: srdjanblagojevic80@gmail.com (S.B.); bojan.kuzmanovic@vs.rs (B.K.)

Received: 30 August 2020; Accepted: 16 September 2020; Published: 26 September 2020

check for updates

Abstract: An important condition for the performance sustainability of organizations, in the public and state sector, is the maintenance and improvement of the employee motivation policy model. Motivation for work has a significant impact on the success of modern organizations; the impact is multidimensional and complex, and it has been confirmed empirically. Motivation is influenced by numerous factors, of different influence and hierarchy, which depend on socio-economic conditions and cultural determinants, but also on the characteristics of employees, and it is necessary to constantly monitor them. The paper presents a sustainable model of employee motivation in public administration and the results of research on the hierarchy and the impact of motivation factors on employees of different demographic characteristics. The research was conducted by an anonymous survey of 2128 respondents (1576 employees in the state administration and 552 persons employed in the Republic of Serbia outside the state administration), in the period March–June 2020. The questionnaire developed for this research has 16 questions on motivating factors with a scale with a high level of internal consistency. Using one-factor analysis of variance (ANOVA), tests of subsequent comparisons, and *t*-test, statistically significant differences are determined for the influence of motivational factors on respondents of different levels of education, age, role in organization, gender, and family status—also, the magnitude of influence ranges from small to large. The influence of motivational factors is greater for older people, for people with a higher level of education, for managers in relation to workers, for women in relation to men, as well as for persons in marriage in relation to persons out of wedlock. By applying multi-criteria analysis, the factors are ranked: the amount of salary and good interpersonal relations are at the top, while criticism and information about work are in the last positions. As a factor that negatively affects motivation, low wages stands out. The difference in the motivation of state administration members was determined by comparing the results of motivational factors' influence with the appropriate sample of employees outside the state administration. Differences were registered for 15 out of 16 factors (only for the factor high wages, there is no difference), and the magnitude of the impact is medium and small.

Keywords: state administration; motivation (psychology); sustainability; politics; organizational performance; motivation factors; hierarchy

1. Introduction

Motivation and cognition of motivated behavior are extremely complex because people's behavior is not always consistent or rational and the level of motivation varies not only between individuals but also within one individual at different times. Motivation is one of the most frequently researched topics of people's behavior in organizations [1]. Individuals differ not only by the level of motivation—how motivated they are—but also by the orientation of that motivation, what motivates them [2]. Interest in motivation and the factors that guide it is as old as the human race and arises from three reasons:

improving productivity, efficiency, and creativity; improving the quality of working life in organizations; and strengthening competitiveness and success [3]. In addition, motivation management creates corporate resources that cannot be copied and help develop a sustainable competitive advantage. Both extrinsic and intrinsic motivation are necessary for this development, and in a knowledge-based society, the focus will be on tasks that require employees to identify with and enjoy their work [4] (p. 284). Motivation for work can be managed, according to Mihailović [5]; it can be challenged and encouraged and the model of managing motivated behavior can be established in the organization. Knowledge of motivational factors is the key to building a system for managing motivated behavior.

The aim of this paper is to determine the hierarchy of motivational factors and formulate a sustainable model of employee motivation policy in public administration, based on the assumptions of the theory of needs and the theory of expectations. The sustainable model assumes that the formulation and changes of motivation policy are based on the results of continuous research, whereby both the research and the instruments that will be applied will be variable, adapt to environmental conditions, and register significant changes in motivation research results. Significant changes in employee motivation will be monitored in relation to the results of previous research and changes in the environment through the application of appropriate analytical techniques and procedures (SWOT, PEST, etc.). The paper will also present the differences in the hierarchy of motivational factors for employees of different demographic characteristics, as well as differences in the impact of certain factors on employees of different characteristics, which is a condition for applying an individual approach to employee motivation and adjustment to needs, interests, and preferences.

Given that virtually all motivational theories have been developed in the USA or Western Europe, where numerous studies have been conducted, and that there are pronounced cultural differences, there is a need to determine the influence of motivational factors in their own cultural environment, current time frame, and specific organizational environment—public administration. Bearing in mind that extensive research on the motivation of public administration employees has not been conducted in Serbia, there was a need to conduct the research and compare the results with previous research and assumed results, which can provide a basis for formulating a sustainable motivation policy.

The paper is organized as follows: after the introduction, the second part presents a literature review, the third part presents the model and procedure of the study, the fourth part presents the results, the fifth part offers a discussion and comparison of the results with previous research, and the sixth part lists the most significant findings and contributions of the study.

2. Literature Review

Numerous authors from the fields of organizational behavior, management, and psychology have defined the concept of motivation, and a brief and concise definition was given by Pinder. According to him, motivation for work is a set of forces inside and outside the personality that initiate work and determine its form, direction, intensity, and duration [6] (p. 11). The three key elements of motivation that are observed are intensity, direction, and persistence. Intensity refers to the effort invested, but high intensity will lead to favorable work results only if the efforts are channeled in a direction that benefits the organization, the persistence dimension measures how long a person can sustain the effort because motivated individuals work long enough to achieve the goal [1].

Due to the importance of motivation, a large number of researchers have tried to unveil the secrets of motivation and numerous theories have been developed. Still, there is no universally accepted way of presenting motivational theories. According to Armstrong [7], three main areas of motivational theories are instrumentality, content, and process. Bahtijarević-Šiber and Mihailović classify the most important theories of motivation into two groups: theories of needs and theories of processes [3,8], while Robbins [1] classifies motivational theories in to need theory, self-determination theory and cognitive evaluation theory, goal settings theory, reinforcement theory, equity theory (theory of organizational justice), and expectancy theory. Pinder, on the other hand [9], considers motivation through a motivational framework by analyzing needs, traits, values, context, person–context fit, cognition, and affect/emotion.

Continuous and long-term interest in motivation by a wide range of researchers is completely understandable because discovering motivational mechanisms and knowledge about factors and mechanisms that are getting employees to work has numerous benefits (reduced fluctuation and absenteeism, positive impact on job satisfaction, positive impact on employees' health). The significance of this research is primarily reflected in the impact of motivation on work performance. Kovach [10] believes that organizations that know what motivates their employees to work are in a better position to stimulate them to work better and improve organizational performance, while Spitzer [11] states that different people respond to different incentives and advises that before investing in reward and recognition systems, organizations should survey their employees and allow employees to choose the specific reward that suits them best. The research that was initiated and realized on the assumptions based on need theories focused on the question "what motivates the worker" and tried to determine the hierarchy of motivational factors (needs). Mihailović [5] (p. 48) defines motivational factors as a hypothetical relationship between many individual, social, and organizational aspects of work and workers with manifestations of motivated behavior.

In addition to the hierarchy of factors, in his research, Kovach analyzed the responses of employees of different characteristics and found differences in the influence of factors depending on gender, age, income level, type of work, and organizational level [10] (p. 60). Additionally, Kovach determined significant differences in the ranking of factors by workers and managers (managers answered the question of what motivates workers). Similar studies have been conducted several times in the United States [12], China [13], Russia [14], Malaysia [15,16], Greece [17,18], Pakistan [19], and the Nordic countries [20]. Motivation in Serbia and surrounding countries has been investigated several times in extensive research conducted by Jezernik in 1960 and 1962, Možina in 1969, Jušić in 1971, and Mihailović in 1988 [5] (pp. 50–52). Moreover, a certain amount of research was realized on appropriate samples, of which the characteristic research of Kujović in 1989 and Mijačika in 1989 stand out [21,22], as well as recent smaller research in Serbia [23] and in Montenegro [24].

The question of the attitude towards money, that is to say, the dilemma of whether money is a motivating factor or not, or under what conditions it can be used as a motivator, is very important in the development of motivational policies. Despite the fact that people are different and have different needs and work values [13], it is noticed that the amount of salary stands out as the most important motivating factor in more research in different time periods and cultures [12,13,17–19,25–32], also in Serbia and in the surrounding countries of Croatia, Slovenia, Bosnia and Herzegovina, and Montenegro [5,21–24]. However, some authors critically consider the impact of money and consider money only as a "hygienic" factor, i.e., that money is not a motivator or not the most important one [33–39]. It provides motivation only under the right conditions, i.e., a poorly designed or applied system of monetary compensation demotivates [7]; in other words, money for workers who are on the edge of existence is only a stimulus, but not a motivation [40]. A significant number of studies show that money reduces intrinsic motivation [4,34,35,41–43], that its influence depends on the type of employees [4], or that it will positively affect the materialist-oriented type of person ("income maximizers"). In addition, monetary rewards decrease performance in executing complex tasks in order to reach difficult goals [44] (p. 264). Furthermore, it is noticed that when people are paid for work, they feel that they want to work less; that is, they work only because they have an obligation [1] (p. 210).

The importance of motivation is also reflected in the influence on performance sustainability. The impact of motivation on performance and sustainability of performance (performance sustainability implies a model of continuous changes and improvements in motivational policies and procedures to preserve or improve the level of employee motivation that ensures maintenance or improvement of existing performance levels) was identified in several studies presented below.

The positive influence on performance by motivation was found in some models of motivation [45,46] such as the social identity model for work motivation and performance [47], as well as in the self-determination theory model, according to which autonomous motivation affects performance [48]. In addition, Lord, in his study [49], concludes that the successful application of motivators improves job satisfaction, which results in increased productivity. Mani finds that the inability to encourage skilled workers leads to undermining the system's effectiveness [50]. The influence of motivation on performance has been found both in previous research [51–55] and in several experimental studies in the last 10 years. Significant impact was found for feedback [56] as well as changes in the payment scheme, when the fixed fee was changed for a fee per piece, the effect increased and vice versa—when the payment per piece was changed for a fixed fee, the effect decreased [57,58]. The studies also found a positive impact of involvement as well as autonomy on efficiency. Workers who had more autonomy at work achieved higher performance compared to workers with less autonomy [59], and involvement (measured through empowerment, teamwork, and development capacity) also led to a higher level of organizational effectiveness [60]. Zhang, in a study conducted in China [61], in addition to finding that motivation has a positive effect on performance, also finds well-internalized extrinsic motivation was more important than intrinsic motivation in predicting work performance and that even small increases in external regulation could lead to an increase in work performance. There are also findings that intrinsic motivation has a positive impact on performance when leadership is socially oriented, which gives us options to improve performance without monetary incentives [62].This is especially interesting for sectors where salary changes are not always under the direct responsibility of management—as is the case in public (state) administration. Igalens [63], in France, finds that under certain conditions "exempt" employees (employees who are not paid for overtime: professionals, administrators, and executives) can be motivated by individualized compensation schemes, and that motivation affects their efforts. Jenkins meta-analyzes 39 studies [64] and underscores that there is a positive link between financial incentives and performance, and that there is no evidence that they destroy intrinsic motivation.

Research around the world over the last 10 years has confirmed the existence of a direct link and a positive impact of motivation on performance in Australia [65], Pakistan [66], Denmark [67], Italy [68], Ghana [69], and Indonesia [70]. In Greece, Manolopoulos [18] finds that there is a positive influence of intrinsic motivation factors of public sector employees on performance. Hong, in Taiwan, finds that benefits have a significant impact on motivation and performance [54]. Wright [71] finds that motivation affects performance but also it has a moderating role in influencing the situation and ability to perform. The influence of motivation on performance was also determined by Mihailović [5] in Serbia, while Armstrong states that motivational strategies aim to create a work environment and develop policies and practices that will ensure a higher level of employee performance [7] (p. 180).

The assumption, in the setting of the motivation policy model, that there are differences that depend on the characteristics of the respondents for the influence of motivational factors is based both on experience and on the results of previous research [5,10,12,16,18–20,22,72–75].

Additionally, Maslow [76] (p. 82) believes that people are much more similar than it seems at first glance and that their goals are much more universal than the paths leading to those goals, because the paths to achieving goals are strongly influenced by society. Hofstede [77] (p. 16) also believes that human characteristics are universal, that all societies have the same problems, but also different solutions. Bahtijarević-Šiber [3] (p. 709) recommends that an individual approach should be developed in motivating employees and that the motivation strategy should be adjusted to the needs, interests, and preferences of employees. One should not disregard the development, changing needs, complexity, and dynamism of motivation. The premise of effective influence on motivation is to constantly analyze and monitor the needs, aspirations, and preferences of employees.

The specificity of public administration as a working environment is noticeable, and the specificity in the hierarchy of motivational factors of civil servants is confirmed by the results of several studies which found differences in motivation and job satisfaction of public and private sector employees in different times and cultural frameworks in Estonia [78], Greece [18], Australia [79], Italy [80], the Czech Republic [81], Taiwan [54], Denmark [72], the USA [28], and Croatia [82]. The results of the research indicate that there are differences among employees in the public and private sectors in their level of motivation [18,72], the motivational potential of jobs [79], and job satisfaction [78,80–82], as well as the hierarchy of motivators [28,54,72,78]. On the other hand, Jurgensen [83] in his extensive longitudinal analysis (for the period from 1949 to 1975) finds no significant differences in job satisfaction, while Wright [71] believes that there is not enough evidence to conclude that the differences in the working environment in the private and public sector are such that they have a great influence on motivation—he finds that the basic problem is to create a valid sample.

Different countries have different backgrounds and contextual influences, so environmental and contextual influences of Serbia may influence the results and they may be different from other countries. According to Hofstede [84], the environment and national culture of the Republic of Serbia have the following characteristics: high power distance, high avoidance of uncertainty, high degree of collectivism, dominant female values, and, in terms of long-term-short-term orientation, Serbia is in the middle of surveyed countries (Appendix A Table A1).

As stated by Janićijević, in relation to the dominant cultural patterns in the world, Serbian national culture is closest in its characteristics to Latin American countries (Chile, Mexico, Venezuela, Colombia) and, to a lesser extent, Latin European countries (France, Italy, Spain). With Latin European and Latin American countries Serbian culture shares a high power distance and uncertainty avoidance but, unlike Latin European individualistic cultures, Serbian national culture is a collectivist one. Serbian culture has female values in opposition to Latin American cultures which value masculinity more. In terms of female values, Serbian culture is similar to Scandinavia, but differs in all other dimensions. Serbian culture shares a high uncertainty avoidance with German national cultures, but differs in a high power distance and collectivism. Ultimately, Serbian culture differs from Anglo-Saxon culture in all dimensions [85] (p. 54).

A section from the literature review used to identify and set aside motivational factors for the list of motivational factors that were used in our research is shown in Table 1.

Table 1. Motivation factors in different research.

Motivation Factors	Authors
Amount of responsibility	Kovach [10]; Harpaz [73]; Wiley [12]; Yang [13]; Brislin et al. [86]; Linc [29]; Huddleston and Good [27]; Elizur et al. [37]
Prizes and awards	Kovach [10]; Herzberg et al. [87]; Islam [75]; Wiley [12]; Yang [13]; Mijacika [22]; Linc [29]
Working time	Harpaz [73]; Brislin et al. [86]; Karland Sutton [28]
Good working conditions	Kujovic [21]; Kovach [10]; Harpaz [73]; Mihailovic [5]; Wiley [12]; Yang [13]; Islam [75]; Mijacika [22]; Brislin et al. [86]; Karl and Sutton [28]; Charles and Marshall [26]; Ismail and Ahmed [16]; Manolopoulos [18]; Worthley et al. [88]; Wright and Kim [89]; Elizur et al. [37]; Silverthorne [39]; Fisher and Yuan [32]; Chitiris [17]; Cukic [90]
Quality of supervision and leadership	Kujovic [21]; Mihailovic [5]; Yang [13]; Islam [75]; Mijacika [22]; Chitiris [17]; Brislin et al. [86]; Karl and Sutton [28]; Charles and Marshall [26]; Ismail and Ahmed [16]; Worthley et al. [88]; Huddleston and Good [27]; Elizur et al. [37]; Cukic [90]
Feeling of being involved or in on things	Mihailovic [5]; Karl and Sutton [28]; Charles and Marshall [26]; Fisher and Yuan [32]; Chitiris [17]; Silverthorne [39]; Elizur et al. [37]
Criticism	Kovach [10]; Wiley [12]; Karl and Sutton [28]; Charles and Marshall [26]; Silverthorne [39]; Fisher and Yuan [32]; Chitiris [17]; Wright and Kim [89]
Interesting work	Kujovic [21]; Kovach [10]; Harpaz [73]; Wiley [12]; Yang [13]; Islam [75]; Mijacika [22]; Brislin et al. [86]; Karl and Sutton [28]; Charles and Marshall [26]; Ismail and Ahmed [16]; Manolopoulos [18]; Worthley et al. [88]; Wright and Kim [89]; Elizur et al. [37]; Fisher and Yuan [32]; Chitiris [17]; Silverthorne [39]; Cukic [90]

Table 1. *Cont.*

Motivation Factors	Authors
Promotion	Kovach [10]; Herzberg et al. [87]; Mihailovic [5]; Wiley [12]; Yang [13]; Islam [75]; Mijacika [22]; Brislin et al. [86]; Karl and Sutton [28]; Charles and Marshall [26]; Ismail and Ahmed [16]; Manolopoulos [18]; Worthley et al. [88]; Wright and Kim [89]; Linc [29]; Huddleston and Good [27]; Chitiris [17]; Cukic [90]
Possibility of development and improvement	Kujovic [21]; Harpaz [73]; Islam [75]; Karl and Sutton [28]; Ismail and Ahmed [16]; Manolopoulos [18]; Worthley et al. [88]; Linc [29]; Huddleston and Good [27]; Silverthorne [39]; Fisher and Yuan [32]; Elizur et al. [37]
Job security	Kovach [10]; Harpaz [73]; Wiley [12]; Yang [13]; Islam [75]; Brislin et al. [86]; Karl and Sutton [28]; Charles and Marshall [26]; Ismail and Ahmed [16]; Manolopoulos [18]; Worthley et al. [88]; Linc [29]; Elizur et al. [37]; Silverthorne [39]; Fisher and Yuan [32]; Chitiris [17]
Good interpersonal relationships	Kujovic [21]; Kovach [10]; Harpaz [73]; Mihailovic [5]; Wiley [12]; Yang [13]; Islam [75]; Mijacika [22]; Brislin et al. [86]; Karl and Sutton [28]; Charles and Marshall [26]; Ismail and Ahmed [16]; Manolopoulos [18]; Worthley et al. [88]; Linc [29]; Huddleston and Good [27]; Elizur et al. [37]; Fisher and Yuan [32]; Silverthorne [39]; Chitiris [17]; Cukic [90]
High wages	Kujovic [21]; Kovach [10]; Harpaz [73]; Mihailovic [5]; Wiley [12]; Yang [13]; Islam [75]; Mijacika [22]; Brislin et al. [86]; Karl and Sutton [28]; Charles and Marshall [26]; Ismail and Ahmed [16]; Manolopoulos [18]; Worthley et al. [88]; Linc [29]; Huddleston and Good [27]; Elizur et al. [37]; Chitiris [17]; Fisher and Yuan [32]; Silverthorne [39]; Cukic [90]
Challenging job and sense of achievement	Mihailovic [5]; Herzberg et al. [87]; Harpaz [73]; Yang [13]; Islam [75]; Brislin et al. [86]; Karl and Sutton [28]; Charles and Marshall [26]; Ismail and Ahmed [16]; Manolopoulos [18]; Worthley et al. [88]; Linc [29]; Huddleston and Good [27]; Elizur et al. [37]; Cukic [90]
Decision-making influence	Mihailovic [5]; Mijacika [22]; Worthley et al. [88]; Wright and Kim [89]
Work/job recognition	Mihailovic [5]; Herzberg et al. [87]; Mijacika [22]; Brislin et al. [86]; Worthley et al. [88]; Wright and Kim [89]; Elizur et al. [37]

The list of factors is adjusted to the environment and time, whereby factors with ideological connotation (self-management), those that are not current in the current time frame (housing), and factors that are not in widespread use in Serbia, especially in government services and the public sector (profit sharing and various types of benefits) were excluded.

Based on the analyzed literature, we set three hypotheses:

Hypothesis 1. *There are significant differences in the influence of motivational factors on employees of different demographic characteristics (gender, age, education, marital status, position).*

Hypothesis 2. *Money is not an important motivational factor for persons employed in public administration.*

Hypothesis 3. *There are significant differences in the influence of motivational factors on employees in the state administration in relation to other employees in the Republic of Serbia.*

3. Materials and Methods

Motivation research was realized in five steps and it was designed as a sustainable, continuous model. The model is shown in Figure 1.

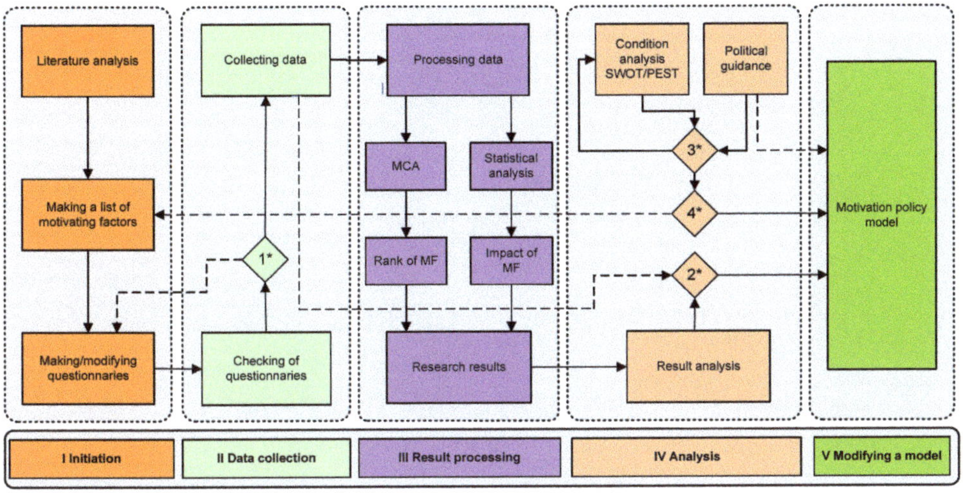

Figure 1. The concept of developing a framework for motivation research.

3.1. Initiation

The initiation stage involved the analysis of literature and conducted research in order to make a list of important motivating factors while adapting to cultural and socio-economic conditions. In this step, the relevant literature was analyzed—realized research in the world [10,12,13,16–18,26–29,32,37,39,73,75,86–89] and also previous research in Serbia and surrounding and neighboring countries [5,21,22,90]. The research questionnaire was made by entering the mentioned 16 factors whose influence on motivation the respondents evaluated using a five-point Likert scale (offered answers about the degree of influence: no influence, little, moderately, considerably, greatly). The questionnaire also has a section related to the ranking of motivational factors (after assessing the impact of 16 factors on motivation, respondents ranked only five factors that have the highest impact on motivation, assigning them ranks from 1 to 5) and factors that negatively affect motivation. There is also the part that collects data on the characteristics of the respondents.

3.2. Data Collection

Data collection was realized in two steps, by conducting a preliminary study on a suitable sample (135 respondents), and then by collecting data on a suitable sample of 2128 persons (1576 employees in the security sector of state administration and a control sample of 552 respondents—employees in the Republic of Serbia; detailed sample structure is shown in Table 2). Based on the pre-research, the obtained results (step 1*: if $\alpha < 0.8$ or response rate <90%; then data collection; else refinement of the questionnaire) and usability of the questionnaire (internal reliability of the scale, number/percentage of unusable/incomplete questionnaires, observed uncertainties that respondents had) were assessed and, if necessary, the instruments were refined, and then the research was realized.

Table 2. Demographic variables of the respondents.

Variable	Category	Number		Percentage	
		N1	N2	N1	N2
Gender	Male	1308	188	83.0	35.9
	Female	267	335	17.0	64.1
Age	<25	139	27	8.8	5.2
	26–35	428	143	27.2	27.6
	36–45	546	190	34.6	36.6
	>46	463	159	29.4	30.6
Education	Elementary	6	3	0.4	0.6
	High school	940	125	59.6	23.9
	College	417	226	26.5	43.3
	MA	142	150	9.0	28.7
	PhD	71	18	4.5	3.5
Marital status	Married	1093	332	69.4	63.5
	Single	483	191	30.6	36.5
Position	Manager	345	117	21.9	22.4
	Non-manager	1231	405	78.1	77.6

Note: N = 2128 (N1 = 1576, N2 = 552, missing 29–33).

3.3. Result Processing

Procedures and techniques of statistical processing and multicriteria analysis were applied. Statistical processing of the collected data was performed by using the IBM SPSS Statistic 23 program (descriptive statistics, analysis of variance, normality tests, *t*-test, and scale reliability). Significance of differences in attitudes about the influence of motivational factors on motivation was determined by analysis of variance and subsequent tests (post hoc tests—Tukey HSD) for multiple comparisons (for categorical variables with three or more subpopulations) and the application of two-tailed *t*-test, for independent samples (for categorical variables with two subpopulations).

To estimate the effect size, we used the eta square (η^2) using Equation (1). Eta square represents the proportion of variance in the variable explained by the independent variable and ranges between 0 and 1, with the influence being small ($\eta^2 = 0.01$), medium ($\eta^2 = 0.06$), and large ($\eta^2 = 0.138$) [91] (p. 24),

$$\eta^2 = \frac{t^2}{t^2 + (N1 + N2 - 2)} \tag{1}$$

N1 and N2 are the sizes of groups, and t is the value of two-tailed *t*-test.

The ranking of motivational factors was realized with the method of arithmetic averaging [92] by applying the Equations (2) and (3), then by normalizing the sum of the ranks and with conversion to aggregated weights,

$$W_j = \frac{\sum_{i=1}^{m} w_{i,j}}{m} \; J = 1, \ldots, n \tag{2}$$

$$W_{i,j} = 100 - s_n\left(r_{i,j} - 1\right) i = 1, \ldots, m, j = 1, \ldots, n \tag{3}$$

3.4. Analysis

Analysis of results and determination of differences in subpopulations and analysis of changes in the environment was done with SWOT/PEST analysis.

Based on the research results (step 2*), policy changes can be formulated in the motivation model. In the following research, if there were no significant differences in relation to the results of the previous research, no corrections of the motivation model would be made and the research would be realized again in the next period of motivation monitoring from step II. In case of significant differences, the necessary changes in motivation policies should be formulated in order to adjust the model.

If there were no significant changes in the results (step 3*) of the environment analysis or conditions that affect the functioning of the organization by using appropriate analytical techniques—SWOT, PEST, etc. (for more on how to implement SWOT/PEST analysis, see references [93–98])—the analysis would be evaluated in the next assessment period, and if there were changes, the next step would be taken (step 4*). Political guidelines in case of rapid changes have a direct impact on changes in motivation models, and in case of regular situations and minor changes, they are taken into account in step 3* and have an impact on the analysis of the situation on changes in the motivation model or on changes in the list of factors for examining motivation.

Significant and pronounced changes (if they exist in step 4*) require direct changes in the motivation model (also in the instruments) and changes of smaller scopes and significance require changes in the data collection instruments (forming new lists of factors).

3.5. Modifying a Model

Based on the previous steps and the obtained results, changes can be realized and the motivation model upgraded by changes and improvements in motivation policies (changes in the amount and type of incentives, changes in the organization of work, changes in the structure of the organization, etc.).

4. Results

The study was conducted in the period 2018–2020. A list of 16 factors was made by analyzing the literature: amount of responsibility, prizes and awards, working time, good working conditions, quality of supervision and leadership, feeling of being involved or in on things, criticism, interesting work, promotion, possibility of development and improvement, job security, good interpersonal relationships, high wages, challenging job and sense of achievement, decision-making influence, and work/job recognition (Table 1, Table A7).

The pre-research which checked the scale-questionnaire in which the mentioned factors were entered was realized in October 2018 on a suitable sample of 135 employees in the state administration. It was determined that the questionnaire is clear to all respondents, and the internal consistency (Cronbach's alpha) is at the required level, $\alpha = 0.877$.

The research was realized during 2020 on a suitable sample of 1576 employees in the state administration. The sample included 7% of all employees in the security sector of state administration. In relation to the structure of employees, the sample has minimal deviations, also in relation to age categories. Deviation in relation to gender is5% (the number of female respondents in the sample is higher than their representation in the population). The research was conducted in 8 cities, in 20 different locations, and in 27 different organizational units. The survey at all locations was conducted by the same researcher, thus excluding the negative impact or approach of the interviewers on the survey results. The detailed structure by gender, age category, education, marital status, and role in the organization is shown in Table 2 (N1). The survey of a control sample with 552 people was conducted through the internet platform soscisurvey.de; respondents accessed the platform via a link which was distributed by the "snowball" method (Table 2, N2).

After processing the data, the values of descriptive statistics for the sample were determined (Table 3), as well as the rank for 16 factors. The metric characteristics of the scale—internal consistency (Cronbach's alpha)—were checked and are at a high level, $\alpha N1 = 0.838$ ($\alpha N2 = 0.895$ for respondents who filled out the survey via the internet platform), which is fully within the recommendations, since Nunnally recommends a minimum level of 0.8 and 0.9 as excellent for basic and applied research [99].

Table 3. Descriptive statistics and rank motivators.

Motivators	Rank	Min	Max	M	SD
Good interpersonal relationships	1	1	5	4.16	0.908
Challenging job and sense of achievement	2	1	5	3.94	0.953
Amount of responsibility	3	1	5	3.92	0.899
High wages	4	1	5	3.88	1.100
Quality of supervision and leadership	5	1	5	3.83	0.938
Working time	6	1	5	3.82	1.109
Job security	7	1	5	3.70	1.232
Interesting work	8	1	5	3.66	0.913
Good working condition	9	1	5	3.61	1.039
Decision-making influence	10	1	5	3.49	0.945
Feeling of being involved or in on things	11	1	5	3.48	0.926
Promotion	12	1	5	3.48	1.286
Prizes and awards	13	1	5	3.43	1.084
Possibility of development and improvement	14	1	5	3.39	1.183
Work/job recognition	15	1	5	3.35	0.983
Criticism	16	1	5	2.78	1.014
∑ *Motivation*		1.94	5.00	3.62	0.561

Note: N = 1576.

Using one-way analysis of variance in the SPSS program (one-way ANOVA procedure), the results of independent categorical variables with three or more subpopulations were compared: age and education. Significant differences in arithmetic mean values in subpopulations were determined, after which tests of subsequent comparisons were applied (post hoc—Tukey HSD).

Tests confirm statistically significant differences for different age groups in the assessment of the impact on motivation for nine factors, with seven factors having a greater impact on older people, and, as expected, two developmental factors have a greater impact on younger employees (promotion, possibility of development and improvement). The most significant differences were found between groups younger than 25 years old and older than 45 years old for amount of responsibility ($\eta^2 = 0.041$) and decision-making influence ($\eta^2 = 0.050$) and between groups from 26 to 35 years old and older than 45 years old for amount of responsibility ($\eta^2 = 0.047$) (Table 4).

Table 4. Age—multiple comparisons.

Motivators	(I)	(J)	Mean Diff. (I–J)	Std. Error	Sig.	η^2	95% CI	
							LL	UL
Interesting work	<25	36–45	−0.273 *	0.086	0.008	0.013	−0.49	−0.05
		>45	−0.391 *	0.088	0.000	0.028	−0.62	−0.17
	26–35	>45	−0.259 *	0.061	0.000	0.019	−0.41	−0.10
Quality of supervision and leadership	<25	36–45	−0.276 *	0.089	0.010	0.014	−0.50	−0.05
		>45	−0.364 *	0.090	0.000	0.026	−0.60	−0.13
	26–35	>45	−0.183 *	0.063	0.019	0.009	−0.34	−0.02
Amount of responsibility	<25	36–45	−0.249 *	0.084	0.016	0.013	−0.47	−0.03
		>45	−0.415 *	0.086	0.000	0.041	−0.64	−0.20
	26–35	36–45	−0.235 *	0.057	0.000	0.016	−0.38	−0.09
		>45	−0.401 *	0.059	0.000	0.047	−0.55	−0.25
	36–45	>45	−0.166 *	0.056	0.016	0.009	−0.31	−0.02
Decision-making influence	<25	26–35	−0.348 *	0.091	0.001	0.022	−0.58	−0.11
		36–45	−0.443 *	0.089	0.000	0.034	−0.67	−0.21
		>45	−0.567 *	0.090	0.000	0.050	−0.80	−0.33
	26–35	>45	−0.218 *	0.063	0.003	0.014	−0.38	−0.06

Table 4. *Cont.*

Motivators	(I)	(J)	Mean Diff. (I–J)	Std. Error	Sig.	η^2	95% CI	
							LL	UL
Feeling of being involved or in on things	>45	<25	0.238 *	0.089	0.039	0.012	0.01	0.47
		26–35	0.210 *	0.062	0.004	0.013	0.05	0.37
Challenging job and sense of achievement	26–35	>45	−0.165 *	0.064	0.048	0.008	−0.33	0.00
Promotion	>45	26–35	−0.309 *	0.086	0.002	0.015	−0.53	−0.09
		36–45	−0.229 *	0.081	0.024	0.008	−0.44	−0.02
Job security	>45	<25	0.379 *	0.119	0.008	0.014	0.07	0.68
		26–35	0.225 *	0.082	0.033	0.008	0.01	0.44
Possibility of development and improvement	>45	<25	−0.460 *	0.114	0.000	0.026	−0.75	−0.17
		26–35	−0.347 *	0.079	0.000	0.021	−0.55	−0.14
		36–45	−0.206 *	0.074	0.029	0.008	−0.40	−0.02

Note: * Difference significant at 0.05.

Numerous differences have been identified in relation to the education level. As the level of education increases, the influence of motivational factors grows (Table 5). Differences in the influence of motivational factors on respondents of different levels of education were found for all analyzed factors except for criticism. The greater the difference in the education level, the greater the difference in the influence of motivational factors. Differences for the group of respondents whose highest level of education was primary school (E) are not shown because the group has only six respondents. Compared to persons with secondary education (high school diploma—H), higher impact on motivation has nine factors on persons with a bachelor's degree (college diploma—C) and with a master's degree (MA), while 12 factors have higher impact son persons with a doctoral degree (PhD). There were also differences in the motivation factors' impacts on persons with a university degree (C) in relation to persons with a master's degree (MA) for three factors and in relation to persons with a doctorate (PhD) for 12 factors. Differences in the impact on persons with a master's degree and a doctorate were determined for four factors. The effect sizes for the identified differences (η^2) range from small to large. The most significant differences were found between the groups with high school (H) and PhD (PhD) educations for the factors quality of supervision and leadership ($\eta^2 = 0.110$), decision-making influence ($\eta^2 = 0.134$), and promotion ($\eta^2 = 0.108$). The group with a bachelor's(C) vs. a doctorate (PhD) differed for the factors quality of supervision and leadership ($\eta^2 = 0.114$) and decision-making influence ($\eta^2 = 0.118$), as well as the groups with master's degree (MA) and doctorate (PhD) for the factors decision-making influence ($\eta^2 = 0.108$) and feeling of being involved or in on things ($\eta^2 = 0.104$).

Table 5. Education level—multiple comparisons.

Motivators	(I)	(J)	Mean Diff. (I–J)	Std. Error	Sig.	η^2	95% CI	
							LL	UL
Interesting work	H	C	−0.222 *	0.052	0.000	0.012	−0.37	−0.08
		MA	−0.457 *	0.080	0.000	0.035	−0.68	−0.24
		PhD	−0.753 *	0.110	0.000	0.074	−1.05	−0.45
	C	PhD	−0.531 *	0.115	0.000	0.069	−0.84	−0.22
Quality of supervisionand leadership	H	C	−0.230 *	0.053	0.000	0.013	−0.38	−0.08
		MA	−0.530 *	0.082	0.000	0.050	−0.75	−0.31
		PhD	−0.945 *	0.112	0.000	0.110	−1.25	−0.64
	C	MA	−0.300 *	0.088	0.006	0.027	−0.54	−0.06
		PhD	−0.716 *	0.117	0.000	0.114	−1.03	−0.40
	MA	PhD	−0.415 *	0.132	0.014	0.069	−0.78	−0.06
Amount of responsibility	H	PhD	−0.339 *	0.110	0.018	0.016	−0.64	−0.04

Table 5. *Cont.*

Motivators	(I)	(J)	Mean Diff. (I–J)	Std. Error	Sig.	η^2	95% CI	
							LL	UL
Good interpersonal relationships	H	C	−0.169 *	0.053	0.012	0.007	−0.31	−0.03
		MA	−0.294 *	0.081	0.003	0.011	−0.51	−0.07
		PhD	−0.632 *	0.110	0.000	0.076	−0.93	−0.33
	C	PhD	−0.462 *	0.115	0.001	0.072	−0.78	−0.15
Decision-making influence	H	C	−0.291 *	0.054	0.000	0.020	−0.44	−0.14
		MA	−0.428 *	0.083	0.000	0.030	−0.65	−0.20
		PhD	−0.900 *	0.113	0.000	0.134	−1.21	−0.59
	C	PhD	−0.609 *	0.118	0.000	0.118	−0.93	−0.29
	MA	PhD	−0.472 *	0.133	0.004	0.108	−0.84	−0.11
Feeling of being involvedor in on things	H	PhD	−0.680 *	0.113	0.000	0.057	−0.99	−0.37
	C	PhD	−0.647 *	0.118	0.000	0.094	−0.97	−0.33
	MA	PhD	−0.556 *	0.133	0.000	0.104	−0.92	−0.19
Good working condition	H	PhD	−0.572 *	0.127	0.000	0.029	−0.92	−0.23
	C	PhD	−0.456 *	0.132	0.005	0.035	−0.82	−0.09
Challenging job and sense of achievement	H	C	−0.248 *	0.055	0.000	0.015	−0.40	−0.10
		MA	−0.333 *	0.084	0.001	0.014	−0.56	−0.10
		PhD	−0.798 *	0.115	0.000	0.085	−1.11	−0.48
	C	PhD	−0.550 *	0.120	0.000	0.074	−0.88	−0.22
	MA	PhD	−0.465 *	0.136	0.006	0.081	−0.84	−0.09
Promotion	H	C	−0.347 *	0.074	0.000	0.016	−0.55	−0.15
		MA	−0.638 *	0.113	0.000	0.036	−0.95	−0.33
		PhD	−1.082 *	0.154	0.000	0.108	−1.50	−0.66
	C	PhD	−0.736 *	0.161	0.000	0.090	−1.18	−0.30
Job security	H	PhD	−0.441 *	0.151	0.030	0.015	−0.85	−0.03
Possibility of development and improvement	H	C	−0.371 *	0.103	0.000	0.021	−0.56	−0.19
		MA	−0.751 *	0.141	0.000	0.054	−1.03	−0.47
		PhD	−0.906 *	0.111	0.006	0.073	−1.29	−0.52
	C	MA	−0.380 *	0.147	0.003	0.024	−0.68	−0.08
		PhD	−0.535 *	0.067	0.000	0.046	−0.94	−0.13
Prizes and awards	C	PhD	−0.449 *	0.139	0.004	0.023	−0.83	−0.07
High wages	PhD	H	0.452 *	0.135	0.004	0.017	0.08	0.82
		C	0.426 *	0.141	0.004	0.027	0.04	0.81
Work/job recognition	H	C	−0.199 *	0.057	0.005	0.009	−0.35	−0.04
		MA	−0.291 *	0.087	0.008	0.010	−0.53	−0.05
		PhD	−0.671 *	0.119	0.000	0.052	−1.00	−0.34
	C	PhD	−0.472 *	0.125	0.001	0.046	−0.81	−0.13
Working time	H	C	−0.295 *	0.064	0.000	0.015	−0.47	−0.12
		MA	−0.530 *	0.099	0.000	0.036	−0.80	−0.26

Note: E—elementary school, H—high school, C—college, MA—master's degree, PhD—doctorate of social or technical sciences. * Difference significant at 0.05.

Differences in relation to the variable position were determined for 12 factors, with a greater impact on the manager than on the non-manager positions, and the differences in the impact are small. For the factor decision-making influence, the magnitude of the impact (η^2 = 0.033) is close to middle levels (Table 6).

Table 6. Position (*t*-test).

Motivators	Manager		Non-Manager		t	P	η^2
	M	SD	M	SD			
Interesting work	3.87	0.896	3.60	0.909	4.909	0.000	0.015
Quality of supervision and leadersh.	4.08	0.913	3.76	0.932	5.811	0.000	0.021
Good interpersonal relationships	4.40	0.721	4.09	0.943	6.597	0.000	0.027
Decision-making influence	3.80	0.867	3.41	0.949	7.339	0.000	0.033
Good working condition	3.71	0.980	3.58	1.053	2.111	0.035	0.003
Challenging job and sense of achiev.	4.12	0.910	3.89	0.958	4.064	0.000	0.010
Promotion	3.84	1.144	3.38	1.306	6.432	0.000	0.026
Job security	3.91	1.012	3.64	1.282	4.142	0.000	0.011
Possibility of development and impr.	3.70	1.087	3.31	1.195	5.757	0.000	0.021
High wages	4.10	0.950	3.82	1.131	4.679	0.000	0.014
Work/job recognition	3.50	1.000	3.31	0.975	3.076	0.002	0.006
Working time	4.05	0.991	3.76	1.132	4.606	0.000	0.013

Note: Manager = 345, non-manager = 1231, df = 1574, Difference significant at 0.05.

Statistically significant differences in relation to the variable gender were found for six factors, to men, the more important factor was good interpersonal relationships while women found feeling of being involved or in on things, good working condition, job security, possibility of development and improvement, and prizes and awards significantly more important. Differences in impact are minimal (Table 7).

Table 7. Gender (*t*-test).

Motivators	Male		Female		t	P	η^2
	M	SD	M	SD			
Good interpersonal relationships	4.18	0.881	4.03	0.937	2.206	0.028	0.003
Feeling of being involved or in on things	3.45	0.932	3.63	0.881	−2.822	0.005	0.005
Good working condition	3.59	1.043	3.74	1.011	−2.162	0.031	0.003
Job security	3.67	1.239	3.86	1.189	−2.441	0.015	0.004
Possibility of development and impr.	3.23	1.169	3.54	1.242	−2.153	0.031	0.003
Prizes and awards	3.40	1.096	3.56	1.015	2.223	0.027	0.003

Note: Male = 1308, female = 267, df = 1573, Difference significant at 0.05.

Differences in relation to the variable marital status were determined for eight factors. Married people had greater influence from interesting work, quality of supervision and leadership, amount of responsibility, decision-making influence, job security, and high wages. Out-of-wedlock people were more influenced by factors related to development and promotion (promotion, possibility of development and improvement), and the differences in influence are small (Table 8).

Table 8. Marital status (*t*-test).

Motivators	Married		Single		t	P	η^2
	M	SD	M	SD			
Interesting work	3.71	0.897	3.56	0.941	2.877	0.004	0.005
Quality of supervision and leadersh.	3.88	0.939	3.70	0.924	3.605	0.000	0.008
Amount of responsibility	3.99	0.870	3.78	0.948	4.247	0.000	0.011
Decision-making influence	3.54	0.916	3.38	1.000	3.213	0.001	0.007
Promotion	3.44	1.322	3.57	1.197	−1.971	0.049	0.002
Job security	3.74	1.203	3.59	1.292	2.207	0.028	0.003
Possibility of development and impr.	3.34	1.179	3.52	1.185	−2.809	0.005	0.005
High wages	3.92	1.004	3.77	1.034	2.440	0.015	0.004

Note: Married = 1093, single = 483, df = 1574, Difference significant at 0.05.

Applying Equations (2) and (3) based on the ranking of factors (respondents after assessing the impact of 16 factors on motivation ranked only 5 factors that have the highest impact on motivation), the rank of factors was obtained and shown by subpopulations (Table A2), also normalized values are shown (Table A3 and Figure A1).

The ranking results show that the motivating factor with the greatest influence is high wages (rank 1; 0.139);in the second position is the factor good interpersonal relationships (0.128).The difference between them and other factors is pronounced (job security, rank 3—0.087; working time, rank 4—0.082; and quality of supervision and leadership, rank 5—0.079). It is evident that the differences in values for other motivational factors ranked from 6–14 positions are not large (in the range 0.068 to 0.030), and that, in the end, there are feeling of being involved or in on things (rank 15; 0.017) and criticism (rank 15; 0.011) whose values stand out as significantly lower than other motivational factors.

The most pronounced deviations from the established ranks are for job security (3), which is ranked by women in the 1st position (0.119) and men in5th (0.082). Managers rank this factor in the 8th position (0.067), and non-managers in the 3rd (0.094). The importance of this factor decreases with the growth of education, so persons with a master's degree ranked it in the 9th position (0.058), and persons with a doctorate in the 12th position (0.035). Quality of supervision and leadership are ranked lower by unmarried people and people under 25 than by married and people from older age groups. The possibility of development and improvement is much more important for younger people (0.091 for people under 25) than for older people (0.026—older than 45). There is also a difference in the significance of achievement (for persons with a high school degree 0.049 and for PhD −0.103).

The rank of the factors that negatively affect motivation was also determined by applying Equations (2) and (3) based on the ranking of the offered five factors (respondents ranked the negative influence on the motivation of five factors). The ranking results confirm the importance of salary because the respondents ranked low wages as the most significant demotivating factor (the most significant factor in 12 out of 15 subpopulations). The biggest differences were found in the group of managers and non-managers. The ranking of factors that negatively affect motivation is shown in tables and graphs by subpopulations (Tables A4 and A5 and Figure A2).

The existence of specificity in the influence of motivational factors of employees in the state administration organizational unit from the security sector, in relation to the general population control sample, was determined by *t*-test of aggregate independent samples (Table 9). Differences in the influences of motivational factors were determined for 15 factors (for all examined factors except high wages), and the effect sizes of the influences range from small to medium. The most significant differences were found for the factors interesting work ($\eta^2 = 0.035$), feeling of being involved or in on things ($\eta^2 = 0.054$), and good working conditions ($\eta^2 = 0.040$).

It was found that the factors interesting work ($\eta^2 = 0.055$) and good working condition ($\eta^2 = 0.030$) were more significant to women in the control sample than to women employed in the surveyed state administration organization, and there were differences in influence in relation to the education level in the manner that feeling of being involved or in on things ($\eta^2 = 0.067$) and good working condition ($\eta^2 = 0.061$) had a greater influence on persons in the control sample with a university degree. Interesting work ($\eta^2 = 0.052$) also had a greater influence on persons with a master's degree in the control sample. The most significant differences were found for persons of PhD level—for the factors quality of supervision and leadership ($\eta^2 = 0.182$) and decision-making influence ($\eta^2 = 0.115$) and the impact was greater on persons employed in public administration (Table A6).

Table 9. Differences in the influences of motivational factors on the sample and the control sample (*t*-test).

Motivators	N2		N1		M Diff.	t	Sig	η^2
	M	SD	M	SD				
Interesting work	4.06	0.933	3.66	0.913	0.40	8.717	0.000	0.035
Quality of supervision and leaders.	3.97	1.016	3.83	0.938	0.14	2.841	0.005	0.004
Amount of responsibility	4.09	0.861	3.92	0.899	0.17	3.946	0.000	0.007
Good interpersonal relationships	4.26	0.877	4.16	0.908	0.10	2.284	0.025	0.002
Decision-making influence	3.78	0.914	3.49	0.945	0.29	6.359	0.000	0.019
Feeling of being involved or in on th.	3.96	0.865	3.48	0.926	0.48	11.01	0.000	0.054
Good working conditions	4.06	0.939	3.61	1.039	0.45	9.419	0.000	0.040
Challenging job and sense of achiev.	4.20	0.902	3.94	0.953	0.26	5.742	0.000	0.015
Promotion	3.83	1.117	3.48	1.286	0.35	5.678	0.000	0.015
Job security	3.99	1.103	3.70	1.232	0.29	4.887	0.000	0.011
Possibility of development and imp.	3.66	1.134	3.39	1.183	0.27	4.664	0.000	0.010
Prizes and awards	3.69	1.168	3.43	1.048	0.26	4.751	0.000	0.011
High wages	3.86	1.026	3.83	0.938	0.02	0.374 (not significant)		
Criticism	2.88	1.021	2.78	1.014	0.10	1.984	0.048	0.002
Work/job recognition	3.46	0.955	3.35	0.983	0.12	2.311	0.021	0.002
Working time	3.95	1.056	3.82	1.109	0.13	2.399	0.017	0.003

Note: N_2—control sample = 552, N_1—state administration = 1576, df = 2126, Difference significant at 0.05.

5. Discussions

The research determined the hierarchy of motivational factors for the work of employees in the security sector of state administration (Table 3, Table A2, Table A3, Figure A1), and established the existence of differences in the influence of motivational factors on different subpopulations of the sample.

The obtained results indicate the importance of money and that employees' incomes are not at the required level and are not in line with the basic principles of the motivational theory of the hierarchy of needs [76], according to which needs are hierarchically organized, and unmet needs of lower levels dominate influencing motivation. On the other hand, the high rank of good interpersonal relationships indicates that this factor is internalized at the level of autonomous motivation—in accordance with the theory of self-determination [48,100,101].

The following had greater impacts on older employees: interesting work, quality of supervision and leadership, amount of responsibility, decision-making influence, feeling of being involved or in on things, job security, and challenging job and sense of achievement. On younger employees, higher influence was exerted by the factors related to promotion and development: promotion and possibility of development and improvement. The effect size of the determined differences is mostly small and ranges up to the limits of the middle level (η^2 = 0.008 to 0.050). Research in the United States had similar results [10], where the importance of interesting work increases with age, and the importance of advancement and personal development decreases with age. In the Scandinavian countries, motivation and intrinsic motivation increase with age [20], and the results of research in several countries [73] find that younger people are more motivated with the possibility for personal development and improvement. In Russia, the opportunity for the development of abilities and skills and advancement is more important for younger people, while for older people, job security and colleagues' respect are more important [29]. In Japan, job security has also been found to be more important to older workers [88]. In Greece, as the manager's age increases, so does his motivation [17].

The influence of motivational factors grows with the increase of education level for all factors, and the magnitude of the influence of the determined differences is the greatest for factors decision-making influence and quality of supervision and leadership. As the level of education increases, the influence of motivational factors also increases, nine factors (interesting work, quality of supervision and leadership, amount of responsibility, decision-making influence, challenging job and sense of achievement, promotion, possibility of development and improvement, work/job recognition, working time) have a greater impact on persons with a university degree and master's degree (C and MA) compared to persons with high school only (H). Moreover, 12 factors (interesting work, quality of supervision and leadership, amount of responsibility, decision-making influence, feeling of being involved or in on things, good working condition, challenging job and sense of achievement, promotion, possibility of development and improvement, prices and awards, high wages, work/job recognition) have a greater impact on doctors of science (PhD) in relation to respondents with a bachelor's degree (C). Furthermore, there are also differences between the respondents with PhD degrees and Master of Science degrees, where four factors (quality of supervision and leadership, decision-making influence, feeling of being involved or in on things, challenging job and sense of achievement) have a greater influence on PhDs. The effect size is in the range from small to large (η^2 = 0.007 to 0.134). Similar results have been found in China [13], as well as in the Scandinavian countries [20] and Greece [18], where people with a higher level of education are more intrinsically motivated. In Russia, as well, college-educated employees value praise and recognition more than employees with high school only [102]. In Malaysia, also, working conditions are more important for people with a doctoral degree than for people with a university degree and a master's degree [75], as well as in Bosnia and Herzegovina [22], where people with a higher level of training are more influenced by a good managerial attitude.

The role in the organization (position) influence is different for 12 factors; managers are more motivated than non-managers by all 12 factors (interesting work, quality of supervision and leadership, good interpersonal relationship, decision-making influence, good working condition, challenging job and sense of achievement, promotion, possibility of development and improvement, high wages, work/job recognition, working time), and the size of the influence is the greatest for decision-making influence, and it is bigger for the managers in relation to the non-managers—the size of the influence is small (η^2 = 0.006 to 0.033). Differences in the influence of motivation factors for different levels have been found in several studies [10,14,73,75,103]. Additionally, non-managers in the public sector value the sense of control over work less [28], and employees in managerial positions are significantly more satisfied than other employees and have a higher level of intrinsic motivation than other employees [20].

The influence of motivational factors in relation to gender is different for six factors. Men are more motivated by good interpersonal relationships, while women were more motivated by all six factors (feeling of being involved or in on things, good working condition, job security, possibility of development and improvement, and prizes and awards). The differences in influence are very small (η^2 = 0.003 to 0.005). Similar results have been found in Denmark, where job security is more important for women in the public sector [72]. Recognition and good working conditions have also been found to be more important for women in the United States [12].

Differences in the influence of motivational factors were determined for eight factors in relation to family status. Married people show greater influence from interesting work, quality of supervision and leadership, amount of responsibility, decision-making influence, job security, and high wages, while single people are more motivated by factors related to development and advancement: promotion and possibility of development and improvement. The differences in impact are small (η^2 = 0.002 to 0.011). In Greece, in the public sector, married people have a higher level of extrinsic motivation, and unmarried people have a higher level of intrinsic motivation [18].

After the analysis of research results, the role of salary amount can be viewed in two ways, as a hygienic factor whose absence creates dissatisfaction [34] (low salary is the most important demotivating factor), but also as a significant extrinsic motivating factor (salary is in fourth place in relation to the influence on motivation).

The specifics of motivation for employees in the security sector of state administration and the justification for the development of a sustainable model of motivation policy for a specific organization was confirmed by establishing differences in the influence of motivational factors for members of the security sector and the control sample of random employees in the Republic of Serbia. The differences were determined for 15 factors (for all factors except high wages), and the magnitude of the impact is in the range of small to medium. Differences were also found in the subpopulations of the two samples, so that the factors interesting work and good working condition are more important to women in the Republic of Serbia than to women employed in the surveyed organization of state administration (security sector). Moreover, the feeling of being involved or in on things and good working condition have greater influence on people with a college education. Interesting work also has a greater impact on persons with a master's degree. The most significant differences were found for persons at the PhD level, for the factors quality of supervision and leadership and decision-making influence, and the impact is greater on persons employed in public administration. Differences in the size of the impact by subpopulations range from small to large ($\eta^2 = 0.002$ to 0.182). Similar results were obtained in Taiwan where employees in private corporations had greater needs for benefits than their counterparts in public organizations [54]. In the Czech Republic, employees working in public/governmental organizations are the least satisfied, and significant differences have been found for all aspects of work, except for the nature of work and communications [81]. In Italy, civil servants differ from private-sector employees in the way they assess the following factors: satisfaction with job security, peer relationships, safety, and health characteristics. Differences between civil servants and private sector employees in Italy do not exist in their assessment of the level of the factors of effort needed for the job and interesting job [80]. In the USA, information about their work is more important to workers in the private sector, but in contrast to our results, public sector workers are more interested in interesting work, while private-sector workers are more interested in their salaries [28].

Hypotheses 1 and 3 are supported by findings (significant differences in the influence of motivational factors were found depending on the demographic characteristics of the respondents, as well as significant differences in the influence of motivational factors on public administration employees and other employees in the Republic of Serbia). Contrary to expectations and hypothesis 2 —money is not a significant motivating factor for public administration employees—findings are not supported by the data used in this study (high wages ranked in the first position, and low wages as a demotivator are also in the first position, which indicates that money in public administration of the Republic of Serbia is a significant motivating factor).

6. Conclusions

The paper presents the results of research conducted in order to develop a sustainable model of motivation for the work of employees in the security sector of public (state) administration. The research identified a hierarchy of 16 factors that have an impact on work motivation: good interpersonal relationships, challenging job and sense of achievement, amount of responsibility, high wages, quality of supervision and leadership, working time, job security, interesting work, good working condition, decision-making influence, feeling of being involved or in on things, promotion, prizes and awards, possibility of development and improvement, work/job recognition, and criticism. The resulting order indicates that good relationships, nature of the profession, and the essence of the work have immense impacts on work motivation (1. Good interpersonal relationships, 2. Challenging job and sense of achievement, 3. Amount of responsibility, 8. Interesting work), as well as the characteristics of leaders (5. Quality of supervision and leadership). The smallest influences on work motivation were work/job recognition (15th) and criticism (16th). The pronounced impact of money as a motivator for work, which can be noticed experientially is not fully confirmed by the resulting hierarchy (4. High wages), however, the importance of the salary for the respondents came up during the ranking of demotivating factors, where the small salary was ranked as the most important demotivating factor.

The existence of differences in the influence of motivational factors for work of the analyzed sample in relation to the control sample was confirmed, as well as differences depending on the characteristics of employees: age, level of training, gender, role in organization, and marital status.

Specificity of the examined sample is reflected in the difference from the general population of employees in 15 factors, and the most significant differences were found for interesting work, feeling of being involved or in on things, and good working condition. Similar differences were found in research in Italy, the Czech Republic, and Taiwan, while, on the other hand, in the USA, different results were obtained in relation to an interesting job which is more important for public sector employees and the level of salaries which is more important for private-sector employees.

We also emphasize the results of ranking, which was realized simultaneously with the assessment of the influence of motivational factors, whereby the ranking method more precisely determines the differences in the hierarchy of motivational factors. It led us to the conclusion that in the security sector of public (state) administration, money is the most influential motivating factor for the employees, because high wages are in the first position ahead of good interpersonal relations.

The obtained results met the goal of the research; they provide a good basis for further research and methodological and practical improvements of data collection instruments. We believe that the results can be applied in improving motivation and developing a sustainable model of motivation policy in public administration, which should be based on increasing salaries, maintaining and developing good interpersonal relationships, and an adequate attitude of managers toward executors, as well as job enrichment.

We also believe that the presented model and organization of motivation research can have a wider use value; it can be applied in other countries and their public companies and state bodies, especially because the research was conducted on a large sample of employees who expressed their views based on perceptions and experience in specific jobs. Bear in mind that the use of results has minor limitations because the sample was not stratified by all attributes of the respondents, which partially reduces the prospects of generalization. Additionally, the influence of a limited number of factors was investigated.

We believe that the following research can be further improved by working on a stratified sample by all attributes and randomization of the order of factors, as well as by using an appropriate computer platform.

Author Contributions: Conceptualization, M.K. and S.B.; methodology, S.B.; validation, M.K., S.B., and B.K.; formal analysis, M.K.; investigation, M.K.; writing—original draft preparation, M.K.; writing—review and editing, S.B. and B.K.; project administration, B.K. All authors have read and agreed to the published version of the manuscript.

Funding: This research received no external funding.

Conflicts of Interest: The authors declare no conflict of interest.

Appendix A

Table A1. Dimensions of national cultures in the world in relation to Serbia.

Culture (State)	N	Min–Max	Serbia		Anglo-Saxon		German		Scandinavian		Latin European		Latin American	
Dimension			Index	Rank	GBR	USA	DEU	AUT	SWE	NOR	ITA	FRA	MEX	VEN
Power distance	76	11–104	86	8	35	40	35	11	31	31	50	68	81	81
Individualism	76	6–91	25	55–56	89	91	67	55	71	69	76	71	30	12
Masculinity	76	5–110	43	47–50	66	62	66	79	5	8	70	43	69	73
Uncertainty avoidance	76	8–112	92	11–13	35	46	65	70	29	50	75	86	82	76
Long-term orientation	93	0–100	52	38–39	51	26	83	60	53	35	61	63	24	16
Indulgence versus restraint	93	0–100	28	70–72	69	68	40	63	78	55	30	48	97	100

Note: GBR—Great Britain, DE—Germany, AU—Austria, SW—Sweden, NO—Norway, IT—Italy, FR—France, ME—Mexico, VE—Venezuela. Source: Adapted from Hofstede, G. (2010).

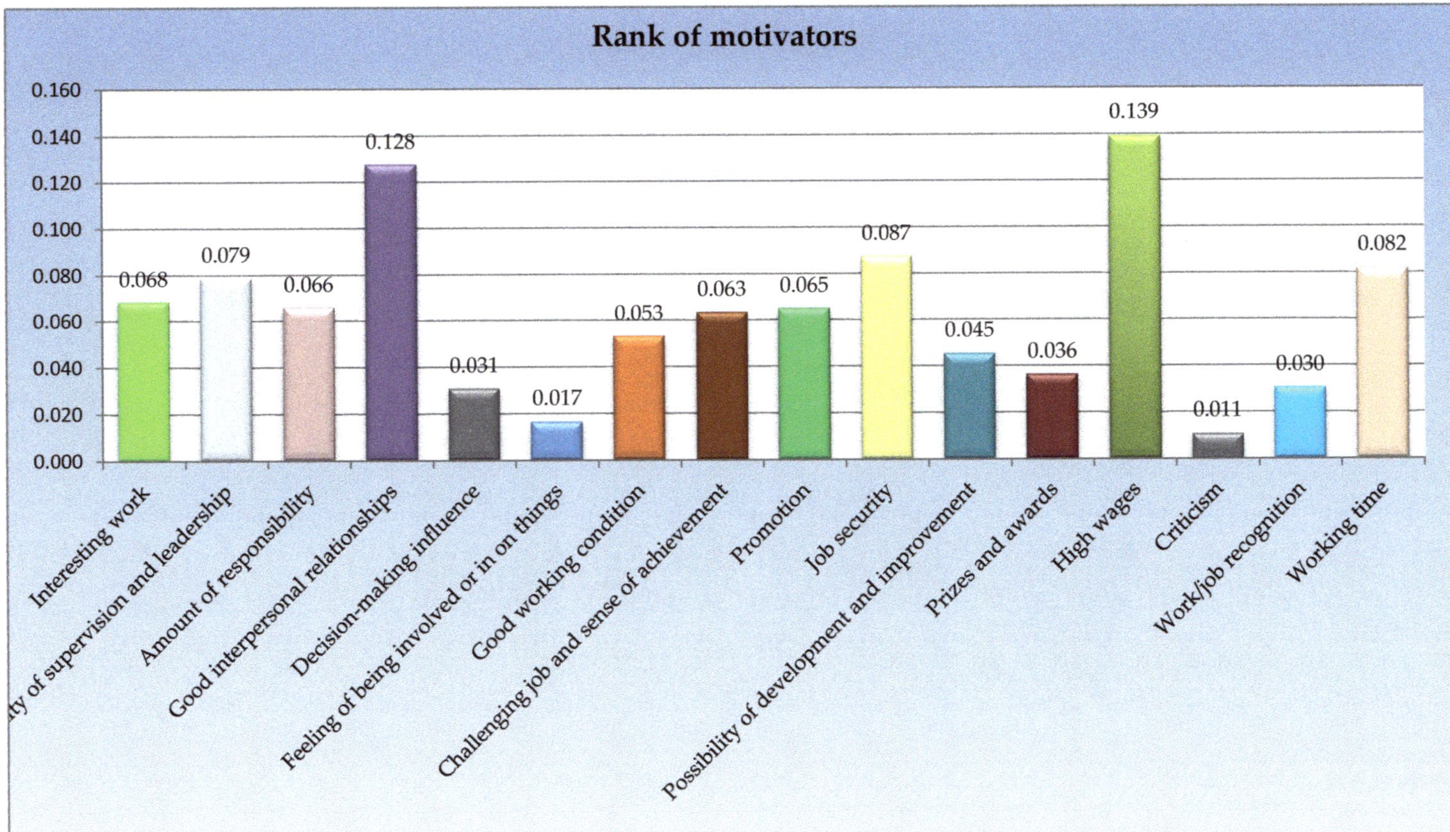

Figure A1. Normalized values of motivating factors—ranking result.

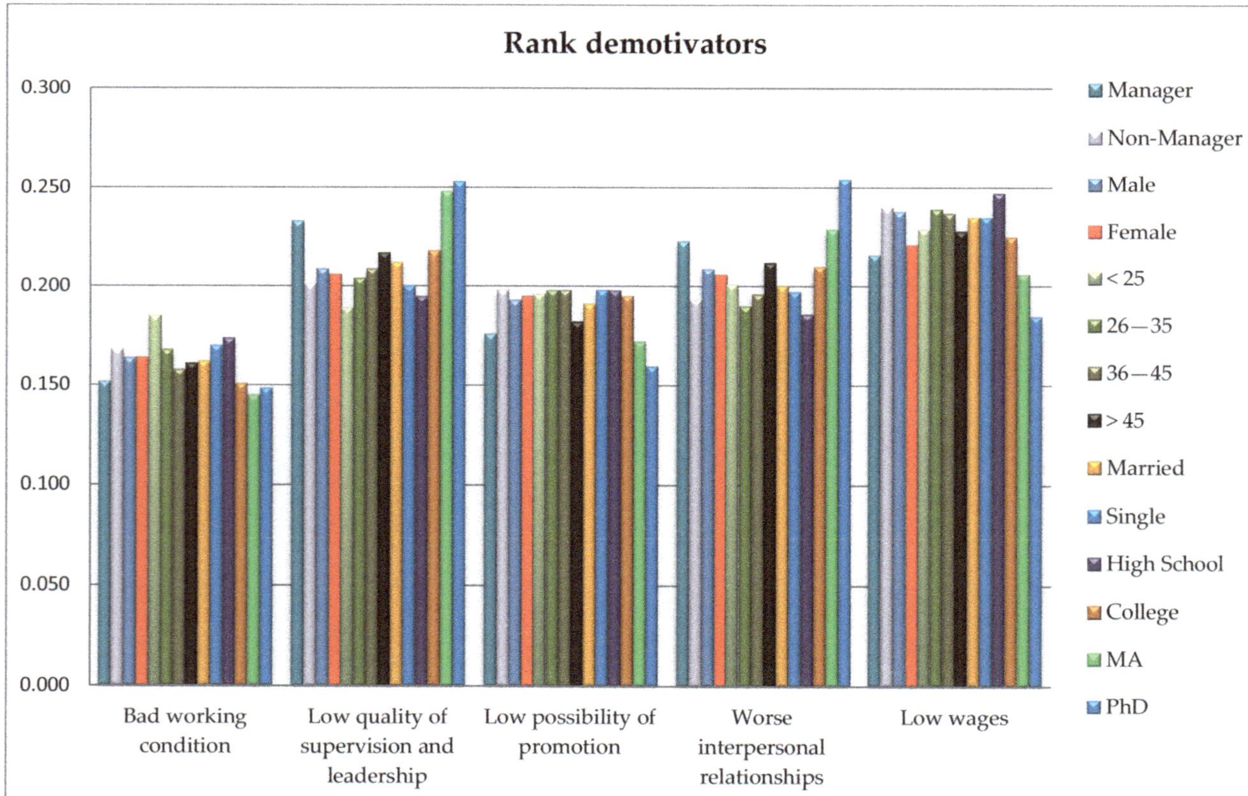

Figure A2. Rank of demotivators.

Table A2. Rank of motivators.

Motivators	Rank	Gender		Position		Age				Education Level			Mar. Stat.	
		M	F	Man	Nom	<25	26-35	36-45	>45	H	C	MA	PhD	Mar	Sin
High wages	1	1	2	2	1	2	1	1	2	1	1	2	5	1	1
Good interpersonal relationships	2	2	3	1	2	1	2	2	1	2	2	1	1	2	2
Job security	3	5	1	8	3	3	4	5	3	3	4	9	12	4	3
Working time	4	3	4	4	4	7	3	3	7	4	3	3	8	5	4
Quality of supervision and leadership	5	4	9	3	5	11	7	4	4	5	5	5	2	3	11
Interesting work	6	6	6	5	7	9	6	7	6	9	6	6	4	7	7
Amount of responsibility	7	7	10	9	6	8	10	8	5	6	9	10	7	6	10
Promotion	8	8	5	6	8	6	5	6	10	7	8	7	6	8	6
Challenging job and sense of achievement	9	9	7	7	9	5	8	10	8	10	7	4	3	9	5
Good working condition	10	10	11	11	10	12	11	9	9	8	11	12	10	10	9
Possibility of development and improvement	11	11	8	10	11	4	9	11	14	12	10	8	10	11	8
Prizes and awards	12	12	12	14	12	10	12	12	13	11	14	15	15	12	12
Decision-making influence	13	13	14	13	13	14	14	14	11	14	13	13	9	13	14
Work/job recognition	14	14	13	12	14	13	13	13	12	13	12	11	13	14	13
Feeling of being involved or in on things	15	15	15	15	15	16	15	15	15	15	15	14	14	15	15
Criticism	16	16	16	16	16	15	16	16	16	16	16	16	16	16	16

Note: N = 1576, M—male, F—female, Man—manager, Nom—Non-manager, H—High school, C—College, MA—master degree, PhD—doctor of social or technical sciences, Mar. stat.—Marital status, Mar—Married, Sin—Single.

Table A3. Rank of motivators, normalized value.

Motivators	Σ	Gender		Position		Age				Education Level Mar. Stat					
		M	F	Man	Nom	<25	26–35	36–45	>45	H	C	MA	PhD	Mar	Sin
High wages	0.139	0.145	0.117	0.135	0.142	0.111	0.148	0.154	0.125	0.150	0.139	0.119	0.079	0.144	0.133
Good interpersonal relationships	0.128	0.131	0.112	145	0.122	0.123	0.114	0.126	0.143	0.119	0.132	0.142	0.184	0.129	0.124
Job security	0.087	0.082	0.119	0.067	0.094	0.092	0.088	0.082	0.093	0.100	0.082	0.058	0.035	0.088	0.089
Working time	0.082	0.084	0.075	0.090	0.080	0.071	0.091	0.088	0.071	0.080	0.087	0.102	0.052	0.083	0.082
Quality of supervision and leadership	0.079	0.082	0.062	0.094	0.074	0.049	0.068	0.085	0.089	0.069	0.080	0.091	0.162	0.088	0.057
Interesting work	0.068	0.067	0.070	0.073	0.066	0.052	0.069	0.064	0.075	0.060	0.077	0.082	0.086	0.068	0.067
Amount of responsibility	0.066	0.067	0.058	0.055	0.068	0.055	0.051	0.060	0.088	0.068	0.062	0.056	0.061	0.069	0.057
Promotion	0.065	0.064	0.073	0.071	0.064	0.085	0.079	0.070	0.042	0.065	0.065	0.070	0.071	0.064	0.069
Challenging job and sense of achievement	0.063	0.062	0.066	0.069	0.061	0.086	0.062	0.052	0.070	0.049	0.076	0.094	0.103	0.059	0.073
Good working condition	0.053	0.053	0.052	0.041	0.056	0.047	0.048	0.056	0.055	0.064	0.038	0.029	0.038	0.050	0.058
Possibility of development and improve.	0.045	0.042	0.064	0.050	0.045	0.091	0.061	0.038	0.026	0.038	0.057	0.066	0.038	0.037	0.064
Prizes and awards	0.036	0.035	0.041	0.024	0.040	0.050	0.038	0.037	0.029	0.047	0.024	0.013	0.007	0.034	0.041
Decision-making influence	0.031	0.030	0.030	0.033	0.029	0.027	0.028	0.028	0.036	0.029	0.031	0.023	0.051	0.032	0.025
Work/job recognition	0.030	0.030	0.035	0.038	0.028	0.035	0.029	0.031	0.030	0.029	0.033	0.034	0.021	0.028	0.035
Feeling of being involved or in on things	0.017	0.016	0.018	0.009	0.018	0.011	0.014	0.017	0.019	0.018	0.014	0.015	0.013	0.016	0.015
Criticism	0.011	0.011	0.009	0.007	0.012	0.014	0.012	0.011	0.008	0.015	0.005	0.006	0.000	0.010	0.012

Note: N = 1576, M—male, F—female, Man—manager, Nom—Non-manager, H—High school, C—College, MA—master degree, PhD—doctor of social or technical sciences, Mar.stat.—Marital status, Mar—Married, Sin—Single.

Table A4. Rank of demotivators, normalized value.

Motivators	Σ	Gender		Position		Age				Education Level Mar. Stat.					
		M	F	Man	Nom	<25	26–35	36–45	>45	H	C	MA	PhD	Mar	Sin
Bad working condition	0.164	0.164	0.164	0.152	0.168	0.185	0.168	0.158	0.161	0.174	0.151	0.145	0.149	0.162	0.170
Low quality of supervision and leaders	0.209	0.209	0.206	0.233	0.201	0.189	0.204	0.209	0.217	0.195	0.218	0.248	0.253	0.212	0.200
Low possibility of promotion	0.193	0.193	0.195	0.176	0.198	0.196	0.198	0.198	0.182	0.198	0.195	0.172	0.160	0.191	0.198
Worse interpersonal relationships	0.200	0.209	0.206	0.223	0.193	0.201	0.190	0.196	0.212	0.186	0.210	0.229	0.254	0.200	0.197
Low wages	0.234	0.238	0.221	0.216	0.240	0.229	0.239	0.237	0.228	0.247	0.225	0.206	0.185	0.235	0.235

Note: N = 1576, M—male, F—female, Man—manager, Nom—Non-manager, H—High school, C—College, MA—master degree, PhD—doctor of social or technical sciences, Mar. stat.—Marital status, Mar—Married, Sin—Single.

Table A5. Rank of demotivators.

Motivators	Rank	Gender		Position		Age				Education Level				Mar. Stat.	
		M	F	Man	Nom	<25	26–35	36–45	>45	H	C	MA	PhD	Mar	Sin
Bad working condition	5	5	5	5	5	5	5	5	5	5	5	5	5	5	5
Low quality of supervision and leadership	2	2	3	2	2	4	2	2	2	3	2	1	2	2	2
Low possibility of promotion	4	4	4	4	3	3	3	3	4	2	4	4	4	4	4
Bad interpersonal relationships	3	3	2	2	4	2	4	4	3	4	3	2	1	3	3
Low wages	1	1	1	3	1	1	1	1	1	1	1	3	3	1	1

Note: N = 1576, M—male, F—female, Man—manager, Nom—Non-manager, H—High school, C—College, MA—master degree, PhD—doctor of social or technical sciences, Mar. stat.—Marital status, Mar—Married, Sin—Single.

Table A6. Differences in motivational factors' impacts between analyzed sample—group and control sample-group (*t*-test).

Motivators	N2/Respondents			N1/Respondents			t	df	Mean Diff.	Sig	μ^2
	M	SD	N	M	SD	N					
Interesting work	4.08	0.919	335 female	3.64	0.887	267 female	5.927	600	0.44	0.000	0.055
	4.06	0.957	128 male	3.67	0.919	1308 male	5.412	1494	0.39	0.000	0.019
	3.71	1.038	125 H	3.53	0.912	940 H	2.038	1063	0.18	0.042	0.004
	4.06	0.912	226 C	3.75	0.915	417 C	4.106	641	0.31	0.000	0.026
	4.35	0.742	150 MA	3.99	0.799	142 MA	3.992	290	0.36	0.000	0.052
Quality of supervision and leadership	4.01	0.985	335 female	3.82	0.845	267 female	2.502	600	0.190	0.013	0.010
	3.72	1.127	18 PhD	4.62	0.663	71 PhD	4.396	87	0.900	0.000	0.182
Good working condition	4.08	0.915	335 female	3.74	1.011	267 female	4.323	600	0.340	0.000	0.030
	4.03	0.973	128 male	3.59	1.043	1308 male	5.453	1494	0.440	0.000	0.020

Table A6. *Cont.*

Motivators	N2/Respondents			N1/Respondents			t	df	Mean Diff.	Sig	μ^2
	M	SD	N	M	SD	N					
	3.90	0.987	125 H	3.54	1.082	940 H	3.530	1063	0.360	0.000	0.012
	4.17	0.880	226 C	3.66	0.995	417 C	6.457	641	0.510	0.000	0.061
Promotion	3.63	1.133	125 H	3.28	1.327	940 H	2.815	1063	0.350	0.005	0.007
	3.80	1.112	335 female	3.54	1.236	267 female	2.721	600	0.260	0.007	0.012
	3.90	1.124	128 male	3.47	1.296	1308 male	4.321	1494	0.430	0.000	0.012
Good interpersonal relationships	4.29	0.878	125 H	4.06	0.941	940 H	2.587	1063	0.270	0.010	0.006
	4.28	0.848	335 female	4.03	1.024	267 female	3.276	600	0.250	0.001	0.018
	4.33	0.767	18 PhD	4.69	0.523	71 PhD	−2.357	87	0.360	0.021	0.060
Decision-making influence	3.61	0.983	125 H	3.34	0.961	940 H	2.943	1063	0.270	0.003	0.008
	3.67	0.767	18 PhD	4.24	0.547	71 PhD	−3.367	87	0.530	0.001	0.115
Feeling of being involved or in on things	3.83	0.940	125 H	3.43	0.936	940 H	3.940	1063	0.400	0.000	0.014
	3.96	0.821	226 C	3.47	0.904	417 C	6.774	641	0.490	0.000	0.067
Challenging job and sense of achievement	4.41	0.813	150 MA	4.14	0.912	142 MA	2.673	290	0.270	0.008	0.024
Job security	4.09	1.088	226 C	3.74	1.137	417 C	3.783	641	0.350	0.000	0.022

Table A7. Motivation factors—meaning.

1	Interesting work	An interesting job, good work organization, well-designed job according to skills or education, a job you love, stimulating job, amusing job, etc.
2	Quality of supervision and leadership	Satisfaction with the leading style, loyalty of the leader, attitude of the leader, correct leadership with respect for subordinates.
3	Amount of responsibility	Independence in work, independence in performing tasks, a sense of "one's own" control over work, that no one "stands over worker's head" while working.
4	Good interpersonal relationships	Good friends at work, healthy work environment without quarrels and conflicts, socializing with colleagues, reputation with colleagues, good associates.
5	Decision-making influence	Participation in decision-making, respect for suggestions and opinions of workers, seeking and respecting the views of workers.
6	Feeling of being involved or in on things	Employees are familiar with the goals/business/realization of tasks, information isnot hidden, and employees are involved and up to date with processes.
7	Good working condition	Safety at work, good working conditions, good organization of work, good microclimatic conditions (temperature, humidity, cleanliness).
8	Challenging job and sense of achievement	Self-actualization—to achieve and be what you can be, your own satisfaction in doing work, the ability to excel in work, respect for the work done, achievement, success in work.
9	Promotion	Possibility of promotion, appointment to a higher position or advances in the profession.
10	Job security	No fear of losing a job, that the job is safe/secure.
11	Possibility of development and improvement	Possibility of professional development, availability of courses, possibility of career development.
12	Prizes and awards	Awards and recognitions for finished jobs/tasks/hard work.
13	High wages	The amount of salary in general—that is, the money received for work—does not refer to satisfaction or dissatisfaction with the current salary, but to the importance of the amount of salary and money received for work.
14	Criticism	Criticism of the mistake made and fear of punishment or consequences.
15	Work/job recognition	Job significance/status, status expert in that field, business contacts that job provides.
16	Working time	That working hours correspond to the needs, that workers can be absent from work if necessary, that vacations are used according to the plan and needs, that workers do not work often overtime.

References

1. Robbins, S.; Judge, T. *Organizational Behavior*, 15th ed.; Prentice Hall: Upper Saddle River, NJ, USA, 2013.
2. Ryan, R.; Deci, E. Intrinsic and Extrinsic Motivations: Classic Definitions and New Directions. *Contemp. Educ. Psychol.* **2000**, *25*, 54–67. [CrossRef] [PubMed]
3. Bahtijarevic-Šiber, F. *Human Resource Management*; Golden Marketing: Zagreb, Croatia, 1999.
4. Frey, B.S.; Osterloh, M. *Successful Management by Motivation, Balancing Intrinsic and Extrinsic Incentives*; Springer: Berlin, Germany, 2002.
5. Mihailovic, D. *Structure of Motivation for Work*; Centar: Belgrade, Serbia, 1988.
6. Pinder, C.C. *Work Motivation in Organizational Behavior*, 2nd ed.; Psychology Press: New York, NY, USA, 2014.
7. Armstrong, M.; Taylor, S. *Handbook of Human Resource Management Practices*, 13th ed.; Kogan Page Publishers Inc.: London, UK, 2014.
8. Mihailovic, D. *Organizational Behavior*; Faculty of Organizational Sciences: Belgrade, Serbia, 2009.
9. Latham, G.P.; Pinder, C.C. Work Motivation Theory and Research at the Dawn of the Twenty-First Century. *Ann. Rev. Psychol.* **2005**, *56*, 485–516. [CrossRef]

10. Kovach, K. What motivates employees? Workers and supervisors give different answers. *Bus. Horiz.* **1987**, *30*, 58–65. [CrossRef]
11. Spitzer, D. Power rewards: Rewards that really motivate. *Manag. Rev.* **1996**, *85*, 45–50.
12. Wiley, C. What motivates employees according to over 40 years of motivation survey. *Int. J. Manpow.* **1997**, *18*, 263–280. [CrossRef]
13. Yang, F. Work, motivation and personal characteristics: And in-depth study of six organizations in Ningbo. *Chin. Manag. Stud.* **2011**, *5*, 272–297. [CrossRef]
14. Katsva, M.; Condrey, S.E. Motivating personnel at Russian nuclear power plants: A case study of motivation theory application. *Public Pers. Manag.* **2005**, *34*, 343–356. [CrossRef]
15. Islam, R.; Burhem, B. Employee motivation: A Comparasion between Malaysia and UAE. *Malays. Manag. Rev.* **2007**, *42*, 101–122.
16. Ismail, A.Z.; Ahmed, S. Employee perceptions on reward/recognition and motivating factors: A comparison between Malaysia and UAE. *Am. J. Econ.* **2015**, *5*, 200–207.
17. Chitiris, L. Who are the work-motivated managers in the hotel industry—An exploratory study. *Int. J. Hosp. Manag.* **1990**, *9*, 293–304. [CrossRef]
18. Manolopoulos, D. An evaluation of employee motivation in the extended public sector in Greece. *Empl. Relat.* **2007**, *30*, 63–85. [CrossRef]
19. Qayyum, A.S. An Empirical Analysis of Employee Motivation and Role of Demographics: The Banking Industry of Pakistan. *Glob. Bus. Manag. Res. Int. J.* **2012**, *1*, 1–14.
20. Eskildsen, J.K.; Kristensen, K.; Westlund, A.H. Work motivation and job satisfaction in the Nordic countries. *Empl. Relat.* **2004**, *26*, 122–136. [CrossRef]
21. Kujović, B. Motivation for the Work of Members of the Police. Master's Thesis, University of Belgrade, Belgrade, Serbia, 1989.
22. Mijcika, N. Labour motivation and survey of some of its aspects. *J. Sociol.* **1989**, *20*, 151–164.
23. Dinić, D.; Miltojević, V. Motivational factors for dedication at work—Case study. *Facta Univ. Ser. Work. Living Environ. Prot.* **2017**, *14*, 207–224. [CrossRef]
24. Rakić, T.; Živković, S.; Veljković, M. Hierarchy of Work Motives and Motivators With the Aim of Forming More Efficient Working Environment. *Facta Univ. Ser. Work. Living Environ. Prot.* **2017**, *14*, 87–98.
25. Uddin, R.; Hogue, N.; Ibrahim, M.; Mamun, A. Work Motivation: A Study on Regular and Part-time Employees of Bangladesh. *Int. J. Manag. Bus. Res.* **2014**, *4*, 235–245.
26. Charles, K.R.; Marshall, L.H. Motivational Preferences of Caribbean Hotel Workers: An Exploratory Study. *Int. J. Contemp. Hosp. Manag.* **1992**, *4*, 25–29. [CrossRef]
27. Huddleston, P.; Good, L.K. Job motivators in Russian and Polish retail firms. *Int. J. Retail. Distrib. Manag.* **1999**, *27*, 383–393. [CrossRef]
28. Karl, K.A.; Sutton, C.L. Job Values in Today's Workforce: A Comparison of Public and Private Sector Employees. *Public Pers. Manag.* **1998**, *27*, 515–527. [CrossRef]
29. Linc, S. Motivating Russian workers: Analysis of age and gender differences. *J. Socio-Econ.* **2004**, *33*, 261–289.
30. Chiu, R.K.; Tang, T.L.; Luk, V.W. Retaining and motivating employees: Compensation preferences in Hong Kong and China. *Pers. Rev.* **2002**, *31*, 402–431. [CrossRef]
31. Giannikis, S.; Mihail, D. Motivation of working women in the Greek retail sector: An empirical analysis. *Int. J. Manpow.* **2010**, *31*, 4–20. [CrossRef]
32. Fisher, C.; Yuan, A. What motivates employees? A comparison of US and Chinese responses. *Int. J. Hum. Resour. Manag.* **1998**, *9*, 516–528. [CrossRef]
33. Cameron, J.; Pierce, W.D. Reinforcement, Reward, and Intrinsic Motivation: A Meta-Analysis. *Rev. Educ. Res.* **1994**, *64*, 363–423. [CrossRef]
34. Herzberg, F. *One More Time: How Do You Motivate Employees?* Harvard Business Review Boston: Boston, MA, USA, 1968; pp. 53–62.
35. Kohn, A. Challenging Behaviorist Dogma: Myths about Money and Motivation. *Compens. Benefits Rev.* **1998**, *30*, 27–37. [CrossRef]
36. Pfeffer, J. Six dangerous myths about pay. *Harv. Bus. Rev.* **1998**, *76*, 108–119.
37. Elizur, D.; Borg, I.; Hunt, R.; Beck, I.M. The structure of work values: Across cultural comparison. *J. Organ. Behav.* **1991**, *12*, 21–38. [CrossRef]

38. Savery, L.K.; Wingham, D.L. Coping with the Career Plateau: Motivators for Directors of Child-care Centres. *Leadersh. Organ. Dev. J.* **1991**, *12*, 17–19. [CrossRef]
39. Silverthorne, C.P. Work Motivation in the United States, Russia, and the Republic of China (Taiwan): A Comparison. *J. Appl. Soc. Psychol.* **1992**, *22*, 1631–1639. [CrossRef]
40. Ilić, J. *Work Motivation*; Draganić: Belgrade, Serbia, 1995.
41. Deci, E.L. Effects of externally mediated rewards on intrinsic motivation. *J. Personal. Soc. Psychol.* **1971**, *18*, 105–115. [CrossRef]
42. Amabile, T.M. How to kill creativity. *Harv. Bus. Rev.* **1998**, *76*, 77–87.
43. Kerr, S. On the folly of rewarding A, while hoping for B. *Acad. Manag. J.* **1975**, *18*, 769–783.
44. Erez, M.; Gopher, D.; Arzi, N. Effects of goal difficulty, self-set goals, and monetary rewards on dual task performance. *Organ. Behav. Hum. Decis. Process.* **1990**, *47*, 247–269. [CrossRef]
45. Lawler, E.E.; Porter, L.W. The Effect of Performance on Job Satisfaction. *Ind. Relat.* **1967**, *7*, 20–28. [CrossRef]
46. Vroom, V.H. *Work and Motivation*; Wiley: New York, NY, USA, 1964.
47. Van Knippenberg, D. Work Motivation and Performance: A Social Identity Perspective. *Appl. Psychol.* **2000**, *49*, 357–371. [CrossRef]
48. Gagne, M.; Deci, E.L. Self-determination theory and work motivation. *J. Organ. Behav.* **2005**, *26*, 331–362. [CrossRef]
49. Lord, R.L. Traditional Motivation Theories and Older Engineers. *Eng. Manag. J.* **2002**, *14*, 3–7. [CrossRef]
50. Mani, B.G. Performance Appraisal Systems, Productivity, and Motivation: A Case Study. *Public Per. Manag.* **2002**, *31*, 141–159. [CrossRef]
51. Pritchard, R.D.; Leonard, D.W.; Von Bergen, C.W.; Kirk, R.J. The effects of varying schedules of reinforcement on human task performance. *Organ. Behav. Hum. Perform.* **1976**, *16*, 205–230. [CrossRef]
52. Terborg, J.R.; Miller, H.E. Motivation, behavior, and performance: A closer examination of goal setting and monetary incentives. *J. Appl. Psychol.* **1978**, *63*, 29–39. [CrossRef]
53. Vallerand, R.J.; Bissonnette, R. Intrinsic, Extrinsic, and A motivational Styles as Predictors of Behavior: A Prospective Study. *J. Personal.* **1992**, *60*, 599–620. [CrossRef]
54. Hong, J.C.; Yang, S.D.; Wang, L.J.; Chiou, E.F.; Su, F.Y.; Huang, S.L. Impact of employee benefits on work motivation and productivity. *Int. J. Career Manag.* **1995**, *7*, 10–14. [CrossRef]
55. Bono, J.; Judge, T. Self-concordance at work: Toward understanding the motivational effects of transformational leaders. *Acad. Manag. J.* **2003**, *46*, 554–571.
56. Kuhnen, C.M.; Tymula, A. Feedback, Self-Esteem, and Performance in Organizations. *Manag. Sci.* **2012**, *58*, 94–113. [CrossRef]
57. Masclet, D.; Peterle, E.; Larribeau, S. Gender differences in tournament and flat-wages chemes: An experimental study. *J. Econ. Psychol.* **2015**, *47*, 103–115. [CrossRef]
58. Benndorf, V.; Rau, H.; Sölch, C. Gender differences in motivational crowding out of work performance. *Econ. Inq.* **2018**, *57*, 1–21. [CrossRef]
59. Kassim, Z.; Yusoff, I.Y.; Fong, N.L. Job Empowerment and Customer Orientation of Bank Employees in Kuching, Malaysia. *Contemp. Manag. Res.* **2012**, *8*, 131–140. [CrossRef]
60. Edminah, A.; Ahiazu, A. Employee Involvement and Organizational Effectiveness. *J. Manag.* **2013**, *32*, 661–674.
61. Zhang, J.; Zhang, Y.; Song, Y.; Gong, Z. The different relations of extrinsic, introjected, identified, regulation and intrinsic motivation on employees' performance: Empirical studies following self-determination theory. *Manag. Decis.* **2016**, *54*, 1–27. [CrossRef]
62. Buch, R.; Kuvaas, B.; Dysvik, A.; Schyns, B. If and when social and economical leader member exchange relationships predict follower work effort. *Leadersh. Organ. Dev. J.* **2014**, *35*, 725–739. [CrossRef]
63. Igalens, J.; Roussel, P. A Study of the relationship between Compensation package, Work motivation and Job satisfaction. *J. Organ.* **1999**, *20*, 1003–1025. [CrossRef]
64. Jenkins, D.G., Jr.; Mitra, A.; Gupta, N.; Shaw, J.D. Are financial incentives related to performance? A meta-analytic review of empirical research. *J. Appl. Psychol.* **1998**, *83*, 777–787. [CrossRef]
65. Taylor, J. Organizational Influences, Public Service Motivation and Work Outcomes: An Australian Study. *Int. Public Manag. J.* **2008**, *11*, 67–88. [CrossRef]
66. Manzoor, Q.-A. Impact of Employees Motivation on Organizational Effectiveness. *Bus. Manag. Strategy* **2011**, *3*, 1–12. [CrossRef]

67. Andersen, L.B.; Heinesen, E.; Pedersen, L.H. How Does Public Service Motivation Among Teachers Affect Student Performance in Schools? *J. Public Adm. Res. Theory* **2014**, *24*, 651–671. [CrossRef]
68. Palma, R. Public service motivation and employee outcomes in the Italian public sector: Testing the mediating effect of person-organization fit. *J. Appl. Quant. Methods* **2016**, *11*, 1–16.
69. Kuranchie-Mensah, B.; Amponsah-Tawiah, K. Employee Motivation and Work Performance: A Comparative Study of Mining Companies in Ghana. *J. Ind. Eng. Manag.* **2016**, *9*, 255–309. [CrossRef]
70. Al-Musadieq, M.; Raharjo, K.; Solimun, S.; Fernandes, A. The mediating effect of work motivation on the influence of job design and organizational culture against HR performance. *J. Manag. Dev.* **2018**, *37*, 452–469. [CrossRef]
71. Wright, B.E. Public-Sector Work Motivation: A Review of the Current Literature and a Revised Conceptual Model. *J. Public Adm. Res. Theory* **2001**, *11*, 559–586. [CrossRef]
72. Cleal, B.; Sell, L. Job Satisfaction, Work Environment, and Rewards: Motivational Theory Revisited. *Labour* **2011**, *25*, 1–23.
73. Harpaz, I. The Importance Of Work Goals: An International Perspective. *J. Int. Bus. Stud.* **1990**, *21*, 75–93. [CrossRef]
74. Heyns, M.M.; Kerr, M.D. Generational differences in work place motivation. *SA J. Hum. Resour. Manag.* **2018**, *16*, 1–10. [CrossRef]
75. Islam, R. Employee motivation: A Malaysian perspective. *Int. J. Commer. Manag.* **2008**, *18*, 344–362. [CrossRef]
76. Maslow, A. *Motivation and Personality*; NOLIT: Beograd, Serbia, 1982.
77. Hofstede, G.; Bond, M.H. The Confucius connection: From cultural roots to economic growth. *Organ. Dyn.* **1988**, *16*, 5–21. [CrossRef]
78. Kamdron, T. Work motivation and job satisfaction of Estonian higher officials. *J. Public Adm.* **2005**, *28*, 1211–1240. [CrossRef]
79. Lee-Ross, D. An exploratory study of work motivation among private and public sector hospital chefs in Australia. *J. Manag. Dev.* **2002**, *21*, 576–588. [CrossRef]
80. Ghinetti, P. The Public-Private Job Satisfaction Differential in Italy. *Labour* **2007**, *21*, 361–388. [CrossRef]
81. Franěk, M.; Večeřa, J. Personal characteristics and job satisfaction. *Ekon. Manag.* **2008**, *4*, 63–76.
82. Sindik, J. Research of the motivation consequences at employees with a different education level. *Econ. Though* **2011**, *1*, 101–119.
83. Jurgensen, C.E. Job preferences (What makes a job good or bad?). *J. Appl. Psychol.* **1978**, *63*, 267–276. [CrossRef]
84. Hofstede, G. *Cultures and Organizations: Software of the Mind*, 3rd ed.; McGraw-Hill: London, UK, 2010.
85. Janićijević, N. The influence of national culture on organizational structure of the company in Serbia. *Econ. Ann.* **2003**, *156*, 45–66.
86. Brislin, R.W.; MacNab, B.; Worthley, R.; Kabigting, F.; Zukis, B. Evolving Perceptions of Japanese Workplace Motivation. *Int. J. Cross Cult. Manag.* **2005**, *5*, 87–104. [CrossRef]
87. Herzberg, F.; Mausner, B.; Snyderman, B. *The Motivation to Work*; John Wiley: New York, NY, USA, 1959.
88. Worthley, R.; MacNab, B.; Brislin, R.; Ito, K.; Rose, L.E. Workforce motivation in Japan: An examination of gender differences and management perceptions. *Int. J. Hum. Resour. Manag.* **2009**, *20*, 1503–1520. [CrossRef]
89. Wright, B.E.; Kim, S. Participation's influence on job satisfaction: The importance of job satisfaction: The importance of job characteristics. *Rev. Public Pers. Adm.* **2004**, *24*, 18–40. [CrossRef]
90. Čukić, V. Motivational potential and organizational balance of employees at the beginning of the economic crisis in Serbia. *Bus. Trends* **2013**, *2*, 1–10. [CrossRef]
91. Cohen, J.W. *Statistical Power Analysis for the Behavioral Sciences*, 2nd ed.; Lawrence Erlbaum Associates: Hillsdale, NJ, USA, 1988.
92. Milićević, M.; Župac, G. Subjective approach to the determination of criteria weights. *Mil. Tech. Cour.* **2012**, *2*, 48–70. [CrossRef]
93. Biswas, S.; Bandyopadhyay, G.; Guha, B.; Bhattacharjee, M. An ensemble approach for portfolio selection in a multi-criteria decision making frame work. *Decis. Mak. Appl. Manag. Eng.* **2019**, *2*, 138–158.
94. Tanackov, I.; Jevtić, Ž.; Stojić, G.; Sinani, F.; Ercegovac, P. Rare Events Queueing System—REQS. *Oper. Res. Eng. Sci. Theory Appl.* **2019**, *2*, 1–11. [CrossRef]
95. Pickton, D.W.; Wright, S. What's SWOT in strategic analysis? *Strateg. Chang.* **1998**, *7*, 101–109. [CrossRef]

96. Helms, M.; Nixon, J. Exploring SWOT analysis—Where are we now? Are view of academic research from the last decade. *J. Strategy Manag.* **2010**, *3*, 215–251. [CrossRef]
97. Kim-Keung, H.J. Formulation of a Systemic PEST Analysis for Strategic Analysis. *Eur. Acad. Res.* **2014**, *2*, 6478–6492.
98. Sammut-Bonnici, T.; Galea, D. PEST analysis. In *Wiley Encyclopedia of Management*; Cooper, C., Ed.; John Wiley & Sons, Ltd.: New York, NY, USA, 2015; Volume 12, pp. 1–7.
99. Nunnally, J.C. *Psychometric Theory*, 2nd ed.; McGraw-Hill: New York, NY, USA, 1978.
100. Ryan, R.; Deci, E. Self-determination theory and the facilitation of intrinsic motivation, social development, and well-being. *Am. Psychol.* **2000**, *55*, 68–78. [CrossRef] [PubMed]
101. Deci, E.; Ryan, R. Facilitating optimal motivation and psychological well-being across life's domains. *Can. Psychol.* **2008**, *49*, 14–23. [CrossRef]
102. Huddleston, P.; Good, L.; Frazier, B. The influence of firm characteristics and demographic variable son Russian workers' work motivation and job attitudes. *Int. Rev. Retail. Distrib. Consum. Res.* **2002**, *12*, 395–421.
103. Takahashi, K. Effects of wage and promotion incentives on the motivation levels of Japanese employees. *Career Dev. Int.* **2006**, *11*, 193–203. [CrossRef]

© 2020 by the authors. Licensee MDPI, Basel, Switzerland. This article is an open access article distributed under the terms and conditions of the Creative Commons Attribution (CC BY) license (http://creativecommons.org/licenses/by/4.0/).

sustainability

MDPI

Article

Development of a Multi-Criteria Model for Sustainable Reorganization of a Healthcare System in an Emergency Situation Caused by the COVID-19 Pandemic

Dragan Pamučar [1,2], Mališa Žižović [3], Dragan Marinković [4,*], Dragan Doljanica [5], Saša Virijević Jovanović [5] and Pavle Brzaković [5]

1 Informetrics Research Group, Ton Duc Thang University, Ho Chi Minh City 758307, Vietnam; dragan.pamucar@tdtu.edu.vn

2 Faculty of Civil Engineering, Ton Duc Thang University, Ho Chi Minh City 758307, Vietnam

3 Faculty of Technical Sciences in Cacak, University of Kragujevac, Svetog Save 65, 32102 Cacak, Serbia; zizovic@gmail.com

4 Berlin Institute of Technology, Department of Structural Analysis, 10115 Berlin, Germany

5 MEF Faculty of Applied Management, Economics and Finance, University Business Academy in Novi Sad, Jevrejska 24, 11000 Belgrade, Serbia; dragan.doljanica@mef.edu.rs (D.D.); sasa.virijevic@mef.edu.rs (S.V.J.); pavle.brzakovic@mef.edu.rs (P.B.)

* Correspondence: dragan.marinkovic@tu-berlin.de

Received: 5 August 2020; Accepted: 10 September 2020; Published: 11 September 2020

check for updates

Abstract: Healthcare systems worldwide are facing problems in providing health care to patients in a pandemic caused by the SARS-CoV-2 virus (COVID-19). The pandemic causes an extreme disease to spread with fluctuating needs among patients, which significantly affect the capacity and overall performance of healthcare systems. In addition, its impact on the sustainability of the entire economic and social system is enormous and certain sustainable management strategies need to be selected. To meet the challenges of the COVID-19 pandemic and ensure sustainable performance, national healthcare systems must adapt to new circumstances. This paper proposes an original multi-criteria methodology for the sustainable selection of strategic guidelines for the reorganization of a healthcare system under the conditions of the COVID-19 pandemic. The selection of an appropriate strategic guideline is made on the basis of defined criteria and depending on infection capacity and pandemic spread risk. The criteria for the evaluation of strategic guidelines were defined on the basis of a survey in which the medical personnel engaged in the crisis response team during the COVID-19 pandemic in the Republic of Serbia participated. The Level-Based Weight Assessment (LBWA) model and Measuring Attractiveness by a Categorical-Based Evaluation Technique (MACBETH) method were used to determine the weight coefficient criteria, while a novel fuzzy Ranking of Alternatives through Functional Mapping of Criterion Subintervals into a Single Interval (RAFSI) model was used to evaluate the strategic guidelines. The proposed multi-criteria methodology was tested in a case study in the Republic of Serbia. The validity of the proposed methodology is shown through the simulation of changes in input parameters of Bonferroni aggregation functions and through a comparison with other multi-criteria methodologies.

Keywords: COVID-19; multi-criteria; level-based weight assessment; fuzzy sets; sustainability

1. Introduction

The World Health Organization (WHO) declared a global pandemic, coronavirus disease 2019 (COVID-19) in March 2020, when the international mission of SARS-CoV-2 virus research began.

The pandemic was caused by a new virus from the coronavirus group, which was first isolated in 1962. Since then, some coronaviruses have been known to infect only some animals, others humans, while some can break the species barrier and cause conditions ranging from mild colds to severe acute respiratory syndrome (SARS). The coronavirus disease (COVID-19) pandemic has brought many surprises and posed many new challenges to healthcare systems worldwide and their sustainable performance. The pandemic has set numerous operational, logistical, organizational and moral–ethical requirements before management, healthcare workers and associates: the organization of specialized training for medical personnel in accordance with emerging conditions; the conversion of facilities for patient accommodation; the procurement of medical equipment; information and communication management; the continuous monitoring of healthcare systems; the availability of health care to specific groups of population in new conditions, etc.

The sustainability of healthcare systems during a pandemic is defined by the Global Health Security Index (GHS Index). The GHS Index assesses the preparedness of healthcare systems to prevent the outbreak and spread of epidemics/pandemics, assesses the preparedness for a rapid response, and also assesses compliance with international protocols in crisis situations, including the COVID-19 pandemic. There are two strategic phases in the fight against the COVID-19 pandemic: (1) Phase I—rapid and aggressive action (rapid reorganization of healthcare systems, restriction of movement, etc.) and (2) Phase II—long-term effort to maintain control over the virus until finding an optimal solution (vaccination, appropriate therapy or increasing the level of community immunity). An integral part of Phase II also includes capacity improvement, forecasting pandemic dynamics, he risk assessment of the further spread of the pandemic and other analyses. The assessment of each parameter reflects structural or other deficiencies in terms of funding, the ability to provide adequate services, management, and workforce and resource management. All of this exposes healthcare systems to great challenges and different outcomes. Due to all of the stated specific characteristics of the COVID-19 pandemic, it is necessary to continuously analyze all weaknesses and strengths of healthcare systems, forming a plan based on adaptive models of healthcare system organization since immediate requirements and the unpredictability of COVID-19 infection require agility and the flexibility of all services to respond in a timely manner, especially qualified medical personnel.

Accordingly, this paper proposes an adaptive model for the organization of a healthcare system in a country, or a region within a country, in cases of emergencies or specific pandemic conditions. The starting point is an assumption that there are (or may be) spas, i.e., spa rehabilitation centers in a given country (region) and that by the interruption of work within the industry, the lives of their users will not be endangered. There are such centers in almost all European countries. In many European countries, spa tourism is very developed, as evidenced by the data shown in Table 1 [1].

Table 1. Overview of spa capacities in several European countries [1].

No.	Country	Number of Spas	Average Number of Visitors	Average Number of Overnight Stays
	Germany	313	>550,000 *	>3,550,000
	Spain	128	>1,000,000	>2,000,000
	France	96	>200,000	>4,000,000
	Austria	81	>500,000	>4,100,000
	Serbia	36	>200,000	>1,000,000
	Czech Republic	34	>300,000	>5,500,000
	Slovakia	21	>100,000	>1,500,000
	Slovenia	15	>200,000	>1,000,000

* The average number of visitors and overnight stays is shown for one year.

It should be emphasized that Table 1 shows data for several European countries since we did not have publicly available data for the remaining European countries. Considering these data, we can conclude that in most European countries there is a great potential of spa centers which can give support to healthcare systems in providing health care to patients in a pandemic.

Hospitals, as healthcare facilities that play a key role in a healthcare system, face problems worldwide in providing health care to patients with different types of diseases. Today, these problems have significantly increased with the occurrence of the COVID-19 pandemic. The pandemic causes an extreme spread of disease with fluctuating needs of patients, which significantly affects the capacity and overall performance of hospitals. In order to cope with the challenges of the COVID-19 pandemic, and remain sustainable under such conditions, national healthcare systems have to adapt to the new circumstances and form special hospitals in which only patients infected with the SARS-CoV-2 virus are treated. This paper proposes a multi-criteria model for the development of strategic guidelines for the adjustment and reorganization of the healthcare system in the Republic of Serbia under the conditions of the COVID-19 pandemic. Four strategic guidelines have been proposed and defined on the basis of a survey conducted with members of the Republic Crisis Response Team in Serbia in the period of March–May 2020. The strategic guidelines are based on the capacity utilization of spa centers and their personnel as a support for the healthcare system of a certain region in the Republic of Serbia. The selection of a strategic guideline that determines the utilization rate of spa centers depends on pandemic infection capacity, which is defined by the WHO. In addition to pandemic infection capacity, the selection of a strategic guideline is also influenced by the prediction of disease spread, which is defined for each country or region individually. Based on pandemic infection capacity and the prediction of disease spread, we determined five levels of risk on the basis of which a multi-criteria model for the selection of strategic guidelines is defined.

Accordingly, the aim of this paper is to develop a multi-criteria model for the evaluation and selection of strategic guidelines that define a strategy for reorganization and adjustment of healthcare systems under the conditions of the COVID-19 pandemic. The proposed multi-criteria model is based on the application of the Level-Based Weight Assessment (LBWA) model [2], Measuring Attractiveness by a Categorical-Based Evaluation Technique (MACBETH) method [3] and the Ranking of Alternatives through Functional Mapping of Criterion Subintervals into a Single Interval (RAFSI) model [4]. The LBWA–MACBETH model was used to determine the weight coefficients of criteria, while the fuzzy RAFSI model was used to evaluate and select strategic guidelines for reorganizing and adjusting the healthcare system. In addition, the proposed model has the following advantages which improve the literature that considers the application of multi-criteria techniques in the medical field:

- The original multi-criteria methodology was developed for the reorganization and adjustment of healthcare systems in an emergency situation caused by the COVID-19 pandemic;
- The multi-criteria methodology allows the evaluation and selection of a strategy for the reorganization of a healthcare system of a region/country depending on a risk level defined on the basis of pandemic infection capacity and the prediction of disease spread;
- Strategic guidelines and criteria for the evaluation of guidelines were defined on the basis of several months of research in which medical experts and other members of the crisis response team in Serbia participated;
- The presented multi-criteria methodology represents a contribution to the decision-making methodology, which is reflected through an original extension of the RAFSI algorithm in a fuzzy environment;
- Since decision making under the conditions of the COVID-19 pandemic is accompanied by numerous uncertainties and dynamic characteristics of input parameters, a special contribution of this paper is an original hybrid fuzzy LBWA–MACBETH–RAFSI model that allows the evaluation and selection of alternative solutions under conditions of uncertainty;
- The fuzzy LBWA–MACBETH–RAFSI multi-criteria methodology is characterized by flexibility and applicability in various areas where strategic decision making is required;
- The proposed multi-criteria methodology provides a new framework for resource management. The model presented in this paper can help decision makers in hospitals and national policy

makers to establish strategic guidelines for the adjustment of healthcare systems depending on the capacity of the COVID-19 pandemic.

The paper is structured into four sections. After the introduction, in which the problem and objectives of the paper are presented, the second section of the paper presents an analysis of the literature that includes studies which consider the application of multi-criteria techniques in the healthcare field. In the third section of the paper, a multi-criteria model for the evaluation of strategic guidelines for the organization of a healthcare system under the conditions of the COVID-19 pandemic is formulated. The fourth section presents the implementation of the multi-criteria model and the validation of the results shown. In the fourth section of the paper, conclusions and guidelines for future research are presented.

2. Literature Review

The assessment of preparedness of healthcare systems to work in crisis situations, such as the COVID-19 pandemic, has been performed in a small number of papers to date. This especially refers to the application of multi-criteria optimization techniques for assessing the degree of organization of healthcare systems, as well as for making strategic decisions. The analysis of the literature shows that most authors apply two main concepts for the analysis of the preparedness of healthcare systems in crisis situations. The first concept implies the application of multi-criteria models, while the second concept is based on the application of statistical tools for data processing and the analysis of their statistical correlation. In their study, Nekoie-Moghadam et al. [5] presented a comprehensive literature review with an analysis of various methodologies used to assess the ability of healthcare institutions to work in crisis situations. Nekoie-Moghadam et al. [5] considered studies that analyzed the level of preparedness of healthcare institutions in terms of logistical capacity, capacity of medical personnel, preparedness of communications for an emergency, management skills, training of medical and non-medical personnel, development of procedures for evacuation of seriously ill people and isolation in case of infectious diseases, disaster recovery, transport capacity, etc. [5–7]. In all analyzed studies, the authors applied different statistical tools for processing data collected by questionnaire surveys [5]. A comprehensive review of the literature that considers the application of statistical tools for the analysis of the preparedness of healthcare institutions to work in crisis situations is also presented in [5–8]. Tabatabaei and Abbasi [9] defined the hospital safety index on the basis of which the risk assessment of healthcare institutions for work in crisis conditions was performed. In their study, Samsuddin et al. [10] showed that human resources, their training and the ability of healthcare institution to adjust in a timely manner are key factors for defining a risk level for work in crisis situations. Marzaleh et al. [11] performed an analysis of the assessment of the preparedness of emergency services in hospitals to work in crisis situations. For the evaluation, they used 31 criteria which were processed using statistical tools and the Delphi technique. A similar methodology was applied by Shabanikiya et al. [12] for hospital capacity analysis. Mulyasari et al. [13] assessed hospital capacity preparedness in Japan using six parameters that consisted of 21 indicators. The study analyzed the preparedness of healthcare systems to work in a crisis situation caused by a tsunami in the Tohoku area.

In addition to the application of statistical tools for assessing the significance of factors for the evaluation of healthcare institutions, there are a number of publications in which the evaluation of healthcare institutions is performed using multi-criteria tools. Hosseini et al. [14] evaluated hospitals using the Technique for Order of Preference by Similarity to Ideal Solution (TOPSIS) multi-criteria technique. The significance of four clusters and 21 factors used to evaluate healthcare facilities was defined based on experts' group preferences. In their study, Ortiz-Barrios et al. [15] demonstrated the application of a complex hybrid multi-criteria technique for the evaluation of hospitals. An integrated Analytic Hierarchy Process (AHP) and Decision-Making Trial and Evaluation Laboratory (DEMATEL) model were used to determine the weight coefficients of factors, while the Technique for Order of Preference by Similarity to Ideal Solution method was used for the evaluation. In order to consider interrelations and define graphically the relationships between the factors used for evaluation,

Ortiz-Barrios et al. [15] used the Decision-Making Trial and Evaluation Laboratory (DEMATEL) method. After defining the significance and interrelation of factors using the DEMATEL method, Ortiz-Barrios et al. [15] showed the TOPSIS methodology as a powerful tool for defining ideal reference values on the basis of which the comparison and evaluation of hospitals were performed. Similar to [15], the advantages of using the DEMATEL technique in their study were shown by Roy et al. [16]. However, in contrast to previous studies that ignored the existence of inaccuracies and uncertainties in the data used to evaluate healthcare systems, in their study, Roy et al. [16] used rough numbers to present inaccuracies in expert preferences.

Given that the COVID-19 pandemic is ongoing and that there are no publications considering the problem of evaluation and assessment of the preparedness of medical institutions to work under the conditions of COVID-19 pandemic, the following section analyzes only a few articles dealing with multi-criteria techniques in the COVID-19 pandemic. Sarkar [17] performed a spatial analysis and mapping of areas susceptible to COVID-19 infection in Bangladesh. A multi-criteria methodology (AHP method), which was implemented in a Geographic Information System (GIS), was used to map the area. In addition to the presented study, a spatial analysis of the risk of spreading COVID-19 infection in urban areas in Italy was performed by Sangiorgio and Parisi [18]. They used the Artificial Neural Network and GIS to create spatial risk maps in 257 city districts. Nardo et al. [19] showed the application of multi-criteria decision analysis to determine weights for eleven criteria in order to prioritize COVID-19 non-critical patients for admission to hospital in healthcare settings with limited resources. Yildirim et al. [20] evaluated the available COVID-19 treatment options in hospitals. For the evaluation, they used a hybrid methodology based on the application of two well-known multi-criteria tools: (1) Preference Ranking Organization Method for Enrichment of Evaluations (PROMETHEE) and (2) Vlsehriterijumska Optimizacija I Kompromisno Resenje (VIKOR) method. The mathematical formulation of both models is presented by fuzzy arithmetic and fuzzy sets.

By analyzing the presented articles, we can conclude that there is a gap in the literature that considers the application of hybrid multi-criteria methods for the evaluation of healthcare systems under the conditions of the COVID-19 pandemic. In general, there are only a few articles in the literature that consider the application of hybrid multi-criteria techniques for the evaluation of hospital capacities in crisis situations [14,15,21]. Despite the fact that decision making under the conditions of the COVID-19 pandemic and crisis situations in general is accompanied by a greater or lesser number of uncertainties, there are only a few studies that apply fuzzy numbers to treat uncertainties [20]. Additionally, the analysis of the literature shows that there is no study that considers the possibility of including spa centers in healthcare systems in order to reduce the pressure on primary hospital capacities under the conditions of COVID-19 pandemic. This study proposes an original multi-criteria methodology (fuzzy LBWA–MACBETH–RAFSI model) that has not been considered in the literature so far. The proposed methodology fills a gap in the literature considering the application of multi-criteria techniques for evaluating healthcare systems under the conditions of COVID-19 pandemic and for evaluation of healthcare systems in general. It should be emphasized that the novel fuzzy LBWA–MACBETH–RAFSI multi-criteria methodology has a universal algorithm that can be applied to solve various real-world problems, which refreshes and improves the field of multi-criteria decision making from a theoretical and methodological perspective.

3. MCDM Methodology for Proposing a Possible Organization of a Healthcare System under the Conditions of the COVID-19 Pandemic

The basic model of organizing healthcare systems in case of emergency is that each healthcare center within a region/country organizes work within its infectious disease ward in cooperation with appropriate infectious disease clinics. Experiences from a period of February–May 2020 have shown that organizing a healthcare system in this way satisfies the needs of users only in the case of low to medium pandemic infection capacity, i.e., in the case of low to medium disease spread risk. In the case of a higher pandemic infection capacity and higher pandemic spread risk, it is necessary to reorganize the healthcare system using secondary healthcare facilities. A large number of countries, including Serbia, in addition to secondary healthcare facilities, have used other facilities of large capacity (e.g., sports halls) for the accommodation and isolation of mildly ill patients.

However, in case of an emergency, spa centers can be successfully re-oriented to providing care for patients referred by appropriate healthcare institutions. Depending on the infection capacity and pandemic spread risk, it is possible to define variants of the inclusion of spa centers for patient care. Each variant of the inclusion of spa centers in healthcare systems is a strategic guideline for a possible reorganization of healthcare systems. In this research, four variants (strategic guidelines) for the reorganization of healthcare systems are defined:

Strategic guideline 1 (*A*1): A healthcare center organizes its work through three levels. The initial level implies that each healthcare center organizes the admission of infected patients within its infectious disease ward in cooperation with infectious disease clinics. When the healthcare center capacity becomes insufficient for the admission and care of patients, the second level of organization is applied. Within the second level, the healthcare center is divided into the main and infectious disease part of the healthcare center. In the infectious disease part, patients who have received positive results after testing for the coronavirus are taken care of. The main part of the healthcare center provides medical services to patients who are not infected. If the second level of organization cannot provide a satisfactory level of health care, the third level of organization is applied. In the third level, the organization of the healthcare center of the second level is maintained, activating additional capacities for the admission of infected patients, such as sports halls, hotels, hostels, etc., which are adapted for the admission and care of positive patients.

Strategic guideline 2 (*A*2): A healthcare center organizes its work through two levels. The initial level, as for *A*1, implies that the healthcare center organizes the admission of infected patients within its infectious disease ward in cooperation with infectious disease clinics. When the healthcare center capacity becomes insufficient for the admission and care of positive patients, the second level of the organization is applied. The second level of organization involves the activation of spa centers to provide medical services to patients who are positive to the coronavirus and have milder symptoms. The infectious disease ward of the healthcare center treats patients who are positive to the coronavirus, who have more severe symptoms and who need artificial lung ventilation. Other units (non-infectious disease units) of the healthcare center provide medical services to patients suffering from other diseases. Here, a spa center represents a recreation and wellness center that is not connected to any healthcare center, but which has staff that as medical personnel can be involved in providing care for positive patients. Patients from superior healthcare institutions are referred to the spa center. In accordance with their protocols, healthcare institutions engage medical personnel to work in spa centers. The defined protocols also apply to the present staff of the spa center, which may, depending on the assessment of the crisis response team, be engaged in treatment.

Strategic guideline 3 (*A*3): A healthcare center organizes its work through two levels: (1) the initial level implies that the healthcare center organizes the admission of infected patients within its infectious disease ward in cooperation with infectious disease clinics and (2) the second level is applied when the healthcare center capacity becomes insufficient for the admission and care of positive patients. The second level involves the use of spa centers to provide medical services to patients infected with the coronavirus. In this strategic guideline, spa centers are rehabilitation centers that are an integral part of the observed healthcare center. In addition to medical personnel from superior healthcare institutions, the personnel from the spa center are directly involved in a work process. For the staff from the spa center, an assessment is made as to whether they are trained to work in emergency situations. For non-trained staff of rehabilitation center, the healthcare center organizes training for work in pandemic conditions.

Strategic guideline 4 (*A*4): The strategic guideline *A*4 represents healthcare center organization through two levels. Within the initial level, the healthcare center organizes the admission and care of all patients. Coronavirus-positive patients are treated within an infectious disease ward in collaboration with infectious disease clinics. Patients who are not positive are treated in the remaining specialized units of the healthcare center. When the healthcare center capacity becomes insufficient for the admission and care of patients, the second level of organization is applied, which includes the use of spa centers. Here, a spa center represents a rehabilitation center that is connected to several healthcare centers that gravitate towards the spa center and it is capable of providing medical care for patients suffering from infectious diseases. The personnel of this center are trained to work under the conditions of a pandemic. In addition to medical personnel from a directly superior healthcare center and trained staff of the spa center, the personnel of other regional centers that gravitate towards the spa center are also involved in a work process of the spa center. At the same time, these regional centers refer their patients for treatment to the spa center. If the spa center has reached its full capacity, other facilities, such as halls, hotels, hostels and others, are used and adapted for the admission and care of positive patients.

The selection of an appropriate strategic guideline for the reorganization of healthcare systems is made depending on infection capacity and pandemic spread risk. The main cause of the epidemic is that one infected person spreads the infection to (on average) p other persons. If $p > 1$, the infection spreads exponentially; if $p < 1$, the infection subsides and disappears [22]. The parameter p can be changed in several ways. One way is to find a number of immune people in the community over time that cannot be infected, which reduces p. The higher the number of immune people, the lower the p, and when p falls below 1, the infection subsides [23]. Based on the value of p and risk assessment (θ), infection capacity and pandemic spread risk (δ), $\delta = p \cdot \theta$, are defined. The value of θ is defined by experts in the crisis response team on the basis of a five-degree scale: 1—very low risk, 2—low risk, 3—medium risk, 4—high risk and 5—very high risk. For the purpose of this study, we defined a scale on the basis of which infection capacity and pandemic spread risk (δ) are determined; see Table 2. The limit values of the scale are defined on the basis of recommendations by experts from the crisis response team in the Republic of Serbia.

Table 2. A scale for defining infection capacity and pandemic spread risk.

No.	Linguistic Value	Numerical
1.	Low infection capacity and risk	$0 \le \delta \le 2.0$
2.	Medium infection capacity and risk	$2.0 < \delta \le 3.0$
3.	Serious infection capacity and risk	$3.0 < \delta \le 5.0$
4.	High infection capacity and risk	$5.0 < \delta \le 7.0$
5.	Very high infection capacity and risk	>7.0

After defining infection capacity and pandemic spread risk, a multi-criteria model is applied to select an appropriate strategic guideline for the reorganization of the healthcare system; see Figure 1.

The multi-criteria model implies the application of the LBWA model for determining the weight coefficients of criteria. The evaluation and selection of an appropriate guideline is completed using the fuzzy RAFSI method. The evaluation of guidelines (alternatives) in relation to defined criteria C_j ($j = 1, 2, \ldots, n$) is performed by experts using a predefined fuzzy linguistic scale. After defining expert correspondent matrices, expert preferences are aggregated into an averaged initial matrix in which uncertainties are represented by triangular fuzzy numbers. Fuzzy linguistic variables were chosen due to their ability to present uncertainties and dilemmas that exist when evaluating strategic guidelines. In the following sections, the mathematical formulation of the fuzzy LBWA–MACBETH–RAFSI multi-criteria methodology is given.

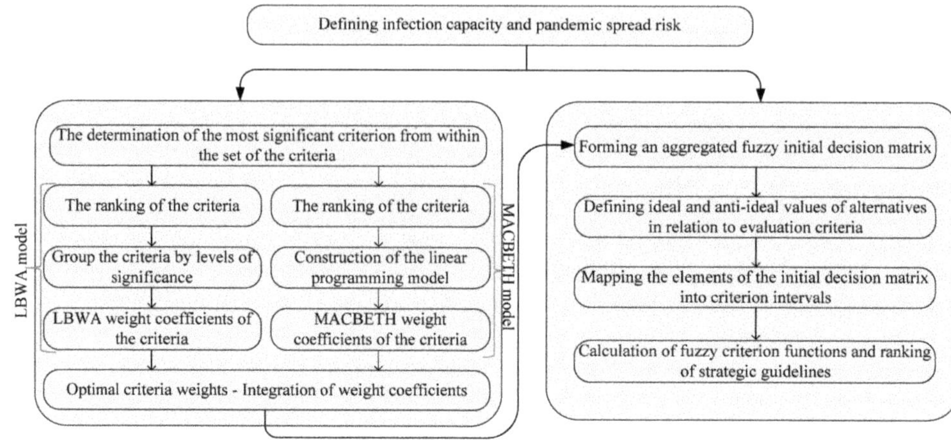

Figure 1. Multi-criteria model for the selection of strategic guidelines for the reorganization of the healthcare system.

3.1. LBWA Model for Determining Weight Coefficients of Criteria

The LBWA method (Figure 2) for determining weight coefficients of criteria is characterized by a small number of pairwise comparisons of criteria and a rational and logical mathematical algorithm [2]. Despite the fact that the LBWA method belongs to a group of multi-criteria methods developed recently, there are already a number of publications that exploit the advantages of the LBWA methodology [24–28].

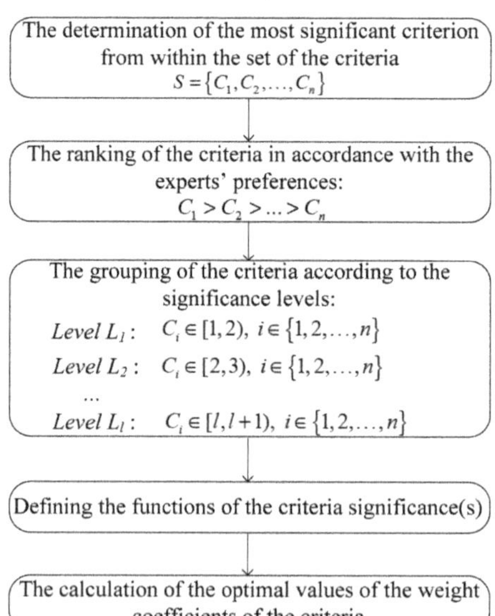

Figure 2. The LBWA model.

From a group of subjective models for determining the weights of criteria, the LBWA methodology is distinguished on the basis of the following advantages: (1) The LBWA model allows the calculation of weight coefficients with a small number of criteria comparisons; (2) The LBWA model algorithm does not become more complex with the increase in the number of criteria, which makes it suitable for use in complex multi-criteria decision making (MCDM) models with a larger number of evaluation criteria; (3) The LBWA model allows decision makers to present their preferences through a logical algorithm when prioritizing criteria; (4) LBWA methodology eliminates inconsistencies in expert preferences since comparisons in the LBWA model are always consistent. Keeping in mind the stated advantages of LBWA, the authors decided to use the LBWA model for determining the weights of the criteria in this paper. The following section presents the algorithm for the LBWA model that includes six steps.

Step 1: Determining the most significant criterion from the set of criteria. Assume that there is a group of k experts who have defined the set of criteria $C = \{C_1, C_2, \ldots, C_n\}$, where n represents the total number of criteria. Experts arbitrarily make a decision on the selection of the most influential criterion C_B from the set C.

Step 2: Grouping criteria by levels of significance. The following algorithm is used to group the criteria by levels of significance:

Level L_1: At the level L_1, group the criteria whose significance is equal to the significance of the criterion C_B or up to twice less than the significance of the criterion C_B, i.e., $C_j \in [1,2),\ j = 1,2,\ldots,n$.

Level L_2: At the level L_2, group the criteria whose significance is exactly twice less than the significance of the criterion C_B or up to three times less than the significance of the criterion C_B, i.e., $C_j \in [2,3),\ j = 1,2,\ldots,n$.

Level L_l: At the level L_l, group the criteria whose significance is exactly l times less than the significance of the criterion C_B or up to $l+1$ times less than the significance of the criterion C_B, i.e., $C_j \in [l,l+1),\ j = 1,2,\ldots,n$.

If the significance of the criterion C_j is denoted by $s(C_j)$, where $j = 1,2,\ldots,n$, then we have $L = L_1 \cup L_2 \cup \cdots \cup L_l$, where for every level $i = 1,2,\ldots,l$, the following applies:

$$L_i = \left\{C_{i_1}, C_{i_2}, \ldots, C_{i_s}\right\} = \left\{C_j \in L\ :\ i \leq s(C_j) < i+1\right\} \tag{1}$$

Step 3: Comparison of criteria by significance. Within the formed levels of significance, each expert compares the criteria by their significance. Each criterion $C_{i_p} \in L_i$ within the subset $C_i = \left\{C_{i_1}, C_{i_2}, \ldots, C_{i_s}\right\}$ is assigned the value I_{i_p} such that the most significant criterion C_B within the first level is assigned the value $I_B = 0$. Moreover, if C_{i_p} is more significant than C_{i_q} then $I_{i_p} < I_{i_q}$, and if C_{i_p} is equivalent to C_{i_q} then $I_{i_p} = I_{i_q}$. The maximum value on the scale for the comparison of criteria is defined by applying Equation (2):

$$\lambda = \max\{|L_1|, |L_2|, \ldots, |L_l|\} \tag{2}$$

Step 4: Defining the elasticity coefficient. Based on the defined maximum value of the scale for the comparison of criteria (λ), Equation (2), the elasticity coefficient $\Delta\lambda$ ($\Delta\lambda > \lambda$) is defined:

$$\Delta\lambda = \max\{|L_1|, |L_2|, \ldots, |L_l|\} \tag{3}$$

Step 5. Calculation of the influence function of criteria. For each criterion C_j within the observed level of significance $C_j \in L_i$, the influence function of the criterion is defined:

$$f(C_j) = \frac{\Delta\lambda}{i \cdot \Delta\lambda + I_{i_p}} \tag{4}$$

where i represents the number of significance level in which the criterion C_j is classified, $\Delta\lambda$ represents the elasticity coefficient, while I_{i_p} represents the value assigned to the criterion C_j within the observed level L_i.

Step 6. Calculation of the optimal values of the weight coefficients of criteria. By applying Equation (5), the weight coefficient of the most influential criterion C_B is calculated:

$$w_B^{LBWA} = \frac{1}{1 + \sum_{\substack{j=1 \\ j \neq B}}^{n} f(C_j)} \tag{5}$$

where w_B represents the value of the weight coefficient of the most influential criterion (C_B), while $f(C_j)$ represents the function of the influence of criteria defined by Equation (4).

The values of the weight coefficients of the remaining criteria are obtained by applying Equation (6):

$$w_j^{LBWA} = f(C_j) \cdot w_B \tag{6}$$

where w_B represents the value of the weight coefficient of the most influential criterion (C_B).

3.2. MACBETH Model for Determining Weight Coefficients of Criteria

The Measuring Attractiveness by a Categorical-Based Evaluation Technique (MACBETH) method [3] is one of the multi-criteria decision making methods that depends upon the qualitative judgments between alternatives and criteria [3]. The algorithm of the MACBETH method is similar to the algorithm of the Analytical Hierarchy Process (AHP) method, since both methods use comparative comparison matrices. The main difference between these two methods is that AHP uses a ratio scale but MACBETH uses an interval scale [3]. The semantic categories used for pairwise comparisons can be seen in Table 3.

Table 3. Semantic Categories for Measuring Attractiveness by a Categorical-Based Evaluation Technique (MACBETH).

Semantic Categories	Scale
No	0
Very weak	1
Weak	2
Moderate	3
Strong	4
Very strong	5
Extreme	6

The MACBETH method uses a linear programming model for finding the importance levels of the alternatives or criteria. In this study, the MACBETH method is used for the importance levels of the criteria. This linear programming model has been constructed as follows (Burgazoğlu, 2015):

$$\begin{aligned} & z_{\min} = \phi(o_1) \\ & o_i > o_j \Rightarrow \phi(o_i) \geq \phi(o_j) + \delta(i,j);\ \forall o_i, o_j, i, j \in \{1, 2, \ldots, n\} \\ & \phi(o_i) - \phi(o_j) \geq \phi(o_k) - \phi(o_l) + \delta(i,j,k,l);\ \forall o_i, o_j, i, j, k, l \in \{1, 2, \ldots, n\} \\ & \phi(o_n) = 1 \end{aligned} \tag{7}$$

After solving linear model (7), the criteria weights have been determined. The next step is to normalize these values. The normalization process can be made as in Equation (8).

$$w_j^{MACBETH} = \frac{\phi(o_j)}{\sum_{j=1}^{n} \phi(o_j)} \tag{8}$$

The final values of the weighting coefficients of the criteria were obtained by fusing the values of the weighting coefficients obtained by the LBWA and MACBETH methodology, Equation (9):

$$w_j = \xi \cdot w_j^{LBWA} + (1-\xi) \cdot w_j^{MACBETH} \tag{9}$$

where w_j $(j = 1, 2, \ldots, n)$ represents the final values of the weighting coefficients, w_j^{LBWA} represents the value of the weighting coefficient obtained using the LBWA methodology, $w_j^{MACBETH}$ represents the value of the weighting coefficient obtained using the MACBETH methodology, while the coefficient $\xi \in [0, 1]$ defines the percentage share of weights criteria in the final decision. It is recommended to define a value $\xi = 0.5$ for the initial ranking of alternatives, since for this value both methodologies equally (50% each) participate in defining the final weights of the criteria.

3.3. *Fuzzy RAFSI Method for the Evaluation of Alternatives*

The Ranking of Alternatives through Functional Mapping of Criterion Subintervals into a Single Interval (RAFSI) method [4] integrates three starting points on which it bases its decision making algorithm: (1) defining referential points—ideal and anti-ideal criteria values; (2) defining the relationship between alternatives in relation to defined ideal/anti-ideal values; (3) defining criterion functions that map criterion sub-intervals into a unique criterion interval. The RAFSI method has three key advantages that recommend it for further application [4]: (*i*) the simple algorithm of the RAFSI method; (*ii*) the fact that the RAFSI method has a new methodology for data standardization that enables the transformation of data from an initial decision matrix into an interval that is suitable for rational decision making; (*iii*) the fact that the mathematical formulation of the RAFSI method eliminates the rank reversal problem, which is one of the significant shortcomings of many traditional MCDM methods.

The following section shows the extension of the RAFSI method to a fuzzy environment (RAFSI-F). By applying fuzzy sets, the RAFSI algorithm has been improved and adapted to handle inaccuracies and uncertainties that arise when solving real-world problems. The algorithm of the RAFSI-F method is realized through six steps:

Step 1: Forming an aggregated fuzzy initial decision matrix. Assume that the evaluation of alternatives from the set A_i $(i = 1, 2, \ldots, m)$ is carried out by k experts. The experts evaluate the alternatives in relation to the defined set of criteria C_j $(j = 1, 2, \ldots, n)$ using a predefined fuzzy linguistic scale. Then, we can present the judgment of k expert as a matrix $X^{(e)} = [\widetilde{\xi}_{ij}^{(e)}]_{m\times n}$, where $1 \le e \le k$.

$$X^{(e)} = \begin{bmatrix} \widetilde{\xi}_{ij}^{(e)} & \widetilde{\xi}_{12}^{(e)} & \cdots & \widetilde{\xi}_{1n}^{(e)} \\ \widetilde{\xi}_{21}^{(e)} & \widetilde{\xi}_{22}^{(e)} & \cdots & \widetilde{\xi}_{2n}^{(e)} \\ \vdots & \vdots & \ddots & \vdots \\ \widetilde{\xi}_{m1}^{(e)} & \widetilde{\xi}_{m2}^{(e)} & \cdots & \widetilde{\xi}_{mn}^{(e)} \end{bmatrix}_{m\times n} ;1 \le i \le m; 1 \le j \le n; 1 \le e \le k \tag{10}$$

where $\widetilde{\xi}_{ij}^{(e)} = \left(\xi_{ij}^{l(e)}, \xi_{ij}^{s(e)}, \xi_{ij}^{u(e)}\right)$; $(i = 1, \ldots, m;\ j = 1, \ldots, n)$ represents a fuzzy value from the fuzzy linguistic scale.

Accordingly, we obtain a matrix $X^{(e)}$ for each expert, i.e., we obtain k expert initial decision matrices $X^{(1)}, X^{(2)}, \ldots, X^{(e)}, \ldots, X^{(k)}$. For each expert matrix $X^{(e)} = \left[\widetilde{\xi}_{ij}^{(e)}\right]_{m\times n}$ at position (i, j), we obtain a fuzzy sequence $\widetilde{\xi}_{ij}^{(e)} = \left(\xi_{ij}^{l(e)}, \xi_{ij}^{s(e)}, \xi_{ij}^{u(e)}\right)$. By applying Equation (11), we obtain an averaged fuzzy number

$\widetilde{\xi}_{ij} = \left(\xi^l_{ij}, \xi^s_{ij}, \xi^u_{ij}\right)$, where ξ^l_{ij} and ξ^u_{ij} represent the lowest and highest limits of the fuzzy number interval, respectively, while ξ^s_{ij} represents the value in which the fuzzy number $\widetilde{\xi}_{ij}$ has a maximum value.

$$\widetilde{\xi}_{ij} = \left(\xi^l_{ij}, \xi^s_{ij}, \xi^u_{ij}\right) = \begin{cases} \xi^l_{ij} = \left(\frac{1}{k(k-1)} \sum\limits_{\substack{i,j=1 \\ i \neq j}}^{k} \xi_i^{lp} \xi_j^{lq} \right)^{\frac{1}{p+q}} \\ \xi^s_{ij} = \left(\frac{1}{k(k-1)} \sum\limits_{\substack{i,j=1 \\ i \neq j}}^{k} \xi_i^{sp} \xi_j^{sq} \right)^{\frac{1}{p+q}} \\ \xi^u_{ij} = \left(\frac{1}{k(k-1)} \sum\limits_{\substack{i,j=1 \\ i \neq j}}^{k} \xi_i^{up} \xi_j^{uq} \right)^{\frac{1}{p+q}} \end{cases} \tag{11}$$

By applying Equation (11), we obtain an averaged fuzzy initial decision matrix

$$X = \begin{bmatrix} \widetilde{\xi}_{11} & \widetilde{\xi}_{12} & \cdots & \widetilde{\xi}_{1n} \\ \widetilde{\xi}_{21} & \widetilde{\xi}_{22} & \cdots & \widetilde{\xi}_{2n} \\ \vdots & \vdots & \ddots & \vdots \\ \widetilde{\xi}_{m1} & \widetilde{\xi}_{m2} & \cdots & \widetilde{\xi}_{mn} \end{bmatrix}_{m \times n} \tag{12}$$

where $\widetilde{\xi}_{ij} = \left(\xi^l_{ij}, \xi^s_{ij}, \xi^u_{ij}\right)$ denotes the value of the i-th alternative for the j-th criterion ($i = 1, 2, \ldots, m$; $j = 1, 2, \ldots, n$). The elements of the matrix X (Equation (12)) represent the fuzzy numbers obtained by averaging the elements from the expert initial decision matrices $X^{(k)}$.

Step 2: Defining ideal and anti-ideal values. For each criterion $C_j (j = 1, 2, \ldots, n)$, a decision maker defines $\widetilde{\xi}_{I_j}$ and $\widetilde{\xi}_{N_j}$, where $\widetilde{\xi}_{I_j}$ represents an ideal value by criterion C_j, while $\widetilde{\xi}_{N_j}$ represents an anti-ideal value by criterion C_j.

Step 3: Mapping the elements of the initial decision matrix into criterion intervals. In the previous step, the criterion intervals $C_j \in \left[\widetilde{\xi}_{N_j}, \widetilde{\xi}_{I_j}\right]$ (for *max* criteria) and $C_j \in \left[\widetilde{\xi}_{I_j}, \widetilde{\xi}_{N_j}\right]$ (for *min* criteria) are defined.

For each alternative from the set A_i ($i = 1, 2, \ldots, m$), we define the function $\widetilde{f}_{A_i}(C_j)$ that maps the criterion intervals from the aggregated initial decision matrix (12) to the criterion interval $[n_1, n_b]$, Equation (13):

$$\widetilde{f}_{A_i}(C_j) = \frac{n_b - n_1}{\widetilde{\xi}_{I_j} - \widetilde{\xi}_{N_j}} \widetilde{\xi}_{ij} + \frac{\widetilde{\xi}_{I_j} \cdot n_1 - \widetilde{\xi}_{N_j} \cdot n_b}{\widetilde{\xi}_{I_j} - \widetilde{\xi}_{N_j}} \tag{13}$$

where n_b and n_1 represent a relation that shows how much an ideal value is better than an anti-ideal value; $\widetilde{\xi}_{I_j}$ and $\widetilde{\xi}_{N_j}$ represent the ideal and anti-ideal value for the criterion C_j, respectively, while $\widetilde{\xi}_{ij}$ denotes the value of the i-th alternative for the j-th criterion from the aggregated initial decision matrix (12). It is recommended that the ideal value is at least six times better than the anti-ideal (barely acceptable values), i.e., that $n_1 = 1$ and $n_b = 6$.

Thus, we obtain a standardized decision matrix $T = \left[\widetilde{\varphi}_{ij}\right]_{m\times n}(i = 1,2,\ldots,m,\ j = 1,2,\ldots,n)$ in which all elements of the matrix are transformed into an interval $\widetilde{\varphi}_{ij} \in [n_1,\ n_b]$.

$$T = \begin{array}{c} \\ A_1 \\ A_2 \\ \vdots \\ A_m \end{array} \begin{array}{c} \begin{array}{cccc} C_1 & C_2 & \ldots & C_n \end{array} \\ \left[\begin{array}{cccc} \widetilde{\varphi}_{11} & \widetilde{\varphi}_{12} & \cdots & \widetilde{\varphi}_{1n} \\ \widetilde{\varphi}_{21} & \widetilde{\varphi}_{22} & \cdots & \widetilde{\varphi}_{2n} \\ \vdots & \vdots & \ddots & \vdots \\ \widetilde{\varphi}_{m1} & \widetilde{\varphi}_{m2} & \cdots & \widetilde{\varphi}_{mn} \end{array}\right] \end{array} \tag{14}$$

where the elements $\widetilde{\varphi}_{ij}$ of the matrix T are obtained by applying Equation (13), i.e., $\widetilde{\varphi}_{ij} = f_{A_i}(C_j)$.

Step 4: Forming a normalized decision matrix $N = \left[\hat{\varphi}_{ij}\right]_{m\times n}$ $(i = 1,2,\ldots,m,\ j = 1,2,\ldots,n)$. By applying Equation (15), the normalization of the element of matrix T is performed.

$$\hat{\varphi}_{ij} = \begin{cases} \frac{\widetilde{\varphi}_{ij}}{2A},\ for\ \max\ criteria \\ \frac{H}{2\widetilde{\varphi}_{ij}},\ for\ \min\ criteria \end{cases} \tag{15}$$

where A and H represent the arithmetic and harmonic means of the elements n_1 and n_b, respectively, i.e.,

$$A = \frac{n_1 + n_b}{2} \tag{16}$$

$$H = \frac{2}{\frac{1}{n_1} + \frac{1}{n_b}} \tag{17}$$

Thus, we form a new normalized decision matrix, Equation (18):

$$N = \begin{array}{c} \\ A_1 \\ A_2 \\ \vdots \\ A_m \end{array} \begin{array}{c} \begin{array}{cccc} C_1 & C_2 & \ldots & C_n \end{array} \\ \left[\begin{array}{cccc} \widetilde{\gamma}_{11} & \widetilde{\gamma}_{12} & \cdots & \widetilde{\gamma}_{1n} \\ \widetilde{\gamma}_{21} & \widetilde{\gamma}_{22} & \cdots & \widetilde{\gamma}_{2n} \\ \vdots & \vdots & \ddots & \vdots \\ \widetilde{\gamma}_{m1} & \widetilde{\gamma}_{m2} & \cdots & \widetilde{\gamma}_{mn} \end{array}\right] \end{array} \tag{18}$$

where $\widetilde{\gamma}_{ij} \in [0,1]$ represent the normalized elements of the matrix N.

Step 5: Calculation of fuzzy criterion functions of alternatives $\widetilde{Q}(A_i)$ and ranking the alternatives. By applying Equation (19), the criterion functions of alternatives $\widetilde{Q}(A_i)$ are calculated and the ranking of alternatives is performed.

$$\widetilde{Q}(A_i) = \sum_{j=1}^{n} w_j \widetilde{\gamma}_{ij} \tag{19}$$

From the considered set of alternatives, the alternative that has a higher value of fuzzy criterion function $\widetilde{Q}(A_i)$ is selected. In order to simplify the ranking of criterion functions $\widetilde{Q}(A_i)$, it is proposed defuzzification using the equation $Q(A_i) = \left(Q(A_i)^l + 4\cdot Q(A_i)^s + Q(A_i)^u\right)/6$.

4. Application of Fuzzy LBWA–MACBETH–RAFSI Multi-Criteria Model

This section presents the application of a proposed multi-criteria methodology for the selection of strategic guidelines for the reorganization of the healthcare system under the conditions of the COVID-19 pandemic in the Republic of Serbia.

For the evaluation of strategic guidelines, five criteria were defined: C1—Assessment of which strategic guideline provides the best results in circumstances considered; C2—Assessment of the quality of services provided for patients treated in an infectious disease ward of a clinical center; C3—Assessment of the quality of services provided in non-infectious disease units of a clinical center; C4—Assessment of the effectiveness of preventing the spread of infection in the considered region/country; C5—Assessment of the ratio of the cost for implementing the strategic guideline to the quality of performance of the healthcare center within the strategic guideline. The criteria were defined on the basis of a questionnaire survey of 30 experts who participated in the work of the Republic Crisis Response Team. Most of the surveyed experts are medical professionals. The values of the weights of the criteria are directly dependent on infection capacity and pandemic spread risk (δ). Depending on infection capacity and pandemic spread risk (δ), the values of weight coefficients of criteria are defined and a strategic guideline for the reorganization of the healthcare system is selected. Based on statistical reports and experts' estimation, the Republic Crisis Response Team have defined the values $p = 1.8$ and $\theta = 3$, on the basis of which we obtain $\delta = 5.4$. Since the value $\delta = 5.4$ is in the interval $5.0 < \delta \leq 7.0$, this implies that there is a high infection capacity and pandemic spread risk (Table 2). Based on the defined value δ, experts of the crisis response team were surveyed and the values of weight coefficients were defined using the LBWA model.

(a) Application of LBWA model

The following section presents the procedure for defining weight coefficients applying the LBWA model:

Step 1: Determining the most significant criterion from the set of criteria $S = \{C_1, C_2, \ldots, C_5\}$. A total of 30 experts from three crisis response teams participated in the research. Since the experts were grouped within the crisis response teams, group decisions were submitted by each crisis response team separately. Thus, in this study, each crisis response team is viewed as a separate expert preference. As the most significant/influential criterion, all three expert groups (*EG*1, *EG*2 and *EG*3) defined the criterion C_1.

Step 2. Grouping criteria within significance levels. In accordance with expert preferences, the criteria are grouped into the following levels:

$$\begin{aligned}&EG1 \text{ and } EG3: \\ &L_1 = \{C_1, C_2, C_3\}, \\ &L_2 = \{C_4\}, \\ &L_3 = \{C_5\}.\end{aligned}$$

$$\begin{aligned}&EG2: \\ &L_1 = \{C_1, C_2, C_3\}, \\ &L_2 = \{C_4\}, \\ &L_3 = \varnothing \\ &L_4 = \{C_5\}.\end{aligned}$$

Step 3. Based on Equation (2), the maximum value of scale (λ) for the evaluation of criteria within the significance levels is defined.

$$EG1 \text{ and } EG3: \quad \left.\begin{aligned}&L_1 = \{C_1, C_2, C_3\}, \\ &L_2 = \{C_4\}, \\ &L_3 = \{C_5\}.\end{aligned}\right\} \Rightarrow \lambda = \max\{|L_1|, |L_2|, |L_3|\} = 3$$

$$EG2: \quad \left.\begin{aligned}&L_1 = \{C_1, C_2, C_3\}, \\ &L_2 = \{C_4\}, \\ &L_3 = \varnothing \\ &L_4 = \{C_5\}.\end{aligned}\right\} \Rightarrow \lambda = \max\{|L_1|, |L_2|, |L_3|, |L_4|\} = 3$$

The scale for the evaluation of criteria is defined for each expert group separately. The maximum number of criteria within the significance levels is the same for all three expert groups. Therefore, the same scale for evaluation of criteria $I \in [0,3]$ was defined. Comparisons of the criteria by expert groups $EG1$, $EG2$ and $EG3$ are shown in Table 4.

Table 4. Comparisons of the criteria within significance levels by $EG1$, $EG2$ and $EG3$.

Criteria	*EG1*	*EG2*	*EG3*
C1	Level 1; I = 0.0	Level 1; I = 0.0	Level 1; I = 0.0
C2	Level 1; I = 1.0	Level 1; I = 1.2	Level 1; I = 1.5
C3	Level 1; I = 1.5	Level 1; I = 1.4	Level 1; I = 1.7
C4	Level 2; I = 0.5	Level 2; I = 1.0	Level 2; I = 0.8
C5	Level 3; I = 0.0	Level 4; I = 0.0	Level 3; I = 0.3

Step 4. Based on the value λ and conditions that the elasticity coefficient is $\Delta\lambda > \lambda$, $\lambda = \max\{|L_1|, |L_2|, \ldots, |L_l|\}$, in this research, the value $\Delta\lambda = 4$ is taken as the elasticity coefficient.

Step 5. Defining the influence function of the criteria. By applying Equation (4), the influence functions of the criteria for each expert group are calculated:

$$EG1: f(C_j) = \begin{matrix} C1 \\ C2 \\ C3 \\ C4 \\ C5 \end{matrix} \begin{Bmatrix} 1.000 \\ 0.800 \\ 0.727 \\ 0.471 \\ 0.333 \end{Bmatrix}; EG2: f(C_j) = \begin{matrix} C1 \\ C2 \\ C3 \\ C4 \\ C5 \end{matrix} \begin{Bmatrix} 1.000 \\ 0.769 \\ 0.741 \\ 0.444 \\ 0.250 \end{Bmatrix}; EG3: f(C_j) = \begin{matrix} C1 \\ C2 \\ C3 \\ C4 \\ C5 \end{matrix} \begin{Bmatrix} 1.000 \\ 0.727 \\ 0.702 \\ 0.455 \\ 0.325 \end{Bmatrix}$$

Thus, for the criterion $C2$ from $EG1$, by applying Equation (4), we obtain the value $f(C_2) = 4/(1 \cdot 4 + 1) = 0.800$. In a similar way, we obtain the remaining values of the influence function of the criteria.

Step 6. Calculation of optimal values of the weight coefficients of the criteria. By applying Equation (5), we obtain the weight coefficient of the best criterion (C1).

$$EG1: w_1 = \frac{1}{1 + 0.800 + 0.727 + 0.471 + 0.333} = 0.300$$

$$EG2: w_1 = \frac{1}{1 + 769 + 0.741 + 0.444 + 0.250} = 0.312$$

$$EG3: w_1 = \frac{1}{1 + 0.727 + 0.702 + 0.455 + 0.325} = 0.312$$

The values of the weight coefficients of the remaining criteria are obtained by applying Equation (6).

$$EG1: w_j = \begin{matrix} w_1 \\ w_2 \\ w_3 \\ w_4 \\ w_5 \end{matrix} \begin{bmatrix} 0.300 \\ 0.240 \\ 0.218 \\ 0.141 \\ 0.100 \end{bmatrix}; EG2: w_j = \begin{matrix} w_1 \\ w_2 \\ w_3 \\ w_4 \\ w_5 \end{matrix} \begin{bmatrix} 0.312 \\ 0.240 \\ 0.231 \\ 0.139 \\ 0.078 \end{bmatrix}; EG3: w_j = \begin{matrix} w_1 \\ w_2 \\ w_3 \\ w_4 \\ w_5 \end{matrix} \begin{bmatrix} 0.312 \\ 0.227 \\ 0.219 \\ 0.142 \\ 0.101 \end{bmatrix}$$

By arithmetic averaging, we obtain the final values of weight coefficients $w_j^{LBWA} = (0.308, 0.236, 0.223, 0.141, 0.093)^T$.

(b) Application of MACBETH model:

The following section presents the procedure for defining weight coefficients applying the MACBETH model.

In the first step, the criteria were discussed with the decision makers in the expert groups and the group decision can be seen in Table 5.

Table 5. Pairwise comparison matrix of expert groups.

Crit.	C2	C3	C4	C5
C1	moderate	strong	moderate	moderate
C2		weak	weak	weak
C3			moderate	moderate
C4				weak

According to the decision makers' evaluations, the linear programming model (7) is constructed as follows:

$$\begin{array}{l} z_{\min} = \phi(o_1) \\ \phi(o_1) \geq \phi(o_2) + 3;\ \phi(o_1) \geq \phi(o_3) + 5;\ \ldots; \\ \phi(o_3) \geq \phi(o_5) + 3;\ \phi(o_4) \geq \phi(o_5) + 2; \\ \phi(o_1) - \phi(o_2) \geq \phi(o_4) - \phi(o_5) + 1; \\ \phi(o_1) - \phi(o_3) \geq \phi(o_4) - \phi(o_5) + 3; \\ \ldots \\ \phi(o_3) - \phi(o_4) \geq \phi(o_4) - \phi(o_5) + 1; \\ \phi(o_3) - \phi(o_5) \geq \phi(o_4) - \phi(o_5) + 1; \\ \phi(o_5) = 1 \\ \forall \phi(o_j) \geq 0;\ j = 1, 2, \ldots, 5;. \end{array}$$

By solving the linear model, we get the criterion values $\phi(o_j) = (11, 8, 6, 3, 1)$. By applying Equation (8), the values $\phi(o_j)$ are normalized, and we obtain the final values of the weighting coefficients according to the MACBETH methodology $w_j^{MACBETH} = (0.379, 0.276, 0.207, 0.103, 0.034)$.

Finally, based on the obtained values of the weights of the criteria according to the LBWA and MACBETH methodology, using Equation (9) we obtain the combined values of the weighting coefficients; see Table 6.

Table 6. Combined values of weight coefficients.

Criteria	LBWA	MACBETH	Final Weights
C1	0.308	0.379	0.344
C2	0.236	0.276	0.256
C3	0.223	0.207	0.215
C4	0.141	0.103	0.122
C5	0.093	0.034	0.064

To obtain aggregated values of weighting coefficients, a value $\xi = 0.5$ was used, whereby both methodologies equally participate in defining the final weights of the criteria.

(C) Application of fuzzy RAFSI model:

After defining the weight coefficients of the criteria, the experts from the crisis response teams evaluated the strategic guidelines using the fuzzy RAFSI model. The following section presents the application of the fuzzy RAFSI model for the evaluation of four strategic guidelines for the reorganization of the healthcare system for the value $\delta = 5.4$. The experts evaluated the strategic guidelines A_i $(i = 1, 2, \ldots, 4)$ in relation to five criteria C_j $(j = 1, 2, \ldots, 5)$ defined in the previous section of the paper. All criteria C_j $(j = 1, 2, \ldots, 5)$ belong to the group of *max* criteria. To evaluate the strategic guidelines (alternatives), the experts used a fuzzy linguistic scale shown in Table 7.

Table 7. Fuzzy linguistic scale for evaluating alternatives [29,30].

Linguistic Terms	Membership Function
Very poor (VP)	(1, 1, 1)
Poor (P)	(1, 2, 3)
Medium poor (MP)	(2, 3, 4)
Medium (M)	(3, 4, 5)
Medium high (MH)	(4, 5, 6)
High (H)	(5, 6, 7)
Very high (VH)	(6, 7, 8)
Extremely high (EH)	(7, 8, 9)
Absolutely high (AH)	(8, 9, 9)

After the evaluation of the strategic guidelines, expert correspondence matrices are obtained, which are presented in the following section:

$$X^{(e)} = \begin{array}{c} \\ A_1 \\ A_2 \\ A_3 \\ A_4 \end{array} \begin{array}{c} \begin{array}{ccccc} C_1 & C_2 & C_3 & C_4 & C_5 \end{array} \\ \begin{bmatrix} \text{H; MP; H} & \text{M; H; H} & \text{MH; MH; H} & \text{H; H; VH} & \text{MH; VH; VH} \\ \text{H; H; M} & \text{M; M; H} & \text{H; MH; M} & \text{MH; H; H} & \text{VH; H; VH} \\ \text{VH; H; VH} & \text{VH; EH; EH} & \text{H; EH; VH} & \text{H; VH; MH} & \text{MH; MH; H} \\ \text{VH; VH; H} & \text{H; VH; VH} & \text{MH; VH; VH} & \text{VH; H; VH} & \text{H; H; MH} \end{bmatrix} \end{array}$$

By applying Equation (11), we obtain an averaged initial decision matrix:

$$X = \begin{array}{c} \\ A_1 \\ A_2 \\ A_3 \\ A_4 \end{array} \begin{array}{c} \begin{array}{ccccc} C_1 & C_2 & C_3 & C_4 & C_5 \end{array} \\ \begin{bmatrix} (3.87, 4.90, 5.92) & (4.28, 5.29, 6.30) & (4.32, 5.32, 6.32) & (5.32, 6.32, 7.33) & (5.29, 6.30, 7.30) \\ (4.28, 5.29, 6.30) & (3.61, 4.62, 5.63) & (3.96, 4.97, 5.97) & (4.65, 5.66, 6.66) & (5.66, 6.66, 7.66) \\ (5.66, 6.66, 7.66) & (6.66, 7.66, 8.66) & (5.97, 6.98, 7.98) & (4.97, 5.97, 6.98) & (4.32, 5.32, 6.32) \\ (5.66, 6.66, 7.66) & (5.66, 6.66, 7.66) & (5.29, 6.30, 7.30) & (5.66, 6.66, 7.66) & (4.65, 5.66, 6.66) \end{bmatrix} \end{array}$$

Step 1: Experts defined an ideal $\widetilde{\xi}_{I_j}$ and anti-ideal $\widetilde{\xi}_{N_j}$ value for each criterion C_j $(i = 1, 2, \ldots, 5)$. For the considered example, by consensus, the decision makers defined the following ideal and anti-ideal points: $\widetilde{\xi}_{I_j} = (10, 10.5, 11.0)$ and $\widetilde{\xi}_{N_j} = (0.5, 0.5, 1.0)$. Since the same linguistic scale have been used to evaluate all criteria, the same ideal and anti-ideal points, $\widetilde{\xi}_{I_j} = (10, 10.5, 11.0)$ and $\widetilde{\xi}_{N_j} = (0.5, 0.5, 1.0)$, apply to all criteria.

Step 2: Based on the defined ideal and anti-ideal points, criterion intervals $C_j \in \left[\widetilde{\xi}_{N_j}, \widetilde{\xi}_{I_j}\right]$ are formed, where $\widetilde{\xi}_{N_j} = (0.5, 0.5, 1.0)$ and $\widetilde{\xi}_{I_j} = (10, 10.5, 11.0)$. In this research, it is defined that the ideal value of the alternative is nine times higher than the anti-ideal value, i.e., that $n_1 = 1$ and $n_s = 9$. By applying Equation (13), we define the functions on the basis of which the standardization of criteria is performed, i.e., the transformation of the values of the initial decision matrix into an interval $[1, 9]$. Thus, we obtain the following fuzzy functions:

$$\widetilde{f}_{A_i}(C_j) = \frac{9-1}{(10, 10.5, 11) - (0.5, 0.5, 1.0)}\widetilde{\xi}_{ij} + \frac{(10, 10.5, 11) \cdot 1 - (0.5, 0.5, 1.0) \cdot 9}{(10, 10.5, 11) - (0.5, 0.5, 1.0)}$$

i.e.,

$$\widetilde{f}_{A_i}(C_j) = \left(f^l_{ij}(C_j), f^s_{ij}(C_j), f^u_{ij}(C_j)\right) = \begin{cases} f^l_{A_i}(C_1) = \frac{9-1}{10-1}\xi^l_{ij} + \frac{10 \cdot 1 - 1 \cdot 9}{10-1} = 0.89 \cdot \xi^l_{ij} + 0.11 \\ f^s_{A_i}(C_1) = \frac{9-1}{10.5-0.5}\xi^s_{ij} + \frac{10 \cdot 1 - 0.5 \cdot 9}{10.5-0.5} = 0.80 \cdot \xi^s_{ij} + 0.60 \\ f^u_{A_i}(C_1) = \frac{9-1}{11-0.5}\xi^u_{ij} + \frac{11 \cdot 1 - 0.5 \cdot 9}{11-0.5} = 0.76 \cdot \xi^s_{ij} + 0.62 \end{cases}$$

where $\widetilde{\xi}_{ij} = \left(\xi^l_{ij}, \xi^s_{ij}, \xi^u_{ij}\right)$ represent elements of the aggregated initial decision matrix.

Since all criteria values in the initial decision matrix cover the same value interval, i.e., they are defined using the same linguistic scale, the same function $\widetilde{f_{A_i}}(C_j)$ is applied to map all criteria C_j $j = 1,2,\ldots,5$).

By applying the function $\widetilde{f_{A_i}}(C_j)$, we obtain a standardized decision matrix ($T = \left[\widetilde{\varphi}_{ij}\right]_{4\times 5}$, $i = 1,2,\ldots,4$, $j = 1,2,\ldots,5$) in which all elements of the matrix are transformed into an interval $[1,9]$.

$$T = \begin{array}{c} \\ A1 \\ A2 \\ A3 \\ A4 \end{array} \begin{array}{c} \begin{array}{ccccc} C1 & C2 & C3 & C4 & C5 \end{array} \\ \begin{bmatrix} (3.55,4.52,5.13) & (3.91,4.83,5.42) & (3.95,4.86,5.44) & (4.84,5.66,6.20) & (4.81,5.64,6.18) \\ (3.91,4.83,5.42) & (3.31,4.30,4.91) & (3.62,4.57,5.17) & (4.24,5.13,5.69) & (5.13,5.93,6.45) \\ (5.13,5.93,6.45) & (6.02,6.73,7.22) & (5.41,6.18,6.70) & (4.52,5.38,5.93) & (3.95,4.86,5.44) \\ (5.13,5.93,6.45) & (5.13,5.93,6.45) & (4.81,5.64,6.18) & (5.13,5.93,6.45) & (4.24,5.13,5.69) \end{bmatrix} \end{array}$$

By substituting the values from the matrix X, Equation (12), into the function $\widetilde{f_{A_i}}(C_j)$, Equation (13), we obtain the remaining values of the elements $\widetilde{\varphi}_{ij}$. Elements at position A_i-C_1 are obtained by applying the function $\widetilde{f_{A_i}}(C_j)$:

$$\widetilde{f_{A_1}}(C_1) = (3.55,4.52,5.13) = \begin{cases} f_{A_1}^{l}(C_1) = 0.89 \cdot \xi_{11}^{l} + 0.11 = 0.89 \cdot 3.87 + 0.11 = 3.55 \\ f_{A_1}^{s}(C_1) = 0.80 \cdot \xi_{11}^{s} + 0.60 = 0.80 \cdot 4.90 + 0.60 = 4.52 \\ f_{A_1}^{u}(C_1) = 0.76 \cdot \xi_{11}^{u} + 0.62 = 0.76 \cdot 5.92 + 0.62 = 5.13 \end{cases}$$

$$\widetilde{f_{A_2}}(C_1) = (3.91,4.83,5.42) = \begin{cases} f_{A_2}^{l}(C_1) = 0.89 \cdot 4.28 + 0.11 = 3.91 \\ f_{A_2}^{s}(C_1) = 0.80 \cdot 5.29 + 0.60 = 4.83 \\ f_{A_2}^{u}(C_1) = 0.76 \cdot 6.30 + 0.62 = 5.42 \end{cases}$$

$$\widetilde{f_{A_3}}(C_1) = (5.13,5.93,6.45) = \begin{cases} f_{A_3}^{l}(C_1) = 0.89 \cdot 5.66 + 0.11 = 5.13 \\ f_{A_3}^{s}(C_1) = 0.80 \cdot 6.66 + 0.60 = 5.93 \\ f_{A_3}^{u}(C_1) = 0.76 \cdot 7.66 + 0.62 = 6.45 \end{cases}$$

$$\widetilde{f_{A_4}}(C_1) = (5.13,5.93,6.45) = \begin{cases} f_{A_3}^{l}(C_1) = 0.89 \cdot 5.66 + 0.11 = 5.13 \\ f_{A_3}^{s}(C_1) = 0.80 \cdot 6.66 + 0.60 = 5.93 \\ f_{A_3}^{u}(C_1) = 0.76 \cdot 7.66 + 0.62 = 6.45 \end{cases}$$

Step 3: We calculate the arithmetic mean and the harmonic mean of the minimum and maximum elements, $n_1 = 1$ and $n_b = 9$:

$$A = (n_1 + n_b)/2 = (1+9)/2 = 5$$

$$H = \frac{2}{\frac{1}{n_1} + \frac{1}{n_b}} = \frac{2}{\frac{1}{9} + \frac{1}{1}} = 1.80$$

So, for $n_1 = 1$ and $n_b = 9$, we obtain that the arithmetic mean is $A = 5.5$, while the harmonic mean is $H = 1.80$.

Step 4: By applying Equation (15), the elements of the matrix T are normalized and transformed depending on whether they belong to the *min* or *max* type of criteria. Since, in our example, all the criteria are of the *max* type, only the arithmetic mean will be used to normalize the elements of the T matrix. So, we obtain a new matrix $N = \left[\widetilde{\gamma}_{ij}\right]_{4\times 5}$ ($i = 1,2,\ldots,4$, $j = 1,2,\ldots,5$), Equation (18).

$$N = \begin{array}{c} \\ A1 \\ A2 \\ A3 \\ A4 \end{array} \begin{array}{c} \begin{array}{ccccc} C1 & C2 & C3 & C4 & C5 \end{array} \\ \begin{bmatrix} (0.35,0.45,0.51) & (0.39,0.48,0.54) & (0.39,0.49,0.54) & (0.48,0.57,0.62) & (0.48,0.56,0.62) \\ (0.39,0.48,0.54) & (0.33,0.43,0.49) & (0.36,0.46,0.52) & (0.42,0.51,0.57) & (0.51,0.59,0.65) \\ (0.51,0.59,0.65) & (0.60,0.67,0.72) & (0.54,0.62,0.67) & (0.45,0.54,0.59) & (0.39,0.49,0.54) \\ (0.51,0.59,0.65) & (0.51,0.59,0.65) & (0.48,0.56,0.62) & (0.51,0.59,0.65) & (0.42,0.51,0.57) \end{bmatrix} \end{array}$$

Thus, for example, for criterion $C1$, at position $A1$-$C1$, we obtain that $\widetilde{\gamma}_{11} = \frac{(3.55,4.52,5.13)}{2 \cdot 5} = \left(\frac{3.55}{10}, \frac{4.52}{10}, \frac{5.13}{10}\right) = (0.355, 0.452, 0.513)$.

Step 5: By applying Equation (19), the criterion functions of alternatives $\widetilde{Q}(A_i)$ are calculated. The alternative is ranked on the basis of value $\widetilde{Q}(A_i)$, so it is more desirable that the alternative has a higher value $\widetilde{Q}(A_i)$.

The ranking of alternatives is shown in Table 8.

Table 8. Criterion functions and final ranking of alternatives.

Alt.	$\widetilde{Q}(A_i)$	Crisp $Q(A_i)$	Rank
$A1$	(0.397, 0.489, 0.547)	0.4829	3
$A2$	(0.382, 0.475, 0.534)	0.4690	4
$A3$	(0.527, 0.605, 0.658)	0.6011	1
$A4$	(0.501, 0.582, 0.635)	0.5771	2

Based on the obtained results, we can single out the strategic guideline A3 as the dominant solution.

5. Validation of the Results

The validation of the results was carried out through two sections. The first section considers the analysis of the influence of changing parameters p and q in the Bonferroni integration operator. The second section presents the validation of the results through comparison with other multi-criteria techniques.

5.1. Influence of Changing Parameters p and q on the Results of Ranking

The Bonferroni integration operator was used to integrate expert preferences into an aggregated initial decision matrix. During the formation of the aggregated initial decision matrix on the basis of which the initial rank was obtained (Table 8), the values of the parameters $p = q = 1$ were adopted. Considering the mathematical equation of the Bonferroni integration operator, it is clear that different values of parameters p and q lead to the transformation of the mathematical Equation (11) [31,32]. Having in mind the stated facts, it is necessary to examine the influence of changing parameters p and q on changing the values in the aggregated decision matrix. Additionally, changes in values in the aggregate decision matrix indirectly lead to changes in the values of criterion functions and potential changes in the ranks of strategic guidelines. Therefore, in the following section, the influence of changing parameters p and q on changing the values of the criterion functions of strategic guidelines is analyzed.

In order to validate the results, the change in parameters p and q was simulated and analyzed in the interval [1, 100] through 300 scenarios. In the first 200 scenarios, the impact of changing only one parameter of the parameters in the interval $[1, 100]$ was analyzed, while the other parameter had the value one. Thus, through the first 100 scenarios, the change in the parameter value $1 \le p \le 100$ was simulated, while in the next 100 scenarios, the change in the parameter $1 \le q \le 100$ was simulated; see Figure 3a. In the last 100 scenarios, the change in both parameters in the interval $1 \le p, q \le 100$ was simulated; see Figure 3b.

By analyzing the results from Figure 3, we notice that for changes in the values of parameters $1 \le p, q \le 100$ through all 300 scenarios, there is no change in the ranks of strategic guidelines and that the initial rank $A4 > A3 > A2 > A1$ is confirmed. We notice that the change in the values of the parameters in the interval $1 \le p, q \le 100$ leads to a change in the values of criterion functions. However, changes in the values of the criterion functions of the strategic guidelines are not large enough to cause changes in the initial rank. In addition, the results presented show that the initial ordering is credible and robust despite drastic changes in input parameters. It is necessary to emphasize that such an analysis is an indispensable step in multi-criteria models in which integration functions are used for the aggregation of input parameters. The integration functions have free parameters whose values are arbitrarily determined by decision makers. This is confirmed by numerous studies that have shown that changes in the values of integration function parameters can often lead to changes in initial ranks [33–35].

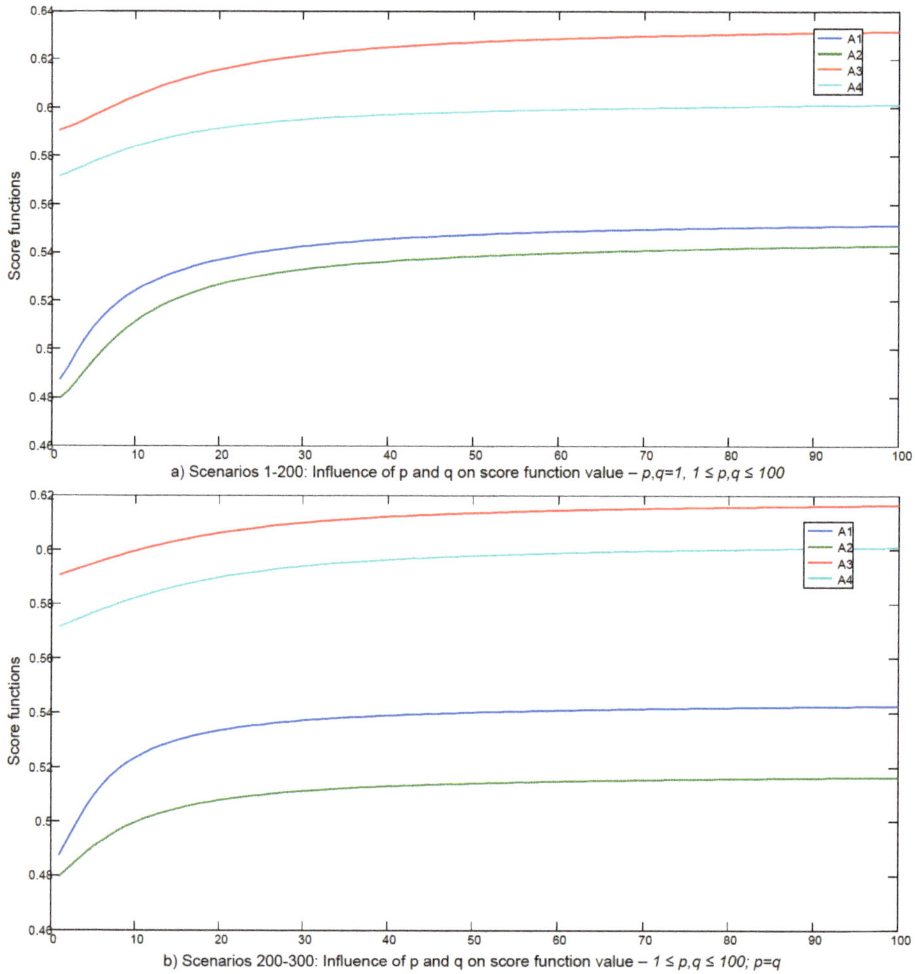

Figure 3. Influence of parameters p and q on changing the utility function of strategic guidelines.

5.2. Computation of Stability of Ranking Based on Different Ranking Methodologies

In order to validate the initial rank, four more multi-criteria techniques were applied under the same conditions (with the same input values) as for the RAFSI method: Compressed Proportional Assessment (COPRAS) [36], Multi-Attributive Ideal-Real Comparative Analysis (MAIRCA) [37], Multi-Attributive Border Approximation area Comparison (MABAC) [38] and Measurement of Alternatives and Ranking According to Compromise Solution (MARCOS) [39]. A comparative overview of the ranks obtained by applying the above multi-criteria techniques is shown in Figure 4.

Based on the comparison of results shown in Figure 4, the order of ranking of the strategic guidelines according to the LBWA–MACBETH–RAFSI model is similar to the order proposed by the multi-criteria techniques developed in [36–39]. The presented analysis proves the validity of the methodology proposed in this paper and that the proposed selection of strategic guideline A3 is credible.

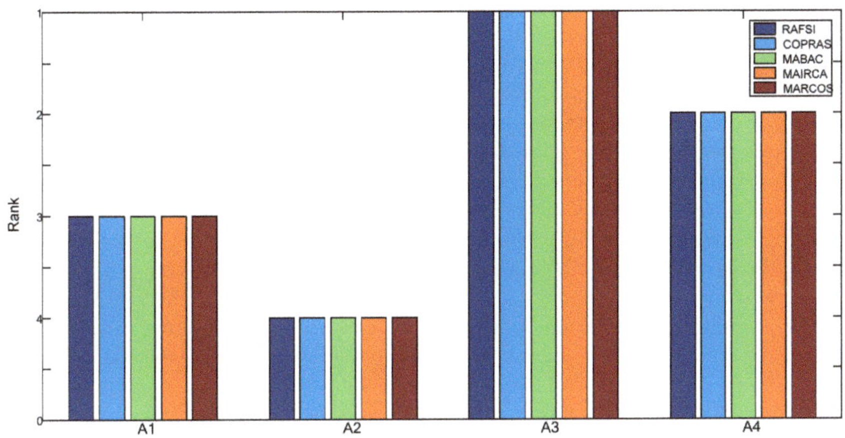

Figure 4. The comparison of different MCDM methodologies.

6. Conclusions

The adaptability of healthcare systems is a key segment that affects the efficiency of the healthcare system of each country. This fact is particularly evident in crisis situations, such as the COVID-19 pandemic, which has affected the whole world. Therefore, it is necessary to efficiently manage the organization of healthcare systems in order to provide adequate medical care to patients and reduce dangers for the population caused by the COVID-19 pandemic to a minimum. This paper presents a multi-criteria model that proposes strategic guidelines for possible adaptation of healthcare system depending on infection capacity and pandemic spread risk. Four strategic guidelines have been defined and five criteria for the evaluation of the strategic guidelines have been determined. The paper proposes a scale on the basis of which infection capacity and pandemic spread risk are defined. Based on the estimation of infection capacity and pandemic spread risk by the crisis response team, the weight coefficients of criteria are defined and a strategic guideline is selected for the possible reorganization of a healthcare system of a region within a country or a whole country. The LBWA and MACBETH methods were proposed to define the values of the weight coefficients of the criteria, while the novel fuzzy RAFSI method was proposed for the evaluation of strategic guidelines and for the proposal of the possible organization of healthcare systems.

The proposed multi-criteria methodology was tested for values of high infection capacity and pandemic spread risk ($1.5 < \delta \leq 2.0$). The application of the fuzzy LBWA–MACBETH–RAFSI methodology was demonstrated and the results were validated. Based on the validation of the results, strategic guideline S3 was proposed as the best solution for the reorganization of the healthcare system under the conditions of a high infection capacity and pandemic spread risk. The testing was performed using the example of the healthcare system in the Republic of Serbia.

The proposed methodology has several contributions that can be observed from two perspectives. The first standpoint refers to the contribution of the proposed methodology within the healthcare filed, where we can highlight the following advantages: (1) an original methodology for reorganization and adjustment of healthcare systems in emergency situations caused by the COVID-19 pandemic has been proposed; (2) a methodology for determining infection capacity and pandemic spread risk has been defined, which is a prerequisite for the application of a multi-criteria methodology; (3) strategic guidelines and criteria for the evaluation of the guidelines were defined on the basis of real research conducted during the COVID-19 pandemic and in which medical experts and other members of the crisis response team in Serbia participated; (4) the proposed methodology provides a new framework for managing the organization of healthcare systems under the dynamic conditions of a pandemic. The model enables national policy makers to establish strategic guidelines for adapting healthcare systems to the capacity of the COVID-19 pandemic; (5) the universality of the proposed multi-criteria methodology allows the application of the proposed methodology for the reorganization of healthcare systems of other countries with minor pre-adjustments of strategic guidelines and criteria depending on the specific characteristics of the country or region in which it is applied.

The second standpoint refers to the contributions of the proposed methodology within the MCDM area, where we can highlight the following contributions: (1) within the proposed multi-criteria methodology, an original extension of the RAFSI method in a fuzzy environment was performed; (2) the fuzzy LBWA–MACBETH–RAFSI model has a flexibility and universality that allow it to be effectively applied to various areas where the application of a rational and robust methodology is required; (3) the hybrid fuzzy LBWA–MACBETH–RAFSI model contributes to the improvement of the literature that considers the theoretical and practical application of multi-criteria techniques; (4) the proposed multi-criteria methodology allows the evaluation of alternatives despite inaccuracies and a lack of quantitative information in a decision-making process and (5) the methodology for the evaluation of healthcare systems in emergency situations is improved through a new approach to treating inaccuracies since the application of this or a similar approach to the healthcare field was not observed in our analysis of the literature.

There are certain limitations that should be considered when interpreting the results of this study. The effectiveness of the proposed fuzzy MCDM tool is evident, but acceptance by other users and decision makers can be a limiting factor. Most users and decision makers readily accept mathematically simple and easy-to-understand decision-making tools. However, tools that require the processing of group information, while respecting the uncertainties and inaccuracies that exist in group preferences, are complex in nature. Therefore, the model used in this paper does not fall into the mathematically simple category of decision-making tools. A particular limitation of this model is the complex mathematical formulations for the fusion of expert preferences into a single group decision. However, integrating the LBWA–MACBETH–RAFSI model into a decision-making system will make it more acceptable for use by heath care management who have to deal with a high degree of uncertainty and inaccuracy during decision making. The proposal of a user-oriented decision support system was made during the preparation of this study and is based on the implementation of Visual Basic and Matlab software packages.

Other limitations of our study relate to the determination of weighting criteria that significantly depend on the subjective characteristics of the available experts participating in the study. Another limitation is the small size of the sample examined and the possible impact of survey formatting on study results. Therefore, it is recommended that further research be organized on a larger sample of respondents familiar with the context of the decision and consider the possibility of increasing the number of criteria that would be grouped into clusters.

The proposed LBWA–MACBETH–RAFSI model is a tool that can be successfully integrated with other MCDM techniques. The development of hybrid multi-criteria models for group decision making, which would be based on the integration of LBWA–MACBETH–RAFSI model and other MCDM tools, is one of the future directions for its application. Another logical step for the future improvement of the LBWA–MACBETH–RAFSI model is its application in other uncertain theories, such as rough, grey, neutrosophic, etc. Numerous linguistic approaches have been developed in recent years, such as extensions of linguistic variables in a neutrosophic environment and an unbalanced linguistic approach. These approaches have attracted a lot of attention in the field of decision making, through the possibility of applying linguistic variables in a decision-making process. Linking these linguistic approaches to the LBWA–MACBETH–RAFSI model and researching the possibility of linguistic modeling of preferences are interesting and promising topics for future research.

Author Contributions: Conceptualization, methodology, validation, writing—original draft preparation, D.P., M.Ž. and D.M.; review and editing, D.D., S.V.J. and P.B. All authors have read and agreed to the published version of the manuscript.

Funding: We acknowledge support by the German Research Foundation and the Open Access Publication Fund of TU Berlin.

Acknowledgments: The authors wish to acknowledge the support received from German Research Foundation and the TU Berlin.

Conflicts of Interest: The authors declare no conflict of interest.

References

1. Stankovic, S.M. *Banje Srbije*; Zavod za Udžbenike: Beograd, Serbia, 2010.
2. Zizovic, M.; Pamucar, D. New model for determining criteria weights: Level Based Weight Assessment (LBWA) model. *Decis. Mak. Appl. Manag. Eng.* **2019**, *2*, 126–137. [CrossRef]
3. Bana, E.; Costa, C.A.; Vansnick, J.C. MACBETH-An Interactive Path Towards The Construction of Cardinal Value Functions. *Int. Trans. Oper. Res.* **1994**, *1*, 489–500. [CrossRef]
4. Zizovic, M.; Pamucar, D.; Albijanic, M.; Chatterjee, P.; Pribicevic, I. Eliminating Rank Reversal Problem Using a New Multi-Attribute Model—The RAFSI Method. *Mathematics* **2020**, *8*, 1015. [CrossRef]
5. Nekoie-Moghadam, M.; Kurland, L.; Moosazadeh, M.; Ingrassia, P.L.; Della Corte, F.; Djalali, A. Tools and checklists used for the evaluation of hospital disaster preparedness: A systematic review. *Disaster Med. Public Health Prep.* **2016**, *10*, 781–788. [CrossRef] [PubMed]
6. Fallah-Aliabadi, S.; Ostadtaghizadeh, A.; Ardalan, A.; Fatemi, F.; Khazai, B.; Mirjalili, M.R. Towards developing a model for the evaluation of hospital disaster resilience: A systematic review. *BMC Health Serv. Res.* **2020**, *20*, 64. [CrossRef]
7. Verheul, M.L.; Dückers, M.L. Defining and operationalizing disaster preparedness in hospitals: A systematic literature review. *Prehosp. Disaster Med.* **2020**, *35*, 61–68. [CrossRef]
8. Alruwaili, A.; Islam, S.; Usher, K. Disaster Preparedness in Hospitals in the Middle East: An Integrative Literature Review. *Disaster Med. Public Health Prep.* **2019**, *13*, 806–816. [CrossRef]
9. Tabatabaei, S.A.N.; Abbasi, S. Risk assessment in social security hospitals of Isfahan Province in case of disasters based on the hospital safety index. *Int. J. Health Syst. Disaster Manag.* **2016**, *4*, 82.
10. Samsuddin, N.M.; Takim, R.; Nawawi, A.H.; Alwee, S.N.A.S. Disaster preparedness attributes and hospital's resilience in Malaysia. *Procedia Eng.* **2018**, *212*, 371–378. [CrossRef]
11. Marzaleh, M.A.; Rezaee, R.; Rezaianzadeh, A.; Rakhshan, M.; Hadadi, G.; Peyravi, M. Emergency Department Preparedness of Hospitals for Radiation, Nuclear Accidents, and Nuclear Terrorism: A Qualitative Study in Iran. *Iran. Red Crescent Med. J.* **2019**, *21*. [CrossRef]
12. Shabanikiya, H.; Jafari, M.; Gorgi, H.A.; Seyedin, H.; Rahimi, A. Developing a practical toolkit for evaluating hospital preparedness for surge capacity in disasters. *Int. J. Disaster Risk Reduct.* **2019**, *34*, 423–428. [CrossRef]
13. Mulyasari, F.; Inoue, S.; Prashar, S.; Isayama, K.; Basu, M.; Srivastava, N.; Shaw, R. Disaster preparedness: Looking through the lens of hospitals in Japan. *Int. J. Disaster Risk Sci.* **2013**, *4*, 89–100. [CrossRef]
14. Hosseini, S.M.; Bahadori, M.; Raadabadi, M.; Ravangard, R. Ranking hospitals based on the disasters preparedness using the TOPSIS technique in western Iran. *Hosp. Top.* **2019**, *97*, 23–31. [CrossRef] [PubMed]
15. Ortiz-Barrios, M.A.; Aleman-Romero, B.A.; Rebolledo-Rudas, J.; Maldonado-Mestre, H.; Montes-Villa, L.; De Felice, F.; Petrillo, A. The analytic decision-making preference model to evaluate the disaster readiness in emergency departments: The ADT model. *J. Multi-Criteria Decis. Anal.* **2017**, *24*, 204–226. [CrossRef]
16. Roy, J.; Adhikary, K.; Kar, S.; Pamucar, D. A rough strength relational DEMATEL model for analysing the key success factors of hospital service quality. *Decis. Mak. Appl. Manag. Eng.* **2018**, *1*, 121–142. [CrossRef]
17. Sarkar, S. COVID-19 Susceptibility Mapping Using Multicriteria Evaluation. *Disaster Med. Public Health Prep.* **2020**, 1–17. [CrossRef]
18. Sangiorgio, V.; Parisi, P. A multicriteria approach for risk assessment of Covid-19 in urban district lockdown. *Saf. Sci.* **2020**, *130*, 104862. [CrossRef]
19. Nardo, P.D.; Gentilotti, E.; Mazzaferri, F.; Cremonini, E.; Hansen, P.; Goossens, H.; Tacconelli, E. Multi-Criteria Decision Analysis to prioritize hospital admission of patients affected by COVID-19 in low-resource settings with hospital-bed shortage. *Int. J. Infect. Dis.* **2020**. [CrossRef]

20. Yildirim, F.S.; Sayan, M.; Sanlidag, T.; Uzun, B.; Uzun, D.; Ozsahin, I. A Clinical Decision Support System for the Treatment of COVID-19 with Multi-Criteria Decision-Making Techniques. *JMIR Med. Inform.* **2020**. [CrossRef]
21. Biswas, T.K.; Das, M.C. Selection of the Barriers of Supply Chain Management in Indian Manufacturing Sectors Due to COVID-19 Impacts. *Oper. Res. Eng. Sci. Theor. Appl.* **2020**, *3*, 1–12.
22. Giordano, G.; Blanchini, F.; Bruno, R.; Colaneri, P.; Di Filippo, A.; Di Matteo, A.; Colaneri, M. Modelling the COVID-19 epidemic and implementation of population-wide interventions in Italy. *Nat. Med.* **2020**, *26*, 855–860. [CrossRef] [PubMed]
23. Kucharski, A.J.; Russell, T.W.; Diamond, C.; Liu, Y.; Edmunds, J.; Funk, S.; Eggo, R.M. Early dynamics of transmission and control of COVID-19: A mathematical modelling study. *PRIME PubMed* **2020**, *20*, 553–558. [CrossRef]
24. Diyaley, S.; Chakraborty, S. Optimization of multi-pass face milling parameters using metaheuristic algorithms. *Facta Univ. Ser. Mech. Eng.* **2019**, *17*, 365–383. [CrossRef]
25. Biswas, S.; Pamucar, D. Facility Location Selection for B-Schools in Indian Context: A Multi-Criteria Group Decision Based Analysis. *Axioms* **2020**, *9*, 77. [CrossRef]
26. Pamucar, D.; Deveci, M.; Canitez, F.; Lukovac, V. Selecting an Airport Ground Access Mode Using Novel Fuzzy LBWA-WASPAS-H Decision Making Model. *Eng. Appl. Artif. Intell.* **2020**, *93*, 103703. [CrossRef]
27. Deveci, M.; Özcan, E.; John, R.; Covrig, C.-F.; Pamucar, D. A Study on Offshore Wind Farm Siting Criteria Using a Novel Interval-valued Fuzzy-rough based Delphi Method. *J. Environ. Manag.* **2020**, *270*, 110916. [CrossRef] [PubMed]
28. Sinani, F.; Erceg, Z.; Vasiljević, M. An evaluation of a third-party logistics provider: The application of the rough Dombi-Hamy mean operator. *Decis. Mak. Appl. Manag. Eng.* **2020**, *3*, 92–107.
29. Petrovic, G.; Mihajlovic, J.; Cojbasic, Z.; Madic, M.; Marinkovic, D. Comparison of three fuzzy MCDM methods for solving the supplier selection problem. *Facta Univ. Ser. Mech. Eng.* **2019**, *17*, 455–469. [CrossRef]
30. Badi, I.; Shetwan, A.; Hemeda, A. A grey-based assessment model to evaluate health-care waste treatment alternatives in Libya. *Oper. Res. Eng. Sci. Theory Appl.* **2019**, *2*, 92–106. [CrossRef]
31. Durmić, E.; Stević, Ž.; Chatterjee, P.; Vasiljević, M.; Tomašević, M. Sustainable supplier selection using combined FUCOM—Rough SAW model. *Rep. Mech. Eng.* **2020**, *1*, 34–43. [CrossRef]
32. Pamucar, D. Normalized weighted Geometric Dombi Bonferoni Mean Operator with interval grey numbers: Application in multicriteria decision making. *Rep. Mech. Eng.* **2020**, *1*, 44–52. [CrossRef]
33. Pamucar, D.; Božanic, D.; Lukovac, V.; Komazec, N. Normalized weighted geometric Bonferroni mean operator of interval rough numbers—Application in interval rough DEMATEL-COPRAS. *Facta Univ. Ser. Mech. Eng.* **2018**, *16*, 171–191. [CrossRef]
34. Lazarevic, I. Using the ELECTRE MLO Multi-Criteria Decision-Making Method in Stepwise Benchmarking—Application in Higher Education. *Oper. Res. Eng. Sci. Theory Appl.* **2019**, *2*, 77–91. [CrossRef]
35. Pamucar, D.; Jankovic, A. The application of the hybrid interval rough weighted Power-Heronian operator in multi-criteria decision making. *Oper. Res. Eng. Sci. Theory Appl.* **2020**, *3*, 54–73. [CrossRef]
36. Zavadskas, E.K.; Kaklauskas, A.; Banaitis, A.; Kvederyte, N. Housing credit access model: The case for Lithuania. *Eur. J. Oper. Res.* **2004**, *155*, 335–352. [CrossRef]
37. Pamucar, D.; Vasin, L.J.; Lukovac, L. Selection of railway level crossings for investing in security equipment using hybrid DEMATEL-MARICA model. In Proceedings of the XVI International Scientific-Expert Conference on Railway, Railcon, Australia, 9–10 October 2014; pp. 89–92.
38. Pamucar, D.; Cirovic, G. The selection of transport and handling resources in logistics centres using Multi-Attributive Border Approximation area Comparison (MABAC). *Expert Syst. Appl.* **2015**, *42*, 3016–3028. [CrossRef]
39. Stevic, Z.; Pamucar, D.; Puska, A.; Chaterjee, P. Sustainable supplier selection in healthcare industries using a new MCDM method: Measurement of Alternatives and Ranking according to COmpromise Solution (MARCOS). *Comput. Ind. Eng.* **2020**, *140*, 106231. [CrossRef]

© 2020 by the authors. Licensee MDPI, Basel, Switzerland. This article is an open access article distributed under the terms and conditions of the Creative Commons Attribution (CC BY) license (http://creativecommons.org/licenses/by/4.0/).

sustainability

MDPI

Article

Dry Port Terminal Location Selection by Applying the Hybrid Grey MCDM Model

Snežana Tadić [1], Mladen Krstić [1], Violeta Roso [2,*] and Nikolina Brnjac [3]

1 Faculty of Transport and Traffic Engineering, University of Belgrade, 11000 Belgrade, Serbia; s.tadic@sf.bg.ac.rs (S.T.); m.krstic@sf.bg.ac.rs (M.K.)

2 Department of Technology Management and Economics, Chalmers University of Technology, 41296 Gothenburg, Sweden

3 Faculty of Transport and Traffic Sciences, University of Zagreb, 10000 Zagreb, Croatia; nikolina.brnjac@fpz.hr

* Correspondence: violeta.roso@chalmers.se; Tel.: +46-739-196775

Received: 28 July 2020; Accepted: 24 August 2020; Published: 27 August 2020

Abstract: Globalization and decentralization of production generate the intensive growth of goods and transport flows, mostly performed by the maritime transport. Ports, as the main nodes in the global logistics networks, are becoming congested, space for their expansion limited, and traffic in their hinterland congested. As a solution to these and many other hinterland-transport-related problems stands out the development of dry port (DP) terminals. Selection of their location is one of the most important strategic decisions on which depends their competitiveness in the market and the functionality of the logistics network. Accordingly, the evaluation and selection of locations for the development of the DP in accordance with the requirements of various stakeholders is performed in this paper, as a prerequisite for the establishment of an ecological, economic, and socially sustainable logistics network in the observed area. To solve this problem, a new hybrid model of multi-criteria decision-making (MCDM) that combines Delphi, AHP (Analytical Hierarchy Process), and CODAS (Combinative Distance-based Assessment) methods in a grey environment is developed. The main contributions of this paper are the defined model, the problem-solving approach based on finding a compromise solution, simultaneous consideration of the environmental, economic, and social sustainability of the DP concept and its implementation in the regional international markets. The applicability of the approach and the defined MCDM model is demonstrated by solving a real-life case study of ranking the potential DP locations in the Western Balkans region. Based on the obtained results, it is concluded that in the current market conditions, it would be most realistic to open three DP terminals, in Zagreb, Ljubljana, and Belgrade.

Keywords: dry port; location; sustainability; MCDM; grey; Delphi; AHP; CODAS

1. Introduction

Trends in the global economy, implying decentralized production and a significant increase in the flow of goods around the world, generate major problems and challenges for the logistics sector [1]. As 80% of the total quantity and 70% of the total value [2] of goods flows are done by maritime transport, seaports as nodes in the logistics networks through which connections are established between two transport systems, maritime and land, become key factors for supply chain efficiency [1]. However, the increasing intensity of activities in ports leads to increased traffic congestion, especially in the immediate vicinity of the port, lack of space for port operations, and environmental pollution [1]. There is also a significant increase in pressure on the transport flows in the port hinterland, especially in terms of sustainability, which is manifested in the form of public opinion pressure or through formal legislation [3]. In addition, the part of the logistics chain in the port hinterland represents as much as 60% of the total logistics costs of the chain [4]. As a sustainable solution to all these problems, in not only environmental, but also economic and social terms, the use of dry port (DP) terminals is advocated by researchers.

When it first appeared, the term DP referred to a land terminal where (or for which) shipping companies could issue a bill of loading [5]. Later, the term DP had different definitions and was equated with terms such as inland clearance depot, inland container depot, intermodal terminal, land freight terminal, land port, etc. [6]. There are different definitions of DP in the literature, but in this paper the term DP is used in accordance with the definition given by Roso et al. [7], according to which the DP is an inland intermodal terminal directly connected by various modes of transport, primarily rail with one or more seaports, in which users can leave/pick up the goods as if directly at a seaport. It is important to emphasize that the DP, being an inland intermodal terminal, can be connected to the other inland terminals (freight centers) in the logistics network in a wider catchment area in the port hinterland by various transport modes.

The main positive effects and, at the same time, the basic motive for the implementation of the DP concept are the reduction of negative impacts generated by transport flows on the environment, efficiency improvement of the logistics activities, and economic development of the region. DP plays a key role in achieving the sustainability of logistics processes in the port hinterland, and its efficiency and ability to reach full potential largely depend on the choice of an adequate location [3]. Terminal location is one of the most important strategic decisions in the implementation of the DP concept. The functionality of the entire logistics network [8] and its competitiveness in the market depend on this decision [9,10], which is why it requires a thorough analysis [1,6,11]. Therefore, the main purpose of this research is to develop a new hybrid MCDM model, combining the Delphi, AHP (Analytical Hierarchy Process), and CODAS (Combinative Distance-based Assessment) methods in a grey environment, for the location of the DP terminal as a way to achieve economic, environmental, and social sustainability of the logistics system. The model provides compromise solutions that take into account the requirements of different stakeholders and was tested in a real-life case study of locating the DP terminals in the Western Balkans. For the observed area, Zagreb, Ljubljana, and Belgrade are obtained as the most favorable locations for the establishment of the DP terminals. An approach to solving this location problem, based on reaching the compromise solution and including the various aspects of the sustainable development, the new hybrid model, and its application in a case study covering the regional international market, is the main contribution of this paper.

The paper is organized as follows. The second section provides an overview of the literature on the application of the DP concept in the function of sustainability, the location of DP terminals, and the methods that make up the hybrid MCDM model. The third section describes the developed MCDM model in detail and gives the steps of its application. The structure of the problem for the observed case study, application of the MCDM model, sensitivity analysis, and the discussion of the obtained results are presented in the fourth section. The last section provides concluding remarks and future research directions.

2. Literature Review

This section provides an overview of the literature in accordance with the basic aspects of observing the defined problem. The first aspect involves the development of the DP terminals as a concept for achieving environmental, economic, and social sustainability. The second aspect refers to the technique of the DP terminal location selection as one of the most important factors for its efficient functioning and achieving the full potential. The third aspect involves an overview of the basic characteristics of the methods that make up the model developed in this paper for locating the DP terminals.

2.1. Dry Port Terminals in the Function of Sustainability

The term "dry port" was first mentioned in scientific papers in 1984 as the "Port of Memphis as a dry port for New Orleans" [6], but more extensive research on DP terminals was done by Beresford and

Dubey on landlocked countries in Africa in 1990 [7,12]. DP is interesting as a subject of research due to numerous advantages generated by its implementation, and some of them are: significant reduction of traffic congestion in the immediate vicinity of the seaport it serves, as well as on the roads connecting the port to the hinterland; more efficient transport and faster movement of goods; improved rail–sea intermodal capacity; port competitiveness improvements through the better market access; significant increase of the market share of railway and inland waterway transport operators; transport cost reduction; expansion of the range of logistics services offered to the customers (shippers and recipients); improvement of the efficiency and reliability of supply chains; reduction of the environmental impacts (air pollution, noise, vibration); development of the regional economy through the creation of new jobs; and attraction of a larger volume of flows, etc. [3,6,13–15].

The DP concept has been the subject of various studies dealing with the geographical location of the terminals (e.g., [16]), hinterland access strategies (e.g., [17]), the selection of efficient terminal types (e.g., [18]), directions of the concept development (e.g., [19]), etc. However, taking into account the role of the DP terminal in logistics networks and some of the basic effects it generates, one of the most important directions of research relates to the analysis of the sustainability of the concept. The impact of the DP terminal concept on the environmental sustainability has been most frequently investigated in the literature, especially the effect of reducing the emissions of carbon dioxide and other carbon-based components [6,20–24], and also other positive effects on the environment, such as reduction of emissions of other gases, particles, noise and vibration, and reduction of traffic congestion, traffic accidents, etc. (e.g., [25]). There are also a significant number of studies that, in addition to the environmental, also take into account the economic or social sustainability (e.g., [6,13,26,27]), while the studies that simultaneously analyze environmental, economic, and social sustainability are very rare [1,28,29]. This paper follows up the previous research in this area and builds upon it by considering the most favorable locations for establishing the concept of the DP terminals based on criteria that include the requirements of all stakeholders in the decision-making process in order to achieve environmental, economic, and social sustainability of the logistics system of the observed area.

2.2. Dry Port Terminal Location Selection

The location of the DP terminal plays a key role in achieving the competitive advantage of the terminal, high efficiency of the terminal operation, and the full potential of the terminal in achieving the goals of sustainable development. In the literature, the problem of the DP terminal location selection has been solved in different ways, i.e., by applying different techniques. A significant group is formed by the studies in which the DP terminals are located by solving the optimization problems, most often formulated as the mixed integer linear programming (MILP) problems (e.g., [30–34]). Various metaheuristics are often used to solve these problems, e.g., greedy algorithm (e.g., [35]), genetic algorithm (e.g., [36]), or heuristics developed exclusively to solve the defined problem (e.g., [37]). There are also examples in the literature where DP terminal locations are selected using the cluster analysis (e.g., [38–40]), simulation models (e.g., [26,41,42]), or by combining different techniques, e.g., geographic information systems (GIS) and heuristics [43], data mining and complex network theory [44], etc. However, the problem with all of the above techniques is that decisions on DP terminal locations are made solely on the basis of one or several parameters, most often investment and operating costs, and neglect the whole set of qualitative factors that have a significant impact on location [11]. Therefore, for solving this type of problem the MCDM methods, able to consider a large number of criteria, stand out in the literature. Some of the MCDM methods applied to locating the DP terminals, either alone or in combination with some other methods, in a conventional form or in the fuzzy environment, are: Analytical Network Process (ANP) [10,45,46], AHP [8,47], a combination of AHP and ELECTRE methods [48], a combination of CFA, MACBETH and PROMETHEE methods [49], etc. However, in most of these studies, the location of the DP terminal was done without considering the requirements and goals of different stakeholders. One of the few studies in which the views of different stakeholders were considered when locating the DP terminals is the work of Nguyen and

Notteboom [11], but even they did not obtain the location as a compromise solution but simply by aggregating the results of the independent potential location rankings by each stakeholder. This is another research gap that this paper covers, since the final ranking of the potential DP locations is obtained as a compromise solution that integrates the requirements of all stakeholders.

Considering the geographical area of research, most papers deal with the DP terminal locations in Asia (e.g., [10,35,36,43,46]) and Europe (e.g., [8,26,33,41]), and considerably fewer with locations in Africa (e.g., [34]), North America (e.g., [50]), and South America (e.g., [47]). The vast majority of papers deal with the DP terminal locations for the area of a state or region within a state, as well as for serving a single port [51]. Studies covering areas of multiple countries and ports, such as the case in this paper, are very rare [30,31,52].

2.3. *An Overview of the Methods that Make Up the MCDM Model*

The paper proposes a new hybrid MCDM model based on a combination of Delphi, AHP, and CODAS methods in a grey environment to solve the defined problem. As the problem has a hierarchical structure with criteria, sub-criteria, and alternatives at different levels, it was justified to apply the AHP method to solve it. However, as the AHP method requires comparison of all pairs of elements (criteria, sub-criteria, alternatives) at the same hierarchical levels, and considering the large scale of the problem (large number of elements), the AHP method was used only to determine the criteria weights. The CODAS method was used for the evaluation of the alternatives and the final ranking, which significantly reduced the complexity of the problem. Since evaluations were performed by the decision makers representing different stakeholders for whom the considered criteria may have different significance, a part of the Delphi method that enabled the unification of the different evaluations was introduced into the model. The entire MCDM model was developed in a grey environment because it enabled the processing of partial data, as well as the combination of unclear and incomplete assessments by the decision makers. Below is a review of the literature on the methods that make up the model.

The Delphi method was developed by Dalkey and Helmer [53], and is generally used to iteratively process the decision makers' opinions until a consensus on the subject of research is reached [54]. The Delphi method is used in situations where there is a risk of subjectivity; when problems are large, complex, and multidisciplinary; when there is uncertainty in the decision-making process and the time frame for decision-making is short; when the opinions of multiple decision makers are sought and anonymity is desirable, etc. [55]. Accordingly, the basic characteristics as well as the advantages of this method are: anonymity, iterativity, controlled feedback, and statistical group response [55]. Some of the disadvantages are the need for multiple repetition of the questionnaire to achieve the convergence of the assessments and the high cost of data collection, especially for large and complex problems [56]. The Delphi method is widely accepted in the literature, and either alone or in combination with some other methods it has been used to solve various problems, such as defining the typical structures of the intermodal terminals [57], selecting the development indicators of the urban sustainability [58], planning of intermodal terminals [59], selecting the maintenance strategy for ship machinery systems [60], locating intermodal terminals [61], etc.

The AHP method was developed by Saaty [62], and is based on defining the hierarchical structure of the problem, i.e., arranging the elements of the problem (criteria, sub-criteria, alternatives) in several levels and their mutual comparison within them. As a result, it gives the weights, i.e., the significance of the elements at each hierarchical level in relation to the decision-making goal. It is used to solve complex problems with a large number of elements that have a strong mutual influence [63]. The AHP method is theoretically easy to understand and simple to apply in practice [64], allows consideration of both quantitative and qualitative parameters [62], and the pairwise comparison of elements, on which it is based, makes it one of the most transparent and technically most appropriate methods for determining the weights (significance) of the problem elements [65]. On the other hand, the main disadvantages of the method are that it requires a large number of comparisons in cases of larger problems, making it very difficult to maintain consistency of the decision makers' assessments [63], as well as that it uses a discrete scale for comparisons, which cannot adequately express ambiguity and vagueness in the decision makers' assessments [66]. The AHP method requires a greater effort to obtain a solution, compared to some other methods that perform direct ranking of elements, such as the SMART or SWARA methods [67], but on the other hand allows decision makers to better understand the relative importance of and interactions between the elements and to perform more precise evaluation, resulting in better solutions [68,69]. In addition, compared to the methods that require complete consistency, e.g., MACBETH or FUCOM [70], the AHP method is more flexible and allows for an increase or decrease in the tolerance threshold according to the decision makers' preferences [71,72]. Because of the above, AHP is one of the most commonly used MCDM methods, especially for determining the criteria weights, and either alone or in combination with some other methods it has found wide application for solving various problems, such as evaluating energy systems [73], selecting intermodal transport chains [74], locating warehouses for humanitarian logistics [75], ranking of logistics systems scenarios [76], locating dry port terminals [77], locating city logistics terminals [78], etc.

The CODAS method was developed by Ghorabaee et al. [79], and is based on determining the overall performance of the elements of the decision problem (usually an alternative), based on the Euclidean and Taxicab distances from the negative ideal point (solution, alternative). Euclidean distances are used as the primary measures of evaluation, and if these distances are too close for some alternatives, Taxicab distances are included in the evaluation to make the alternatives easier to compare. The degree of closeness (similarity) of Euclidean distances is determined on the basis of the closeness threshold parameter. The CODAS method, as well as methods such as TOPSIS, VIKOR, EDAS, COPRAS, etc., belongs to the group of distance-based methods. The main advantage of these, in relation to some other methods such as AHP, ANP, BWM, etc., is that they do not require comparison of all pairs of elements (criteria, sub-criteria, alternatives), which makes them much more efficient, especially when considering problems with a large number of elements. On the other hand, the advantage of the CODAS method over the other methods from the same group is the ability to evaluate the elements of the problem by applying more than one type of distance, which significantly increases the accuracy of results and allows fine differentiation between closely ranked alternatives [79]. CODAS is a relatively new method, but in a short period of time it has been applied, either alone or in combination with some other methods, to solve problems in various areas, such as locating dismantling centers in reverse logistics [80], evaluating organizational culture in the field of innovation and sustainable development [81], selecting staff [82], selecting suppliers [83,84], selecting renewable energy sources [85], etc.

Most of the shortcomings of the individual methods that make up the model can be solved by applying intuitive or interval sets (e.g., fuzzy, rough, grey). Fuzzy [86], rough [87], and grey [88] sets are the different ways of representing uncertainty in data sets [89]. They are introduced into the MCDM with the aim of more efficient problem-solving in situations where elements of the problem are assessed by the decision makers, who very often give inaccurate, incomplete, or ambiguous evaluations. Although there are some overlaps between them, these sets differ significantly. Grey sets, unlike fuzzy and rough sets, allow better processing of partial data as well as combining obscure and incomplete data into one model [89]. In addition, grey sets can present situations that cannot be represented by either fuzzy or rough sets [89]. Therefore, in this paper, the previously described MCDM methods are combined in a grey environment. Grey extension of the Delphi method was performed by Lingkan [90] and has since been applied to solve various problems, either alone or in combination with other methods [91–93], while grey extension of the AHP method was first performed by Liu et al. [94] and has since been recorded in different variants of the AHP method applications in the grey environment for solving various problems, either alone or in combination with some other methods [95–97]. There have been no examples in the literature so far of extending the CODAS method in the grey environment, nor of combining these three methods in any form (conventional or in an environment of intuitive or interval assessments). This is another research gap covered by this paper, i.e., another of the main contributions of the paper.

3. Hybrid Grey MCDM Model

The hybrid MCDM model proposed in this paper for the DP terminal location selection involved the application of the Grey Delphi-based grey AHP (GDAHP) method for determining criteria and sub-criteria weights and the Grey CODAS (GCODAS) method for ranking and selecting the most favorable alternatives. The steps for applying the model are explained in detail below, and a general overview of the model structure is given in Figure 1.

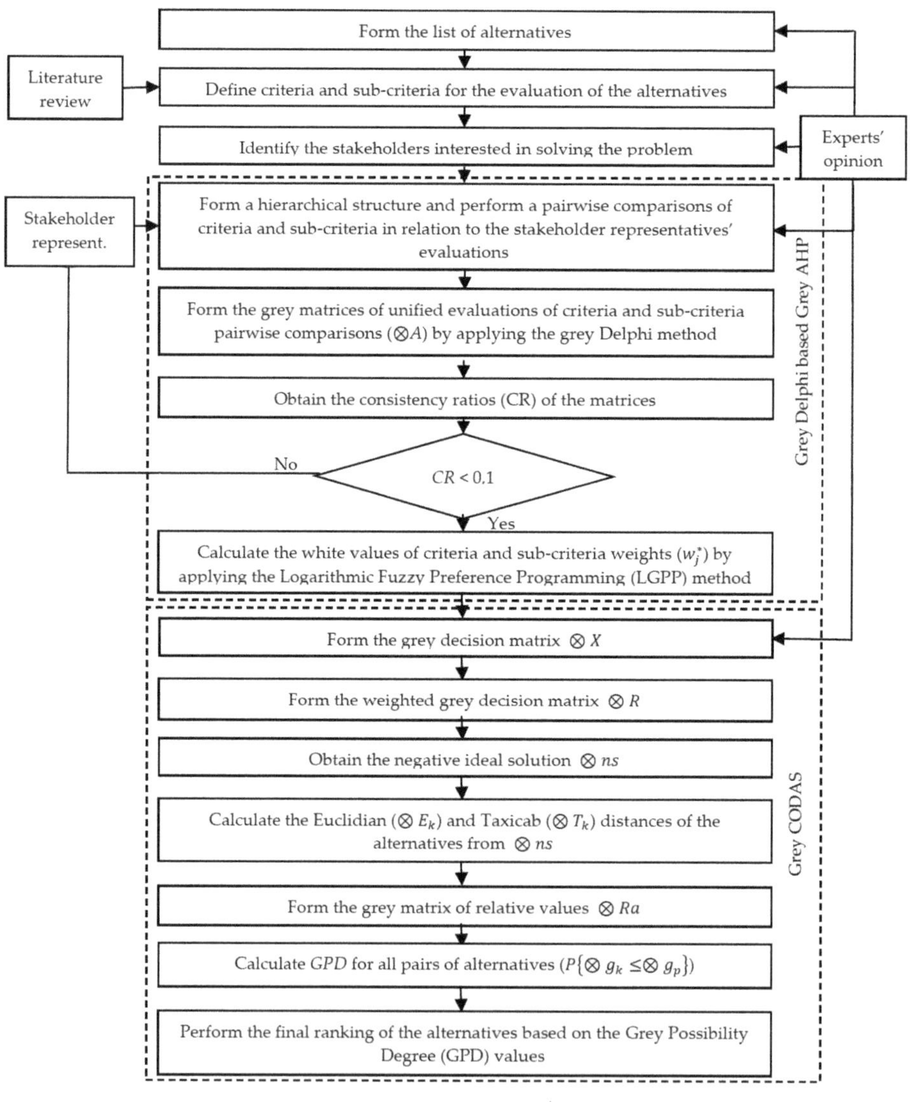

Figure 1. Structure of the proposed hybrid multi-criteria decision-making (MCDM) model.

Step 1: Define the problem structure, i.e., form the sets of alternatives, criteria, and sub-criteria for their evaluation and identify the stakeholders.

Step 2: Define the grey scale for the evaluation of the criteria, sub-criteria, and the alternatives by the decision makers. Linguistic terms and the corresponding grey values are given in Table 1.

Table 1. Grey scale for the evaluations.

Linguistic Term	Abbreviation	Grey Scale
None	N	[0, 2]
Very low	VL	[1, 3]
Low	L	[2, 4]
Moderately low	ML	[3, 5]
Medium	M	[4, 6]
Moderately high	MH	[5, 7]
High	H	[6, 8]
Very high	VH	[7, 9]
Extremely high	EH	[8, 10]

Step 3: Obtain the criteria and sub-criteria weights by applying the GDAHP method. Decision makers belonging to the different stakeholders perform the pairwise comparisons of the elements (criteria and sub-criteria) using the linguistic terms, which can be transformed into grey values according to the relations given in Table 1. These values indicate the relative importance, i.e., the elements' preference in relation to the other elements.

Step 3.1: Unify the decision makers' evaluations by applying the following equations:

$$\otimes\ \alpha_{ij} = \left[\underline{\alpha}_{ij}, \overline{\alpha}_{ij}\right],\ i = j = 1, \ldots, n, \tag{1}$$

$$\underline{\alpha}_{ij} = \left(\prod\nolimits_{h=1}^{o} \underline{a}_{ijh}\right)^{1/o}, \tag{2}$$

$$\overline{\alpha}_{ij} = \left(\prod_{h=1}^{o} \overline{a}_{ijh}\right)^{1/o}, \tag{3}$$

where $\underline{\alpha}_{ij}$ and $\overline{\alpha}_{ij}$ are the lower and upper values of the unified grey evaluation $\otimes\ \alpha_{ij}$, respectively, and stands $\underline{\alpha}_{ij} \leq \overline{\alpha}_{ij}$. $\underline{a}_{ijh}$ and $\overline{a}_{ijh}$ are the lower and upper values of the grey evaluation $\otimes\ a_{ij}$, indicating the comparison of the element (criteria, sub-criteria) i in relation to the element j by the decision maker belonging to the stakeholder h. n is the number of criteria/sub-criteria, and o is the number of stakeholders involved in the process.

Step 3.2: Form the grey decision matrix ($\otimes\ A$) elements of which are the unified values of the criteria/sub-criteria comparisons $\otimes\ \alpha_{ij}$:

$$\otimes\ A = \begin{bmatrix} \otimes\ \alpha_{11} & \otimes\ \alpha_{21} & \cdots & \otimes\ \alpha_{n1} \\ \otimes\ \alpha_{21} & \otimes\ \alpha_{22} & \cdots & \otimes\ \alpha_{n2} \\ \vdots & \vdots & \ddots & \vdots \\ \otimes\ \alpha_{n1} & \otimes\ \alpha_{n2} & \cdots & \otimes\ \alpha_{nn} \end{bmatrix}. \tag{4}$$

Step 3.3: Calculate the relative weights of the elements (criteria, sub-criteria). For each matrix of the pairwise comparisons, it is necessary to obtain the priority vector. For obtaining the priority vector the Logarithmic Grey Preference Programming (LGPP) method is developed, based on the Logarithmic Fuzzy Preference Programming (LFPP) method developed by Wang and Chin [98]. The elements of the grey decision matrix $\otimes A$ are approximated by the logarithms of the grey evaluations $\otimes\ \alpha_{ij}$ by applying the following equation:

$$\ln \otimes \alpha_{ij} \approx \left[\ln \underline{\alpha}_{ij}, \ln \overline{\alpha}_{ij}\right];\ i = j = 1, \ldots, n. \tag{5}$$

For obtaining the elements' priority values (w_j) it is necessary to solve the following non-linear priority model:

$$\min J = (1-\lambda)^2 + M\sum_{i=1}^{n-1}\sum_{j=i+1}^{n}\left(\varepsilon_{ij}^2 + \eta_{ij}^2\right), \tag{6}$$

subject to:

$$y_i - y_j - \lambda \ln\left(\frac{\overline{\alpha}_{ij}}{\underline{\alpha}_{ij}}\right) + \varepsilon_{ij} \geq \ln \underline{\alpha}_{ij};\ i = 1,\ldots,n-1;\ j = i+1,\ldots,n, \tag{7}$$

$$-y_i + y_j - \lambda \ln\left(\frac{\overline{\alpha}_{ij}}{\underline{\alpha}_{ij}}\right) + \eta_{ij} \geq -\ln \overline{\alpha}_{ij};\ i = 1,\ldots,n-1;\ j = i+1,\ldots,n, \tag{8}$$

$$\lambda,\ y_{i,j} \geq 0; i = j = 1,\ldots,n, \tag{9}$$

$$\varepsilon_{ij}^2,\ \eta_{ij}^2 \geq 0; i = 1,\ldots,n-1; j = i+1,\ldots,n, \tag{10}$$

where λ is the minimum membership degree obtained in the following way:

$$\lambda = \min\left(\ln\left(w_i/w_j\right)\right),\ i = 1,\ldots,n-1; j = i+1,\ldots,n. \tag{11}$$

J denotes the objective function, which should be minimized, $w_{i,j}$ are the white values of the priority vector, $y_{i,j} = \ln w_{i,j}$ for $i = 1,\ldots,n-1; j = i+1,\ldots,n$, and M is the large enough constant such as $M = 10^3$. ε_{ij}^2 and η_{ij}^2 for $i = 1,\ldots,n-1; j = i+1,\ldots,n$ are the non-negative deviation variables introduced in order to avoid λ from taking a negative value. It is most desirable that the values of the deviation variables are as small as possible, and that they meet the following inequalities:

$$\ln w_i - \ln w_j - \lambda \ln\left(\frac{\overline{\alpha}_{ij}}{\underline{\alpha}_{ij}}\right) + \varepsilon_{ij} \geq \ln \underline{\alpha}_{ij};\ i = 1,\ldots,n-1;\ j = i+1,\ldots,n,$$

$$-\ln w_i + \ln w_j - \lambda \ln\left(\frac{\overline{\alpha}_{ij}}{\underline{\alpha}_{ij}}\right) + \eta_{ij} \geq -\ln \overline{\alpha}_{ij};\ i = 1,\ldots,n-1;\ j = i+1,\ldots,n.$$

Let $y_j^*(j = 1,\ldots n)$ be the optimal solution to the model (6)–(11). Normalized criteria values for the grey decision matrix $\otimes A = \left[\otimes \alpha_{ij}\right]_{n\times n}$ are obtained in the following way:

$$w_j^* = \frac{\exp\left(y_j^*\right)}{\sum_{j=1}^{n}\exp\left(y_j^*\right)},\ j = 1,\ldots,n, \tag{12}$$

where exp() is the exponential function, namely, $\exp\left(y_j^*\right) = e^{y_j^*}$ for $j = 1,\ldots,n$. This method results in white normalized weights.

In order to control the result of the method, the Consistency Ratio (CR) for each matrix is calculated as follows [98]:

$$CR = CI/RI, \tag{13}$$

where CI is the Consistency Index and is calculated as follows:

$$CI = \frac{\lambda_{max} - n}{n-1}. \tag{14}$$

λ_{max} is the eigenvalue of the matrix $\otimes A$. RI the Random Index, whose values for matrices of various sizes are contained in Saaty [99]. The CR values should be less than 0.10 in order for comparisons to be acceptable.

Step 4: Perform the alternatives evaluation by applying the GCODAS method developed in this paper as the grey extension of the conventional CODAS method developed by Ghorabaee et al. [79].

Step 4.1: Form the grey decision matrix ($\otimes X$) in the following way:

$$\otimes X = \left[\otimes x_{kj}\right]_{m\times n} = \begin{bmatrix} \otimes x_{11} & \otimes x_{12} & \cdots & \otimes x_{1n} \\ \otimes x_{21} & \otimes x_{22} & \cdots & \otimes x_{2n} \\ \vdots & \vdots & \ddots & \vdots \\ \otimes x_{m1} & \otimes x_{m2} & \cdots & \otimes x_{mn} \end{bmatrix}, \tag{15}$$

where $\otimes x_{kj} = \left[\underline{x}_{kj}, \overline{x}_{kj}\right]$ represents the grey evaluation of the alternative k ($k = 1, \ldots, m$) in relation to the criterion j ($j = 1, \ldots, n$), while $\underline{x}_{kj}$ and $\overline{x}_{kj}$ represent the lower and upper values of the grey number $\otimes x_{kj}$.

Step 4.2: Form the weighted grey decision matrix ($\otimes R$) in the following way:

$$\otimes R = \left[\otimes r_{kj}\right]_{m\times n}, \tag{16}$$

$$\otimes r_{kj} = \left[\underline{r}_{kj}, \overline{r}_{kj}\right] = w_j^* \times \otimes x_{kj}, \tag{17}$$

where w_j^* represents the values indicating the weight of the criterion j.

Step 4.3: Define the negative ideal solution $\otimes ns$ in the following way:

$$\otimes ns = \left[\otimes ns_j\right]_{1\times m}, \tag{18}$$

$$\otimes ns_j = \left[\underline{ns}_j, \overline{ns}_j\right] = \left[\min_k \underline{r}_{kj}, \min_k \overline{r}_{kj}\right]. \tag{19}$$

Step 4.4: Calculate the grey Euclidian distances ($\otimes E_k$) of the alternatives from the negative ideal solution in the following way:

$$\otimes E_k = \left[\underline{E}_k, \overline{E}_k\right], \tag{20}$$

$$\underline{E}_k = \sqrt{\sum_{j=1}^{m} \max\left(0, \min\left(\underline{e}_{kj} \times \underline{e}_{kj}, \underline{e}_{kj} \times \overline{e}_{kj}, \overline{e}_{kj} \times \underline{e}_{kj}, \overline{e}_{kj} \times \overline{e}_{kj}\right)\right)}, \tag{21}$$

$$\overline{E}_k = \sqrt{\sum_{j=1}^{m} \max\left(\underline{e}_{kj} \times \underline{e}_{kj}, \underline{e}_{kj} \times \overline{e}_{kj}, \overline{e}_{kj} \times \underline{e}_{kj}, \overline{e}_{kj} \times \overline{e}_{kj}\right)}, \tag{22}$$

where $\underline{e}_{kj}$ and $\overline{e}_{kj}$ are the lower and upper values of the grey number $\otimes e_{kj}$ obtained in the following way:

$$\otimes e_{kj} = \left[\underline{e}_{kj}, \overline{e}_{kj}\right] = \otimes r_{kj} - \otimes ns_j, \tag{23}$$

$$\underline{e}_{kj} = \underline{r}_{kj} - \overline{ns}_j, \tag{24}$$

$$\overline{e}_{kj} = \overline{r}_{kj} - \underline{ns}_j. \tag{25}$$

Step 4.5: Calculate the grey Taxicab distances ($\otimes T_k$) of the alternatives from the negative ideal solution in the following way:

$$\otimes T_k = \left[\underline{T}_k, \overline{T}_k\right], \tag{26}$$

$$\underline{T}_k = \sum\nolimits_{j=1}^{m} \left|\underline{e}_{kj}\right|, \tag{27}$$

$$\overline{T}_k = \sum\nolimits_{j=1}^{m} \left|\overline{e}_{kj}\right|. \tag{28}$$

Step 4.6: Form the grey matrix ($\otimes Ra$) of the relative values of comparisons of the alternatives' distances from the negative ideal solution ($\otimes g_{kp}$) in the following way:

$$\otimes Ra = \left[\otimes g_{kp}\right]_{m \times m}, \tag{29}$$

$$\otimes g_{kp} = \left[\underline{g}_{kp}, \overline{g}_{kp}\right], \tag{30}$$

where $\underline{g}_{kp}$ and $\overline{g}_{kp}$ are the lower and upper values of the grey number $\otimes g_{kp}$ obtained in the following way:

$$\underline{g}_{kp} = \left(\underline{E}_k - \overline{E}_p\right) + \left(\psi\left(\underline{E}_k - \overline{E}_p\right) \times \left(\underline{T}_k - \overline{T}_p\right)\right), \tag{31}$$

$$\overline{g}_{kp} = \left(\overline{E}_k - \underline{E}_p\right) + \left(\psi\left(\overline{E}_k - \underline{E}_p\right) \times \left(\overline{T}_k - \underline{T}_p\right)\right), \tag{32}$$

where $p = 1, \ldots, m$ is the index of the alternative with which all the other alternatives k are compared (all pairs of alternatives are compared), and ψ represents the function that determines the equality threshold of the Euclidian distances of any two alternatives, and it is obtained in the following way:

$$\psi = \begin{cases} 1\ if \left|white(\otimes E_k) - white\left(\otimes E_p\right)\right| \geq \tau \\ 0\ if \left|white(\otimes E_k) - white\left(\otimes E_p\right)\right| < \tau \end{cases}, \tag{33}$$

where $white(\otimes E_{k,p})$ is obtained by applying the following equation [100]:

$$White\left(\otimes E_{k,p}\right) = \left(\underline{E}_{k,p} + \overline{E}_{k,p}\right)/2. \tag{34}$$

In the Function (33) τ represents the threshold parameter defined by the decision maker. It is recommended that this parameter take the value between 0.1 and 0.5. Introduction of this parameter ensures that if the difference of the Euclidian distances of any two alternatives is greater than τ, the alternatives are compared by the values, including the Taxicab distances, as well.

Step 4.7: Calculate the Grey Possibility Degrees (GPDs) for all pairs of alternatives in the following way [101]:

$$GPD\left\{\otimes q_k \leq \otimes q_p\right\} = \frac{\max\left(0, L(\otimes q_k) + L\left(\otimes q_p\right) - \max\left(0, \overline{q}_k - \underline{q}_p\right)\right)}{L(\otimes q_k) + L\left(\otimes q_p\right)}, \tag{35}$$

$$L\left(\otimes q_{k,p}\right) = \left|\overline{q}_{k,p} - \underline{q}_{k,p}\right|, \tag{36}$$

where L represents the span of the grey value obtained as the difference between the lower and upper values of a grey number, $\otimes q_k$ and $\otimes q_p$ are the grey values of the unified distances of the alternatives from the negative ideal solution, while $\overline{q}_k$ and $\underline{q}_k$, i.e., $\overline{q}_p$ and $\underline{q}_p$, are their lower and upper values, respectively, obtained in the following way:

$$\underline{q}_{k,p} = \sum\nolimits_{k,p=1}^{m} \underline{g}_{k,p}, \tag{37}$$

$$\overline{q}_{k,p} = \sum\nolimits_{k,p=1}^{m} \overline{g}_{k,p}. \tag{38}$$

The defined GPDs have the following characteristics:

If $\otimes q_k = \otimes q_p$, then $GPD\left\{\otimes q_k \leq \otimes q_p\right\} = 0.5$,

If $\underline{q}_p > \overline{q}_k$, then $GPD\left\{\otimes q_k \leq \otimes q_p\right\} = 1$,

If $\overline{q}_p < \underline{q}_k$, then $GPD\left\{\otimes q_k \leq \otimes q_p\right\} = 0$.

When there is an intersection between $\otimes q_k$ and $\otimes q_p$, if $GPD\{\otimes q_k \leq \otimes q_p\} < 0.5$ then $\otimes q_p$ is less than $\otimes q_k$, and if $GPD\{\otimes q_k \leq \otimes q_p\} > 0.5$ then $\otimes q_k$ is less than $\otimes q_p$.

Step 4.8: Perform the final ranking of the alternatives based on the *GPD* values. The best alternative is the one that has the $GPD < 0.5$ in most of the alternative pairs comparisons.

4. Case Study: Locating a Dry Port in the Western Balkans

The case study discussed in this section of the paper implies the location selection for the DP terminal in the Western Balkans region, in a geographical and not a political sense, which includes the territories of the following countries: Slovenia, Croatia, Serbia, Bosnia and Herzegovina, Montenegro, North Macedonia, and Albania. This region is of great importance for the transport flows through Europe. It is a gateway for a significant part of the goods flows entering Europe, primarily from Asia, through the ports that gravitate towards it: Trieste, Koper, Rijeka, Split, Ploče, Bar, Durres, and Thessaloniki [102,103]. In addition, it is very well-connected by land with both Central and Western Europe via Pan-European [102] and TENtec corridors [103], and with Asia via Eurasian corridors [104]. All this makes the region of the Western Balkans ideal for the development of the DP concept. However, not all countries in this region are equally suitable for the development of the concept. Apart from obvious differences in geographical terms (position, access to the sea, inland waterways, etc.), the countries of the Western Balkans also differ significantly in terms of political, economic, and technological development as a consequence of different cultural and historical heritage. There are also significant differences in the transport sector. Slovenia long ago understood the importance of adequate planning and development of this area through institutionalization and definition of financing models, with the developed system of public bodies, institutions and economic organizations, and related documents defined in accordance with the needs of the country [105]. Croatia and Serbia lag behind, so even though they have defined strategies and plans, they are not updated regularly and do not have clearly defined dynamics, priority goals, and measures [105]. The problem in these countries is the lack, or insufficient development, of the systems that interlock the economy and science, state bodies, and economic entities at the national level, so groups within a sector of the ministry responsible for traffic or transportation deal with transport [105]. Other countries, Montenegro, Bosnia and Herzegovina, Macedonia, and Albania, above all, have inadequately developed plans that contain clumsily inserted parts of planning documents of other countries. Because of the above, the independent development of the DP concept in each of the countries of the region is not justified, but it is necessary to observe a broader, regional aspect that would imply the opening of the DP terminals for the markets transcending the national borders. Accordingly, a comprehensive approach that takes into account different aspects of the problem is necessary to locate them. The structure of the problem is defined below, i.e., the sets of potential locations, as well as the criteria and sub-criteria for their evaluation.

4.1. Potential Locations and the Criteria for Their Evaluation

The problem considered in this paper was structured as follows. The capitals of the Western Balkan countries, being the economic, administrative, political, traffic, etc., centers of the region, were defined as potential locations (PLs) for the establishment of the DP. These were Ljubljana (PL_1), Zagreb (PL_2), Belgrade (PL_3), Sarajevo (PL_4), Podgorica (PL_5), Skopje (PL_6), and Tirana (PL_7). For the evaluation of potential DP locations, 20 sub-criteria were defined that could be classified into one of the four groups of criteria: environmental (EN), economic (EC), infrastructure (IN), and socio-political (SP). Environmental, economic, and socio-political criteria in the process of selecting the DP locations directly included the principles of the basic pillars of sustainability (environmental, economic, and social). On the other hand, the infrastructure criteria indirectly considered the environmental and economic sustainability of the DP locations. The criteria were defined based on an extensive review of the literature on locating the DP terminals using the MCDM methods [8,10,11,45,46,48,49,106]. An overview of the defined criteria and sub-criteria, as well as their description, are given in Table 2.

Table 2. Criteria and sub-criteria for the evaluation of the potential dry port (DP) locations.

Criteria	Sub-Criteria	Description
Environmental criteria (EN)	Energy consumption (C_1)	The amount of energy required to realize the transport flows between ports and potential DP locations, as well as the total energy savings as a result of the DP terminal implementation.
	Harmful emissions (C_2)	The amount of harmful gasses, particles, and noise emission as a result of the transport flows realization between ports and potential DP locations, as well as the total emission reduction as a result of the DP terminal implementation.
	Traffic congestion (C_3)	The degree of traffic congestion as a result of activating the DP location as well as the overall reduction of traffic congestion (primarily on roads) as a result of the DP terminal implementation.
	Traffic accidents (C_4)	The risk of traffic accidents as a result of more intensive transport flows due to the activation of the DP location, as well as the overall reduction of accidents (primarily on roads) as a result of the DP terminal implementation.
Economic criteria (EC)	Economic development (C_5)	The economic development of the country in which the potential DP location is located, observed through the GDP, production indices, trade, state of the market, etc., as well as the impact of the DP terminal implementation on the economic growth.
	Volume of flows (C_6)	Potential volume of flows that the DP could attract at the designated location.
	Investment costs (C_7)	Costs of activating a potential DP location that depend on the land price, costs of building and equipping the terminal, etc.
	Exploitation costs (C_8)	The operating costs of the DP terminal, which largely depend on the labor price in the countries of potential locations, but also the price of energy, consumables, tax policy, utilities, etc.
	Transportation costs (C_9)	Freight transport costs depending on the mode of transport, as well as the costs of establishing and maintaining lines in rail and inland waterway transport.
	Transport time (C_{10})	The time required for the realization of transport, depending on the mode of transport, established lines, quality of infrastructure, etc.

Table 2. *Cont.*

Criteria	Sub-Criteria	Description
Infrastructure criteria (IN)	Distances (C_{11})	Distances of potential DP locations from the ports they are supposed to serve on one side, and potential markets (primarily Central and Western European countries) on the other.
	Transport infrastructure (C_{12})	The degree of development and condition of the transport infrastructure of all modes of transport used in connecting the ports, potential DP locations and markets.
	Multimodality (C_{13})	Possibility of using multiple transport modes in connecting the ports and potential DP locations, and especially in connecting the DP locations with the markets in Europe.
	Network status (C_{14})	Connection of potential DP locations with the TENec network, development of the logistics network in the surrounding area (presence of logistics centers, logistics providers, etc.), the possibility of achieving a significant status in the network.
	Information infrastructure (C_{15})	The development of information technologies in the surroundings of the potential location, application of newer generation networks, internet speed, etc.
Socio-political criteria (SP)	Political stability (C_{16})	Stability of the political situation in the country where the potential DP location is situated, which includes corruption control, government efficiency, absence of violence, rule of law, freedom of speech, etc.
	EU integration (C_{17})	Status of the country/region in which the potential DP location is located in relation to the EU (member, accession negotiations are underway, candidate, potential candidate).
	Strategies and planes (C_{18})	The development and implementation of development strategies and plans (economy, logistics and transport, etc.) at the national, regional, EU level.
	Regulation (C_{19})	Existence and the degree of application of laws in the field of logistics, transport, traffic, international trade, customs, finance, insurance, etc.
	Labor force (C_{20})	Employment rate and availability of labor force, primarily qualified experts in the field of logistics.

4.2. *Application of the Hybrid MCDM Model for Ranking the Potential DP Locations*

The first step in applying the model, in addition to defining a set of alternatives and criteria and sub-criteria for their evaluation, involved identifying stakeholders interested in solving the problem. Three stakeholders were identified: investors/owners and operators (Inv.), users (Use.), and government/administration and residents (A&R). The main goal of investors and operators was to maximize profit, i.e., financial benefits, as a result of successful operation of the terminal. A large number of secondary goals contributed to the realization of this primary goal, such as minimization of logistics costs, optimization of position in the logistics chain, synergy with existing activities, synergy with external strategic initiatives, maximization of expansion possibilities, and utilization of the available infrastructure capacity, etc. The main goal of the users was to get quality service at a reasonable price. Regardless of whether the logistics and organization of transport were the primary activities of users, service quality was crucial for the efficiency of their business, while the price directly affected the possibility of making a profit. The goals of administration and residents were divided into economic and strategic. The main economic goal was the economic development of the surrounding area (city, region, state) in which the terminal was located. The main strategic goal was the ecological and social responsibility through preserving the environment, improving mobility and accessibility, reducing traffic congestion, reducing traffic accidents, etc.

For the purpose of this research, the representatives belonging to each group of defined stakeholders, who had the noticeable knowledge and experience in the investigated field, were selected. They were given by the questioners the instructions to neglect their personal territorial, political, cultural, etc., preferences in order to provide an attitude that was as objective as possible of the stakeholder group they represented, for the observed area. The questions were designed to be as simple as possible in order to capture the attitudes of the stakeholders' representatives by the defined criteria. For example, they were asked: "In your opinion, how much more important is energy consumption (C_1) than harmful emissions (C_2)?" In the same manner the questioner followed all remaining criteria and sub-criteria. In accordance with their preferences, the stakeholders' representatives evaluated the significance of the criteria with linguistic evaluations, which were then converted into the grey evaluations using the relations given in Table 1. An example of an evaluation of the EN criteria by the stakeholders' representatives is given in Table 3.

Table 3. Evaluation of the EN criteria by the stakeholders' representatives.

	C_1			C_2			C_3			C_4		
	Inv.	**Use.**	**A&R**	**Inv.**	**Use.**	**A&R**	**Inv.**	**Use.**	**A&R**	**Inv.**	**Use.**	**A&R**
C_1	/	/	/	M	VL					VL		
C_2			S	/	/	/			VL			L
C_3	VL	L	ML	MH	ML		/	/	/	L	VL	VL
C_4		VL	L	ML	L					/	/	/

Inv.—investors/owners and operators, Use.—users, A&R—government/administration and residents.

By applying the Equations (1)–(3), their evaluations were re-unified and the grey decision matrix (4) was formed. For each matrix of comparison of criteria, and then the sub-criteria, the nonlinear priority model (6) was solved, subject to (7)–(11), and the optimal solutions were obtained and normalized by using the Equation (12). In this way, the white weights of criteria and sub-criteria were obtained. For each matrix of criteria or sub-criteria comparison, the consistency ratios were calculated by the Equations (13) and (14), and since all the values were less than 0.10, it was concluded that the evaluations were valid. The final sub-criteria weights for the potential locations evaluation were obtained by multiplying the weights of criteria with the weights of the belonging sub-criteria. These weights are shown in Table 4.

Then, the evaluation of alternatives in relation to the sub-criteria was performed according to the reports and indicators of the development of logistics, transport, economy, and political situation in each country [107], statistical data [108], environmental indicators of the transport chain operations [109], the results of previous research [30–32,110], and the experience of the decision makers. Linguistic evaluations of the alternatives in relation to the sub-criteria, given in Table 5, were then converted into the grey evaluations by applying the relations given in Table 1, and thus a grey decision Matrix (15) was formed.

Table 4. Criteria and sub-criteria weights.

Criterion	Weight	Sub-Criterion	Weight	Final Weight
EN	0.230	C_1	0.088	**0.020**
		C_2	0.048	**0.011**
		C_3	0.663	**0.153**
		C_4	0.201	**0.046**
EC	0.427	C_5	0.204	**0.087**
		C_6	0.491	**0.210**
		C_7	0.111	**0.047**
		C_8	0.056	**0.024**
		C_9	0.111	**0.047**
		C_{10}	0.028	**0.012**
IN	0.197	C_{11}	0.013	**0.003**
		C_{12}	0.131	**0.026**
		C_{13}	0.585	**0.115**
		C_{14}	0.237	**0.047**
		C_{15}	0.033	**0.007**
SP	0.145	C_{16}	0.564	**0.082**
		C_{17}	0.017	**0.002**
		C_{18}	0.096	**0.014**
		C_{19}	0.288	**0.042**
		C_{20}	0.035	**0.005**

Table 5. Evaluations of the alternatives in relation to the sub-criteria.

	PL_1 (Ljubljana)	**PL_2 (Zagreb)**	**PL_3 (Belgrade)**	**PL_4 (Sarajevo)**	**PL_5 (Podgorica)**	**PL_6 (Skopje)**	**PL_7 (Tirana)**
C_1	VH	EH	H	MH	M	VL	ML
C_2	VH	EH	M	MH	H	VL	ML
C_3	M	ML	L	MH	EH	H	VH
C_4	ML	ML	MH	EH	H	M	VH
C_5	EH	VH	H	M	MH	ML	L
C_6	VH	EH	H	MH	M	ML	L
C_7	ML	H	EH	H	MH	VH	H
C_8	L	ML	MH	EH	MH	VH	H
C_9	VH	EH	EH	H	MH	ML	M
C_{10}	MH	H	EH	VH	M	ML	L
C_{11}	H	EH	VH	M	L	N	VL
C_{12}	EH	VH	H	MH	M	MH	ML
C_{13}	M	MH	EH	M	ML	ML	ML
C_{14}	EH	VH	H	L	N	M	VL
C_{15}	EH	VH	VH	H	MH	M	ML
C_{16}	EH	VH	ML	L	H	M	MH
C_{17}	EH	EH	VH	MH	VH	H	H
C_{18}	EH	H	H	MH	MH	ML	ML
C_{19}	EH	VH	H	M	MH	ML	L
C_{20}	VH	VH	EH	MH	M	ML	L

By applying the Equations (16) and (17), the weighted grey decision matrix was formed, and then by applying the Equations (18) and (19), the negative ideal solution was defined. By applying the Equations (20)–(25), the grey Euclidian distances were obtained, and by applying the Equations (26)–(28), the grey Taxicab distances of the alternatives from the negative ideal solution were obtained. Afterwards, by applying the Equations (29)–(34), the grey matrix of the relative values of the alternatives was obtained and is presented in Table 6.

Table 6. Grey matrix of relative values of the alternatives.

	PL_1	PL_2	PL_3	PL_4	PL_5	PL_6	PL_7
PL_1	[−1.1, 1.1]	[−3.2, 5.4]	[−1.3, 0.9]	[−5.6, 2.8]	[−5.3, 2.9]	[−6.5, 0.8]	[−5.6, 1]
PL_2	[−5.4, 3.2]	[−1, 1]	[−6.1, 2.9]	[−6.9, 1.8]	[−6.5, 2]	[−7.8, −0.1]	[−6.9, 0.1]
PL_3	[−0.9, 1.3]	[−2.9, 6.1]	[−1.1, 1.1]	[−1.2, 0.9]	[−1, 0.9]	[−6.2, 1.5]	[−5.2, 1.7]
PL_4	[−2.8, 5.6]	[−1.8, 6.9]	[−0.9, 1.2]	[−0.9, 0.9]	[−0.8, 1]	[−5.2, 2.3]	[−4.2, 2.5]
PL_5	[−2.9, 5.3]	[−2, 6.5]	[−0.9, 1]	[−1, 0.8]	[−0.8, 0.8]	[−5.3, 1.9]	[−4.4, 2.1]
PL_6	[−0.8, 6.5]	[0.1, 7.8]	[−1.5, 6.2]	[−2.3, 5.2]	[−1.9, 5.3]	[−0.9, 0.9]	[−0.6, 1]
PL_7	[−1, 5.6]	[−0.1, 6.9]	[−1.7, 5.2]	[−2.5, 4.2]	[−2.1, 4.4]	[−1, 0.6]	[−0.7, 0.7]

By applying the Equations (35)–(38), the GPD values were calculated, based on which the final ranking of the alternatives, i.e., potential locations, was obtained (Table 7).

Table 7. Grey Possibility Degree (GPD) values of the alternatives' comparisons and the final ranking.

GPD (PLk ≤ PLk)	PL_1	PL_2	PL_3	PL_4	PL_5	PL_6	PL_7	Rank
PL_1	/	0.583	0.443	0.391	0.407	0.271	0.3	2
PL_2	0.417	/	0.353	0.311	0.324	0.205	0.228	1
PL_3	0.557	0.647	/	0.435	0.454	0.294	0.329	3
PL_4	0.609	0.689	0.565	/	0.519	0.365	0.401	5
PL_5	0.593	0.676	0.546	0.481	/	0.345	0.381	4
PL_6	0.729	0.795	0.706	0.635	0.655	/	0.541	7
PL_7	0.7	0.772	0.671	0.599	0.619	0.459	/	6

4.3. Sensitivity Analysis

In order to examine the stability of the obtained solution, a sensitivity analysis was performed in which the influence of changes in certain parameters on the result was examined. For this purpose, six additional scenarios were defined, the first of which (Sc.1) implied the equalization of the importance of all criteria, while the other scenarios implied the exclusion of one of the five most important criteria, i.e., the criteria with the greatest weight and thus the strongest influence on the potential location ranking. Criteria C_6 (Sc.2), C_3 (Sc.3), C_{13} (Sc.4), C_5 (Sc.5), and C_{16} (Sc.6), respectively, were excluded in the scenarios. Changes in the ranking of potential locations in additional scenarios compared to the initial (basic) scenario are shown in Table 8 and Figure 2. As it can be seen from the conducted sensitivity analysis, in all scenarios except Sc.2, PL_2 was ranked first and PL_1 second. PL_3 was ranked third in all scenarios except Sc.2 and Sc.4. The remaining potential locations were mostly ranked between fourth and seventh place in all scenarios, and the only step forward was made by PL_5, which was ranked second in Sc.2 and third in Sc.4, respectively. Taking everything into account, it can be concluded that the obtained solution is sufficiently stable and that Zagreb (PL_2), Ljubljana (PL_1), and Belgrade (PL_3) stand out as the most favorable potential locations, respectively. The worst locations for the opening of the DP terminal are Skopje (PL_6) and Tirana (PL_7), respectively.

Table 8. Sensitivity analysis.

	Basic	Sc.1	Sc.2	Sc.3	Sc.4	Sc.5	Sc.6
PL_1	2	2	1	2	2	2	2
PL_2	1	1	3	1	1	1	1
PL_3	3	3	4	3	4	3	3
PL_4	5	4	7	4	5	5	5
PL_5	4	5	2	5	3	4	4
PL_6	7	6	6	6	7	7	7
PL_7	6	7	5	7	6	6	6

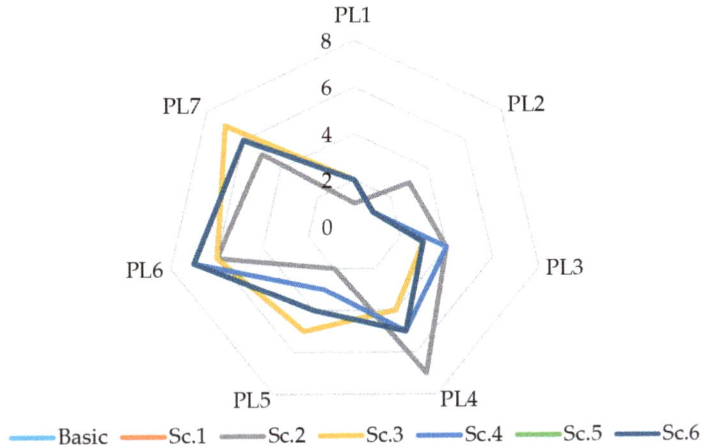

Figure 2. Sensitivity analysis.

4.4. Discussion of the Obtained Results and Analysis of the Proposed Approach

Zagreb (PL_2) is the best ranked potential location for the opening of the DP terminal in the Western Balkans region. As well as in the case of Ljubljana (PL_1), which is the second-ranked potential location, proximity and good transport connections with the ports of Koper, Trieste, Rijeka, Split, and Ploče, as well as the fact that the countries whose capitals are these potential locations are EU member states, mostly contributed to their good ranking. As such, they have more developed economies, they are more politically stable and part of a single customs zone, they have higher levels of development of planning and regulatory documents, etc., which result in better potential to attract more goods and better connectivity. On the other hand, Belgrade (PL_3), as the capital of Serbia, is ranked third, with a significant advantage over the remaining PLs, despite the fact that Serbia is not an EU member and is the only country in the Western Balkans, besides North Macedonia, that has no access to the sea. Belgrade has a very good geographical position at the crossroads between Western Europe and Asia. It is the only PL that has the possibility of using inland waterways, and although no seaport is close to it, all ports are within a radius of 700 km, which gives it a significant advantage in relation to all the other PLs and makes it the only one with a realistic possibility to achieve a significant connection with all observed ports. Cheaper labor and lower investment and operating costs are other factors that make this PL better than PL_2 and PL_1. On the other hand, although plans and regulations are not at a satisfactory level, they are significantly more developed than in the other countries, whose capitals are ranked lower. The lack and insufficient development of planning and regulatory documents, as well as political instability, lack of professional staff, and underdeveloped economy are some of the main reasons for the weaker ranking of the other PLs. In addition, these PLs mainly gravitate to the local ports (in the immediate vicinity), they are not located, or are partially located, on major transport routes, and it is not realistic to expect that they could attract a larger volume of flows and achieve more significant status in the network.

The implications of the obtained results of potential location rankings are such that when considering the Western Balkans region for the opening of a DP terminal, it is not possible to talk about the opening of a single terminal. None of the defined PLs would fully satisfy the requirements of the market, goods flows between the observed ports and their wider catchment areas, i.e., the markets of Central and Western Europe. Accordingly, the end result of the analysis is the establishment of three DP terminals, in Zagreb, Ljubljana, and Belgrade, since this would be the only way to achieve the full effect of redirecting flows from the observed ports and to create preconditions for the development of a sustainable logistics network in the Western Balkans and its integration into a comprehensive European network. This does not mean that DP terminals should never be considered for the remaining PLs, but it means that in the current market conditions the emphasis should be placed on establishing the selected locations, while the remaining locations could be activated in later stages of the DP terminal network development. These PLs would most likely be used for the establishment of the close- and mid-range DP terminals that would enable the connection of the ports from their immediate surroundings with the rest of the network.

The applicability of the proposed hybrid model, which combines Delphi, AHP, and CODAS methods in a grey environment, was proven by solving a real-life case study. Since the problem was structured as a hierarchy, with the ultimate goal at the top, groups of criteria and sub-criteria, and the alternatives, i.e., the potential locations, at the bottom, the AHP method imposed itself as a logical technique for solving it. However, considering that when applying only the AHP method the dimensions of the problem would be very large due to the need to compare all pairs of alternatives in relation to each of the sub-criteria, and that its solution would require significant time, financial and other resources, the CODAS method was introduced in the MCDM model. Since, on the other side, the criteria did not have the same significance for all stakeholders involved in the problem, a part of the Delphi method was introduced in the model in order to obtain unified assessments of the criteria. The model was developed in a grey environment because the elements of the problem were evaluated by the decision makers who very often gave inaccurate, incomplete, or ambiguous assessments. The model is universally applicable and after certain adjustments it can be used to solve various problems, in both this and other areas. The main advantages of the defined model are simplicity of application, provision of quality results with reasonable consumption of resources, and the possibility of fine differentiation between mutually similar alternatives. The limitations of the defined model are reflected in the inability to consider possible cause-and-effect relationships between the criteria themselves. This can be solved by applying the ANP (Analytical Network Process) method [99], but in that case one should take into account the number of criteria and sub-criteria that are considered, because the problem dimensions grow exponentially with their increase. The model also does not take into account the different influence that stakeholders have on problem solving, but views them all as equals, which is not usually the case in real-life situations, but this aspect can easily be included in the model by assigning the different weights to the stakeholders.

5. Conclusions

The subject of the paper was the selection and evaluation of locations for the DP terminals with the aim of establishing an environmentally, economically, and socially sustainable logistics network in the hinterlands of the ports, taking into account the requirements of different stakeholders. To solve the defined problem, a new hybrid MCDM model, which combined Delphi, AHP, and CODAS methods in a grey environment, was developed. The applicability of the developed model was successfully demonstrated by solving a real-life case study of ranking the potential locations for the establishment of the DP terminals in the Western Balkans region. As the result, the most realistic scenario would be the opening of three DP terminals, in Zagreb, Ljubljana, and Belgrade.

The main contributions of the paper are defining a new approach for the ranking and selection of the DP terminal locations based on finding a compromise solution that integrates the requirements of all stakeholders, as well as the extension of the CODAS method in a grey environment and development of the new hybrid MCDM model. In addition, the paper makes a significant contribution and upgrade to the previous research in the field of sustainability analysis of the DP concept and its implementation for the needs of regional international markets.

In future research, the defined MCDM model can be extended to include cause-and-effect relationships between the criteria and sub-criteria, e.g., by including the ANP method, as well as the different influences of stakeholders through the assignment of weighting coefficients to their grades. The number of stakeholders' representatives could also be increased in order to capture the fine differentiations in the characteristics of the regions within the observed area. Since the defined MCDM model is universally applicable and can be applied to solve various problems after certain adjustments, one of the directions of future research can be its application to solve problems with a similar structure, in this or any other area. In addition, in future research, some new, different approaches can be developed to solve the defined problem, and the problem itself can be expanded to include a wider area, a larger number of ports, and more potential locations. It would be interesting to define models and plans for the future development of the logistics network and to explore the possibilities and conditions for activating the other potential DP locations, both existing and the newly defined.

Author Contributions: Conceptualization, S.T., M.K., V.R. and N.B.; Formal analysis, S.T., M.K., V.R. and N.B.; Methodology, S.T. and M.K.; Writing—original draft, S.T., M.K., V.R. and N.B. All authors have read and agreed to the published version of the manuscript.

Funding: This research received no external funding.

Conflicts of Interest: The authors declare no conflict of interest.

References

1. Awad-Núñez, S.; González-Cancelas, N.; Soler-Flores, F.; Camarero-Orive, A. How Should The Sustainability of The Location of Dry Ports Be Measured? A Proposed Methodology Using Bayesian Networks and Multi-Criteria Decision Analysis. *Transport* **2015**, *30*, 312–319. [CrossRef]
2. UNCTAD—United Nations Conference on Trade and Development. *Review of Maritime Transportation*; United Nations Publication, UNCTAD: Geneva, Switzerland, 2017; Available online: https://unctad.org/en/PublicationsLibrary/rmt2017_en.pdf (accessed on 16 July 2020).
3. Khaslavskaya, A.; Roso, V. Outcome-Driven Supply Chain Perspective on Dry Ports. *Sustainability* **2019**, *11*, 1492. [CrossRef]
4. Beresford, A.; Pettit, S.; Xu, Q.; Williams, S. A study of dry port development in China. *Marit. Econ. Logist.* **2012**, *14*, 73–98. [CrossRef]
5. UNCTAD—United Nations Conference on Trade and Development. *Multimodal Transport and Containerisation. TD/B/C.4/238/Supplement 1, Part Five: Ports and Container Depots*; United Nations Publication, UNCTAD: Geneva, Switzerland, 1982.
6. Khaslavskaya, A.; Roso, V. Dry ports: Research outcomes, trends, and future implications. *Marit. Econ. Logist.* **2020**, *22*, 265–292. [CrossRef]
7. Roso, V.; Woxenius, J.; Lumsden, K. The dry port concept: Connecting container seaports with the hinterland. *J. Transp. Geogr.* **2009**, *17*, 338–345. [CrossRef]
8. Roso, V.; Brnjac, N.; Abramovic, B. Inland Intermodal Terminals Location Criteria Evaluation: The Case of Croatia. *Transp. J.* **2015**, *54*, 496. [CrossRef]
9. Wang, C.; Chen, Q.; Huang, R. Locating dry ports on a network: A case study on Tianjin Port. *Marit. Policy Manag.* **2018**, *45*, 71–88. [CrossRef]
10. Wei, J.; Sun, A.; Zhuang, J. The Selection of Dry Port Location with the Method of Fuzzy-ANP. In *Lecture Notes in Electrical Engineering*; LNEE 72; Luo, Q., Ed.; Springer: Berlin/Heidelberg, Germany, 2010; pp. 265–273. [CrossRef]
11. Nguyen, L.C.; Notteboom, T. A Multi-Criteria Approach to Dry Port Location in Developing Economies with Application to Vietnam. *Asian J. Shipp. Logist.* **2016**, *32*, 23–32. [CrossRef]
12. Beresford, A.K.C.; Dubey, R.C. *Handbook on the Management and Operation of Dry Ports*; United Nations Publication: Geneva, Switzerland, 1990; Available online: https://unctad.org/en/Docs/rdpldc7_en.pdf (accessed on 16 July 2020).
13. Henttu, V.; Hilmola, O.-P. Financial and environmental impacts of hypothetical Finnish dry port structure. *Res. Transp. Econ.* **2011**, *33*, 35–41. [CrossRef]
14. Hui, F.; Aye, L.; Duffield, C. Engaging Employees with Good Sustainability: Key Performance Indicators for Dry Ports. *Sustainability* **2019**, *11*, 2967. [CrossRef]
15. Veenstra, A.; Zuidwijk, R.; Van Asperen, E. The extended gate concept for container terminals: Expanding the notion of dry ports. *Marit. Econ. Logist.* **2012**, *14*, 14–32. [CrossRef]
16. Roso, V.; Russell, D.; Ruamsook, K.; Stefansson, G. Inland port services for seaport's competitive advantage. *World Rev. Intermodal Transp. Res.* **2015**, *5*, 263–280.

17. Monios, J. The role of inland terminal development in the hinterland access strategies of Spanish ports. *Res. Transp. Econ.* **2011**, *33*, 59–66. [CrossRef]
18. Tadić, S.; Krstić, M.; Brnjac, N. Selection of efficient types of inland intermodal terminals. *J. Transp. Geogr.* **2019**, *78*, 170–180. [CrossRef]
19. Bask, A.; Roso, V.; Andersson, D.; Hämäläinen, E. Development of seaport–dry port dyads: Two cases from Northern Europe. *J. Transp. Geogr.* **2014**, *39*, 85–95. [CrossRef]
20. Facchini, F.; Digiesi, S.; Mossa, G. Optimal dry port configuration for container terminals: A non-linear model for sustainable decision making. *Int. J. Prod. Econ.* **2020**, *219*, 164–178. [CrossRef]
21. Li, W.; Hilmola, O.-P.; Panova, Y. Container Sea Ports and Dry Ports: Future CO2 Emission Reduction Potential in China. *Sustainability* **2019**, *11*, 1515. [CrossRef]
22. Kurtulus, E.; Cetin, I.B. Assessing the Environmental Benefits of Dry Port Usage: A Case of Inland Container Transport in Turkey. *Sustainability* **2019**, *11*, 6793. [CrossRef]
23. Tao, X.; Wu, Q.; Zhu, L. Mitigation potential of CO_2 emissions from modal shift induced by subsidy in hinterland container transport. *Energy Policy* **2017**, *101*, 265–273. [CrossRef]
24. Regmi, M.B.; Hanaoka, S. Assessment of Modal Shift and Emissions along a Freight Transport Corridor Between Laos and Thailand. *Int. J. Sustain. Transp.* **2013**, *9*, 192–202. [CrossRef]
25. Roso, V. Evaluation of the dry port concept from an environmental perspective: A note. *Transp. Res. Part D Transp. Environ.* **2007**, *12*, 523–527. [CrossRef]
26. Lattila, L.; Henttu, V.; Hilmola, O.-P. Hinterland operations of sea ports do matter: Dry port usage effects on transportation costs and CO_2 emissions. *Transp. Res. Part E Logist. Transp. Rev.* **2013**, *55*, 23–42. [CrossRef]
27. Cezar-Gabriel, C. Performance assessment in operating dry ports. *Ann. Fac. Econ.* **2010**, *1*, 934–938.
28. Tsao, Y.-C.; Thanh, V.-V. A multi-objective mixed robust possibilistic flexible programming approach for sustainable seaport-dry port network design under an uncertain environment. *Transp. Res. Part E Logist. Transp. Rev.* **2019**, *124*, 13–39. [CrossRef]
29. Awad-Núñez, S.; Cancelas, N.G.; Orive, A.C. Setting of factors considered when determining the location area of a dry port and its hierarchies through a Delphi methodology. *Rev. Transp. Territ.* **2015**, *13*, 100–121. [CrossRef]
30. Krstić, M.; Kovač, M.; Tadić, S. Dry port location selection: Case study for the Adriatic ports. In Proceedings of the XLVI International Symposium of Operational Research, SYM-OP-IS 2019, Kladovo, Serbia, 15–18 September 2019; pp. 303–308. (In Serbian).
31. Tadić, S.; Kovač, M.; Zečević, S.; Krstić, M. Implementation of the dry port concept in the West Balkans region. In Proceedings of the VII International symposium: New Horizons of Transport and Communications, Doboj, Republic of Srpska, 29–30 November 2019; pp. 422–427. (In Serbian).
32. Tadić, S.; Krstić, M.; Kovač, M. Location of dry port terminals: Case study for the territory of Republic of Serbia. In Proceedings of the fifth scientific conference with international participation "Politehnika", Belgrade, Serbia, 13 December 2019; pp. 558–563. (In Serbian).
33. Ambrosino, D.; Sciomachen, A. Location of Mid-range Dry Ports in Multimodal Logistic Networks. *Procedia Soc. Behav. Sci.* **2014**, *108*, 118–128. [CrossRef]
34. Mwemezi, J.J. PInland container depot integration into logistics networks based on network flow model: The Tanzanian perspective. *Afr. J. Bus. Manag.* **2012**, *6*, 7149–7157. [CrossRef]
35. Feng, X.; Zhang, Y.; Li, Y.; Wang, W. A Location-Allocation Model for Seaport-Dry Port System Optimization. *Discret. Dyn. Nat. Soc.* **2013**, *2013*, 1–9. [CrossRef]
36. Chang, Z.; Notteboom, T.; Lu, J. A two-phase model for dry port location with an application to the port of Dalian in China. *Transp. Plan. Technol.* **2015**, *38*, 442–464. [CrossRef]
37. Ng, A.K.Y.; Gujar, G.C. The Spatial Characteristics of Dry Ports in India. *Trans. Commun. Bull. Asia Pac.* **2009**, *78*, 102–111.
38. Li, F.; Shi, X.; Hu, H. Location Selection of Dry Port Based on AP Clustering—The Case of Southwest China. *J. Syst. Manag. Sci.* **2011**, *1*, 79–88.
39. Wang, Y.; Wang, J. The optimal location of dry port: A case study of the hinterland of Western Side of the Taiwan Straits Port Group. In Proceedings of the IEEE 17th International Conference on Industrial Engineering and Engineering Management, Xiamen, China, 29–31 October 2010; pp. 1864–1868.
40. Zhang, Z. Applying fuzzy C-clustering to location planning of dry port. *J. Shanghai Marit. Univ.* **2008**, *29*, 34–37.

41. Muravev, D.; Rakhmangulov, A. Environmental Factors' Consideration at Industrial Transportation Organization in the «Seaport—Dry port» System. *Open Eng.* **2016**, *6*, 476–484. [CrossRef]
42. Henttu, V.; Lättilä, L.; Hilmola, O.P. *Financial and Environmental Impacts of a Dry Port to Support Two Major Finnish Transit Seaports*; Department of Industrial Management, Faculty of Technology Management, Lappeenranta University of Technology: Lappeenranta, Finland, 2010.
43. Abbasi, M.; Pishvaee, M.S. A two-stage GIS-based optimization model for the dry port location problem: A case study of Iran. *J.Ind. Syst. Eng.* **2018**, *11*, 50–73.
44. Van Nguyen, T.; Zhang, J.; Zhou, L.; Meng, M.; He, Y. A data-driven optimization of large-scale dry port location using the hybrid approach of data mining and complex network theory. *Transp. Res. Part E Logist. Transp. Rev.* **2020**, *134*, 101816. [CrossRef]
45. Abdoulkarim, H.T.; Fatouma, S.H.; Kalgora, B. The Selection of Dry Port Location by Analytic Network Process Model: A Case Study of Dosso-Niger. *J. Transp. Technol.* **2019**, *9*, 146–155. [CrossRef]
46. Wang, C.-H.; Wei, J.Y. Research on the Dry Port Location of Tianjin Port Based on Analytic Network Process. In Proceedings of the 2008 International Seminar on Business and Information Management, Wuhan, China, 19 December 2008; pp. 75–78.
47. Gonçalves, W.; De Freitas, R.R.; Chiquieri, J.; Júnior, L.O.R.; Mattos, A.L.; Zatta, F.N. Regional development and Organization of space: An Approach to the Dry Port location using the Hybrid Method Fuzzy-AHP. *Int. J. Adv. Eng. Res. Sci.* **2019**, *6*, 521–530. [CrossRef]
48. Ka, B. Application of Fuzzy AHP and ELECTRE to China Dry Port Location Selection. *Asian J. Shipp. Logist.* **2011**, *27*, 331–353. [CrossRef]
49. Komchornrit, K. The Selection of Dry Port Location by a Hybrid CFA-MACBETH-PROMETHEE Method: A Case Study of Southern Thailand. *Asian J. Shipp. Logist.* **2017**, *33*, 141–153. [CrossRef]
50. Rahimi, M.; Asef-Vaziri, A.; Harrison, R. An Inland Port Location-Allocation Model for a Regional Intermodal Goods Movement System. *Marit. Econ. Logist.* **2008**, *10*, 362–379. [CrossRef]
51. Flämig, H.; Hesse, M. Placing dryports. Port regionalization as a planning challenge—The case of Hamburg, Germany, and the Süderelbe. *Res. Transp. Econ.* **2011**, *33*, 42–50. [CrossRef]
52. Do, N.-H.; Nam, K.-C.; Le, Q.-L.N. A consideration for developing a dry port system in Indochina area. *Marit. Policy Manag.* **2011**, *38*, 1–9. [CrossRef]
53. Dalkey, N.; Helmer, O. An Experimental Application of the DELPHI Method to the Use of Experts. *Manag. Sci.* **1963**, *9*, 458–467. [CrossRef]
54. Delbecq, A.L.; Van De Ven, A.H.; Gustafson, D.H. *Group Techniques for Program Planning: A Guide to Nominal Group and Delphi Processes*; Scott Foresman: Glenview, IL, USA, 1975.
55. Safari, H.; Soufi, M.; Aghasi, E. Select a Hypermarket Location Based on Fuzzy Multi Criteria Decision Making (F-MCDM) Techniques (Hybrid of F-Delphi, F-AHP, F-LLSM and F-PROMTHEE). *Kuwait Chapter Arab. J. Bus. Manag. Rev.* **2014**, *4*, 76–95. [CrossRef]
56. Ishikawa, A.; Amagasa, M.; Shiga, T.; Tomizawa, G.; Tatsuta, R.; Mieno, H. The max-min Delphi method and fuzzy Delphi method via fuzzy integration. *Fuzzy Sets Syst.* **1993**, *55*, 241–253. [CrossRef]
57. Tadić, S.; Krstić, M.; Zečević, S. Defining the Typical Structures of the Intermodal Terminals. In *Quantitative Methods in Logistics*; Vidović, M., Kilibarda, M., Zečević, S., Radivojević, G., Eds.; Faculty of Transport and Traffic Engineering, University of Belgrade: Beograd, Serbia, 2020; pp. 67–86.
58. Chan, P.; Lee, M.-H. Developing Sustainable City Indicators for Cambodia through Delphi Processes of Panel Surveys. *Sustainability* **2019**, *11*, 3166. [CrossRef]
59. Tadić, S.; Krstić, M.; Roso, V.; Brnjac, N. Planning an Intermodal Terminal for the Sustainable Transport Networks. *Sustainability* **2019**, *11*, 4102. [CrossRef]
60. Emovon, I.; Norman, R.; Murphy, A.J. Hybrid MCDM based methodology for selecting the optimum maintenance strategy for ship machinery systems. *J. Intell. Manuf.* **2015**, *29*, 519–531. [CrossRef]
61. Zečević, S.; Tadić, S.; Krstić, M. Intermodal Transport Terminal Location Selection Using a Novel Hybrid MCDM Model. *Int. J. Uncertain. Fuzziness Knowl.-Based Syst.* **2017**, *25*, 853–876. [CrossRef]
62. Saaty, T.L. *The Analytic Hierarchy Process: Planning, Priority Setting, Resource Allocation*; McGraw-Hill International Book Co.: New York, NY, USA, 1980.
63. Calabrese, A.; Costa, R.; Levialdi, N.; Menichini, T. Integrating sustainability into strategic decision-making: A fuzzy AHP method for the selection of relevant sustainability issues. *Technol. Forecast. Soc. Chang.* **2019**, *139*, 155–168. [CrossRef]

64. Wang, J.; Jing, Y.-Y.; Zhang, C.-F.; Zhao, J.-H. Review on multi-criteria decision analysis aid in sustainable energy decision-making. *Renew. Sustain. Energy Rev.* **2009**, *13*, 2263–2278. [CrossRef]
65. Zardari, N.H.; Ahmed, K.; Shirazi, S.; Bin Yusop, Z. Literature review. In *Weighting Methods and Their Effects on Multi-Criteria Decision Making Model Outcomes in Water Resources Management*; Springer: Berlin, Germany, 2015; pp. 7–67.
66. Tesfamariam, S.; Sadiq, R. Risk-based environmental decision-making using fuzzy analytic hierarchy process (F-AHP). *Stoch. Environ. Res. Risk Assess.* **2006**, *21*, 35–50. [CrossRef]
67. Hajkowicz, S.A.; McDonald, G.T.; Smith, P.N. An Evaluation of Multiple Objective Decision Support Weighting Techniques in Natural Resource Management. *J. Environ. Plan. Manag.* **2000**, *43*, 505–518. [CrossRef]
68. Ishizaka, A.; Siraj, S. Are multi-criteria decision-making tools useful? An experimental comparative study of three methods. *Eur. J. Oper. Res.* **2018**, *264*, 462–471. [CrossRef]
69. Konidari, P.; Mavrakis, D. Multi-criteria evaluation of climate policy interactions. *J. Multi-Criteria Decis. Anal.* **2006**, *14*, 35–53. [CrossRef]
70. Durmić, E. The Evaluation of the Criteria for Sustainable Supplier Selection by Using the FUCOM Method. *Oper. Res. Eng. Sci. Theory Appl.* **2019**, *2*, 91–107. [CrossRef]
71. Ishizaka, A.; Nemery, P. *Multi-criteria Decision Analysis: Methods and Software*; John Wiley & Sons: Hoboken, NJ, USA, 2013.
72. Alonso, J.A.; Lamata, M.T. Consistency in the analytic hierarchy process: A new approach. *Int. J. Uncertain. Fuzz.* **2006**, *14*, 445–459. [CrossRef]
73. Taylan, O.; AlAmoudi, R.; Kabli, M.; Aljifri, A.; Ramzi, F.; Herrera-Viedma, E. Assessment of Energy Systems Using Extended Fuzzy AHP, Fuzzy VIKOR, and TOPSIS Approaches to Manage Non-Cooperative Opinions. *Sustainability* **2020**, *12*, 2745. [CrossRef]
74. Zečević, S.; Tadić, S.; Krstić, M. Selection of the intermodal transport chain variant using AHP method. In *Proceedings of the 3rd International Conference on Traffic and Transport Engineering-ICTTE, Guimarães, Portugal, 4–7 September 2016*; Scientific Research Center Ltd.: Belgrade, Serbia, 2016; pp. 339–345.
75. Roh, S.Y.; Jang, H.M.; Han, C.H. Warehouse Location Decision Factors in Humanitarian Relief Logistics. *Asian J. Shipp. Logist.* **2013**, *29*, 103–120. [CrossRef]
76. Tadić, S.; Zečević, S.; Krstić, M. Ranking of Logistics System Scenarios for Central Business District. *Promet Traffic Transp.* **2014**, *26*, 159–167. [CrossRef]
77. Kashani, A.H.; Nodehi, A.N.; Bandpey, S. Prioritization of Dry Ports by Using Pair Comparison Technique to Enhance AHP. *J. Int. Bus. Stud.* **2014**, *3*, 34–45.
78. Tadić, S.; Zečević, S.; Krstić, M. City logistics terminal location selection using combined fuzzy AHP and fuzzy TOPSIS analysis. In Proceedings of the 1st International Conference on Traffic and Transport Engineering, ICTTE, Belgrade, Serbia, 29–30 November 2012; pp. 345–358.
79. Ghorabaee, M.K.; Zavadskas, E.K.; Turskis, Z.; Antucheviciene, J. A new Combinative Distance-Based Assessment (CODAS) method for multi-criteria decision-making. *Econ. Comput. Econ. Cybern.* **2016**, *50*, 25–44.
80. Karagoz, S.; Deveci, M.; Simic, V.; Aydin, N.; Bolukbas, U. A novel intuitionistic fuzzy MCDM-based CODAS approach for locating an authorized dismantling center: A case study of Istanbul. *Waste Manag. Res.* **2020**, *38*, 660–672. [CrossRef] [PubMed]
81. Sansabas-Villalpando, V.; Olguín, I.J.C.P.; Pérez-Domínguez, L.; Rodríguez-Picón, L.A.; Méndez-González, L.C. CODAS HFLTS Method to Appraise Organizational Culture of Innovation and Complex Technological Changes Environments. *Sustainability* **2019**, *11*, 7045. [CrossRef]
82. Yeni, F.B.; Özçelik, G. Interval-Valued Atanassov Intuitionistic Fuzzy CODAS Method for Multi Criteria Group Decision Making Problems. *Group Decis. Negot.* **2018**, *28*, 433–452. [CrossRef]
83. Bolturk, E. Pythagorean fuzzy CODAS and its application to supplier selection in a manufacturing firm. *J. Enterp. Inf. Manag.* **2018**, *31*, 550–564. [CrossRef]
84. Badi, I.; Abdulshahed, A.M.; Shetwan, A. A case study of supplier selection for a steelmaking company in Libya by using the Combinative Distance-based Assessment (CODAS) model. *Decis. Mak. Appl. Manag. Eng.* **2018**, *1*, 1–12. [CrossRef]

85. Bolturk, E.; Karasan, A. Interval valued neutrosophic CODAS method for renewable energy selection. In *Data Science and Knowledge Engineering for Sensing Decision Support*; Liu, J., Lu, J., Xu, Y., Martinez, L., Kerreiçinde, E., Eds.; World Scientific Publishers: London, UK, 2018; pp. 1026–1033. [CrossRef]
86. Zadeh, L. Fuzzy sets. *Inf. Control.* **1965**, *8*, 338–353. [CrossRef]
87. Pawlak, Z. Rough sets. *Int. J. Comput. Inf. Sci.* **1982**, *11*, 341–356. [CrossRef]
88. Ju-Long, D. Control problems of grey systems. *Syst. Control. Lett.* **1982**, *1*, 288–294. [CrossRef]
89. Yang, Y.; John, R. Grey sets and greyness. *Inf. Sci.* **2012**, *185*, 249–264. [CrossRef]
90. Lingkan, Y. Analyzing the main factors in the forming of torrential rain debris flow—Applying Grey Systems Theory in Dealing with the Findings of an Investigation by the Delphi Method. *B. Soil Water Conserv.* **1987**, *2*. Available online: https://en.cnki.com.cn/Article_en/CJFDTOTAL-STTB198702005.htm (accessed on 26 August 2020).
91. Lü, X.; Fan, H.-T.; Xin, L. Study on the risk assessment approach of port facility security based on a comprehensive model of Delphi method, analytic hierarchy process, grey theory and fuzzy evaluation method. In Proceedings of the 2015 International Conference on Transportation Information and Safety (ICTIS), Wuhan, China, 25–28 June 2015; pp. 628–632.
92. Chen, Y.J.; Duan, L.; Xiao, Y. Constructing driving behavior indicators using principal component analysis and combination of fuzzy and grey Delphi method. *J. Saf. Environ.* **2012**, *12*, 228–232.
93. Ma, Z.; Shao, C.; Ma, S.; Ye, Z. Constructing road safety performance indicators using Fuzzy Delphi Method and Grey Delphi Method. *Expert Syst. Appl.* **2011**, *38*, 1509–1514. [CrossRef]
94. Liu, G.; Zhou, Z.C.; Shen, Y.W. Warfare Efficiency Evaluation of Surface Warships Formation Based on AHP and Grey Evaluation Method. *Command Control Simul.* **2007**, *29*, 69–73.
95. Sahoo, S.; Dhar, A.; Debsarkar, A.; Kar, A. Future Scenarios of Environmental Vulnerability Mapping Using Grey Analytic Hierarchy Process. *Nat. Resour. Res.* **2019**, *28*, 1461–1483. [CrossRef]
96. Guru, S.; Mahalik, D.K. A comparative study on performance measurement of Indian public sector banks using AHP-TOPSIS and AHP-grey relational analysis. *Opsearch* **2019**, *56*, 1213–1239. [CrossRef]
97. Çelikbilek, Y. Using an Integrated Grey AHP–MOORA Approach for Personnel Selection: An Application on Manager Selection in the Health Industry. *Alphanumeric J.* **2018**, *6*, 69. [CrossRef]
98. Wang, Y.; Chin, K.-S. Fuzzy analytic hierarchy process: A logarithmic fuzzy preference programming methodology. *Int. J. Approx. Reason.* **2011**, *52*, 541–553. [CrossRef]
99. Saaty, T.L.; Vargas, L.G. *The Analytic Network Process*; RWS Publications: Pittsburgh, PA, USA, 1996.
100. Osati, M.; Omidvari, M. Performance measurement of electricity suppliers using PROMETHEE and balance scorecard. *Manag. Sci. Lett.* **2016**, *6*, 387–394. [CrossRef]
101. Mukherjee, S.; Ghosh, B. Application of Grey Possibility Degree in Comparing Poverty. International Journal of Innovative Research in Science. *Eng. Technol.* **2015**, *4*, 4698–4703. [CrossRef]
102. Pan-European Corridors to Major Transport Axes. Available online: https://www.unece.org/trans/main/wp5/wp5_workshop2.html (accessed on 17 July 2020).
103. EC—European Commission. *The Core Network Corridors: Trans European Transport Network*; Directorate General for Mobility and Transport: Brussels, Belgium, 2013.
104. UN—United Nations. *Joint Study on Developing Euro-Asia Transport Linkages, Prepared under the United Nations Development Account Project on "Capacity Building through Cooperation in Developing Land and Land-Sea Interregional Transport Linkages"*; United Nations: Geneva, Switzerland, 2008.
105. Tadić, S.; Zečević, S.; Milenković, D. Intermodal transport treatment in developed and developing countries. *Tehnika* **2017**, *67*, 897–902. (In Serbian) [CrossRef]
106. Black, J.; Roso, V.; Marušić, E.; Brnjac, N. Issues in Dry Port Location and Implementation in Metropolitan Areas: The Case of Sydney, Australia. *Trans. Marit. Sci.* **2018**, *7*, 41–50. [CrossRef]
107. The World Bank. Global Rankings. 2018. Available online: https://lpi.worldbank.org/international/global (accessed on 17 July 2020).
108. Eurostat. Main Tables. Available online: https://ec.europa.eu/eurostat/web/national-accounts/data/main-tables (accessed on 17 July 2020).

109. EcoTransit. Calculation Parameters. Available online: https://www.ecotransit.org/calculation.en.html (accessed on 17 July 2020).
110. Tadić, S.; Zečević, S.; Milenković, D. Problems regarding intermodal transport in the Danube Region. In *Proceedings of the 4th International Conference on Traffic and Transport Engineering, ICTTE 2018, Belgrade, Serbia, 27–28 September 2018*; City Net Scientific Research Center Ltd.: Belgrade, Serbia, 2018; pp. 483–489.

© 2020 by the authors. Licensee MDPI, Basel, Switzerland. This article is an open access article distributed under the terms and conditions of the Creative Commons Attribution (CC BY) license (http://creativecommons.org/licenses/by/4.0/).

sustainability

MDPI

Article

Evaluation of Process Orientation Dimensions in the Apparel Industry

Andrea Dobrosavljević [1], Snežana Urošević [1], Milovan Vuković [1], Miroslav Talijan [2] and Dragan Marinković [3,*]

1 Technical Faculty in Bor, University of Belgrade, Vojske Jugoslavije 12, 19210 Bor, Serbia; an.dobrosavljevic@gmail.com (A.D.); surosevic@tfbor.bg.ac.rs (S.U.); mvukovic@tfbor.bg.ac.rs (M.V.)

2 Military Academy, University of Defence, Pavla Jurišića Šturma 33, 11000 Belgrade, Serbia; mtalijan@gmail.com

3 Faculty of Mechanical Engineering and Traffic Systems, TU Berlin, Str. d. 17. Juni 135, 10623 Berlin, Germany

* Correspondence: dragan.marinkovic@tu-berlin.de

Received: 8 April 2020; Accepted: 15 May 2020; Published: 19 May 2020

check for updates

Abstract: The dimensions that influence the establishment of business process management (BPM) practices and the progression to higher levels of process maturity derive from exploring the dimensions of process orientation of organizations. Small and medium-sized clothing enterprises (SME's) are characterized by various specifics that can affect the degree of process orientation adoption and the pace of transition from lower to higher levels of process maturity. According to these specifics, the acceptance of the process approach may be differently affected. For the purpose of adequate evaluation and prioritization of the most influential dimensions, a new integrated multicriteria decision-making (MCDM) model that combines classical and fuzzy theory was developed. First, the full consistency method (FUCOM) method was applied, followed by the fuzzy pivot pairwise relative criteria importance assessment (fuzzy PIPRECIA) method to obtain more accurate criteria values. Prioritization of the most influential BPM dimension contributes to highlighting the area of business that needs to be primarily strengthened by appropriate actions for successful establishment of BPM in apparel industry SMEs. Within this research, the prioritized dimension refers to human resource management in accordance with the specific aspects of business within the apparel industry.

Keywords: business process orientation (BPO); business process management (BPM); FUCOM; fuzzy PIPRECIA; apparel industry; SMEs

1. Introduction

In the contemporary business environment, organizations are becoming aware of the need to implement adequately defined and controlled processes, which will contribute not only to acquiring the status of a process-mature organization, but also to improving the methods of management, development, and product quality [1]. Organizational orientation towards business processes and managing them ensures gaining knowledge about the possibilities of adopting best business practices and differentiating from the competition [2]. In the case of textile manufacturing organizations in countries such as Serbia, Bulgaria, and North Macedonia, several specific aspects make the whole approach to the implementation of business process management (BPM) relevant because it can contribute to the generation of new insights important for the improvement of business processes of these types of organizations. There were textile giants in these countries, in the past, that employed thousands of workers. The past and present picture of this sector is quite different. Today, there are mainly garment manufacturing small and medium-sized clothing enterprises (SMEs) [3].

The evaluation of the influence of different dimensions of process orientation is partly stimulated, among other specific aspects of the apparel industry, by the statement of author Chong [4], who stated that although process-oriented research is mainly focused on the operations of large organizations, this concept applies to the business of small and medium-sized enterprises (SMEs).

Consideration of the apparel industry SMEs' business appears through the literature precisely because of the particularities that accompany their business, among which the labor-intensive character and the dominant presence of the female workforce can be emphasized [5–7]. Besides, specific aspects of the apparel industry are highlighted in terms of constant model changes, short deadlines due to the change in fashion trends, and grading, which signifies the production of items by size. Then, there are the particularities of the textile material in terms of impermanence in shape, solidity, creasing, and pattern. All of these specific aspects make the production process complex [8]. The apparel industry is creative; however, design as a strategic resource is underutilized due to limited human and financial resources, as well as the informal process of developing and innovating apparel within SMEs [9,10]. The managerial role of the owner, as well as informality of interpersonal relationships and communication processes in SMEs, can, also, be added to this set of specific aspects [11]. The problems encountered by SMEs in the apparel industry in the surveyed countries are reflected in the lack of skilled personnel, batch size, type of assortment, production subcontracting or "lohn" production, cheap labor, and poor information, among others [12].

The fact is that in the modern business world, only those organizations that are ready to adapt quickly and effectively to generate new, more appropriate, and flexible concepts of business and development manage to survive [13,14]. In such organizations, the issue of adopting and implementing certain business practices based on strengthening certain elements or dimensions of process orientation arises.

2. Motivation and Research Aim

The business process consists of a set of activities performed in coordination with an organizational environment, with each element of the activity being subject to effective management [15]. BPM is implemented through the application of methods, techniques, and software for design, adoption, control, and analysis of operational processes that encompass people, organization, implementation, documentation, and other sources of information [16]. However, some authors have emphasized that in the SMEs sector there is a lack of understanding and interest for the proper implementation of BPM practice with respect to the principles of the concept [17,18]. Key elements in applying a BPM approach relate to a clear explanation of intentions to carry out this practice, the links between BPM and strategic programs that an organization develops, and the acquisition of process competencies, skills, and knowledge, among others [19]. An array of elements, such as structure, focus, measurement, responsibility, and consumer orientation, make up an integral part or basis of the business process orientation [20,21].

Various authors have researched the elements that influence the adoption of process orientation and BPM, and in the next section, this issue will be considered through a literature review. Considering the dimensions as important parts for understanding the overall picture of the establishment of BPM practices, the thought of the multidimensionality of this practice develops. Each of the elements can be categorized as an element of a particular process dimension. This statement was the driver of the research question within this paper. To what extent do each of the identified categories of process orientation adoption elements realistically contribute to the establishment of BPM practice and the progression towards a state of stability and high degree of process maturity within the SMEs of the apparel industry?

The aim of this paper was, therefore, reflected in the evaluation of the adopted process dimensions, while prioritizing the most influential one considering the apparel industry sector and its specific aspects. In this way we sought to identify which link in the business process management chain makes a more pronounced contribution to enhancing the process maturity of apparel industry SMEs, in comparison to other links. The findings of such a nature, except the scientific contributions, can provide a practical contribution in terms of advising organizations in which direction they should focus their activities and in what areas they need to make improvements in order to achieve higher process maturity.

In the next section, a literature review of the study of the elements and characteristic dimensions of process orientation is presented, with an emphasis on the recognized importance of each. The application of different multicriteria decision-making methods in the textile and clothing industry is also highlighted in this section. The research part of the paper begins with a description of the methodological framework and structure of the research. The presented methodology, which includes the application of the FUCOM and fuzzy PIPRECIA methods as elements of an integrated multicriteria model, was implemented to evaluate the process orientation dimensions of the apparel industry organizations. Finally, in addition to concluding considerations, guidelines for further research in this area are provided.

3. Literature Review of Research on the Process Orientation Dimensions

BPM is driven by a few basic rules concerning the appropriate way of mapping and documenting process activities, a focus on consumers, and a reliance on systems and documentation procedures to maintain discipline, consistency, and repeatability of performance quality. Also, BPM is concerned with the measurement and processing activities and business process management with continuous optimization, as pointed out by [22]. According to this author, BPM represents the approach of cultural changes, and it is not only about the establishment of good systems and the right kind of structure. It should be inspired by the best practices to ensure the achievement of superior competitiveness. Within each of these rules, it is possible to notice the existence of dimensions that influence the establishment of BPM. Adoption of the process view may depend on elements such as alignment of the process with strategic goals, established culture, human resources, management style, application of methods, and development of information systems, according to authors [23]. Another author [24] highlighted the success factors of BPM. These are strategy alignment, level of IT investment, performance measurement, level of employee specialization, organizational change, the appointment of process owner, implementation of proposed changes, implementation of continuous improvement systems, process standardization, computerization, automation, training, and employee empowerment. These elements and factors are also considered in the literature as interconnected areas of BPM capability [25,26].

3.1. *Significance of Process Orientation Dimensions for Establishment of BPM*

An organization's strategic program can be more easily implemented by applying BPM practices [27]. There is an assumption within the literature that process activities support the organization's strategy [28,29]. The chosen strategy must rely on processes that are coordinated and designed to simplify its successful implementation [30]. Harmonization of operational and strategic priorities is the basis for gaining competitiveness, and involvement of top management is important to ensure that the right priorities are set [31,32].

The requirements of the business are expressed through a description of the business processes, which are related to the strategy, and the explanation of the organizational goals, which answers the questions "why" and "what" is being done. These requirements also motivate the definition of specific processes, that is, the answer to the question "how" [33]. BPM relies on documented procedures to ensure discipline, consistency, and repeatability of performance quality [22]. An organization that does not, in any way, define or document its processes, inevitably remains at the lowest level of process maturity [34].

Business process performance reflects in overall organizational performance [35,36]. BPM relies on the measurement of activities to evaluate the performance of each process, to set goals, and to deliver results at a level that reaches corporate goals [22]. Performance measurement and evaluation help to evaluate the effectiveness and efficiency of performance, as well as the performance potential of different objects within an organization, so determining specific performance measures or performance indicators is essential [37,38]. Considering that consumer orientation is one of the core values that the BPM culture concept emphasizes, process performance measures based on consumer expectations are of great importance [11,39,40].

A process-oriented organization tries to organize responsibilities as horizontally as possible, as opposed to the traditional vertical hierarchical structure [41]. The business processes of all partners in the network should be conceptually modeled, focusing on documenting the flow and structure of interorganizational process activities [42]. Communication in the service of process execution, as well as the application of adequate techniques, are integrated within the organizational structure [43].

In a process-oriented organization, human resources are the drivers that enable the execution of a process. Process quality depends on the competencies of the human resources that work on its realization [44,45]. The application of BPM has a significant impact on the people in the organization. Adopting a process orientation affects employees because they have to accept responsibility for the outcomes of their processes. Accordingly, employees get new and different roles [41], like the role of the process owner responsible for overseeing functioning and process performance; the match between organizational information, management systems, and process needs; and the quality of the measures used in the process to measure process performance [46]. Management should focus on educating employees about process implementation and the benefits of process-based performance management [47]. It is of great importance to train the employees and make them able to work in a changing business environment [48]. Process-mature organizations encourage their employees to be process-oriented, enabling them to participate actively in process definition and process improvement activities. Rewards and incentives for the efforts of the employees should exist. Employees in such an environment develop a belief in the process, and when errors occur, they are focused on improvements rather than the measures of punishment. The value and importance of good staff are recognizable in process-mature organizations, therefore the knowledge and skills of employees are nurtured because modern business operations and the presence of new technologies require the upgrading of specialized skills and knowledge of human resources [49,50].

Business processes are conducted in a group environment, where the effectiveness of their realization depends on elements such as communication, coordination, and cooperation [51]. Organizational culture plays a key role during any change initiative [52]. Research [53,54] has identified four distinctive values that define the concept of culture within business process management. These values include customer orientation, excellence, responsibility, and teamwork.

Market orientation enhances organizations' motivation to explore consumer and competitor reactions, as it is marked by the degree of information generated from the external environment. Then, it is used to share information with the external environment and to respond to consumer needs and demands [55,56]. The application of BPM practices provides the opportunity to gain a competitive advantage and secure the market position of the organization [57]. Besides, careful BPM can help clients gain extraordinary experience in consuming products and services marketed by an organization [58].

The desire to meet the requirements of both internal and all external stakeholders, including suppliers, lies in the core of the BPM [59]. The term "cross-organizational business processes" is referred to in the literature, with the notion that contemporary businesses meet the wave of cross-organizational BPM among partners [60,61]. This term refers to the process collaboration that exceeds the limits of the organization while maintaining the autonomy of the organization in terms of the freedom of modification of the internal operation to achieve organizational objectives while achieving the common partnership goals [62].

The application of information and communication technologies to support BPM in companies and administration is quite an important element. Adequate techniques and software tools to support activities from design to control, as well as analysis of operating business processes, are being put into practice to facilitate value creation [63]. Information technology typically provides automatic support for business processes and inter-process connections [36]. Besides, they enable business process innovation in line with industry best practices and IT trends [64].

All of the business dimensions can reflect the state of the process orientation by how they perform. Using a holistic approach and evaluating each one of them might be beneficial. Some of the authors [65–68] highlighted some of these dimensions as factors of success or dimensions for analysis of process maturity within organizations operating in various industries. Authors [69] applied the fuzzy Delphi method to isolate dimensions by which BPM practice adoption can be assessed in a labor-intensive business environment, from a wider list of factors from the literature. The final list of dimensions consists of: (1) compliance of process and strategic goals; (2) process identification, documentation, and standardization; (3) performance measurement and business process improvement; (4) process structure; (5) human resource management; (6) process organizational culture; (7) focus on consumers and competitors; (8) supplier relations processes; and (9) process-based information systems development.

3.2. Application of Multicriteria Decision-Making Methods in Textile and Apparel Industries

Multicriteria decision-making methods [70,71] take into account expert evaluation of the criteria. Experts rank alternatives according to specified standards in the most objective way possible [72]. Each of the multicriteria methods has its advantages and disadvantages, as well as its application areas. In order to solve complex decision problems, a combination of several methods is also possible [73]. These methods have found application in the field of BPM through the evaluation and selection of business processes [74], evaluation of the most appropriate business process outsourcing decision [75], assessment of process factors when developing a methodology for analyzing business processes weaknesses and possible improvements [76], and BPM decision-making support [77], among others.

The analytic hierarchy process (AHP) method was applied to formulate a decision support system for determining the potential for clustering in the textile and clothing sectors [78]. The AHP method for calculation of weight coefficients of each criterion was applied to solve the problems of ranking and selection of suppliers within the textile industry in China and India, and then using the calculated weights, the ranking of suppliers was performed using the TOPSIS method [79]. Authors [80] grouped suppliers using the AHP method and cluster analysis. The AHP method was used to make a prototype selection of fashion items by [81]. Fuzzy AHP was used for the selection of adequate manufacturing technology for the textile laminating process [82].

In addition to the AHP method, which is present in a large number of papers related to the apparel and textile industries, the combination of PROMETHEE II and V methods has been applied in the development of a multicriteria optimization model for the optimal blending of cotton [83]. Ranking of workplaces according to the severity of working conditions in the textile industry was done by [84] by applying the ELECTRA method.

Some of the newer MCDM methods, such as the full consistency method (FUCOM) [85] and pivot pairwise relative criteria importance assessment (PIPRECIA) [86] might, as well as the ones previously mentioned, make their contribution to the research in the textile and apparel industry. The successful implementation of these methods is notable in the examples of sustainable supplier selection [87,88].

4. Methodology

Within the apparel industry in Serbia, Bulgaria, and North Macedonia, as well as in the other European countries, the most common business entities are small and medium-sized enterprises (SMEs). SMEs are interesting for research on BPM, and the sector within which they operate makes them even more interesting because business processes are creative and labor-intensive. The initial assumption of this research was that the establishment of BPM in apparel industry organizations can be considered by taking different process orientation dimensions into account. Each of these dimensions can have a different degree of contribution according to the conditions of business in a particular industry. With that in mind, this research attempted to evaluate and prioritize different process orientation dimensions according to their influence on BPM establishment in the apparel industry based on experts' opinions, and to discuss the practical contribution of the prioritized dimensions to the establishment of efficient, effective, and continuous BPM in the apparel industry organizations. Figure 1 illustrates the proposed steps to conduct this research in order to fulfil the stated research aim.

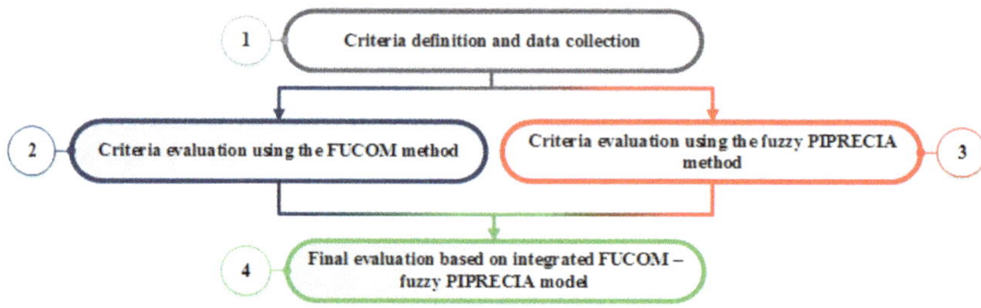

Figure 1. Proposed methodology for evaluation of process orientation dimensions' influence in the apparel industry using an integrated FUCOM–fuzzy PIPRECIA model.

The proposed research methodology for conducting an evaluation of process orientation dimensions in the apparel industry consisted of four major research phases, as presented in Figure 1. The initial phase of this research focused on criteria definition and data collection. Considering the dimensions that influence the establishment of BPM, described in detail within the literature review, nine criteria stood out. The optimal size of expert panel was investigated by [89], who came to the conclusion that final calculation scores are stable when more than 15 experts provide their estimations, and with 30 experts involved the stable mean can be reached. They further stated that having a large panel of experts is good, but expert panels of smaller sizes can, as well, provide a reliable evaluation of criteria. Within this research, a panel of 24 experts from apparel industry was formed. By contacting them, a database of criteria evaluations was collected. Collected data, in the form of expert evaluations, presented the input for the second and the third phase.

The second phase relied on the application of the full consistency (FUCOM) method to calculate the weight coefficients of dimensions based on the evaluations of each expert. One of the new methods based on the principles of pairwise comparison and validation of results through deviation for maximum consistency is the full consistency method (FUCOM) [85]. The decisive advantages of applying the FUCOM method are the small number of pairwise comparisons of criteria ($n - 1$ comparisons only), the ability to validate results by determining deviations from the maximum consistency (DMC) of comparisons, and to appreciate the transitivity in pairwise comparisons of criteria. This is particularly true of the first and second steps of the FUCOM method, in which decision makers rank the criteria according to their personal preferences and make comparisons of pairs of ranking criteria. However, unlike other subjective models, FUCOM showed smaller deviations in the criteria weight values obtained relative to the optimal value [85,90–92]. In addition, the methodological procedure of the FUCOM method eliminates the problem of redundancy of comparison of criteria pairs, which exists in some subjective models for determining the weight coefficients of criteria. The assumption is that there are n evaluation criteria in the multi-criteria model denoted by wj, $j = 1, 2, \ldots, n$, and that their weight coefficients should be determined. Subjective weighting models based on pairwise comparisons require the decision maker to determine the degree of influence of the criterion i and on the criterion j [93].

The third phase used the fuzzy PIPRECIA and the inverse fuzzy PIPRECIA to calculate weight coefficients of the same dimensions using the fuzzy logic. The fuzzy logic in PIPRECIA method was introduced by [94]. Inclusion of fuzzy logic in decision-making increases the reliability of the decision by minimization of misjudgment possibility [95,96]. The main advantage of the PIPRECIA method is that it allows criteria evaluation without sorting by relevance [97].

Phases two and three provided the input for final evaluation. The main focus within the fourth phase was on the comparison and final calculation of weight coefficients of considered process orientation dimensions in the business environment of the apparel industry using an integrated FUCOM–fuzzy PIPRECIA model.

5. Evaluation of Process Orientation Dimensions Priority within the Apparel Industry

5.1. Criteria Definition and Data Collection

By reducing the initial list of 13 factors, through which it is possible to observe the level of process orientation of organizations in labor-intensive activities, a list of nine factors was obtained within the research [69]. As labor-intensive character appears to be one of the particularities within apparel industry organizations, along with the dominant participation of the female workforce, the creative character, and the need to acquire specific knowledge, the possibility of evaluating the influence of these dimensions in the apparel industry has been noticed. Table 1 presents the complete list of these dimensions, codified in order to be used in the integrated FUCOM–fuzzy PIPRECIA multicriteria decision-making model as criteria. A description of each dimension is, as well, provided.

Table 1. Codification of the criteria used in the questionnaire.

Code	Criteria	Description of the Criteria
C1	Compliance of process and strategic goals	Business processes support the strategy and are aligned and designed to simplify its execution.
C2	Process identification, documentation, and standardization	The well-defined and documented procedures ensure discipline, consistency, and repeatability of performance quality.
C3	Performance measurement and business process improvement	Measuring the performance of each individual process is important for goal setting, delivering excellent results, and continuity in improvements.
C4	Process structure	The process organizational structure integrates elements in the service of execution of business processes, with a horizontal organization of responsibilities.
C5	Human resources management	Adopting a process orientation affects employees because they have to accept responsibility for the outcomes of their processes.
C6	Process organizational culture	Organizational culture plays a key role in adopting values aligned with BPM practices.
C7	Focus on consumers and competitors	The possibility of achieving competitive advantage and securing the organization's market position arises under the influence of adequate BPM in relations with consumers and competitors.
C8	Supplier relations processes	Impact of managing cross-organizational business processes with suppliers.
C9	Process-based information systems development	Adequate techniques and software tools to support activities from design to control, as well as analysis of operating business processes are being put into practice to facilitate value creation.

Source: [22,30,41,43,53,54,57,62,63,69].

Experts evaluated defined criteria using a seven-point scale given in the questionnaire for experts' assessments. By comparing the criteria, experts measured their preferences using scores from 1 to 7, where 1 stood for an almost equal value of two compared criteria, 2 for slightly more significant criteria, 3 for moderately more significant criteria, 4 for more significant criteria, 5 for much more significant criteria, 6 for dominantly more significant criteria, and 7 for absolutely more significant criteria [97].

Objective evaluation of the criteria was carried out by 24 experts from Serbia, Bulgaria, and North Macedonia in the period from September to October 2019. Experts who evaluated the criteria gained practical experience by investigating organizations' aspirations to adopt business process management practices or by being responsible for process execution within the textile and apparel industry. This group included experts in managerial positions in the procurement, production, quality control, and sales processes, as well as those engaged in educational and scientific research in the textile and apparel industries.

5.2. Criteria Evaluation Using the FUCOM Method

Table 2 shows the evaluation of the criteria by all decision makers, which is essentially the first and second step of the FUCOM method.

Table 2. The first two steps of the FUCOM method—ranking and rating.

DM1	C4	C5	C6	C2	C3	C7	C8	C1	C9
	1	1	1	1.5	1.5	1.5	1.8	1.8	1.8
DM2	C1	C5	C2	C8	C9	C7	C4	C3	C6
	1	1	1.5	1.5	1.5	1.8	1.8	1.8	2.7
DM3	C5	C9	C6	C3	C2	C4	C1	C7	C8
	1	1	1.5	1.5	1.8	1.8	2.7	4	5
DM4	C3	C2	C5	C9	C1	C7	C8	C6	C4
	1	1.5	1.5	1.5	1.8	2.7	2.7	4	4
DM5	C3	C9	C5	C2	C1	C4	C6	C7	C8
	1	1	1	1.5	1.8	1.8	1.8	2.7	2.7
...									
DM20	C5	C9	C8	C6	C4	C1	C2	C3	C7
	1	1	1.5	1.5	1.5	1.5	1.8	1.8	1.8
DM21	C7	C8	C9	C6	C5	C4	C3	C2	C1
	1	1	1.5	1.5	1.5	1.5	1.8	1.8	1.8
DM22	C9	C7	C5	C3	C2	C1	C4	C6	C8
	1	1.5	1.5	1.5	1.5	1.8	1.8	1.8	2.7
DM23	C5	C7	C9	C1	C4	C6	C2	C3	C8
	1	1.5	1.5	1.5	1.5	1.8	1.8	1.8	2.7
DM24	C1	C3	C7	C9	C8	C6	C5	C4	C2
	1	1	1	1	1.5	1.5	1.5	1.5	1.5

The following is an example of a FUCOM method calculation for a third decision maker. As can be seen from Table 2, a ranking and rating was performed showing that the best criterion was C5 and the worst was C8. Based on the obtained criteria importance, the comparative values of the criteria importance for each decision-maker were calculated:

$$\varphi_{C_5/C_9} = 1/1 = 1,\ \varphi_{C_9/C_6} = 1.5/1 = 1.5,\ \varphi_{C_6/C_3} = 1.5/1.5 = 1,\ \varphi_{C_3/C_2} = 1.8/1.5 = 1.2,$$
$$\varphi_{C_2/C_4} = 1.8/1.8 = 1,\ \varphi_{\frac{C_4}{C_1}} = 2.7/1.8 = 1.5,\ \varphi_{\frac{C_1}{C_7}} = 4/2.7 = 1.48,\ \varphi_{\frac{C_7}{C_8}} = 5/4 = 1.25.$$

In the third step, the final values of the weighting coefficients of the evaluation criteria were calculated, which needed to satisfy two conditions:

(1) The final values of the weighting coefficients should satisfy the condition where:

$$w_5/w_9 = 1,\ w_9/w_6 = 1.5,\ w_6/w_3 = 1,\ w_3/w_2 = 1.2,\ w_2/w_4 = 1,\ w_4/w_1 = 1.5,$$
$$w_1/w_7 = 1.48,\ w_7/w_8 = 1.25.$$

(2) In addition to the defined relations, the final values of the weighting coefficients should satisfy the condition of mathematical transitivity:

$$\varphi_{C_5/C_6} = 1 \times 1.5 = 1.5,\ \varphi_{C_9/C_3} = 1.5 \times 1 = 1.5,\ \varphi_{C_6/C_2} = 1 \times 1.2 = 1.2,\ \varphi_{C_3/C_4} = 1.2 \times 1 = 1.2,$$
$$\varphi_{C_2/C_1} = 1 \times 1.5 = 1.5,\ \varphi_{C_4/C_7} = 1.5 \times 1.48 = 2.22,\ \varphi_{C_1/C_8} = 1.48 \times 1.25 = 1.85.$$

$$w_5/w_6 = 1.5,\ w_9/w_3 = 1.5,\ w_6/w_2 = 1.2,\ w_3/w_4 = 1.2,\ w_2/w_1 = 1.5,\ w_4/w_7 = 2.22,$$
$$w_1/w_8 = 1.85.$$

The final model for determining the final values of the weighting coefficients of the evaluation criteria can be defined by applying the model:

$$\min \chi$$
$$s.t.\begin{cases} \left|\frac{w_5}{w_9} - 1\right| = \chi,\ \left|\frac{w_9}{w_6} - 1.5\right| = \chi,\ \left|\frac{w_6}{w_3} - 1\right| = \chi,\ \left|\frac{w_3}{w_2} - 1.2\right| = \chi,\ \left|\frac{w_2}{w_4} - 1\right| = \chi, \\ \left|\frac{w_4}{w_1} - 1.5\right| = \chi,\ \left|\frac{w_1}{w_7} - 1.48\right| = \chi,\ \left|\frac{w_7}{w_8} - 1.25\right| = \chi, \\ \left|\frac{w_5}{w_6} - 1.5\right| = \chi,\ \left|\frac{w_9}{w_3} - 1.5\right| = \chi,\ \left|\frac{w_6}{w_2} - 1.2\right| = \chi,\ \left|\frac{w_3}{w_4} - 1.2\right| = \chi, \\ \left|\frac{w_2}{w_1} - 1.5\right| = \chi,\ \left|\frac{w_4}{w_7} - 2.22\right| = \chi,\ \left|\frac{w_1}{w_8} - 1.85\right| = \chi, \\ \sum_{j=1}^{9} w_j = 1,\ w_j \geq 0, \forall j \end{cases}$$

Criteria values according to the third decision maker were 0.70, 0.106, 0.127, 0.106, 0.190, 0.127, 0.047, 0.038, and 0.190.

The calculation for the other decision makers was performed in an identical manner and the results are presented in Figure 2. After finding the mean of all values for each criterion, taking into account the values of all experts, the final values of the weights of the criteria for the FUCOM method were obtained.

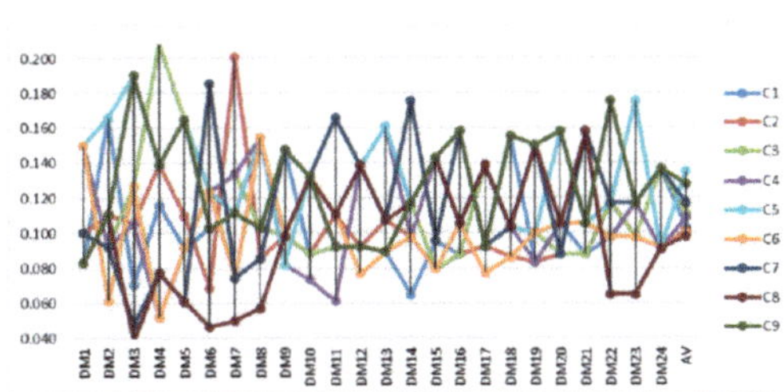

Figure 2. Criteria weights for each of the decision makers and final criteria values obtained by the FUCOM method.

The final values of the criteria obtained by the FUCOM method are indicated by AV, which stands for average value, in Figure 2. Using the FUCOM method calculation, the following values of the criteria were obtained: C1 = 0.103, C2 = 0.101, C3 = 0.111, C4 = 0.109, C5 = 0.135, C6 = 0.099, C7 = 0.117, C8 = 0.098, C9 = 0.128. The final values can be ranked from highest to lowest. At the same time, it was concluded that criteria C5, called human resource management, was the best-rated criteria by experts, while the lowest-rated and ninth-ranked criteria, C8, is called supplier relations processes.

5.3. Criteria Evaluation Using the Fuzzy PIPRECIA Method

The evaluation was performed using a linguistic scale that involves quantification into fuzzy triangular numbers. Table 3 shows the evaluation of the criteria for fuzzy PIPRECIA and inverse fuzzy PIPRECIA by the decision makers and the geometric mean (GM), based on which values were further calculated.

Table 3. Criteria evaluation by decision makers for fuzzy PIPRECIA and inverse fuzzy PIPRECIA method.

PIPR.	C2			C3			...	C8			C9		
DM1	1.100	1.150	1.200	1.000	1.000	1.000		0.500	0.667	1.000	1.000	1.000	1.000
DM2	0.500	0.667	1.000	0.500	0.667	1.000		1.100	1.150	1.200	1.000	1.000	1.000
DM3	1.100	1.150	1.200	1.100	1.150	1.200		0.500	0.667	1.000	1.500	1.750	1.800
DM4	1.100	1.150	1.200	1.100	1.150	1.200		1.000	1.000	1.000	1.200	1.300	1.350
DM5	1.100	1.150	1.200	1.100	1.150	1.200		1.000	1.000	1.000	1.300	1.450	1.500
		...					...				...		
DM20	0.500	0.667	1.000	1.000	1.000	1.000		1.100	1.150	1.200	1.100	1.150	1.200
DM21	1.000	1.000	1.000	1.000	1.000	1.000		1.000	1.000	1.000	0.500	0.667	1.000
DM22	1.100	1.150	1.200	1.000	1.000	1.000		0.400	0.500	0.667	1.300	1.450	1.500
DM23	0.500	0.667	1.000	1.000	1.000	1.000		0.400	0.500	0.667	1.200	1.300	1.350
DM24	0.500	0.667	1.000	1.100	1.150	1.200		0.500	0.667	1.000	1.100	1.150	1.200
GM	0.795	0.912	1.095	0.969	1.015	1.076		0.689	0.793	0.956	1.082	1.185	1.267
PIPR-I	C8			C7				C2			C1		
DM1	1.000	1.000	1.000	1.100	1.150	1.000		1.200	1.000	1.000	1.000	0.500	0.667
DM2	1.000	1.000	1.000	0.500	0.667	1.000		1.000	1.100	1.150	1.200	1.100	1.150
DM3	0.250	0.286	0.333	1.100	1.150	0.250		1.200	0.500	0.667	1.000	0.500	0.667
DM4	0.400	0.500	0.667	1.000	1.000	0.400		1.000	0.500	0.667	1.000	0.500	0.667
DM5	0.333	0.400	0.500	1.000	1.000	0.333		1.000	0.500	0.667	1.000	0.500	0.667
		...					...				...		
DM20	0.500	0.667	1.000	0.500	0.667	0.500		1.000	1.000	1.000	1.000	1.100	1.150
DM21	1.100	1.150	1.200	1.000	1.000	1.100		1.000	1.000	1.000	1.000	1.000	1.000
DM22	0.333	0.400	0.500	1.200	1.300	0.333		1.350	1.000	1.000	1.000	0.500	0.667
DM23	0.400	0.500	0.667	1.200	1.300	0.400		1.350	1.000	1.000	1.000	1.100	1.150
DM24	0.500	0.667	1.000	1.100	1.150	0.500		1.200	0.500	0.667	1.000	1.100	1.150
GM	0.511	0.598	0.732	0.929	1.013	0.511		1.122	0.765	0.862	1.018	0.735	0.863

Based on the evaluation of the criteria and the geometric mean, a matrix *sj* was formed. Following the rules of operations with fuzzy numbers, the k_j matrices were obtained as follows:

$$\overline{k_1} = (1.000, 1.000, 1.000),$$
$$\overline{k_2} = (2 - 1.058,\ 2\ \ -0.899,\ 2 - 0.795) = (0.942, 1.101, 1.205).$$

Fuzzy weights, denoted by q_j, were calculated in the following manner:

$$\overline{q_1} = (1.000, 1.000, 1.000),\ \overline{q_2} = \left(\frac{1.000}{1.205}, \frac{1.000}{1.101}, \frac{1.000}{0.942}\right) = (0.830, 0.908, 1.062).$$

The relative weight (w_j) of the criterion was then determined:

$$\overline{w_1} = \left(\frac{1.000}{11.086}, \frac{1.000}{6.910}, \frac{1.000}{5.377}\right) = (0.090, 0.145, 0.186)$$
$$\overline{w_2} = \left(\frac{0.830}{11.086}, \frac{0.908}{6.910}, \frac{1.062}{5.377}\right) = (0.075, 0.131, 0.197).$$

In order to determine the final weights of the criteria, the inverse methodology of fuzzy PIPRECIA method was applied. The following calculations refer to determination of coefficient $k_j{}'$; fuzzy weights for inverse fuzzy PIPRECIA, denoted by $q_j{}'$; and relative weights for the inverse fuzzy PIPRECIA, denoted by $w_j{}'$.

$$\overline{k_9{}'} = (1.000, 1.000, 1.000),$$
$$\overline{k_8{}'} = (2 - 0.861,\ \ 2 - 0.720,\ 2 - 0.628) = (1.139, 1.280, 1.372)$$

$$\overline{q_9{}'} = (1.000, 1.000, 1.000),\ \overline{q_8{}'} = \left(\frac{1.000}{1.372}, \frac{1.000}{1.280}, \frac{1.000}{1.139}\right) = (0.729, 0.781, 0.878)$$

$$\overline{w_9{}'} = \left(\frac{1.000}{9.844}, \frac{1.000}{6.019}, \frac{1.000}{4.702}\right) = (0.102, 0.166, 0.213)$$

$$\overline{w_8{}'} = \left(\frac{0.729}{9.844}, \frac{0.781}{6.019}, \frac{0.878}{4.702}\right) = (0.074, 0.130, 0.187).$$

The complete calculation and results of the inverse fuzzy PIPRECIA method are presented in Table 4 and illustrated in Figure 3. The last column shows the deficient values of the relative weights of the criteria.

Table 4. Calculation and results of fuzzy PIPRECIA and inverse fuzzy PIPRECIA.

P.	sj			kj			qj			wj			DF
C1				1.000	1.000	1.000	1.000	1.000	1.000	0.090	0.145	0.186	0.143
C2	0.795	0.899	1.058	0.942	1.101	1.205	0.830	0.908	1.062	0.075	0.131	0.197	0.133
C3	0.930	0.996	1.084	0.916	1.004	1.070	0.776	0.904	1.159	0.070	0.131	0.216	0.135
C4	0.757	0.863	1.023	0.977	1.137	1.243	0.624	0.796	1.187	0.056	0.115	0.221	0.123
C5	1.041	1.086	1.125	0.875	0.914	0.959	0.650	0.870	1.356	0.059	0.126	0.252	0.136
C6	0.600	0.703	0.870	1.130	1.297	1.400	0.465	0.671	1.200	0.042	0.097	0.223	0.109
C7	0.879	0.969	1.089	0.911	1.031	1.121	0.414	0.651	1.317	0.037	0.094	0.245	0.110
C8	0.689	0.793	0.963	1.037	1.207	1.311	0.316	0.539	1.270	0.029	0.078	0.236	0.096
C9	0.953	1.055	1.172	0.828	0.945	1.047	0.302	0.570	1.535	0.027	0.083	0.285	0.107
SUM							5.377	6.910	11.086				
P-I	sj′			kj′			qj′			wj′			DF
C1	0.795	0.899	1.058	0.942	1.101	1.205	0.222	0.427	1.257	0.023	0.071	0.267	0.096
C2	0.730	0.835	1.006	0.994	1.165	1.270	0.267	0.470	1.185	0.027	0.078	0.252	0.099
C3	0.798	0.906	1.062	0.938	1.094	1.202	0.339	0.548	1.178	0.034	0.091	0.250	0.108
C4	0.681	0.763	0.889	1.111	1.237	1.319	0.408	0.599	1.104	0.041	0.100	0.235	0.112
C5	1.065	1.123	1.171	0.829	0.877	0.935	0.538	0.741	1.227	0.055	0.123	0.261	0.135
C6	0.619	0.765	1.019	0.981	1.235	1.381	0.503	0.650	1.017	0.051	0.108	0.216	0.117
C7	0.951	1.027	1.121	0.879	0.973	1.049	0.695	0.803	0.999	0.071	0.133	0.212	0.136
C8	0.628	0.720	0.861	1.139	1.280	1.372	0.729	0.781	0.878	0.074	0.130	0.187	0.130
C9				1.000	1.000	1.000	1.000	1.000	1.000	0.102	0.166	0.213	0.163
SUM							4.702	6.019	9.844				

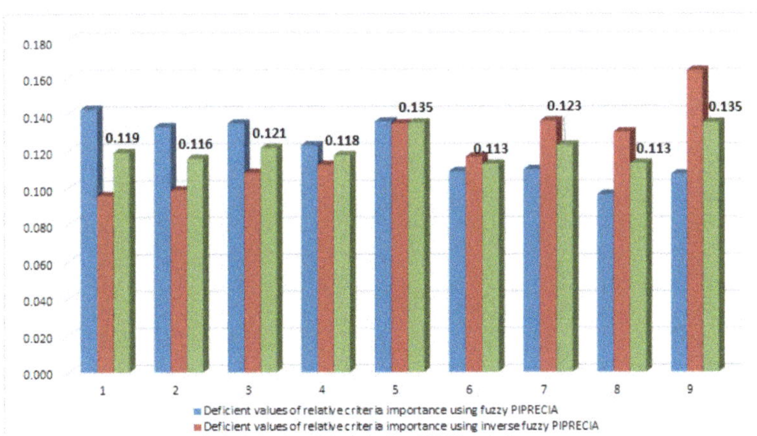

Figure 3. Final values of the criteria obtained using the fuzzy PIPRECIA method.

5.4. Final Evaluation Based on Integrated FUCOM–Fuzzy PIPRECIA Model

The integrated FUCOM–fuzzy PIPRECIA model enables the final evaluation of the process orientation dimensions of the clothing industry organizations. Figure 4 shows the values obtained by the FUCOM method marked by columns, the values obtained by the fuzzy PIPRECIA method marked with the red line, and the final values, expressed as the mean of the FUCOM and fuzzy PIPRECIA criteria, indicated by the green line.

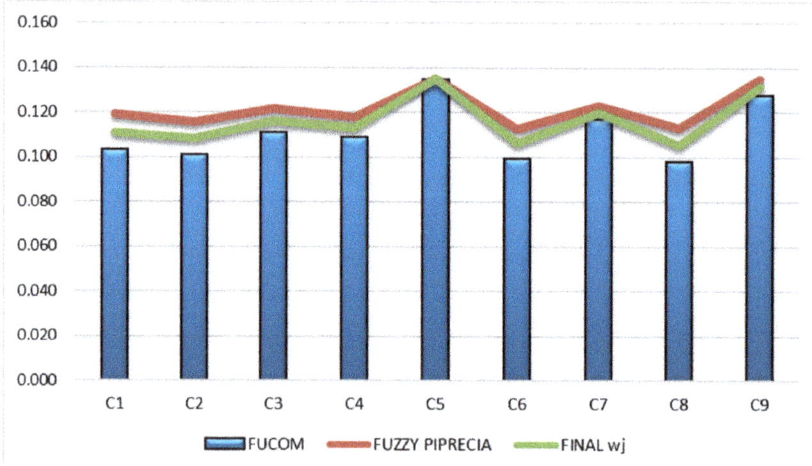

Figure 4. Final criteria values obtained using the integrated FUCOM–fuzzy PIPRECIA model.

Therefore, the best-ranked criteria was human resource management (0.135), followed by the process-based information systems development (0.131), followed by a focus on consumers and competitors (0.120). In fourth place was the criteria of performance measurement and business process improvement (0.116), fifth was the process structure (0.113), and sixth was the compliance of process and strategic goals (0.113). The lowest on the impact scale were process identification, documentation, and standardization (0.108), ranked seventh, followed by process organizational culture (0.106), eighth, and finally supplier relations processes (0.106), ninth. It should be emphasized that the criteria of process organizational culture and supplier relations processes share the same weight value, both in the result of the fuzzy PIPRECIA method and in the final result of the integrated multicriteria model.

6. Discussion

For SMEs in the apparel industry to adequately apply the principles of BPM and ensure growth and development based on the application of this practice, it is necessary to satisfy several elements of process orientation. Earlier research has noted the importance of each of the dimensions examined, but within this paper, the need to evaluate the strength of their individual impact on the adoption of process orientation and the establishment of BPM practice in SMEs in the apparel industry in Serbia, Bulgaria, and North Macedonia has been recognized.

The integrated FUCOM–fuzzy PIPRECIA model made it possible to evaluate and prioritize process orientation dimensions based on the evaluations of the 24 experts, taking into account the SMEs of the apparel industry and the specificities that occur in the businesses of these organizations. Ranking the criteria based on the final values of the weight coefficients of the implemented integrated multicriteria model provided the answer to the basic research question that motivated this research, which read: "To what extent do each of the identified categories of process orientation adoption elements realistically contribute to the establishment of BPM practices and the progression towards a state of stability and a high degree of process maturity within the SMEs of the apparel industry?"

The final values of the overall integrated model shown in Figure 4 indicate the most influential contribution of the human resource management element in comparison to the other eight considered elements. Routine tasks can be automated, however, within labor-intensive industries such as the apparel industry, which involves a large number of manual operations in the manufacturing processes, especially of uniquely designed garments; human resources are required [24,98]. Managing them is also necessary.

Process-based information systems development follows the previous element in its potential impact. The apparel industry, as well as others in terms of business modernization, is accompanied by the automatization and implementation of information technology [61]. As [99] stated, manual cutting of materials is still present in the production of apparel because of lower production costs, however, this makes the production process labor-intensive and reduces productivity. The impact of this element is increasing due to the recognized need to use computer numerical control (CNC) machines in production with the implementation of specialized tailoring software.

The criterion of focus on consumers and competitors was third-ranked within the integrated FUCOM–fuzzy PIPRECIA model. A system that responds quickly and adequately to the needs and demands of consumers attracts a large number of consumers and ensures their loyalty while gaining a more favorable competitive position [100].

The dimensions related to performance measurement and business process improvement, process structure, compliance of process and strategy goals, and process identification, documentation, and standardization were ranked fourth to seventh. Measuring the performance of the creative processes within the apparel industry can be of great importance, as regularly measuring and setting improvement goals would ensure the delivery of added value to consumers. The lack of human and financial resources in the SMEs of the apparel industry is an obstacle to carrying out a large number of activities that strengthen the process orientation. A large number of activities within the work practices of these companies are performed by one person, and the responsibility for performing multiple processes and multiple process roles is also up to one person.

The criterion of process organizational culture is estimated to have a lower contribution and was, accordingly, ranked eighth. The value of the weight coefficient was the same for the dimension relating to processes in supplier relations, which was ranked ninth, out of the nine evaluated criteria. Therefore, it is noted that this criterion has the weakest impact compared to the previously discussed ones on the establishment of BPM practices in apparel SMEs.

7. Conclusions

BPM is a comprehensive approach, dependent on strategic and operational elements, the use of modern techniques and tools, the involvement of human resources, and a horizontal focus that enables the delivery of added value to end consumers [22].

The textile and apparel industries in Serbia, Bulgaria, and North Macedonia represent a fruitful area of research, taking into account the past and the present picture of business within these industries in these areas. Based on the review of available research regarding the application of multicriteria methods in the textile, but especially the apparel industry, no research has been observed focusing on the ranking of process orientation dimensions; that is, elements that influence the higher level of establishment of BPM practices within apparel industry SMEs. The considered elements have implications for establishing BPM practice within different areas, however, the conducted research provides an opportunity to see the impact of different process dimensions in accordance with the specificities of this sector.

A new integrated FUCOM–fuzzy PIPRECIA model has been developed in order to achieve the research objective related to the evaluation of adopted process orientation dimensions, prioritizing the most influential ones in accordance with the research problem. Through various phases within this model, the calculation of the weight coefficients of each of the considered process orientation dimensions was performed based on the responses of 24 experts.

The final values of the integrated model highlighted the human resource management (HRM) element as the most influential link in the chain of influence on the establishment of BPM practice in apparel industry SMEs in Serbia, Bulgaria, and North Macedonia. Clearly, in the labor-intensive sector with a majority share of female workers, this element is recognized as the most influential. In terms of impact, this element is accompanied by process-based information systems development and a focus on consumers and competitors. Other dimensions provide a weaker contribution to the establishment of BPM practice and the adoption of process orientation, while the weakest contribution was attributed to the dimension of supplier relations processes. Although the impacts of all the dimensions considered are different according to the sector and size of the organizations, each of them has an impact, and none of the dimensions were excluded by this research.

To properly manage business processes and build a mature organization, all the elements need to be considered. For SMEs within the apparel industry to achieve this, they need to focus their initial steps on human resource management activities, then the development of process-based information systems, in line with the rankings obtained from the implementation of the integrated FUCOM–fuzzy PIPRECIA model.

The practical significance of conducted prioritization within this research is reflected in highlighting the importance and priority of particular dimensions according to the specific aspects of business in that domain. Decision makers in the domain can use these results to develop a process-based development strategy.

Based on this result, there is a need for further research within the HRM dimension to provide a more comprehensive understanding of the activities that need to be undertaken for this dimension to have its full impact on the establishment of BPM practices in the apparel industry SMEs.

Future research will be devoted to a deeper understanding of the problem by applying calculated weights of process orientation dimensions in future evaluations to generate solutions for improvements. An evaluation of the attitudes of managers and employees regarding the adoption of BPM practices, considering the role of process owners within the creative processes of the apparel industry, will also be done. Particular attention will be paid to assessing the degree of process maturity within small and medium-sized enterprises in the apparel industry based on the evaluations of managers and employees.

Author Contributions: Conceptualization, A.D. and S.U.; methodology, A.D. and D.M; validation, S.U. and M.T.; formal analysis, D.M.; investigation, A.D.; writing—original draft preparation, A.D.; writing—review and editing, M.V. and M.T.; supervision, S.U. All authors have read and agreed to the published version of the manuscript.

Funding: We acknowledge support by the German Research Foundation and the Open Access Publication Fund of TU Berlin.

Acknowledgments: The authors wish to acknowledge the support received from German Research Foundation and the TU Berlin.

Conflicts of Interest: The authors declare no conflict of interest.

References

1. Meidan, A.; García-García, J.A.; Escalona, M.J.; Ramos, I. A survey on business processes management suites. *Comput. Stand. Interfaces* **2017**, *51*, 71–86. [CrossRef]
2. Škrinjar, R.; Vukšić, V.; Štemberger, M. Adoption of Business Process Orientation Practices: Slovenian and Croatian Survey. *Bus. Syst. Res.* **2010**, *1*, 5–19. [CrossRef]
3. Stojanović, Đ.; Bašić, A. Extended Supply Chain and Small and Medum Retail Enterprises in Serbian Garment Industry. *Fibres Text. East. Eur.* **2013**, *21*, 13–17.
4. Chong, S. Business process management for SMEs: An exploratory study of implementation factors for the Australian wine industry. *J. Inf. Syst. Small Bus.* **2007**, *1*, 41–58.
5. Dimitrijević, D.; Spaić, O.; Urošević, S.; Adamović, Ž.; Đurić, Ž. Korelacija specifičnih performansi MSP odevne industrije država u razvoju i implementacija IKT tehnologija. *Tekst. Ind.* **2018**, *66*, 42–54.
6. Moniz, A.B.; Paulos, M.R. The globalization in the clothing sector and its implications for work organisation: A view from the Portuguese case. In Proceedings of the International Sociological Association First ISA Forum of Sociology, Barcelona, Spain, 5–8 September 2008; pp. 1–16. Available online: https://www.researchgate.net/publication/23646278_The_globalisation_in_the_clothing_sector_and_its_implications_for_work_organisation_a_view_from_the_Portuguese_case (accessed on 17 May 2020).

7. Bair, J.; Gereffi, G. Local Clusters in Global Chains: The Causes and Consequences of Export Dynamism in Torreon's Blue Jeans Industry. *World Dev.* **2001**, *29*, 1885–1903. [CrossRef]
8. Urošević, S.; Dobrosavljević, A. Conditions of Development of Small and Medium Enterprises in Clothing Industry. In *How to Prevent SMEs Failure (Actions Based on Comparative Analysis in Višegrad Countries and Serbia)*; Technical Faculty in Bor: Bor, Serbia, 2019; pp. 583–627.
9. Lan, T. Development Countermeasure of Clothing Creativity and Brand Construction from the Perspective of Cultural and Creative Industries. *Argos* **2019**, *36*, 72.
10. Ceptureanu, S.I.; Ceptureanu, E.G.; Melinte, C.S.; Borisov, D. Capabilities of SMEs in Romanian clothing industry. *Ind. Text.* **2016**, *67*, 265–269.
11. Dobrosavljević, A.; Urošević, S. Evaluation of orientation on performances and process improvements in manufacturing organizations. *Tehnika* **2019**, *74*, 287–294. [CrossRef]
12. Dimitrijević, D.; Milenković, L.; Antić, S.; Ilić, A.; Krstić, D. Uticaj primene CAD sistema u malim preduzećima odevne industrije zemalja u razvoju. *Tekst. Ind.* **2015**, *63*, 23–29.
13. Sorak, M.; Urošević, S.; Dragić, M.; Sorak, L. Improvement methodology of important clothing characteristics, by applying quality tools. *Ind. Text.* **2015**, *66*, 283–288.
14. Dimitrijević, D.; Adamović, Ž.; Đurić, Ž. Proces generisanja modela MSP—Implementacija efikasnih istraživačkih metoda - 1. deo. *Tekst. Ind.* **2018**, *66*, 85–95.
15. Lemańska-Majdzik, A.; Okręglicka, M. Identification of Business Processes in an Enterprise Management. *Procedia Econ. Financ.* **2015**, *27*, 394–403. [CrossRef]
16. Pyon, C.U.; Woo, J.Y.; Park, S.C. Service improvement by business process management using customer complaints in financial service industry. *Expert Syst. Appl.* **2011**, *38*, 3267–3279. [CrossRef]
17. Hewitt, S. Business excellence: Does it work for small companies? *Tqm Mag.* **1997**, *9*, 76–82. [CrossRef]
18. Rolinek, L.; Plevny, M.; Kubecova, J.; Kopta, D.; Rost, M.; Vrchota, J.; Marikova, M. Level of Process Management Implementation in SMEs and Some Related Implications. *Transform. Bus. Econ.* **2015**, *14*, 360–377.
19. Pritchard, J.; Armistead, C. Business process management – lessons from European business. *Bus. Process Manag. J.* **1999**, *5*, 10–35. [CrossRef]
20. Reijers, H.A. Implementing BPM systems: The role of process orientation. *Bus. Process Manag. J.* **2006**, *12*, 389–409. [CrossRef]
21. Vom Brocke, J.; Rosemann, M. *Handbook on Business Process Management*, 2nd ed.; Springer: Berlin, Germany, 2015.
22. Zairi, M. Business process management: A boundaryless approach to modern competitiveness. *Bus. Process Manag. J.* **1997**, *3*, 64–80. [CrossRef]
23. De Bruin, T.; Rosemann, M. Towards a Business Process Management Maturity Model. In Proceedings of the 13th European Conference on Information Systems, Information Systems in a Rapidly Changing Economy, Regensburg, Germany, 26–28 May 2005; Verlag and the London School of Economics: London, UK, 2005; pp. 521–532.
24. Trkman, P. The critical success factors of business process management. *Int. J. Inf. Manag.* **2010**, *30*, 125–134. [CrossRef]
25. Niehaves, B.; Plattfaut, R.; Becker, J. Business process management capabilities in local governments: A multi-method study. *Gov. Inf. Q.* **2013**, *30*, 217–225. [CrossRef]
26. Plesa, S.; Prostean, G. Business Process Management for Model Based Design Automotive Projects. *Procedia Soc. Behav. Sci.* **2018**, *238*, 313–322. [CrossRef]
27. Macedo de Morais, R.; Kazan, S.; Inês Dallavalle de Pádua, S.; Lucirton Costa, A. An analysis of BPM lifecycles: From a literature review to a framework proposal. *Bus. Process Manag. J.* **2014**, *20*, 412–432. [CrossRef]
28. Huff, A.S.; Reger, R.K. A Review of Strategic Process Research. *J. Manag.* **1987**, *13*, 211–236. [CrossRef]
29. Tallon, P.P. A Process-Oriented Perspective on the Alignment of Information Technology and Business Strategy. *J. Manag. Inf. Syst.* **2007**, *24*, 227–268. [CrossRef]
30. Neely, A.D.; Adams, C.; Kennerley, M. *The Performance Prism: The Scorecard for Measuring and Managing Business Success*; Prentice Hall, Financial Times: London, UK, 2002.
31. Carpinetti, L.C.R.; Buosi, T.; Gerólamo, M.C. Quality management and improvement: A framework and a business-process reference model. *Bus. Process Manag. J.* **2003**, *9*, 543–554. [CrossRef]

32. Lawler, E.E. Total Quality Management and employee involvement: Are they compatible? *AMP* **1994**, *8*, 68–76. [CrossRef]
33. Kazhamiakin, R.; Pistore, M.; Roveri, M. A Framework for Integrating Business Processes and Business Requirements. In Proceedings of the Eighth IEEE International Enterprise Distributed Object Computing Conference, Monterey, CA, USA, 24 September 2004; IEEE: Piscataway, NJ, USA, 2004; pp. 9–20.
34. Vlahović, N.; Milanović, L.; Škrinjar, R. Turning Points in Business Process Orientation Maturity Model: An East European Survey. *WSEAS Trans. Bus. Econ.* **2010**, *7*, 22–32.
35. Lockamy, A.; McCormack, K. The development of a supply chain management process maturity model using the concepts of business process orientation. *Supply Chain Manag.* **2004**, *9*, 272–278. [CrossRef]
36. Elbashir, M.Z.; Collier, P.A.; Davern, M.J. Measuring the effects of business intelligence systems: The relationship between business process and organizational performance. *Int. J. Account. Inf. Syst.* **2008**, *9*, 135–153. [CrossRef]
37. Zamecnik, R.; Rajnoha, R. Business Process Performance Measurement Under Conditions of Business Practice. *Procedia Econ. Financ.* **2015**, *26*, 742–749. [CrossRef]
38. Van Looy, A.; Shafagatova, A. Business process performance measurement: A structured literature review of indicators, measures and metrics. *Springer Plus* **2016**, *5*, 1797. [CrossRef] [PubMed]
39. Schmiedel, T.; vom Brocke, J.; Recker, J. Culture in Business Process Management: How Cultural Values Determine BPM Success. In *Handbook on Business Process Management 2*; vom Brocke, J., Rosemann, M., Eds.; Springer: Berlin/Heidelberg, Germany, 2015; pp. 649–663.
40. Trkman, P.; Mertens, W.; Viaene, S.; Gemmel, P. From business process management to customer process management. *Bus. Process Manag. J.* **2015**, *21*, 250–266. [CrossRef]
41. Willaert, P.; Van den Bergh, J.; Willems, J.; Deschoolmeester, D. The Process-Oriented Organisation: A Holistic View Developing a Framework for Business Process Orientation Maturity. In *Business Process Management*; Alonso, G., Dadam, P., Rosemann, M., Eds.; Lecture Notes in Computer Science; Springer: Berlin/Heidelberg, Germany, 2007; Volume 4714, pp. 1–15.
42. Albani, A.; Dietz, J.L.G. Current trends in modeling inter-organizational cooperation. *J. Enterp. Inf. Manag.* **2009**, *22*, 275–297. [CrossRef]
43. Kwak, Y.H.; Ibbs, C.W. Project Management Process Maturity (PM)2 Model. *J. Manag. Eng.* **2002**, *18*, 150–155. [CrossRef]
44. Vanhaverbeke, W.; Torremans, H. Organizational structure in process-based organizations. *Knowl. Process Manag.* **1999**, *6*, 41–52. [CrossRef]
45. Heravizadeh, M.; Mendling, J.; Rosemann, M. Dimensions of Business Processes Quality (QoBP). In *Business Process Management Workshops*; Ardagna, D., Mecella, M., Yang, J., Eds.; Lecture Notes in Business Information Processing; Springer: Berlin, Heidelberg, 2009; Volume 17, pp. 80–91.
46. Hammer, M. The Process Audit. *Harv. Bus. Rev.* **2007**, *85*, 1–16.
47. Fleacă, E.; Fleacă, B. The Business Process Management Map – an Effective Means for Managing the Enterprise Value Chain. *Procedia Technol.* **2016**, *22*, 954–960. [CrossRef]
48. Krysińska, J.; Janaszkiewicz, P.; Prys, M.; Różewski, P. Knowledge Resources Development Process in Business Process Outsourcing (BPO) Organizations. *Procedia Comput. Sci.* **2018**, *126*, 1145–1153. [CrossRef]
49. Paulk, M.C. Practices of High Maturity Organizations. In Proceedings of the 11th Software Engineering Process Group (SEPG) Conference, Atlanta, GA, USA, 23 February 1999; Software Engineering Institute: Pittsburgh, PA, USA, 1999; pp. 8–11.
50. Cuevas, G.; Calvo-Manzano, J.A.; García, I. Some Key Topics to be Considered in Software Process Improvement. In *Computer Systems and Software Engineering: Concepts, Methodologies, Tools, and Applications*; IGI Global: Hershey, PA, USA, 2018; pp. 134–160.
51. Haake, J.M.; Wang, W. Flexible support for business processes: Extending cooperative hypermedia with process support. *Inf. Softw. Technol.* **1999**, *41*, 355–366. [CrossRef]
52. Škerlavaj, M.; Štemberger, M.I.; Škrinjar, R.; Dimovski, V. Organizational learning culture—The missing link between business process change and organizational performance. *Int. J. Prod. Econ.* **2007**, *106*, 346–367. [CrossRef]
53. Schmiedel, T.; vom Brocke, J.; Recker, J. Which cultural values matter to business process management? Results from a global Delphi study. *Bus. Process Manag. J.* **2013**, *19*, 292–317. [CrossRef]

54. Schmiedel, T.; vom Brocke, J.; Recker, J. Development and validation of an instrument to measure organizational cultures' support of Business Process Management. *Inf. Manag.* **2014**, *51*, 43–56. [CrossRef]
55. Qi, L.; Wang, K.; Wang, X.; Zhang, F. Research on the relationship among market orientation, customer relationship management, customer knowledge management and business performance. *Manag. Sci. Eng.* **2008**, *2*, 31–37.
56. Dong, X.; Zhang, Z.; Hinsch, C.A.; Zou, S. Reconceptualizing the elements of market orientation: A process-based view. *Ind. Mark. Manag.* **2016**, *56*, 130–142. [CrossRef]
57. Matei, G. SOA and BPM, a Partnership for Successful Organizations. *Inform. Econ.* **2011**, *15*, 39–54.
58. Viaene, S.; Van den Bergh, J.; Schröder-Pander, F.; Mertens, W. BPM—Quo Vadis? *BPTrends* **2010**. Available online: https://www.bptrends.com/bpt/wp-content/publicationfiles/TWO%2009-14-10-BPM-Quo%20Vadis-van%20den%20Berghet%20al1.pdf (accessed on 17 May 2020).
59. Burlton, R. BPM Critical Success Factors Lessons Learned from Successful BPM Organizations. *BPTrends* **2011**. Available online: https://www.bptrends.com/publicationfiles/10-04-2011-ART-BPM%20Critical%20Success%20Factors-Burlton.pdf (accessed on 17 May 2020).
60. Norta, A.; Grefen, P.; Narendra, N.C. A reference architecture for managing dynamic inter-organizational business processes. *Data Knowl. Eng.* **2014**, *91*, 52–89. [CrossRef]
61. Hoyer, V. Modeling Collaborative e-Business Processes in SME environments. *J. Inf. Sci. Technol.* **2008**, *5*, 46–63.
62. Yongchareon, S.; Liu, C.; Zhao, X. An Artifact-Centric View-Based Approach to Modeling Inter-organizational Business Processes. In *Web Information System Engineering—WISE 2011*; Bouguettaya, A., Hauswirth, M., Liu, L., Eds.; Lecture Notes in Computer Science; Springer: Berlin/Heidelberg, Germany, 2011; Volume 6997, pp. 273–281.
63. Houy, C.; Fettke, P.; Loos, P.; van der Aalst, W.M.P.; Krogstie, J. BPM-in-the-Large—Towards a Higher Level of Abstraction in Business Process Management. In *E-Government, E-Services and Global Processes*; Janssen, M., Lamersdorf, W., Pries-Heje, J., Rosemann, M., Eds.; IFIP Advances in Information and Communication Technology; Springer: Berlin/Heidelberg, Germany, 2010; Volume 334, pp. 233–244.
64. Rahimi, F.; Møller, C.; Hvam, L. Business process management and IT management: The missing integration. *Int. J. Inf. Manag.* **2016**, *36*, 142–154. [CrossRef]
65. Škrinjar, R.; Trkman, P. Increasing process orientation with business process management: Critical practices'. *Int. J. Inf. Manag.* **2013**, *33*, 48–60. [CrossRef]
66. McCormack, K. Business Process Orientation: Do You Have It? *Qual. Prog.* **2001**, *34*, 51–58.
67. Bosilj Vukšić, V.; Hernaus, T.; Kovačić, A. *Upravljanje Poslovnim Procesima—Organizacijski i Informacijski Aspekti*; Školska knjiga: Zagreb, Croatia, 2008.
68. Milanović Glavan, L. Upravljanje procesima u poduzećima Republike Hrvatske. *Zbornik radova—Sarajevo Business and Economics Review (SBER)* **2011**, *31*, 105–125.
69. Dobrosavljević, A.; Urošević, S. Separation of factors important for the adoption of business process management practices in labor-intensive activities. *Tehnika* **2019**, *74*, 861–867. [CrossRef]
70. Pamučar, D.; Božanić, D.; Lukovac, V.; Komazec, N. Normalized weighted geometric bonferroni mean operator of interval rough numbers–application in interval rough dematel-copras model. *Facta Univ. Ser. Mech. Eng.* **2018**, *16*, 171–191. [CrossRef]
71. Petrović, G.; Mihajlović, J.; Ćojbašić, Ž.; Madić, M.; Marinković, D. Comparison of three fuzzy MCDM methods for solving the supplier selection problem. *Facta Univ. Ser. Mech. Eng.* **2019**, *17*, 455–469.
72. Morente-Molinera, J.A.; Wu, X.; Morfeq, A.; Al-Hmouz, R.; Herrera-Viedma, E. A novel multi-criteria group decision-making method for heterogeneous and dynamic contexts using multi-granular fuzzy linguistic modelling and consensus measures. *Inf. Fusion* **2020**, *53*, 240–250. [CrossRef]
73. Lee, H.-C.; Chang, C.-T. Comparative analysis of MCDM methods for ranking renewable energy sources in Taiwan. *Renew. Sustain. Energy Rev.* **2018**, *92*, 883–896. [CrossRef]
74. Cho, C.; Lee, S. A study on process evaluation and selection model for business process management. *Expert Syst. Appl.* **2011**, *38*, 6339–6350. [CrossRef]
75. Perçin, S. Fuzzy multi-criteria risk-benefit analysis of business process outsourcing (BPO). *Inf. Manag. Comput. Secur.* **2008**, *16*, 213–234. [CrossRef]
76. Coskun, S.; Basligil, H.; Baracli, H. A weakness determination and analysis model for business process improvement. *Bus. Process Manag. J.* **2008**, *14*, 243–261. [CrossRef]

77. Campos, A.C.S.M.; de Almeida, A.T. Using multiple criteria decision analysis for supporting decisions of Business Process Management. In Proceedings of the 2010 IEEE International Conference on Industrial Engineering and Engineering Management, Macao, China, 7–10 December 2010; IEEE: Piscataway, NJ, USA, 2010; pp. 52–56.
78. Aluftekin, N.; Tas, A.; Yuksel, O.; Cakar, G.E.; Bayraktar, F. Assessment of cluster potential and decision making criteria in the textile and apparel sector using the Analytic Hierarchy Process (AHP). *Afr. J. Bus. Manag.* **2011**, *5*, 9125–9136.
79. Sasi, J.C.; Digalwar, A.K. Application of AHP and TOPSIS Method for Supplier Selection between India & China in Textile Industry. *Int. Res. J. Eng. Technol.* **2015**, *2*, 1730–1738.
80. Stojanov, T.; Ding, X. Supplier Selection for Mixed-Model Production: A Case Study from the Apparel Industry. *Fibres Text. East. Eur.* **2015**, *23*, 8–12.
81. Moretti, I.C.; Braghini Junior, A.; Colmenero, J.C. Using the analytic hierarchy process for selecting prototypes in the development process of fashion garment products. *Acta Sci. Technol.* **2017**, *39*, 367. [CrossRef]
82. Mastrocinque, E.; Coronado Mondragon, A.E.; Hogg, P.J. Manufacturing technology selection in the supply chain context by means of fuzzy AHP: A case in the high performance textile industry. *ARPN J. Eng. Appl. Sci.* **2016**, *11*, 240–246.
83. Chakraborty, S.; Ramakrishnan, K.R.; Mitra, A. A multi-criteria decision support model for optimal cotton fibre blending. *J. Text. Inst.* **2018**, *109*, 1482–1492. [CrossRef]
84. Urošević, S.; Radosavljević, D.; Stefanović, V.; Đorđević, D.; Kokeza, G. Multicriteria ranking of a job positions by ELECTRA methods in order to improve the analysis and conditions at work in companies in the textile industry. *Ind. Text.* **2017**, *68*, 388–395. [CrossRef]
85. Pamučar, D.; Stević, Ž.; Sremac, S. A New Model for Determining Weight Coefficients of Criteria in MCDM Models: Full Consistency Method (FUCOM). *Symmetry* **2018**, *10*, 393. [CrossRef]
86. Stanujkić, D.; Zavadskas, E.K.; Karabašević, D.; Smarandache, F.; Turskis, Z. The Use of Pivot Pairwise Relative Criteria Importance Assessment Method for Determining the Weights of Criteria. *Rom. J. Econ. Forecast.* **2017**, *20*, 116–133.
87. Durmić, E. The evaluation of the criteria for sustainable supplier selection by using the FUCOM method. *Oper. Res. Eng. Sci. Theory Appl.* **2019**, *2*, 91–107. [CrossRef]
88. You, S.Y.; Zhang, L.J.; Xu, X.G.; Liu, H.C. A New Integrated Multi-Criteria Decision Making Multi-Objective Programming Model for Sustainable Supplier Selection and Order Allocation. *Symmetry* **2020**, *12*, 302. [CrossRef]
89. Campagne, C.S.; Roche, P.; Gosselin, F.; Tschanz, L.; Tatoni, T. Expert-based ecosystem services capacity matrices: Dealing with scoring variability. *Ecol. Indic.* **2017**, *79*, 63–72. [CrossRef]
90. Nenadić, D. Ranking dangerous sections of the road using MCDM model. *Decis. Mak. Appl. Manag. Eng.* **2019**, *2*, 115–131. [CrossRef]
91. Puška, A.; Stojanović, I.; Maksimović, A. Evaluation of sustainable rural tourism potential in Brcko district of Bosnia and Herzegovina using multi-criteria analysis. *Oper. Res. Eng. Sci. Theory Appl.* **2019**, *2*, 40–54. [CrossRef]
92. Fazlollahtabar, H.; Smailbašić, A.; Stević, Ž. FUCOM method in group decision-making: Selection of forklift in a warehouse. *Decis. Mak. Appl. Manag. Eng.* **2019**, *2*, 49–65. [CrossRef]
93. Stević, Ž.; Durmić, E.; Gajić, M.; Pamučar, D.; Puška, A. A Novel Multi-Criteria Decision-Making Model: Interval Rough SAW Method for Sustainable Supplier Selection. *Information* **2019**, *10*, 292. [CrossRef]
94. Stević, Ž.; Stjepanović, Ž.; Božičković, Z.; Das, D.; Stanujkić, D. Assessment of Conditions for Implementing Information Technology in a Warehouse System: A Novel Fuzzy PIPRECIA Method. *Symmetry* **2018**, *10*, 586. [CrossRef]
95. Albu, A.; Precup, R.E.; Teban, T.A. Results and challenges of artificial neural networks used for decision-making and control in medical applications. *Facta Univ. Ser. Mech. Eng.* **2019**, *17*, 285–308. [CrossRef]
96. Popović, G.; Đorđević, B.; Milanović, D. Multiple Criteria Approach in the Mining Method Selection. *Industrija* **2019**, *47*, 47–62. [CrossRef]
97. Vesković, S.; Milinković, S.; Abramović, B.; Ljubaj, I. Determining criteria significance in selecting reach stackers by applying the fuzzy PIPRECIA method. *Oper. Res. Eng. Sci. Theory Appl.* **2020**, *3*, 72–88. [CrossRef]

98. Güner, M.G.; Ünal, C. Line Balancing in the Apparel Industry Using Simulation Techniques. *Fibres Text. East. Eur.* **2008**, *16*, 75–78.
99. Nemeša, I. Automatizovani sistemi za krojenje tekstilnih materijala. *Tekst. Ind.* **2017**, *65*, 24–31.
100. De Toni, A.; Meneghetti, A. The production planning process for a network of firms in the textile-apparel industry. *Int. J. Prod. Econ.* **2000**, *65*, 17–32. [CrossRef]

© 2020 by the authors. Licensee MDPI, Basel, Switzerland. This article is an open access article distributed under the terms and conditions of the Creative Commons Attribution (CC BY) license (http://creativecommons.org/licenses/by/4.0/).

sustainability

MDPI

Article

A New Comprehensive Approach for Efficient Road Vehicle Procurement Using Hybrid DANP-TOPSIS Method

Marko Stokic *, **Davor Vujanovic** and **Dragan Sekulic**

University of Belgrade, Faculty of Transport and Traffic Engineering, Vojvode Stepe 305, 11000 Belgrade, Serbia; d.vujanovic@sf.bg.ac.rs (D.V.); d.sekulic@sf.bg.ac.rs (D.S.)

* Correspondence: m.stokic@sf.bg.ac.rs; Tel.: +381-64-006-0007

Received: 29 April 2020; Accepted: 11 May 2020; Published: 14 May 2020

check for updates

Abstract: The efficient vehicle procurement is an important business segment of different companies with their own vehicle fleet. It has a significant influence on reducing transport and maintenance costs and on increasing the fleet's energy efficiency. It is indispensable that managers consider various criteria from several aspects when procuring a vehicle. In that sense, we defined 13 relevant criteria and divided them into four multidisciplinary aspects: Construction-technical, financial, operational, and environmental. Decision-Making Trial and Evaluation Laboratory-Based Analytic Network Process (DANP) method was applied to evaluate the significance of defined criteria and aspects and their interdependency. It is established that the three most important criteria for vehicle procurement are vehicle price, vehicle maintenance, and vehicle selling price. The most important aspect is construction technical aspect, while the aspect of the environment is the least important. The Technique for Order Preference by Similarity to Ideal Solution (TOPSIS) method was used to rank eight different vehicles, which were considered by vehicle fleet manager at the observed company. This model assists fleet managers in the selection of the most suitable vehicle for procurement, while significantly reducing decision-making time and simultaneously observing all necessary criteria and their weights. Moreover, we have considered 10 different scenarios to establish whether and how the rank of the observed alternatives would change.

Keywords: vehicle procurement; MCDM; fleet energy efficiency; transport costs; maintenance costs; alternative fuel vehicles

1. Introduction

The vehicle procurement represents one of the significant activities of the fleet managers [1]. Transport and logistics companies often have a heterogeneous vehicle fleet, composed of different vehicle types and different construction-operation characteristics [2]. One of the main goals of the observed companies is to reduce transport and maintenance costs, as well as to increase the fleet's energy efficiency in order to increase profit [3]. This paper analyzes the road vehicles procurement as one of the factors which contribute to the achievement of the defined goal. Besides choosing the right vehicle type during the procurement, an additional potential for achieving the defined goal is to choose the appropriate vehicle model within the selected type. However, the problem is that the managers during vehicle procurement often observe only certain criteria of certain aspects that they consider important, while other criteria and aspects that they insufficiently know about are neglected and considered insignificant. An additional problem is that managers evaluate the importance of observed criteria according to their (subjective) perceptions. All of the above leads to higher transport and maintenance costs and lower fleet's energy efficiency, and thereby, to lower profit.

Vehicle procurement has been often covered in the literature as part of solving the fleet renewal or fleet replacement problem [4–6]. Some authors propose various models of integer programming to achieve efficient road vehicle fleet renewal or replacement [7–9]. Although the authors observed many different criteria, there is a lack of criteria that indirectly have an impact on transport and maintenance costs and energy efficiency. Moreover, the interdependent influences between criteria are not taken into account even though they exist in managers' decision-making during the vehicle procurement process. The task of fleet renewal consists of defining the number and type of vehicles that are added to the fleet by procurement or that are removed from the fleet by selling [10], but there are only a few papers that deal with the decision which type and model of vehicle to procure. This is especially important for road vehicles since there are many different models of the same vehicle type (e.g., various models of van with gross vehicle weight up to 3.5 t) where each model has different characteristics which have an impact on transport and maintenance costs and energy efficiency. Besides that, although it gives good results, the integer programming is not commonly used by vehicle fleet managers, and it is not how they make a decision during the road vehicle procurement process. Considering the above-mentioned, the observed problem belongs to the Multi-Criteria Decision Making (MCDM).

In order to determine criteria weights, we used the Decision-Making Trial and Evaluation Laboratory-Based Analytic Network Process (DANP) method because the interdependent effects of the vehicle procurement criteria are non-reciprocal values. DANP method calculates non-weighted supermatrices by applying pairwise comparisons to the total impact matrix within the whole system, which better describes the real system [11]. The integration of the Decision-Making Trial and Evaluation Laboratory (DEMATEL) and Analytic Network Process (ANP) method is still widely used for solving different problems. The authors of [12] used integration of these two methods to establish the relationship between six important performance criteria and eighteen key performance indicators that describe the performance measurement system of any food supply chain, as well as to determine their ranking. Gudiel Pineda et al. [13] used the DANP method to obtain relative weights of the criteria important to evaluate the financial and operational performance of airlines. Hsu et al. [14], in their paper, observe multiple criteria regarding the quality of service for bike-sharing systems, as well as the relationships between them in order to better reflect the real-world situation and thus they used DANP method to determine the observed criteria relative weights. The authors of [15] have researched the inhibitors in the intermodal freight transport system, and they used grey-DEMATEL and ANP (gDANP) method to establish relationships between these inhibitors, as well as their relative weights. DANP method was used in [16] in order to determine relative weights and interdependency of 11 criteria from four clusters with an aim to support vehicle fleet managers in decision making on the selection of vehicle service centers. After that, we used the well-known Technique for Order Preference by Similarity to Ideal Solution (TOPSIS) method to rank the observed vehicles. This method is one of the best in addressing the rank reversal problem [17]. The TOPSIS method as MCDM method was applied in [18], where authors used this method to integrate environmental, economic, and social dimensions in order to assess the sustainability of battery electric vehicles and to compare it with the sustainability of internal combustion engine vehicles. TOPSIS method was used in [19] for problem-solving in urban public transportation by ranking observed alternatives (electric buses) and evaluating them through seven different criteria. Additionally it was used in the selection of Brazilian furniture industry suppliers [20], an egg supplier for a Thai restaurant [21], evaluation of different fuel characteristics [22], vehicle manufacturer evaluation [23], performance evaluation of seven Indian industries based on financial ratio analysis taking into account 15 criteria [24], selection of the best railroad container terminal location out of 11 potentional ones observing seven different criteria [25], etc.

We are not aware of a model that selects the most suitable road vehicle while taking into account construction-technical, financial, operational, and environmental aspects of road vehicle, simultaneously from three points of view: Transport costs, maintenance costs, and energy efficiency. Thus, the main contribution of this paper is such a developed model for efficient vehicle procurement.

This model is easily understandable for vehicle fleet managers since it describes better their perception in case there are multiple interdependent criteria. It significantly reduces decision making time, observing all necessary criteria and their weights. The aim of the model is to increase companies' profit through a comprehensive and multidisciplinary approach. Additionally, we have conducted a literature review and engaged 50 experts to define relevant criteria for road vehicle procurement and their relative weights. Finally, this model is applicable for the selection of alternative fuel vehicles (AFVs) as well as for conventional (diesel) vehicles.

The paper is structured as follows. Section 2 comprises a literature review with the aim of defining the relevant vehicle procurement criteria. The developed model methodology is presented step-by-step in Section 3. The model implementation, obtained results, and discussion are presented in Sections 4 and 5, while the conclusions are given in Section 6.

2. Vehicle Procurement Criteria

In this chapter, a literature review was conducted to identify significant aspects and criteria for vehicle procurement within the developed model that have a direct and indirect impact on transport and maintenance costs and fleet's energy efficiency.

Many authors, while researching efficient fleet renewal/replacement, consider criteria such as the existence of new vehicle technologies, which includes vehicle equipment regarding safety and technical systems [26]. The same author points out the vehicle comfort and vehicle technical condition (age) as relevant criteria in order to increase the competitiveness of companies in the market. The older vehicles contribute to increased transport and maintenance costs [27], as well as to increased environmental costs [2]. The vehicle technical condition concerning the vehicle mileage is important in the calculation of the expected costs [6]. In order to increase a vehicle's energy efficiency and reduce CO_2 emission, vehicle manufacturers endeavor to reduce engine capacity [28,29]. It is established that the homogeneous fleet in air traffic has an influence on reducing transport and maintenance costs, pilots and technicians training costs, spare parts stock costs, and other costs [30]. Since this can be applied for road vehicles, the criterion compatibility with the existing vehicle fleet is also significant for our model. Vehicle price is considered in solving strategic fleet renewal in shipping and depends on the size, age and vessel technical condition [27]. This criterion is also used to calculate the life cycle costs of vehicles in [18]. For example, the electric commercial vehicles (ECVs) have approximately three times higher procurement costs in relation to conventional vehicles, and because of that, the fleet managers should not plan ECV procurement if the vehicle life cycle is shorter than 12 years [8]. Thus, government subsidies as a form of financing options when buying a vehicle are necessary for the electric vehicles to become more competitive [7]. The national and local financial subsidies were observed in taken into account in [18]. The criterion financing options when buying a vehicle in the form of vehicle leasing was taken into consideration in the developed model for fleet renewal with the aim to minimize AFVs' expected and risk costs [9]. The discounted value of vehicle price, also as a form of financing options, should be contained within economic factors in the model for fleet replacement [8]. As the vehicle ages and its' technical condition degrades, vehicle selling price decreases as well, while transport and maintenance costs tend to increase [31]. One of the factors that influence transport (operating) costs is fuel/energy consumption [18]. Fuel/energy consumption is the key parameter in stochastic processes because it depends on vehicle operation conditions and drivers' skills [6]. With a decrease in fuel efficiency with diesel-fueled commercial vehicles, ECVs become much more competitive [8]. The fuel (energy) costs are about 62% lower for electric vans, about 37% lower for compressed natural gas (CNG) vans, and about 2% higher for biodiesel B30 vans, compared to diesel vans [32]. Some vehicle manufacturers and distributors, with a longer period of vehicle maintenance at their service centers, provide certain discounts on vehicle price [33]. In this way, fleet managers obtain lower initial vehicle procurement costs, but the overall maintenance costs are higher (due to the contracted extended maintenance in authorized service centers). While considering the AFVs, it is important to take into account the preventive maintenance costs since the electric vehicles have lower maintenance costs than conventional [8]. The maintenance costs are about 30% lower for electric vans and about 11% higher for biodiesel B30 and CNG vans compared to diesel vans [32]. This criterion is also observed in [18] for comparing battery electric vehicles and internal combustion engine vehicles. Besides all, legislation requirements, competition, and clients themselves play a remarkable role when procuring vehicles [33]. In that sense, in order to be in compliance with legislation requirements and to conduct efficient vehicle procurement, it is necessary to consider vehicle's CO_2 emissions. The CO_2 emissions depend directly on the vehicle age and weight, so older or heavier vehicles emit more CO_2 than newer or lighter vehicles [34]. The AFVs, such as electric vehicles, can significantly reduce harmful gas emissions, but only if electricity is generated from renewable energy sources [35]. CO_2 emission is 50%-70% lower for electric vans, 28% lower for biodiesel B30, and about 5% lower for CNG vans compared to diesel vans [32]. The vehicle dimensions and weight sometimes represent a limitation for vehicle access to the transportation network, especially in dense urban environments [2]. In this sense, it is also necessary to take into account the criterion of vehicle's compliance with the road-traffic infrastructure requirements on the transport network. This criterion can be quantified and measured through the vehicle turning diameter, which is of great importance in urban areas where diameters of road curves are smaller, and traffic lanes' width is usually narrower. A potential to increase the fleet's energy efficiency is much greater if managers optimize vehicle utilization and choose the right size and type of vehicle according to transport operations [36]. In that sense, we will observe vehicle's compliance with the customers' transport requirements as a criterion in our model.

Based on the literature review and authors' own experience, we defined 13 relevant criteria and divided them into the four multidisciplinary aspects: Construction-technical, financial, operational, and environment (Table 1).

Table 1. Preview of relevant criteria and aspects related to vehicle procurement.

Aspects/Criteria	Criteria Description	Criteria Parameter
	CONSTRUCTION-TECHNICAL ASPECT (*CT*)	
CT1–Vehicle equipment regarding safety and technical systems	Existence of advanced onboard systems: ABA, Cruise Control with Variable Speed Limiter, Active Distance Assist Distronic, Blind Spot Assist Mirror, Rear View Camera, Stop&Start System, TSR, LGS, TPMS, HSA.	Number of safety-technical systems: $CT1_p$
CT2–Engine characteristics	Engine capacity	Engine capacity (cm^3): $CT2_p$
CT3–Vehicle comfort	Existence of further devices and equipment on the vehicle: Automatic air condition, driver's armrest, Electric windows, Rain sensors, LCD display with navigation, Bluetooth connection, Automatic transmission, Seat heating, Seat cooling, Seat massagers	Number of on-board devices/equipment: $CT3_p$
CT4–Vehicle's compatibility with the existing vehicle fleet	Number of the observed vehicle model in the existing vehicle fleet	Percentage of the observed vehicle model (%): $CT4_p$
CT5–Vehicle's technical condition	$1+\frac{\text{Vehicle current mileage at the moment of procurement}}{\text{Average vehicle annual mileage in observed company}}$	Vehicle mileage coefficient: $CT5_p$
	FINANCIAL ASPECT (*F*)	
F1–Vehicle price	The price of the observed vehicle model on the market	Vehicle price (£): $F1_p$
F2–Financing options when buying a vehicle	Existence of further options: Leasing, government subsidies, discounts on the vehicle price	Number of options: $F2_p$
F3–Vehicle selling price	The residual value of the observed vehicle model on the market with the certain number of years and miles traveled	Vehicle selling price (£): $F3_p$
	OPERATIONAL ASPECT (*O*)	
O1–Fuel/energy costs	Specific fuel (energy) consumption multiplied by fuel/energy price	Fuel (energy) costs (£/100 km): $O1p$

Table 1. *Cont.*

Aspects/Criteria	Criteria Description	Criteria Parameter
O2–Vehicle maintenance	Price of preventive maintenance multiplied by the number of obligatory annual preventive maintenance services (vehicle manufacturer requirements during warranty period)	Preventive maintenance costs (£/year): $O2_p$
	ASPECT OF THE ENVIRONMENT (*E*)	
E1–Environmental protection requirements	Amount of CO_2 emission from the observed vehicle model	CO_2 emission (g/km): $E1_p$
E2–Vehicle's compliance with the road-traffic infrastructure requirements on the transport network	Compliance of vehicle turning diameter with the characteristics of the road infrastructure on the transport network (road curve diameter, curves frequency, traffic lane width, etc.)	Vehicle turning diameter "curb to curb" (m): $E2_p$
E3–Vehicle's compliance with customers' transport requirements	Compliance of the vehicle's payload capacity with the cargo amount that should be carried	Utilization of vehicle's payload capacity (%): $E3_p$

Note: ABA: Active Brake Assist; TSR: Traffic Sign Recognition; LGS: Lane Guard System; TPMS: Tire Pressure Monitoring System; HSA: Hill Start Assist.

3. Building an Evaluation Model for Vehicle Procurement

The procedure of the developed model for efficient vehicle procurement is shown in Figure 1. There are six phases that are described in the following subchapters.

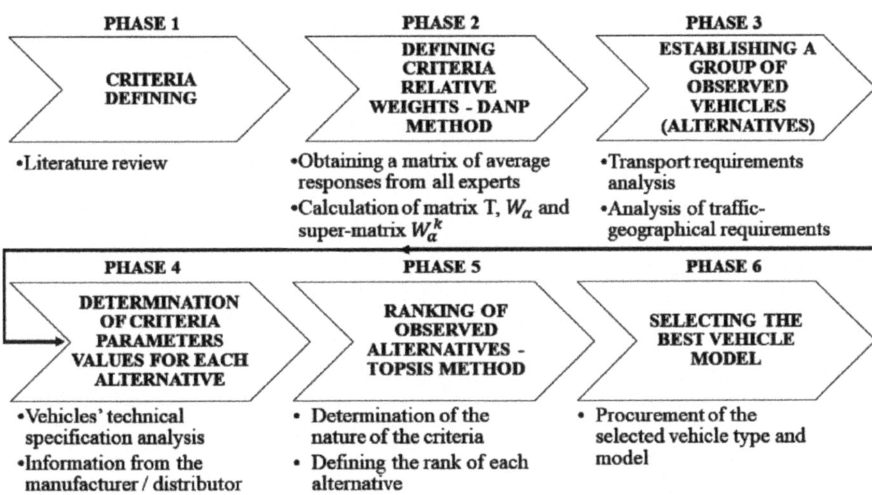

Figure 1. The procedure of the Multi-Criteria Decision Making (MCDM) model for efficient vehicle procurement.

3.1. *Defining Criteria Relative Weights with DANP Method*

DANP was used to determine the intensity of relevant criteria interdependencies, as well as to calculate the criteria significance to attaining the main goal of the observed companies. DANP shows the criteria interdependent effects as non-reciprocal values and calculates non-weighted supermatrices by applying pairwise comparisons (comparison of impact pairs) to the total impact matrix within the whole system, which better describes the real system [11]. The basic steps of DANP are [11]:

Step 1: Defining criteria (factors) and aspects (dimensions). In order to define the relevant criteria and aspects, it is necessary to conduct a scientific literature review in the considered area and conduct a survey.

Step 2: Calculation of direct-relation matrix-**B**. Matrix **B** represents the matrix of average responses of all H experts as follows:

$$\mathbf{B}=[\mathbf{b}_{ij}]_{n\times n}=\frac{1}{\mathbf{H}}\sum_{\mathbf{k}=1}^{\mathbf{H}}[\mathbf{p}_{ij}^{k}];\ 1 \leq i,j \leq n\ ;1 \leq k\ \leq H \tag{1}$$

where $P^k = \left[\mathbf{p}_{ij}^k\right]_{n\times n}$ represents the answers of the expert k on the interaction of the ith and j^{th} factors in the form of integer non-negative numbers within the $n \times n$ rank matrix.

Step 3: Calculation of normalized direct-relation matrix-$\boldsymbol{X}$. Matrix X is calculated as follows:

$$\mathbf{X}=[\mathbf{x}_{ij}]_{n\times n}=\mathbf{g}\cdot\mathbf{B}\ ;\ 0 \leq x_{ij} \leq 1;\ 1 \leq i,j \leq n \tag{2}$$

where $\boldsymbol{g}$ is the normalization factor which is calculated as follows:

$$\mathbf{g}=\mathbf{min}\left(\frac{1}{\mathbf{max}_{1\leq i\leq n}\sum_{j=1}^{n}\mathbf{b}_{ij}},\frac{1}{\mathbf{max}_{1\leq j\leq n}\sum_{i=1}^{n}\mathbf{b}_{ij}}\right) \tag{3}$$

Step 4: Determining the total impact matrix-$\boldsymbol{T}$. In order to determine the total impact matrix $\boldsymbol{T}$, it is used the identity matrix $\boldsymbol{I}$ of rank $n \times n$, as follows:

$$\mathbf{T}=[\mathbf{t}_{ij}]_{n\times n}=\mathbf{X}(\mathbf{I}-\mathbf{X})^{-1};\ 1 \leq i,j \leq n; \lim_{k\to\infty} X^k = [0]_{n\times n} \tag{4}$$

Step 5: Adjusting a threshold value-$\boldsymbol{q}$. According to expert opinion, the threshold value $\boldsymbol{q}$ can be adjusted in order to disregard the criteria relationships with small impacts in the matrix $\boldsymbol{T}$.

Step 6: Calculating the criteria influence and making the Influence Relation Map–IRM. On the basis of adopted threshold value q, the row and column sum values of the matrix T are calculated as follows

$$\mathbf{r}=[\mathbf{r}_i]_{n\times 1}=\left[\sum\nolimits_{j=1}^{n}\mathbf{t}_{ij}\right]_{n\times 1}';\ 1 \leq i \leq n;\ \mathbf{s}=[\mathbf{s}_j]_{1\times n}'=\left[\sum\nolimits_{i=1}^{n}\mathbf{t}_{ij}\right]_{1\times n}';\ 1 \leq j \leq n \tag{5}$$

where a row sum values $\boldsymbol{r}$ indicates the overall influence of a given criterion on other criteria, and a column sum values $\mathbf{s}$ indicates the overall influence of other criteria on a given criterion. A sign $'$ indicates a transposed matrix. The expression $(r + s)$ shows the importance of criteria and aspects. The expression $(r - s)$ sorts the criteria and aspects into cause and receive groups. With positive $(r - s)$ value, a criterion (or aspect) is into the cause group. Otherwise, criterion (or cluster) is into the receive group.

Step 7: Obtaining the unweighted super-matrix-$\boldsymbol{W}$. For this purpose, the total impact matrix-$\boldsymbol{T}$ is used. The matrix $\boldsymbol{T}$ actually represents the total impact matrix for criteria-$\mathbf{T_c}$. The normalized total impact matrix for criteria $\mathbf{T}_c^{\alpha}$ is calculated, according to Expression (6):

$$\mathbf{T}_c^{\alpha}= \begin{array}{cc} & \begin{array}{cccc} \mathbf{D}_1 & \mathbf{D}_2 & \cdots & \mathbf{D}_m \\ c_{11}\cdots c_{1n_1} & c_{21}\cdots c_{2n_2} & \cdots & c_{m1}\cdots c_{mn_m} \end{array} \\ \begin{array}{cc} \mathbf{D}_1 & \begin{array}{c} c_{11} \\ c_{12} \\ \vdots \\ c_{1n_1} \end{array} \\ \vdots & \vdots \\ \mathbf{D}_2 & \begin{array}{c} c_{21} \\ c_{22} \\ \vdots \\ c_{2n_2} \end{array} \\ \vdots & \vdots \\ \mathbf{D}_m & \begin{array}{c} c_{m1} \\ c_{m2} \\ \vdots \\ c_{mn_m} \end{array} \end{array} & \begin{bmatrix} \mathbf{T}_c^{\alpha 11} & \mathbf{T}_c^{\alpha 12} & \cdots & \mathbf{T}_c^{\alpha 1m} \\ \mathbf{T}_c^{\alpha 21} & \mathbf{T}_c^{\alpha 22} & \cdots & \mathbf{T}_c^{\alpha 2m} \\ \vdots & \vdots & \ddots & \vdots \\ \mathbf{T}_c^{\alpha m1} & \mathbf{T}_c^{\alpha m2} & \cdots & \mathbf{T}_c^{\alpha mm} \end{bmatrix} \end{array} \tag{6}$$

where D_m indicates the mth aspect and c_{mn_m} marks the nth criterion in the mth aspect. The matrix $\mathbf{T}_c^{\alpha_{11}}$ shows the normalized sub-matrix of the matrix $\mathbf{T}_c^{\alpha}$ with criteria impacts in the aspect D_1.

The super-matrix **W** is calculated, as shown in Expression (7):

$$W=(T_c^{\alpha})'=\begin{array}{c c} & \begin{array}{cccc} D_1 & D_2 & \cdots & D_m \\ c_{11}\cdots c_{1n_1} & c_{21}\cdots c_{2n_2} & \cdots & c_{m1}\cdots c_{mn_m} \end{array} \\ \begin{array}{cc} D_1 & \begin{array}{c} c_{11} \\ c_{12} \\ \vdots \\ c_{1n_1} \end{array} \\ \vdots & \vdots \\ D_2 & \begin{array}{c} c_{21} \\ c_{22} \\ \vdots \\ c_{2n_2} \end{array} \\ \vdots & \vdots \\ D_m & \begin{array}{c} c_{m1} \\ c_{m2} \\ \vdots \\ c_{mn_m} \end{array} \end{array} & \begin{bmatrix} W^{11} & W^{12} & \cdots & W^{1m} \\ W^{21} & W^{22} & \cdots & W^{2m} \\ \vdots & \vdots & \ddots & \vdots \\ W^{m1} & W^{m2} & \cdots & W^{mm} \end{bmatrix} \end{array} \quad (7)$$

where $W^{ij}=(\mathbf{T}_c^{\alpha_{ji}})'$ indicates a transposition of the sub-matrix $\mathbf{T}_c^{\alpha_{ji}}$ in the normalized total impact matrix for criteria $\mathbf{T}_c^{\alpha}$, while $I = 1,2, \ldots m; j = 1,2, \ldots m$.

Step 8: Calculating the weighted super-matrix-$\mathbf{W}_{\alpha}$. For this purpose, the total impact matrix for aspects $\mathbf{T_D}$ is used. The matrix $\mathbf{T_D}$ represents the sum of all impacts of every aspect in the matrix $\mathbf{T_c}$. The normalized total impact matrix for aspects $\mathbf{T}_D^{\alpha}$ is calculated, as shown in Expression (8)

$$T_D^{\alpha}=\begin{bmatrix} t_D^{11}/d_1 & \cdots & t_D^{1j}/d_1 & \cdots & t_D^{1m}/d_1 \\ \vdots & & \vdots & & \vdots \\ t_D^{i1}/d_i & \cdots & t_D^{ij}/d_i & \cdots & t_D^{im}/d_i \\ \vdots & & \vdots & & \vdots \\ t_D^{m1}/d_m & \cdots & t_D^{mj}/d_m & \cdots & t_D^{mm}/d_m \end{bmatrix}=\begin{bmatrix} t_D^{\alpha 11} & \cdots & t_D^{\alpha 1j} & \cdots & t_D^{\alpha 1m} \\ \vdots & & \vdots & & \vdots \\ t_D^{\alpha i1} & \cdots & t_D^{\alpha ij} & \cdots & t_D^{\alpha im} \\ \vdots & & \vdots & & \vdots \\ t_D^{\alpha m1} & \cdots & t_D^{\alpha mj} & \cdots & t_D^{\alpha mm} \end{bmatrix} \quad (8)$$

where t_D^{ij} indicates the sum of all impacts from the sub-matrix $\mathbf{T}_c^{ij}$ of the total impact matrix for criteria and where d_i denotes the sum of ith row in the super-matrix for aspects $\mathbf{T_D}$, while $i = 1,2, \ldots ,m$.

The super-matrix $\mathbf{W}_{\alpha}$ is calculated according to Expression (9):

$$W_{\alpha}=T_D^{\alpha}\cdot W=\begin{bmatrix} t_D^{\alpha 11}\cdot W^{11} & t_D^{\alpha 21}\cdot W^{12} & \cdots & \cdots & t_D^{\alpha m1}\cdot W^{1m} \\ t_D^{\alpha 12}\cdot W^{21} & t_D^{\alpha 22}\cdot W^{22} & \vdots & & \vdots \\ \vdots & \cdots & t_D^{\alpha ji}\cdot W^{ij} & \cdots & t_D^{\alpha mi}\cdot W^{im} \\ \vdots & & \vdots & & \vdots \\ t_D^{\alpha 1m}\cdot W^{m1} & t_D^{\alpha 2m}\cdot W^{m2} & \cdots & \cdots & t_D^{\alpha mm}\cdot W^{mm} \end{bmatrix} \quad (9)$$

Step 9: Obtaining the criteria relative weights. It is necessary to multiply the super-matrix $\mathbf{W}_\alpha$ by itself multiple times in order to calculate the limited weighted super-matrix $\mathbf{W}_\alpha^k$ with a stable convergence value, as shown in Expression (10):

$$\lim_{k\to\infty} (\mathbf{W}_\alpha)^k = \mathbf{W}_\alpha^k \quad (10)$$

where the number k denotes a positive integer number, which tends to infinity. Every vector of the limited super-matrix $\mathbf{W}_\alpha^k$ represents the relative weight of each criterion in relation to the defined goal.

3.2. Establishing a Group of Observed Vehicles–Defining Alternatives

Within this phase, the suitable vehicle types and models necessary for the realization of the transport process are defined. Since the observed companies are different regarding their characteristics, when defining a set of considered vehicles (alternatives) for procurement, it is necessary to take into account the following elements: The available company budget, the transport network specifics, the characteristics of the customers' transport requirements, and legislation requirements.

Road characteristics (road curve diameter, traffic lanes width, existing traffic regulations, etc.) can be limiting factors for certain vehicle types and models. This is especially the case in urban areas with small road curve diameters and lane widths, with traffic restrictions for certain vehicles, such as trucks over 7.5 t gross vehicle weight (GVW) and vehicles with lower Euro standard than it is prescribed. Based on the analysis of customers' transport requirements, the quantities and types of cargo most often transported by the company have to be determined, and therefore, the suitable types and models of vehicles are established (by type of vehicle's cargo compartment, vehicle's payload capacity). Legislation requirements (vehicle emission, noise, etc.) also represent limiting factors for certain vehicle categories when defining the vehicles set, especially for vehicles that operate in urban areas.

3.3. Determination of Criteria Parameters Values for Each Alternative

After defining a set of considered vehicles, it is necessary to determine for each vehicle the parameter values that quantitatively express each of the defined criteria, as noted in Table 1. These values of parameters are determined on the basis of the vehicles' technical and operational data using vehicle specifications, available scientific and professional literature, information obtained from vehicle distributors, as well as from vehicle service centers. After determining the parameter value of the defined criteria for each vehicle, it is necessary to rank the considered vehicles by applying the TOPSIS method.

3.4. Ranking of Observed Alternatives with TOPSIS Method

TOPSIS method [37] was applied in order to rank the considered vehicles (alternatives) within the procuring process. The basic steps of the TOPSIS method are [37]:

Step 1: Construct the normalized decision matrix. Starting matrix $\mathbf{X_T} = \left[x_{ij}\right]_{m\times n}$ is transformed into a normalized matrix $\mathbf{R_T} = \left[r_{ij}\right]_{m\times n}$, where r_{ij} are normalized values obtained by normalization. i represents i^{th} alternative considered A_i $(1 \le i \le m)$ and j represents j^{th} criterion K_j $(1 \le j \le n)$. Normalization of starting matrix is carried out by vector normalization based on Equations (11) and (12):

$$r_{ij} = \frac{x_{ij}}{\sqrt{\sum_{i=1}^{m} x_{ij}^2}};\ \forall K_j \in K';\ \text{for benefit criteria} \quad (11)$$

$$r_{ij} = \frac{\frac{1}{x_{ij}}}{\sqrt{\sum_{i=1}^{m} \left(\frac{1}{x_{ij}}\right)^2}};\ \forall K_j \in K'';\ \text{for cost criteria} \quad (12)$$

Step 2: Construct the weighted normalized decision matrix. This step involves incorporating the relative weights of the criteria $\mathbf{W_j} = [w_j]_{1\times n}$ obtained with the DANP method on the basis of $\mathbf{W^k_\alpha}$ matrix, where *j* represents j^{th} criterion and *j* = 1, ... ,*n*. Criteria relative weights are included in the decision matrix, which transforms the matrix $\mathbf{R_T}$ into a weighted matrix $\mathbf{V_T}$ based on Equation (13):

$$\mathbf{V_T} = [\mathbf{v_{ij}}]_{m\times n} = [\mathbf{W_j} \cdot \mathbf{r_{ij}}]_{m\times n}; \; i = 1,\ldots,m; \; j = 1,\ldots,n \quad (13)$$

Step 3: Determine the ideal and anti-ideal solution. Ideal $(\mathbf{A}^+)$ and anti-ideal $(\mathbf{A}^-)$ solutions are vectors that are calculated, as shown in Equations (14) and (15):

$$\mathbf{A}^+ = \left(\left(\max_i \mathbf{v_{ij}} \middle| \forall \mathbf{K_j} \in \mathbf{K'}\right), \left(\min_i \mathbf{v_{ij}} \middle| \forall \mathbf{K_j} \in \mathbf{K''}\right)\right) = \left(\mathbf{v}_1^+, \mathbf{v}_2^+, \ldots, \mathbf{v}_j^+, \ldots, \mathbf{v}_n^+\right), \; i = 1,\ldots,m; i = 1,\ldots,n \quad (14)$$

$$\mathbf{A}^- = \left(\left(\min_i \mathbf{v_{ij}} \middle| \forall \mathbf{K_j} \in \mathbf{K'}\right), \left(\max_i \mathbf{v_{ij}} \middle| \forall \mathbf{K_j} \in \mathbf{K''}\right)\right) = \left(\mathbf{v}_1^-, \mathbf{v}_2^-, \ldots, \mathbf{v}_j^-, \ldots, \mathbf{v}_n^-\right), \; i = 1,\ldots,m; i = 1,\ldots,n \quad (15)$$

Step 4: Calculate the separation measure. The alternative distance from ideal $(\mathbf{S}_i^+)$ and anti-ideal $(\mathbf{S}_i^-)$ solution is calculated by Euclidean distance in n-dimensional space of criteria as shown in Equation (16):

$$\mathbf{S}_i^+ = \sqrt{\sum_{j=1}^{n} (\mathbf{v_{ij}} - \mathbf{v}_j^+)^2}\,; \; \mathbf{S}_i^- = \sqrt{\sum_{j=1}^{n} (\mathbf{v_{ij}} - \mathbf{v}_j^-)^2}; \; i = 1, \ldots, m; \; j = 1,\ldots,n \quad (16)$$

Step 5: Calculate the relative closeness to the ideal solution. This is calculated as follows (17):

$$\mathbf{C_i} = \frac{\mathbf{S}_i^-}{\mathbf{S}_i^- + \mathbf{S}_i^+}, \; i = 1,\ldots,m \quad (17)$$

where $0 \leq \mathbf{C_i} \leq 1$. The closer $\mathbf{C_i}$ is to 1 the alternative is relatively closer to $\mathbf{A}^+$.

Step 6: Rank the preference order. A compromise solution is one or more alternatives that make the set $\mathbf{A}^*$ determined in accordance with (18):

$$\mathbf{C_i} = \frac{\mathbf{S}_i^-}{\mathbf{S}_i^- + \mathbf{S}_i^+}, \; i = 1,\ldots,m \quad (18)$$

where the rank of alternatives $\mathbf{A_i}$ (from best to worst) corresponds to the order of $\mathbf{C_i}$ values ranked in descending order.

3.5. Selecting The Best Vehicle Model

After obtaining the best vehicle model for procurement, an analysis of the obtained results is carried out according to defined criteria. It is considered by which criteria the chosen alternative takes precedence over the other alternatives. In addition, the expected effects and losses of other alternatives are analyzed. Fleet managers consider the obtained results and make a decision on the choice of an appropriate vehicle for procurement. The presented model is a support for fleet managers in the road vehicle procurement process in order to reduce the transport and maintenance costs and increase the fleet's energy efficiency.

4. Model Application

The developed model was applied at a company that owns a vehicle fleet for automotive spare parts distribution. According to the model's second phase (Figure 1), the relative weights of the vehicle procurement criteria were calculated. In accordance with the other phases of the model, a set of vehicles (alternatives) was determined, as well as values of the observed parameters. Afterwards, the observed alternatives were ranked.

4.1. Calculating Criteria Relative Weights

Within Phase 2 of the developed model, a survey was conducted to gather experts' opinions on the interdependencies between the defined criteria. The survey included 50 opinions of experts in this field. The demographic information of experts is shown in Table 2. The survey was conducted during the period from January to March 2019. The experts were informed about the model idea and therefore tasked to determine the interdependencies of 13 defined criteria with a score from 0 to 4 (from "no impact" to "very strong impact"). The inconsistent rate of 50 responses was less than 5%.

Table 2. Demographic information of survey participants.

Demographic Information		**Frequency**	**Percentage**
Age	25–34	10	20%
	35–44	26	52%
	45–54	12	24%
	55–65	2	4%
Education level	High school	1	2%
	College	5	10%
	Bachelor of Science	7	14%
	Master of Science	16	32%
	Doctor of Science	21	42%

Based on the results of the survey and Equations (1)–(4), the total impact matrix-T was calculated, which simultaneously represents the total impact matrix for criteria $-T_c$ (Table 3). As can be seen from Table 3, the more important criteria are those with higher value of (r + c): Vehicle procurement price (F1), Vehicle maintenance (O2), and Engine characteristics (CT2). The criterion that influences the other criteria the most has the highest value of (r − s), and that is Environmental protection requirements (E1). The criterion with the lowest value of (r − s) is influenced the most by the other criteria, and that is Vehicle procurement price (F1).

Table 3. Total impact matrix of criteria T_c and effects.

Criteria.	**CT1**	**CT2**	**CT3**	**CT4**	**CT5**	**F1**	**F2**	**F3**	**O1**	**O2**	**E1**	**E2**	**E3**
CT1	0.0477	0.0850	0.0699	0.0874	0.1196	0.2012	0.0712	0.1582	0.1463	0.1732	0.0591	0.0784	0.0876
CT2	0.0790	0.0724	0.0773	0.1055	0.1165	0.2133	0.0865	0.1603	0.1800	0.1709	0.0978	0.1184	0.1486
CT3	0.0614	0.0673	0.0301	0.0639	0.0621	0.1554	0.0575	0.1175	0.1140	0.1130	0.0525	0.0675	0.0745
CT4	0.0800	0.1090	0.0600	0.0456	0.0800	0.1345	0.0641	0.0888	0.1077	0.1539	0.0514	0.0920	0.1038
CT5	0.0987	0.1078	0.0955	0.0773	0.0644	0.2096	0.0829	0.1548	0.1599	0.1737	0.0888	0.0747	0.0880
F1	0.0959	0.0969	0.0812	0.0745	0.0981	0.0928	0.0949	0.1293	0.1050	0.1222	0.0621	0.0697	0.0869
F2	0.0757	0.0757	0.0587	0.0690	0.0648	0.1596	0.0315	0.0877	0.0629	0.0985	0.0436	0.0438	0.0592
F3	0.0552	0.0782	0.0526	0.0586	0.0857	0.1450	0.0733	0.0566	0.0955	0.1078	0.0540	0.0420	0.0714
O1	0.0483	0.0812	0.0339	0.0492	0.0693	0.1204	0.0555	0.1169	0.0548	0.0903	0.0771	0.0618	0.0851
O2	0.0764	0.0927	0.0565	0.0910	0.1133	0.1585	0.0695	0.1427	0.1325	0.0741	0.0788	0.0510	0.0759
E1	0.1010	0.1469	0.0901	0.1033	0.1513	0.2068	0.0927	0.1605	0.1557	0.1775	0.0510	0.1004	0.1152
E2	0.0922	0.1222	0.0740	0.1200	0.0977	0.1881	0.0675	0.1351	0.1682	0.1494	0.0717	0.0492	0.1314
E3	0.0522	0.1088	0.0591	0.1134	0.0966	0.1704	0.0591	0.1202	0.1586	0.1418	0.0575	0.0878	0.0548
(r + s)	2.3484	2.8704	1.8756	2.2296	2.6957	3.3649	1.8370	2.6045	2.5846	2.9593	2.4977	2.4035	2.4628
(r − s)	0.4211	0.3824	0.1979	0.1120	0.2567	−0.9462	0.0246	−0.6528	−0.6973	−0.5334	0.8070	0.5299	0.0981

In order to show the interdependencies of aspects, the total impact matrix of aspects $-T_D$ was calculated (Table 4). As can be seen in Table 4, the Construction-technical aspect (CT) is very important. The aspect of the environment (E) has the most influence on other aspects, while the Financial aspect (F) is influenced the most by the other aspects.

Table 4. Total impact matrix of aspects T_D and effects.

Aspects	Construction-Technical CT	Financial F	Operational O	Environment E
Construction technical-CT	1.9634	1.9558	1.4926	1.2831
Financial-F	1.1208	0.8707	0.5918	0.5327
Operational-O	0.7118	0.6635	0.3516	0.4297
Environment-E	1.5288	1.2004	0.9513	0.7189
(r + s)	12.020	7.8064	5.5439	7.3639
(r − s)	1.3701	−1.5744	−1.2307	1.4350

Based on the calculated total caused and received effects, Figure 2 shows the Influence Relation Map (IRM), in order to clarify the relationships between defined criteria and aspects. As can be seen in Figure 2, aspect E has the highest value of (r − s), so it affects all other aspects. Aspect CT influences aspects O and F, while aspect F has the most received effects from the other aspects (Figure 2). When considering the impacts of criteria within one aspect such as e.g., CT aspect, the CT1 criterion affects all other criteria within CT aspect because it has the highest value of (r − c), while the CT4 criterion with the lowest value of (r − c) receives the most effects from the other criteria of CT aspect.

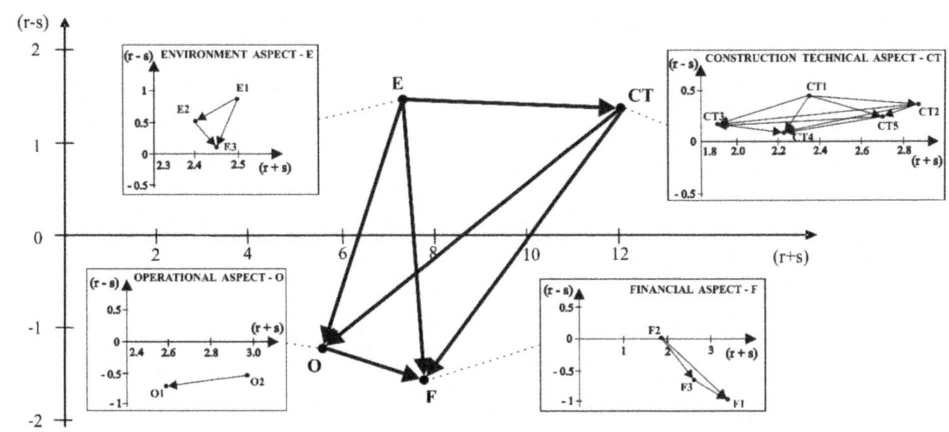

Figure 2. Influence Relation Map (IRM) for vehicle procurement.

Based on Equation (10), a limited weighted super-matrix W^k_α was calculated (Table 5). The vectors in super-matrix W^k_α represent the relative weights of vehicle procurement criteria W_j. As it is shown in Table 5, the criterion Vehicle procurement price (F1) has the highest relative weight, i.e., has the highest importance for the procurement of vehicles, in achieving the defined goal. Other significant criteria for vehicle procurement are: Vehicle maintenance (O2), Vehicle selling price (F3), and Fuel/energy Cost (O1). The most important vehicle procurement aspect is construction-technical aspect (CT), followed by financial (F), operational (O) and aspect of the environment (E), which has the least significance for the vehicle procurement process (Table 5).

Table 5. Relative weights of criteria and aspects.

Criteria	*CT1*	*CT2*	*CT3*	*CT4*	*CT5*	*F1*	*F2*	*F3*	*O1*	*O2*	*E1*	*E2*	*E3*
Relative weights-W_j	0.0603	0.0773	0.0516	0.0643	0.0762	0.1291	0.0588	0.1002	0.0963	0.1039	0.0544	0.0562	0.0715
Rank	9	5	13	8	6	1	10	3	4	2	12	11	7
Aspects	Construction technical *CT*					Financial *F*			Operational *O*		Environment *E*		
Local weights	0.3296					0.2881			0.2002		0.1821		
Rank	1					2			3		4		

4.2. Establishing and Ranking the Vehicles and Selecting the Best Vehicle for Procurement

The observed company (Delmax Ltd.) has a heterogeneous fleet of 66 vehicles, with small pickup vehicles (GVW up to 2.5 t), vans (GVW up to 3.5 t), and trucks (GVW up to 15 t). The analysis of the operation data shows that the vehicles' average annual mileage is 55,000 km. Based on the analysis of transport routes and history of customers' transport requirements in the observed company, it was found that more than 90% of the transport routes pass through urban areas, and the weight of the transported cargo is between 800 and 1300 kg in more than 80% cases. According to the obtained results of conducted analyses, it was determined that possible vehicle types that can meet transportation requirements are: Vans with GVW up to 3.5 t and trucks with GVW up to 7.5 t. Regarding this, different types and models of vehicles from different manufacturers were considered in order to determine which vehicle model contributes the most to achieving the company's goal. For the purpose of calculating the rank of alternatives, we have used the adjusted program code written in Python 3.7 [38].

Alternative A5 is a truck with a GVW up to 7.5 t, and the other alternatives are vans with a GVW up to 3.5 t (Table 6). Alternatives A1, A2, A3, and A4 are conventional (diesel) vans from different manufacturers, while the alternative A6 is an electric van. The alternatives A7 and A8 are a variation of the alternative A2 that use different alternative fuel CNG and biodiesel B30, respectively. Criteria values of the observed alternatives were determined based on the brochures from vehicle manufacturers' websites and websites that advertise used vehicles, as well as from authorized maintenance services and vehicle dealers. These values were also determined based on the characteristics of the observed companies' vehicle fleet and their transport requirements, considered the price of diesel is £1.08/lit and the price of electricity is £0.11/kWh. After determining the nature of each criterion (min./max.) the relative closeness of the observed alternatives to the ideal solution (C_i) was calculated on the basis of Equations (11)–(18), and the rank of each alternative was determined (Table 6).

Table 6. Criteria values and rank of alternatives.

Criteria	*CT1*	*CT2*	*CT3*	*CT4*	*CT5*	*F1*	*F2*	*F3*	*O1*	*O2*	*E1*	*E2*	*E3*	C_i	*Rank*
A1 diesel	6	2179	5	10.61	1	38801	1	7600	10.42	495	159	12.6	84	0.485	**4**
A2 diesel	2	2300	4	24.24	1	36693	1	8000	8.21	243	186	12.8	84	0.570	**1**
A3 diesel	1	2298	4	12.12	1	38331	1	7500	7.45	248	182	13.6	89	0.489	**3**
A4 diesel	3	2143	1	3.03	1	39852	1	11500	8.75	288	213	13.6	100	0.459	**5**
A5 diesel	2	4580	1	0	1	59605	1	21150	15.34	471	375	13.2	33	0.315	**8**
A6 electric	1	0	4	0	1	66373	1	11445	3.63	174	55	13.6	94	0.522	**2**
A7 CNG	2	3000	4	0	1	43901	1	5832	5.18	270	177	12.8	84	0.413	**7**
A8 B30	2	2300	4	0	1	36693	1	8000	8.40	269	134	12.8	84	0.432	**6**
max/min	max	min	max	max	min	min	max	max	min	min	min	min	max		
Crit. Wj	0.0603	0.0773	0.0516	0.0643	0.0762	0.1291	0.0588	0.1002	0.0963	0.1039	0.0544	0.0562	0.0715		

5. Results and Discussion

As it is shown in Table 6 the A2 (diesel van) is the first ranked alternative, while the A6 (electric van) is the second one. This vehicle has the lowest procurement price (about £30,000 lower than A6 and more than £1600 lower than the third ranked alternative, i.e., A3), which is the most important criterion. If a period of five years in operation (275,000 km) is observed, it can be seen that the sum of fuel/energy and maintenance costs is the lowest for the observed electric van (about £10,800), which is almost £13,000 less than for A2, about £11,000 less than for A3 and £4700 for CNG van. Regarding CO_2 emission, it can be seen that A6 is the best alternative since it emits about 15 t of CO_2 during a period of five years (well to wheel emission), which is 36 t less than the first ranked alternative and 88 t less than A5. However, if we observe overall transport and maintenance costs (sum of vehicle price, fuel/energy, and maintenance costs minus vehicle selling price) we can see that A2 has about £13,300 lower costs than A6. Besides the higher procurement price of alternative A3, the first ranked vehicle has one additional safety/technical system, twice as good compatibility with the existing vehicle fleet, higher vehicle selling price, as well as smaller vehicle turning diameter (which is very important in dense urban environments). Despite the fact that alternative A1 has the most safety-technical systems and equipment regarding vehicle comfort, it has very high maintenance costs (£495/year), which is the second most important criterion. Some of the reasons why the alternative A5 is the worst ranked are very high procurement, fuel/energy, and maintenance costs, as well as low payload utilization (33% of payload utilization means that this vehicle has 67% of free space while transporting goods, which means about 725,800 tonne-kilometer of loss compared to full payload utilization observed over a period of five years and 275,000 km traveled). It can be seen that electric van (A6) is second ranked mostly due to its high procurement price, zero compatibility with the existing vehicle fleet, and very low importance of criterion related to CO_2 emission. Thus, we have considered different scenarios to establish whether and how the rank of the alternatives would change.

Scenario S-0 represents the existing scenario, while the S-1 represents the ranking of observed alternatives if the electric power is produced from renewable sources (zero emission of CO_2 from well to wheel). S-2 and S-3 represent scenarios where there are subsidies of £15,000 and £20,000, whereby the use of renewable sources is significantly increased, respectively. Additionally, we have observed scenarios where the relative weights of criteria vehicle price (F1) and CO_2 emission (E1) are the same (S-4), they have the same weights but the electric van emission is zero (S-5), where the importance of E1 criterion is twice as high as the importance of F1 (S-6), and where the relative weights of criteria F1 and E1 switch places (S-7). We have researched the rank of alternatives when the compatibility with the existing vehicle fleet of all alternatives is the same (S-8). Scenario S-9 represents the case where the number of safety technical systems is the same for all vehicles. Since the maintenance costs for alternative A1 are very high, in scenario S-10 we examined what would happen if these costs were the same as for the alternative A2. The ranks of alternatives for all of these scenarios are shown in Figure 3.

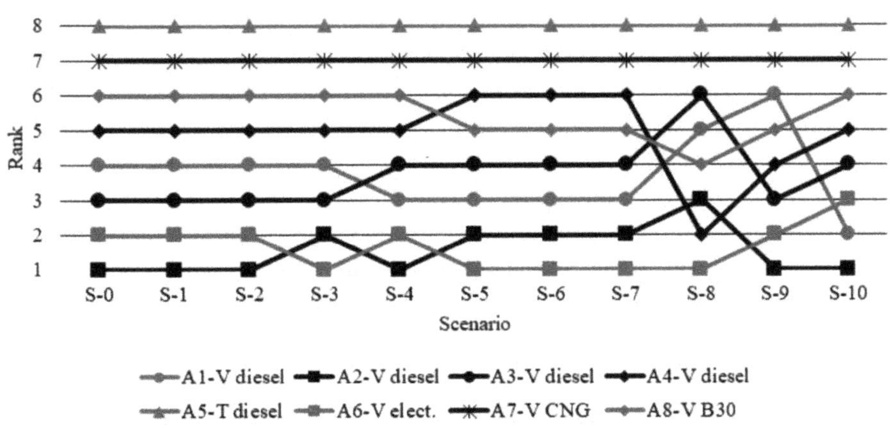

Figure 3. Ranks of alternatives in different scenarios.

As can be seen from Figure 3, in our case study, scenarios S-1 and S-2 are the same as the basic scenario. If there are subsidies of £20,000 (about 30% of new electric vehicle price) and the electric energy production is mostly from renewable sources (S-3), A6 is the first ranked alternative. In scenario

S-4 the only change is between alternatives A1 and A3. Scenarios S-5, S-6, and S-7 show that there are changes in rank between alternatives A2 and A6 (which is first ranked alternative in this case), between A1 and A3, as well as between A8 and A4. If the compatibility with the existing vehicle fleet of all alternatives is the same (S-8), then electric van is the first ranked alternative, and the second one is A4, while the first ranked alternative from the basic scenario (A2) is now the third ranked. The rank of alternatives A4 and A8 in scenario S-9 has improved, while the rank of alternative A1 has gone down from fourth to sixth. In the final scenario (S-10) it can be seen that the alternative A1 would be second ranked if its' maintenance costs were lower, such as for the alternative A2. The rank of alternatives A5 and A7 is the same for all scenarios.

6. Conclusions

Vehicle procurement as a factor that contributes to making more profit in companies with their own vehicle fleet has been researched in this paper. We developed a model that selects the most suitable road vehicle while taking into account construction-technical, financial, operational, and environmental aspects of a road vehicle, simultaneously from three points of view: Transport costs, maintenance costs, and energy efficiency. Using the DANP method and engaging 50 experts from this field, we determined the criteria relative weights and their interdependency, which was ignored in most previous studies. The most important criteria for vehicle procurement are vehicle price (F1), vehicle maintenance (O2), vehicle selling price (F3), and fuel/energy costs (O1), while environmental protection requirements (E1) and vehicle comfort (CT3) are the least significant criteria. The TOPSIS method was used to rank the observed vehicles by their suitability. Besides the basic scenario, we have considered 10 more different scenarios to establish whether and how the rank of the observed alternatives would change.

On the basis of collected criteria parameter values observing five–year period and annual mileage of 55,000 km, the first ranked alternative (diesel van) has lower overall transport and maintenance costs (fuel/energy costs, maintenance costs, vehicle price and vehicle selling price) for £13,300 than second ranked alternative (electric van). In our case (observed company) the electric vehicle price should be about 30% lower (there should be subsidies) in order for this alternative to become the first ranked. We think there should be government subsidies, as well as the increase in usage of renewable energy sources, which will together encourage vehicle fleet managers to procure electric vehicles. For example, this difference in the vehicle price can be used to train the maintenance technicians and drivers, as well as to buy some spare parts and equipment to maintain electric vehicles. We can see that the sum of fuel/energy costs and vehicle maintenance costs of electric van are almost £15,000 lower, on average, than for diesel vans (observing five-year period). However, if overall transport and maintenance costs of electric van are observed, it can be seen that these costs are about £10,000 higher, on average, for five years in operation than for diesel vans.

This model is evaluated by road vehicle fleet managers as easy to understand and operate, while it gives extremely satisfying results in a short period of time. The managers said that without the model they were not aware of all criteria, and they did not observe all of them in previous vehicle procurements. Moreover, it was almost impossible for them to simultaneously observe all relevant criteria necessary for vehicle procurement, as well as their weights and parameters (vehicle data). With the model, the only thing managers should do is to provide vehicles' data and import them into the model. Managers in the observed company during the vehicle procurement process have replaced two old, less suitable vans, with two new ones of the same manufacturer and the same model as the *A2* alternative, based on the obtained results of the model. By choosing this alternative instead of second ranked alternative, this company saved about 20% of overall transport and maintenance costs. Moreover, this alternative has proven to be a better choice than all of the other ones as well, taking into account transport and maintenance costs and energy efficiency.

Since the criteria values of the observed alternatives should be determined based on the brochures from vehicle manufacturers' websites and websites that advertise used vehicles, as well as from authorized maintenance services and vehicle dealers, the accuracy of the obtained information could affect the accuracy of derived results. Additionally, the limitation could represent the non-existence of parameter information that is needed for the model (e.g., vehicle CO_2 emission), as well as a little longer time to collect all necessary data compared to the time when managers observed only a few criteria.

Author Contributions: Conceptualization, D.V.; formal analysis, M.S.; investigation, M.S.; methodology, D.V.; software, M.S.; supervision, D.V.; validation, D.S.; visualization, D.S.; writing—original draft, M.S. All authors have read and agreed to the published version of the manuscript.

Funding: This paper has been realized within the project "Development of the Model for Managing the Vehicle Technical Condition in Order to Increase its Energy Efficiency and Reduce Exhaust Emissions" (TR36010), supported by the Ministry of Education, Science and Technological Development of the Republic of Serbia.

Acknowledgments: The authors wish to express their gratitude to all managers in the company Delmax Ltd. and vehicle service centers for providing all necessary information for this research.

Conflicts of Interest: The authors declare no conflict of interest.

References

1. Alvarez, J.F.; Tsilingiris, P.; Engebrethsen, E.S.; Kakalis, N.M.P. Robust fleet sizing and deployment for industrial and independent bulk ocean shipping companies. *INFOR* **2011**, *49*, 93–107. [CrossRef]
2. Hoff, A.; Andersson, H.; Christiansen, M.; Hasle, G.; Løkketangen, A. Industrial aspects and literature survey: Fleet composition and routing. *Comput. Oper. Res.* **2010**, *37*, 2041–2061. [CrossRef]
3. Vujanović, D.; Momcilović, V.; Medar, O. Influence of an integrated maintenance management on the vehicle fleet energy efficiency. *Therm. Sci.* **2018**, *22*, 1525–1536. [CrossRef]
4. Mørch, O.; Fagerholt, K.; Pantuso, G.; Rakke, J. Maximizing the rate of return on the capital employed in shipping capacity renewal. *Omega* **2017**, *67*, 42–53. [CrossRef]
5. Patricksson, Ø.S.; Fagerholt, K.; Rakke, J.G. The fleet renewal problem with regional emission limitations: Case study from Roll-on/Roll-off shipping. *Transp. Res. Part C Emerg. Technol.* **2015**, *56*, 346–358. [CrossRef]
6. Ansaripoor, A.H.; Oliveira, F.S.; Liret, A. A risk management system for sustainable fleet replacement. *Eur. J. Oper. Res.* **2014**, *237*, 701–712. [CrossRef]
7. Figliozzi, M.A.; Boudart, J.A.; Feng, W. Economic and environmental optimization of vehicle fleets. *Transp. Res. Rec. J. Transp. Res. Board* **2011**, *2252*, 1–6. [CrossRef]
8. Feng, W.; Figliozzi, M. An economic and technological analysis of the key factors affecting the competitiveness of electric commercial vehicles: A case study from the USA market. *Transp. Res. Part C Emerg. Technol.* **2013**, *26*, 135–145. [CrossRef]
9. Ansaripoor, A.H.; Oliveira, F.S.; Liret, A. Recursive expected conditional value at risk in the fleet renewal problem with alternative fuel vehicles. *Transp. Res. Part C Emerg. Technol.* **2016**, *65*, 156–171. [CrossRef]
10. Pantuso, G.; Fagerholt, K.; Wallace, S.W. Uncertainty in Fleet Renewal: A Case from Maritime Transportation. *Transp. Sci.* **2016**, *50*, 390–407. [CrossRef]
11. Lee, W.S.; Huang, A.Y.; Chang, Y.Y.; Cheng, C.M. Analysis of decision making factors for equity investment by DEMATEL and Analytic Network Process. *Expert Syst. Appl.* **2011**, *38*, 8375–8383. [CrossRef]
12. Sufiyan, M.; Haleem, A.; Khan, S.; Khan, M.I. Evaluating food supply chain performance using hybrid fuzzy MCDM technique. *Sustain. Prod. Consum.* **2019**, *20*, 40–57. [CrossRef]
13. Gudiel Pineda, P.J.; Liou, J.J.H.; Hsu, C.C.; Chuang, Y.C. An integrated MCDM model for improving airline operational and financial performance. *J. Air Transp. Manag.* **2018**, *68*, 103–117. [CrossRef]
14. Hsu, C.C.; Liou, J.J.H.; Lo, H.W.; Wang, Y.C. Using a hybrid method for evaluating and improving the service quality of public bike-sharing systems. *J. Clean. Prod.* **2018**, *202*, 1131–1144. [CrossRef]
15. Kumar, A.; Anbanandam, R. Analyzing interrelationships and prioritising the factors influencing sustainable intermodal freight transport system: A grey-DANP approach. *J. Clean. Prod.* **2020**, *252*, 119769. [CrossRef]
16. Vujanović, D.B.; Momcilović, V.M.; Vasić, M.B. A hybrid multi-criteria decision making model for the vehicle service center selection with the aim to increase the vehicle fleet energy efficiency. *Therm. Sci.* **2018**, *22*, 1549–1561. [CrossRef]
17. Sakthivel, G.; Ilangkumaran, M.; Gaikwad, A. A hybrid multi-criteria decision modeling approach for the best biodiesel blend selection based on ANP-TOPSIS analysis. *Ain Shams Eng. J.* **2015**, *6*, 239–256. [CrossRef]

18. Wang, Y.; Zhou, G.; Li, T.; Wei, X. Comprehensive evaluation of the sustainable development of battery electric vehicles in China. *Sustainability* **2019**, *11*, 5635. [CrossRef]
19. Hamurcu, M.; Eren, T. Electric Bus Selection with Multicriteria Decision Analysis for Green Transportation. *Sustainability* **2020**, *12*, 2777. [CrossRef]
20. dos Santos, B.M.; Godoy, L.P.; Campos, L.M.S. Performance evaluation of green suppliers using entropy-TOPSIS-F. *J. Clean. Prod.* **2019**, *207*, 498–509. [CrossRef]
21. Sureeyatanapas, P.; Sriwattananusart, K.; Niyamosoth, T.; Sessomboon, W.; Arunyanart, S. Supplier selection towards uncertain and unavailable information: An extension of TOPSIS method. *Oper. Res. Perspect.* **2018**, *5*, 69–79. [CrossRef]
22. Hoseinpour, M.; Sadrnia, H.; Tabasizadeh, M.; Ghobadian, B. Evaluation of the effect of gasoline fumigation on performance and emission characteristics of a diesel engine fueled with B20 using an experimental investigation and TOPSIS method. *Fuel* **2018**, *223*, 277–285. [CrossRef]
23. Sabaghi, M.; Mascle, C.; Baptiste, P. Application of DOE-TOPSIS technique in decision-making problems. *IFAC-PapersOnLine* **2015**, *48*, 773–777. [CrossRef]
24. Anthony, P.; Behnoee, B.; Hassanpour, M.; Pamucar, D. Financial performance evaluation of seven Indian chemical companies. *Decis. Mak. Appl. Manag. Eng.* **2019**, *2*, 19–37. [CrossRef]
25. Milosavljević, M.; Bursać, M.; Tričković, G. The selection of the railroad container terminal in Serbia based on multi criteria decision making methods. *Decis. Mak. Appl. Manag. Eng.* **2018**, *1*, 1–15. [CrossRef]
26. Karagülle, A.Ö. The Evaluation of Fleet Structures in Turkish Aviation Industry from Strategic Management Point of View. *Procedia–Soc. Behav. Sci.* **2012**, *58*, 93–97. [CrossRef]
27. Bakkehaug, R.; Eidem, E.S.; Fagerholt, K.; Hvattum, L.M. A stochastic programming formulation for strategic fleet renewal in shipping. *Transp. Res. Part E Logist. Transp. Rev.* **2014**, *72*, 60–76. [CrossRef]
28. de Salvo Junior, O.; Vaz de Almeida, F.G. Influence of technologies on energy efficiency results of official Brazilian tests of vehicle energy consumption. *Appl. Energy* **2019**, *241*, 98–112. [CrossRef]
29. Jiang, C.; Xu, H.; Srivastava, D.; Ma, X.; Dearn, K.; Cracknell, R.; Krueger-Venus, J. Effect of fuel injector deposit on spray characteristics, gaseous emissions and particulate matter in a gasoline direct injection engine. *Appl. Energy* **2017**, *203*, 390–402. [CrossRef]
30. Merkert, R.; Hensher, D.A. The impact of strategic management and fleet planning on airline efficiency–a random effects tobit model based on dea efficiency scores. *Transp. Res. Part A Policy Pract.* **2011**, *45*, 686–695. [CrossRef]
31. Riechi, J.; Mácian, V.; Tormos, B.; Avila, C. Optimal fleet replacement: A case study on a Spanish urban transport fleet. *J. Oper. Res. Soc.* **2017**, *68*, 886–894. [CrossRef]
32. *Low Carbon Vehicle Partnership, The Low Emission Van Guide*; Guide for van operators; Cenex: Leicestershire, UK, 2019; pp. 5–27.
33. Keles, P.; Hartman, J.C. Case study: Bus fleet replacement. *Eng. Econ. A J. Devoted to Probl. Cap. Invest.* **2004**, *49*, 253–278. [CrossRef]
34. Mijailović, R. The optimal lifetime of passenger cars based on minimization of CO2 emission. *Energy* **2013**, *55*, 869–878. [CrossRef]
35. Degirmenci, K.; Breitner, M.H. Consumer purchase intentions for electric vehicles: Is green more important than price and range? *Transp. Res. Part D Transp. Environ.* **2017**, *51*, 250–260. [CrossRef]
36. Liimatainen, H.; Pöllänen, M. Trends of energy efficiency in Finnish road freight transport 1995–2009 and forecast to 2016. *Energy Policy* **2010**, *38*, 7676–7686. [CrossRef]
37. Hwang, C.-L.; Yoon, K. *Multiple Attribute Decision Making*; Lecture Notes in Economics and Mathematical Systems; Springer: Heidelberg, Germany, 1981; Volume 186, pp. 128–140. ISBN 978-3-642-48318-9.
38. Papathanasiou, J.; Ploskas, N. TOPSIS. In *Multiple Criteria Decision Aid*; Pardalos, M., Du, D.-Z., Eds.; Springer Nature: Cham, Switzerland, 2018; Volume 136, pp. 1–30. ISBN 978-3-319-91646-0.

© 2020 by the authors. Licensee MDPI, Basel, Switzerland. This article is an open access article distributed under the terms and conditions of the Creative Commons Attribution (CC BY) license (http://creativecommons.org/licenses/by/4.0/).

sustainability

MDPI

Article

Evaluation of Criteria for the Implementation of High-Performance Computing (HPC) in Danube Region Countries Using Fuzzy PIPRECIA Method

Milovan Tomašević [1], Lucija Lapuh [1], Željko Stević [2,*], Dragiša Stanujkić [3] and Darjan Karabašević [4]

1 Faculty of Information Studies in Novo Mesto, Ljubljanska cesta 31a, 8000 Novo Mesto, Slovenia; milovan.tomasevic@fis.unm.si (M.T.); lucija.lapuh@fis.unm.si (L.L.)

2 Faculty of Transport and Traffic Engineering, University of East Sarajevo, Vojvode Mišića 52, 74000 Doboj, Bosnia and Herzegovina

3 Department of Management, Technical Faculty in Bor, University of Belgrade, Vojske Jugoslavije 12, 19210 Bor, Serbia; dstanujkic@tfbor.bg.ac.rs

4 Faculty of Applied Management, Economics and Finance, University Business Academy in Novi Sad, Jevrejska 24, 11000 Belgrade, Serbia; darjan.karabasevic@mef.edu.rs

* Correspondence: zeljkostevic88@yahoo.com or zeljko.stevic@sf.ues.rs.ba

Received: 12 March 2020; Accepted: 7 April 2020; Published: 9 April 2020

check for updates

Abstract: The use of computers with outstanding performance has become a real necessity in order to achieve greater efficiency and sustainability for the accomplishment of various tasks. Therefore, with the development of information technology and increasing dynamism in the business environment, it is expected that these computers will be more intensively deployed. In this paper, research was conducted in Danube region countries: Austria, Bosnia and Herzegovina, Bulgaria, Croatia, Czech Republic, Germany, Hungary, Moldova, Montenegro, Romania, Serbia, Slovakia, Slovenia, and Ukraine. The aim of the research was to determine what criteria are most significant for the introduction of high-performance computing and the real situation in each of the countries. In addition, the aim was to establish the infrastructure needed to implement such a system. In order to determine the partial significance of each criterion and thus the possibility of implementing high-performance computing, a multi-criteria model in a fuzzy environment was applied. The weights of criteria and their rankings were performed using the Fuzzy PIvot Pairwise RElative Criteria Importance Assessment—fuzzy PIPRECIA method. The results indicate different values depend on decision-makers (DMs) in the countries. Spearman's and Pearson's correlation coefficients were calculated to verify the results obtained.

Keywords: supercomputers; Fuzzy PIPRECIA; Danube region countries; high-performance computing (HPC)

1. Introduction

The use of supercomputers is increasing today, both in science and in economy. Advances in algorithms and software technologies at all levels are essential for further progress in problem-solving with the use of supercomputers. A supercomputer represents computer architecture of high-capacity, i.e., high performance, capable of processing large amounts of data in a very short time. Such computers are usually considered high-performance computers whose construction is not based primarily on von Neumann architecture. Typically, it enables the distribution and parallelization of computer processes. Although today's desktop computers are more powerful than supercomputers developed just a decade ago, it can be said that common characteristics of supercomputers, regardless of the period in which they occurred, are as follows: maximum available processing speed, maximum possible memory size, the largest physical dimensions, and the highest price, compared to other computers.

Supercomputers are used to solve a variety of problems, including intensive calculations [1], e.g., in inventory management [2], military intelligence [3], climate forecasting [4], earthquake modeling [5], transport [6], production [7], human health and safety [8], medical applications [9], i.e., practically in every field of science or business. The role of supercomputers in all these fields is becoming increasingly important, and supercomputers are increasingly influencing future progress. Clustering of European supercomputers was initiated in 2002, by the project Distributed European Infrastructure for Supercomputing (DEISA), when 11 of the largest supercomputing centers from seven European countries were connected to a 10 Gbit/s network [10]. The next frontier in achieving performance for supercomputers is 1018 instructions per second and is expected to be reached around 2021–2022 [11].

The idea that computing power can be provided to customers as a useful service is not new, as it has been dating since 1966 [12]. Despite that, new technologies, as well as customer demands for high-quality services, come daily, and all of that obviously leads to day-to-day, new competition. In recent decades, the economy and industry have faced a large inflow of new and intensively data-oriented services, including data collection, storage, analysis, and use [13]. Search engines [14], social networks, and business intelligence [15] are just a few examples of such trends [16].

The amount of digital data in our world has grown enormously, with an exponential trend. The size of the digital universe is estimated to grow from 4.4 zettabytes in 2013 to 44 zettabytes by 2020 [17,18]. The vast amount of data being produced, commonly referred to as Big Data [19], provides great potential in the form of undiscovered structures and relationships [15]. In order for this potential to be exploited, new knowledge acquired, and business value gained, the produced data must be accessed, then processed, analyzed, and visualized in the most efficient way. To this end, the application of specialized architectures based on the use of supercomputers, often referred to as high-performance computing (HPC) architectures, becomes a necessity in practice [20–22]. This will certainly also mean the rapid transition from small, closed computers and storage architectures to large, open, and service-oriented infrastructures [23]. Since introducing early supercomputers in the 1980s, the high-performance industry has been striving for continuous performance growth. This can be noticed from the trends expressed by the list of top 500 supercomputers [24], which shows an exponential increase in computing power over the last twenty years. Traditionally, the computing power provided by HPC is mainly used in enterprises, as well as research institutes and academic organizations, which benefit from the rich offerings given by HPC. For example, from the arrangement of complex sequences in DNA to complex simulations in meteorological phenomena, HPC has been proven to be the primary basis for providing procedures for solving complex problems that require very high computational performance [25–28].

Science and industry face extensive challenges to cope with very large complex data sets. These big data needed to be processed. A collaborative exchange between scientists and data analysis experts is essential to provide insights and solutions for a specific challenge [29].

HPC providers appreciate cooperation in HPC-related aspects among national and foreign research centers, as well as mainly with national enterprises. They believe that cooperation with an industry is important for the future development of their organization and that they can help enterprises to meet their needs through HPC. On the other hand, enterprises believe that cooperation with science or industry could foster the HPC usage and their organization development [30]. HPC intermediaries, which provide consultancy or advisory services in the field of HPC, are important for connecting users of HPC services and HPC centers, as many companies and public research centers lack technical knowledge about HPC. In Europe there is a strong presence of independent software vendors whose are successfully working with research institutions. They struggle to expand their businesses successfully due to their difficulties in raising financing. Commercial banks are generally engaged in the financing of private and commercially oriented HPC centers [31].

HPC is, according to the European Commission [32], a strategic resource for Europe's future. Creating awareness among the students about the importance of HPC topics is very important to educate a workforce to continue work in this area. Students can be introduced to HPC topics through courses, projects, and summer internship programs. Quality of education can be, according to the experiences from India [33], enhanced with better outcome in undergraduate computer science and engineering programs with the introduction of HPC courses into their curriculum.

Taking into account the increased requirements for performing complex data analyses in research and the application of HPC technologies to support those analyses, there is an objective need for an adequate evaluation and selection of an appropriate HPC architecture, which presents a challenge to the engineering community in the field of computing and informatics.

Supercomputer software, algorithms, and hardware are closely coupled. As architectures change, new software solutions are needed. If a selection of architectures is made without considering software and algorithms, the results of such a decision may be unsatisfactory.

Educated and qualified people are an important part of a supercomputing system. Professionals, i.e., multidisciplinary expert teams for supercomputers, need a set of specialized knowledge of the applications they work with, as well as of various supercomputing technologies. This further implies that both forms of support are necessary for a supercomputing system to operate effectively. In addition to the long implementation and required customizations, one of the problems that occurs when deciding to select some HPC architecture is the possibility of various types of risks occurring. As the biggest, a financial risk arises, and together with it a business risk. It implies that in addition to the resources that need to be allocated for the analysis when selecting some HPC architecture, it is also necessary to provide the resources for the architecture itself, as well as the standards that the architecture requires in order to work. In addition, a team of experts should be provided, which is also necessary to support the implementation of the HPC architecture, in order to achieve positive business results, particularly in the case of specialized problems or projects and the creation of space for future projects. Different risks can lead to the failure of the HPC architecture to be implemented, most often when it cannot meet specific requirements or needs.

In order to avoid the risk of failure of implementation and application of some HPC architecture, it is necessary to clearly define all the undertakings in the entire life cycle of the HPC architecture, starting with the analysis of the architecture selection and throughout its lifetime, in order to achieve positive business results.

A typical process for selecting some architecture could be implemented in several steps. In the first step, an assessment of the current state is important, when it is decided whether architecture changes are needed and if so, why and under what conditions. Then, the next steps would be to analyze plans and budgets, taking into consideration requirements and needs, i.e., assessing the needs for a new solution. Following these information and constraints, we explore current technologies and their capabilities. The next step is deciding which solution is appropriate for the desired changes. The whole process results in decisions that represent the basis for negotiating and creating contracts for the procurement and implementation of the desired architecture.

In such a process, it is necessary to have as much information as possible before making a decision, since when establishing some architecture in an organization, wrong decisions can be made by not paying attention to each criterion and its domain in the required way, so such an approach will not produce the desired results. Failure to achieve the desired results can lead to failure in the implementation and application of the desired HPC architecture. Therefore, it is necessary to provide a multidimensional database model and analysis procedures over such a model, in order to provide opportunities for decision simulations according to specific criteria.

Despite cost reductions and ease of access to high-performance computing platforms, they are still unavailable to most institutions and companies. Supercomputing systems consume a significant amount of energy, leading to high operating costs, reduced reliability, and waste of natural resources.

This fact clearly indicates the sensitivity of a decision-making process regarding the selection and implementation of such an architecture.

2. Methods

2.1. Operations on Fuzzy Numbers

A fuzzy number [34] $\overline{A}$ on R can be a triangular fuzzy number (TFN) if its membership function $\mu_{\overline{A}}(x)$: R→[0,1] is equal to following Equation (1) [35,36]:

$$\mu_{\overline{A}}(x) = \begin{cases} \frac{x-l}{m-l} & l \le x \le m \\ \frac{u-x}{u-m} & m \le x \le u \\ 0 & otherwise \end{cases}, \quad (1)$$

From Equation (1), l and u mean the lower and upper bounds of the fuzzy number $\overline{A}$, and m is the modal value for $\overline{A}$. The TFN can be denoted by $\overline{A} = (l, m, u)$.

The operational laws of TFN $\overline{A}_1 = (l_1, m_1, u_1)$ and $\overline{A}_2 = (l_2, m_2, u_2)$ are displayed as the following equations [37,38].

Addition:

$$\overline{A_1} + \overline{A_2} = (l_1, m_1, u_1) + (l_2, m_2, u_2) = (l_1 + l_2, m_1 + m_2, u_1 + u_2) \quad (2)$$

Multiplication:

$$\overline{A_1} \times \overline{A_2} = (l_1, m_1, u_1) \times (l_2, m_2, u_2) = (l_1 \times l_2, m_1 \times m_2, u_1 \times u_2) \quad (3)$$

Subtraction:

$$\overline{A_1} - \overline{A_2} = (l_1, m_1, u_1) - (l_2, m_2, u_2) = (l_1 - u_2, m_1 - m_2, u_1 - l_2) \quad (4)$$

Division:

$$\frac{\overline{A_1}}{\overline{A_2}} = \frac{(l_1, m_1, u_1)}{(l_2, m_2, u_2)} = \left(\frac{l_1}{u_2}, \frac{m_1}{m_2}, \frac{u_1}{l_2}\right) \quad (5)$$

Reciprocal:

$$\overline{A_1}^{-1} = (l_1, m_1, u_1)^{-1} = \left(\frac{1}{u_1}, \frac{1}{m_1}, \frac{1}{l_1}\right) \quad (6)$$

2.2. Fuzzy PIvot Pairwise RElative Criteria Importance Assessment–Fuzzy (PIPRECIA) Method

The fuzzy PIPRECIA method [39] consists of 11 steps that are shown below, and so far, it has been used in a few studies. Stanković et al. [38] used fuzzy PIPRECIA in integration with the newly developed Fuzzy Measurement Alternatives and Ranking according to the COmpromise Solution (fuzzy MARCOS) for road traffic risk analysis. The significance of the six criteria for evaluating road sections was evaluated using Fuzzy PIPRECIA. The results showed that the most important criterion in their study was the number of access points on each section, i.e., the second criterion. The original study [39] in which fuzzy PIPRECIA was developed dealt with the assessment of conditions for the implementation of information technology in the warehouse. This method was integrated into Strengths, Weaknesses, Opportunities, and Threats (SWOT) analysis to obtain values of each SWOT dimension. Marković et al. [40] showed that fuzzy PIPRECIA as a subjective Multi-Criteria Decision-Making (MCDM) method can be very successfully applied in integration with other objective methods, such as Criteria Importance Through Intercriteria Correlation (CRITIC). They used an integrated model for the ranking banks in order to achieve business excellence and sustainability. Fuzzy PIPRECIA in the study in [41] was used for determining criteria weights for the evaluation of green suppliers. This method was successfully applied in integration with Rough Simple Additive Weighting (Rough SAW) method.

Step 1. Form the required benchmarking set of criteria and form a decision-making team. Sort the criteria according to marks from the first to the last, and this means that they need to be sorted unclassified. Therefore, in this step, their significance does not play any role.

Step 2. In order to determine the relative importance of criteria, each decision-maker individually evaluates pre-sorted criteria by starting from the second criterion, Equation (7):

$$\overline{s_j^r} = \begin{cases} > \overline{1} & if \quad C_j > C_{j-1} \\ = \overline{1} & if \quad C_j = C_{j-1} \\ < \overline{1} & if \quad C_j < C_{j-1} \end{cases}, \tag{7}$$

where $\overline{s_j^r}$ denotes the assessment of criteria by a decision-maker r.

In order to obtain a matrix $\overline{s_j}$, it is necessary to perform the averaging of matrix $\overline{s_j^r}$ using a geometric mean. Decision-makers evaluate criteria by applying the defined scales in Tables 1 and 2.

Table 1. Scale 1–2 for the assessment of criteria [22].

			l	m	u	DFV
Almost equal value		1	1.000	1.000	1.050	1.008
Slightly more significant		2	1.100	1.150	1.200	1.150
Moderately more significant		3	1.200	1.300	1.350	1.292
More significant	**Scale 1–2**	4	1.300	1.450	1.500	1.433
Much more significant		5	1.400	1.600	1.650	1.575
Dominantly more significant		6	1.500	1.750	1.800	1.717
Absolutely more significant		7	1.600	1.900	1.950	1.858

Table 2. Scale 0–1 for the assessment of criteria [22].

	l	m	u	DFV	
	0.667	1.000	1.000	0.944	weakly less significant
	0.500	0.667	1.000	0.694	moderately less significant
	0.400	0.500	0.667	0.511	less significant
Scale 0–1	0.333	0.400	0.500	0.406	really less significant
	0.286	0.333	0.400	0.337	much less significant
	0.250	0.286	0.333	0.288	dominantly less significant
		0.250	0.286	0.251	absolutely less significant

When the criterion is of greater importance in relation to the previous one, assessment is made using the above scale in Table 1. In order to make it easier for decision-makers to evaluate the criteria, the table shows the defuzzified value (DFV) for each comparison.

When the criterion is of less importance compared to the previous one, assessment is made using the above-mentioned scale in Table 2.

Step 3. Determine the coefficient $\overline{k_j}$:

$$\overline{k_j} = \begin{cases} = \overline{1} & if \quad j = 1 \\ 2 - \overline{s_j} & if \quad j > 1 \end{cases}. \tag{8}$$

Step 4. Determine the fuzzy weight $\overline{q_j}$:

$$\overline{q_j} = \begin{cases} = \overline{1} & if \quad j = 1 \\ \frac{\overline{q_{j-1}}}{\overline{k_j}} & if \quad j > 1 \end{cases}. \tag{9}$$

Step 5. Determine the relative weight of the criterion $\overline{w_j}$:

$$\overline{w_j} = \frac{\overline{q_j}}{\sum_{j=1}^{n} \overline{q_j}}. \tag{10}$$

In the following steps, the inverse methodology of fuzzy PIPRECIA method (fuzzy PIPRECIA-I) needs to be applied.

Step 6. Perform the assessment of above-defined applying scale, but this time starting from a penultimate criterion:

$$\overline{s_j^{r}}' = \begin{cases} > \overline{1} & if \quad C_j > C_{j+1} \\ = \overline{1} & if \quad C_j = C_{j+1} \\ < \overline{1} & if \quad C_j < C_{j+1} \end{cases}, \tag{11}$$

where $\overline{s_j^{r}}'$ denotes the assessment of criteria by a decision-maker r.

It is again necessary to perform the averaging of matrix $\overline{s_j^{r}}$ by applying a geometric mean.

Step 7. Determine the coefficient $\overline{k_j}'$:

$$\overline{k_j}' = \begin{cases} = \overline{1} & if \quad j = n \\ 2 - \overline{s_j}' & if \quad j > n \end{cases}, \tag{12}$$

where n denotes a total number of criteria. Specifically, in this case, it means that the value of the last criterion is equal to the fuzzy number one.

Step 8. Determine the fuzzy weight $\overline{q_j}'$:

$$\overline{q_j}' = \begin{cases} = \overline{1} & if \quad j = n \\ \frac{\overline{q_{j+1}'}}{\overline{k_j'}} & if \quad j > n \end{cases}. \tag{13}$$

Step 9. Determine the relative weight of the criterion $\overline{w_j}'$:

$$\overline{w_j}' = \frac{\overline{q_j}'}{\sum_{j=1}^{n} \overline{q_j}'}. \tag{14}$$

Step 10. In order to determine the final weights of criteria, it is first necessary to perform the defuzzification of the fuzzy values $\overline{w_j}$ and $\overline{w_j}'$:

$$\overline{w_j}'' = \frac{1}{2}(w_j + w_j'). \tag{15}$$

Step 11. Check the results obtained by applying Spearman and Pearson correlation coefficients.

3. Input Parameters for the Evaluation of Criteria for the Implementation of HPC in Danube Region Countries

Determining the significance of the criteria relevant to the implementation of high-performance computing (HPC) was carried out in 14 countries. Countries that participated in this research were: Austria (AT), Bosnia and Herzegovina (BA), Bulgaria (BG), Croatia (HR), Czech Republic (CZ), Germany (DE), Hungary (HU), Moldova (MD), Montenegro (ME), Romania (RO), Serbia (RS), Slovakia (SK), Slovenia (SL), and Ukraine (UA). All countries were countries of Central and South-East Europe, forming the so-called Danube region. Challenges of the Danube region are, according to

Coscodaru et al. [42], its backwardness and substantial disparities between its well-off westernmost parts and the rest of the region. Most HPC infrastructure and knowledge are located in the West of the Danube region. Those countries are either EU member states, pre-accession countries, or EU neighboring countries. Regional nuances were considered in this paper. Countries, as well as the number of respondents and the number of correctly completed questionnaires, are shown in Figure 1.

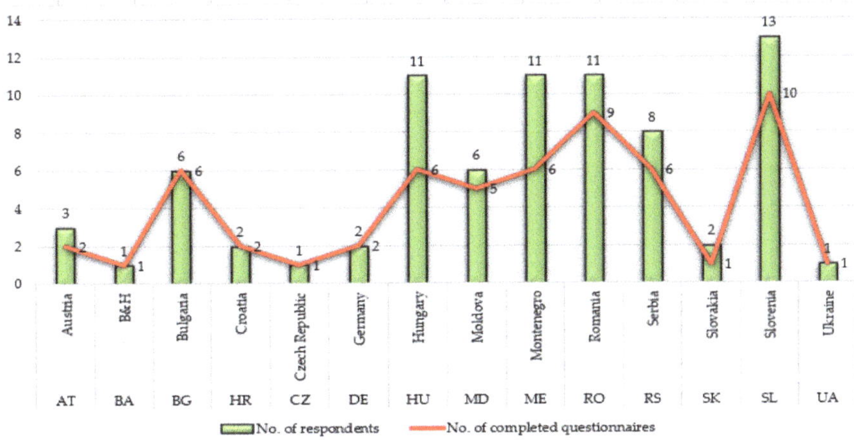

Figure 1. Countries, respondents, and correctly completed questionnaires.

The total number of questionnaires sent to the experts for evaluation was 78. Since it is a very complex area still insufficiently researched, the total number of correctly completed questionnaires was 58, which represents 74.36% of the total number of questionnaires sent. The largest number of decision-makers to whom the questionnaire was forwarded was from Slovenia (13), then from Hungary, Montenegro, Romania, 11 each. For other countries, the number of questionnaires forwarded was fewer, which can be seen in Figure 1. The data for the study were gained by the online survey by the InnoHPC [43] database conducted in 2017 and by Lapuh postdoctoral research in 2018. Experts using HPC employed in the research institutions and HPC providers were part of the survey. Research institutions were chosen by the convenience, which was based on the literature review, the authors' expert analysis on the potential usage of HPC in research organizations, and the online search. The experts were found on the basis of organizations' website research or their published or presented scientific works in the field of HPC. The snowball sampling was used in parallel: participants were asked to suggest partners from other institutions dealing with HPC.

Data in this section represent input parameters in the model. Figure 1 shows the total number of respondents (decision-makers) per each of 14 Danube region countries. The number of respondents was various and depended on the number of experts in each country (marked with green bar). Some of the decision-makers did not properly fulfil surveys or just gave equal assessments for all criteria. Such surveys were not useful for computation in our methodology, so we have excluded them. The number of correctly completed questionnaires is presented in Figure 1 and marked with a red line. For example, in Austria, three respondents were included in the research, but one of them did not fulfil the questionnaire in the proper way, so two questionnaires were entered in the further model.

Figure 2 shows that the performance of correctly completed questionnaires was 100% in the following countries: Bosnia and Herzegovina, Bulgaria, Croatia, Czech Republic, Germany, and Ukraine. Moldova, Romania, Slovenia, and Serbia had slightly lower percentages, 83%, 82%, 77%, and 75%, respectively. Countries with 50% or more correctly completed questionnaires were Slovakia (50%), Hungary and Montenegro (55%), and Austria (67%).

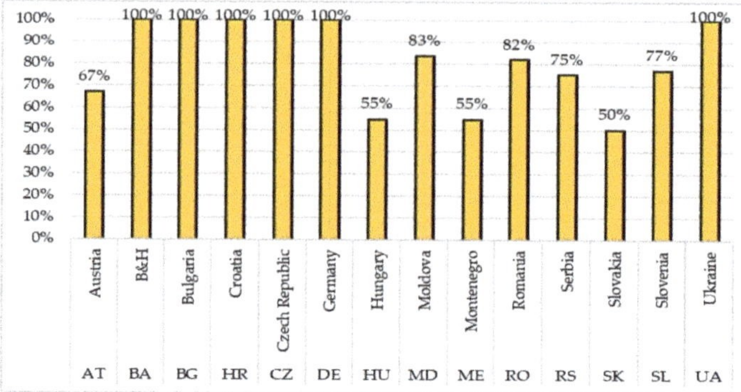

Figure 2. Percentage share of correctly completed answers.

The criteria, shown in Table 3, were evaluated using fuzzy PIPRECIA and fuzzy PIPRECIA-I.

Table 3. Criteria for evaluating the implementation of high-performance computing (HPC).

Designation	Criteria
C_1	Availability of free HPC infrastructure (e.g., having a sort of public funding)
C_2	Availability of commercial HPC infrastructure (where you have to pay for using it)
C_3	Availability of skilled human resources
C_4	Degree to which universities equip students with the necessary knowledge to work in HPC
C_5	Availability of competitive public funding (e.g., direct public funding, grants, awards, baseline funding)
C_6	Availability of private funding for R&D related to HPC
C_7	Degree of awareness about HPC benefits
C_8	Degree of science-industry cooperation related to HPC
C_9	Degree of industry-public authorities' cooperation related to HPC
C_{10}	Degree of science-public authorities' cooperation related to HPC
C_{11}	HPC prioritization in legislative documents and strategies

The point of interest of this study was how the experts employed in the research institutions or by the HPC providers dealing with HPC perceive the general degree of HPC development in the country they are working in. The HPC experts were asked to evaluate the HPC situation in the individual countries, regarding the HPC infrastructure, HPC competences, if curriculums contain gaining HPC skills, about the existence of project calls on gaining funding for HPC usage, about their perspective on the general awareness on advantages of using HPC in the country, if and how organizations in the selected individual countries cooperate related to HPC, and if individual countries encourage HPC usage in the legislative documents or strategies.

4. Results

A detailed calculation procedure performed on two respondents from Austria is shown in Tables 4–8. Attitudes of respondents using linguistic scales (step 2) shown in Tables 1 and 2 are presented in Table 4. It is important to note that when evaluating for fuzzy PIPRECIA, it was performed by Equation (7) and the initial criterion was the second one; therefore, cell C_1 in Table 4 is empty. The evaluation for the inverse fuzzy PIPRECIA method was performed by Equation (11), starting from the penultimate criterion C_{10}.

Table 4. Respondents' attitudes using a linguistic scale for fuzzy PIPRECIA and inverse fuzzy PIPRECIA.

	Fuzzy PIPRECIA						Inverse Fuzzy PIPRECIA					
	DM$_1$			DM$_2$			DM$_1$			DM$_2$		
Criterion	*l*	*m*	*u*	*l*	*m*	*u*	*l*	*m*	*u*	*l*	*m*	*u*
C$_1$							1.100	1.150	1.200	1.000	1.000	1.000
C$_2$	0.500	0.667	1.000	1.000	1.000	1.000	0.400	0.500	0.667	0.500	0.667	1.000
C$_3$	1.200	1.300	1.350	1.100	1.150	1.200	1.100	1.150	1.200	0.667	1.000	1.000
C$_4$	0.500	0.667	1.000	1.000	1.000	1.050	0.400	0.500	0.667	1.100	1.150	1.200
C$_5$	1.200	1.300	1.350	0.500	0.667	1.000	1.300	1.450	1.500	1.200	1.300	1.350
C$_6$	0.333	0.400	0.500	0.400	0.500	0.667	0.400	0.500	0.667	1.000	1.000	1.000
C$_7$	1.200	1.300	1.350	1.000	1.000	1.000	1.100	1.150	1.200	0.400	0.500	0.667
C$_8$	0.500	0.667	1.000	1.200	1.300	1.350	1.100	1.150	1.200	1.000	1.000	1.000
C$_9$	0.500	0.667	1.000	1.000	1.000	1.000	0.400	0.500	0.667	0.500	0.667	1.000
C$_{10}$	1.200	1.300	1.350	1.100	1.150	1.200	1.300	1.450	1.500	1.100	1.150	1.200
C$_{11}$	0.333	0.400	0.500	0.500	0.667	1.000						

Aggregating the values of decision-makers using the average mean yielded the values of s_j shown in Table 5. Subsequently, using Equation (8) in the third step, the values of k_j were obtained as follows:

$$\overline{k_1} = (1.000,\ 1.000,\ 1.000),$$

$$\overline{k_2} = (2 - 1.000,\ 2 - 0.833,\ 2 - 0.750) = (1.000,\ 1.167,\ 1.250).$$

In order to obtain the values of q_j, it was necessary to apply the fourth step, i.e., Equation (9).

$$\overline{q_1} = (1.000, 1.000, 1.000)$$

$$\overline{q_2} = \left(\frac{1.000}{1.250}, \frac{1.000}{1.167}, \frac{1.000}{1.000}\right) = (0.800, 0.857, 1.000)$$

$$\overline{q_3} = \left(\frac{0.800}{0.850}, \frac{0.857}{0.775}, \frac{1.000}{0.725}\right) = (0.941, 1.106, 1.379)$$

In order to obtain the value of w_j, the fifth step, i.e., Equation (10), is applied. The sum previously calculated for the values of q_j was $(6.287, 8.741, 17.158)$, obtained in the following way:

$$\begin{aligned}
&\sum_{j=1}^{n} \overline{q_j} = (6.287, 8.741, 17.158) \\
&= (1.000 + 0.800 + 0.941 + 0.753 + 0.655 + 0.401 + 0.455 + 0.387 + 0.310 + 0.365 + 0.230) = 6.287 \\
&= (1.000 + 0.857 + 1.106 + 0.948 + 0.932 + 0.602 + 0.708 + 0.696 + 0.597 + 0.770 + 0.525) = 8.741 \\
&= (1.000 + 1.000 + 1.379 + 1.415 + 1.715 + 1.210 + 1.467 + 1.778 + 1.778 + 2.453 + 1.962) = 17.158
\end{aligned}$$

$$\overline{w_1} = \left(\frac{1.000}{17.158}, \frac{1.000}{8.741}, \frac{1.000}{6.287}\right) = (0.058, 0.114, 0.159).$$

The following equation, $df_{crisp} = \frac{l+4m+u}{6}$, was then applied to perform the defuzzification of the values, as shown in the penultimate column of Table 5.

$$w_{1crisp} = \frac{0.058 + 4 \times 0.114 + 0.159}{6} = 0.112$$

Finally, the ranks for the obtained criterion values are shown (Table 5), which were further used to determine Spearman's correlation coefficient and determine the final ranks of the criteria.

Table 5. Details of the calculation carried out using fuzzy PIPRECIA.

	s_j			k_j			q_j			w_j			DF	Rank
	l	*m*	*u*	*l*	*m*	*u*	*l*	*m*	*u*	*l*	*m*	*u*		
C_1				1.000	1.000	1.000	1.000	1.000	1.000	0.058	0.114	0.159	0.112	5
C_2	0.750	0.833	1.000	1.000	1.167	1.250	0.800	0.857	1.000	0.047	0.098	0.159	0.100	7
C_3	1.150	1.225	1.275	0.725	0.775	0.850	0.941	1.106	1.379	0.055	0.127	0.219	0.130	1
C_4	0.750	0.833	1.025	0.975	1.167	1.250	0.753	0.948	1.415	0.044	0.108	0.225	0.117	4
C_5	0.850	0.983	1.175	0.825	1.017	1.150	0.655	0.932	1.715	0.038	0.107	0.273	0.123	3
C_6	0.367	0.450	0.583	1.417	1.550	1.633	0.401	0.602	1.210	0.023	0.069	0.193	0.082	11
C_7	1.100	1.150	1.175	0.825	0.850	0.900	0.445	0.708	1.467	0.026	0.081	0.233	0.097	8
C_8	0.850	0.983	1.175	0.825	1.017	1.150	0.387	0.696	1.778	0.023	0.080	0.283	0.104	6
C_9	0.750	0.833	1.000	1.000	1.167	1.250	0.310	0.597	1.778	0.018	0.068	0.283	0.096	9
C_{10}	1.150	1.225	1.275	0.725	0.775	0.850	0.365	0.770	2.453	0.021	0.088	0.390	0.127	2
C_{11}	0.417	0.533	0.750	1.250	1.467	1.583	0.230	0.525	1.962	0.013	0.060	0.312	0.094	10
Σ							6.287	8.741	17.158					

Using the methodology of inverse fuzzy PIPRECIA, the values shown in Table 6 were obtained. It is important to note that in the same way as previously described, these values were obtained by applying Equations (11)–(14). Assessment by decision-makers from the sixth step was previously shown in Table 4.

Step 7. Determining the coefficient $\overline{k_j}'$:

$$\overline{k_{11}}' = (1.000,\ 1.000,\ 1.000)$$

$$\overline{k_{10}}' = (2 - 1.350,\ 2 - 1.300,\ 2 - 1.200) = (0.650,\ 0.700,\ 0.800).$$

Step 8. Determining the fuzzy weight $\overline{q_j}'$:

$$\overline{q_{11}}' = (1.000, 1.000, 1.000)$$

$$\overline{q_{10}}' = \left(\frac{1.000}{0.800}, \frac{1.000}{0.700}, \frac{1.000}{0.650}\right) = (1.250, 1.429, 1.538)$$

$$\overline{q_9}' = \left(\frac{1.250}{1.550}, \frac{1.429}{1.417}, \frac{1.538}{1.167}\right) = (0.806, 1.008, 1.319).$$

In order to obtain the value of w_j, the ninth step, i.e., Equation (14), was applied. The sum previously calculated for the values of q_j was obtained in the following way:

$$\sum_{j=1}^{n} \overline{q_j}' = (7.520, 11.093, 17.833)$$
$$= (1.000 + 1.250 + 0.806 + 0.849 + 0.679 + 0.522 + 0.697 + 0.557 + 0.499 + 0.322 + 0.339) = 7.520$$
$$= (1.000 + 1.429 + 1.008 + 1.090 + 0.928 + 0.742 + 1.188 + 1.011 + 1.093 + 0.771 + 0.834) = 11.093$$
$$= (1.000 + 1.538 + 1.319 + 1.465 + 1.374 + 1.177 + 2.048 + 1.920 + 2.133 + 1.828 + 2.031) = 17.833$$

$$\overline{w_{11}}' = \left(\frac{1.000}{17.833}, \frac{1.000}{11.093}, \frac{1.000}{7.520}\right) = (0.056, 0.090, 0.133).$$

The following equation, $df_{crisp} = \frac{l+4m+u}{6}$, was then applied to perform the defuzzification of the values.

$$w_{11}'_{crisp} = \frac{0.056 + 4 \times 0.090 + 0.133}{6} = 0.092$$

Step 10. In order to determine the final weights of criteria, it was necessary to apply Equation (15). For example:

$$\overline{w_1}'' = \frac{1}{2}(0.112 + 0.098) = 0.105.$$

Table 6. Details of the calculation carried out using fuzzy PIPRECIA-I.

	$s_{j'}$			$k_{j'}$			$q_{j'}$			$w_{j'}$			DF	Rank
	l	*m*	*u*	*l*	*m*	*u*	*l*	*m*	*u*	*l*	*m*	*u*		
C1	1.050	1.075	1.100	0.900	0.925	0.950	0.339	0.834	2.031	0.019	0.075	0.270	0.098	6
C2	0.450	0.583	0.833	1.167	1.417	1.550	0.322	0.771	1.828	0.018	0.070	0.243	0.090	10
C3	0.883	1.075	1.100	0.900	0.925	1.117	0.499	1.093	2.133	0.028	0.098	0.284	0.118	3
C4	0.750	0.825	0.933	1.067	1.175	1.250	0.557	1.011	1.920	0.031	0.091	0.255	0.108	4
C5	1.250	1.375	1.425	0.575	0.625	0.750	0.697	1.188	2.048	0.039	0.107	0.272	0.123	2
C6	0.700	0.750	0.833	1.167	1.250	1.300	0.522	0.742	1.177	0.029	0.067	0.157	0.076	11
C7	0.750	0.825	0.933	1.067	1.175	1.250	0.679	0.928	1.374	0.038	0.084	0.183	0.093	8
C8	1.050	1.075	1.100	0.900	0.925	0.950	0.849	1.090	1.465	0.048	0.098	0.195	0.106	5
C9	0.450	0.583	0.833	1.167	1.417	1.550	0.806	1.008	1.319	0.045	0.091	0.175	0.097	7
C10	1.200	1.300	1.350	0.650	0.700	0.800	1.250	1.429	1.538	0.070	0.129	0.205	0.132	1
C11				1.000	1.000	1.000	1.000	1.000	1.000	0.056	0.090	0.133	0.092	9
Σ							7.520	11.093	17.833					

The columns labeled DF in Tables 5 and 6 contain the defuzzified weights of the criteria. On the basis of these values, the final weight of the criteria was calculated using Equation (15), as shown in Table 7 and Figure 3.

Table 7. The final weights of criteria obtained from Austrian respondents.

	wp	wp-i	Final wj	Rank
C1	0.112	0.098	0.105	5
C2	0.100	0.090	0.095	9
C3	0.130	0.118	0.124	2
C4	0.117	0.108	0.113	4
C5	0.123	0.123	0.123	3
C6	0.082	0.076	0.079	11
C7	0.097	0.093	0.095	8
C8	0.104	0.106	0.105	6
C9	0.096	0.097	0.097	7
C10	0.127	0.132	0.129	1
C11	0.094	0.092	0.093	10

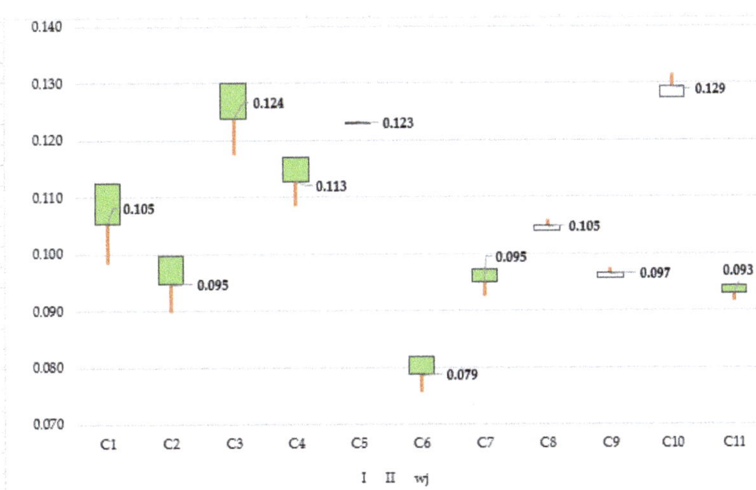

Figure 3. Final criteria weights according to DMs for Austria.

Spearman's correlation coefficient was 0.900, while Pearson's correlation coefficient was 0.957, which is a very high correlation of both the ranks and values of the criteria obtained by fuzzy and inverse fuzzy PIPRECIA.

As can be seen in Figure 3, the tenth criterion was the most significant—the degree of science-public authorities' cooperation related to HPC, which also had a minor variation in weights considering both approaches. A slightly lower value was obtained for C_3—availability of skilled human resources—which was in the second position, with a variation of 0.012. The third most significant criterion was C_5—availability of competitive public funding (e.g., direct public funding, grants, awards, baseline funding)—with a value of 0.123, which indicates that it had almost the same significance as C_3. Its variation or deviation was equal to zero, since it had an identical value by applying both approaches. In the fourth position is C_4—degree to which universities equip students with the necessary knowledge to work in HPC—whose value was 0.113 and a deviation of 0.09. The fifth most significant criterion was C_1—availability of free HPC infrastructure (e.g., having some sort of public funding)—and it had a value of 0.105. Other criteria were less significant, with values below 0.100.

The determination of criteria weights from the remaining 13 countries was carried out in a similar way. The obtained criteria weights according to countries are shown in Table 8 and Figure 4.

Table 8. Criteria weights according to countries.

	C1	C2	C3	C4	C5	C6	C7	C8	C9	C10	C11
AT	0.105	0.095	0.124	0.113	0.123	0.079	0.095	0.105	0.097	0.129	0.093
BA	0.180	0.084	0.142	0.117	0.097	0.082	0.099	0.099	0.099	0.121	0.085
BG	0.125	0.102	0.125	0.113	0.126	0.092	0.107	0.101	0.099	0.108	0.110
HR	0.116	0.114	0.103	0.101	0.110	0.091	0.100	0.099	0.098	0.115	0.114
CZ	0.127	0.128	0.148	0.101	0.102	0.083	0.098	0.100	0.101	0.122	0.085
DE	0.092	0.092	0.092	0.092	0.092	0.092	0.092	0.083	0.095	0.086	0.104
HU	0.144	0.106	0.116	0.105	0.105	0.094	0.109	0.104	0.094	0.113	0.113
MD	0.097	0.092	0.122	0.121	0.111	0.100	0.114	0.107	0.104	0.109	0.100
ME	0.111	0.112	0.107	0.113	0.107	0.104	0.101	0.107	0.105	0.111	0.108
RO	0.108	0.117	0.115	0.116	0.108	0.095	0.108	0.112	0.097	0.103	0.102
RS	0.119	0.114	0.124	0.133	0.104	0.092	0.097	0.097	0.094	0.109	0.095
SK	0.091	0.130	0.106	0.107	0.109	0.090	0.107	0.109	0.111	0.114	0.117
SL	0.119	0.125	0.108	0.098	0.103	0.095	0.106	0.108	0.105	0.123	0.121
UA	0.110	0.128	0.104	0.124	0.085	0.086	0.088	0.127	0.088	0.129	0.090

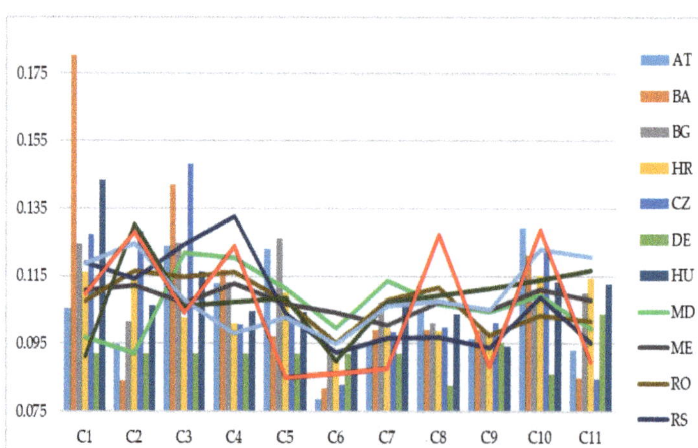

Figure 4. Criteria weights according to countries.

The significance and ranks of criteria according to countries are shown in Table 9 and Figure 5.

Table 9. Ranks of criteria according to countries.

	C_1	C_2	C_3	C_4	C_5	C_6	C_7	C_8	C_9	C_{10}	C_{11}
AT	5	9	2	4	3	11	8	6	7	1	10
BA	1	10	2	4	8	11	5	5	5	3	9
BG	3	8	2	4	1	11	7	9	10	6	5
HR	1	3	6	7	5	11	8	9	10	2	4
CZ	3	2	1	7	5	11	9	8	6	4	10
DE	3	3	3	3	3	3	3	11	2	10	1
HU	1	6	2	7	8	11	5	9	10	3	4
MD	10	11	1	2	4	8	3	6	7	5	9
ME	4	2	7	1	8	10	11	6	9	3	5
RO	7	1	3	2	6	11	5	4	10	8	9
RS	3	4	2	1	6	11	8	7	10	5	9
SK	10	1	9	8	6	11	7	5	4	3	2
SL	4	1	5	10	9	11	7	6	8	2	3
UA	5	2	6	4	11	10	9	3	8	1	7

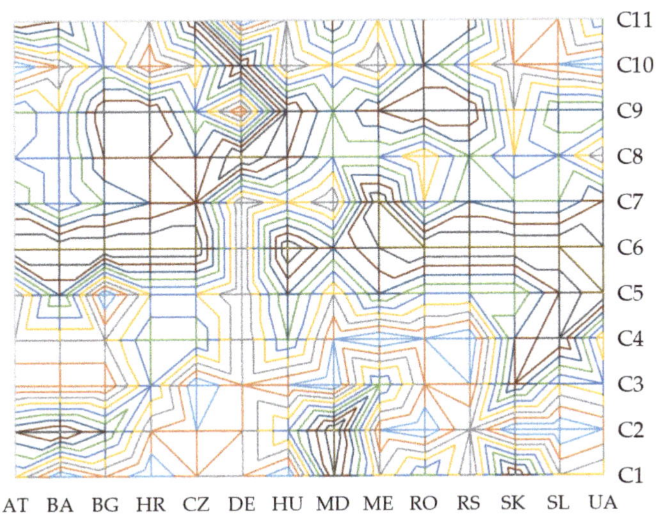

Figure 5. Significance of criteria according to countries.

As can be seen from Tables 8 and 9, and Figures 4 and 5, there was a big difference in the importance of the criteria according to the attitudes of respondents from different countries.

Based on the above, it can be seen that there were noticeable differences regarding the importance of the criteria in different countries. According to the Austrian respondents, four criteria, namely, C_3, C_4, C_5, and C_{10}, were the most influential, where criterion C_{10}—degree of science-public authorities' cooperation related to HPC—had the highest weight, that is, 0.13. It is also significant that the weights of these criteria were approximately the same, that is, the difference between weights of criteria C_{10} and C_4 was very small, at 0.01. According to the opinions of respondents from Bosnia and Herzegovina, criteria C_1, C_3, C_4, and C_4 were identified as the most influential, where criterion C_1—availability of free HPC infrastructure (e.g., having sort of public funding)—has the highest weight, that is, 0.18.

It should be noted that a higher difference between the weight of the most influential criterion and the weight of the least influent criterion was observed in this case, at 0.01. In the case of Bulgaria, five criteria were identified as the most significant, namely C_1—availability of free HPC infrastructure (e.g., having some sort of public funding), C_3—availability of skilled human resources, C_4—degree to which universities equip students with the necessary knowledge to work in HPC, C_5—availability of competitive public funding (e.g., direct public funding, grants, awards, baseline funding), and C_{11}—HPC prioritization in legislative documents and strategies.

Based on the research conducted in Croatia, the five most influential criteria could be identified, namely C_1, C_2, C_5, C_{10}, and C_{11}. The criteria C_1—availability of free HPC infrastructure (e.g., having some sort of public funding), and C_{10}—degree of science-public authorities' cooperation related to HPC, had approximately the same weights, while the weights of the criteria C_2, C_5, and C_{11} were slightly smaller. In the case of the Czech Republic, the criteria C_3—availability of skilled human resources, C_2—availability of commercial HPC infrastructure (where you have to pay for using it), C_1—availability of free HPC infrastructure (e.g., having sort of public funding), and C_{10}—degree of science-public authorities' cooperation related to HPC, were recognized as the most significant ones. In the case of Germany, the difference between the weight of the most and least influential criterion was very low at only 0.02, which is why all the criteria had approximately the same significance.

According to the respondents from Hungary, almost 50% of the total importance belonged to criteria C_1, C_3, C_{10}, and C_{11}, while according to the respondents from Moldova six criteria, namely C_3, C_4, C_5, C_7, C_8, and C_{10}, were singled out as the most influential. In the case of Montenegro respondents, the difference between the weights of the best and worst placed criteria was only 0.01, which is why almost all the criteria had approximately the same significance, and the most significant were the criteria C_4, C_2, C_1, C_9, and C_{11}, to which more than 60% of the weight belonged. According to Romanian respondents, the most important criteria were C_2—availability of commercial HPC infrastructure (where you have to pay for using it), C_4—degree to which universities equip students with the necessary knowledge to work in HPC, and C_3—availability of skilled human resources.

In the case of Serbia, the most important criteria were C_4, C_3, C_1, and C_2, while in the case of Slovakia, the most important criteria were C_2, C_{11}, and C_{10}. According to the opinion of the respondents from Slovenia, the most important criteria were C_2, C_1, C_{11}, and C_1, while according to the opinion of the respondents from Ukraine, the most important were criteria C_{10}, C_2, C_8, and C_4.

The number of occurrences of criteria from the first to eleventh position in the rankings is shown in Figure 6.

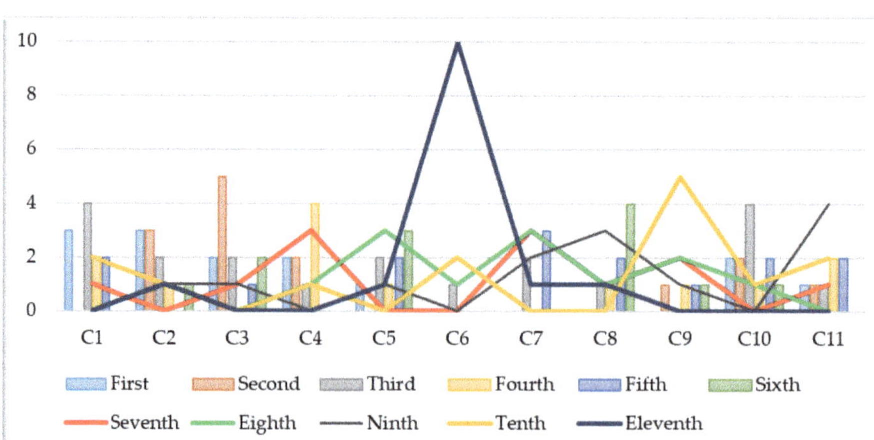

Figure 6. Number of occurrences of criteria from the first to eleventh position in the rankings.

From the above table, it can be seen that criteria C_1—availability of free HPC infrastructure (e.g., having some sort of public funding), C_2—availability of commercial HPC infrastructure (where you have to pay for using it), C_3—availability of skilled human resources, C_4—degree to which universities equip students with the necessary knowledge to work in HPC, and C_{10}—degree of science-public authorities' cooperation related to HPC were often highly ranked. Their importance was also confirmed by the high mean values of their weights.

Criteria C_1—availability of free HPC infrastructure (e.g., having some sort of public funding) and C_2—availability of commercial HPC infrastructure (where you have to pay for using it) had the highest number of appearances in the first position, three times each, while criteria C_3—availability of skilled human resources, C_4—degree to which universities equip students with the necessary knowledge to work in HPC, and C_{10}—degree of science-public authorities' cooperation related to HPC, each had two appearances in the first position. Criterion C_2—availability of commercial HPC infrastructure (where you have to pay for using it) is also interesting because it was once identified as the least important criterion. Criterion C_6—availability of private funding for R&D related to HPC, can be mentioned as the least influential criterion because it was identified as the least important criterion based on the attitudes of respondents from nine countries.

5. Conclusions

A supercomputer represents computer architecture of high performance, capable of processing large amounts of data in a very short time. Supercomputers can be used for solving a variety of very complex problems, including intensive calculations. HPC has a very important role in computer science and until now has been used for solving a variety of computationally intensive tasks in different areas, such as quantum mechanics, molecular modeling, and physical simulations. In addition, HPC has become indispensable in the field of cryptanalysis. Therefore, the role of supercomputers is becoming increasingly important. Considering the importance of the use of supercomputers, which is evident from this research, it can be concluded that it can have a significant impact on increasing sustainability from the cost aspect. This is achieved through the higher speed of solving complex problems and greater efficiency in executing all processes, as well as decision-making based on previously implemented algorithms. The speed and success of the application of information technology will become the basic factor of the strength and usability [44].

In this paper, research was carried out regarding the evaluation of criteria for the implementation of HPC in Danube region countries by using the fuzzy PIPRECIA method. Therefore, the determination of the significance of the criteria relevant to the implementation of high-performance computing was carried out in Danube region countries. The significance of the 11 criteria was determined using the fuzzy PIPRECIA method, based on the views of 58 successfully interviewed experts from 14 Danube region countries.

Main findings and results of the study indicate that the criterion named "degree of science-public authorities' cooperation related to HPC" was recognized as the most important, and its weight was 0.129. The second and the third influential criteria were the "availability of skilled human resources" and the "availability of competitive public funding", whose weights were 0.124 and 0.123, respectively. Finally, the less influential was criterion "availability of private funding for R&D related to HPC", whose weight was 0.079.

Author Contributions: Conceptualization, M.T. and L.L.; methodology, Ž.S. and D.S.; validation, D.K. and Ž.S.; investigation, M.T. and L.L.; data curation, M.T. and L.L.; writing—original draft preparation, Ž.S. and D.K.; writing—review and editing, D.S. All authors have read and agreed to the published version of the manuscript.

Funding: This article is based on the research funded by the Republic of Slovenia and the European Union under the European Regional Development Fund (Grant reference C3330-17-529007).

Conflicts of Interest: The authors declare no conflict of interest.

References

1. Ayres, D.L.; Cummings, M.P. Heterogeneous Hardware Support in BEAGLE, a High-Performance Computing Library for Statistical Phylogenetics. In Proceedings of the 46th International Conference on Parallel Processing Workshops (ICPPW), Bristol, UK, 14–17 August 2017; pp. 23–32.
2. Kessentini, M.; Narjès Bellamine, B.S.; Sami, S. Agent-Based Modeling and Simulation of Inventory Disruption Management in Supply Chain. In Proceedings of the International Conference on High Performance Computing Simulation (HPCS), Orleans, France, 16–20 July 2018; pp. 1008–1014.
3. Molyakov, A. China Net: Military and Special Supercomputer Centers. *J. Electr. Electron. Eng.* **2019**, *7*, 95–100. [CrossRef]
4. Wang, Y.; Jiang, J.; Zhang, J.; He, J.; Zhang, H.; Chi, X.; Yue, T. An efficient parallel algorithm for the coupling of global climate models and regional climate models on a large-scale multi-core cluster. *J. Supercomput.* **2018**, *74*, 3999–4018. [CrossRef]
5. Rodgers, A.J.; Pitarka, A.; Petersson, N.A.; Sjögreen, B.; McCallen, D.B. Broadband (0-4 Hz) Ground Motions for a Magnitude 7.0 Hayward Fault Earthquake with Three-Dimensional Structure and Topography. *Geophys. Res. Lett.* **2018**, *45*, 739–747. [CrossRef]
6. Volokhov, V.M.; Varlamov, D.A.; Zyubina, T.S.; Zyubin, A.S.; Volokhov, A.V.; Amosova, E.S. supercomputer simulation of nanocomposite components and transport processes in the Li-ion power sources of new types. In *Russian Supercomputing Days*; Springer: Cham, Germany, 2017; pp. 299–312.
7. Ezell, S.J.; Atkinson, R.D. The Vital Importance of High-Performance Computing to US Competitiveness. Information Technology and Innovation Foundation, 2016. Available online: http://www2.itif.org/2016-high-performance-computing.pdf?_ga=2.70059347.218455825.1581320442-155308587.1581320442 (accessed on 10 February 2020).
8. Cockrell, C.; An, G. Sepsis reconsidered: Identifying novel metrics for behavioral landscape characterization with a high-performance computing implementation of an agent-based model. *J. Theor. Biol.* **2017**, *430*, 157–168. [CrossRef] [PubMed]
9. Albu, A.; Precup, R.E.; Teban, T.A. Results and challenges of artificial neural networks used for decision-making and control in medical applications. *Facta Univ. Ser. Mech. Eng.* **2019**, *17*, 285–308. [CrossRef]
10. Gentzsch, W.; Lederer, H. *DEISA Mini-Symposium on Extreme Computing in an Advanced Supercomputing Environment, Parallel Computing: From Multicores and GPU's to Petascale 477 B*; Barbara Chapman University of Houston, USA Frédéric Desprez INRIA, France Gerhard R. Joubert TU Clausthal, Germany Alain Lichnewsky GENCI, France Frans Peters Philips Research, The Netherlands and Thierry Priol INRIA, France, Eds.; IOS Press: Amsterdam, The Netherlands, 2010.
11. High-Performance Computing and EuroHPC Initiative. Available online: http://europa.eu/rapid/press-release_MEMO-18-5901_en.htm (accessed on 27 June 2019).
12. Fayyad, U.; Piatetsky-Shapiro, G.; Smyth, P. The KDD Process for Extracting Useful Knowledge from Volumes of Data. *Commun. Acm* **1996**, *39*, 27–34. [CrossRef]
13. Chen, M.; Mao, S.; Liu, Y. Big Data: A Survey. *Mob. Netw. Appl.* **2014**, *19*, 171–209. [CrossRef]
14. Wu, X.; Zhu, X.; Wu, G.Q.; Ding, W. Data Mining with Big Data. *IEEE Trans. Knowl. Data Eng.* **2014**, *26*, 97–107.
15. Chen, H.; Chiang, R.; Storey, V. Business Intelligence and Analytics: From Big Data to Big Impact. *MIS Quarterly* **2012**, 1165–1188. [CrossRef]
16. Götz, M. Scalable Data Analysis in High Performance Computing. Ph.D. Thesis, Faculty of Industrial Engineering, Mechanical Engineering and Computer Science, Reykjavík, Iceland, December 2017.
17. Zahid, F. Network Optimization for High Performance Cloud Computing. Ph.D. Thesis, Faculty of Mathematics and Natural Sciences at the University of Oslo, Oslo, Norway, August 2017.
18. Turner, V.; Gantz, J.F.; Reinsel, D.; Minton, S. The digital universe of opportunities: Rich data and the increasing value of the internet of things. *Idc Anal. Future* **2014**, *16*.
19. De Mauro, A.; Greco, M.; Grimaldi, M.; Giannakopoulos, G.; Sakas, D.P.; Kyriaki-Manessi, D. What is big data? A consensual definition and a review of key research topics. In Proceedings of the AIP Conference Proceedings, Madrid, Spain, 5–8 September 2014; Volume 1644, pp. 97–104.
20. Arora, R. *Conquering Big Data with High Performance Computing*; Springer: Basel, Switzerland, 2016.

21. HPC Wire—Since 1987—Covering the Fastest Computers in the World and the People Who Run Them. Available online: https://www.hpcwire.com/ (accessed on 23 July 2019).
22. Fox, G.; Qiu, J.; Jha, S.; Ekanayake, S.; Kamburugmuve, S. Big data, simulations and hpc convergence. In *Workshop on Big Data Bench-marks*; Springer: Basel, Switzerland, 2015; pp. 3–17.
23. Rappa, M.A. The utility business model and the future of computing services. *Ibm Syst. J.* **2004**, *43*, 32. [CrossRef]
24. Top 500 Super Computer Sites. Available online: http://www.top500.org/ (accessed on 23 July 2019).
25. Collins, J.R.; Stephens, R.M.; Gold, B.; Long, B.; Dean, M.; Burt, S.K. An exhaustive DNA micro-satellite map of the human genome using high performance computing. *Genomics* **2003**, *82*, 10–19. [CrossRef]
26. Zheng, X.; Levine, D.; Shen, J.; Gogarten, S.M.; Laurie, C.; Weir, B.S. A high-performance computing toolset for relatedness and principal component analysis of SNP data. *Bioinformatics* **2012**, *28*, 3326–3328. [CrossRef]
27. Michalakes, J.; Dudhia, J.; Gill, D.; Henderson, T.; Klemp, J.; Skamarock, W.; Wang, W. The weather research and forecast model: Software architecture and performance. In Proceedings of the Eleventh ECMWF Workshop on the Use of High Performance Computing in Meteorology, Reading, UK, 25–29 October 2004; World Scientific: Singapore; pp. 156–168.
28. Sanbonmatsu, K.Y.; Tung, C.S. High performance computing in biology: Multimillion atom simulations of nanoscale systems. *J. Struct. Biol.* **2007**, *157*, 470–480. [CrossRef] [PubMed]
29. Jäkel, R.; Peukert, E.; Nagel, W.E.; Rahm, E. ScaDS Dresden/Leipzig—A competence center for collaborative big data research. *Inf. Technol.* **2018**, *60*, 327–333.
30. Suklan, J. Gap Analysis: HPC Supply and Demand. In *Go with the Flow: High Performance Computing and Innovations in the Danube Region*; Roncevic, B., Coscodaru, R., Fric, U., Eds.; Vega Press Ltd.: London, UK; Budapest, Hungary; Ljubljana, Slovenia, 2019; pp. 47–59.
31. Gigler, B.-S.; Casorati, A.; Verbeek, A. Financing the Future of Supercomputing, How to Increase Investment in High Performance Computing in Europe. European Investment Bank, 2018. Available online: https://www.eib.org/attachments/pj/financing_the_future_of_supercomputing_en.pdf (accessed on 2 April 2020).
32. European Commission, Horizon 2020. The EU Framework Programme for Research and Innovation: High-Performance Computing (HPC). Available online: https://ec.europa.eu/programmes/horizon2020/en/h2020-section/high-performance-computing-hpc (accessed on 2 April 2020).
33. Neelima, B. High Performance Computing education in an Indian engineering institute. *J. Parallel Distrib. Comput.* **2017**, *105*, 73–82. [CrossRef]
34. Volosencu, C. Properties of Fuzzy Systems. *WSEAS Trans. Syst.* **2009**, *8*, 210–228.
35. Chatterjee, P.; Stević, Ž. A two-phase fuzzy AHP-fuzzy TOPSIS model for supplier evaluation in manufacturing environment. *Oper. Res. Eng. Sci. Theory Appl.* **2019**, *2*, 72–90. [CrossRef]
36. Bozanic, D.; Tešić, D.; Milićević, J. A hybrid fuzzy AHP-MABAC model: Application in the Serbian Army–The selection of the location for deep wading as a technique of crossing the river by tanks. *Decis. Mak. Appl. Manag. Eng.* **2018**, *1*, 143–164. [CrossRef]
37. Volosencu, C. *Fuzzy Logic*; IntechOpen Ltd.: London, UK, 2020.
38. Stanković, M.; Stević, Ž.; Das, D.K.; Subotić, M.; Pamučar, D. A New Fuzzy MARCOS Method for Road Traffic Risk Analysis. *Mathematics* **2020**, *8*, 457. [CrossRef]
39. Stević, Ž.; Stjepanović, Ž.; Božičković, Z.; Das, D.K.; Stanujkić, D. Assessment of conditions for implementing information technology in a warehouse system: A novel fuzzy piprecia method. *Symmetry* **2018**, *10*, 586. [CrossRef]
40. Marković, V.; Stajić, L.; Stević, Ž.; Mitrović, G.; Novarlić, B.; Radojičić, Z. A Novel Integrated Subjective-Objective MCDM Model for Alternative Ranking in Order to Achieve Business Excellence and Sustainability. *Symmetry* **2020**, *12*, 164. [CrossRef]
41. Đalić, I.; Stević, Ž.; Karamasa, C.; Puška, A. A novel integrated fuzzy PIPRECIA–interval rough SAW model: Green supplier selection. *Decis. Mak. Appl. Manag. Eng.* **2020**, *3*, 126–145. [CrossRef]
42. Coscodaru, R.; Modic, D.; Roncevic, B. High-Performance Computing as a Tool of Transnational Innovation Policy. In *Go with the Flow: High Performance Computing and Innovations in the Danube Region*; Roncevic, B., Coscodaru, R., Fric, U., Eds.; Vega Press Ltd.: London, UK; Budapest, Hungary; Ljubljana, Slovenia, 2019; pp. 5–20.

43. InnoHPC Project. Available online: http://www.interreg-danube.eu/approved-projects/innohpc (accessed on 14 October 2019).
44. Tomašević, M. Application solution to the stage of aggregation method for assessing the quality of service provided. *Oper. Res. Eng. Sci. Theory Appl.* **2019**, *2*, 86–100. [CrossRef]

© 2020 by the authors. Licensee MDPI, Basel, Switzerland. This article is an open access article distributed under the terms and conditions of the Creative Commons Attribution (CC BY) license (http://creativecommons.org/licenses/by/4.0/).

sustainability

MDPI

Article

Sustainability of the Optimum Pavement Model of Reclaimed Asphalt from a Used Pavement Structure

Edis Softić [1], Veljko Radičević [2], Marko Subotić [3], Željko Stević [3,*], Zlatan Talić [4] and Dragan Pamučar [5]

[1] Faculty of Technical Sciences, University of Bihać, dr Irfana Ljubijankica bb, 77000 Bihać, Bosnia and Herzegovina; edis.softic@bih.net.ba
[2] Technical College of Applied Sciences, Uroševac (Leposavić), 24. Novembar bb, 38218 Leposavić, Serbia; veljkoradi@gmail.com
[3] Faculty of Transport and Traffic Engineering Doboj, University of East Sarajevo, Vojvode Mišića 52, 74000 Doboj, Bosnia and Herzegovina; marko.subotic@sf.ues.rs.ba
[4] Faculty of Polytechnic, University of Zenica, Fakultetska 1, 72000 Zenica, Bosnia and Herzegovina; zlatan.talic@divel.ba
[5] Military Academy, Department of Logistics, University of Defence, Pavla Jurisica Sturma 33, 11000 Belgrade, Serbia; dragan.pamucar@va.mod.gov.rs
* Correspondence: zeljkostevic88@yahoo.com or zeljko.stevic@sf.ues.rs.ba; Tel.: +387-66-795-413

Received: 13 February 2020; Accepted: 28 February 2020; Published: 3 March 2020

Abstract: This paper demonstrates and provides additional findings and instructions to produce new cold-recycled layers of pavement structures spatially and temporally sustainable. At the same time, recycled pavement structures have been enhanced with optimum amounts of new stone materials and binders made of cement and foamed bitumen. The subject of the research is based on the examination of recycled asphalt from surface and bituminous base courses of pavement structures for use on higher-type roads. The aim of the research is to model the process of producing recycled asphalt by cold recycling to optimize the process of influential parameters. In addition, one of the primary goals of the research is to demonstrate a sustainable way of producing new cold-recycled layers of pavement structures. The obtained results indicated the inevitability of the use of recycled material from pavement structures with the possibility of applying secondary and tertiary crushing of recycled mass, which depends on the type of layer for which the recycled material would be used. The research resulted in an optimum mixture variant of the stabilization layer of pavement structure that consists mainly of recycled material from a worn pavement structure improved with a relatively small amount of new aggregate with the addition of minimal stabilizers made of cement and foamed bitumen. The results showed that the optimum mixture variant of the stabilization layer is spatially and temporally stable. Additionally, the presented optimum variant of the stabilization layer enables sustainable development of road networks with minimum consumption of new natural resources.

Keywords: cold recycling; indirect tensile strength; sustainable binder materials

1. Introduction

In many developed countries, the period of rapid expansion of new road networks has reached its peak, existing road infrastructure is outdated, and a significant number of roads are approaching the end of their exploitation life. Pavement damages are the result of many different impacts on roads and occur in varying degrees of intensity. In many Western European and US countries, binding materials with improved viscoelastic range should play a leading role in sustainable road construction concerning the environmental aspects [1].

Cold recycling technology for the reconstruction of pavement structures is a rational and environmentally advanced approach in the system of modern road exploitation. As a raw material to produce new pavement layers of extremely good mechanical characteristics, the damaged layers of existing pavement structures are used, with the amount of binder lower than in conventional asphalt mixtures. It significantly reduces the need for the exploitation of natural resources and the use of new materials to develop a sustainable road network [2–4]. Compared to conventional pavement rehabilitation procedures, the usage of recycled layers reduces energy consumption during construction and indirectly reduces emission of greenhouse gases. In addition, by this procedure, the time required to execute the works was shortened, too. With all the above, the activities of removing, transporting, and depositing old pavement material are minimized or do not exist at all [2]. The recycling of asphalt obtained by the process of scraping surface layers of pavement structures consists of the selection of waste material, crushing of asphalt layers, removal of impurities, and mixing with natural aggregate. All built-in components must be removed and prevented from mixing with other materials such as concrete, earth, clay, glass, plaster, paper, or wood. The plants for recycling aggregates are not significantly different from plants that produce crushed aggregate from natural sites. Any variant that promotes the reuse of construction components and the recycling of scraps of construction materials would transform the method by which the construction sector currently operates and would create a vision for new business development [5].

The paper presents the current and improved state of recycling of current flexible pavement structures under different conditions with the aid of experimental research and mathematical modeling in a function of parameters related to the production of recycled asphalt. The scope of the research included recycling materials from existing flexible pavement structures on both main roads and roads with low traffic loads. The materials tested were obtained from road maintenance, reconstruction, and demolition.

2. Background of the Research

Testing and development of methods and equipment suitable for recycling materials from pavement layers is ongoing constantly. New technologies allow an increasing number of pavement materials to be reused. The best technologies allow nearly 100% of the extracted asphalt to be used to produce new asphalt concrete. However, this extremely efficient process has not become widely used yet. Still, it is a challenge for researchers and innovators to continue developing technological improvements to increase the overall efficiency of recycling methods [6–8].

As a useful strategy for preserving material in the road economy is a so-called "scale concept", which is widespread. According to this concept, illustrated in Figure 1, the efforts should concentrate on the use of the most durable pavement materials, for example, the construction of roads with long design life [6,7].

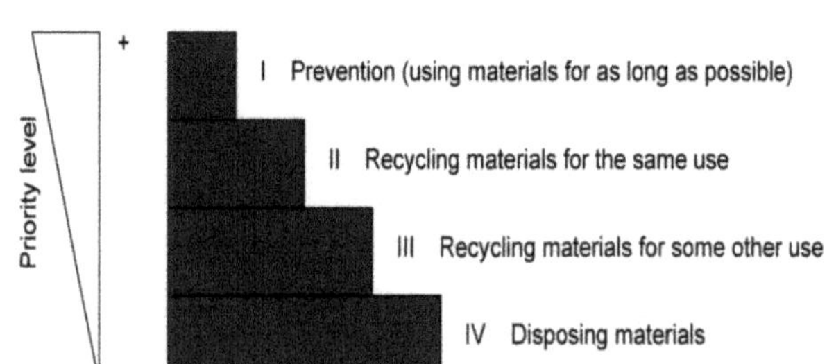

Figure 1. Scale concept.

However, individual components can be modified to obtain an optimally designed asphalt mixture. Road sections that would be recycled would result in minimal traffic disruptions, so that the road would be open for traffic as works progress. Additionally, a thorough examination of pavement material that is the subject of recycling is necessary, with the identified deficiencies of the recycled mixture being adjusted by certain additives in the mixture. Consequently, pavement materials have been studied for many years [9]. In a study [10], an experimental research of recycled asphalt with high percentages used in a new pavement structure is presented. The study characterizes the mechanical behavior of bituminous mixtures. Two semi-dense mixtures of 12 and 20 mm were analyzed as the maximum aggregate size containing 40% and 60% of the recycled mixture, i.e., (S-12 and S 20, in accordance with Spanish specifications). The authors [11] worked on an analytical approach to evaluate the performance of cold asphalt mixture of recycled layers. The experimental research was conducted by producing in parallel two series of cold-prepared asphalt concrete, one with new mineral aggregates and the other with recycled aggregate from a pavement structure. The study identified the influence of different variables on mixtures of both types of samples. Researchers [12] conducted research focused on the use of recycled roofing shingles in pavement layers. Tests showed that a percentage of 4–6% of shingles is the optimum proportion that fully meets the qualitative and quantitative characteristics of pavement structure. Author [13] implemented and tested the mechanical properties of foam-stabilized mixes for a recycled pavement mixture in detail. He experimentally presented and compared the mechanical properties of foamed bitumen created by two different gradations and eight types of bitumen from six different sources. The foam-stabilized mixes provided reasonable moisture resistance as the index of retained strength (IRS) exceeded 80 to 90% after 5 days of soaking. In addition, the same author in [14] analyzed the characterization of foamed bitumen, the quality and mechanical properties of mixes. In his research, the quality of foamed bitumen was assessed by two empirical parameters, namely half lifetime and expansion ratio. New methods were developed to characterize the foam quality that was specifically developed in that study. This researcher proposed a method that would use a so-called brook field.

For roads with high traffic loads, maintenance and reconstruction activities can lead to significant release of asphalt concrete, so recycling of extracted material becomes a high priority. For roads with low traffic loads, relatively small amounts of extracted asphalt are produced, and it is more important to properly recycle granular materials, as well as unbound or bonded substrate layers. Recycling of asphalt into new asphalt layers is of less importance due to the limited quantity and quality of the material so that its reuse as an unbound aggregate in the substrate or subbase course is more widespread [15–17]. Some other authors in [18] provided the analysis of results obtained by measuring biggest rutting depths determined on asphalt samples of different composition. The resistance to rutting was measured by a small-size device according to procedure B as specified in EN 12697-22 [19], and the composition of the sample was determined by a spatial model in which components were expressed as volume concentrations. Statistical analysis of the research results has revealed dependence of rutting depth on asphalt sample composition, which is valid for all test samples. In experimental work done by the authors [20] described and analyzed the results obtained by measuring tensile strength of asphalt samples of different composition, which discovered the dependence of the tensile strength on the asphalt sample composition. This test was measured by an indirect method, and the asphalt sample composition was defined according to a spatial model in which components were expressed by volume concentrations. The statistical analysis of the measurement results revealed the functional dependence of the asphalt sample tensile strength on its composition. Optimization of foaming and bitumen properties have been extensively elaborated by experts at study [15], where foamed bitumen was produced after adding water under regulated pressure (water pressure at a gauge of 1 Bar to be greater than air pressure) and the required temperature of bitumen.

3. Materials and Methods

3.1. Materials

Due to the specificity of recycling machinery and simplification of recycling process, the techniques of recycling have been defined at the international level. The recycling techniques involve defining sites and materials in flexible pavement structures, the way of recycling current pavement structures and conditions for using the same structure for new roads. Based on the set of rules, recycled asphalt can be used as:

1. cold-formed on site;
2. hot-formed on site;
3. milled and transported to a landfill with refining.

For cold recycling, granular material can be directly embedded, but with certain limitations in quantity due to inappropriate accuracy in its formula and stabilizers, which are determined by laboratory. The amount used on site is determined by practical tests and its percentage varies within the range of 20–50%.

Primary raw material in a cold recycling process is a recycled aggregate that is produced by mechanical treatment (milling and, if necessary, further crushing) of existing pavement layers in an on-site recycling process, optionally mixing with new stone material that is brought to the site of installation and added in order to improve grain-size distribution of whole recycled aggregate mixture. Stone aggregates in a mixture of milled recycled aggregate originally used to produce current pavement layers meet certain quality parameters. When recycling, it is not necessary to re-prove some of aggregate properties, such as shape index, crushing resistance, total sulfur content, and freezing and thawing resistance. The above does not apply if the aggregate is from another source or its origin is unknown [6,10].

3.2. Hypothesis

The initial hypothesis of the study Equation (1) is that the influence of milled asphalt in the process of producing bituminous base courses of pavement is the objective function of input variables:

$$Fc = optF = f\ (a_1, a_2, b_1, b_2, s_z, d_z, \rho, p_a, c) \qquad (1)$$

where

a_1 represents a recycled aggregate from asphalt layers,
a_2 represents a new aggregate to be added to the asphalt mixture,
b_1 represents bitumen extracted from the recycled asphalt mixture,
b_2 represents new bitumen to be added to the asphalt mixture,
s_z represents the indirect tensile strength of the specimen,
d_z represents the indirect tensile strength in the wet and dry state of the specimen,
ρ represents the density of the specimen,
p_a represents the characteristics of the asphalt surface,
c represents cement in a new added mixture.

By applying the method of optimization, the optimum solution of the process of making recycled asphalt mass by cold process will be determined. As a result, testing recycled material in road construction must be shifted from a simple testing of mixture components to a mixture performance test.

The process of producing bituminous asphalt base of road construction made of milled asphalt enhanced with hydraulic cement binders and cold-foamed bitumen is a complex technological process conditioned by several parameters, some parameters are variable and some are of constant size.

Their significance depends on the characteristics of the observed output size. Due to the specific composition of asphalt mixture, which is not common in practice and can be classified as rigid and flexible road surfacing in a certain way, it is practically impossible to determine reliably and theoretically its composition using analytical models. Accordingly, the application of experimental analysis of the investigated parameters is of great importance for improving the technological procedure for processing and installing the asphalt mixture made of recycled asphalt that is produced by cold process, with the aim of shortening the production process and its control, i.e., reducing production costs while maintaining the quality of asphalt mixture and installation in accordance with required standards.

The aim of the experimental research is to investigate influential parameters, such as: maximum percentage of milled asphalt from a pavement structure, cement binder admixture, foamed bitumen binder admixture, and the analysis of experimental data in terms of achieving the maximum tensile strength ratio of wet and dry specimens, maximum density ratio of dry and wet specimens without compromising quality and by reducing the overall costs of the production process. The purpose of the experiment plan is to generate mathematical models, a relevant equation (Second-degree polynomials) that describes the process. If parameters studied in the experiment are really those that affect the process, and the data obtained by the experiment of acceptable accuracy and precision, then it is possible to develop a model that describes the process in a credible way.

3.3. Sample Preparation

In this case, the main goal is to create a model that can reliably predict the amount of cement required (percentage) if the amount of bitumen is known (also percentage), the required tensile strength of the dry specimen and the ratio of tensile strength of the wet and dry specimen. By identifying influential factors, it determines the influential sizes and their changes significantly affect the desired (output) size. In this regard, the basic sizes that enter the process (mathematical model) are:

- Percentage amount of bitumen
- Percentage amount of cement

The quality of the recycled mixture produced by a cold recycling process is determined by the Indirect Tensile Strength (ITS) measured by a standard procedure on a laboratory-prepared test object. The laboratory test object is prepared according to the Marshall method in a mold of 100.6 mm (4 inch) with 2 × 75 strokes on a standard Marshall rammer. Thereafter, according to the procedure, a specimen is left for 24 h in a mold at room temperature (20–25 °C). The specimen is then removed from the mold and conditioned for 72 h at a constant temperature of 60 °C in an air circulation thermostat. The indirect tensile strength is measured on specimens thus prepared in dry (ITSs) and wet (ITSv) conditions. To test specimens in dry conditions, laboratory test objects prepared as described above are conditioned at a temperature of 25 ± 1 °C for at least one and not more than two hours prior to the test.

The following section provides equations used to calculate the properties of the laboratory test object prepared by compaction according to a modified Proctor method in the process of determining the quality of recycled mixture prepared by cold recycling procedure.

- The density of the wet specimen is defined by the Equation (2):

$$\rho_{AU(saturated)} = \frac{4 \cdot M_{AU}}{\pi \cdot d^2 \cdot h} \cdot 1000 \tag{2}$$

  where h represents the mean height of the specimen [m], while d represents the diameter of the specimen [m].
- The density of the dry specimen is defined by the following Equation (3):

$$\rho_{AU(dry)} = \frac{100}{\% \frac{masH_2O}{AU} + 100} \cdot \frac{4 \cdot M_{AU}}{\pi \cdot d^2 \cdot h} \cdot 1000 \tag{3}$$

where $\%_{mas}H_2O$ represents the mass percentage of water in the specimen, AU represents the surface of the asphalt specimen [m^2], h represents the mean height of the specimen [m], and d represents the diameter of the specimen [m].

The following section presents equations used to calculate the properties of the laboratory test object prepared by compaction according to the Marshall (2 × 75 strokes at 25 °C) method in the process of determining the quality of asphalt mixture prepared by cold recycling procedure.

The ITS of the dry and water-saturated specimen is obtained by the following Equation (4):

$$IVC = \frac{2 \cdot P}{\pi \cdot h \cdot d} \tag{4}$$

where h represents the mean height of the specimen [m], d represents the diameter of the specimen [m], P represents the fracture force [N].

- The retained tensile strength is obtained by the Equation (5):

$$ZVC = \frac{IVC_{(water-saturated\ specimen)}}{IVC_{(dry\ specimen)}} \tag{5}$$

The experiment examined the following parameters of producing laboratory specimens with the content of 85% of recycled aggregate in five levels:

- bitumen admixture B = 0.5; 1.0; 1.5; 2.0; 2.5. (%)
- cement admixture C = 0; 1.0; 1.5; 2.0. (%)

The parameters were tested in different combinations with different percentage amounts of cement, foamed bitumen, and fresh aggregate according to a standard procedure.

3.4. The Laboratory Testing Procedure

The laboratory testing procedure in the process of designing cold recycling consists of the following phases.

3.4.1. Preparation of Raw Materials for Designing the Composition of Recycled Mixture:

- For the purpose of sampling the pavement structure, determining the thickness and composition of individual pavement layers, testing grain-size distribution, plasticity index, moisture content, composition and properties of milled asphalt,
- Defining the composition design; combination of different aggregates (mixing ratio of milled asphalt to stone material, depth of intervention, type of binder),
- Testing the optimum moisture content by MPP (modified Proctor procedure) (4) and (5),
- Determining the binder content in milled asphalt by extraction of binder from heated milled asphalt mixture,
- Determining penetration on bituminous binder separated from milled asphalt mixture,
- Testing how much cement/lime (depending on plasticity index) is needed to be added,
- Testing how much bitumen binder (depending on physical and mechanical properties) is needed to be added.

3.4.2. Determination of the Grain-Size Distribution of Recycled Mixture (RM) as Specified in EN 933-1/A1:

- Determining the grain-size distribution of milled asphalt as specified in EN 933-1/A1 [21]
- Determining the grain-size distribution of a milled mechanically compacted unbound layer under asphalt layers as specified in EN 933-1/A1 [21]

- Determining the number of added fractions to improve the granulometry of a mixture of milled asphalt and unbound layers below the asphalt
- Determining the plasticity index for a mixture of recycled material as specified in ASTM D 422
- Determining optimum moisture by MPP as specified in EN 1097-5 and AASHTO T-180. The method for determining optimum moisture is defined in detail in a composition design procedure [22]

3.4.3. Determining Dry or Laboratory Reference Density and Optimum Humidity

Mixtures consisting of an unbound bituminous layer and a coarse aggregate are generally less sensitive to moisture in terms of determining their optimum humidity during compaction. Therefore, the optimum moisture content of these mixtures is determined by manual mixing, which can be used to deduce cohesion and workability of the mixture.

Optimum humidity is used to evaluate the behavior of hand-mixed specimens, while the dry laboratory reference volume is determined using the Proctor test as specified in EN 13286-2 at optimum humidity on one test specimen. In mixtures that are moisture sensitive when compacting, the optimum humidity is determined by the standard procedure as specified in EN 13286-2 [23].

The optimum moisture content can also be determined during the compaction of test specimens. If there is water release from the mold during the production of test specimens, without further increase in density and with water beginning to flow from the specimen, the humidity in the prepared specimen is close to optimum. When adding water to the mixture, the presence of water in bitumen emulsion must be taken into account. The real humidity of the mixture is determined according to EN 1097 [22].

3.4.4. Determining Binder Characteristics and Optimum Binder Amounts in Recycled Mixtures

For recycled mixtures stabilized with foamed bitumen, it is necessary to determine the optimum content of foamed bitumen and foaming bitumen characteristics (expansion coefficient and degradation half-life, optimum bitumen foaming temperature, and optimum amount of water for bitumen foaming). For designing the composition, we select the bitumen content at which the highest value of the ITS of wet specimen group is obtained. For recycled mixtures stabilized with bituminous emulsion: determination of the optimum content and characteristics of bitumen emulsion and stability if bitumen emulsion is used as a binder in combination with cement. For recycled mixtures stabilized with cement or other hydraulic binders: determination of the optimum content of cement binder [24]. The calculation of the cement proportion of the mixture is carried out using the Equation (6):

$$\frac{C}{mixture} = \frac{MC}{\rho_{ZS} \cdot D_{ZS}} \cdot 100\% \tag{6}$$

where % *C/mixture* represents the proportion of cement in an embedded, fully compacted base course [% (m/m)], *MC* represents the mass of cement spreading over m^2 expressed in [kg/ m^2], ρ_{ZS} represents the density (dry Proctor density) of the fully compacted base course expressed in [kg], whereas D_{ZS} represents the thickness of the base course made of recycled cement-stabilized pavement structure expressed in [m].

If, in laboratory, a certain optimum content of cement is expressed as a percentage, then the sum of cement and recycled material must be 100% by definition. The unit used to express the proportion of cement in the mixture is [% (m/m)]. If, in laboratory, a certain content of cement is expressed as a fraction (C/100RM) per 100 parts of recycled material, then the proportion of cement in the mixture (C/100RM) is calculated by the Equation (7) [16]:

$$\frac{C}{100 \cdot RM} = \frac{\%C/mixture}{1 - \frac{\%C/mixture}{100}} [kgC/100kgRM] \tag{7}$$

where *C/100RM* represents the proportion of cement per 100 parts of recycled aggregate; while % *C* represents the percentage of cement in the mixture.

3.4.5. Designing Test Specimens and Measuring the ITS and Water Resistance

Producing test specimens for measuring the ITS and water resistance. Six test specimens are prepared for one RM. Three of them are used to test the tensile strength of cross-section after seven days and three to test the decrease of ITS after seven days of air storage and seven days of storage in water. The preparatory specimen for testing the mass is made using the Equation (8): [6]

$$m = V \cdot p \cdot \left(1 + \frac{w}{100}\right) [g] \tag{8}$$

where V represents the volume of specimen in [cm^3]; p represents the volume density of dry soil in [g/cm^3]; w represents the humidity of the mixture in [%].

In a mixture with a cement binder or other hydraulic binders, bitumen emulsion with a cement binder or foamed bitumen, the specimen is stored in a mold at a temperature of (20 ± 2) °C for a period of (24 ± 6) h. The compacted specimen is then weighed with an accuracy of ± 1 [g] and its average height (average of four measurements) ± 1 mm is determined. The compacted specimen should have a height of (125 ± 20) mm. Holding it at humidity of 90% to 100% can be replaced by using airtight containers. The nurturing of test specimens and their testing is performed as follows:

- Seven days after being produced, three specimens are tested for cross-section tensile strength
- Seven days after being produced, three specimens are placed in water for further seven days at a temperature of (20 ± 2) °C, after which they are tested for water resistance.

When using slow-sticking binders, maturation time is prolonged.

3.4.6. Laboratory Testing of Specimen Density

Two specimens need to be tested as specified in EN 12697-6+A1, procedure D. The EN12697-6 +A1 standard is intended for asphalt produced by hot processing. Regarding the standard, weighing is possible with an accuracy of ± 1 g [25].

3.4.7. Determining the Maximum Density of Recycled Mixture

Determining the maximum volume of the mixture is carried out on two test specimens in water as specified in EN 12697-5+A1, procedure A. The EN12697-5+A1 standard is intended for asphalt produced by hot processing. Regarding the standard, weighing is possible with an accuracy of ± 1 [g] [25].

3.4.8. Calculation of Void Proportion

The proportion of voids is determined by the maximum density of the mixture and specimen density by the Equation (9):

$$V_m = \frac{P_m}{P_t} \tag{9}$$

where V_m represents the proportion of voids, p_m represents the maximum density of the mixture; p_t represents the specimen density.

3.4.9. Determining the ITS and Water Resistance

At this stage, there is a procedure as specified in EN 13286-42 satisfying the following conditions: (1) the specimen should be air-tempered; (2) the specimen should stand for four hours at a temperature of (15 ± 1) °C; (3) the pressure bar has an average width of 0.1 cm; (4) the specimen is tested with a constant pressure displacement of (50 ± 1) mm/min [26]. The results of testing specimen strength by indirect tension are calculated using the Equation (10):

$$R_t = \frac{2 \cdot F}{n \cdot H \cdot D} \tag{10}$$

where R_t represents the strength of the specimen at cross-section [kN/m]; F represents the maximum load force at specimen cracking [F]; H represents the specimen height [mm]; D represents the specimen diameter [mm].

The result is determined by an average of three specimens. If one of the values differs by more than 20% from the average value, the specimen must be discarded and the average is calculated based on two remaining values.

3.4.10. Determining the Compressive Strength and Resistance to Freezing and Water (EN 14227-1)

The maximum time from the moment of producing a fresh mixture to preparing a test specimen is one hour (in case of using foamed bitumen, three hours). If it is not possible to ensure timely delivery of fresh mixture to the laboratory for testing, the production of on-site test specimens must be ensured.

In mixtures stuck with cement binders and other hydraulic binders, the following features are controlled: (1) humidity, (2) compressive strength and freezing and water resistance (EN 14227-1) [27], (3) ITS, and (4) water resistance. In mixtures stuck with cement binders and bituminous emulsion or foamed bitumen, the following features are controlled: (1) humidity and (2) ITS and water resistance. In mixtures stuck with bitumen emulsion or foamed bitumen, the following features are controlled: (1) humidity, (2) ITS and water resistance, (3) void proportion.

3.4.11. Determining the Expansion Coefficient and Degradation Half-Life of Foamed Bitumen

The purpose of this test is to determine the characteristics and properties of foamed bitumen to ensure good-quality wrapping of aggregate grain. The experimental space is defined by the area of changes in basic factors. For the purposes of the experiment, i.e., modeling, tests have been carried out with factor changes that can be found in practical applications. It means that the percentage of bitumen changes at an interval of 0.5%–2.5%, and the percentage of cement changes at an interval of 0%–2%. For a degree model, one of the values (mean or one of boundary values) is determined by interpolation, i.e., extrapolation since tabular experimental data can hardly reach the condition that $fsr^2 = f_{max}xf_{min}$ required by this model. Based on the previous experience and research, for defining the dependence of achieved tensile strength for a given composition on a wet and dry specimen, the starting point is the first-order model that will be extended to a model with two-factor interaction. All input sizes (factors) are varied at two levels, implying minimum, maximum, and central-point values that may occur at the interval of interest. The mathematical model is developed for a two-level two-factor 2^k plan of the experiment that requires a minimum of $N = 2^k = 2^2 = 4$ (Equation (11)) of experimental point. The plan will be expanded with additional points (repeated measurements) at the center of the interval (central point). If such a model proves to be inadequate, there is a switch to a higher-order model, i.e., the previous model is extended by four symmetric points on the central axes ($\pm\alpha$) to improve accuracy. This central composite rotatable plan has in total:

$$N = 2^k + 2{\cdot}k + n_0 \tag{11}$$

i.e., $2^2 + 4 + 5 = 13$ experimental points where N represents the total number of experiments, k represents the number of varied factors, $k = 2$, while n_0 represents the number of repetitions of the experiment at the central point, for this model $n_0 = 5$.

The distance from the central point of supplementary measurement points ($\pm\alpha$) is selected in a way to maintain the symmetry and orthogonality of the model. For this case, the number of repeated measurements n_o and the value $\pm\alpha$ are interdependent and for $n_o = 5$, follows that $\alpha = 1.4142$. The values of the output size for that value of input sizes are obtained by extrapolation from the tables of experimental data. The specified distance value of supplementary experimental points from the central point of the plan (α) ensures that the plan remains orthogonal and the distribution of points remains optimum. Extension in the number of measurements at the central axes (axial points) and at the central point is an efficient replacement for the 3^k model, which requires more experiments is presented in Figure 2.

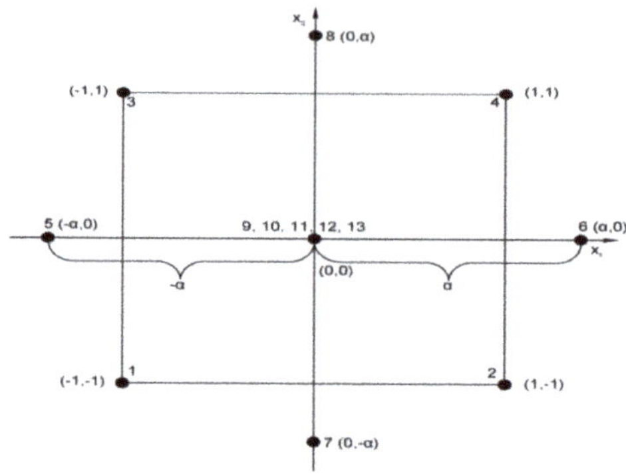

Figure 2. Position of experimental points in space.

Input sizes present the amount of bitumen B (%) and the amount of cement C (%), while the output size presents the ITS of dry specimens I (MPa). The state function of the process is defined as $z = f(B, C)$. A graphical representation of input-output sizes is given in Figure 3.

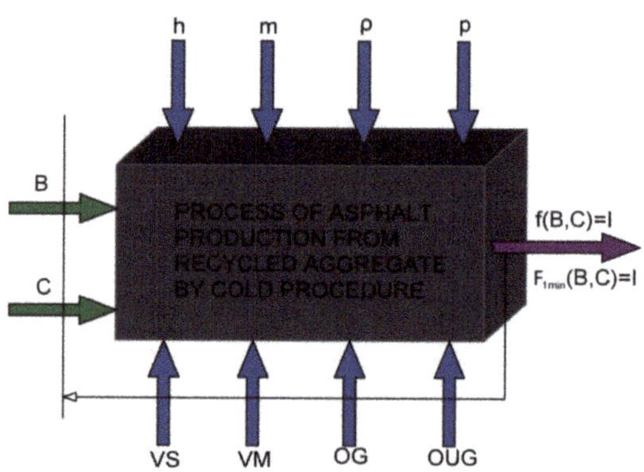

Figure 3. Input-output sizes of the technological process for the production of bituminous base asphalt mixture from recycled aggregate enhanced by hydraulic binders.

The independently variable sizes presented in Figure 3 are B (bitumen expressed in %) and C (cement expressed in %). The dependent variable shown in Figure 3 is I, which represents the ITS of dry specimens (MPa). Other sizes are specimen height (h), specimen mass (m), specimen density (ρ), fracture force (P), ITS of dry specimen (VS), wet/dry specimen strength ratio (VM), density ratio of dry to wet specimen (OG), the dependence of total specimen density on bitumen proportion (OUG). The influential factors are: B– x1; C – x2 and $\alpha = 1.4142$.

3.5. Experimental Plan

The space of coded coordinates is an area that is necessary for the design of an experiment matrix plan in which the coding of basic factors is required. Independent variables (x_1, x_2) in stochastic process modeling are usually transformed into coded forms (X_1, X_2), Equation (12):

$$-1 \leq X_i \leq 1; i = 1, 2, 3, \ldots \quad (12)$$

where: $Xi_{min} = -1$, for the value $f_i = f_{i\,min}$; $Xi_{sr} = 0$; for the value $f_i = f_{isr}$; $Xi_{min} = 1$; for the value $f_i = f_i max$.

The coded values of variables, regardless of their physical measuring unit, are expressed by two values, +1 and −1, which are obtained by using coding equations, so that the upper (maximum) level takes on the value +1 and the lower (minimum) −1. When there is an intermediate level, the encoded value is zero. In this case, the coding is performed using the Equation (13):

$$X_i = \frac{x_i - x_{0i}}{\frac{x_{i\,max} - x_{i\,min}}{2}} = \frac{x_i - x_{0i}}{\Delta x_i} \tag{13}$$

The medium part of the physical value is obtained by the Equation (14):

$$X_{0i} = \frac{x_{i\,min} + x_{i\,min}}{2} \tag{14}$$

where X_i represents the coded value of independent variables, i represents the number of independent variables (i = 1,2,3, ...), x_i represents the physical value of independent variables at the upper or bottom level, x_{0i} represents the physical value of independent variables at the center of the plan, i.e., zero-mean value, and Δx_i represents the boundary interval of the physical value of the variables from the midpoint to the maximum, i.e., minimum value of the variable.

The variation interval of factor I is half of the difference between the maximum (f_{imax}) and (f_{imin}) values of the *i-th* parameter (i = 1,2,3) and can be defined by Equation (15) as follows:

$$X_i = \frac{f_i - f_{0i}}{I_i} \frac{f_{i-} f_{0i}}{I_i};\ i = 1, 2, 3 \ldots \tag{15}$$

where:

$f_{i0} = \frac{f_{i\,max} + f_{i\,min}}{2}$ represents the mean value of the *i-th* factor;

$I_i = \frac{f_{i\,max} + f_{i\,min}}{2}$ represents the variation interval.

A desirable feature of each experiment is mutual independence of the evaluation of main effects and their interactions, which can be achieved by orthogonality and rotatability of the experiment. It is an orthogonal experiment if the sum of products of coded states for any two columns in the experiment matrix is equal to zero. A rotatable plan is a special form of central compositional plan that is very often applied in modeling and adaptive control in multi-variable processes. In addition to application properties, these plans also have optimality features that are suitable for optimizing selected processes and technologies. The rotatability of the central compositional plan of the experiment is achieved by adding the state of the experiment so that all states are equidistant from the center of the experiment, i.e., the rotatability depends on the axial distance α (the distance of the states at the axes from the center). The experiment is said to be rotatable if $\alpha = \sqrt[4]{F}$, where F represents the number of factor states (F = 2k in the case of a complete factor experiment).

In Table 1 are shown lists the physical and coded values of the percentages of bitumen and cement that added to the mix of recycled asphalt. Minimal values are coded as "−α", maximal values are coded with "+α", while average values are denoted as −1, 0 and +1.

Table 1. Levels and variations of input factors in the space of physical and coded coordinates.

Coded Values of Input Factors	Xi	−α	−1	0	+1	+α
Physical values	Bitumen [%]	0.75	1.0	1.6	2.2.	2.45
	Cement [%]	0.793	1.0	1.5	2.0	2.207

In Table 2 are shown the physical and coded values of the percentages of bitumen and cement and the variation of compressive strengths depending on the additive of binder (bitumen and cement). The number of performed experiments is 13 that shown in Table 2, also.

Table 2. Distribution of experimental points in the space of coded coordinates.

No. of Experiment a N	Coded Values								Results for Tensile Strength [MPa]	Results for Tensile Strength [MPa]	
	X_0	X_1	X_2	X_1X_2	X_1^2	X_2^2	%B	%C	σ_S	σ_M	
1	+1	−1	−1	+1	+1	+1	1.0	1	0.313	0.269	
2	+1	+1	−1	−1	+1	+1	2.2	1	0.325	0.300	
3	+1	−1	+1	−1	+1	+1	1.0	2	0.343	0.386	
4	+1	+1	+1	+1	+1	+1	2.2	2	0.380	0.457	
5	+1	$-\alpha$	0	0	α^2	0	0.75	1.5	0.359	0.289	
6	+1	$+\alpha$	0	0	α^2	0	2.45	1.5	0.408	0.357	
7	+1	0	$-\alpha$	0	0	α^2	1.6	0.79	0.272	0.292	Coded Coordinates
8	+1	0	$+\alpha$	0	0	α^2	1.6	2.21	0.346	0.480	
9	+1	0	0	0	0	0	1.6	1.5	0.376	0.326	
10	+1	0	0	0	0	0	1.6	1.5	0.413	0.348	
11	+1	0	0	0	0	0	1.6	1.5	0.374	0.358	
12	+1	0	0	0	0	0	1.6	1.5	0.421	0.313	
13	+1	0	0	0	0	0	1.6	1.5	0.396	0.336	

SQUARE MODEL: $y = b_0 + b_1x_1 + b_2x_2 + b_{11}x_{11} + b_{22}x_{22} + b_{12}x_1x_2$

4. Results and Discussion

4.1. Decoding a Dry Specimen Model and Wet Specimen Model

Table 3 shows the combined percentage of bitumen and cement carried out in the experimental study. For given percentage values of cement and bitumen, values of medium density and ITS and their mutual relations in the dry and wet state were obtained. The percentage value of bitumen is given by random selection of the experiment.

Table 3. Combined percentage of bitumen and cement carried out in the experimental study.

Percentage of Cement (%)	Percentage of Bitumen (%)	Average Density	ITS of the Specimen		
			Dry (MPa)	Wet (MPa)	Ratio Dry/Wet (%)
0.0%	0.5%	2.136	0.158	0.075	47.0
	1.0%	2.128	0.166	0.067	40.5
	1.5%	2.130	0.169	0.079	46.7
	2.0%	2.137	0.164	0.091	55.3
	2.5%	2.108	0.163	0.089	55.0
1.0%	0.5%	2.157	0.293	0.235	80.2
	1.0%	2.146	0.313	0.269	85.9
	1.5%	2.147	0.327	0.290	88.7
	2.0%	2.132	0.289	0.276	95.5
	2.5%	2.125	0.322	0.295	91.5
1.5%	0.6%	2.156	0.364	0.321	88.1
	1.1%	2.134	0.387	0.339	87.6
	1.6%	2.147	0.396	0.336	84.9
	2.1%	2.161	0.395	0.364	92.1
	2.6%	2.128	0.390	0.362	92.8
2.0%	0.7%	2.158	0.320	0.340	106.5
	1.2%	2.167	0.345	0.409	118.3
	1.7%	2.158	0.381	0.457	120.2
	2.2%	2.148	0.380	0.457	120.4
	2.7%	2.144	0.416	0.446	107.3

After the coefficients of the mathematical model have been determined in coded coordinates and the significance and adequacy of the model have been confirmed, we return to the space of physical coordinates by decoding. When the above Equations (13)–(15) are applied to the input factors in the opposite direction, returning from the coded to the physical coordinate system are obtained:

$$x_1 = \%B\ (bitumen\ percentage)\cdot\Delta x_1 = \frac{2.2-1.0}{2} = 0.6;\ x_{01} = \frac{2.2+1}{2} = 1.6;\ x_1 = \frac{x_1-1.6}{0.6}$$

$$x_2 = \%C\ (cement\ percentage)\cdot\Delta x_2 = \frac{2.0-1.0}{2} = 0.5;\ x_{02} = \frac{2+1}{2} = 1.5\cdot x_2 = \frac{x_2-1.5}{0.5}$$

Substituting the bold values (x_1 and x_2) in the mathematical model yields a mathematical model in a system of physical coordinates x_1, x_2 corresponding to the percentage of bitumen and cement, respectively. Inserting the numerical values, the following is obtained:

$$Y(X_1,X_2) = 0.396 + 0.0147\ X_1 + 0.0238\ X_2 + 0.0061\ X_1X_2 - 0.0077\ X_1^2 - 0.0449X_2^2$$

$$y(x_1,x_2) = 0.396 + 0.0147\left(\frac{x_1-1.6}{0.6}\right) + 0.0238\left(\frac{x_2-1.5}{0.5}\right) + 0.0061\left(\frac{x_1-1.6}{0.6} * \frac{x_2-1.5}{0.5}\right) - 0.0077\left(\frac{x_1-1.6}{0.6}\right)^2 + 0.0449\left(\frac{x_2-1.5}{0.5}\right)^2$$

After arranging, the mathematical model takes the following form:

$$y(x_1,x_2) = -0.125 + 0.0628x_1 + 0.5539x_2 + 0.02033x_1x_2 - 0.02147x_1^2 - 0.1796x_2^2$$

On the other hand, considering only the significant coefficients, we obtain that:

$$y(x_1,x_2) = -0.125 + 0.5539x_2 - 0.1796\,x_2^2$$

where x_1 represents the percentage of bitumen (%B), while x_2 represents the percentage of cement (%C).

The quality of the approximation is shown in the 3D representation of the model and experimental points. Figure 4 presents the curvature of the surface (functions of two independent variables) representing the second-order model, as well as the position of points in the experiment space on the basis of which the coefficients of the model have been calculated are clearly visible.

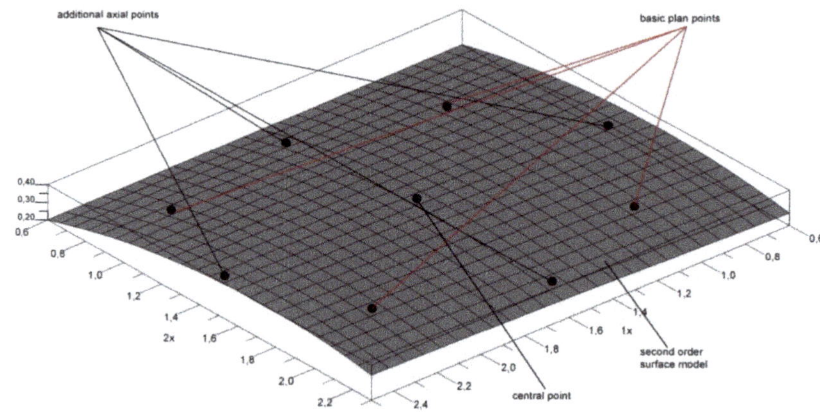

Figure 4. A second-order model for a dry specimen with the position of experimental points.

When encoding patterns are applied to the input factors in the opposite direction, equations for decoding, i.e., returning from the encoded to the physical coordinate system are obtained:

$$x_1 = \%B\ (bitumen\ percentage){\cdot}\Delta x_1 = \frac{2.2-1.0}{2} = 0.6;\ x_{01} = \frac{2.2+1}{2} = 1.6;\ x_1 = \frac{x_1-1.6}{0.6}$$

$$x_2 = \%C\ (cement\ percentage){\cdot}\Delta x_2 = \frac{2.0-1.0}{2} = 0.5;\ x_{02} = \frac{2+1}{2} = 1.5{\cdot}x_2 = \frac{x_2-1.5}{0.5}$$

Inserting the numerical values, the following is obtained:

$Y(X_1,X_2) = 0.3362 + 0.02477\ X_1 + 0.06748\ X_2 + 0.01\ X_1X_2 - 0.006975\ X_1{}^2 + 0.02452X_2{}^2$

$$y(x_1,x_2) = 0.3362 + 0.02477\left(\frac{x_1-1.6}{0.6}\right) + 0.06748\left(\frac{x_2-1.5}{0.5}\right) + 0.01\left(\frac{x_1-1.6}{0.6} * \frac{x_2-1.5}{0.5}\right) - 0.006975\left(\frac{x_1-1.6}{0.6}\right)^2 + 0.0245\left(\frac{x_2-1.5}{0.5}\right)^2$$

After arranging, the mathematical model takes the following form:

$$y(x_1,\ x_2) = 0.3188 + 0.0533x_1 - 0.21266x_2 + 0.0333x_1x_2 - 0.0194x_2^1 - 0.098x_2^2$$

On the other hand, considering only the significant coefficients, the following is obtained:

$$y(x_1,\ x_2) = 0.3188 + 0.0533x_1 - 0.21266x_2 - 0.098x_2^2$$

where x_1 represents the percentage of bitumen (%B), while x_2 represents the percentage of cement (%C). The representation of the previous model in 3D in the space of physical coordinates is given in Figure 5. It is graphically confirmed that the model adequately approximates the experimental data, Figure 5.

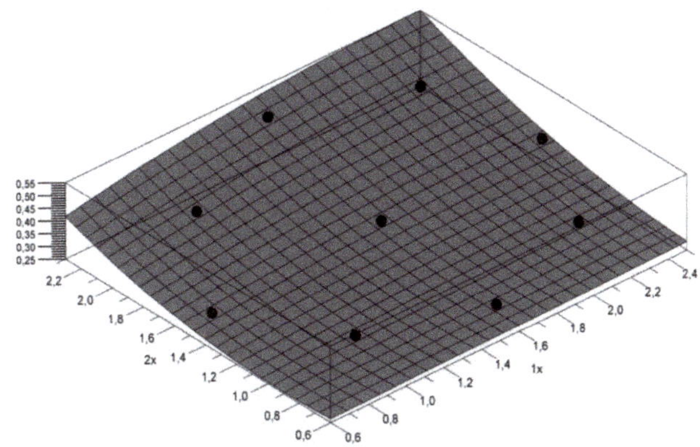

Figure 5. The 3D model representation of a wet specimen in the space of physical coordinates.

In addition to the above test, the multiple regression coefficient can be checked as a supplementary indicator to evaluate the adequacy of the model. In this case, applying the pattern to the results given in the previous tables, the following is obtained *R = 0.6552*:

The high value of the regression coefficient for the dry specimen is an indication that the previously developed model describes experimental data properly, i.e., confirms its adequacy. The coefficient of determination (R^2) confirms that a high percentage (95.5%) of the variability of the output size (% I) can be justified by changes in the input sizes.

Model prediction can be graphically compared with experimental data if for each performed experiment, the real (measured) value and the value obtained by computing is presented, Figure 6.

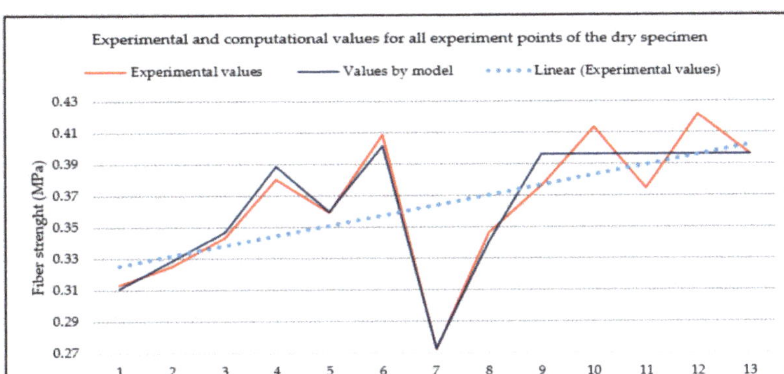

Figure 6. Comparison of experimental and computational values for all experiment points of the dry specimen.

A graphical representation of the relationship between experimental and model points is shown in Figure 6. The multiple regression coefficient in the case of a wet specimen has the following value *R = 0.986*.

The high value of the regression coefficient presented in Figure 7 is an indication that the previously developed model describes experimental data properly, i.e., its adequacy is confirmed.

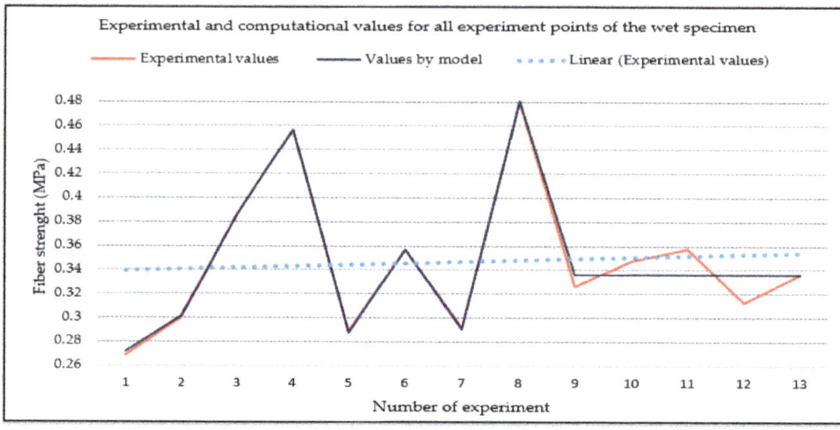

Figure 7. Comparison of experimental and computational values for all experiment points of the wet specimen.

4.2. Discussion on the Results of Testing the Optimum Content of Bitumen and Cement in a Stabilization Layer

It should be emphasized that in our area, asphalt layers (bituminous base and wearing) are used almost exclusively in all layers of pavement structures. In support of this fact, and based on a review of the total number of constructed roads at the Cantonal and Federal Roads Directorates, as well as the Public Enterprise "*Putevi Republike Srpske*", not a single section of road was constructed from recycled layers of existing and dilapidated pavement structures by cold recycling. In addition, it is important to note that most of regional and main roads, in terms of standard and technical regulations, do not meet the criteria of bearing and high-quality drainage since most of them are made with a bituminous base course, mainly of asphalt of limestone origin. According to the results obtained by testing the optimum content of bitumen and cement binders, it can be concluded that the condition on tested specimens is satisfactory according to standard parameters. The diversity of the total number of specimens leads us to the following discussion:

- The highest values of ITS of wet and dry specimens are provided by recycled asphalt specimens with a percentage of cement of 2% and a percentage of bitumen of: (0.7; 1.2; 1.7; 2.2; 2.7)%. In this specific case, the highest value of ITS is provided by a specimen made with a percentage of cement content of 2%, and bitumen of 2.2%, which have a strength ratio of 120.4 which is 50% higher than an allowed value of 80.

- The lowest value of ITS is given by the specimen made of recycled layer with a percentage of cement content of 0.5% and bitumen of 1%, which are 50% less than the allowed value.

- Recycled masses of pavement structure which contain a higher percentage of asphalt concrete (AB 11, AB 16) than an average specimen provides extremely high values of ITS with lower percentages of cement and bitumen binders.

- It has been clearly established that different types of recycled asphalt stabilization mixture produced from the same RM, with different values of grain-size distribution provide different values of ITS, as well as density of dry and wet specimens. At the same time, recycled stabilization mixtures of fine-grained structure provide better results.

- The type of binder used in a recycled stabilization asphalt mixture is crucial and affects the value of ITS when it comes to RMs of the same type. It is especially evident in a more detailed analysis of specimens with a cement content of 0.5–2% and a bitumen content of 0.5–2.5%.

- By the extraction of the obtained grain-size distribution of individually separated asphalt concrete layers that have to be recycled, it can be confirmed that there is no significant difference between the improved RM compared to the mixture originally sampled without improvement.

- Stabilization layers of recycled mixture improved with a certain percentage of cement and bitumen have a much rougher structure than basic asphalt layers (AB 16, BNS 16, BNHS 16, BNS 22, BNHS 22, BNS 32, etc.) since it has a large number of grains larger than 16 mm.
- These specimens show a significant increase in the value of slip resistance coefficient with respect to the basic types of asphalt layers and referring to this type of mixture, as well as the type and position of the base course.
- The value of slip resistance coefficient, as well as the roughness of the stabilization layer obtained by milling a worn pavement structure enhanced with cement and foamed bitumen binders, rises with increasing binders.
- The nature of aggregates in a recycled pavement layer greatly influences the value of the ITS of wet and dry specimens, their interrelationship with the same percentages of binder materials. This fact is significant and is based on studies that show that most deformations of pavement structures arise from longitudinal and transverse cracks of flattened ruts that indirectly depend on the type and origin of aggregates used in the structure. The values of ITS of stabilized layers of recycled pavement structure are on average 30% lower for limestone rocks compared to aggregates of eruptive origin that represent the basis of recycled pavement material.
- The selection of the type and proportion of hydraulic stabilizing agent determined by experimental testing based on plasticity index (IP < 10), cement with a percentage of 1% was selected.
- The search for the optimum content of bitumen in stabilized recycled material was performed by experimental testing during the formula preparation for cement and bitumen of stabilized granular material in the range of bitumen content of 0.5%–2.5%. For this prepared series of specimens, the ITS of dry and water-saturated specimens was measured. The following optimum proportions of foamed bitumen were identified to be in a range of 2.2%–2.4%.

4.3. Discussion on the Results on Formula Preparation with Different Content Percentages of Added Aggregate in the Total Mass of Recycled Mixture

The specificity of the milled material from the worn pavement structure intended to produce recycled stabilization mixture in bituminous base courses of the pavement is its limited use.

Besides the stated, the milled pavement material can be successfully used in surfacing. This application limits production spectrum to obtaining precisely defined percentages of aggregates 0–32 mm, which in a technological sense significantly influences the choice of reuse. Analyzing the grain-size distribution as the most important qualitative parameter, the plants in the world that are used exclusively in technological schemes for processing milled material from flexible pavement structure have faced with certain limitations when it comes to the selection of processing technology and economic justification of particular pavement layers. According to this analysis, and based on the test results, it has been determined that the amount and type of binder and grain-size distribution is a key and prevalent factor and parameter for a clear selection of an optimum formula with different content percentages of aggregate added to the total mass of the RM.

When searching for the optimum grain-size mixture of milled material, it was difficult to find such a grain-size mixture that would best meet the boundary curves defined by the Wirtgen Cold Recycling Manual for pavement structures.

It has been found and experimentally determined that a mixture of milled material from the pavement structure and added fraction of stone material have to be mixed in a ratio of 85:15 in both designed compositions. Such a mixture does not satisfy fully conditioned granulometry, but largely meets the recommendations for materials intended to be mixed with cement and foamed bitumen.

In a procedure for making a formula, the optimum moisture content of mixtures of grain-size material, which ranged from 4.8–5%, was experimentally determined.

4.4. Discussion on the Selection of Technology for Processing Recycled Mixtures from Pavement Structures in Terms of Economic Justification

In addition to the grain-size distribution of recycled aggregate from the pavement structure, the shape of aggregate grain, the addition of binder stabilizers and foamed bitumen, the main feature when selecting the technology for processing recycled stone from the pavement structure in terms of its technical soundness and usability in new base courses of pavement is the cost-effectiveness of the selected system. This is due to the fact that as a rule, good stone resistance to wear implies increased compressive strength and rock hardness.

As a result, more energy is required to process a recycled mass from the pavement structure, since the crush resistance is directly dependent on the physical and mechanical characteristics and binder of the recycled material from the pavement structure.

However, what is indicative and decisive in the assessment of technology selection when it comes to recycled pavement material is the question of the value of unit costs and costs due to the processing of worn-out pavement structures into a recycled mass that would be wholly or largely used in bituminous base courses of new pavement structures.

It can be very easily checked by analysis determined by direct measurement of specific propulsion energy consumption, which in this case has been obtained from the energy consumed and the quantities of usable recycled material obtained from the worn-out pavement structure. Specific propulsion energy consumption during the production of recycled material from the pavement structure on a Wirtgen WR 2500 S machine ranges from 30–40% of the price compared to the price of a new aggregate.

Accordingly, it was found that the specific consumption of propulsion energy in crushing and processing of worn-out pavement structures compared to the use of a new aggregate is two to three times lower. This indicator is directly dependent on the type of rock, the crushing plant used, its working principle, the type and origin of the material from the pavement structure that needs to be recycled and other parameters.

5. Conclusions

The mathematical model presented in the paper showed dependence, i.e., interaction between the retained and added amount of binder with regard to the percentage of milled (used) aggregate and the added new aggregate that was necessary in order to adjust defined limits of grain-size distribution curve. The paper presents that the improved model is optimum and not overly complex for implementation since it has been defined and tested in actual, realistic, and practical, conditions. There is always concern that many of laboratory tests used on traditional materials cannot predict a true effect of recycled materials.

Multiple regression coefficients for a dry and wet specimen are $Rs = 0.9552$, $Rm = 0.986$ proving good and mutual dependence of factors in the mathematical model, showing that this mathematical model describes the process with sufficiently proper accuracy. It should be remembered that the amount of resistance obtained is primarily related to the amount of cement binder added to the recycled material. In addition, tests have shown that adding more binding material does not mean better specimen characteristics since recycled materials treated with a cement binder tend to become stiffer and at the same time very prone to cracking. An increase in the amount of cement binder increases the stiffness of RM, which is associated with a decrease in fatigue resistance and an increase in shrinkage and consequent cracking. The amount of cracking resulting from the shrinkage associated with cement binders can be mitigated by the following procedures: (1) adding the amount of cement binder as closely as possible to the designed content in the previous mixture, (2) compacting the recycled mixture below the optimum humidity content or saturation of 75%, (3) by controlled drying of recycled material layer.

The results presented in this study have shown that recycled materials stabilized with bitumen provide a flexible material with better fatigue resistance properties when compared to those treated with cement stabilizing agents. Bitumen-stabilized mixtures are resistant to cracking due to shrinkage and can be opened for traffic before those mixtures treated only with cement. Additionally, FDR with a bitumen-stabilizing binder, using bitumen, makes a stabilized material rather than an asphalt layer, so it has a higher content of voids of between 10% and 20%. Bitumen-stabilized material tends to act partly as unbound material (stone aggregate) and partly as viscous-elastic material capable of withstanding repeated tensile stresses. Based on the results obtained, we can conclude that the initial hypothesis is fully confirmed.

Based on the newly acquired knowledge, experience, and the results obtained in this paper, it may be recommended to continue researching on the detection of quantitative and qualitative characteristics of materials from flexible pavement worn-out structures that have to be recycled and further improved with binders in order to use the same material in new layers of bituminous base and wearing pavement courses. It is also possible to improve further the technology of milling and processing of worn-out layers of flexible pavements in accordance with the indicators of the technical road parameters. In this regard, the direction of scientific research should be specifically focused on:

- the impact of the degree of grain wear and its suitability for fabrication in recycled pavement layers with respect to the exploitation lifetime of the road, depending on the quantity of newly added binder material
- the impact of binder (cement-foamed bitumen-filler-new fraction) on increasing use of recycled aggregate from worn-out pavement structures
- state of roads and study of economic and technical aspects of the development of transport network in Bosnia and Herzegovina
- the impact of grain size and shape on adding, i.e., consuming new binders in a recycled mass of pavement structure, as well as the impact of technology on processing and the shape of recycled aggregate
- combining other binder materials (cement, lime, fly ash, bituminous emulsion, geocrete, etc.) with the aim of maximizing the use of recycled materials from a pavement structure while meeting technical regulations and standards in the field of road construction
- dependence of tire consumption on the type of newly constructed road from a recycled mass if used for producing wearing courses of flexible pavement structure
- the possibility of using a recycled mass from the pavement structure to improve the substrate and ballast prism on railway tracks if it is economically and technically proven that the material does not meet the minimum conditions for reuse for road construction.

Author Contributions: Conceptualization, E.S. and V.R.; methodology, E.S. and Z.T.; validation, M.S., Ž.S.; formal analysis, E.S., V.R., and M.S.; writing—original draft preparation, E.S. and Z.T.; writing—review and editing, Ž.S. and D.P.; supervision, D.P. All authors have read and agreed to the published version of the manuscript.

Funding: This research received no external funding.

Conflicts of Interest: The authors declare no conflict of interest.

References

1. Nazarko, J.; Radziszewski, P.; Dębkowska, K.; Ejdys, J.; Gudanowska, A.; Halicka, K.; Kilon, J.; Kononiuk, A.; Kowalski, K.J.; Król, J.B.; et al. Foresight Study of Road Pavement Technologies. *Proced. Eng.* **2015**, *122*, 129–136. [CrossRef]
2. Puodziukas, V.; Svarpliene, A.; Braga, A. Measures for Sustainable Development of Road Network. *Trans. Res. Proced.* **2016**, *14*, 965–972. [CrossRef]
3. Kehagia, F. The implementation of sustainability in highway projects. *Int. J. Sustain. Dev. Plan.* **2009**, *4*, 61–69. [CrossRef]
4. Alkilani, S.G.; Jupp, J.R. Paving the Road for Sustainable Construction in Developing Countries: A Study of the Jordanian Construction Industry. *Australas. J. Constr. Econ. Build. Conf. Ser.* **2012**, *12*, 84–93. [CrossRef]
5. Iacovidou, E.; Purnell, P. Mining the physical infrastructure: Opportunities, barriers and interventions in promoting structural components reuse. *Sci. Total Environ.* **2016**, *557–558*, 791–807. [CrossRef] [PubMed]

6. Noesler, I.; Beckedahl, H. Adhesion Between Aggregates and Bitumen—Performance Testing of Compacted Asphalt Specimens by Means of the Dynamic Indirect Tensile Test. In Proceedings of the 3th Eurasphalt & Eurobitume Congress, Barcelona, Spain, 20–22 September 2000.
7. Martinez, F.; Angelone, S. Determination of Fracture Parameters of Asphalt Mixes by the Repeated Indirect Tensile Test. In Proceedings of the Sixth RILEM Symposium, Zürich, Switzerland, 14–16 April 2003; pp. 387–393.
8. De Freitas, E.; Pereira, P.; Picado-Santos, L.; Papagiannakis, A. Effect of Construction Quality, Temperature, and Rutting on Initiation of Top-Down Cracking. *Trans. Res. Rec. J. Trans. Res. Board* **2005**, *1929*, 174–182. [CrossRef]
9. Elizondo-Martinez, E.J.; Andres-Valeri, V.C.; Rodriguez-Hernandez, J.; Castro-Fresno, D. Proposal of a New Porous Concrete Dosage Methodology for Pavements. *Materials* **2019**, *12*, 3100. [CrossRef] [PubMed]
10. Valdés, G.; Pérez-Jiménez, F.; Miró, R.; Martínez, A.; Botella, R. Experimental study of recycled asphalt mixtures with high percentages of reclaimed asphalt pavement (RAP). *Constr. Build. Mater.* **2011**, *25*, 1289–1297. [CrossRef]
11. Giuliani, F.; Rastelli, S. An analytical approach to evaluate the performance of cold recycled asphalt mixtures. In Proceedings of the International RILEM Conference on the Use of Recycled Materials in Buildings and Structures, Barcelona, Spain, 8–11 November 2004.
12. Uzarowski, L.; Prilesky, H.; Berube, E.; Henderson, V.; Rizvi, R. Laboratory Testing of Vancuver HMA Mixes Containing Recycled Asphalt Shingles. In *Proceedings of the Development of New Technologies for Classification of Materials Session*; Soils and Materials Standing Committee: Vancouver, BC, Canada, 2010.
13. Saleh, M. Effect of Aggregate Gradation, Mineral Fillers, Bitumen Grade, and Source on Mechanical Properties of Foamed Bitumen-Stabilized Mixes. *Trans. Res. Rec. J. Trans. Res. Board* **2006**, *1952*, 90–100. [CrossRef]
14. Saleh, M. *Characterisation of From Bitumen Quality and the Mechanical Properties of Foam Stabilised Mixes*; University California at Davisi: Davisi, CA, USA, 2008.
15. *Wirtgen Cold Recycling Manual*, 1st ed.; Withgen GmbH: Windhagen, Germany, 2012.
16. Ramljak, Z.; Pejnović, V. Practical Mechanical Test for the Design and Control of Asphaltic Mixes. In *Yugoslav National Report*; Testing of Hydrocarbon Binders and Materials; RILEM Third International Symposium: Belgrade, Serbia, 1983; pp. 1–27.
17. Poulikakos, L.D.; Takahashi, S.; Partl, M.N. A Comparison of Swiss and Japanese Porous Asphalt Through Various Mechanical Tests. In Proceedings of the 4th Swiss Transport Research Conference, Ascona, Switzerland, 25–26 March 2004.
18. Strineka, A.; Putrić Brkić, J.; Sekulić, D. Utjecaj sastava na deformabilnost asfalta. *Građevinar* **2010**, *62*, 803–811.
19. EN 12697-22. *Bituminous Mixtures. Test Methods for Hot Mix Asphalt—Part 22: Wheel Tracking*; Asociación Española de Normalización y Certificación (AENOR): Madrid, Spain, 2008.
20. Ramljak, Z.; Strineka, A.; Šafran, K. Dependance of the Asphalt Sample Composition on its Tensile Strength. *Građevinar* **2005**, *57*, 141–150.
21. EN, BS. 933-1. *Tests for Geometrical Properties of Aggregates, Determination of Particle Size Ddistribution. Sieving Method*; European Standard: Brussels, Belgium, 2012.
22. EN 1097-5. *Tests for Mechanical and Physical Properties of Aggregates. Determination of the Water Content by Drying in a Ventilated Oven*; NSAI: Dublin, Ireland, 2000.
23. EN 13286-2. *Unbound and Hydraulically Bound Mixtures—Part 2: Test Methods for Laboratory Reference Density and Water Content—Proctor Compaction*; NSAI: Dublin, Ireland, 2010.
24. Al-Khateeb, G.; Buttlar, W.G. *Hollow-Cylinder Tensile Tester for Asphaltic Paving Mixtures. Mid Continent Transportation Symposium*; Center for Transportation Research and Education, Iowa State University: Ames, IA, USA, 2000; pp. 14–19.
25. EN 12697-6. *Bituminous Mixtures. Test Methods for Hot Mix Asphalt. Part 6: Determination of Bulk Density of Bituminous Specimens*; Spanish Association for Standardization and Certification (AENOR): Madrid, Spain, 2012.

26. EN 13286-42. *Unbound and Hydraulically Bound Mixtures—Part 42: Test Method for the Determination of the Indirect Tensile Strength of Hydraulically Bound Mixtures*; European Comity of Normalization: Brussels, Belgium, 2003.
27. EN 14227-1. *Hydraulically Bound Mixtures—Specifications—Part 1: Cement Bound Granular Mixtures*; British Standards Institute: London, UK, 2004.

© 2020 by the authors. Licensee MDPI, Basel, Switzerland. This article is an open access article distributed under the terms and conditions of the Creative Commons Attribution (CC BY) license (http://creativecommons.org/licenses/by/4.0/).

sustainability

MDPI

Article

Dynamic Gaming Case of the R-Interdiction Median Problem with Fortification and an MILP-Based Solution Approach

Yiyong Xiao, Pei Yang, Siyue Zhang, Shenghan Zhou *, Wenbing Chang and Yue Zhang

School of Reliability and System Engineering, Beihang University, Beijing 100191, China; xiaoyiyong@buaa.edu.cn (Y.X.); yangpei@buaa.edu.cn (P.Y.); zhang_sy@buaa.edu.cn (S.Z.); changwenbing@buaa.edu.cn (W.C.); Zhangyue1127@buaa.edu.cn (Y.Z.)

* Correspondence: zhoush@buaa.edu.cn; Tel.: +86-010-82316003

Received: 26 November 2019; Accepted: 7 January 2020; Published: 13 January 2020

check for updates

Abstract: This paper studies the cyclic dynamic gaming case of the *r*-interdiction median problem with fortification (CDGC-RIMF), which is important for strengthening a facility's reliability and invulnerability under various possible attacks. We formulated the CDGC-RIMF as a bi-objective mixed-integer linear programming (MILP) model with two opposing goals to minimize/maximize the loss from both the designer (leader) and attacker (follower) sides. The first goal was to identify the most cost-effective plan to build and fortify the facility considering minimum loss, whereas the attacker followed the designer to seek the most destructive way of attacking to cause maximum loss. We found that the two sides could not reach a static equilibrium with a single pair of confrontational plans in an ordinary case, but were able to reach a dynamically cyclic equilibrium when the plan involved multiple pairs. The proposed bi-objective model aimed to discover the optimal cyclic plans for both sides to reach a dynamic equilibrium. To solve this problem, we first started from the designer's side with a design and fortification plan, and then the attacker was able to generate their worst attack plan based on that design. After that, the designer changed their plan again based on the attacker's plan in order to minimize loss, and the attacker correspondingly modified their plan to achieve maximum loss. This game looped until, finally, a cyclic equilibrium was reached. This equilibrium was deemed to be optimal for both sides because there was always more loss for either side if they left the equilibrium first. This game falls into the subgame of a perfect Nash equilibrium—a kind of complete game. The proposed bi-objective model was directly solved by the CPLEX solver to achieve optimal solutions for small-sized problems and near-optimal feasible solutions for larger-sized problems. Furthermore, for large-scale problems, we developed a heuristic algorithm that implemented dynamic iterative partial optimization alongside MILP (DIPO-MILP), which showed better performance compared with the CPLEX solver when solving large-scale problems.

Keywords: facility location; *r*-interdiction median problem with fortification (RIMF); cyclic dynamic equilibrium game; mixed-integer linear programming (MILP)

1. Introduction

The reliability and security of critical infrastructures is vital to the stability of a country's social system. The growing threat of international terrorism may imperil facilities and cause massive loss to people's lives and properties. Such man-made disasters often seriously damage facilities, resulting in general or partial impacts on the normal order of society, leading to large-scale supply shortages of products or service interruptions. Some attacks will affect the sustainability of key facilities, making it difficult to ensure the normal supply, so it is necessary to consider the location of facilities under attack. In practice, there may be multiple rounds of confrontation between the attackers and defenders, so dynamic attack defense issues need to be considered to ensure the long-term sustainability of the facilities.

Terrorists always seek out the most critical facilities to conduct attacks and maximize the loss of the social system. To strengthen the security and reliability of critical facilities, facility locations should be well planned and fortified, and coping measures, such as backup systems and service reassignments, should be prepared for in advance of an attack. Our problems fall into the class of an incapacitated facility location, which is a variant of the p-median problem first proposed by German scholar Alfred Weber [1]. In the paper of Huizhen [2], they further study the mathematical nature of the semiLagrangian relaxation (SLR) applied to solve the un-capacitated facility location (UFL). On this basis, SLR for UFL is improved in theory, and the way to improve its computing capability is discussed. Monabbati [3] use their proposed sub-additive dual ascent procedure to find an optimal sub-additive dual function based on Klabjan's generator sub-additive function to solve the so called uncapacitated facility location problem (UFLP). Glover [4] proposes a simple multi-wave algorithm for solving the uncapacitated facility location problem (UFLP) to minimize the combined costs of selecting facilities to be opened and assigning each customer to an opened facility in order to meet the customers' demands. On the basis of an incapacitated facility location problem with interdiction and fortification, the r-interdiction median problem with fortification (RIMF) occurs. In the existing literature, node interdictions have been divided into two groups, i.e., r-interdiction median models (RIM) and r-interdiction covering models (RIC). The RIM model was originally formulated by Church and Scaparra [5], where interdiction strategies seek p facilities that have already been located. The model determines a subset of r facilities, and if these facilities are lost, the impact on the average service distance or the total weighted distance from the customer is at its greatest. In the RIC model, the objective is to specify r facilities of p existing facilities, which, when removed, results in a maximum drop in coverage. Church and Scaparra [6] then put forward the interdiction median problem with fortification (IMF) model, which involves finding a subset of q facilities in p different locations of the supply or service facility; when these q facilities harden, they provide the best protection from a subsequent optimal r-interdiction strike. Church and Scaparra [7] propose the r-interdiction median problem with fortification (RIMF) model, where the leader and follower respectively conduct interdiction and fortification, assuming that one side has complete information about the other side. Figure 1 below shows the study process from the p-median problem to the RIMF problem.

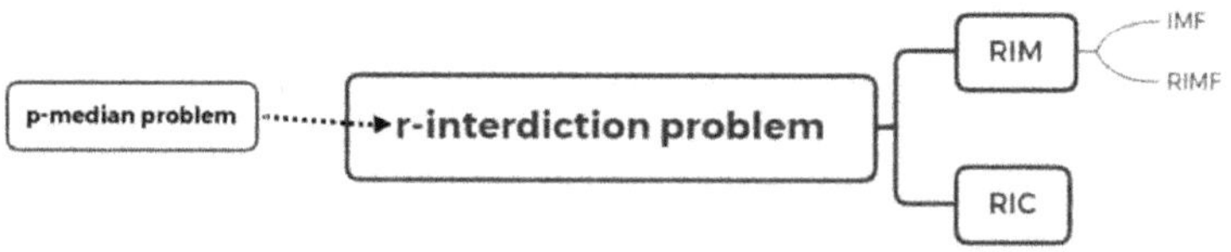

Figure 1. The study process from the p-median problem to the r-interdiction median problem with fortification (RIMF) problem.

A number of studies on the r-interdiction problem in the literature have strived to develop more reliable and invulnerable systems. Snyder [8] presents a stochastic location model with risk pooling (LMRP) to optimize location, inventory, and allocation decisions under random parameters described by discrete scenes. Scaparra et al. [9] present an optimization modeling approach to allocate protection resources among a system of facilities so that the disruptive effects of possible intentional attacks to the system are minimized. In the work of Aksen et al. [10], the authors elaborate on a budget-constrained extension of the RIMF model, where the objective function is to find the optimal allocation of protection resources to a given service system consisting of p facilities such that the disruptive effects of r possible attacks are minimized. Liberatore et al. [11] present the stochastic r-interdiction median problem with fortification (S-RIMF), where the model takes into consideration a random number of possible losses. Zhu et al. [12] studies the r-interdiction median problem with probabilistic protection, assuming that defensive resources are allocated based on the degree of importance of the service facility. Medal et al. [13] integrate the facility location and the facility reinforcement decisions into the formulation and assume that the decision-maker is risk averse and thus interested in mitigating against the facility disruption scenario with the greatest consequences. Mahmoodjanloo [14] puts forward a tri-level model under the defender–attacker–defender framework based on leader–follower games for facility protection against disturbance in an r-interdiction median problem. Mahmoodjanloo [14] focuses on reducing the effect of intentional attacks, in which facilities are located and strengthened within a limited budget. Sadeghi [15] presents a new formulation and solution method for the partial fortification and interdiction of a tri-level shortest path problem which extends the existing network interdiction models to a more practical environment. Zheng et al. [16] present an exact approach to solve the r-interdiction median problem with fortification. Their methods include a greedy heuristic and an iterative algorithm, solving a set cover problem iteratively to ensure the best solution upon termination. Khanduzi [17] implements two novel approaches to solve the multi-period interdiction problem with fortification. Roboredo [18] proposes a branch-and-cut algorithm for the RIMF problem, which is the best exact algorithm found. Furthermore, Roboredo [19] also put forward a branch-and-cut algorithm for the r-interdiction covering problem with fortification (RICF) problem, which is faster in solving large instances compared with the exact method found in the literature. Biswas and Pal [20] propose an interesting hybrid goal programing model and genetic algorithm in a fuzzy environment. Barma et al. [21] present a novel linear programming(LP) model with antlion optimization algorithm for multi-depot vehicle routing problem(VRP). In conclusion, in the previous literatures, there is almost no multi-round attack defense confrontation. Most papers generally consider a bi-objective function for at most a three level RIMF problem. In this paper we present, for the first time, a cyclic equilibrium gaming case of the r-interdiction median problem with fortification based on the study of Dong [22]. We consider it a dynamic Stackelberg game [23], involving two non-cooperative, fully rational players to protect (or attack) the facility as much as possible. Each side makes the most optimal decision based on the other's decision. We present a bi-objective mixed integer linear programming model to formulate the cyclic dynamic gaming case of the r-interdiction median problem with fortification (CDGC-RIMF). Using the cyclic algorithm, the computer generate two decision packages for both the attacker and defender until the equilibrium was cyclically reached. For the large-sized problem, we developed a heuristic algorithm based on the partial optimization strategy [24,25] to efficiently solve the problem with near-optimal solutions. Contributions of the paper are outlined as follows.

- This article first brings up the CDGC-RIMF problem, which considers cyclic decision periods/phases in large-scale scenes with two opposite objectives.
- This article constructs the mixed-integer linear programming (MILP) model so as to solve it by an efficient and easy way through the CPLEX solver.
- For the large-size scenes, we developed a heuristic algorithm that implemented dynamic iterative partial optimization alongside MILP (DIPO-MILP).
- Our model put the wise decision of the attacker into consideration. The objective of attacker was to consider the optimized rearrangement of the defender faced with the destruction to achieve the worst loss possible.
- Our algorithm generated two decision packages for both the attacker and defender, thereby reaching an optimal cyclic equilibrium.
- In the cyclic confrontation process, each side made the most optimal confrontation plan until the balance of the cyclic confrontation was reached. Through this model, the decision package of the two sides could be produced. For the protector, it could predict the other side's decision in advance (the protector knew that the attacker would make the most optimal confrontation plan) and then make the decision for the next step. It could help the facility manager make wise

decisions for site selection and defense in advance, so as to ensure the long-term sustainability of the facility.

The remaining paper is organized as follows. In Section 2, we formally describe the CDGC-RIMF model. In Section 3, we present the notation, the property of the cyclic bi-objective MILP model, and a partial optimization algorithm for large-scale problems. In Section 4, we present some computational experiments, including the small- and large-scale problems, and analyze the protection effect to make up a fortified facility. Finally, in Section 5, we summarize our work and present suggestions for future works.

2. Problem Description and Formulation

The problem involved a general service supply system composed of several facilities and a number of demand nodes that received goods or services from their nearest service sites. A number p of facilities were built out of a set of potential locations sites, denoted as F ($j \in F$). Notation N, indexed by i, represents the set of demand nodes. Each demand node i, $i \in N$ was associated with a demand a_i. The distances between the facilities and the demand nodes were known parameters, denoted as d_{ij}, where $i \in N$ and $j \in F$. The service cost was expressed as the sum of total demands weighted by the distances to their closest facilities. Each candidate site was characterized with a fixed cost f_j to build a facility and a fixed fortification cost g_j to protect the built facility. The attacker (follower) intended to carry out the most devastating attack possible and was assumed to have the ability to destroy the maximum r number of facilities after considering all possible design plans from the designer. The designer, who was able to protect g facilities from being attacked, tried to minimize the establishment and protective cost as well as system loss with consideration of all possible attacks by the attacker.

We formulated the cyclic dynamic gaming case of the r-interdiction median problem as a bi-objective MILP model, as shown below. The sets, parameters, and decision variables used in the mathematical formulation were introduced in advance, as follows.

Sets:

N: Set of demands;

F: Set of candidate sites.

Parameters:

i: Index of demand that $i \in N$;

j: Index representing candidate site that $j \in F$;

a_i: Demand of node i;

m: Number of potential sites;

p: Number of built (located) facilities;

r: Number of interdicted facilities;

q: Number of defense facilities, $q < p$, $r < p-q$;

g_j: Setup cost at site j;

h_j: Fortification cost at site j;

d_{ij}: Distance from demand i to site j;

M: A large number.

Variables:

x_j: $\begin{cases} 1, \text{if site j is selected to operate a facility} \\ 0, \text{otherwise} \end{cases}$;

y_{ij}: $\begin{cases} 1, \text{if customer i is served by the facility at site j} \\ 0, \text{otherwise} \end{cases}$;

s_j: $\begin{cases} 1, \text{if the facility at site i is interdicted} \\ 0, \text{otherwise} \end{cases}$;

z_j: $\begin{cases} 1, \text{if the facility at site j is fortified} \\ 0, \text{otherwise} \end{cases}$;

b_j: $\begin{cases} \text{1, if sit j has a valid facility} \\ \text{0,otherwise} \end{cases}$;

X: Set for facility location, $X = \{x_j\}$;

Y: Set for service assignment, $Y = \{y_{ij}\}$;

S: Set for interdicted facility, $S = \{s_j\}$;

Z: Set for defense facility, $Z = \{z_j\}$.

The main goal of the designer was to design a service network with the objective of minimizing the sum of fixed and variable costs of a system facing disturbance. The goal of the attacker was to maximize the objective of the designer.

Objective 1 (from the designer):

$$\text{Minimize}: \ \min f(X,Z) = \sum_{i \in N, j \in F} a_i d_{ij} y_{ij} + \sum_{j \in F} \left(g_j x_j + h_j z_j\right). \quad (1)$$

Objective 1 (from the attacker):

$$\text{Maximize}: \ \max f(S) = \sum_{i \in N, j \in F} a_i d_{ij} y_{ij}. \quad (2)$$

Subject to:

(1) $\sum_{j \in F} x_j \le p \qquad \forall j \in F,$

(2) $\sum_{j \in F} s_j \le r \qquad \forall j \in F,$

(3) $\sum_{j \in F} z_j \le q \qquad \forall j \in F,$

(4) $\sum_{j \in F} y_{ij} = 1 \qquad \forall i \in N,$

(5) $\begin{cases} y_{ij} \le x_j \\ y_{ij} \le 1 - s_j + z_j \end{cases} \qquad \forall i \in N, j \in F,$

(6) $z_j \le x_j \qquad \forall j \in F,$

(7) $\begin{cases} b_j \le x_j \\ b_j \le 1 - s_j + z_j \\ b_j \ge x_j - s_j \\ b_j \ge x_j + z_j - 1 \end{cases} \qquad \forall j \in F,$

(8) $y_{ij} \cdot d_{ij} \le d_{ij'} + M \cdot \left(1 - b_{j'}\right) \qquad \forall i \in N; j, j' \in F,$

(9) $x_j \in \{0,1\}, y_{ij} \in \{0,1\}, z_j \in \{0,1\}, s_J \in \{0,1\}, b_j \in \{0,1\} \qquad \forall i \in N, j \in F.$

In the above model, Constraint (1) indicates that the designer built p facilities out of the set of candidate facilities. Constraint (2) ensures that, at most, r facilities were to be interdicted. Constraint (3) stipulates that the designer could maximally protect q facilities. Constraint (4) makes sure that each demand was serviced by one facility. Constraint (5) states that only the facilities that were set up and not attacked at the same time (or were attacked but protected at the same time) could serve the demands. Constraint (6) indicates that only the built facilities were allowed to be protected. Constraint (7) determines whether a site had a valid facility, i.e., $b_j = 1$ if site j had a valid facility or $b_j = 0$ if otherwise. Constraint (8) is a new constraint for the closest assignment, a different version from the related constraints devised by Church [6] and Liberatore et al. [11]. This constraint normally used the set of existing valid facilities (not including j) that was further than j from demand i. However, these sets had to be recalculated when the fortification or interdiction changed; a phenomenon associated with considerable computational complexity. Constraint (9) states that all of the decision variables were binary.

3. Solution Approaches for Cyclic Dynamic Equilibrium

3.1. Solution Definitions and Analysis

Below are the notations used for the definitions and analysis.

U: Set of all possible design plans;
V: Set of all possible attack plans;
D: A design plan and $D \in U$;
A: An attack plan and $A \in V$;
A|D: The optimal attack plan for the given design plan;
D|A: The optimal design plan for the given attack plan;
$O_{D/A}$: The optimal minimum objective cost of the design plan for the given attack plan;
$O_{A/D}$: The optimal maximum objective cost of the attack plan for the given design plan.

Definition 1. *General optimal design (GOD) plan: The general optimal design plan was a plan by the designer that was optimized for a given attack plan.*

(1) *ATTACKER:* $A = \{A_i / A_i \in V\}$;
(2) *DESIGNER:* $D = \{D_i / D_i$ *satisfies* $O_{D/A}, A_i \in V, D_i \in U\}$.

When solving the objective of the designer, we fixed the variable s_j *in S with given values, then searched for the minimum cost of all of the designer's plans to thwart the opponent, as Figure 2 shows.*

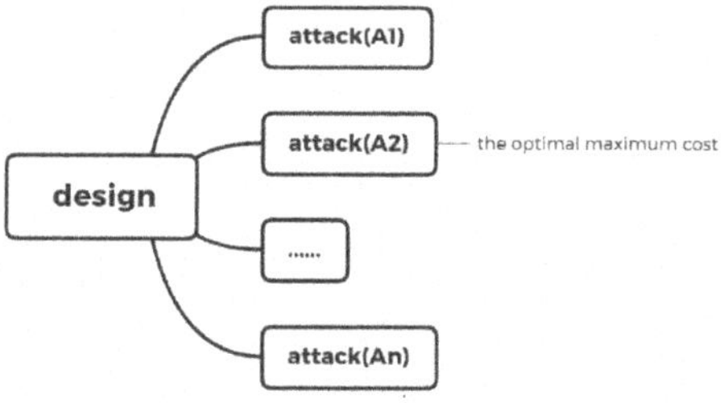

Figure 2. The optimal attack strategy.

Definition 2. *General optimal attack (GOA) plan: The general optimal attack was the plan by the designer that was optimized for a given design plan.*

(1) *DESIGNER:* $D = \{Di / D_i \in U\}$;
(2) *ATTACKER:* $A = \{Ai/Ai$ *satisfies* $O_{A/D}, A_i \in V, D_i \in U\}$.

When solving the objective of the attacker, we fixed the variable x_j, z_j *in S with given values, then searched for the maximum cost of all of the attacker's plans to thwart the opponent, as Figure 3 shows.*

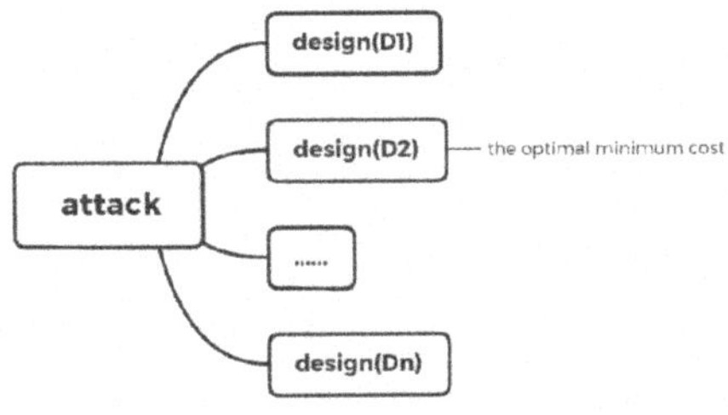

Figure 3. The optimal design strategy.

Definition 3. *The ideal design plan (IDP): The ideal design plan had the lowest objective cost of all of the design plans, occurring when there was no interdiction. The ideal design plan was a special case of GOD.*

(1) *ATTACKER:* $A = \{\theta\}$;
(2) *DESIGNER:* $D = \{D_i/D_i$ *satisfies* $O_{D/A}, A_i \in V, D_i \in U\}$.

In this situation of the fortification set $Z = \{\theta\}$, *it was obvious that the designer did not need to set up the fortification facility because there was no interdiction from the attacker. The ideal design plan was a special case of GOD.*

Definition 4. *The ideal attack plan (IAP): The ideal attack plan had the highest objective cost of all of the attack plans, occurring when there was no fortification. The ideal attack plan was a special case of GOA.*

(1) *DESIGNER:* $D = \{D_i/ D_i \in U\}$;
(2) *ATTACERK:* $A = \{A_i/A_i$ *satisfies* $O_{A/D}, A_i \in V, D_i \in U\}$.

This situation, where $Z = \{\theta\}$, *was very useful for the attacker. Attacking the facility unprotected would lead to huge damage.*

Definition 5. *The cycled confrontation strategy (CCS): The cycled confrontation strategy was expressed as {A1/D1, D1/A2, A2/D3, D3/A4, A4/D1}. There were two sets of plans from two sides, where each plan of one set was the optimal plan when responding to the opposite plan from the other side. The cycle stopped when the end plan was the same as the existing plan, as shown in Figure 4.*

Definition 6. *The round of confrontation (RC): Every time one side carried out a strategy, the other side found a strategy to use to fight back. This was one round of confrontation. In the example shown in Figure 4, we demonstrate that there were nine rounds from the beginning to the end of the opposite strategies.*

Definition 7. *The round of cycled confrontation (RCC): In the example shown in Figure 4, we demonstrate that there were six rounds from the beginning of the cycle (D2) to the end of the cycle when D2 existed again.*

Definition 8. *The evaluation indicators for two kinds of solution methods, as shown in Table 1.*

Table 1. The evaluation indicators.

	Index	Name
1	$\overline{C_A}$	Average attack cost
2	$\overline{C_D}$	Average design cost
3	$\overline{C_A}-\overline{C_D}$	The optimality of solutions

Property 1. *The cycled confrontation strategy began with the IDP and IAP.*

Property 2. *The cycled confrontation strategy existed because the plans from both sides were finite.*

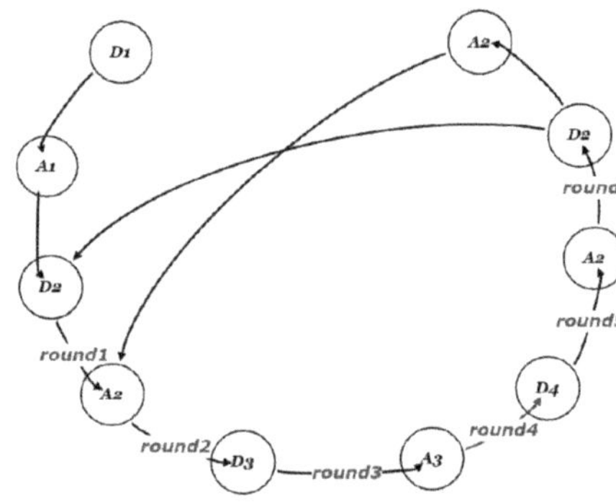

Figure 4. The optimal attack strategy.

3.2. *A Bi-Objective Solution Framework*

We solved CDGC-RIMF with a MILP-based solution approach. Our approach started by creating an initial design plan D, then fixed D_i (fix x_i, z_i) to achieve the attack plan A_i (fix s_i) using the MIP solver CPLEX. After the attack plan, we fixed the attack plan A_i (fix s_i) to get the next design plan. The solution process continued until the repeated attack plan existed, as shown in Figure 5.

1) Let i = 1.
2) Find an initial design plan D, and let $D_i \leftarrow D$
3) Repeat
   4) Fix D_i as a given design plan and use the MIP solver to find am attack plan, noted as A_i, forming $A_i // D_i$
   Partial Optimization Loop
   5) Fix A_i as a given attack plan and use the MIP solver to find a design plan, noted as D_{i+1}, forming $D_{i+1} // A_i$
   Partial Optimization Loop
   6) Let i = i + 1
7) Until $D_i = D_{ii}$ exists where $1 \le D_{ii} \le i-1$
8) Output the equilibrium method

Figure 5. The bi-objective solution approach.

3.3. *A Heuristic Partial Optimization Algorithm*

The large-sized complex problem with many possible decision variables could not be optimized completely within an acceptable Central Processing Unit (CPU) time. Therefore, some heuristic algorithms were put forward to reduce the solution space so as to save time and improve the calculation efficiency, in particular, the heuristic concentration algorithm. As Rosing and Hodgson [26] illustrated in their paper, the heuristic concentration procedure mainly consists of two stages. Firstly, it was necessary to find a concentration set according to some rules, and secondly, the optimal solution in the concentration set could be solved every time by generating a concentration set. Based on Rosing and Hodgson's heuristic concentration procedure, the MILP-based neighborhood searching algorithms by Xiao et al. [24,25] and You et al. [27,28] were introduced, and the concept of the partial set was also presented. For this paper, we developed a heuristic algorithm for large-sized problems following the steps showed above. First, we fixed all of the nodes, then generated iteratively different partial sets that were subsequently used to unfix the nodes in the partial set, and then optimized the unfixed nodes. A rule used to generate the partial set was that nodes selected to be in the partial set more times should have a lower possibility of being selected in the next round. The main framework of the proposed heuristic algorithm is outlined in Figure 6. The partial optimization forced the optimization to stay within only a small range of the variables so it could be achieved in a short computational time. An iterative implement of partial optimization on different selected ranges continuously improved the solution until it was good enough or until the time limit was reached.

As shown in Figure 6, the partial optimization algorithm (POA) procedure started with an initial random attacker plan (x_0, z_0) and a designer plan (s_0), obtained by randomly generating the number of q interdictions and p facilities. Then (in Step 1,2), the confrontation cyclic started with the initial attacker and the designer plan. The first confrontation plan was from the defender. The index P suggested an improvement for every round of partial optimization. The parameter W indicated the number of to-be-selected nodes for partial optimization and was assigned an initial value in Step 3. Our experiments showed that parameter W was sensitive and considerably influenced the algorithm efficiency. The appropriate value interval for W was between 35 and 45, according to our experimental comparisons. After that, the partial optimization loop started at Step 4, containing Steps 4.1–4.11. In Step 4.2–4.5, all variable instances of x_j were first fixed, and then a number W of instances were selected to be unfixed. Each time, we selected a number W of x_j following the selection rule, that nodes chosen more often had a lower possibility of being chosen. After that, the CPLEX solver was applied to optimize the selected nodes (in Step 4.6). In Steps 4.7 and 4.8, the new solution was judged regarding whether it improved upon the incumbent solution. If yes, then it was accepted as the new incumbent solution and the non-improvement index P was reset to 0; otherwise, the P was increased by one. The loop stopped if there was no further improvement even after P_{max} continuous attempts, or if the total CPU time was longer than the set limited time T_{max}, or if all of the nodes were optimized at the same time (in Step 4.6). In steps 4.10–4.12, the number W was adjusted dynamically to ensure that the solution time solved through CPLEX was between t_{min} and t_{max}. The number W increased by 5 if the last CPU time was less than t_{min} or reduced by 5 (Δ = 5) if the last CPU time was more than t_{max}. Finally, the obtained C was the best solution found and became the output of the algorithm. The attacker adopted similar partial optimization as the partial optimization in the defender's plan to develop its coping plan.

Procedure: The partial optimization algorithm (POA)

Input parameters: T_{max}, P_{max}, t_{min}, t_{max}, W_{min}, Δ, W_0; output result: x_j, z_j, s_j, $a.obj$, $d.obj$

1) For the designer: Let x_j, $z_j \leftarrow$ Initialization: Random generation of a designer strategy

2) For the attacker: Let $s_j \leftarrow$ Initialization: Random generation of an attacker strategy

3) Let x_j, z_j, $s_j \leftarrow x_0$, z_0, s_0, $P \leftarrow 0$, and $W \leftarrow W_0$

4) **Iterative Neighborhood Search (INS) Loop Begins**

4.1) **For the designer:**

4.2) Let $f \leftarrow$ null

4.3) Apply a rule to select a number W of nodes from F into f

4.4) For j in F, fix all x_j decision variables

Figure 6. *Cont.*

4.5) For j in f, unfix x_j decision variables

4.6) Call the CPLEX solver to get a new solution X',Z', d.obj' and record the used CPU time t

4.7) **If** $d.obj'$ improved upon $d.obj$, **then** let $d.obj \leftarrow d.obj'$ and $P \leftarrow 0$

4.8) **Otherwise,** let $P \leftarrow P + 1$

4.9) **If** $P \geq P_{max}$ **or** total CPU time $\geq T_{max}$ or $W \geq$ number of CFs **then** break the loop and stop

4.10) **If** $t \leq t_{min}$, **then** let $W \leftarrow W + \Delta$

4.11) **if** $t \geq t_{max}$, **then** let $W \leftarrow W - \Delta$

4.12) **if** $W < W_{min}$, **then** let $W \leftarrow W_{min}$

4.13) **For the attacker: The process was similar to 4.2 – 4.12:**

Just substitute x_j **with** s_j **and substitute** $d.obj$ **with** $a.obj$

4) **Loop End**

Figure 6. The framework of the POA algorithm.

4. Computational Tests

The goal of the computational experiments was to validate the model. The codes by AMPL were run on a Linux PC server with two 2.30 GHz Intel Xeon (R) CPUs and 128 GB RAM. We used the MIP solver AMPL/CPLEX (version 12.6.0.1) to solve the tested instances.

4.1. The Generation of the Tested Instances

The data sets and the parameter settings used in the experiments were generated randomly, as shown in Table 2. In this paper, we used small-, medium-, and large-scale problems to validate our model.

Table 2. The parameters for different scale problems.

		Nodes of Demands CU	Candidate Facilities CF	Facilities to be Built F(p)	Number of Attacks (r)	Number of Fortification $z(q)$
Small-Scale	**1**	25	10	5	3	1
	2	50	20	10	5	3
	3	75	30	15	8	4
	4	100	40	20	10	5
Medium-Scale	**5**	250	50	25	13	6
	6	300	60	30	15	8
	7	350	70	35	18	9
	8	400	80	40	20	10
Large-Scale	**9**	900	90	45	23	11
	10	1000	100	50	25	13
	11	1100	110	55	28	14
	12	1200	120	60	30	15

Note: candidate facility (CF); client (CU).

For the small-scale problem, we set a number of 10, 20, 30, or 40 candidate facilities (CFs), respectively, within a square region of 100 × 100 distance units. The actual limited number of facilities (P) was set by CF/2. Every selected facility served five clients, so the clients (CU) assigned to each facility were about 25, 50, 75, or 100. The number of fortified facilities and interdiction facilities were as shown in column 6–7. The number of fortified facilities (r) was calculated by $p/2$, while the number of interdiction facilities was calculated by $r/2$. The coordinates of the facilities and the clients in the small-scale problem were generated between 0 and 100. For the medium problem, the coordinates of the facilities and clients were in the square region of 200 × 200 distance units, with the CFs listed as 50, 60, 70, or 80. The number of CUs, rs, and qs were calculated in a similar way, as in the small-scale problem. In the large-scale problem, these values were in the square region of 400 × 400 distance units. The numbers relating to CF, p, CU, r, and q are shown in columns 3–7.

4.2. Analysis of the Tested Results

Tables 3 and 4 display the computation time, RC, and RCC. The first six columns show the parameters CF, F, CU, r, and q. The seventh and eighth columns, respectively, contain cumulative computing times and independent run times. The last two rows display the round of confrontation and the round of cycled confrontation. As shown in Table 3, (r > q), RCC 4 remained unchanged as the CF increased, while Table 3 (r < q) shows that there was a different RCC 6 for the 50 CFs. The solution time increased as the parameters of CF, F, CU, r, and q increased. For the small- and medium-scale problems, calculations were finished in 500 s. However, it took a long time to solve the large-scale problem; therefore, it is necessary to solve the large-scale problem using another, more efficient, approach.

Table 3. The results for small- and medium-scale problems using CPLEX (r > q). Round of confrontation (RC); round of cycled confrontation (RCC); candidate facility (CF); client (CU).

	CF	F(p)	CU	s(r)	z(q)	time(s)	RC	RCC
1	10	5	25	3	1	0	5	4
2	20	10	50	5	3	1	9	4
3	30	15	75	8	4	3	9	4
4	40	20	100	10	5	5	5	4
5	50	25	250	13	6	62	7	4
6	60	30	300	15	8	170	9	4
7	70	35	350	18	9	267	7	4
8	80	40	400	20	10	527	7	4
9	90	45	900	23	11	5019	11	4
10	100	50	1000	25	13	6297	9	4
11	110	55	1100	28	14	12,030	11	4
12	120	60	1200	30	15	14,687	9	4

Table 4. The results for the small- and medium-scale problems (q > r).

	CF	F(p)	CU	s(r)	z(q)	time(s)	RC	RCC
1	10	5	25	1	3	0	5	4
2	20	10	50	3	5	1	7	4
3	30	15	75	4	8	2	7	4
4	40	20	100	5	10	8	7	4
5	50	25	250	6	13	84	9	6
6	60	30	300	8	15	145	7	4
7	70	35	350	9	18	289	7	4
8	80	40	400	10	20	530	7	4
9	90	45	900	11	23	3379	7	4
10	100	50	1000	13	25	6219	9	4
11	110	55	1100	14	28	7219	7	4
12	120	60	1200	15	30	141,119	11	4

We chose one example to illustrate the concrete game process between the two sides. We took the 10 × 25 problem as an example to demonstrate the confrontation process. As shown in Table 5, the design plans, attack plans, and the cost were listed until the same attack plan existed. Figure 7a–f gives the model solution of Table 5 in graphical form.

Table 5. The strategy of both sides (for data: 10 × 25, p = 6, r = 3, q = 2).

	Decision Maker	Designer		Attacker	Cost
		Location	Fortification		
1	design	1,5,8,10			47,251
1	attack	1,5,8,10		1,5,10	89,291
2	design	3,4,5	5	1,5,10	50,119
2	attack	3,4,5	5	3,4	76,437
3	design	1,5,8,10		3,4	47,251
3	attack	1,5,8,10		1,5,10	89,291

As shown in Figure 7a–f, points selected to be used for facilities were colored green and the outline of the fortified facilities was black. The lines in each figure described assignments of demand to selected facilities. The attack points were displayed as red triangles. In the first stage, when i = 1, the designer selected P (P = 5) facilities from potential locations considering the minimum weighted distance. Faced with the designer's strategy, the attacker adopted an optimal method to ensure maximum destruction. In the second stage, when i = 2, the designer decided how to protect and establish the most reliable facilities to reduce loss. Every time when the attackers carried out the attack strategy, this model provided the protector with a most effective way to maintain the supply as much as possible. Wise decisions helped the facility designer to keep the facilities sustainable and reliable against the multi-confrontation between two sides. The confrontation process continued until the same attack plan existed., and then the final equilibrium was reached.

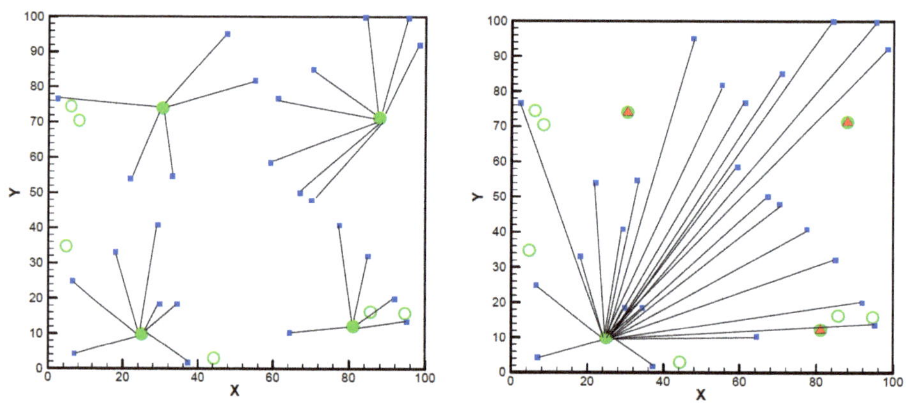

(**a**) (**b**)

Figure 7. *Cont.*

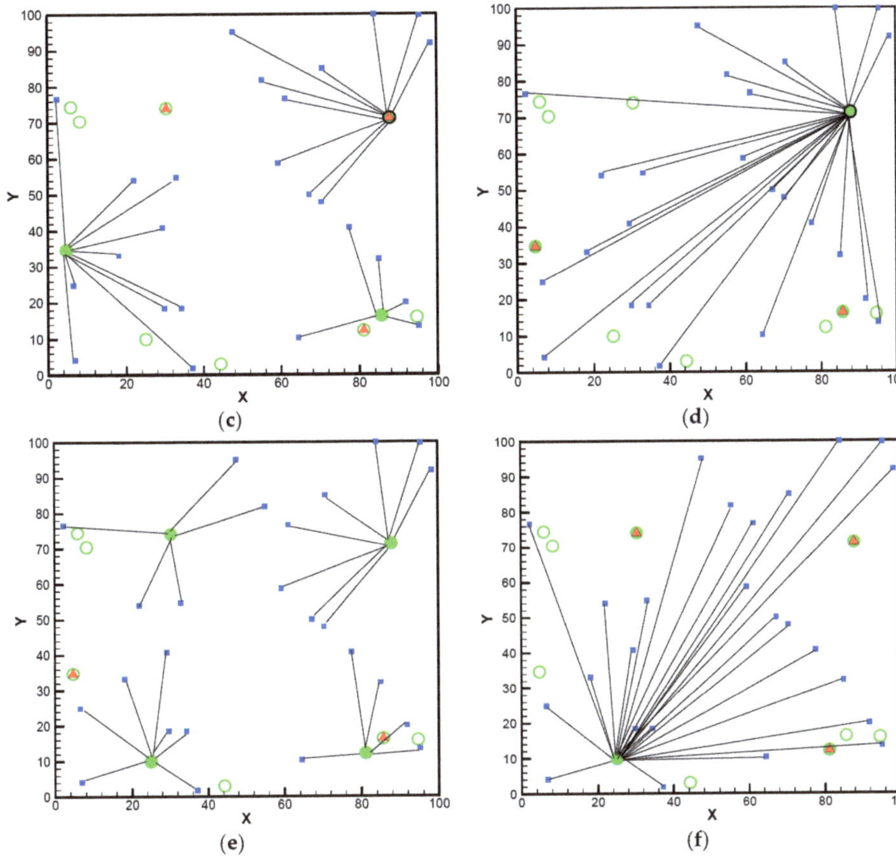

Figure 7. (**a**) $i = 1$, the optimal designer plan; (**b**) $i = 1$, the optimal attacker plan; (**c**) $i = 2$, the optimal designer plan; (**d**) $i = 2$, the optimal attacker plan; (**e**) $i = 3$, the optimal designer plan; and (**f**) $i = 3$, the optimal designer plan. —the potential locations of facilities; —interdiction on the facilities; —facilities in use; —fortified facilities in use, —demands.

For the large-scale problem, we implemented the POA and CPLEX to solve the model at sizes of 90 × 900, 100 × 1000 and 110 × 1100. This experiment showed that the POA had good performance when solving large-scale instances. As shown in Table 6, the POA was much better than CPLEX in terms of CPU time, and the average CPU time of the POA was about 85.1% shorter than that of CPLEX for size 90 × 900, 95.9% for size 100 × 1000 and 96.2% for size 110 × 1100. These results were obtained as averages of the CPU time from algorithms in 10 replications.

Table 6. The comparison of POA and CPLEX for large-scale problems.)

CF	f(p)	cu	s(r)	z(q)	CPLEX		POA		Time Comparison
					$\overline{C_A} - \overline{C_D}$	Time(s)	$\overline{C_A} - \overline{C_D}$	Time(s)	CP_t- PO_t/CP_t
90	45	900	23	11	2,360,771	3890	2,368,109	578	85.1%
100	50	1000	25	13	2,471,750	8615	2,484,027	350	95.9%
110	55	1100	28	14	2,883,380	15,374	2,908,162	581	96.2%

Note: CPU time of CPLEX (CP_t); CPU time of POA (PO_t)

In our model, we considered the cycle of the confrontation strategy, so RC and RCC were not stable when we used the partial optimal algorithm to solve this problem. The RC and RCC numbers influenced the results and CPU times. Even in this situation, we found better optimal results through adjusting the parameter of the partial optimal algorithm and repeated the experiments many times. As Table 6 shows, the POA achieved better results in a shorter CPU time when CF = 90, 100, or 110.

When CF = 120, as the iteration increased, RC and RCC became even more unstable; therefore, finding better average results was more difficult. Therefore, we deleted the cycle to test the single confrontation (RC = 1); in this situation, we found that POA operated better without being influenced by the cycle confrontation. Furthermore, we performed similar experiments when CF = 90, 100, 110, and 120. The results proved that the POA performed even better in the single confrontation. When two approaches produced the same optimal results (the optimal decisions), the time taken by the POA was less than that taken by CPLEX. Among all the large-sized instances, the average CPU time of the POA was 90.3% at most and 79.7% at least, which was shorter than that of CPLEX, as shown in Table 7.

Table 7. The comparison of POA and CPLEX in a single confrontation (ATTACK ONCE).

CF	f(p)	cu	s(r)	z(q)	CPLEX		POA		Time Comparison
					$\overline{C_A}$	Time(s)	$\overline{C_A}$	Time(s)	CP_t- PO_t/CP_t
90	45	900	23	11	4,972,729	370	4,972,729	75	79.7%
100	50	1000	25	13	5,832,329	638	5,832,329	91	85.7%
110	55	1100	28	14	6,437,748	2061	6,437,748	215	89.6%
120	60	1200	30	15	6,402,308	1742	6,402,308	169	90.3%

Note: CPU time of CPLEX (CP_t); CPU time of POA (PO_t)

We fixed the casual attack strategy, then used CPLEX and the POA to achieve the design plan faced with the same attack strategy. The POA achieved the same optimal result in less time, as Table 8 shows. The average reduced CPU time of the POA compared with CPLEX varied between 89.6% for the smallest instance (CF = 90) and 96.2% for the biggest instance (CF = 120). It was obvious that the efficiency had been improved greatly.

Table 8. The comparison of the POA and CPLEX in a single confrontation (DEFENCE ONCE).

CF	f(p)	cu	s(r)	z(q)	CPLEX		POA		Time Comparison
					$\overline{C_D}$	Time(s)	CP_t- PO_t/ CP_t	Time(s)	CP_t- PO_t/CP_t
90	45	900	23	11	1,308,912	1025	1,308,912	107	89.6%
100	50	1000	25	13	1,480,763	1757	1,480,763	137	92.2%
110	55	1100	28	14	1,466,200	2156	1,566,200	169	92.2%
120	60	1200	30	15	1,586,774	2881	1,586,774	109	96.2%

Note: CPU time of CPLEX (CP_t); CPU time of POA (PO_t)

We also discussed the effect of protective resources (q) on the designer's efficiency. Table 9 shows that the number of fortified facilities increased with the cost of changing designer. Columns 1–5 illustrate the parameters of candidate facilities, limited facilities, serving demands, number of interdiction facilities, and number of fortification facilities. For the instances CF = 10, 20, 30, and 40 listed here, the number of fortification facilities (q) varied between 1 and 10, which was no more than the number of interdiction facilities. Columns 6–7 show the RC and RCC, which almost remained the same as q increased.

Table 9. The sensitivity of parameter q (for the small-scale problem).

CF	f(p)	cu	s(r)	z(q)	RC	RCC	$\overline{C_A}$
10	5	25	3	1	6	4	48,207.00
				2 . . . 10	6	4	. . .
20	5	25	3	1	10	4	63,059.00
				2	10	4	52,318.50
				3 . . . 10	. . .	. . .	. . .
30	15	75	8	1	8	4	85,995.00
				2	8	4	85,189.20
				3 . . . 10	. . .	. . .	. . .
40	20	100	10	1	6	4	101,512.33
				2	6	4	101,354.67
				3	6	4	101,301.67
				4 . . . 10	. . .	. . .	. . .

Note: " . . . " represents that the cost remained the same as the last iteration.

Often, most protection benefits were achieved through the first two or three fortifications, and subsequent security investments gradually reduced in efficiency. Overall, the fortification of the second facility still played an important role regarding improvement of the results. For instance, in problems where CF = 20 operating facilities, when the number of q exceeds 2, there was no improvement regarding the cost of the design.

5. Conclusions

In this paper, a bi-objective cyclic dynamic equilibrium gaming model of the *r*-interdiction median problem with fortification was presented. In this model, the two sides were wise to make a decision according to the operational experience. The system designer (defender) could decide regarding the interdiction. The attacker could decide to destroy essential facilities after considering the possible reassignment of the designer faced with one interdiction, which was the worst-case loss. The model could give two operational strategy packages when the two sides achieved equilibrium. To solve the large-scale problem, the partial optimization algorithm was used, and we made the comparison between the exact method using CPLEX and the partial optimization algorithm. Compared with the previous work, the strategy packages can help the facility manager to make decisions in advance against the attackers and better keep the facility in long-term sustainability. Additionally, we found that the partial optimization algorithm yielded better solutions with higher search efficiency.

In our analysis, attacks on a given facility were always successful. Defenders could only perform passive defenses and defenders could not interrupt or even destroy attackers. New models should be developed to include the defender's positive defense and the success rate of interdiction. Our model was defined on a network of nodes and arcs where each node was assumed to represent a local area of demand, as well as represent a potential position for a facility. In future work, we could choose the optimal position in a continuous network with no potential facilities. Regarding the partial optimization parameter, we could find a way to achieve self-adjustment rather than by manual adjustment.

Author Contributions: This manuscript was written by P.Y. under the supervision of Y.X. and Y.Z. The modeling, data analysis, and algorithm process were executed by S.Z. (Siyue Zhang) and P.Y.; S.Z. (Shenghan Zhou) and W.C. were responsible for the data acquisition and model design. All authors have read and agreed to the published version of the manuscript.

Funding: This work is supported by the National Natural Science Foundation of China (Grant Nos. 71871003, 71971013 and 71971009).

Acknowledgments: Thanks for the support of my friend: Meng You.

Conflicts of Interest: The authors declare no conflict of interest.

References

1. Friedrich, C.J. Alfred Weber's theory of the location of industries. *Am. J. Sociol.* **1929**, *35*, 853.
2. Huizhen, Z.; Xin, W.; Liang, M.A. A new semi-Lagrangian relaxation method to solve the un-capacitated facility location problem. *Oper. Res. Trans.* **2015**, *19*, 37–47.
3. Monabbati, E.; Kakhki, H.T. On a class of subadditive duals for the uncapacitated facility location problem. *Appl. Math. Comput.* **2015**, *251*, 118–131. [CrossRef]
4. Glover, F.; Hanafi, S.; Guemri, O. A simple multi-wave algorithm for the uncapacitated facility location problem. *Front. Eng. Manag.* **2018**, *5*, 41–55. [CrossRef]
5. Church, R.L.; Scaparra, P.M.; Middleton, R. Identifying Critical Infrastructure: The Median and Covering Facility Interdiction Problems. *Ann. Assoc. Am. Geogr.* **2004**, *94*, 491–502. [CrossRef]
6. Church, R.L.; Scaparra, P.M. Protecting critical assets: The r-interdiction median problem with fortification. *Geogr. Anal.* **2007**, *39*, 129–146. [CrossRef]
7. Scaparra, P.M.; Church, R.L. A bilevel mixed-integer program for critical infrastructure protection planning. *Comput. Oper. Res.* **2008**, *35*, 1905–1923. [CrossRef]
8. Snyder, L.V.; Scaparra, M.P.; Daskin, M.S.; Church, R.L. Planning for Disruption in Supply Chain Networks. In *Models, Methods, and Applications for Innovative Decision Making*; Informs: Catonsville, MD, USA, 2016. [CrossRef]
9. Scaparra, P.M.; Church, R.L. An exact solution approach for the interdiction median problem with fortification. *Eur. J. Oper. Res.* **2008**, *189*, 76–92. [CrossRef]
10. Aksen, D.; Piyade, N.; Aras, N. The budget constrained r-interdiction median problem with capacity expansion. *Cent. Eur. J. Oper. Res.* **2010**, *18*, 269–291. [CrossRef]
11. Liberatore, F.; Scaparra, P.M.; Daskin, S.M. Analysis of facility protection strategies against an uncertain number of attacks: The stochastic R-interdiction median problem with fortification. *Comput. Oper. Res.* **2011**, *38*, 357–366. [CrossRef]
12. Zhu, Y.; Zheng, Z.; Zhang, X.; Cai, K. The r-interdiction median problem with probabilistic protection and its solution algorithm. *Comput. Oper. Res.* **2013**, *40*, 451–462. [CrossRef]
13. Medal, R.H.; Pohl, A.E.; Rossetti, D.M. A multi-objective integrated facility location-hardening model: Analyzing the pre- and post-disruption trade off. *Eur. J. Oper. Res.* **2014**, *237*, 257–270. [CrossRef]
14. Mahmoodjanloo, M.; Parvasi, P.S.; Ramezanian, R. A tri-level covering fortification model for facility protection against disturbance in r-interdiction median problem. *Comput. Ind. Eng.* **2016**, *102*, 219–232. [CrossRef]
15. Sadeghi, S.; Seifi, A.; Azizi, Z. Tri-level shortest path network interdiction with partial fortification. *Comput. Ind. Eng.* **2017**, *106*, 400–411. [CrossRef]
16. Zheng, K.; Albert, A.L. An exact algorithm for solving the bi-level facility interdiction and fortification problem. *Oper. Res. Lett.* **2018**, *46*, 573–578. [CrossRef]
17. Khanduzi, R.; Maleki, R.H.; Akbari, R. Two novel combined approaches based on TLBO and PSO for a partial interdiction/fortification problem using capacitated facilities and budget constraint. *Soft Comput.* **2018**, *22*, 5901–5919. [CrossRef]
18. Roboredo, M.C.; Pessoa, A.A.; Aizemberg, L. An exact approach for the r-interdiction median problem with fortification. *Rairo-Oper. Res.* **2019**, *53*, 505–516. [CrossRef]
19. Roboredo, M.C.; Aizemberg, L.; Pessoa, A.A. An exact approach for the r-interdiction covering problem with fortification. *Cent. Eur. J. Oper. Res.* **2019**, *27*, 111–131. [CrossRef]
20. Biswas, P.; Pal, B.B. A fuzzy goal programming method to solve congestion management problem using genetic algorithm. *Decis. Mak. Appl. Manag. Eng.* **2019**, *2*, 36–53. [CrossRef]
21. Barma, P.S.; Dutta, J.; Mukherjee, A. A 2-opt guided discrete antlion optimization algorithm for multi-depot vehicle routing problem. *Decis. Mak. Appl. Manag. Eng.* **2019**, *2*, 112–125.
22. Dong, J.R.; Huang, C.H.; Zuo, X.R.; Xiao, Y.Y. A bi-objective MIP model for the dynamic gaming case of the Capacitated R-interdiction median problem with fortification. In Proceedings of the 4th International Symposium on Project Management Conference, Wuhan, China, 18–19 June 2016.
23. Simaan, M.; Cruz, B.J. Additional Aspects of the Stackelberg Strategy in Nonzero-Sum Games. *J. Optim. Theoryapplications* **1973**, *11*, 613–626. [CrossRef]

24. Xiao, Y.; Kaku, I.; Zhao, Q.; Zhang, R. Neighborhood search techniques for solving uncapacitated multilevel lot-sizing problems. *Comput. Oper. Res.* **2012**, *39*, 647–658. [CrossRef]
25. Xiao, Y.; Konak, A. A genetic algorithm with exact dynamic programming for the green vehicle routing & scheduling problem. *J. Clean. Prod.* **2017**, *167*, 1450–1463.
26. Rosing, K.E.; Hodgson, J.M. Heuristic concentration for the p-median: An example demonstrating how and why it works. *Comput. Oper. Res.* **2002**, *29*, 1317–1330. [CrossRef]
27. You, M.; Xiao, Y.Y.; Zhang, S.Y.; Yang, P.; Zhou, S.H. Optimal mathematical programming for the warehouse location problem with Euclidean distance linearization. *Comput. Ind. Eng.* **2019**, *136*, 70–79. [CrossRef]
28. You, M.; Xiao, Y.Y.; Zhang, S.Y.; Zhou, S.Y.; Yang, P.; Pan, X. Modeling the Capacitated Multi-Level Lot-Sizing Problem under Time-Varying Environments and a Fix-and-Optimize Solution Approach. *Entropy* **2019**, *21*, 377. [CrossRef]

© 2020 by the authors. Licensee MDPI, Basel, Switzerland. This article is an open access article distributed under the terms and conditions of the Creative Commons Attribution (CC BY) license (http://creativecommons.org/licenses/by/4.0/).

sustainability

MDPI

Article

Robust Scheduling of Waste Wood Processing Plants with Uncertain Delivery Sources and Quality

Balázs Dávid [1,2], Olivér Ősz [3] and Máté Hegyháti [3,4,*]

1 InnoRenew CoE, 6310 Izola, Slovenia; balazs.david@innorenew.eu or balazs.david@famnit.upr.si
2 Department of Information Sciences and Technologies, University of Primorska, 6000 Koper, Slovenia;
3 Department of Information Technology, Széchenyi István University, Széchenyi István University, 9026 Győr, Hungary; osz.oliver@sze.hu
4 Institute of Informatics and Mathematics, University of Sopron, 9400 Sopron, Hungary
* Correspondence: hegyhati@sze.hu or hegyhati.mate@uni-sopron.hu

Abstract: While the study of reverse wood value chains has become an important topic recently, optimization-focused studies usually consider network-level problems and decisions, and do not address the individual processes in the network. In the case of waste wood, one such important process is the scheduling of the various machines in a waste wood processing facility to treat incoming wood deliveries with multiple sources and varying quality. This paper proposes a robust multi-objective mixed-integer linear programming model for the optimization of this process that considers the uncertain origins and compositions of the incoming deliveries, while aiming to minimize both lateness and energy consumption. An exhaustive study is performed on instance sets of different sizes and structures to show the efficiency and the limits of the proposed model both in single- and multi-objective cases.

Keywords: reverse wood value chain; scheduling; waste-wood processing

check for updates

Citation: Dávid, B.; Ősz, O.; Hegyháti, M. Robust Scheduling of Waste Wood Processing Plants with Uncertain Delivery Sources and Quality. *Sustainability* **2021**, *13*, 5007. https://doi.org/10.3390/su13095007

Academic Editors: Dragan Pamucar and Željko Stević

Received: 11 April 2021
Accepted: 28 April 2021
Published: 29 April 2021

Publisher's Note: MDPI stays neutral with regard to jurisdictional claims in published maps and institutional affiliations.

Copyright: © 2021 by the authors. Licensee MDPI, Basel, Switzerland. This article is an open access article distributed under the terms and conditions of the Creative Commons Attribution (CC BY) license (https://creativecommons.org/licenses/by/4.0/).

1. Introduction

The topics of recycling and reuse are getting an increased recognition due to the growing importance of sustainability and environmental consciousness. Adding values to residues and wastes not only helps industries meet their commitments towards various EU- and country-level regulations, but also provides an environmentally friendly alternative to other, more traditional resource streams. Still, biomass wastes are mainly considered for energy production [1], which leads to scientific studies of biomass value chains concentrating mostly on efficient use for energy as well [2]. However, there are certain biomass types where viable alternative uses also exist. Waste wood is a good example for this: it is a versatile material that can be reused and recycled after its original function becomes obsolete, offering a more sustainable option than energy production [3].

Reverse logistics is the field that considers the logistics network of return product flows, putting an emphasis on the end-of-life recovery and the reuse of resources [4]. The three major processes of a reverse logistics network are distribution, production planning and inventory management [5]. Naturally, this may change or become more specialized for certain industrial fields. While the processes remain the same for biomass supply chains, extra emphasis should be put on processing, conditioning and intermediate production, as these can influence the attributes of the transported, stored and utilized biomass products [6]. However, the majority of papers in literature only study inventory and product return management and network design [7], and the optimization of production processes is rarely considered. This is especially true for waste wood value chains. While the traditional forward resource flows of wood are well-studied [8,9] and facility-level processes like sawmilling [10] or cutting pattern optimization [11] are also considered, the available literature on the reverse logistics processes of waste wood is significantly scarcer. Apart from studies about the use of waste wood for energy [12,13], the few existing papers revolve around variations of network design [14–17], but the optimization problems arising in the facilities of these networks are not addressed.

This paper studies the arising scheduling problems in a waste wood processing facility over a planning horizon, and presents an optimization approach for their solution. These facilities are considered as an integral and intermediate process of a waste wood reverse logistics network where various types of wood waste are first collected from the available sources of the network, then transported to such a facility for processing, and finally delivered to customers to satisfy their demands. The wood deliveries arriving to a facility go through a series of transformation processes, and the resulting product is then transported to the customer in its required form. As the incoming deliveries to such a facility can have different origins (e.g., building/industrial or household waste) and characteristics (e.g., coated or uncoated) [18], the exact processing steps that have to be taken might vary between deliveries based on these features. Because of these characteristics, there is an uncertain element to the problem that also has to be taken into account during optimization. Moreover, each delivery has a priority and a deadline that also have to be considered during scheduling.

A mixed-integer linear programming model (MILP) is presented for scheduling the wood deliveries on the various machines of the processing facility in the required order over a fixed planning horizon of several days/weeks. The model considers the robustness of the problem by taking into account the uncertainty that arises because of the varying delivery sources and quality. Optimization is done using two different objective functions; either ensuring timely completion of the deliveries by considering their priority-weighted lateness, or minimizing the energy consumption of the available machines at the facility. This allows the generation of different solutions for the same problem scenario using multi-objective optimization, which can aid the decision-making process of production planning at the facility.

The efficiency of the model is presented on input instances of varying sizes and structures. As the acquisition of accurate and exact real-world data is difficult on this level of a supply chain, these instances are randomly generated. The characteristics (origin and type) of incoming deliveries are determined based on real-world distributions published in the literature, while the available machines of the facility have the characteristics of actual existing waste wood processing machinery.

The outline of this paper is the following. First, the arising scheduling problem is defined in detail, describing all constraints that are be taken into account, and a multi-objective MILP model is formalized for the optimization of this problem. Then, the results of this model are presented on various input instances, showing its capabilities and limits. Finally, the paper is concluded with the presentation of future extension possibilities of the current work.

2. Problem Definition

2.1. General Process and Infrastructure

The approach proposed in this paper tackles the mid-term scheduling of stochastic jobs in a waste wood processing facility. The time horizon of interest spans over several days, starting from the earliest delivery of input materials until the latest deadline. For the scope of this study, time frames of one and two weeks were both examined.

The processing facility receives waste wood deliveries from various sources and has to produce shredded wood out of them by pre-determined deadlines. The production process of a delivery will be referred to as a job. There is a one-to-one relation between jobs and deliveries, thus the two terms will often be used interchangeably.

Incoming waste wood deliveries are categorized according to the information in [18]. Deliveries can belong to multiple categories based on their origin: construction waste, industrial waste, household waste, and waste from collection centers. Due to their similar characteristics, construction and industrials wastes will be merged under the same category for the remainder of this paper. The same will be true for wastes from households and collection centers. Waste wood of the former category usually arrives directly from the sites, while the latter can come either directly from households or regional collections centers. Independently of their origin, deliveries can also be categorized by the type of the delivered material: they consist of either solid wood or derived timber products.

A key feature of a delivery is the possibility and percentage of contaminations. On top of the above characteristics, a part of each delivery is coated wood that has to go through a removal process. The ratio of coated wood in a delivery depends on its other features. Derived timber products from households or collection centers are coated in the majority of the cases, while this is not true for industrial and building waste, which is only partly coated. A more detailed classification of deliveries would be possible, however, from the aspect of the current investigation, jobs sharing the same delivery category are assumed to have similar properties. Furthermore, collecting accurate real-world data for finer classification may be practically infeasible.

The processing of each job consists of six or seven steps. Five of these steps, in processing order, are: inspection, coating removal, shredding, screening, and re-shredding. The sixth step, metal separation can be done either manually between inspection and coating removal, or in an automated fashion by a (magnetic) separator after shredding. For solid wood deliveries from industrial sites, a seventh step, pre-shredding is also required to make the materials suitable for the shredding equipment. The interconnection of the above processes can be seen in Figure 1.

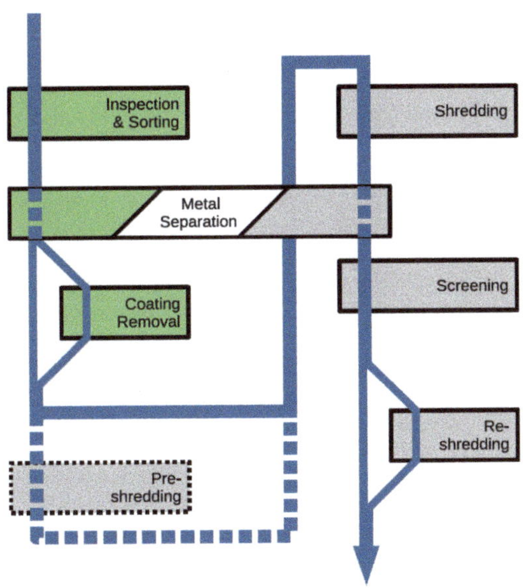

Figure 1. The possible flow of waste wood through the different transformation processes (green: manual steps, gray: machine steps).

In Figure 1, each box corresponds to one of the processing steps of the facility, while the blue line represents the flow of the materials. The green-colored steps must be carried out by dedicated crews, gray-colored steps are performed by machines. Coating removal is performed only for the coated part of each delivery, while pre-shredding is only done for solid industrial wood deliveries. The two metal separation steps have a mutually excluding relation: exactly one of them has to be included in the transformation process of every delivery. Similarly to coating removal, re-shredding is only performed on a portion of all materials.

The facility has two types of resources to carry out these tasks. Shredding, pre-shredding, automated metal separation, screening, and re-shredding is done by high-value machines with limited availability. There is no overlap among the machines suitable for these steps, except for shredding and re-shredding, which use the same machines. The rest of the tasks, i.e., inspection, coating removal, and manual metal separation are carried out by dedicated crews. Both machines and the crews have an estimated processing speed, expressed in the processed mass over time. Several suitable machines may be assigned to work in parallel on the same delivery. However, the processing steps are non-preemptive: they may not be interrupted and resumed later on the same machine. The only interruption occurs at the end of shifts, and the next day processing continues with the same job. It is assumed that setup and shutdown operations are done before and after the shifts, respectively. Specialized crews for inspection and removal tasks work on a single job at a time, and—similarly to machines—do not switch to other jobs until they are finished with the current one.

While the paper considers shredded wood as the end-product produced by the facility, different wood processing techniques can also be handled in a similar fashion. The proposed modeling and solution methodology is flexible enough to accommodate different processes as well, as long as their relations can be defined in a similar way.

2.2. Uncertainty and Objective

The uncertainty of the problem comes from two steps: coating removal and re-shredding. The percentage of coated raw material is not known in advance, it is determined during the inspection step. Similarly, the amount of shredded material, that is not fine enough and requires re-shredding is only revealed during screening. There is, however, statistical data available for both of these for each of the four categories. Since the outcome of the inspection/screening influences the amount of raw material needing coating removal/re-shredding, the time required for these steps is stochastic.

Unless the worst-case scenario is considered for all of the deliveries, the feasibility (i.e., adhering to the deadlines) of a schedule cannot be ensured. However, if processing plans were tailored for the worst-case scenario, the real-life operation would have too much idle time and an oversized machine capacity would be required.

Thus, a different approach and objective is proposed. Two scenarios are modeled simultaneously: worst considered case, and robust case. The theoretical worst case would be when 100% of a delivery requires both coating removal and re-shredding. The probability of this is very low. Instead of this, the worst considered case assumes ratios needing coating removal and re-shredding, such that the probability that the actual ratios will be lower, is 95%. Similarly, the robust case assumes ratios that are higher or equal to the actual ratios 80% of the time, based on the statistics. These percentages can be set to different values by the experts responsible for daily operation planning, to manage the robustness of the solution.

The cumulative probabilities can be translated into material ratios, based on the probability density functions constructed by estimating the distribution from statistical data. Figure 2 shows such a probability density function for coating removal. Aiming for 95% and 80% probabilities, this translates to 81%, and 62% of materials needing coating removal in the worst considered and robust cases, respectively. A separate function is created for each delivery category, and for both coating removal and re-shredding.

By fixing uncertain parameters to constant values in each scenario, the same modeling techniques can be applied as in deterministic scheduling problems. Robustness can be specified through the chosen probability values, and more scenarios can be analyzed by rerunning the solution procedure with different values. A possible future research topic is to incorporate uncertainty into the scheduling model in the form of stochastic variables, fuzzy numbers, or by other techniques.

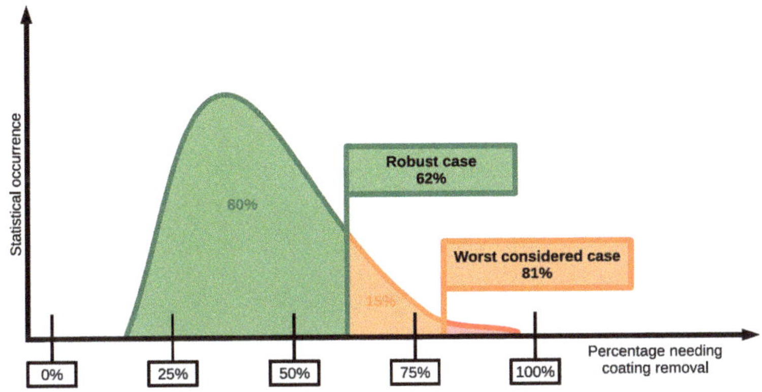

Figure 2. Probability density function for the ratio of raw material needing coating removal based on statistical data.

Scheduling decisions are the same for both scenarios but some constraints and objectives are only connected to one of them.

Delivery deadlines impose hard constraints for the robust case. So a schedule is only feasible if no lateness occurs in this scenario. It is important to note that even if some ratios turn out to be over the specified robust threshold, deadlines may still be satisfied due to reserve machine times, which are either inherently present in the schedule, or resulted from lower ratios for other jobs.

For given robustness thresholds, plenty of schedules may be feasible. To select the best of them, two different objectives are proposed in this paper.

The first objective is to minimize deadline violations for the worst considered case. Deliveries may have different associated penalties, liquidated damages, or importance based on the client. These factors are modeled via a scalar priority, and the exact objective is the priority-weighted sum of lateness. Simply put, the reported solution will be a robust schedule with the lowest possible penalty in the worst case.

Processing machines require a huge amount of electricity to operate. Machine assignment to jobs can influence the total energy consumption of the facility, and as a result, its environmental footprint. The second objective is to provide a robust schedule that minimizes this footprint, coming from the energy needs of setup, shutdown, and operation of the machines. Energy need for setup and shutdown is provided for each machine as is. For operation, two different power values are given per machine: working and idle power. It is assumed that machines are turned off for the night, and only turned off before the shifts end, if they are not required later on the same day.

2.3. Formal Problem Definition

The proposed problem class can be defined formally by the sets and parameters listed below.

2.3.1. Job Related Data

J is the finite set of jobs/deliveries.

$J^{\mathrm{B}}, J^{\mathrm{H}}$ are the sets of jobs with building/industrial and household deliveries, respectively. $J^{\mathrm{B}} \cap J^{\mathrm{H}} = \emptyset$ and $J^{\mathrm{B}} \cup J^{\mathrm{H}} = J$

$J^{\mathrm{S}}, J^{\mathrm{D}}$ are the sets of jobs with solid and derived wood deliveries, respectively. $J^{\mathrm{S}} \cap J^{\mathrm{D}} = \emptyset$ and $J^{\mathrm{S}} \cup J^{\mathrm{D}} = J$

$d_j^a \in \mathbb{Z}^{0,+}$ is the day of arrival of the delivery for job $j \in J$. $[day]$

$d_j^s \in \mathbb{Z}^{0,+}$ is the day of shipping of the product for job $j \in J$. $[day]$

$p_j \in \mathbb{R}^{0,+}$ is the priority of job $j \in J$. $[-]$

$m_j \in \mathbb{R}^{0,+}$ is the total mass of the delivery for job $j \in J$. $[t]$

2.3.2. Infrastructure Related Data

$c^{\mathtt{IS}}, c^{\mathtt{CR}}, c^{\mathtt{MS}} \in \mathbb{R}^{0,+}$ are the throughput capacities for inspection and separation, coating removal, and manual metal separation at the facility, respectively. $[t/h]$

M is the finite set of machines, the union of the following pairwise disjoint sets:

$M^{\mathtt{MS}}$ is the finite set of machines for automated metal separation.

$M^{\mathtt{PS}}$ is the finite set of machines for pre-shredding.

$M^{\mathtt{SH}} = M^{\mathtt{RS}}$ is the finite set of machines for shredding and re-shredding.

$M^{\mathtt{SC}}$ is the finite set of machines for screening.

$c_{m,j} \in \mathbb{R}^{0,+}$ is the throughput capacity of a machine $m \in M$ for job $j \in J$. $[t/h]$

$e_m \in \mathbb{R}^{0,+}$ is the electrical power consumption of a machine $m \in M$. $[kW]$

$w_m \in \mathbb{R}^{0,+}$ is the total amount of electrical energy needed to shutdown and then start up machine $m \in M$. $[kWh]$

$s \in \mathbb{R}^{0,+}$ is the length of the shifts. $[h]$

2.3.3. Uncertain Data

$p^{\mathtt{CR}}_{X,Y}, p^{\mathtt{RS}}_{X,Y} \in [0,1]$ are the percentages requiring coating removal and re-shredding for jobs with origins $X \in \{\mathtt{B},\mathtt{H}\}$ and material type $Y \in \{\mathtt{S},\mathtt{D}\}$ in the robust case. $[-]$

$\bar{p}^{\mathtt{CR}}_{X,Y}, \bar{p}^{\mathtt{RS}}_{X,Y} \in [0,1]$ are the percentages requiring coating removal and re-shredding for jobs with origins $X \in \{\mathtt{B},\mathtt{H}\}$ and material type $Y \in \{\mathtt{S},\mathtt{D}\}$ in the worst-case scenario. $[-]$

3. Proposed Approach

In this section we propose a family of Mixed-Integer Linear Programming (MILP) models to tackle the aforementioned optimization problem. All of these models rely on the same basis, and differ in objective functions and/or several constraints. The models can be applied in various ways, as illustrated in Section 4: solving them provides the globally optimal solution, or for large problem instances, they may report a close-optimal solution within a short amount of time. Having two objective functions, one can derive a Pareto-curve with iterative execution of the optimizer. The models are developed for offline scheduling, however, by fixing variables that represent already applied decisions, the models may be used in a reactive fashion as well.

The operation of the facility is not continuous, s represents the total length of work shifts on a day. However, since tasks can be interrupted at the end of shifts and resumed the next day, the active time of the facility can be considered as a continuous time horizon. If something happens at time $t \geq 0$ in the model described below, it refers to the $t - s \cdot \lfloor \frac{t}{s} \rfloor$-th hour on the $\lceil \frac{t}{s} \rceil$-th day.

3.1. General Structure and Derived Sets

As detailed later, the model follows a precedence-based representation of scheduling decisions. To ease the formulation, several sets are derived from the input data.

As already used for the c, p and $\bar{p}$ parameters and the M sets above, the following symbolic names represent the processing steps:

`IS` Inspection and separation

`CR` Coating removal

`MS` Metal separation

`PS` Pre-shredding

`SH` Shredding

`RS` Re-shredding

SC Screening

Various subsets of these will be used for simpler formalization of the model:

$$S^{\mathrm{C}} = \{\mathtt{IS}, \mathtt{CR}, \mathtt{MS}\} \tag{1}$$

$$S^{\mathrm{M}} = \{\mathtt{MS}, \mathtt{PS}, \mathtt{SH}, \mathtt{RS}, \mathtt{SC}\} \tag{2}$$

$$S = S^{\mathrm{C}} \cup S^{\mathrm{M}} \tag{3}$$

$$\bar{S} = \{\mathtt{CR}, \mathtt{RS}\} \tag{4}$$

S^{C} contains the steps that could be done manually by dedicated crews, while S^{M} contains the steps that can be carried out by machines. S is simply the set of all step types, and $\bar{S}$ contains the steps with uncertain load.

The mandatory precedence relations between these steps are stored as pairs in the P set:

$$P = \{(\mathtt{IS}, \mathtt{CR}), (\mathtt{CR}, \mathtt{PS}), (\mathtt{CR}, \mathtt{SH}), (\mathtt{PS}, \mathtt{SH}), (\mathtt{SH}, \mathtt{SC}), (\mathtt{SC}, \mathtt{RS})\} \tag{5}$$

A key derived set for the formulation is T, the set of all processing steps for all jobs, which will be called the set of tasks and can formally be defined as:

$$T = \{(j, s) \in J \times S \mid s \neq \mathtt{PS} \vee j \in J^{\mathrm{B}} \cap J^{\mathrm{S}}\} \tag{6}$$

Another useful derived set is the set of possible machine-task assignments:

$$A = \{((j, s), m) \in T \times M \mid s \in S^{\mathrm{M}} \wedge m \in M^{s}\} \tag{7}$$

3.2. Basic Scheduling Variables

As mentioned before, the model follows a precedence based logic. Each task is assigned two non-negative continuous variables indicating its start and completion time:

$$t^{s}_{j,s}, t^{c}_{j,s} \in [s \cdot d^{a}_{j}, s \cdot d^{s}_{j}] \qquad \forall (j, s) \in T \tag{8}$$

The decision whether to do metal separation manually or by a machine is represented by the following binary variable:

$$d_j \in \{0, 1\} \qquad \forall j \in J \tag{9}$$

$d_j = 1$ means manual separation, and $d_j = 0$ represents separation by machines.

Binary variables represent assignment decisions as well, 1 referring to a machine assigned to a task, and 0 if it is not:

$$a_{j,s,m} \in \{0, 1\} \qquad \forall ((j, s), m) \in A \tag{10}$$

As parallel execution is allowed, and several machines may be assigned to the same task (e.g., shredding of a single delivery is performed on two or more machines), a continuous variable represents the quantity distribution for each assignment:

$$q_{j,s,m} \in [0, m_j] \qquad \forall ((j, s), m) \in A \tag{11}$$

There are tasks that may share one or more machines (e.g., shredding is done on the same machine for two different deliveries). The set of these (ordered) task pairs is given in the following set:

$$C^{\mathrm{M}} = \{((j_1, s_1), (j_2, s_2)) \in T \times T \mid s_1, s_2 \in S^{\mathrm{M}} \wedge M^{s_1} \cap M^{s_2} \neq \emptyset \wedge (j_1, s_1) \neq (j_2, s_2)\} \tag{12}$$

Note, that as most steps have their separate machines, $s_1 \neq s_2$ only occurs, if one of them is SH and the other is RS.

If such tasks are both assigned to at least one shared machine, they have to be sequenced. The following variable takes the value of 1, if (j_1, s_1) must precede (j_2, s_2) if at least one shared machine is assigned to both of them. The value 0 indicates the opposite direction.

$$b^{\mathrm{M}}_{j_1,s_1,j_2,s_2} \in \{0,1\} \qquad \forall((j_1,s_1),(j_2,s_2)) \in A \tag{13}$$

For manual jobs, each step has a dedicated crew, and thus assignment collision is there for all job pairs. A similar set to C^{M} is defined for manual jobs, however, in this case it only contains triplets, as the step has to be the same for colliding tasks:

$$C^{\mathrm{C}} = \{(j_1,s,j_2) \in J \times S^{\mathrm{C}} \times J \mid j_1 \neq j_2\} \tag{14}$$

And a similar binary variable represents sequencing:

$$b^{\mathrm{C}}_{j_1,s,j_2} \in \{0,1\} \qquad \forall(j_1,s,j_2) \in C^{\mathrm{C}} \tag{15}$$

Finally, **M** will denote a sufficiently large number for relaxations in big-M constraints.

3.3. *Constraints*

3.3.1. Logical and Balance Constraints

There are some logical dependencies between the decisions represented by the introduced binary variables, that can be expressed by the following constraints.

$$a_{j,\mathtt{MS},m} \leq 1 - d_j \qquad \forall((j,\mathtt{MS}),m) \in A \tag{16}$$

$$b^{\mathrm{C}}_{j_1,s,j_2} + b^{\mathrm{C}}_{j_2,s,j_1} = 1 \qquad \forall(j_1,s,j_2) \in C^{\mathrm{C}} \setminus \{\mathtt{MS}\} \tag{17}$$

$$b^{\mathrm{C}}_{j_1,\mathtt{MS},j_2} + b^{\mathrm{C}}_{j_2,\mathtt{MS},j_1} \leq \frac{d_{j_1} + d_{j_2}}{2} \qquad \forall(j_1,\mathtt{MS},j_2) \in C^{\mathrm{C}} \tag{18}$$

$$b^{\mathrm{C}}_{j_1,\mathtt{MS},j_2} + b^{\mathrm{C}}_{j_2,\mathtt{MS},j_1} \geq d_{j_1} + d_{j_2} - 1 \qquad \forall(j_1,\mathtt{MS},j_2) \in C^{\mathrm{C}} \tag{19}$$

$$b^{\mathrm{M}}_{j_1,s_1,j_2,s_2} + b^{\mathrm{M}}_{j_2,s_2,j_1,s_1} = 1 \qquad \forall((j_1,s_1),(j_2,s_2)) \in C^{\mathrm{M}} : s_1 \neq \mathtt{MS} \tag{20}$$

$$b^{\mathrm{M}}_{j_1,\mathtt{MS},j_2,\mathtt{MS}} + b^{\mathrm{M}}_{j_2,\mathtt{MS},j_1,\mathtt{MS}} \leq 1 - \frac{d_{j_1} + d_{j_2}}{2} \qquad \forall((j_1,\mathtt{MS}),(j_2,\mathtt{MS})) \in C^{\mathrm{M}} \tag{21}$$

$$b^{\mathrm{M}}_{j_1,\mathtt{MS},j_2,\mathtt{MS}} + b^{\mathrm{M}}_{j_2,\mathtt{MS},j_1,\mathtt{MS}} \geq 1 - d_{j_1} - d_{j_2} \qquad \forall((j_1,\mathtt{MS}),(j_2,\mathtt{MS})) \in C^{\mathrm{M}} \tag{22}$$

Equation (16) ensures that no machine is assigned to metal separation if it is decided to be done manually. Equation (17) states that for every job pair and manual step, one of the jobs should precede the other for that step. Metal separation is handled separately in Equations (18) and (19) to address the possibility of automated metal separation. Equations (20)–(22) do the same for steps carried out by machines, addressing `MS` separately again.

$$\sum_{(j,s,m) \in A} q_{j,s,m} = m_j \qquad \forall(j,s) \in T : s \in \{\mathtt{PS},\mathtt{SH},\mathtt{SC}\} \tag{23}$$

$$\sum_{(j,\mathtt{MS},m) \in A} q_{j,\mathtt{MS},m} = m_j \cdot (1 - d_j) \qquad \forall j \in J \tag{24}$$

$$\sum_{(j,\mathtt{RS},m) \in A} q_{j,s,m} = p^{\mathtt{RS}}_{\mathrm{X},\mathrm{Y}} \cdot m_j \qquad \forall X \in \{\mathtt{B},\mathtt{H}\}, Y \in \{\mathtt{S},\mathtt{D}\}, j \in J_X \cap J_Y \tag{25}$$

$$q_{j,s,m} \leq m_j \cdot a_{j,s,m} \qquad \forall((j,s),m) \in A \tag{26}$$

Equations (23)–(25) enforce that the sum of the quantities assigned to machines at a step should add up to the total load. Besides metal separation, re-shredding needs a separate constraint as well, due to its uncertain nature. Equation (26) allows only assigned units to process a non-zero quantity for a task.

3.3.2. Processing Time Constraints

The following constraints set the relation between the starting and completion times of tasks.

$$t^c_{j,\mathtt{IS}} = t^s_{j,\mathtt{IS}} + \frac{m_j}{c^{\mathtt{IS}}} \qquad \forall j \in J \tag{27}$$

$$t^c_{j,\mathtt{MS}} \geq t^s_{j,\mathtt{MS}} + \frac{m_j}{c^{\mathtt{MS}}} - \mathbf{M} \cdot (1 - d_j) \qquad \forall j \in J \tag{28}$$

$$t^c_{j,\mathtt{MS}} \leq t^s_{j,\mathtt{MS}} + \frac{m_j}{c^{\mathtt{MS}}} + \mathbf{M} \cdot (1 - d_j) \qquad \forall j \in J \tag{29}$$

$$t^c_{j,\mathtt{CR}} = t^s_{j,\mathtt{CR}} + p^{\mathtt{CR}}_{X,Y} \cdot \frac{m_j}{c^{\mathtt{CR}}} \qquad \forall X \in \{\mathtt{B},\mathtt{H}\}, Y \in \{\mathtt{S},\mathtt{D}\}, j \in J_X \cap J_Y \tag{30}$$

$$t^c_{j,s,c} \geq t^s_{j,s,c} + \frac{q_{j,s,m}}{c_{m,j}} \qquad \forall((j,s),m) \in A \tag{31}$$

Inspection and separation is the simplest to address (27), metal separation needs special attention due to possible automated execution (28), and coating removal has its own equation as well due to its uncertain nature (30).

A single constraint, Equation (31) sets the difference between completion and starting time for all tasks that are carried out by machines.

3.3.3. Production Precedence Constraints

The following 5 equations express mandatory production sequencing due to material flows.

$$t^s_{j,s_n} \geq t^c_{j,s_p} \qquad \forall j \in J, (s_p, s_n) \in P : (j, s_p) \in T \wedge (j, s_n) \in T \tag{32}$$

Equation (32) enforces all mandatory production precedence relations in P for all jobs. Namely, if s_p is a predecessor step of s_n, then the starting time of s_n must be at earliest the completion time of s_p.

$$t^s_{j,\mathtt{MS}} \geq t^c_{j,\mathtt{IS}} - \mathbf{M} \cdot (1 - d_j) \qquad \forall j \in J \tag{33}$$

$$t^s_{j,\mathtt{CR}} \geq t^c_{j,\mathtt{MS}} - \mathbf{M} \cdot (1 - d_j) \qquad \forall j \in J \tag{34}$$

$$t^s_{j,\mathtt{MS}} \geq t^c_{j,\mathtt{SH}} - \mathbf{M} \cdot d_j \qquad \forall j \in J \tag{35}$$

$$t^s_{j,\mathtt{SC}} \geq t^c_{j,\mathtt{MS}} - \mathbf{M} \cdot d_j \qquad \forall j \in J \tag{36}$$

Equations (33) and (34) handle the case when metal separation is done by a dedicated crew, and it must be carried out after inspection and before coating removal. Equations (35) and (36) address the other case, when the metal separation is done by a machine, then it must follow shredding and precede screening.

3.3.4. Scheduling Precedence Constraints

The following two constraints express timing constraints that are the results of sequencing decisions made by $b^{\mathtt{C}}$ and $b^{\mathtt{M}}$ binary variables.

$$t^s_{j_2,s} \geq t^c_{j_1,s} - \mathbf{M} \cdot (1 - b^{\mathtt{C}}_{j_1,s,j_2}) \qquad \forall (j_1, s, j_2) \in C^{\mathtt{C}} \tag{37}$$

$$\begin{gathered} t^s_{j_2,s_2} \geq t^c_{j_1,s_1} - \mathbf{M} \cdot (3 - a_{j_1,s_1,m} - a_{j_2,s_2,m} - b^{\mathtt{M}}_{j_1,s_1,j_2,s_2}) \\ \forall((j_1,s_1),(j_2,s_2)) \in C^{\mathtt{M}}, m \in M^{s_1} \cap M^{s_2} \end{gathered} \tag{38}$$

Equation (37) handles sequencing of dedicated crews, while Equation (38) addresses machine sequencing.

3.4. Objective Functions

As discussed before, two different objective functions are considered here, and each requires additional variables and constraints to be calculated. Most of these variables are fixed or computed, e.g., they do not introduce additional decisions into the model, rather they are the consequences of the decisions made by the preexisting variables.

3.4.1. Priority Weighted Lateness Minimization

As this objective function intends to minimize the total weighted lateness in the worst considered case, an alternate timing for each task needs to be calculated. In order to achieve that, a worst considered scenario versions of t^s, t^c and q variables need to be defined over the same domains: $\bar{t}^s$, $\bar{t}^c$, and $\bar{q}$. The only difference is that the upper limit of $\bar{t}^s$ and $\bar{t}^c$ is infinity instead of $s \cdot d_j^s$. These variables represent the starting and completion times, and the load of machines when coating removal and re-shredding takes longer in the worst considered case.

To set these variables, Equations (23)–(38) need to be duplicated with the worst considered case variables instead of the robust case ones.

An additional set of continuous variables, the number of late days of each job needs to be introduced:

$$\bar{l}_j \in \mathbb{Z}^{0,+} \qquad \forall j \in J \tag{39}$$

Equation (40) calculates the values for these variables:

$$\bar{t}^c_{j,\mathrm{RS}} \le (d_j^s + \bar{l}_j) \cdot s \qquad \forall j \in J \tag{40}$$

Then, the objective function can be defined as:

$$\sum_{j \in J} p_j \cdot \bar{l}_j \to \min \tag{41}$$

3.4.2. Electrical Footprint Minimization

To ease the formulation of this objective function, a set for the considered days (in the robust case) is defined as:

$$H_j = \{d_j^a - 1, \ldots, d_j^s + 1\} \qquad \forall j \in J \tag{42}$$

$$H = \cup_{j \in J} H_j \tag{43}$$

An additional binary variable is introduced to indicate whether a machine is turned on or not on a particular day:

$$u_{m,d} \in \{0,1\} \qquad \forall m \in M, d \in H \tag{44}$$

The total energy consumption is the sum of electricity used for processing tasks and the energy used to start up and shut down machines on active days. This objective can be expressed as:

$$\sum_{(j,s,m) \in A} \frac{q_{j,s,m} \cdot e_m}{c_{m,j}} + \sum_{m \in M, d \in H} u_{m,d} \cdot w_m \to \min \tag{45}$$

However, u needs to be computed from t and a variables. In order to do that, an auxiliary computed binary variable is introduced to indicate if a machine works on a task at a given day:

$$y_{j,s,m,d} \in \{0,1\} \qquad ((j,s),m) \in A, d \in H_j \tag{46}$$

If a machine works on at least one task on a day, it must be turned on, as expressed by Equation (47):

$$u_{m,d} \ge y_{j,s,m,d} \qquad \forall ((j,s),m) \in A, d \in H_j \tag{47}$$

A machine must work on a task at least as many days as many is needed to process the assigned quantity:

$$\sum_{d \in H} y_{j,s,m,d} \geq \frac{q_{j,s,m}}{c_{m,j} \cdot s} \qquad \forall((j,s),m) \in A \tag{48}$$

The following constraints ensure that a task does not start on an earlier day or end on a later one than indicated by the corresponding y variables:

$$t^s_{j,s} \geq d \cdot s - \mathbf{M} \cdot (1 - y_{j,s,m,d} + y_{j,s,m,d-1}) \qquad \forall((j,s),m) \in A, d \in H_j : d > d^a_j \tag{49}$$

$$t^c_{j,s} \leq (d+1) \cdot s + \mathbf{M} \cdot (1 - y_{j,s,m,d} + y_{j,s,m,d+1}) \qquad \forall((j,s),m) \in A, d \in H_j : d < d^s_j \tag{50}$$

$$t^s_{j,s} \leq (d+1) \cdot s + \mathbf{M} \cdot (1 - y_{j,s,m,d}) \qquad \forall((j,s),m) \in A, d \in H_j : d \geq d^s_j \tag{51}$$

$$t^c_{j,s} \geq d \cdot s - \mathbf{M} \cdot (1 - y_{j,s,m,d}) \qquad \forall((j,s),m) \in A, d \in H_j : d \leq d^s_j \tag{52}$$

3.5. Overview of the Mathematical Model

A structural overview of the sets, parameters, and variables used in the proposed model is illustrated in Table 1. The table details the sets and parameters that are provided as input data, and the ones that are derived from others. Moreover, it also highlights the parameters, sets, and variables that are required for the base model or one of the objective functions.

Table 1. Overview of model notations.

	BASE MODEL	
Input data	sets parameters	$J^{\mathrm{B}}, J^{\mathrm{H}}, J^{\mathrm{S}}, J^{\mathrm{D}}, M^{\mathrm{MS}}, M^{\mathrm{PS}}, M^{\mathrm{SH}}, M^{\mathrm{SC}}$ $d^a_j, d^s_j, m_j, c^{\mathrm{IS}}, c^{\mathrm{CR}}, c^{\mathrm{MS}}, c_{m,j}, s, p^{\mathrm{CR}}_{X,Y}, p^{\mathrm{RS}}_{X,Y}$
Derived sets		$J, M, M^{\mathrm{RS}}, S^{\mathrm{C}}, S^{\mathrm{M}}, S, \bar{S}, P, T, A, C^{\mathrm{M}}, C^{\mathrm{C}}$
Decision variables	continuous binary	$t^s_{j,s}, t^c_{j,s}, q_{j,s,m}$ $d_j, a_{j,s,m}, b^{\mathrm{M}}_{j_1,s_1,j_2,s_2}, b^{\mathrm{C}}_{j_1,s,j_2}$
	PRIORITY WEIGHTED LATENESS MINIMIZATION	
Additional input data		$p_j, \bar{p}^{\mathrm{CR}}_{X,Y}, \bar{p}^{\mathrm{RS}}_{X,Y}$
Additional variables	continuous discrete	$\bar{t}^s_{j,s}, \bar{t}^c_{j,s}, \bar{q}_{j,s,m}$ $\bar{l}_j$
Objective function		$\sum_{j \in J} p_j \cdot \bar{l}_j$
	ELECTRICAL FOOTPRINT MINIMIZATION	
Additional input data		H_j, H, e_m, w_m
Additional variables	binary	$u_{m,d}, y_{j,s,m,d}$
Objective function		$\sum_{(j,s,m) \in A} \frac{q_{j,s,m} \cdot e_m}{c_{m,j}} + \sum_{m \in M, d \in H} u_{m,d} \cdot w_m$

4. Numerical Results

In order to properly test the efficiency of the proposed model, a large number of tests had to be carried out on instances of varying sizes and structures. While acquiring single instances can be manageable by monitoring a collection center over a given period, solving only a handful of these does not show the general usefulness of the model, rather its functionality on a couple of dedicated use-cases. Existing large-scale information about waste generation, collection and processing is mostly published as statistical data, aggregated usually in a yearly fashion [19]. Such data cannot be decomposed into actual real-life scenarios, but information about the general structure of real-world instances can be derived from them. Because of this limitation, instances used for testing the model were randomly generated based on studies about the characteristics and distributions of wood waste. Randomly generated instances can turn out to be infeasible, usually in cases where a greater amount of deliveries arriving around the same time cannot be processed by their given deadline due to the limited capacity of the available machines. However, such inputs are also useful for testing purposes to see how efficiently infeasibility can be reported. Proving the infeasibility of scenarios is not always a trivial task, but it should be within the capabilities of the model.

Features of the deliveries such as their origin (building or industrial waste, and wastes from households and collection centers) and type (solid or derived) were determined based on statistics in [18]. While that study discusses the four origins independently of each other, we merged them into two groups based on their similarities for the sake of simplicity and easier presentation. Naturally, the model would function the same way with four different origins.

The total mass of deliveries was determined based on the potential payloads of biomass transport trucks [20]. Two different payload ranges were considered. Instances with a small payload contained deliveries of 19–23 t, that could hypothetically be processed on their arrival day, while the deliveries of instances with large, 31–49 t payloads could take more than one day to process. Information of the machines (throughput and energy use) for the different tasks was based on real machinery [21]. Three machines with different properties were available for every machine task, each extra option offering higher throughput in exchange for more energy use. Only a single crew was provided for each manual step of the process, and their throughput was intentionally set significantly lower than that of the machine steps. This resulted in inspection and sorting being the bottleneck of the scheduling process, as every delivery had to go through this step, and there are no machine alternatives for it. For this reason, tests were run both with the original 10 t/h throughput, and with doubling this efficiency.

A large number of tests were performed with optimizing for either one of the objective functions, or solving the combined bi-objective optimization problem. The proposed model was solved using Gurobi 9.1 solver for every instance. Tests were run on a PC with an Intel Core i7-5820K 3.30 GHz CPU and 32 GB memory.

4.1. Single Objective Optimization

The first set of input instances were generated over a one-week horizon. The arrival day of the deliveries was chosen in a uniformly random way from this period, and the deadlines were also uniformly random: 1–2 days from arrival for small deliveries and 2–3 days for large deliveries. Various instance sets were generated with delivery numbers between 5 and 20, and the average running times of instances of the same set are presented in Table 2 for the slower inspection throughput, and Table 3 for the faster one. Each row of the tables gives the number of deliveries and delivery size (S—small, L—large) of the instances group, and provides an average optimization time for a set of 10 randomized instances in the case of minimizing their lateness or energy use. The optimization time corresponds to the time needed to either achieve an optimal solution or determine the infeasibility of the instance under the current parameter settings. A running time limit of one hour (3600 s) was enforced on the solution process. In some cases, non-optimal solutions were found within the time limit, these will be explained in detail when discussing the corresponding tables. Such occurrences are marked with an * next to the average running time.

In the case of the one-week horizon and a slower inspection throughput, the running time limit was reached for two inputs of the (15,L) instance set while minimizing energy use. The instances were infeasible in both cases. However, this infeasibility was found in under 10 s when optimizing for minimal lateness, so these running times were used when calculating the energy use average. Results for the (20,L) instances are not provided, as they were infeasible in every case, because the planning horizon was too small to schedule every delivery on the available machines.

Table 2. Average results of a one-week horizon and slower inspection and sorting process.

Deliveries	Size	Lateness Sol. Time (s)	Energy Sol. Time (s)
5	S	0.10	0.15
5	L	0.15	0.55
10	S	1.62	2.21
10	L	1.06	7.23
15	S	2.73	23.20
15	L	9.70	154.24
20	S	2.86	339.56
20	L	-	-

Doubling the speed of the inspection and sorting step shows a more efficient solution process, the results of which are presented in Table 3.

Table 3. Average results of a one-week horizon and faster inspection and sorting process.

Deliveries	Size	Lateness Sol. Time (s)	Energy Sol. Time (s)
5	S	0.09	0.13
5	L	0.13	0.49
10	S	0.30	0.46
10	L	0.54	8.62
15	S	1.46	5.35
15	L	38.36	743.70
20	S	7.45	86.17
20	L	979.11 *	1243.16 *

Optimal solutions were found for all instances for the sizes of 5, 10 and 15 deliveries except for one: in the (15,L) instance set, a single solution reached the time limit, and produced a non-optimal result with a 2.18% gap. However, while solutions were found for all instances of the (20,L) instance set, only 5 were optimal when optimizing for lateness, and 7 were optimal when optimizing for energy. The suboptimal solutions found for the remaining instances had a greater than 50% optimality gap in every case. As a result, these are not considered in the averages of Table 3. As bad-quality suboptimal solutions were found for multiple inputs in this (20,L) instance set, this seems to be the limit of the model considering the given time limit. Efficient (optimal, or close-to-optimal) solutions were found for all other problem classes with a short average running time.

The model was tested for larger delivery numbers as well. However, as slower inspection resulted in infeasible solutions for the (20,L) instances, these larger inputs were tested over a two-week planning horizon. The instance sets were generated under the same conditions as the one-week sets. Their average running times can be seen in Table 4 in the case of slower inspection throughput, and Table 5 in the faster case.

Table 4. Average results of a two-week horizon and slower inspection and sorting process.

Deliveries	Size	Lateness Sol. Time (s)	Energy Sol. Time (s)
20	S	3.59	72.37
20	L	21.59 *	494.32 *
25	S	10.02	113.25
25	L	431.46 *	1655.49 *
30	S	17.23 *	549.61 *
30	L	-	-

When using the slower throughput for inspection and sorting, two instances in the (20,L) set did not yield any result in the one hour limit when optimizing for energy.

However, these instances were shown to be infeasible under 60 s when optimizing for lateness, so these values were used for calculating the average runtime. Another instance did not provide an optimal solution in one hour, and the best feasible solution had a 3.79% gap. The (25,L) set had one instance where no optimal solution was found for either objective function in an hour, but the optimality gap of the best solution was 83.88% and 17.29% respectively. Another three instances only gave optimal solutions for minimizing lateness, but could only find near-optimal solutions for minimizing energy with 3.65%, 5.69% and 2.57% gaps. One such instance was also present for (30,S), where only a solution with a 2.33% gap was found in one hour. Two inputs of this set did not yield any results in the one hour running time limit for either objective function. These inputs were not considered in the averages. Minimizing lateness for the same instances yielded an optimal result below 60 s. The instances of the (30,L) set, however, were either all infeasible, of no solution was found in the one hour limit, so their results are not presented in the table.

Again, transitioning to a faster inspection and sorting throughput provides more efficient results, which can be seen in Table 5.

Table 5. Average results of a two-week horizon and faster inspection and sorting process.

Deliveries	Size	Lateness Sol. Time (s)	Energy Sol. Time (s)
20	S	1.68	4.80
20	L	9.10	329.92
25	S	5.28	36.56
25	L	144.57 *	2445.34 *
30	S	17.45	776.60
30	L	-	-

It can be seen from the table that optimal solutions were found for all instances scheduling short deliveries regardless of their problem sizes. In the case of the (25,L) instances, time limit was reached for a single input when optimizing for lateness. The suboptimal solution had a 44.06% optimality gap, and it is not included in the average. Time limit was reached in five cases when minimizing energy consumption for the same instance set. However, these solutions were much closer to the optimum (with respective gaps of 4.74%, 1.82%, 5.31%, 2.87% and 0.93%), and their running time was included in the average. Solution of the (30,L) instance set inputs, however, reached the time limit on every occasion, and while solutions were found, their optimality gap was above 50% in every case. This clearly shows that this instance set is not solvable in the given time limit, and results are not presented for this reason.

If the goal is the optimization of a single objective, it can be seen from the above results that the model can efficiently schedule a big number of smaller deliveries over both a one- and two-week horizon, and has no problem with larger deliveries up to a certain problem size. While solutions could not be acquired for some instances in the given time limit, or their quality was not good enough, this can be remedied with the increase of available running time for the solution process. The solution of the model is significantly easier when optimizing for lateness, which was expected due to the added complexity of the large number of extra binary decision variables in the case of the energy minimization objective.

4.2. Bi-Objective Optimization

In the case of considering both objectives at the same time, a bi-objective optimization problem has to be solved. One option for this is the augmented ε-constraint method introduced in [22]. This method yields multiple non-dominated solutions for the problem, meaning that there is no obvious best solution among them. Such a set of solutions where one cannot find an improved alternative to any of them is called a Pareto front.

The solutions of this front are achieved by solving a series of optimization problems based on the original model. First, the lexicographic method is applied: the objectives are assigned a hierarchical ordering, and the model is solved considering all objectives in this order. Two hierarchical optimization problems are solved, one having the lateness objective on top of the hierarchy, while the other having the energy objective. Using the objective values of these solutions, the possible value range can be determined for each objective. This value range is then divided along multiple grid points, and an optimization problem is solved for every region of this division. The number of problems to be solved depends on the chosen number of grid points, which acts as a parameter of the ε-constraint method. Using G grid points will result in $G + 1$ regions.

Based on the experiences from the single objective cases, bi-objective optimization was carried out for problem sizes of 5–25 deliveries. Inputs with 5–15 deliveries were generated over a one-week horizon, while inputs with 20 and 25 deliveries were generated over a two-week horizon. For each delivery number, 10-10 instances were generated with small and large deliveries. Bi-objective optimization was carried out for all of these instances using both 5 and 10 grid points, resulting in 6 or 11 problems to be solved. A time limit of 30 min (1800 s) was set for the individual solution processes. Instance sets were generated with both slower and faster inspection and sorting throughput similarly to the single-objective case. Aggregated results can be seen in Table 6 for the slower, and Table 7 for the faster throughput. Both tables present the number and size of deliveries in the given instance set, as well as the number of instances where the solution process was terminated due to reaching the time limit. For instances where solutions were achieved, the average number of solutions in the Pareto front and the average required solution time is presented both for the 5 and 10 grid point divisions.

Table 6. Average bi-objective results with slower inspection and sorting process.

Deliveries	Size	Limit Reached	Division 5		Division 10	
			Front	Time (s)	Front	Time (s)
5	S	0	1.20	0.72	1.20	0.89
5	L	0	1.00	1.13	1.00	1.07
10	S	0	1.60	7.43	1.70	11.32
10	L	0	2.10	172.48	2.10	303.07
15	S	0	3.10	46.61	3.20	80.23
15	L	3	3.71	1321.81	3.86	2079.09
20	S	0	3.00	109.42	3.20	181.13
20	L	4	2.50	1331.42	2.67	1981.28
25	S	0	3.10	593.51	3.50	1043.97
25	L	10	-	-	-	-

Table 7. Average bi-objective results with faster inspection and sorting process.

Deliveries	Size	Limit Reached	Division 5		Division 10	
			Front	Time (s)	Front	Time (s)
5	S	0	1.10	0.62	1.10	0.70
5	L	0	1.00	0.92	1.00	0.90
10	S	0	1.80	12.44	1.80	18.75
10	L	0	1.40	185.86	1.40	316.93
15	S	0	2.60	50.88	2.80	85.67
15	L	4	3.67	2314.70	4.33	3952.80
20	S	0	2.50	75.03	2.60	118.81
20	L	4	3.83	1809.51	4.17	3213.87
25	S	0	3.40	631.90	3.70	896.69
25	L	10	-	-	-	-

In the case of using the slower inspection throughput, the finer division of 10 grid points usually produced the same number of solutions for the front than the 5-grid point division, or provided at most one additional. However, having 10 grid points leads to an average of 63% increase in running time for the bigger instances. Instances of the (25,L) set were not solvable in the given time limit, and solutions are not presented for them as a result.

Instances with faster inspection throughput behaved similarly to the previous instance sets. The finer 10-grid point division again produced at most one additional solution compared to the 5-grid point division with an average of 64% increase in running time for the bigger instances. There was a notable exception for one instance in the (15,S) set, where the 10-point division resulted in 9 solutions as opposed to the 6 solutions of the 5-point division. Instances of the (25,L) set were not solvable in the given time limit, and their solutions are not presented.

It can be seen from the above tables that efficient bi-objective optimization of the model is also possible. Multiple non-dominated solutions can be found for the problems in an acceptable time. Results show that increasing the size of the division from 5 to 10 grid points usually results in the same number of solutions, or provides only one additional result. However, this comes at the cost of a significant increase in running time. There was only a single instance where the finer division provided three additional solutions to the front.

5. Conclusions

This paper studied the problem of scheduling the machines in a waste wood processing facility where the incoming deliveries can come from various sources and have uncertain compositions. A multi-objective mixed-integer linear programming model was presented for the problem to provide robust solutions that minimize lateness and energy consumption. The efficiency of the method was shown on instances with various sizes, which were randomly generated based on real-world distribution data. Computational tests showed that the approach can provide mid-term (1–2 weeks) schedules for 25–30 deliveries in under an hour. The bi-objective model allows the consideration of alternative solutions based on both economic and sustainability factors.

The topic has possibilities for future research. In the current model, a simplified energy cost function was considered. The approach can be extended with a more detailed energy model, for example, considering the idle power of machines between tasks during a day, possibly with additional decisions about when to turn them off. Another possible extension is to model different operation profiles for machines, which allow more energy-efficient processing with the expense of slower throughput. To achieve better overall throughput, the scheduling model could be improved to allow overlap between subsequent processing steps. Some extensions can be too computationally intensive to solve with an exact MILP solver, so a metaheuristic solution approach may also be beneficial for future research.

Author Contributions: Conceptualization, B.D., O.Ő. and M.H.; methodology, B.D., O.Ő. and M.H.; software, B.D.; validation, B.D., O.Ő.; formal analysis, B.D. and M.H.; investigation, B.D.; resources, B.D.; writing—original draft preparation, B.D., O.Ő. and M.H.; writing—review and editing, B.D., O.Ő. and M.H.; visualization, M.H. All authors have read and agreed to the published version of the manuscript.

Funding: This research was supported by the National Research, Development and Innovation Office, NKFIH grant no. 129178. Balázs Dávid gratefully acknowledges the European Commission for funding the InnoRenew CoE project (Grant Agreement #739574) under the Horizon2020 Widespread-Teaming program, and the Republic of Slovenia (Investment funding of the Republic of Slovenia and the European Union of the European Regional Development Fund), and he is also grateful for the support of the Slovenian ARRS grant N1-0093. The authors would also like to thank the Slovenian Research Agency (ARRS) for funding infrastructure program IO-0035.

Acknowledgments: Balázs Dávid acknowledges the Slovenian Research Agency (ARRS) for the bilateral project BI-AT/20-21-014. The Authors would like to acknowledge the work of Endre Bodó, who was helping the research with testing the developed method. He was supported by "Integrated program for training new generation of scientists in the fields of computer science", no.

EFOP-3.6.3-VEKOP-16-2017-0002 (supported by the European Union and co-funded by the European Social Fund).

Conflicts of Interest: The authors declare no conflict of interest.

References

1. Tripathi, N.; Hills, C.; Singh, R.; Atkinson, C. Biomass waste utilisation in low-carbon products: Harnessing a major potential resource. *NPJ Clim. Atmos. Sci.* **2019**, *2*, 35, doi:10.1038/s41612-019-0093-5.
2. Nunes, L.; Causer, T.; Ciolkosz, D. Biomass for energy: A review on supply chain management models. *Renew. Sustain. Energy Rev.* **2020**, *120*, 109658, doi:10.1016/j.rser.2019.109658.
3. Garcia, C.A.; Hora, G. State-of-the-art of waste wood supply chain in Germany and selected European countries. *Waste Manag.* **2017**, *70*, 189–197, doi:10.1016/j.wasman.2017.09.025.
4. Dekker, R.; Fleischmann, M.; Inderfurth, K.; Wassenhove, L.N.V. (Eds.) *Reverse Logistics*; Springer: Berlin/Heidelberg, Germany, 2004; doi:10.1007/978-3-540-24803-3.
5. Salema, M.I.G.; Barbosa-Povoa, A.P.; Novais, A.Q. An optimization model for the design of a capacitated multi-product reverse logistics network with uncertainty. *Eur. J. Oper. Res.* **2007**, *179*, 1063–1077, doi:10.1016/j.ejor.2005.05.032.
6. Fiedler, P.; Lange, M.; Schultze, M. Supply logistics for the industrialized use of biomass—Principles and planning approach. In Proceedings of the 2007 International Symposium on Logistics and Industrial Informatics, Wildau, Germany, 13–15 September 2007; pp. 41–46, doi:10.1109/LINDI.2007.4343510.
7. Kazemi, N.; Modak, N.M.; Govindan, K. A review of reverse logistics and closed loop supply chain management studies published in IJPR: A bibliometric and content analysis. *Int. J. Prod. Res.* **2019**, *57*, 4937–4960, doi:10.1080/00207543.2018.1471244.
8. Troncoso, J.J.; Garrido, R.A. Forestry production and logistics planning: An analysis using mixed-integer programming. *For. Policy Econ.* **2005**, *7*, 625–633, doi:10.1016/j.forpol.2003.12.002.
9. Kaltenbrunner, M.; Huka, M.A.; Gronalt, M. Adaptive Model Building Framework for Production Planning in the Primary Wood Industry. *Forests* **2020**, *11*, 1256, doi:10.3390/f11121256.
10. Maturana, S.; Pizani, E.; Vera, J. Scheduling production for a sawmill: A comparison of a mathematical model versus a heuristic. *Comput. Ind. Eng.* **2010**, *59*, 667–674, doi:10.1016/j.cie.2010.07.016.
11. Koch, S.; König, S.; Wäscher, G. Integer linear programming for a cutting problem in the wood-processing industry: A case study. *Int. Trans. Oper. Res.* **2009**, *16*, 715–726, doi:10.1111/j.1475-3995.2009.00704.x.
12. Greinert, A.; Mrówczyńska, M.; Szefner, W. The Use of Waste Biomass from the Wood Industry and Municipal Sources for Energy Production. *Sustainability* **2019**, *11*, 3083, doi:10.3390/su11113083.
13. Marchenko, O.; Solomin, S.; Kozlov, A.; Shamanskiy, V.; Donskoy, I. Economic Efficiency Assessment of Using Wood Waste in Cogeneration Plants with Multi-Stage Gasification. *Appl. Sci.* **2020**, *10*, 7600, doi:10.3390/app10217600.
14. Devjak, S.; Tratnik, M.; Merzelj, F. Model of Optimization of Wood Waste Processing in Slovenia. In *Operations Research '93*; Bachem, A., Derigs, U., Jünger, M., Schrader, R., Eds.; Physica: Heidelberg, Germany, 1994; pp. 103–107.
15. Burnard, M.; Tavzes, Č.; Tošić, A.; Brodnik, A.; Kutnar, A. The Role of Reverse Logistics in Recycling of Wood Products. In *Environmental Implications of Recycling and Recycled Products*; Springer: Singapore, 2015; pp. 1–30, doi:10.1007/978-981-287-643-0_1.
16. Trochu, J.; Chaabane, A.; Ouhimmou, M. Reverse logistics network redesign under uncertainty for wood waste in the CRD industry. *Resour. Conserv. Recycl.* **2018**, *128*, 32–47, doi:10.1016/j.resconrec.2017.09.011.
17. Egri, P.; Dávid, B.; Kis, T.; Krész, M. Robust Reverse Logistics Network Design. In *Logistics Operations and Management for Recycling and Reuse*; Golinska-Dawson, P., Ed.; Springer: Berlin/Heidelberg, Germany, 2020; pp. 37–53, doi:10.1007/978-3-642-33857-1_3.
18. Kharazipour, A.; Kües, U. Recycling of Wood Composites and Solid Wood Products. In *Wood Production, Wood Technology, and Biotechnological Impacts*; Universitätsverlag: Göttingen, Germany, 2007; pp. 509–533.
19. Cocchi, M.; Vargas, M.; Tokacova, K. *European Wood Waste Statistics Report for Recipient and Model Regions*; Technical Report; Absorbing the Potential of Wood Waste in EU Regions and Industrial Bio-based Ecosystems (BioReg): 2019. Available online: https://www.bioreg.eu/assets/delivrables/BIOREG%20D1.1%20EU%20Wood%20Waste%20Statistics%20Report.pdf (accessed on 6 April 2021).
20. Laitila, J.; Asikainen, A.; Ranta, T. Cost analysis of transporting forest chips and forest industry by-products with large truck-trailers in Finland. *Biomass Bioenergy* **2016**, *90*, 252–261, doi:10.1016/j.biombioe.2016.04.011.
21. Komptech. Stationary Machines. Available online: https://komptech.ca/wp-content/uploads/2020/09/Komptech-Stationary-Machines_2018_E.pdf (accessed on 6 April 2021).
22. Mavrotas, G. Effective implementation of the ε-constraint method in Multi-Objective Mathematical Programming problems. *Appl. Math. Comput.* **2009**, *213*, 455–465, doi:10.1016/j.amc.2009.03.037.

sustainability

MDPI

Article

SmartISM: Implementation and Assessment of Interpretive Structural Modeling

Naim Ahmad * and Ayman Qahmash

Department of Information Systems, King Khalid University, Abha 62529, Saudi Arabia; a.qahmash@kku.edu.sa
* Correspondence: nagqadir@kku.edu.sa

Abstract: Interpretive Structural Modeling (ISM) is a technique to establish the interrelationships between elements of interest in a specific domain through experts' knowledge of the context of the elements. This technique has been applied in numerous domains and the list continues to grow due to its simplistic concept, while sustainability has taken the lead. The partially automated or manual application of this technique has been prone to errors as witnessed in the literature due to a series of mathematical steps of higher-order computing complexity. Therefore, this work proposes to develop an end-to-end graphical software, SmartISM, to implement ISM technique and MICMAC (Matrice d'Impacts Croisés Multiplication Appliquée á un Classement (cross-impact matrix multiplication applied to classification)), generally applied along with ISM to classify variables. Further, a scoping review has been conducted to study the applications of ISM in the previous studies using Denyer and Tranfield's (2009) framework and newly developed SmartISM. For the development of SmartISM, Microsoft Excel software has been used, and relevant algorithms and VBA (Visual Basic for Applications) functions have been illustrated. For the transitivity calculation the Warshall algorithm has been used and a new algorithm reduced conical matrix has been introduced to remove edges while retaining the reachability of variables and structure of digraph in the final model. The scoping review results demonstrate 21 different domains such as sustainability, supply chain and logistics, information technology, energy, human resource, marketing, and operations among others; numerous types of constructs such as enablers, barriers, critical success factors, strategies, practices, among others, and their numbers varied from 5 to 32; number of decision makers ranged between 2 to 120 with a median value of 11, and belong to academia, industry, and/or government; and usage of multiple techniques of discourse and survey for decision making and data collection. Furthermore, the SmartISM reproduced results show that only 29 out of 77 studies selected have a correct application of ISM after discounting the generalized transitivity incorporation. The outcome of this work will help in more informed applications of this technique in newer domains and utilization of SmartISM to efficiently model the interrelationships among variables.

Keywords: interpretive structural modeling (ISM); MICMAC (Matrice d'Impacts Croisés Multiplication Appliquée á un Classement (cross-impact matrix multiplication applied to classification)); Microsoft Excel; SmartISM; VBA (Visual Basic for Applications); reduced conical matrix

check for updates

Citation: Ahmad, N.; Qahmash, A. SmartISM: Implementation and Assessment of Interpretive Structural Modeling. *Sustainability* **2021**, *13*, 8801. https://doi.org/10.3390/su13168801

Academic Editors: Dragan Pamucar and Marc A. Rosen

Received: 5 June 2021
Accepted: 4 August 2021
Published: 6 August 2021

Publisher's Note: MDPI stays neutral with regard to jurisdictional claims in published maps and institutional affiliations.

Copyright: © 2021 by the authors. Licensee MDPI, Basel, Switzerland. This article is an open access article distributed under the terms and conditions of the Creative Commons Attribution (CC BY) license (https://creativecommons.org/licenses/by/4.0/).

1. Introduction

Every discipline is expanding its frontier and multiple disciplinary approaches have become essential to solve complex problems. This leads to the study of a large number of constructs of interests simultaneously. These constructs may have been identified in theory or practice. Warfield [1–4] in the 1970's developed a technique to establish an interrelationship model between variables known as interpretive structural modeling (ISM). The holistic picture of important constructs in the structured form derived from ISM technique helps the practitioners to solve the problem effectively. This technique is widely used due to its simplistic procedure and profound value addition in problem solving in different domains.

ISM helps in representing partial, fragmented, and distributed knowledge into integrated, interactive, and actionable knowledge. This technique is therefore particularly useful for the areas that are inherently multidisciplinary, such as sustainability. The discipline of sustainability ensures the performance in three areas: economic, social, and environmental, termed as triple bottom line (TBL) [5], while the world undergoes development. Additionally, the literature shows the maximum number of applications of this technique in the area of sustainability.

The search with the quoted keywords of "interpretive structural modeling" on the single database of Scopus yielded 5184 documents. There is an exponential growth in the usage of this technique from 2007 onward; prior to this year articles are around 10 each year starting from 1974. For the year 2007, 46 documents are listed and the numbers are exponentially increasing each successive year to 1200 documents in the year of 2020. With around 36% contribution in articles, India is leading the application of ISM, followed by China, USA, UK, and Iran. Together these five countries contribute around 71% of total articles. This technique is being used in many disciplines in decreasing order, namely business, engineering, computer science, decision science, environmental science, social science, and others.

ISM helps in modeling the variables and brings out the existing interrelationship structure among them. It helps a group of people or decision makers to debate and share their knowledge and achieve consensus on the relationships among the variables. The participants can share their views without any knowledge of mathematical complexity involved in the underlying steps. A computerized program may automate all the graphical and algebraic computation and convert their inputs into a pictorial model consisting of variables along with the relationships among them. The ISM process does not add any information [6] but brings in structural value [7].

In the same time period of the 1970's, another technique known as MICMAC (Matrice d'Impacts Croisés Multiplication Appliquée á un Classement (cross-impact matrix multiplication applied to classification)) was developed by J. C. Duperrin and M. Godet [8]. MICMAC helps in classification of the variables into one of the four categories, namely dependent, independent, linkage, and autonomous variables. ISM coupled with MICMAC becomes a strong tool to visualize the structure of variables along with the interrelationships between them. ISM is also used in several multi criteria decision making (MCDM) techniques such as analytical hierarchy process (AHP) [9], analytic network process (ANP) [10–12], technique for order of preference by similarity to ideal solution (TOPSIS) [13,14], decision-making trial, evaluation laboratory (DEMATEL) [15,16], and others.

Implementation of this technique and conduction of brainstorming sessions with experts in previous studies [17,18] led to identification of some key challenges such as variables' identification, selection of decision makers and method of decision making, and unavailability of end-to-end software for ISM and MICMAC. Furthermore, the literature shows erroneous applications of steps of ISM such as wrong reachability matrix [19–22], wrong transitivity calculations [9,13,16,23–37], incorrect level partitioning and wrong structure of the model [31,38–42], and incorrect addition [11,14,43–48] or reduction [49–52] of edges affecting the reachability of variables. An error in an earlier step generally leads to an error in subsequent steps. Similarly, the wrong calculation of transitivity leads to wrong MICMAC diagrams. Therefore, there exists some important issues in implementation of this technique, namely identification of variables, decision makers, expertise and experience of decision makers, method of decision making, and computerization of the steps of ISM. Previous ISM reviews [53–55] don't critically analyze the steps of ISM applications in the articles. Similarly, although some automation of the ISM technique has been provided earlier [56,57], there does not exist any end-to-end graphical software that may help in applying this technique and allow the decision makers to focus on sharing knowledge and iterate the ISM technique until a high-confidence consensus model is arrived at. These challenges set the objectives of this research as follows:

- Development of SmartISM, a software tool for ISM and MICMAC using Microsoft Excel and VBA.
- Scoping review of applications of ISM on existing studies to identify application domains, types and numbers of variables studied, composition of decision makers, decision making and data collection techniques, and accuracy of ISM application using SmartISM.

The remainder of the paper is organized into the following sections: literature review, research methodology, development of SmartISM using Microsoft Excel, results, discussion, and conclusion.

2. Literature Review

There are numerous studies in the literature that illustrate the ISM technique. They can be summarized into the seven steps approach with an additional eighth step for MICMAC analysis, as given in the following subsection. The next subsection illustrates the existing available automation of the ISM. The last subsection presents some studies that have reviewed the implementations of ISM.

2.1. ISM and MICMAC Techniques

The interpretive structural modelling (ISM) can be defined as constructs' directional structuring technique based on contextual interrelationships defined by domain experts, utilizing computerized conversion of relations into a pictorial model using matrix algebra and graph theory. It may be explained in the series of steps as follows, which will assist in automating all the processes of the ISM technique.

2.1.1. Elements or Constructs or Variables

Identification of elements or constructs of the subject being studied is the most important of all activities. Similarly, the establishment of their definition along with the theoretical boundaries or scope is very critical. Elements must be explained with the details of their definition, objectives, and possible indications or measurements. These elements are generally identified by literature review, expert opinions, and/or surveys. Some of the unique approaches have been use of thematic analysis [58], upper echelon theory [11], contingency theory [59], content analysis [52], strengths, weaknesses, opportunities, and threats (SWOT) analysis [30], idea engineering workshop [40,60], and Delphi technique [37,61]. One study [42] has defined the source, understanding, and interpretation for each variable.

2.1.2. Decision Makers (DMs)

DMs play a very significant role in ISM as the whole process and outcome are dependent upon their input. There are three important aspects for the selection of a group of DMs such as size, expertise, and diversity. The group of DMs should be representative of all of the stakeholders in the domain of the problem. They should have sound experience of domain and expert level knowledge of variables being studied. The literature shows the number of DMs ranging between 2 [62,63] to 120 [64] with a median value of 11, and very few studies [16,30,41,65] have taken DMs from academia, industry, and government together.

2.1.3. Structural Self-Interaction Matrix (SSIM)

Elements or constructs are interrelated with one of the four relations such as x influences y, y influences x, x and y mutually influence each other, or x and y are unrelated. These relations are almost universally represented by 'V', 'A', 'X', and 'O' characters respectively in the SSIM. These relationships are assigned by DMs based on contextual relationships during pairwise comparison on variables. The number of comparisons is nC2 (mathematical combination), where n is the number of variables in the domain of study. Finally, an n by n matrix is formed with nC2 cells filled with A, V, X, and O symbols and the remaining cells are blank. Most studies have used these standard symbols except few such as [35]. As this is the basic matrix and required for all other steps therefore has been documented in most of the studies except few such as [15,66].

2.1.4. Reachability Matrix (RM) and Final Reachability Matrix (FRM)

RM is the representation of SSIM in binary form. V, A, X, and O symbols of SSIM are replaced with 1, 0, 1, and 0, digits respectively. At their transposed positions by row with column and column with row, 0, 1, 1, and 0 digits are placed, respectively. The constructs are assumed to influence self, so ones are placed at the diagonal positions. The resultant RM is checked for transitive relations. Transitivity is the basic assumption in the ISM such as if variable x influences y and y influences z then x will influence z transitively. This is second-order or two-hop transitivity whereas generalized transitivity means x is related to z through one or more variables. The transitive relations hence identified are represented in the RM with 1*s to distinguish from original 1s and the resulting matrix is known as FRM. FRM also consists of driving and dependence powers of each variable by counting 1s and 1*s in rows and columns respectively. Very few studies mention usage of some software for transitivity calculations such as [56,57]. However, one of the most frequent reasons for incorrect ISM calculations have been wrong transitivity calculations, such as in studies [9,13,16,23–37]. Therefore, this study proposes the use of an established Warshall algorithm [67] for transitivity calculations.

2.1.5. Level Partitioning

This is a very important step to develop the hierarchical directional structure among the variables. Reachability, antecedent, and intersection sets are derived for all the variables from the FRM. For a specific variable, a reachability set consists of itself and all the variables it influences, and an antecedent set consists of itself and all the variables influencing it. Thereafter, the intersection set of reachability and antecedent set is calculated. Variables having the same reachability and intersection sets are given the top rank and are removed for the next iteration and the process is repeated until all variables are ranked. Some studies such as [31,38–42] in the literature had incorrect leveling for variables.

2.1.6. Conical Matrix (CM) and Digraph

CM is row and column wise ordered FRM based on ranks or levels of variables identified in the level partitioning step. Further the levels of each variable are also recorded at the end of row and column in CM. This matrix helps in drawing the digraph to get the first visual output of the hierarchical directional structure of variables. Circular nodes are drawn with variable numbers. Further they are connected with directional edges based upon 1s or 1*s in the CM between pairs of variables. Fewer studies have mentioned CM and digraph [12,20,27,35,65,68–70], as the digraph resembles the final model with a lesser number of edges. The importance of the digraph further goes down in automatic calculation of transitivity.

2.1.7. Reduced Conical Matrix (RCM) and Final ISM Model

Digraph is converted into a final model by replacing the node numbers with names of the variables and representing nodes in the rectangular shapes. Moreover, efforts are made to remove maximum edges from digraph while maintaining the levels and structure of variables and reachability of variables. This is done to improve the readability of the final model. Several studies have committed mistakes at this step either by adding extra edges [11,14,43–48] or omitting edges [49–52] that have affected the reachability of the variables. Therefore, a new algorithm, reduced conical matrix (RCM), has been devised to remove maximum possible edges without affecting structure and reachability of variables, as explained in the fourth section. This RCM is used for making the final ISM model. The final model may further be subjected to validations by different means such as review by DMs, interviews from different sets of participants, or statistical validations.

2.1.8. MICMAC

MICMAC (Matrice d'Impacts Croisés Multiplication Appliquée á un Classement (cross-impact matrix multiplication applied to classification)) in the simplest terms is a variable classification technique. Variables are mapped onto a two-dimensional grid based on their dependence and driving power values, represented on horizontal and vertical axes respectively. The range of these values is between 1 and total number of variables and the axes are bifurcated at mid-points, resulting in four quadrants numbered anti clockwise. These quadrants classify variables into autonomous, dependent, linkage, and independent categories. The autonomous variables are not connected with the remaining system of variables whereas linkage variables are sensitive and strongly connected with independent and dependent variables. The final hierarchical ISM model coupled with the MICMAC analysis greatly improves the understanding of variables. Therefore, most studies have carried out MICMAC analysis except few such as [19,39,47,71,72].

2.2. Implementation of ISM

As originally proposed by Warfield [1–4], the ISM requires its steps to be executed with the assistance of a computer [6]. Some of the more recent studies demonstrate specialized software or routines being developed for ISM. The article [56] mentions the development of the ISM software package in R software. This software package takes the SSIM input in the comma separated (.csv) excel file and provides two outputs in excel file format, namely, "ISM_Matrix" for FRM step to incorporate transitivity calculations and "ISM_Output" for partitioning step to identify the levels of the variables. Similarly, some studies such as [57] have used MATLAB software to calculate the FRM and partitioning steps. The previous studies have attempted to automate FRM and partitioning steps, leading to partial automation of ISM. As pointed out earlier in absence of automation, the final model may introduce errors in edges regardless of correct FRM and leveling, leading to wrong reachability of variables. Further, having all the steps being carried out automatically shows the prompt results to researchers and decision makers for further possible iterations. Therefore, there exists a need to develop an end-to-end graphical software to implement ISM and MICMAC and identify the required algorithms for it.

2.3. Assessment of ISM Applications

The ISM technique is being applied in a range of domains [53–55]. The review article [54] provides 10 different application domains for ISM. It further provides additional parameters such as integration with other MCDM approaches. Similarly the review article [55] identified ISM applications in 14 domains without industry or organizations, 20 industrial sectors, and 4 other areas. Furthermore, among other characteristics, it mentions integration with other MCDM approaches, and the presence of constructs for cost and/or quality. These reviews haven't focused on operationalization of ISM technique. Therefore, there exists a gap to identify the methodology of steps of applications of ISM in the existing articles such as nature and number of variables, compositions of DMs, decision making and data collection techniques, and accuracy of ISM results.

3. Research Methodology

This study addresses two objectives, firstly the development of SmartISM for the implementation of ISM, as explained in the following section. The second objective is the scoping review of literature to identify the scope of ISM and MICMAC applications and the assessment of applications of ISM using SmartISM tool. For the scoping review the five-step framework of Denyer and Tranfield (2009) [73] has been adopted as explained in the following paragraphs, Figure 1. The review process also generated the data necessary for the assessment of application of ISM using SmartISM.

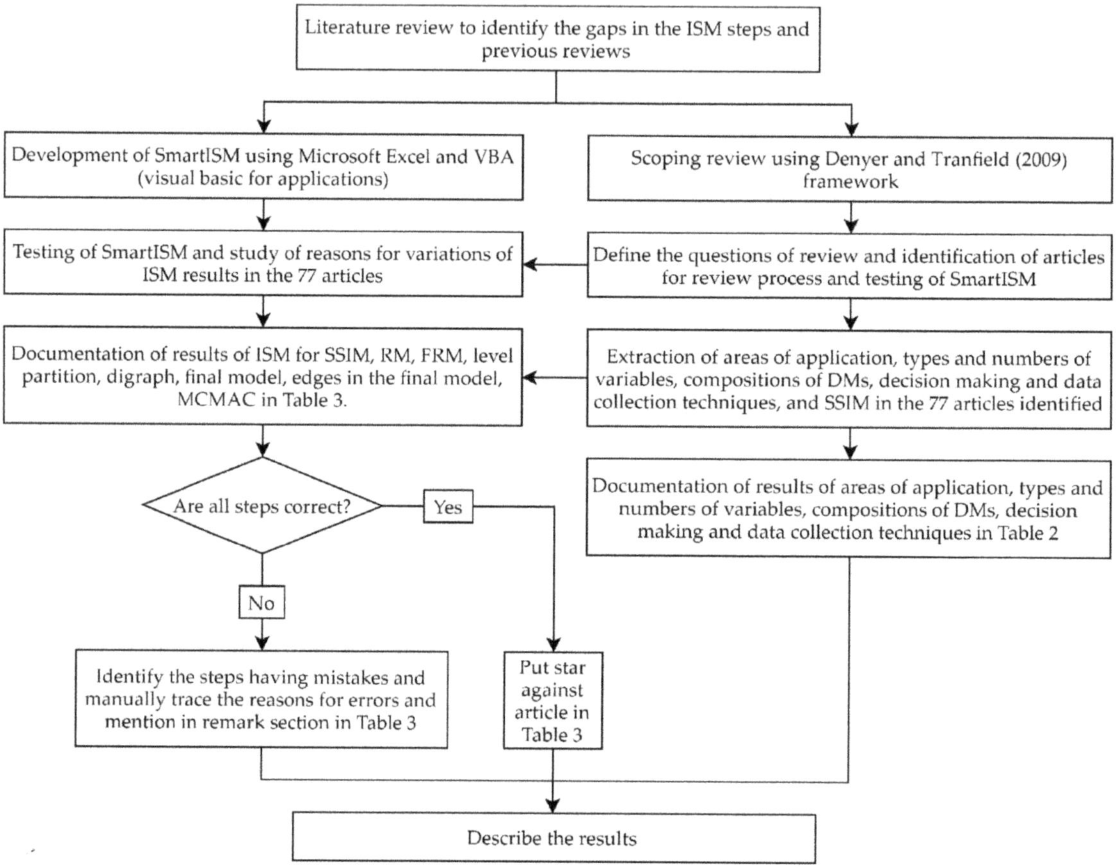

Figure 1. Research Methodology.

Step I: Question formulation: Formulating questions requires identification of context, interventions, mechanisms, and outcomes. In this study, context is considered to be domain, and decision makers; interventions are variables of interest in problem domain; mechanisms are techniques for data collection from DMs; and outcomes are the ISM outputs and MICMAC diagram. In essence there will be the following research questions that will help in addressing the second objective of this study.

- What are the different domains and sub-domains of ISM applications?
- What are the different types and numbers of variables being studied?
- What are the compositions of DMs in different studies?
- What are the different decision making and data collection techniques?
- How accurate has been the application of ISM technique, using SmartISM?

Step II: Locating Studies: As the ISM based studies are huge, only quality sources were considered, rather than an exhaustive search. As per the objectives, articles that had significant discussion and documentation on ISM application as mentioned in Step I were needed. As defined earlier, the steps of ISM are structural self-interaction matrix, initial and final reachability matrix, level partitioning, conical matrix, digraph, and final model. It was observed that an article going into the details of level partitioning had sufficient demonstration of ISM. Therefore, "Interpretive Structural Modeling" + partitioning keywords were used on ScienceDirect database and it resulted in 300 articles up to the year of 2021, of which 291 belonged to review and research articles.

Step III: Study selection and evaluation: These articles were further perused for the relevance to present study and classified into different groups such as definition only, other techniques, no partitioning, non-related, incomplete outputs, and desired study. As the articles were growing nonlinearly each year, therefore, after the year of 2017, a random selection of five articles was preferred to keep the dataset manageable. It resulted in 77 articles in the desired study group that were considered further in this study.

Step IV: Analysis and Synthesis: This step has two components: first the analysis of articles for context, interventions, and mechanisms was performed, as explained in step one, by extracting relevant information as shown in Table 2. Second was the extraction of SSIM from the 77 selected articles to reimplement the ISM technique using SmartISM. The results from the SmartISM were compared with the outcomes illustrated in the article for SSIM, RM, FRM, LP, CM, digraph, final model and edges in the final model, and MICMAC and the variations are summarized in Table 3.

Step V: Reporting and using the results: Results of analysis and synthesis are reported in results and discussion section. They have been provided in a fashion that will assist in informed-adoption and application of ISM and MICMAC, and utilization of SmartISM for academicians, practitioners, and policy makers alike.

Articles' Details

Articles' publication years range from 2005 to 2021. As the articles are increasing non-linearly, therefore 2017 onwards only five articles were randomly chosen for each year. The publication sources having two or more articles have been shown in Table 1. Journals in the area of sustainability have the maximum number of articles. Journal of cleaner production had published 18 articles out of 77 selected articles.

Table 1. Publication sources for two or more articles.

Publication Title	2005	2007	2008	2009	2011	2012	2013	2014	2015	2016	2017	2018	2019	2020	2021	Total
Journal of Cleaner Production							2	2		5	3	1	3	1	1	18
Renewable & Sustainable Energy Reviews							1	1		1	2					5
Resources, Conservation and Recycling				1	1		1	1		1						5
Resources Policy									3	1						4
Procedia Social and Behavioral Sciences									3	1						4
Procedia Engineering								1			1					2
Journal of Environmental Management											1	1				2
Computers & Industrial Engineering											1			1		2
Telematics and Informatics											1	1				2
International Journal of Production Economics						1		1								2
Sustainable production and consumption											1	1				2
Technological Forecasting and Social Change	1										1					2
Miscellaneous Sources with 1 article		1	2		1		1	3	5	1	3	1	2	3	4	27
Total	1	1	2	1	2	1	5	9	11	10	14	5	5	5	5	77

4. Development of SmartISM Using Microsoft Excel

Microsoft Excel provides an excellent environment for graphical representation and modelling of virtually any conceptual framework of any discipline. It has some important features such as cellular addressable input sheets, interactive output, vector graphic objects, integral atomic access of data in multiple ways, many inbuilt data processing functions, backend VBA (Visual Basic for Applications) interface to code any logic or algorithm, mechanisms for development of event driven interfaces, ubiquitous tool and ease of use, and widespread ecosystem of support and training. Hence it makes a natural choice for practitioners, decision makers, and researchers to develop their problem-solving models in Microsoft Excel. Its applications in business statistics and decision making are widely documented. Following are some advanced applications of MS Excel in different domains such as Genetic Analysis [74], Finite Element Analysis in Engineering [75], and Pharmacokinetic Pharmacodynamic fields of Pharmacology [76]. On the flip side it has a drawback to support multiple real-time concurrent users. This section explains the functions and features of VBA to develop SmartISM, an end-to-end graphical software to automate processes of

ISM and MICMAC with the help of pseudo codes. Additionally, the demonstration video for SmartISM has been attached as a Supplementary Material, see Video S1.

Firstly, the SSIM matrix defined by DMs is entered in Excel, and serves as the basic input for other steps of ISM. For n variables, the size of SSIM will be n by n. DMs will compare n(n + 1)/2 or $^{n}C_2$ unique pairs of variables and assign one of the relationships using symbols V, A, X, or O, as explained earlier. Thereafter, eight VBA macros will derive matrices of RM, FRM, CM, and RCM; level partitioning; and draw diagrams of digraph, final model, and MICMAC. RM is a binary form of SSIM using conversion rules for V, A, X, and O as explained earlier and keeping 1s at the diagonal positions of the matrix, as described in the following pseudo code. RM also contains the driving and dependence powers for each variable.

```
Function RM
//copy the content of SSIM into RM
RM ← SSIM
//loops to replace V, A, X and O symbols with 1, 0, 1 and 0, digits; and putting 0,1, 1 //and 0
digits at their transposed positions; and keeping the diagonal elements as 1
For i = 1 To n    //n is the total number of elements
    For j = 1 To n
        If i = j
            RM[i][j] ← 1
        If RM[i][j] = 'V'
            RM[i][j] ← 1
            RM[j][i] ← 0
        If RM[i][j] = 'A'
            RM[i][j] ← 0
            RM[j][i] ← 1
        If RM[i][j] = 'X'
            RM[i][j] ← 1
            RM[j][i] ← 1
        If RM[i][j] = 'O'
            RM[i][j] ← 0
            RM[j][i] ← 0
//count non-zero elements in rows and columns and append to show the driving and
//dependence powers
For i = 1 To n
    RM[i][n + 1] ← Countif(RM[i][] != 0)
    RM[n + 1][i] ← Countif(RM[][i] != 0)
```

The second function FRM requires calculation of transitive relations among variables. For manual calculation, RM can be visualized as a digraph with variables representing nodes and 1s in the RM representing the directed edges. By tracing different paths, transitive relations can be identified. For a large number of variables the process would be tedious and leads to errors, whereas a simple Warshall algorithm [67] for transitive closure can be used to automate it. This algorithm results in generalized transitivity if applied in-place, otherwise it will give second-order or two-hop transitivity. Transitive relations are marked with 1* in FRM, see the pseudo code for main logic in the following paragraph. Moreover, the 1s and 1*s are counted in rows and columns to calculate the driving and dependence powers respectively for each variable.

```
Function FRM
//copy the content of RM into FRM
FRM ← RM
//block for generalized transitivity (all levels) Warshall algorithm in-place
//start three level nested loop to parse through FRM
For k = 1 To n    //n is the total number of elements
    For i = 1 To n
        For j = 1 To n
            If FRM[i][k] = 1 And FRM[k][j] = 1
                FRM[i][j] ← 1
//putting 1* to differentiate between transitive links identified and links in RM
    For i = 1 To n
        For j = 1 To n
            If FRM[i][j] != RM[i][j]
                FRM [i][j] ← *1
//block for second-order transitivity (up to second level only) Warshall algorithm
//start three level nested loop to parse through FRM
For k = 1 To n
    For i = 1 To n
        For j = 1 To n
            If RM[i][k] = 1 And RM[k][j] = 1 And FRM[i][j] = 0
                FRM[i][j] ← 1*
//recount non-zero elements in rows and columns and append to show the driving and
//dependence powers
For i = 1 To n
    RM[i][n + 1] ← Countif(RM[i][] != 0)
    RM[n + 1][i] ← Countif(RM[][i] != 0)
```

The next step is to calculate the ranks of the variables through level partitioning. A new matrix LP is defined with five columns namely elements (Mi), reachability set R(Mi), antecedent set A(Mi), intersection set R(Mi)∩A(Mi) and level, and n rows. For a specific variable Mi in FRM, non-zero cells in the row comprise its reachability set and their corresponding identifiers are kept in the LP row of the same variable Mi. Similarly, non-zero cells in the column comprise its antecedent set and their corresponding identifiers are kept in the LP row of the same variable Mi. The intersection sets are calculated for all variables and variables having the same reachability and intersection sets are given first rank. In the next iteration, identifiers of all the ranked variables are removed from reachability, antecedent, and intersection sets. Again, variables having the same reachability and intersection sets are given the second rank and iteration continues until all the variables are ranked. The iteration results may be copied in one Microsoft Excel Sheet.

```
Function LP
//initiate a matrix LP of size n by 5 to keep element number, reachability set, antecedent //set,
intersection set and levels for each of the n elements; levels will remain empty
For i = 1 To n
    LP[i][1] ← i
    //reachability set R for ith element
    For j = 1 To n
        If FRM[i][j] != 0
            Append jth element to R
    LP[i][2] ← R
    //antecedent set A for ith element
    For j = 1 To n
        If FRM[j][i] != 0
            Append jth element to A
    LP[i][3] ← A
    LP[i][4] ← R∩A
```

```
//iteration for level calculations, where levels j can go up to n
For j = 1 To n
    //remove elements that have levels
    For i = 1 To n
        If LP[i][5] != Null
            For k = 1 To n
                Remove ith element from LP[k][2], LP[k][3], LP[k][4]
    //assign jth level to elements that have equal reachability and intersection sets
    For i = 1 To n
        If LP[i][2] = LP[i][4] And LP[i][5] = Null
            LP[i][5] ← j
    //print the jth iteration results
    Print LP
```

Once the variables are ranked, a digraph can be developed easily by positioning the variables as per their ranks with the help of CM. CM is row and column wise sorted FRM as per variables' ranks or levels. Directed edges can be drawn between variables as per non-zero cells in the CM. Two shape objects Oval and Connector are needed to automate the drawing of digraph. Positing of ovals needs to be carefully assigned, as there can be multiple ovals in one level. The simplest way to identify the needed objects in drawing is to auto record a macro and draw a sample. Afterwards, the macro can be manually edited and static names of the objects can be made dynamic for easy handling in the loop structures of VBA. The pseudo codes for the functions for CM and digraph is as follows.

```
Function CM
//copy the content of FRM into CM
CM ← FRM
//add levels from LP to CM for each element at the end of rows and columns
For i = 1 To n   //n is the total number of elements
    CM[i][n+2] ← LP[i][5]
    CM[n+2][i] ← LP[i][5]
CM.Sort key1: = Range(LP[n + 2][])
CM.Sort key1: = Range(LP[][n + 2]), Orientation: = xlLeftToRight
Function Digraph
//create ovals of size s to represent numbered nodes for each element
For i = 1 To n   //n is the total number of elements
        Ovals[i] ← Shapes.AddShape(msoShapeOval, 0, 0, s, s)
        Ovals[i].TextFrame.Characters.Text ← i
        //define and calculate the position arrays v and h of each rectangle based on
        //drawing canvas size, required interspacing between elements as per number of
        //elements, elements in each level and any offset needed
        Ovals[i].Top ← v[i]
        Ovals[i].Left ← h[i]
//add directed arrows between elements based on edges in CM
For i = 1 To n
        For j = 1 To n
            If CM[i][j] != 0 And i != j
                Shapes.Range(Ovals[i], Ovals[j]).Select
                Shapes.AddConnector(msoConnectorStraight, 0, 0, 0, 0).Select
                Selection.ShapeRange.Line.EndArrowheadStyle ← msoArrowheadOpen
                Selection.ShapeRange.ConnectorFormat.BeginConnect Ovals[i], 1
                Selection.ShapeRange.ConnectorFormat.EndConnect Ovals[j], 5
```

The final model represents variable names in the rectangular boxes in place of their identifiers in ovals and tries to remove maximum possible transitive links from the digraph. Transitive reduction is a technique to reduce the number of transitive links. Transitive reduction is complicated, specifically for the directed cyclic graphs, and the algorithm may even distort the structure of the digraph. Therefore, an algorithm was designed to develop a reduced conical matrix (RCM) that removes maximum links without changing the structure of digraph and reachability of elements. The main logic is to remove incoming links from second lower-level variables from the CM and results in RCM, see the pseudo code for the main logic in the following paragraph. RCM was used to draw automated final ISM model using Rectangle and Connector shape objects, as in the following pseudo code.

```
Function RCM
//copy the content of CM into RCM
RCM ← CM
//start loop to parse through columns
For i = 1 To n   //n is the total number of elements
    //start loop to parse through row of specific column for lower triangular matrix
    For j = i To n
    //search for first non-zero row cell whose level is greater than the level of that
    //column element
            If (RCM[j + 1][i] = 1 Or RCM[j + 1][i] = "1*") And RCM[j + 1][n +2] > RCM[i][n + 2]
                //set the L one higher than the level identified
                L ← RCM[j + 1] [n + 2] + 1
                Break For
    //identify the row that has level equal to L
    For j = i To n
            If RCM[j][n + 2] = L
                Break For
    //set all the rows starting from identified row in preceding step and below up to n
    //as 0
    For j = j To n
            RCM [j][i] ← 0
Function FinalModel
//create rectangles of size s to represent each element with variable text kept in names //array
For i = 1 To n   //n is the total number of elements
    Rects[i] ← Shapes.AddShape(msoShapeRectangle, 0, 0, s, s)
    Rects[i].TextFrame.Characters.Text ← names[i]
    //define and calculate the position arrays v and h of each rectangle based on
    //drawing canvas size, required interspacing between elements as per number of
    //elements, elements in each level and any offset needed
    Rects [i].Top ← v[i]
    Rects [i].Left ← h[i]
//add directed arrows between elements based on edges in RCM
For i = 1 To n
    For j = 1 To n
        If RCM[i][j] != 0 And i != j
            Shapes.Range(Rects [i], Rects [j]).Select
            Shapes.AddConnector(msoConnectorStraight, 0, 0, 0, 0).Select
            Selection.ShapeRange.Line.EndArrowheadStyle ← msoArrowheadOpen
            Selection.ShapeRange.ConnectorFormat.BeginConnect Rects [i], 1
            Selection.ShapeRange.ConnectorFormat.EndConnect Rects [j], 3
```

Lastly, a macro was written to draw a MICMAC diagram. The basic input for this diagram was the dependence and driving powers of variables from FRM. This was the longest macro as it required many shape objects such as Line, Connector, Rectangle, Oval, and Textbox. However, it didn't require any special algorithm to be used. Nevertheless, logic to initiate, aggregate, and draw different objects based on number of variables, and dependence and driving powers in a specified space, required careful arrangement.

```
Function MICMAC
//draw n + 1 horizontal and vertical lines where n is total number of elements spaced at
//s as per canvas size and number of elements, offset has been skipped for simplification
//and add numbered labels for each line
For i = 1 To n + 1
    Shapes.AddLine(0, i*s, n*s, i*s).Line //horizontal lines
    Shapes.AddLine(i*s, 0, i*s, n*s).Line //vertical lines
    With Shapes.AddTextbox(msoTextOrientationHorizontal, i*s, n*s, 30, 20)
        .TextFrame.Characters.Text ← i-1
    With Shapes.AddTextbox(msoTextOrientationHorizontal, 0, i*s, 30, 20)
        .TextFrame.Characters.Text ← i-1
//draw middle horizontal and vertical lines that may be of higher weight
With Shapes.AddLine(0, n*s/2, n*s, n*s).Line //horizontal
    .Weight ← 3
With Shapes.AddLine(n*s/2, 0, n*s/2, n*s).Line //vertical
    .Weight ← 3
//draw horizontal and vertical arrows to demarcate dependence and driving powers
Shapes.AddConnector(msoConnectorStraight, 0, n*s, n*s, n*s).Select
Selection.ShapeRange.Line.EndArrowheadStyle ← msoArrowheadOpen
With Shapes.AddTextbox(msoTextOrientationHorizontal, n*s/2, n*s, 110, 20)
    .TextFrame.Characters.Text ← "Dependence Power"
Shapes.AddConnector(msoConnectorStraight, 0, n*s, 0, s).Select
Selection.ShapeRange.Line.EndArrowheadStyle ← msoArrowheadOpen
With Shapes.AddTextbox(msoTextOrientationUpward, 0, n*s/2, 20, 80)
    .TextFrame.Characters.Text ← "Driving Power"
//write labels for each quadrant such as autonomous (I), dependent (II), linkage (III),
//and independent (IV) variables
With Shapes.AddTextbox(msoTextOrientationHorizontal, 0, s*n, 130, 40)
    .TextFrame.Characters.Text←"I-Autonomous Variables III-Linkage Variables"
With Shapes.AddTextbox(msoTextOrientationHorizontal, s*n, s*n, 135, 40)
    .TextFrame.Characters.Text←"II-Dependent Variables IV-Independent Variables"
//place the I to IV (Roman[]) in quadrants in appropriate positions x[] and y[]
For i = 1 To 4
    With Shapes.AddTextbox(msoTextOrientationHorizontal, x[i], y[i], 20, 20)
        .TextFrame.Characters.Text ← Roman[i]
//set dependence and driving in 2-dimensional arrays x and y
For i = 1 To n
    x[i][1] ← FRM[i][n + 1]
For i = 1 To n
    y[1][i] ← FRM[n + 1][i]
//aggregate elements with same dependence and driving powers in a 2-dimensional
//array E
For i = 1 To n
    If x[1][i], y[i][1] = x[1][i], y[i][1]
        Append i^th element at E[x[1][i]][y[i][1]]
//place ovals of size o and elements on the grid, offsets have been ignored
For i = 1 To nVar
    For j = 1 To nVar
        If E[i][j] != Null
            Shapes.AddShape(msoShapeOval, i*s, j*s, o, o)
            With Shapes.AddShape(msoShapeRectangle, i*s, j*s, 0, 15)
                .TextFrame.Characters.Text = E[i][j]
```

The SmartISM software was extensively tested on studies available in the literature. For any discrepancy between the reported results in the study and the SmartISM, steps were manually verified to validate the results of SmartISM, as shown in Table 3. The sample results of SmartISM for one of the previous studies [77] that had no discrepancy are shown in Figures 2–5.

All Steps	LA(1)	LETM(2)	LFFD(3)	LPE(4)	LRTP(5)	LPP(6)	IDI(7)	RIR(8)	OVIG(9)	IIF(10)	LSP(11)	LAPT(12)	LIS(13)	Driving Power
LA(1)	1	1*	1	1	1	1	1	1	1	1	1	1	1	13
LETM(2)	0	1	0	0	0	0	0	0	0	0	0	1	0	2
LFFD(3)	0	1	1	0	1	1	1	1	1	1*	0	1	1	10
LPE(4)	0	1	1	1	1	1	1	1	1	1	0	1	1	11
LRTP(5)	0	1	0	0	1	1	1	1	1	1	0	1	0	8
LPP(6)	0	1*	0	0	0	1	0	0	0	1	0	1*	0	4
IDI(7)	0	1	0	0	1	1	1	1	1	1	0	1	0	8
RIR(8)	0	1	0	0	0	0	0	1	0	1	0	1	0	4
OVIG(9)	0	1*	0	0	0	0	0	0	1	1	0	1*	0	4
IIF(10)	0	1	0	0	0	0	0	0	0	1	0	1	0	3
LSP(11)	0	1	1	1	1	1	1	1	1	1	1	1	1	12
LAPT(12)	0	1	0	0	0	0	0	0	0	0	0	1	0	2
LIS(13)	0	1	0	0	0	1	0	1	1	1	0	1	1	7
Dependence Power	1	13	4	3	6	8	6	8	8	11	2	13	5	

Figure 2. Final Reachability Matrix (FRM).

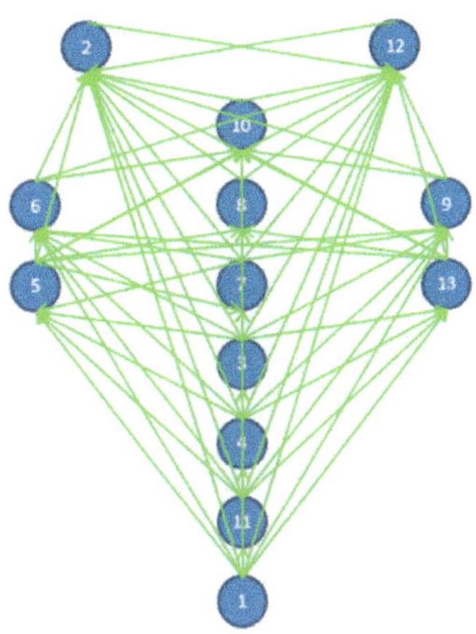

Figure 3. Digraph.

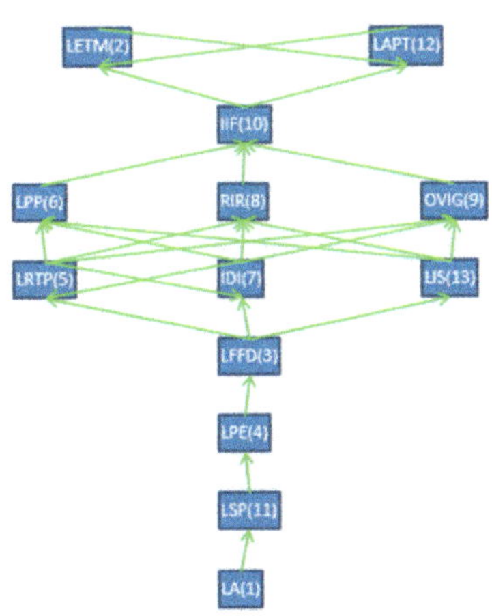

Figure 4. Final ISM Model.

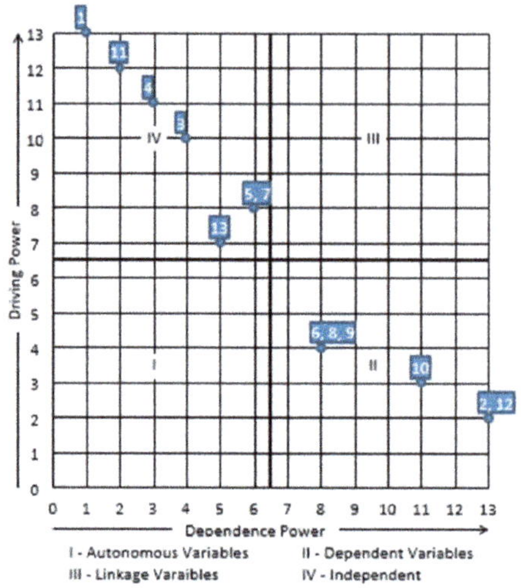

Figure 5. MICMAC Diagram.

5. Results

This section presents the scoping review answers to the questions described in the research methodology section with respect to domain of ISM applications, variables of study, composition of decision makers, decision making, and data collection techniques, as summarized in Table 2. Furthermore, the results of the assessment of ISM technique using SmartISM on the selected 77 papers are summarized in Table 3.

Table 2. Articles' domain, variables, decision makers and techniques, and ISM and other MCDM approaches.

S. No.	Reference	Domain (Sub-Domain)	Variables Description	Method of Decision Making	Number of Decision Makers	Techniques
1	[43]	Sustainability (Eco-Design)	6 barriers	workshops, seminars, telephonic enquiries, and individual and consensus questionnaire	4 excluding authors	ISM and MCDM
2	[15]	Sustainability (Manufacturing)	10 barriers and 10 enablers	two-phase questionnaire survey and online survey	10 (5 academic and 5 industry)	DEMATEL-MMDE-SEM-ISM Separate analysis for academic and industry
3	[49]	Sustainability (GSCM)	11 drivers	decision team	Including experts from the industry	ISM
4	[78]	Sustainability (GSCM)	26 CFS	personal interview, Likert scale questionnaire, decision team	10 (9 industry and 1 government)	ISM
5	[16]	Sustainability (E-Waste recycling)	10 barriers	brainstorming, and interviews	10 (6 industry, 1 NGO, 2 government and 1 academic)	DEMATEL-ISM
6	[13]	Sustainability (Healthcare-Waste)	8 factors	group decision making	industry, academic and government, 9 for AHP	ISM, fuzzy AHP and fuzzy TOPSIS.
7	[14]	Logistics (Reverse logistics Provider)	7 factors	expert opinion	5 for TOPSIS	ISM and fuzzy TOPSIS
8	[60]	Energy (Smart grid technologies)	12 barriers	idea engineering workshop	Industry and academia	ISM and MICMAC
9	[38]	Sustainability (GSCM)	18 factors	face to face interview using questionnaire	15, (OEM 3, tier-1 5, tier-2 4, tier-3 3)	ÌPA, ISM, MICMAC
10	[31]	Sustainability (Landfill Community)	12 barriers	expert opinions, interviews and group discussions	information not available	ISM MICMAC
11	[79]	Energy (Solar Power)	13 barriers	workshop	6 (4 industry and 2 academic)	ISM MICMAC

Table 2. *Cont.*

S. No.	Reference	Domain (Sub-Domain)	Variables Description	Method of Decision Making	Number of Decision Makers	Techniques
12	[32]	SCM (Food-Waste)	9 challenges	consultation	5 (industry and 2 academic)	Exploratory Factor Analysis, ISM MICMAC
13	[80]	SCM (Humanitarian SC)	12 CSF	consultation	7 (3 industry and 4 academic)	ISM MICMAC
14	[81]	Sustainability (SSCM)	13 enablers	one day workshop	15 or 25 industry	ISM MICMAC
15	[82]	Business (Offshoring)	9 elements	consultation and approval of interviewed experts	30 (6 each for 5 cases' development)	ISM MICMAC
16	[62]	Logistics (Reverse logistics)	11 barriers	consultation	2 (1 industry and 1 academic)	ISM MICMAC
17	[77]	Energy	13 barriers	questionnaire and communication	4 for questionnaire and 3 for agreement through communication (mix)	ISM MICMAC
18	[83]	HR (OSH)	8 factors	discussion for 72 h	5 industry	ranking order clustering ROC, ISM
19	[63]	Sustainability (GSCM)	26 barriers identification by more than 5 experts	questionnaire and discussion	2 (1 industry and 1 academic)	ISM MICMAC
20	[33]	Sustainability (GSCM)	20 barriers	brainstorming	5 (3 industry and 2 academic)	ISM MICMAC
21	[84]	Logistics (3rd party Reverse logistics provider)	7 attributes	consultation	information not available	ISM MICMAC
22	[66]	Tourism (Tour Value)	12 enablers	brainstorming	7 (4 industry and 3 academic)	ISM MICMAC
23	[39]	Marketing (Motivation)	17 motivations	laddering interview	52 respondents through content analysis of interviews	Content analysis ISM
24	[44]	Supplier Selection (Corporate social responsibility)	9 CSR issues	direct meetings, telephonic inquiries, and email, and questionnaire	150 firework industrial managers from SMEs and relevant field experts for variable identification	ISM MICMAC
25	[85]	Sustainability (SSCM)	25 SSCM practices	questionnaire and judgement group	45 or more survey response, consolidation judgement group of 15	ISM MICMAC
26	[10]	Sustainability (Production)	20 barriers	workshop	23 experts	ISM ANP
27	[58]	Sustainability (Green Business Model in Construction)	5 barriers	thematic analysis, semi-structured interviews	19	ISM MICMAC
28	[34]	Marketing (Retail Brand)	10 factors	brainstorming sessions	9 (4 industry and 5 academic)	ISM MICMAC
29	[86]	Sustainability (Neighborhoods)	19 barriers	interview	15 (10 industry and 5 academic) Divided into 2 nominal groups for validation	ISM MICMAC
30	[40]	Sustainability (Construction)	8 CSFs for roads and bridges, and 10 CSFs for embarkment	idea engineering workshop	3 (2 industry, 1 academia)	ISM MICMAC
31	[19]	IT (E-Commerce Security)	10 factors	13 semi structured interview	13 industry	Content analysis, ISM
32	[35]	Operations (Maintenance)	8 Maintenance tools and techniques	consultation	5	ISM MICMAC
33	[36]	Operations (Maintenance)	24 factors	discussion	information not available	ISM MICMAC
34	[45]	IT (ERP)	16 CSFs	face to face communication, questionnaire surveys since difficult to gather them	14 industry	ISM MICMA Fuzzy cognitive maps

Table 2. *Cont.*

S. No.	Reference	Domain (Sub-Domain)	Variables Description	Method of Decision Making	Number of Decision Makers	Techniques
35	[61]	Innovation (Manufacturing)	11 innovation enablers (competitiveness)	Delphi technique	industry and academia	ISM Fuzzy MICMAC
36	[37]	IT (Cloud)	16 CSFs	Delphi technique from 20 to 16	14 (11 industry and 3 academic)	ISM MICMAC
37	[87]	Sustainability (SSCM)	10 enablers	27 survey response followed by the telephonic interviews	27 industry survey based on Fawcett et al. (2014)	TISM MICMAC
38	[88]	HR (Team)	11 factors	expert opinions	14 (10 industry and 4 academic)	ISM
39	[89]	HR (OHSAS)	9 factors influencing safety	expert opinions	25 to 16 industry responded	ISM MICMAC
40	[90]	Sustainability (SSCM)	10 hurdles	decision team rated on 5-point Likert scale and brainstorming for SSIM	7 (4 industry, 3 academic)	ISM MICMAC
41	[91]	Energy (Solar)	16 barriers	LR and expert opinion	None	ISM Fuzzy MICMAC
42	[41]	National (National Infrastructure)	13 Indian Critical infrastructure sectors	LR, brainstorming, face to face interview	30 experts (7 academic, 13 industry, 10 gov.)	ISM MICMAC
43	[68]	Energy (Biodiesel)	5 risk factors for ISM	expert opinions through group discussions	research group of senior university professors	ISM Bayesian network
44	[92]	Sustainability (Emission trading system)	15 factors	semi structured interviews, factors selected on 4 to 3 votes, same rule of majority gives way to minority for relationship	58 to 10 to 7 experts	ISM MICMAC
45	[71]	Sustainability (Remanufacturing)	14 obstacles	consultation	industry and academia	ISM
46	[23]	SCM (SCM)	15 factors	2 brainstorming sessions and verification by recirculation of model, 179 survey response to develop correlation matrix	5 industry	ISM MICMAC
47	[46]	Manufacturing	15 factors	340 to 480 self-administered questionnaires	experts from industry and academic	questionnaire-based survey; ISM MICMAC approach; EFA; CFA and GTMA
48	[24]	SCM (Food SCM)	14 Critical causal factors	Input from experts	13 Indian academics in phase 1, 12 industry	ISM MICMAC
49	[64]	Sustainability (Production)	21 criteria	linguistic preferences and that concurrently applies perception judgment-pretesting of survey-purposive sampling method-120 replies	10 researchers then 5 researchers for pre-testing, For ISM 120 responses	fuzzy set theory and ISM
50	[25]	SCM (Food SCM)	16 causes of food wastages	Semi structured interview Brainstorming	6 (4 industry 2 academic)	TISM fuzzy MICMAC
51	[11]	Sustainability (Green IT/IS)	13 psychological drivers of motivation	Upper Echelon Theory (UET) Focus group discussions Survey for ANP	10 (2 industry 8 academic)	Hybrid ISM -ANP
52	[93]	IT (Building Information Modelling)	15 capability factors for BIM adoption	the semi-structured interviews and focus groups were conducted and LR	7 industry	ISM MICMAC
53	[69]	Sustainability (Reverse logistics)	7 grouped barriers	Phone consultation and face to face structured interview	4 (2 industry 2 academic)	ISM MICMAC

Table 2. *Cont.*

S. No.	Reference	Domain (Sub-Domain)	Variables Description	Method of Decision Making	Number of Decision Makers	Techniques
54	[65]	Sustainability (GSCM)	12 behavioral factors	Opinion of experts	15 (10 industry, 3 academic, 2 government)	ISM MICMAC
55	[70]	Sustainability (GHRM)	7 factors	Ability Motivation Opportunity theoretical lens Questionnaire and personal interview Pilot study from 97 respondents to 7 factors ISM on 7 factors SEM to validate model	11 experts listed 42 variables EFA 42 to 32 variables to 7 factors CFA 32 to 24 variables Same 11 experts for ISM	LR, expert opinion to EFA to CFA to ISM MICMAC to ISEM
56	[12]	Operations (Lean Manufacturing)	9 antecedents and critical success factors	Group discussion for ANP and pairwise comparisons, Also, for identifying 9 factors and ISM	7 industry	ANP -horizontal integration of human ISM- vertical integration of human
57	[94]	IT (SCM)	11 technological capabilities	Brainstorming	3 industry	TISM
58	[9]	Supplier relationship management (SRM)	10 factors	library survey method-questionnaire-research method in this survey is descriptive and analytical	managers and the production heads	leveling the factors using ISM model MICMAC, fuzzy TOPSIS-AHP- ranking suppliers
59	[72]	SCM (E-Procurement)	14 barriers and 15 benefits	Five-point scale survey followed by brainstorming session	8 industry for variable reduction + 4 academic for ISM	ISM SEM
60	[20]	Sustainability (SSCM)	32 CSFs of motivation and encouragement	32 CSFs were identified from LR and opinions of academicians and industry practitioners	consultation with industry experts and academic	ISM MICMAC
61	[95]	IT (Social networking sites (SNS))	12 Factors of abandoning usage of SNS	survey on 60 PG students who have abandoned the SNS at least once and validation by LR	60 students as experts	ISM MICMAC
62	[59]	Sustainability (GSCM)	19 influential factors on the implementation	LR, and invoking the contingency theory perspective to get 19 factors group discussion for ISM (4 sessions)	13 Industry	ISM MICMAC
63	[47]	Sustainability (recycling 3D printing waste)	22 barriers	Interview for identification of barriers and 3 experts consulted for relationships between barriers	13 (12 industry and 1 academic)	ISM
64	[26]	Sustainability (Green Lean Six Sigma)	12 enablers	LR, survey and expert opinion for identification of enablers	Not mentioned	ISM MICMAC
65	[21]	IT (Building Information Modelling)	12 barriers	LR and survey for identification of enablers and semi-structured interview for relationships between barriers	5 industry	ISM MICMAC
66	[22]	Sustainability (Oil and gas SSCM)	7 determinants	LR and expert opinion for identification of determinants, individual expert opinion for relationships among determinants	18 (13 industry 5 academic)	ISM MICMAC and SEM

Table 2. *Cont.*

S. No.	Reference	Domain (Sub-Domain)	Variables Description	Method of Decision Making	Number of Decision Makers	Techniques
67	[96]	IT (m-commerce)	10 barriers	LR and survey for identification of determinants, survey for relationships among determinants	Number of participants not mentioned	ISM MICMAC
68	[50]	Building energy performance	16 factors	LR and interview for identification of factors, and interview for relationships	12 industry	ISM MICMAC
69	[51]	Quality of Passenger interaction process	15 enablers	Report, passengers' opinion including authors and observations for identification of enablers, and also for relationships	Authors opinions	ISM MICMAC
70	[52]	IT (Cloud computing)	7 determinants	Content analysis on interview transcripts of 58 industry people for identification of determinants, Group discussion for relations	Not mentioned	ISM MICMAC, AHP and TOPSIS
71	[27]	SC (Lean Sis Sigma)	10 barriers	LR and 10 consultation sessions with 35 experts using brainstorming and focused group method to identify barriers and relationships	35 (industry and academic)	ISM MICMAC
72	[42]	Sustainability (CSR)	12 enablers	LR, survey and interviews for identification of enablers and experts' opinions for relationships	20 (industry and academic)	ISM MICMAC
73	[28]	Automobile (Electric Vehicle)	12 barriers	LR, expert opinion and best worst method to select barriers	10 (7 industry and 3 academic)	ISM MICMAC and Best worst method
74	[97]	Marketing (app-based retailing)	20 factors	LR and survey from expert for factors identification and relationships	33 (30 industry and 3 academic)	ISM MICMAC
75	[48]	SC (Food loss and waste)	8 factors	LR and consultation with experts for factors selection and relationships	11 (5 industry and 6 academic)	ISM MICMAC
76	[29]	New product development (time to market)	19 capabilities and 5 drivers	LR and expert interview for identification of capabilities and drivers and relationships	14 (8 industry and 6 academic)	ISM fuzzy MICMAC
77	[30]	Sustainability (Renewable energy)	13 strategies	LR and SWOT analysis for strategies identification and experts' opinion for relationships	5 (2 industry, 2 academic and 1 government)	SWOT ISM MICMAC

SC: Supply chain, SCM: SC management, SSCM: Sustainable SCM, GSCM: Green SCM, SRM: Supplier relationship management, OSH: Occupational safety and health, IT: Information technology, IS: Information Systems, ERP: Enterprise resource planning, HR: Human resources, GHRM: Green HR Management, OHSAS: Occupational health and safety assessment series, NGO: Non-governmental organization, SME: Small and medium scale enterprises, ISM: Interpretive structural modelling, SSIM: Structural self-interaction matrix, AHP: Analytical hierarchy process, ANP: Analytic network process, TOPSIS: Technique for order of preference by similarity to ideal solution, LR: Literature review, CSF: Critical success factors, SEM: Structural equation modeling, EFA: Exploratory factor analysis, CFA: Confirmatory factor analysis; CSR: Corporate social responsibility; SWOT: Strengths, weaknesses, opportunities, and threats.

Table 3. Assessment of application of ISM using Smart ISM.

S. No.	References	SSIM	RM	FRM	Partitioning	Transitivity	Digraph	Model	Edges	MICMAC	Remarks
1	[43]	Y	Y	Y	Y	2nd Level	N	Y	Varied	Y	Added extra edge from variable 4 to variable 5
2	[15]	N	N	N	N	N	N	N	N	N	No SSIM
3	[49]	Y	Y	Y	Y	2nd Level	N	Y	Varied	Y	Some edges are missed such as between variable 9 and variable 1
4 *	[78]	Y	Y	Y	Y	2nd Level	N	Y	Y	Y	
5	[16]	Y	Y	N	N	N	N	N	N	Y	Wrong calculation for transitivity
6	[13]	Y	Y	N	N	No Level	N	N	N	Y	Transitivity not incorporated though mentioned in the methodology
7	[14]	Y	Y	Y	Y	2nd Level	N	Y	Varied	Y	Added extra edge from R to I
8 *	[60]	Y	Y	Y	Y	All Levels	N	Y	Y	Y	
9	[38]	Y	Y	Y	N	2nd Level	N	N	Y	Y	At 8th level one variable is wrong in model
10	[31]	Y	Y	N	N	No Level	N	N	N	N	Didn't incorporate transitivity though the concept is mentioned
11 *	[79]	Y	Y	Y	Y	All Levels	N	Y	Y	Y	
12	[32]	Y	Y	N	N	All Levels	N	N	N	N	Wrong calculation for transitivity, only two transitive links but shown more
13 *	[80]	Y	Y	Y	Y	2nd Level	N	Y	Y	Y	No transitive links
14 *	[81]	Y	Y	Y	Y	2nd Level	N	Y	Y	Y	
15 *	[82]	Y	Y	Y	Y	2nd Level	N	Y	Y	Y	Retained some transitive links in the model
16 *	[62]	Y	Y	Y	Y	2nd Level	N	Y	Y	Y	
17 *	[77]	Y	Y	Y	Y	2nd Level	N	Y	Y	Y	
18 *	[83]	Y	Y	Y	Y	2nd Level	N	Y	Y	Y	Model upside down
19 *	[63]	Y	Y	Y	Y	2nd Level	N	Y	Y	Y	
20	[33]	Y	Y	N	N	N	N	N	N	Y	Wrong calculation for transitivity and for generalized transitivity; all will be having same dependence and driving powers with max value
21 *	[84]	Y	Y	Y	Y	2nd Level	N	Y	Y	Y	MICMAC axis interchanged and for generalized transitivity all will be having same dependence and driving powers with max value
22	[66]	N	N	N	N	N	N	N	N	N	No SSIM and directly starts with RM
23	[39]	Y	Y	Y	N	All Levels	N	N	N	N	No MICMAC
24	[44]	Y	Y	Y	Y	2nd Level	N	Y	Varied	Y	Added extra edge from variable 6 to variable 9
25 *	[85]	Y	Y	Y	Y	2nd Level	N	Y	Y	Y	
26 *	[10]	Y	Y	Y	Y	2nd Level	N	Y	Y	N	B2 element wrong on MICMAC
27 *	[58]	Y	Y	Y	Y	2nd Level	N	Y	Y	Y	
28	[34]	Y	Y	N	N	N	N	N	N	N	RM and FRM together, one transitive link skipped between elements EM and BN
29 *	[86]	Y	Y	Y	Y	All Levels	N	Y	Y	Y	
30	[40]	Y	Y	Y	N	2nd Level	N	N	N	N	Wrong leveling, one element missing on MICMAC
31	[19]	Y	N	N	N	2nd Level	N	N	N	N	No MICMAC, one extra wrong self-relation
32	[35]	Y	Y	N	N	No Level	Y	N	N	N	Transitivity not incorporated though mentioned in the methodology, non-standard relation symbols in SSIM such as F, R, FR, and X
33	[36]	Y	Y	N	N	2nd Level	N	N	N	N	Wrong transitivity calculation
34	[45]	Y	Y	Y	Y	All Levels	N	Y	Varied	Y	Added wrong edges for F12 and F15
35 *	[61]	Y	Y	Y	Y	2nd Level	N	Y	Y	Y	No transitivity relations found
36	[37]	Y	Y	N	N	2nd Level	Y	N	N	N	Wrong transitivity calculation
37 *	[87]	Y	Y	N	Y	2nd Level	N	Y	Y	N	Skipped one transitive relation between 8 and 9, same results for second-order and generalized transitivity
38 *	[88]	Y	Y	N	Y	No Level	N	Y	Y	Y	Transitivity mentioned but not included; rest of all the things correct
39	[89]	Y	N	N	N	2nd Level	N	N	N	N	Wrong transitivity calculation
40 *	[90]	Y	Y	Y	Y	All Levels	N	Y	Y	Y	Same results for second-order and generalized transitivity
41 *	[91]	Y	Y	Y	Y	All Levels	N	Y	Y	Y	
42	[41]	Y	Y	Y	N	2nd Level	N	N	N	Y	No transitive links
43 *	[68]	Y	Y	N	Y	2nd Level	Y	Y	Y	N	
44	[92]	N	N	N	N	N	N	N	N	N	Non-standard SSIM
45 *	[71]	Y	Y	N	Y	No Level	N	Y	Y	N	Transitivity not calculated and no MICMAC
46	[23]	Y	Y	N	N	2nd Level	N	N	N	N	

Table 3. *Cont.*

S. No.	References	SSIM	RM	FRM	Partitioning	Transitivity	Digraph	Model	Edges	MICMAC	Remarks
47	[46]	Y	Y	Y	Y	2nd Level	N	Y	Varied	Y	Extra edges added
48	[24]	Y	Y	N	N	2nd Level	N	N	N	N	
49	[64]	N	N	N	N	N	N	N	N	N	Non-standard SSIM, fuzzy ISM
50	[25]	Y	Y	N	N	2nd Level	N	N	N	N	TISM
51	[11]	Y	Y	Y	Y	2nd Level	N	Y	Varied	N	One extra edge, MICMAC calculation is correct but drawn wrong
52	[93]	Y	N	N	N	2nd Level	N	N	N	N	For self-relation X shown in SSIM whereas RM and FRM not given
53 *	[69]	Y	Y	Y	Y	2nd Level	Y	Y	Y	Y	
54 *	[65]	Y	Y	Y	Y	2nd Level	Y	Y	Y	Y	For self-relation X shown in SSIM
55 *	[70]	Y	Y	Y	Y	No Level	Y	Y	Y	Y	Transitivity not included
56 *	[12]	Y	Y	Y	Y	2nd Level	Y	Y	Y	Y	
57	[94]	N	N	N	N	N	N	N	N	N	Non-standard SSIM
58	[9]	Y	Y	N	N	2nd Level	N	N	N	N	Many mistakes in transitivity calculation
59	[72]	Y	Y	Y	Y	2nd Level	N	N	N	N	No MICMAC, extra edges
60	[20]	Y	N	N	N	2nd Level	N	N	N	N	RM onwards wrong calculations
61 *	[95]	Y	Y	Y	Y	All Levels	N	Y	Y	N	MICMAC explained in detail
62	[59]	Y	N	N	N	2nd Level	N	N	N	N	Partitioning in simulation didn't give levels, for self-relation X shown in SSIM
63	[47]	Y	Y	Y	Y	2nd Level	N	Y	Varied	N	Some extra edges drawn in final model, no MICMAC
64	[26]	Y	Y	N	N	N	N	N	N	N	Wrong transitivity calculations
65	[21]	Y	N	N	N	All Levels	N	N	N	N	Non-standard SSIM, wrong conversion for relationship between elements B10 and B08 in RM
66	[22]	Y	N	N	Y	All Levels	N	Y	Y	N	in SSIM for CGLC, self-influence is considered wrong and assigned 0
67 *	[96]	Y	Y	Y	Y	2nd Level	N	Y	N	Y	Digraph is not mentioned
68	[50]	N	N	Y	Y	All levels	N	Y	Varied	Y	Adjacency matrix to represent RM, one edge F6 to F1 from adjacency matrix is not drawn
69	[51]	Y	Y	Y	Y	All Levels	N	Y	Varied	Y	Some edges are missed such as between NWFOB and AEPE
70	[52]	Y	Y	N	Y	No Level	N	Y	Varied	Y	1 in SSIM for self-relation, transitivity not included although concept is discussed; some reachability edges missed such as C6 to C1
71	[27]	Y	Y	N	Y	2nd Level	Y	Y	Y	N	5 transitive links missing in 2nd level transitivity and 6 missing from all levels
72	[42]	Y	Y	Y	N	2nd Level	N	N	N	Y	First partitioning iteration is wrong and incorrectly assigns level 1 to E4
73	[28]	Y	Y	N	Y	All levels	N	Y	N	N	One transitive link between B1 and B12 is missed out; B1 is wrongly placed in MICMAC diagram
74 *	[97]	Y	Y	N	Y	No Level	N	Y	Y	N	All 3 transitive links missing, although the concept is mentioned; Wrong MICMAC
75	[48]	Y	Y	Y	Y	All levels	N	Y	Varied	Y	One extra edge between F1 and F4 that changes the reachability
76	[29]	Y	Y	N	N	2nd Level	N	N	N	N	X for self-relation in SSIM, missing transitive link between T14 and T19, self-interaction between for T17 missing, Wrong transitive link given between T19 and T6
77	[30]	Y	Y	N	N	All Levels	N	N	N	N	Many mistakes in transitivity calculations such as between S5 and S13

* All SmartISM reproduced results similar to articles' ISM process results with varied transitivity considerations. Y: SmartISM reproduced results similar to articles' ISM process results, N: SmartISM reproduced results not similar to articles' ISM process results, Varied: edges in the final model varied, although the hierarchical structure of variables may or may not be correct. ISM: Interpretive structural modelling, TISM: Total ISM, SSIM: Structural self-interaction matrix, RM: Reachability Matrix, FRM: Final RM, MICMAC (Matrice d'Impacts Croisés Multiplication Appliquée á un Classement (cross-impact matrix multiplication applied to classification)). Some variables code such as B2, EM, and BN have been given in the table for their description, refer to the articles. Symbol X in SSIM means both variables mutually influence each other.

5.1. Domain of Study

ISM is being applied in numerous fields such as sustainability, social sciences, management, engineering, and information technology. The results show 21 different domains, with highest studies in sustainability (32), supply chain and logistics (13), information technology (9), energy (5), human resource (3), marketing (3), and operations (3) (Table 2). Within sustainability, the highest studies are in the area of green supply chain management (GSCM) and sustainable supply chain management (SSCM) with seven studies each followed by two studies in construction [40,58] and several other areas such as e-waste recycling [16], healthcare waste [13], recycling 3D printing waste [47], green IT/IS [11], among others. In the area of supply chain and logistics, studies have focused on supplier relationship [9] and selection [44], food SCM [24,25], e-procurement [72], and reverse logistics [14,62,84], among others. In the field of information technology studies are conducted in the areas of building information modeling [21,93], cloud computing [37,52], e-commerce security [19], m-commerce [96], enterprise resource planning [45], supply chain management [94], and social networking sites [95]. Energy domain studies were in the area of bio-diesel [68], smart grid technologies [60], and solar energy [79]. For the human resource domain, two studies were in the area of occupational health and safety [83,89] and one in team performance [88]. The studies in the marketing area focused on motivation [39], retail brand [39], and app-based retailing [97]. Furthermore, the articles in the area of operations focused on maintenance [35,36] and lean manufacturing [12]. Some of the innovative areas were landfill communities [31], emission trading system [92], tour value [66], and quality of passenger interaction process [51].

5.2. Variables of Study

Most studies focus on enablers or drivers, challenges or barriers, critical success factors, and influencing or significant elements in the domain of research (Table 2). Other studies have tried to explore different sets of variables. For example, article [84] has studied seven attributes of third party reverse logistics; article [39] has studied 17 motivational factors in the marketing area; article [44] has used 9 corporate social responsibility factors in the area supplier selection; article [85] has studied 25 SSCM practices; article [35] interrelated maintenance tools and technique; article [11] explored 13 psychological drivers of motivation in the area of green IT/IS; article [95] studied the interrelationships between 12 factors for abandoning social networking sites; article [29] studied capabilities and drivers for new product development; and article [30] studied 13 strategies for renewable energy. The number of variables being studied ranged from 5 [58] to 32 [20]. Additionally, some studies explored two types of factors such as 10 barriers and 10 enablers [15] and 14 barriers and 15 benefits [72]. One study gave variables in two applications such as 8 CSFs for roads and bridges and 10 CSFs for embarkment [40]. These variables are identified mostly through literature review, experts' opinions, and/or survey. Some of the unique approaches used to identify variables are thematic analysis [58], upper echelon theory [11], contingency theory [59], content analysis [52], best worst method [28], SWOT (strength, weakness, opportunity, and threat) analysis [30], idea engineering workshop [40,60], and Delphi technique [37,61].

5.3. Domain Experts or Decision Makers

This is the most crucial step as it provides the input for further steps. There are two important aspects, namely, selection of decision makers (DM) and method of information gathering from them. There are three different sets of DMs in the sample studies participants from industry, academia, and government. Participants varied from 120 through survey [64] to 2 [62,63] through group discussion and consultation. The median value of the total number of participants was 10. Only four studies [16,30,41,65] had taken participants from all three sectors: industry, academia, and government. While 56 studies had DMs from industry including others and two studies [51,72] had only academic DMs, 17 studies didn't mention the number of DMs.

5.4. Decision Making and Data Collection Techniques

There are two approaches followed for decision making, namely, discourse and survey. For discourse many techniques have adopted such as idea engineering workshops [40,60], telephonic enquiries [43,44], group decision making [13,52], personal interview [70,78], brainstorming [16,33,34,66], laddering interview [39], direct meetings [44], semi-structured interview [21,58,93] and structured interview [69], Delphi technique [37,61], and focus group discussion [11,27]. Similarly, for surveys different techniques are as follows: individual and consensus questionnaire [43], Likert scale questionnaire [78,90], email questionnaire [44], library survey method [9], and self-administered questionnaire [46].

5.5. Assessment of Application of ISM Technique Using SmartISM

The SSIM matrices from all the 77 articles were entered into the developed SmartISM software and resulted in 77 excel files. Thereafter, for each article results were reproduced in the SmartISM software buy running the macros in 77 excel files. Variations between the reported results in the articles and corresponding SmartISM reproduced results were studied. Due to differences in transitivity incorporation FRM was checked for firstly non-incorporation of transitivity, followed by two-hop transitivity (second-order) and lastly generalized transitivity (all levels). In some cases, second order and generalized transitivity could be same. Furthermore, in case of inconsistency the digraph was manually built and transitivity was traced before reporting the results. Similarly, the complete process was analyzed to identify the reasons for the errors in any of the steps. As ISM technique is sequential, error in one step will cause subsequent steps to be erroneous specially if error exists in the steps of RM, FRM, and level partitioning.

The results of the assessments are summarized in Table 3 where 'Y' means the articles' calculations match with that of SmartISM and 'N' means different results. For each article SSIM, RM, FRM, level partitioning, digraph, final model, edges in the final model, and MICMAC diagrams are verified. Three studies [64,92,94] didn't report standard SSIM and two studies [15,66] had no SSIM, therefore results were not reproduced for them. Out of 72 remaining studies, 29 studies came out correct on all the parameters and their serial numbers (S. No.) are marked with stars '*'. Of these 29 studies, 25 had either second level (two-hop or second-order transitivity) or all levels (generalized transitivity), and four [70,71,88,97] had no transitivity calculations.

The remaining 43 studies with different results from the SmartISM outputs were further analyzed starting from first the step of SSIM and moving on until the variations were identified. One study [19] had one wrong self-relation in RM. Three studies [20–22] had incorrect RM and one article [93] did not provide RM and FRM. Two studies [59,89] didn't provide RM, and their FRM didn't match with the SmartISM output due to wrong transitivity calculations. Furthermore, eighteen studies [9,13,16,23–37] had incorrect transitivity calculations leading to variations in FRM. Five studies [38–42] had variations at the fourth step of level partitioning. One study [31] was checked for level partitioning without transitivity as it was not considered, and came out wrong in assigning levels to variables. Finally, 12 studies had accurate hierarchical structures of variables in the final ISM model but variations in the edges that distorted the reachability of variables. Of these 12 studies, eight studies [11,14,43–48] had some extra edges and four studies [49–52] had some missing edges in the final model.

In the documentation of application of ISM, only eight studies [12,20,27,35,65,68–70] reported digraph. MICMAC analysis has been used by all studies except five [19,39,47,71,72] to explain grouping of the variables. Five studies [27,29,59,65,93] have explicitly mentioned X and one study [52] has mentioned 1 in the SSIM for variables to represent mutual self-relation, although it is a basic assumption in ISM therefore, other studies have not mentioned it.

6. Discussion

The operationalization of the ISM is best to be conducted through software, as there are tedious calculations such as transitivity, level partitioning, and graphical displays of digraph, final ISM model, and MICMAC diagram. Moreover, these calculations and displays need to be iterated until the high confidence model is approved by the experts. Therefore, this study has explained the methodology to develop MS Excel and VBA based, end-to-end software, SmartISM for ISM and MICMAC with the help of pseudo codes. For incorporating transitivity in FRM, the Warshall algorithm has been used, and a new algorithm RCM has been introduced for removing edges from variables' second lower level onwards without affecting reachability and digraph structure. Further, the demonstration video of SmartISM has been added as a Supplementary Material, and this tool has been extensively tested on the existing studies and applied successfully in some of the studies [98–100].

Furthermore, the scoping review shows that the ISM and MICMAC techniques are being applied in different domains of social sciences, management, engineering, and technology such as sustainability, SCM, operations, manufacturing, human resource, information technology, and many other innovative areas. This technique is also employed in different multi criteria decision making techniques such as AHP, ANP, TOPSIS, DEMATEL, etc. There are four important issues that need to be addressed such as variables and their context, decision makers' experience and numbers, decision making and data collection techniques, and utilization of software tools. The nature of variables has been enablers or drivers, challenges or barriers, critical success factors, strategies, capabilities and drivers, and influencing or significant elements in the area of study. Their numbers have varied from 5 to 32 and they have been identified through domain specific literature review, experts' opinions, and/or survey. Furthermore, techniques such as thematic analysis, upper echelon theory, contingency theory, content analysis, best worst method, SWOT analysis, idea engineering workshop, and Delphi technique are used for variables' identification. Similarly, the variables have been explained well to establish the contextual meaning.

Another important aspect is the experts or decision makers, as the whole analysis is dependent upon their knowledge and experience. There should be representation from all stakeholders of the domain being studied. In the best-case, experts from academia, industry, government and regulatory bodies should be selected in the panel of DMs. There are very few studies such as four in the sample of articles that have had DMs from all the stakeholders. The number of DMs varied from 2 to 120, whereas in most of the studies they were 11. Two approaches have been utilized for extracting information from DMs namely discourse and surveys. The discourse techniques are idea engineering workshops, telephonic enquiries, group decision making, personal interview, brainstorming, laddering interview, direct meetings, semi-structured and structured interview, Delphi technique, and focus group discussion. Survey techniques have used individual and consensus questionnaires, Likert scale questionnaire, email questionnaire, library survey method, and self-administered questionnaire.

The SmartISM reproduced results, on the existing studies selected in scoping review, show that only 29 out of 77 studies had correct calculations with varied transitivity incorporation such as no transitivity, second order transitivity, or generalized transitivity. Wrong transitivity calculation has been the most frequent reason for incorrect ISM results followed by variations in drawing edges in the final model that affects the reachability of the variables.

Lastly, five studies didn't report standard SSIM, which is essential to reproduce the calculations. Therefore, as a standard practice some minimum outputs must be reported namely SSIM, FRM, level partitioning (final after all iterations), and final ISM model. Similarly, MICMAC analysis is also an important and indispensable part of ISM, as all studies except five have used it for classifying variables into one of the four groups, namely, dependent, independent, linkage, and autonomous.

7. Conclusions

Human decisions play a very important role in any social or technical system development. Domain experts have intricate knowledge on the system and can predict the contextual interrelationships between the variables of interest in the particular domain. The interpretive structural modelling technique can assemble their tacit knowledge into a tangible hierarchical model leading to an enhanced understanding of the subject. This study has developed a software tool such as SmartISM to implement ISM in an error-free, user-friendly, and graphical style. In addition to automation of existing routines of ISM, the Warshall algorithm is used for transitivity calculations and a new algorithm, reduced conical matrix, has been introduced to convert the digraph into a final model with lesser edges while retaining the digraph structure and reachability of variables. Furthermore, the scoping review of this research will guide practitioners, policy makers, and academicians in applying this technique in different disciplines in an informed way. It will help in managing ISM configuration settings such as variables' selection, composition of decision makers, decision making, and data collection techniques. The poor results of assessment of application of ISM technique in the previous studies necessitate the utilization of an end-to-end software, such as SmartISM, to produce a high confidence final model, explaining the interrelationships between important constructs in the applied domain. To limit the number of articles in the review process only the ScienceDirect database was used, and for the last four years articles were randomly selected; therefore, results should be interpreted accordingly. The future studies will focus on the development of software tools to apply ISM in conjunction with other MCDM techniques such as AHP, ANP, TOPSIS, and DEMATEL.

Supplementary Materials: The following is available online at https://www.mdpi.com/article/10.3390/su13168801/s1, Video S1: The SmartISM software demonstration.

Author Contributions: Conceptualization, N.A. and A.Q.; methodology, N.A.; software, N.A.; validation, N.A. and A.Q.; formal analysis, N.A.; investigation, A.Q.; resources, A.Q.; writing—original draft preparation, N.A.; writing—review and editing, N.A. and A.Q. All authors have read and agreed to the published version of the manuscript.

Funding: This research received no external funding.

Institutional Review Board Statement: Not applicable.

Informed Consent Statement: Not applicable.

Data Availability Statement: The data presented in this study are available on request from the corresponding author.

Acknowledgments: Authors are highly thankful to the King Khalid University, Kingdom of Saudi Arabia for all the support provided for this research work.

Conflicts of Interest: The authors declare no conflict of interest.

References

1. Warfield, J.N. *An Assault on Complexity*; Office of Corporate Communications: Battelle, FI, USA, 1973.
2. Warfield, J.N. Developing subsystem matrices in structural modeling. *IEEE Trans. Syst. Man Cybern.* **1974**, *4*, 74–80. [CrossRef]
3. Warfield, J.N. Developing interconnection matrices in structural modeling. *IEEE Trans. Syst. Man Cybern.* **1974**, *4*, 81–87. [CrossRef]
4. Warfield, J.N. *Societal Systems: Planning, Policy and Complexity*; John Wiley & Sons: New York, NY, USA, 1976.
5. Elkington, J. 25 Years Ago I Coined the Phrase "Triple Bottom Line." Here's Why it's Time to Rethink it. *Harvard Business Review*, 25 June 2018. Available online: https://hbr.org/2018/06/25-years-ago-i-coined-the-phrase-triple-bottom-line-heres-why-im-giving-up-on-it (accessed on 7 April 2021).
6. Farris, D.R.; Sage, A.P. On the use of interpretive structural modeling for worth assessment. *Comput. Electr. Eng.* **1975**, *2*, 149–174. [CrossRef]
7. Warfield, J.N. *Structuring Complex Systems, (A Battelle Monograph No. 4)*; Battelle Meml. Inst: Columbus, OH, USA, 1974.
8. Godet, M. Introduction to la prospective: Seven key ideas and one scenario method. *Futures* **1986**, *18*, 134–157. [CrossRef]
9. Beikkhakhian, Y.; Javanmardi, M.; Karbasian, M.; Khayambashi, B. The application of ISM model in evaluating agile suppliers selection criteria and ranking suppliers using fuzzy TOPSIS-AHP methods. *Expert Syst. Appl.* **2015**, *42*, 6224–6236. [CrossRef]
10. Govindan, K.; Madan Shankar, K.; Kannan, D. Application of fuzzy analytic network process for barrier evaluation in automotive parts remanufacturing towards cleaner production—A study in an Indian scenario. *J. Clean. Prod.* **2016**, *114*, 199–213. [CrossRef]
11. Dalvi-Esfahani, M.; Ramayah, T.; Nilashi, M. Modelling upper echelons' behavioural drivers of Green IT/IS adoption using an integrated Interpretive Structural Modelling—Analytic Network Process approach. *Telemat. Inform.* **2017**, *34*, 583–603. [CrossRef]
12. Wong, W.P.; Wong, K.Y. Synergizing an ecosphere of lean for sustainable operations. *J. Clean. Prod.* **2014**, *85*, 51–66. [CrossRef]
13. Chauhan, A.; Singh, A. A hybrid multi-criteria decision making method approach for selecting a sustainable location of healthcare waste disposal facility. *J. Clean. Prod.* **2016**, *139*, 1001–1010. [CrossRef]

14. Kannan, G.; Pokharel, S.; Kumar, P.S. A hybrid approach using ISM and fuzzy TOPSIS for the selection of reverse logistics provider. *Resour. Conserv. Recycl.* **2009**, *54*, 28–36. [CrossRef]
15. Bhanot, N.; Rao, P.V.; Deshmukh, S.G. An integrated approach for analysing the enablers and barriers of sustainable manufacturing. *J. Clean. Prod.* **2017**, *142*, 4412–4439. [CrossRef]
16. Kumar, A.; Dixit, G. An analysis of barriers affecting the implementation of e-waste management practices in India: A novel ISM-DEMATEL approach. *Sustain. Prod. Consum.* **2018**, *14*, 36–52. [CrossRef]
17. Ahmad, N.; Quadri, N.; Qureshi, M.; Alam, M. Relationship Modeling of Critical Success Factors for Enhancing Sustainability and Performance in E-Learning. *Sustainability* **2018**, *10*, 4776. [CrossRef]
18. Naveed, Q.N.; Qureshi, M.R.N.; Alsayed, A.O.; Ahmad, N.; Sanober, S.; Shah, A. Assimilating E-Learning barriers using an interpretive structural modeling (ISM). In Proceedings of the 2017 4th IEEE International Conference on Engineering Technologies and Applied Sciences (ICETAS), Salmabad, Bahrain, 29 November–1 December 2017; pp. 1–7.
19. Tan, F.T.C.; Guo, Z.; Cahalane, M.; Cheng, D. Developing business analytic capabilities for combating e-commerce identity fraud: A study of Trustev's digital verification solution. *Inf. Manag.* **2016**, *53*, 878–891. [CrossRef]
20. Raut, R.D.; Narkhede, B.; Gardas, B.B. To identify the critical success factors of sustainable supply chain management practices in the context of oil and gas industries: ISM approach. *Renew. Sustain. Energy Rev.* **2017**, *68*, 33–47. [CrossRef]
21. Tan, T.; Chen, K.; Xue, F.; Lu, W. Barriers to Building Information Modeling (BIM) implementation in China's prefabricated construction: An interpretive structural modeling (ISM) approach. *J. Clean. Prod.* **2019**, *219*, 949–959. [CrossRef]
22. Gardas, B.B.; Raut, R.D.; Narkhede, B. Determinants of sustainable supply chain management: A case study from the oil and gas supply chain. *Sustain. Prod. Consum.* **2019**, *17*, 241–253. [CrossRef]
23. Chandramowli, S.; Transue, M.; Felder, F.A. Analysis of barriers to development in landfill communities using interpretive structural modeling. *Habitat Int.* **2011**, *35*, 246–253. [CrossRef]
24. Gokarn, S.; Kuthambalayan, T.S. Analysis of challenges inhibiting the reduction of waste in food supply chain. *J. Clean. Prod.* **2017**, *168*, 595–604. [CrossRef]
25. Jayant, A.; Azhar, M. Analysis of the Barriers for Implementing Green Supply Chain Management (GSCM) Practices: An Interpretive Structural Modeling (ISM) Approach. *Procedia Eng.* **2014**, *97*, 2157–2166. [CrossRef]
26. Khan, I.; Rahman, Z. Brand experience anatomy in retailing: An interpretive structural modeling approach. *J. Retail. Consum. Serv.* **2015**, *24*, 60–69. [CrossRef]
27. Patidar, L.; Soni, V.K.; Kumar Soni, P. Development of a Framework for Implementation of Maintenance Tools and Techniques Using Interpretive Structural Modeling. *Mater. Today Proc.* **2017**, *4*, 8158–8166. [CrossRef]
28. Mishra, R.P.; Kodali, R.B.; Gupta, G.; Mundra, N. Development of a Framework for Implementation of World-class Maintenance Systems Using Interpretive Structural Modeling Approach. *Procedia CIRP* **2015**, *26*, 424–429. [CrossRef]
29. Raut, R.D.; Gardas, B.B.; Jha, M.K.; Priyadarshinee, P. Examining the critical success factors of cloud computing adoption in the MSMEs by using ISM model. *J. High Technol. Manag. Res.* **2017**, *28*, 125–141. [CrossRef]
30. Agarwal, A.; Shankar, R.; Tiwari, M.K. Modeling agility of supply chain. *Ind. Mark. Manag.* **2007**, *36*, 443–457. [CrossRef]
31. Gardas, B.B.; Raut, R.D.; Narkhede, B. Modeling causal factors of post-harvesting losses in vegetable and fruit supply chain: An Indian perspective. *Renew. Sustain. Energy Rev.* **2017**, *80*, 1355–1371. [CrossRef]
32. Balaji, M.; Arshinder, K. Modeling the causes of food wastage in Indian perishable food supply chain. *Resour. Conserv. Recycl.* **2016**, *114*, 153–167. [CrossRef]
33. Kaswan, M.S.; Rathi, R. Analysis and modeling the enablers of Green Lean Six Sigma implementation using Interpretive Structural Modeling. *J. Clean. Prod.* **2019**, *231*, 1182–1191. [CrossRef]
34. Ali, S.M.; Hossen, M.A.; Mahtab, Z.; Kabir, G.; Paul, S.K.; ul Haq Adnan, Z. Barriers to lean six sigma implementation in the supply chain: An ISM model. *Comput. Ind. Eng.* **2020**, *149*, 106843. [CrossRef]
35. Tarei, P.K.; Chand, P.; Gupta, H. Barriers to the adoption of electric vehicles: Evidence from India. *J. Clean. Prod.* **2021**, *291*, 125847. [CrossRef]
36. de Oliveira Mota, R.; Godinho Filho, M.; Osiro, L.; Ganga, G.M.D.; de Sousa Mendes, G.H. Unveiling the relationship between drivers and capabilities for reduced time-to-market in start-ups: A multi-method approach. *Int. J. Prod. Econ.* **2021**, *233*, 108018. [CrossRef]
37. Mukeshimana, M.C.; Zhao, Z.-Y.; Nshimiyimana, J.P. Evaluating strategies for renewable energy development in Rwanda: An integrated SWOT—ISM analysis. *Renew. Energy* **2021**, *176*, 402–414. [CrossRef]
38. Ruiz-Benitez, R.; López, C.; Real, J.C. Environmental benefits of lean, green and resilient supply chain management: The case of the aerospace sector. *J. Clean. Prod.* **2018**, *167*, 850–862. [CrossRef]
39. Xiao, L. Analyzing consumer online group buying motivations: An interpretive structural modeling approach. *Telemat. Inform.* **2018**, *35*, 629–642. [CrossRef]
40. Haleem, A.; Luthra, S.; Mannan, B.; Khurana, S.; Kumar, S.; Ahmad, S. Critical factors for the successful usage of fly ash in roads & bridges and embankments: Analyzing indian perspective. *Resour. Policy* **2016**, *49*, 334–348. [CrossRef]
41. Singh, A.N.; Gupta, M.P.; Ojha, A. Identifying critical infrastructure sectors and their dependencies: An Indian scenario. *Int. J. Crit. Infrastruct. Prot.* **2014**, *7*, 71–85. [CrossRef]
42. Kanji, R.; Agrawal, R. Exploring the use of corporate social responsibility in building disaster resilience through sustainable development in India: An interpretive structural modelling approach. *Prog. Disaster Sci.* **2020**, *6*, 100089. [CrossRef]

43. Jabbour, C.J.C.; Jugend, D.; de Sousa Jabbour, A.B.L.; Govindan, K.; Kannan, D.; Leal Filho, W. "There is no carnival without samba": Revealing barriers hampering biodiversity-based R&D and eco-design in Brazil. *J. Environ. Manag.* **2018**, *206*, 236–245. [CrossRef]
44. Kumar, D.T.; Palaniappan, M.; Kannan, D.; Shankar, K.M.; Thresh Kumar, D.; Palaniappan, M.; Kannan, D.; Shankar, K.M. Analyzing the CSR issues behind the supplier selection process using ISM approach. *Resour. Conserv. Recycl.* **2014**, *92*, 268–278. [CrossRef]
45. Baykasoğlu, A.; Gölcük, İ. Development of a two-phase structural model for evaluating ERP critical success factors along with a case study. *Comput. Ind. Eng.* **2017**, *106*, 256–274. [CrossRef]
46. Jain, V.; Raj, T. Modelling and analysis of FMS productivity variables by ISM, SEM and GTMA approach. *Front. Mech. Eng.* **2014**, *9*, 218–232. [CrossRef]
47. Peeters, B.; Kiratli, N.; Semeijn, J. A barrier analysis for distributed recycling of 3D printing waste: Taking the maker movement perspective. *J. Clean. Prod.* **2019**, *241*, 118313. [CrossRef]
48. Gokarn, S.; Choudhary, A. Modeling the key factors influencing the reduction of food loss and waste in fresh produce supply chains. *J. Environ. Manag.* **2021**, *294*, 113063. [CrossRef] [PubMed]
49. Diabat, A.; Govindan, K. An analysis of the drivers affecting the implementation of green supply chain management. *Resour. Conserv. Recycl.* **2011**, *55*, 659–667. [CrossRef]
50. Xu, X.; Zou, P.X.W. Analysis of factors and their hierarchical relationships influencing building energy performance using interpretive structural modelling (ISM) approach. *J. Clean. Prod.* **2020**, *272*, 122650. [CrossRef]
51. Kadam, S.; Bandyopadhyay, P.K. Modelling passenger interaction process (PIP) framework using ISM and MICMAC approach. *J. Rail Transp. Plan. Manag.* **2020**, *14*, 100171. [CrossRef]
52. Sharma, M.; Sehrawat, R. A hybrid multi-criteria decision-making method for cloud adoption: Evidence from the healthcare sector. *Technol. Soc.* **2020**, *61*, 101258. [CrossRef]
53. Attri, R.; Dev, N.; Sharma, V. Interpretive structural modelling (ISM) approach: An overview. *Res. J. Manag. Sci.* **2013**, *2319*, 1171.
54. Attri, R. Interpretive structural modelling: A comprehensive literature review on applications. *Int. J. Six Sigma Compet. Advant.* **2017**, *10*, 258–331. [CrossRef]
55. Gardas, B.B.; Raut, R.D.; Narkhede, B.E. A state-of the-art survey of interpretive structural modelling methodologies and applications. *Int. J. Bus. Excell.* **2017**, *11*, 505–560. [CrossRef]
56. Anand, A.; Bansal, G. Interpretive structural modelling for attributes of software quality. *J. Adv. Manag. Res.* **2017**, *14*, 256–269. [CrossRef]
57. Guan, L.; Abbasi, A.; Ryan, M.J. Analyzing green building project risk interdependencies using Interpretive Structural Modeling. *J. Clean. Prod.* **2020**, *256*, 120372. [CrossRef]
58. Abuzeinab, A.; Arif, M.; Qadri, M.A. Barriers to MNEs green business models in the UK construction sector: An ISM analysis. *J. Clean. Prod.* **2017**, *160*, 27–37. [CrossRef]
59. Agi, M.A.N.; Nishant, R. Understanding influential factors on implementing green supply chain management practices: An interpretive structural modelling analysis. *J. Environ. Manag.* **2017**, *188*, 351–363. [CrossRef]
60. Luthra, S.; Kumar, S.; Kharb, R.; Ansari, M.F.; Shimmi, S.L. Adoption of smart grid technologies: An analysis of interactions among barriers. *Renew. Sustain. Energy Rev.* **2014**, *33*, 554–565. [CrossRef]
61. Dewangan, D.K.; Agrawal, R.; Sharma, V. Enablers for Competitiveness of Indian Manufacturing Sector: An ISM-Fuzzy MICMAC Analysis. *Procedia Soc. Behav. Sci.* **2015**, *189*, 416–432. [CrossRef]
62. Raci, V.; Shankar, R. Analysis of interactions among the barriers of reverse logistics. *Technol. Forecast. Soc. Chang.* **2005**, *72*, 1011–1029. [CrossRef]
63. Mathiyazhagan, K.; Govindan, K.; NoorulHaq, A.; Geng, Y. An ISM approach for the barrier analysis in implementing green supply chain management. *J. Clean. Prod.* **2013**, *47*, 283–297. [CrossRef]
64. Tseng, M.-L. Modeling sustainable production indicators with linguistic preferences. *J. Clean. Prod.* **2013**, *40*, 46–56. [CrossRef]
65. Muduli, K.; Govindan, K.; Barve, A.; Kannan, D.; Geng, Y. Role of behavioural factors in green supply chain management implementation in Indian mining industries. *Resour. Conserv. Recycl.* **2013**, *76*, 50–60. [CrossRef]
66. Lin, L.-Z.; Yeh, H.-R. Analysis of tour values to develop enablers using an interpretive hierarchy-based model in Taiwan. *Tour. Manag.* **2013**, *34*, 133–144. [CrossRef]
67. Warshall, S. A theorem on boolean matrices. *JACM* **1962**, *9*, 11–12. [CrossRef]
68. Sajid, Z.; Khan, F.; Zhang, Y. Integration of interpretive structural modelling with Bayesian network for biodiesel performance analysis. *Renew. Energy* **2017**, *107*, 194–203. [CrossRef]
69. Bouzon, M.; Govindan, K.; Rodriguez, C.M.T. Reducing the extraction of minerals: Reverse logistics in the machinery manufacturing industry sector in Brazil using ISM approach. *Resour. Policy* **2015**, *46*, 27–36. [CrossRef]
70. Gholami, H.; Rezaei, G.; Saman, M.Z.M.; Sharif, S.; Zakuan, N. State-of-the-art Green HRM System: Sustainability in the sports center in Malaysia using a multi-methods approach and opportunities for future research. *J. Clean. Prod.* **2016**, *124*, 142–163. [CrossRef]
71. Singhal, D.; Tripathy, S.; Jena, S.K.; Nayak, K.K.; Dash, A. Interpretive structural modelling (ISM) of obstacles hindering the remanufacturing practices in India. *Procedia Manuf.* **2018**, *20*, 452–457. [CrossRef]

72. Toktaş-Palut, P.; Baylav, E.; Teoman, S.; Altunbey, M. The impact of barriers and benefits of e-procurement on its adoption decision: An empirical analysis. *Int. J. Prod. Econ.* **2014**, *158*, 77–90. [CrossRef]
73. Denyer, D.; Tranfield, D. Producing a Systematic Review. *SAGE Handb. Organ. Res. Methods* **2009**, *39*, 672–688. [CrossRef]
74. Peakall, R.O.D.; Smouse, P.E. GENALEX 6: Genetic analysis in Excel. Population genetic software for teaching and research. *Mol. Ecol. Notes* **2006**, *6*, 288–295. [CrossRef]
75. Chandrupatla, T.R.; Belegundu, A.D.; Ramesh, T.; Ray, C. *Introduction to Finite Elements in Engineering*; Prentice Hall: Upper Saddle River, NJ, USA, 2002; Volume 2.
76. Zhang, Y.; Huo, M.; Zhou, J.; Xie, S. PKSolver: An add-in program for pharmacokinetic and pharmacodynamic data analysis in Microsoft Excel. *Comput. Methods Programs Biomed.* **2010**, *99*, 306–314. [CrossRef] [PubMed]
77. Wang, G.; Wang, Y.; Zhao, T. Analysis of interactions among the barriers to energy saving in China. *Energy Policy* **2008**, *36*, 1879–1889. [CrossRef]
78. Luthra, S.; Garg, D.; Haleem, A. An analysis of interactions among critical success factors to implement green supply chain management towards sustainability: An Indian perspective. *Resour. Policy* **2015**, *46*, 37–50. [CrossRef]
79. Ansari, M.F.; Kharb, R.K.; Luthra, S.; Shimmi, S.L.; Chatterji, S. Analysis of barriers to implement solar power installations in India using interpretive structural modeling technique. *Renew. Sustain. Energy Rev.* **2013**, *27*, 163–174. [CrossRef]
80. Yadav, D.K.; Barve, A. Analysis of critical success factors of humanitarian supply chain: An application of Interpretive Structural Modeling. *Int. J. Disaster Risk Reduct.* **2015**, *12*, 213–225. [CrossRef]
81. Diabat, A.; Kannan, D.; Mathiyazhagan, K. Analysis of enablers for implementation of sustainable supply chain management—A textile case. *J. Clean. Prod.* **2014**, *83*, 391–403. [CrossRef]
82. D. Vivek, S.; Banwet, D.K.; Shankar, R. Analysis of interactions among core, transaction and relationship-specific investments: The case of offshoring. *J. Oper. Manag.* **2008**, *26*, 180–197. [CrossRef]
83. Cagno, E.; Micheli, G.J.L.; Jacinto, C.; Masi, D. An interpretive model of occupational safety performance for Small- and Medium-sized Enterprises. *Int. J. Ind. Ergon.* **2014**, *44*, 60–74. [CrossRef]
84. Govindan, K.; Palaniappan, M.; Zhu, Q.; Kannan, D. Analysis of third party reverse logistics provider using interpretive structural modeling. *Int. J. Prod. Econ.* **2012**, *140*, 204–211. [CrossRef]
85. Jia, P.; Diabat, A.; Mathiyazhagan, K. Analyzing the SSCM practices in the mining and mineral industry by ISM approach. *Resour. Policy* **2015**, *46*, 76–85. [CrossRef]
86. Shi, Q.; Yu, T.; Zuo, J.; Lai, X. Challenges of developing sustainable neighborhoods in China. *J. Clean. Prod.* **2016**, *135*, 972–983. [CrossRef]
87. Shibin, K.T.; Gunasekaran, A.; Dubey, R. Explaining sustainable supply chain performance using a total interpretive structural modeling approach. *Sustain. Prod. Consum.* **2017**, *12*, 104–118. [CrossRef]
88. Sağ, S.; Kaynak, R.; Sezen, B. Factors Affecting Multinational Team Performance. *Procedia Soc. Behav. Sci.* **2016**, *235*, 60–69. [CrossRef]
89. Rajaprasad, S.V.S.; Chalapathi, P.V. Factors Influencing Implementation of OHSAS 18001 in Indian Construction Organizations: Interpretive Structural Modeling Approach. *Saf. Health Work* **2015**, *6*, 200–205. [CrossRef] [PubMed]
90. Luthra, S.; Haleem, A. Hurdles in Implementing Sustainable Supply Chain Management: An Analysis of Indian Automobile Sector. *Procedia Soc. Behav. Sci.* **2015**, *189*, 175–183. [CrossRef]
91. Sindhu, S.; Nehra, V.; Luthra, S. Identification and analysis of barriers in implementation of solar energy in Indian rural sector using integrated ISM and fuzzy MICMAC approach. *Renew. Sustain. Energy Rev.* **2016**, *62*, 70–88. [CrossRef]
92. Shen, L.; Song, X.; Wu, Y.; Liao, S.; Zhang, X. Interpretive Structural Modeling based factor analysis on the implementation of Emission Trading System in the Chinese building sector. *J. Clean. Prod.* **2016**, *127*, 214–227. [CrossRef]
93. Ahuja, R.; Sawhney, A.; Arif, M. Prioritizing BIM Capabilities of an Organization: An Interpretive Structural Modeling Analysis. *Procedia Eng.* **2017**, *196*, 2–10. [CrossRef]
94. Rajesh, R. Technological capabilities and supply chain resilience of firms: A relational analysis using Total Interpretive Structural Modeling (TISM). *Technol. Forecast. Soc. Chang.* **2017**, *118*, 161–169. [CrossRef]
95. Jothimani, D.; Bhadani, A.K.; Shankar, R. Towards Understanding the Cynicism of Social Networking Sites: An Operations Management Perspective. *Procedia Soc. Behav. Sci.* **2015**, *189*, 117–132. [CrossRef]
96. Rana, N.P.; Barnard, D.J.; Baabdullah, A.M.A.; Rees, D.; Roderick, S. Exploring barriers of m-commerce adoption in SMEs in the UK: Developing a framework using ISM. *Int. J. Inf. Manag.* **2019**, *44*, 141–153. [CrossRef]
97. Mondal, J.; Chakrabarti, S. Insights and anatomy of brand experience in app-based retailing (eRBX): Critical play of physical evidence and enjoyment. *J. Retail. Consum. Serv.* **2021**, *60*, 102484. [CrossRef]
98. Naveed, Q.N.; Ahmad, N.; Qamar, S.; Khan, N.; Naim, A.; Hussain, M.R.; Qureshi, M.R.N.; Alsayed, A.O.; Mohiuddin, K. Relationship modeling for OSN-based E-Learning Deployment. In Proceedings of the 2019 IEEE 6th International Conference on Engineering Technologies and Applied Sciences (ICETAS), Kuala Lumpur, Malaysia, 20–21 December 2019; pp. 1–7.
99. Ahmad, N.; Hoda, N.; Alahmari, F. Developing a Cloud-Based Mobile Learning Adoption Model to Promote Sustainable Education. *Sustainability* **2020**, *12*, 3126. [CrossRef]
100. Ahmad, N. The Structural Modeling of Significant Factors for Sustainable Cloud Migration. *Int. J. Intell. Eng. Syst.* **2021**, *14*, 1–10. [CrossRef]

sustainability

MDPI

Article

Extended Maximal Covering Location and Vehicle Routing Problems in Designing Smartphone Waste Collection Channels: A Case Study of Yogyakarta Province, Indonesia

Diana Puspita Sari [1,2,*], Nur Aini Masruroh [1,*] and Anna Maria Sri Asih [1]

1 Department of Mechanical and Industrial Engineering, Universitas Gadjah Mada, Yogyakarta 55281, Indonesia; amsriasih@ugm.ac.id
2 Department of Industrial Engineering, Diponegoro University, Semarang 50275, Indonesia
* Correspondence: dianapuspitasari@lecturer.undip.ac.id (D.P.S.); aini@ugm.ac.id (N.A.M.)

Abstract: Most people will store smartphone waste or give it to others; this is due to inadequate waste collection facilities in all cities/regencies in Indonesia. In Yogyakarta Province, there is no electronic waste collection facility. Therefore, an e-waste collection network is needed to cover all potential e-waste in the province of Yogyakarta. This study aims to design a collection network to provide easy access to facilities for smartphone users, which includes the number and location of each collection center and the route of transporting smartphone waste to the final disposal site. We proposed an extended maximal covering location problem to determine the number and location of collection centers. Nearest neighbor and tabu search are used in forming transportation routes. The nearest neighbor is used for initial solution search, and tabu search is used for final solution search. The study results indicate that to facilitate all potential smartphone waste with a maximum distance of 11.2 km, the number of collection centers that must be established is 30 units with three pick-up routes. This research is the starting point of the smartphone waste management process, with further study needed for sorting, recycling, repairing, or remanufacturing after the waste has been collected.

Keywords: smartphone waste; collection center; extended maximal covering location problem; transportation route; mathematical model; tabu search

check for updates

Citation: Sari, D.P.; Masruroh, N.A.; Asih, A.M.S. Extended Maximal Covering Location and Vehicle Routing Problems in Designing Smartphone Waste Collection Channels: A Case Study of Yogyakarta Province, Indonesia. *Sustainability* **2021**, *13*, 8896. https://doi.org/10.3390/su13168896

Academic Editors: Željko Stević, Dragan Pamucar and Marc A. Rosen

Received: 28 June 2021
Accepted: 5 August 2021
Published: 9 August 2021

Publisher's Note: MDPI stays neutral with regard to jurisdictional claims in published maps and institutional affiliations.

Copyright: © 2021 by the authors. Licensee MDPI, Basel, Switzerland. This article is an open access article distributed under the terms and conditions of the Creative Commons Attribution (CC BY) license (https://creativecommons.org/licenses/by/4.0/).

1. Introduction

Developing countries such as Indonesia currently have the problem of handling large volumes of electronic waste (e-waste) [1]. It is associated with rapid technological and economic developments, leading to the production of a wider selection of electronic products at more affordable prices [2], thereby increasing public consumption and potential for electronic waste. The Global E-waste Monitor 2017 Quantities, Flows, and Resources ranked Indonesia ninth among the global producers of electronic waste, with smartphones being observed to contribute significantly. It is, however, important to note that the use of smartphones started increasing in 2020 due to the emergence of coronavirus, which prompted people to work and learn from home using online platforms. Records show smartphones are the technological devices with the highest consumption rate (70%), followed by laptops and personal computers [3], but there is no appropriate waste management process [4]. This is indicated by the absence of regulations for the collection and transportation of electronic wastes in Indonesia, with those implemented observed to be limited to informal initiatives. This, therefore, led to the low ranking of the country in the waste management system by the United Nations University. This is one of the major differences between Indonesia and developed countries [5]. A previous study also showed that improper handling of waste is dangerous for environmental sustainability [6].

About 80% of the materials composing smartphones can be recycled effectively [7]. Smartphones contain valuable materials, such as gold, silver, and palladium [8]. Metals in electronic waste, especially smartphones, are present in higher concentrations than in primary ore found in the ground. As an illustration, 300–350 g of secondary gold can be extracted from one ton of smartphones, while every ton of soil in ordinary gold mines only produces 5 g of primary gold [9]. Resource extraction from e-waste is more economical than extracting metal ores from the ground [10]. Thus, smartphone recycling is done because the economic benefits outweigh the costs [11]. Proper management of e-waste is necessary to reduce the problem of metal scarcity [8]. The potential for smartphone waste in Indonesia is quite significant. The total population of Indonesia in 2020 was 270,203,917 people [12]. If 63.53% are smartphone users [13], then the total number of smartphone users is 171,660,549 people. With an average smartphone lifetime of 4.7 years [14], these users produce 36,523,521 units of smartphone waste per year. When this waste is appropriately managed, in addition to minimizing the environmental impact, it can also provide significant economic benefits by producing 5.48–6.39 tons of secondary gold and saving natural resources.

However, so far, the amount of secondary metal recovered through e-waste recycling has been limited [15]; this is due to the limited supply of e-waste. A preliminary study conducted on smartphone users in Indonesia showed that 59% save non-functioning smartphones; 21% dispose of them; and the rest give them out to other people, sell them, and use them in other ways. This is because the public does not know what to do with these items. Meanwhile, Yogyakarta is one of the barometer provinces in Indonesia with an improper electronic waste management system through the formal channel. According to previous studies, government drivers are the factor with the most influence on consumers' intentions to participate in smartphone waste collection programs, followed by facility accessibility [16]. This means that the government needs to develop and implement a formal e-waste management system, starting with the e-waste collection process. One of the alternative electronic waste collection programs applicable to Indonesia is the use of a dropbox [17], but Yogyakarta Province does not currently have any collection points for smartphone waste. Therefore, there is the need to provide a convenient collection channel for the consumers, which is expected to be a major starting point for a formal channel to waste management in the area.

This study aimed to design a collection channel by determining the number and location of the collection center facilities followed by a transportation route from the collection center to the final disposal site. Facility location is related to the finding of a solution that covers customers using a number of facilities. It is, however, important to note that covering problems are fundamental facility location problems [18], which are often categorized as location set covering problems (LSCPs) and maximal covering location problems (MCLPs). The classic MCLP involves looking for the location of several facilities on the network in such a way that the population covered can be maximized [19]. Church and ReVelle first introduced this model in 1973 at a North American Regional Science Council [20]. The purpose was to maximize the demand covered by a particular service distance by placing a certain number of distribution facilities [21]. Therefore, customers or clients are declared covered when they are within a certain coverage distance from at least one facility [22]. The model is also important in the decision-making of the supply chain process, making it relatively important for practical use [23].

Several previous studies have used MCLP to design models and approaches in determining locations. MCLP is used in both the public and private sectors. In the public sector, it has been applied to determine the spread of an ambulance in emergency services [24], the location of emergency warning sirens [25], the location of medical equipment supply centers [26], the location of treatment centers in the event of a disease outbreak [27], appropriate locations for shelters for those temporarily displaced by floods [28], and the location of a waste cooking oil collection center [29]. Its use in the private sector involves determining the location of bank branches [30]. Several researchers have developed MCLPs. For example, Davari et al. [31] developed a MCLP with fuzzy travel times; Arana-Jiménez et al. [32] developed a fuzzy MCLP; Vatsa and Jayaswal ([33,34]) modeled a capacitated multiperiod MCLP with server uncertainty; and Cordeau et al. [9] introduced the MCLP algorithm to determine a subset of facilities, maximizing customer requests by considering budget constraints. A continuous MCLP was also developed by Yang et al. [35] to optimize a continuous location of the cellular network's communication centers for natural disaster rescue. ReVelle et al. [36] also solved the MCLP with heuristic concentration which is used to determine a prominent case solution to maximum coverage locations with a high coverage percentage. Ibarra-Rojas et al. [37] developed a MCLP with accessibility indicators for when facilities have limited service areas, while Alizadeh and Nishi [38] used the hybrid covering location problem for strategic and tactical decisions. Alizadeh and Nishi [39] also developed a multiperiod maximal coverage location problem with different facility configurations as an extension of the classic MCLP. Zhang et al. [40] addressed the issue of locating multimodal facilities in emergency medical rescue.

The classical MCLP is used to determine the minimum number of facilities to maximize the demand covered by a given service distance. The model does not consider costs; it assumes that the number of facilities is minimal and that the investment costs are also minimal. Because each alternative location is assumed to have the same investment costs, it is necessary to develop a model that considers the difference in investment costs between potential locations. In this study, the collection center to be built is an intermediary facility, so it is also necessary to consider transportation costs to the final facility. Therefore, in this study, we develop the MCLP method by considering the investment and transportation costs, hereinafter referred to as an extended maximal covering location problem (e-MCLP). With this development, in addition to minimizing the number of facilities, it will also minimize total costs, including investment costs and transportation costs. Thus, the developed model is expected to provide an affordable facility location to consumers with a minimum total investment and transportation costs from the collection center to the final disposal facility.

The selection of the number and location of the collection centers was followed by the transportation route scheduling plan from the collection center to the final disposal site to determine the optimal route for efficient product distribution. It is defined as the route with the shortest distance and is considered important due to its ability to reduce transportation costs [41]. The vehicle route optimization problem, however, is known as the vehicle routing problem (VRP), which was introduced by Dantzig and Ramser in 1959 to solve the problem of gasoline distribution [42]. VRP is a common discrete optimization problem in transportation and logistics [43]. It is generally an integral part of the vehicle route with the exact delivery location visited once while all the routes start and end at the warehouse [44]. VRP focuses on the distribution of goods from the company's depot to customers and aims to minimize global transportation costs related to distance, fixed costs associated with vehicles and balance routes, and the number of vehicles required to serve consumers [45]. There are three methods of solving VRPs: the exact, heuristic, and metaheuristic methods [46]. However, the exact method is not applicable to a problem with a large input size and a limited time.

The methods used in this study include the heuristic and metaheuristic methods. The heuristic method involved the application of the nearest neighbor (NN) method, which has been widely used to solve VRP. Solomon introduced it in 1987 based on the idea of visiting the closest location from every other location visited [47], and it has been observed to be significantly better and to have more realistic performance in route formation than other methods [48]. This led to its wide application in solving the traveling salesman problem [49], determining routes from one city to another [50], designing waste transportation routes [51], and minimizing travel time and fuel consumption for transportation of agricultural products [52]. The nearest neighbor method is quite effective in its application due to its ability to look for consumers based on the closest distance from the vehicle's last location. It is, important to note that the nearest neighbor method produces the route with the shortest distance compared to other heuristic methods [41]. It is also easy to implement and execute the algorithm, but it does not guarantee the best resulting solution [53], so in this study, nearest neighbor was used to determine the initial solution.

This research applied the tabu search (TS) method, an algorithm considered to have the ability to produce an optimal solution. It was first introduced by Glover [54] based on the idea that allowing uphill motion helps to prevent the solution from becoming stuck in local optimal conditions [55]. The strength of this method lies in its flexible memory structure [54]. This makes its solutions very similar every time it is applied and makes it better than the other methods, such as simulated annealing and genetic algorithm [54]. Several studies have used tabu searches to solve VRP [56], classical VRP, periodic VRP, multidepot VRP, site-dependent VRP [57], heterogeneous fleet VRP [58], VRP with discrete split deliveries and pickups [59], multicompartment VRP [60], heterogeneous multitype fleet VRP with time windows and an incompatible loading constraint [61], multidepot open VRP [62], VRP with cross docks and split deliveries [63], VRP with private fleet and common carrier [57], time-dependent VRP with time windows on a road network [64], consistent VRP [65], and heterogeneous VRP on a multigraph [66]. Shi et al. [67] also used the heuristic solution method for the problem of multidepot vehicle routing-based waste collection and compared the results with the tabu search. Khan et al. [68] presented a sustainable closed-loop supply chain framework that uses a metaheuristic approach, tabu search, and simulated annealing. Tebaldi et al. [69] determined the best route to visit a set of customers, considering vehicle capacity and time constraints. This result underlies the use of the nearest neighbor approach to obtain an initial solution and the use of the metaheuristic tabu search approach to determine the final solution.

2. Materials and Methods

This research was conducted in two main stages: determining the number and location of collection centers and determining the smartphone waste transportation route. The location, number, and capacity of collection centers were determined by developing a maximal covering location problem hereinafter referred to as the extended maximal covering location problem (e-MCLP). The focus of the MCLP is to minimize the number of facilities while ensuring all consumers are covered, but the e-MCLP was developed to consider the costs involved. The model's objective was, therefore, to minimize the total costs, including those associated with investment and transportation from the collection facility to the final disposal site. The costs associated with collecting smartphones are not as high as those for other large volumes of e-waste, but the developed model can be used for other types of waste. The reason for choosing this type of waste is because it has a higher economic value (containing precious metals such as gold, silver, and palladium) than others, with components that allow up to 80% recycling and a large potential for smartphone waste. Meanwhile, for now, informal actors dominate the practice of recycling smartphone waste, which harms the environment. The low collection cost and high economic and environmental benefits are expected to motivate the government to implement the proposed scenario.

The development scenario involves two levels of collection center (CC) facilities, namely the primary collection center (PCC) and the secondary collection center (SCC). Consumers collect their waste at PCC. Instead, local governments carry out transportation from PCC to SCC. Transportation routes are needed in this study because smartphones are products with small volumes, so the capacity of the collection center is not as large as vehicle capacity. If one trip only picks up from one PCC, it becomes inefficient because the vehicle's utility is low, and transportation costs will be higher due to many trips being needed. For this reason, it is necessary to consider the route determination in this study. Routing is expected to increase vehicle utilization and save transportation costs. The output of determining the transportation route is expected to be an input for local governments to schedule waste collection.

Yogyakarta, one of the provinces in Indonesia, is located on Java Island and has an area of 3178.79 km^2. It has a municipality and four regencies: Yogyakarta city and

Gunung Kidul, Bantul, Sleman, and Kulon Progo regencies, with respective areas of 32.5, 1485.36, 506.85, 574.82, and 579.26 km^2. These areas contain 14, 18, 17, 17, and 12 districts, respectively [70], as indicated in Appendix A, for a total of 78 districts. These districts were used as candidates for primary collection centers (PCCs) in this study. The parameters used as input in the mathematical model include the distance between the PCCs, the distance expected by consumers, and the distance from PCC to SCC.

Yogyakarta Province currently has 3 locations serving as final disposal sites (TPAs). The first is the Regional TPA, commonly called Piyungan TPA in Ngablak, Sitimulyo Village, Piyungan District, Bantul Regency. It is an integrated waste disposal site created to serve Yogyakarta City, Bantul Regency, and Sleman Regency [71]. The second location is Wonosari TPA in Wukirsari, Baleharjo, Wonosari, Gunung Kidul Regency, and the third is the Banyuroto TPA in Dlingo, Banyuroto, Nanggulan, Kulon Progo Regency. The Piyungan TPA has the largest capacity and most strategic location among the three, and this makes it suitable to be used as the secondary collection center (SCC). The candidates for the PCCs are district offices, which means the distances between PCCs are the same as those between district offices, and the distance from the PCC to SCC is the distance from the district office to the TPA Piyungan.

The PCC is provided by the government for consumers in the form of a dropbox, while SCC is a waste collection point for all the PCCs in a province. For this research, one SCC was located at the final disposal site in one province while the PCCs were built at the minimum number required to minimize investment costs incurred but with the ability to reach all consumers. Further, a survey conducted on smartphone users, with a total of 325 valid questionnaires, showed the consumers are willing to bring their smartphone waste to a collection facility with a maximum distance of 11.2 km. This means the PCC to be established is based on the number of districts to accommodate the interests of the consumers. Meanwhile, the PCC with the closest distance to the SCC was selected for this research to accommodate government interests by minimizing transportation costs. The PCC is located in the district office, a government-owned facility, and this means it does not require large investment costs since there is no need to procure land and a building, as only the dropbox needs to be prepared. This collection center has the capacity to accommodate all the smartphone waste supplies in the area due to the small product volume. It is important to determine the transportation route to optimize vehicle utility due to the relatively small volume of waste.

The location and capacity of the PCC were used to determine the transportation routes by joining the nearest neighbor approach and the tabu search model (NN-TB). The application of the NN was initiated from the starting point, which is the depot/SCC, and directed towards the PCC with the closest distance, which has not been visited due to several restrictions. The solution obtained at this stage is limited to determining the best route and the consumers to be served next based on the nearest point to the vehicle's last location [72]. It has been previously stated that the nearest neighbor algorithm is easy to implement and execute but does not guarantee the maximum resulting solution [53], and this was the reason it was used in this study to determine only the initial solution. Afterward, the tabu search method was used to search for the optimal route. The metaheuristic method is usually applied to solve combinatorial optimization problems, where the combinations are usually used to calculate the number of exchanges to be made in each iteration [73]. The tabu search algorithm is also a mathematical optimization method that guides the iterative search for solutions by providing tabu status for solutions found [74].

2.1. Collection Center Determination Steps

The parameters used as input in the mathematical model are the distance between PCC candidates, the distance expected by consumers, and the PCC candidate's distance to the final disposal site or secondary collection center (SCC). The distance between PCC candidates and distances between each PCC and SCC were based on Google Maps. The distance matrices between PCC candidates and from the PCC candidates to the SCC are shown in Appendix B. The distance value is essential to determine the number and location of PCCs to be built in the area.

The notation used in the mathematical model of e-MCLP is as follows:

Z	Total cost
IC	Investment cost
TC	Transportation cost
m	The number of the district ($m = 1, 2, \ldots, \mid m \mid$)
k	The number of SCC ($k = 1$)
X_i	Supply point i
X_j	Point j is selected or not as a PCC
X_j	$\begin{cases} 1, & j \text{ become } CCP\ \forall j \in V \\ 0, & \text{otherwise} \end{cases}$
a_{ij}	Distance requirements (fulfilled or not)
a_{ij}	$\begin{cases} 1, & \text{distance } i \text{ to } j \leq D\ \forall i,j \in V \\ 0, & \text{otherwise} \end{cases}$
D	Range (km)
Q_j	The capacity of PCC at point j
Y_i	Coverage of smartphone waste supply at point i (covered or not)
Y_i	$\begin{cases} 1, & \text{the point is covered in the PCC at point } j\ \forall i \in V \\ 0, & \text{otherwise} \end{cases}$
S_i	Supply of smartphone waste at point i

The basic model was developed from the MCLP [75] in the form of e-MCLP, and its functional objective was to minimize the total cost of the number of facilities to be established within the range wanted by the consumer, as shown in Equation (1). The costs considered include those associated with the investment and transportation from PCC to SCC. Furthermore, the PCC was established in a district office, a government facility, which means there was no need to invest money in land acquisition. Therefore, the only investment needed was the procurement of dropbox, and the value is the same for all candidate locations. It is important to note that the PCC locations selected were those with the lowest investment costs and closer to the SCC. The decision variable X_j has a value of 1 or 0, where a value of 1 indicates the point j is selected as a PCC and a value of 0 indicates the point j is not selected as a PC. Dropbox procurement costs are USD 350.37 (USD 1 is equivalent to IDR 14,270.75), and the dropbox service life is 5 years; using the straight-line depreciation method, the annual depreciation cost is USD 70.07 per dropbox. Thus, the investment cost per year is USD 70.07 per dropbox. The vehicle's fuel consumption is 10 km/L at USD 0.67 per liter; therefore the transportation cost is USD 0.067 per kilometer.

Equations (2)–(6) are constraint functions. Equation (2) is a limiting function that requires $a_{ij}X_j$ to be 1, and this means a minimum of one PCC needs to be established within the range of the consumers' point. Meanwhile, Equations (3)–(5) state that X_j, a_{ij}, and Y_i are binary, while Equation (6) states that the PCC capacity at point j is the accumulation of the waste supply multiplication at point i by 1 or 0, where 1 means the waste supply at point i is covered and 0 means it is not covered. Smartphones are, however, usually in small volume and not too large a supply due to the estimation of lifespan at two years. Therefore, the PCC capacity value used in this research is 1, which indicates that the entire waste supply was accommodated.

$$Min\ Z = \sum_{j=1}^{m} IC_j X_j + \sum_{j=1}^{m} \sum_{k}^{1} TC_{jk} X_j \quad (1)$$

$$\sum_{i,j=1}^{m} a_{ij} X_i \geq 1 \quad (2)$$

$$X_j \in [0,1]\ \forall j \in V \quad (3)$$

$$a_{ij} \in [0,1]\ \forall i,j \in V \quad (4)$$

$$Y_i \in [0,1]\ \forall i \in V \quad (5)$$

$$Q_j = \sum_{i=1}^{m} Y_i S_i, \text{ for every } j \quad (6)$$

2.2. Steps to Determine the Transportation Route

The method used in this research was the nearest neighbor and tabu search (NN-TS) method, where the results obtained from the nearest neighbor were used as input in the tabu search. It is important to note that the tabu search was initiated by approaching a local minimum and noting recent movements in a tabu list that forms an adaptive memory to explore better solutions, with its size indicating the degree of diversification and intensification [76]. The mathematical model was, however, first determined before the calculations, and this was based on several assumptions and limitations, which include the following: (1) the vehicle has enough capacity to accommodate smartphone waste; (2) the distance from location $j(a)$ to $j(b)$ is the same as the distance from location $j(b)$ to $j(a)$ due to symmetry; (3) collection activities to PCCs start from 08:00–16:00 WIB with a rest time of 1 h, and this means the planning time horizon for a day is 7 h; (4) one vehicle visits more than one PCC but each PCC is only visited by one vehicle; (5) the average vehicle speed is 45 km/h; (6) the loading time at a PCC is 10 min; (7) the unloading and administration time at the SCC is 30 min. The notation used in the mathematical model of VRP is as follows:

V	The set of all vertices with 0 is a SCC $\{0, 1, 2, \ldots, v\}$
P	The set of PCC $\{1, 2, \ldots, p\}$
E	The set of directed ribs $\{(j(a),j(b)) \mid j(a),j(b) \in V, j(a) \neq j(b)\}$
T	The set of trip $\{1, 2, \ldots, t\}$
C	Vehicles $\{1, 2, \ldots, c\}$
J	Total mileage (km)
$D_{j(a)j(b)}$	Distance from PCC at point $j(a)$ to $j(b)$ (km)
$X^t_{j(a)j(b)c}$	There is a trip from PCC at point $j(a)$ to $j(b)$ on trip t or not
$X^t_{j(a)j(b)c}$	$\begin{cases} 1, \text{ vehicle } c \text{ travels from point } j(a) \text{ to } j(b) \text{ on the trip } t \\ 0, \text{ otherwise} \end{cases}$
$d_{j(a)}$	PCC capacity at PCC at point $j(a)$
Q	Load capacity on a route
$T_{j(a)j(b)}$	Travel time from PCC at point $j(a)$ to $j(b)$
S^t_c	Service time (loading–unloading)
$Y^t_{j(a)c}$	There is a load on PCC at point $j(a)$ carried by vehicle c on trip t or not
$Y^t_{j(a)c}$	$\begin{cases} 1, \text{ there is a load at point } j(a) \text{ carried by vehicle } c \text{ on the trip } t \\ 0, \text{ otherwise} \end{cases}$

The objective function of the VRP mathematical model is to minimize the total distance traveled from the route as shown in Equation (7). The decision variable $X^t_{j(a)j(b)c}$ has a value of 1 or 0; 1 indicates the selected route when vehicle c travels from PCC at point $j(a)$ to $j(b)$ on the trip t, and 0 indicates when the situation is otherwise. Equations (8)–(14) are constraint functions, with Equations (8) and (9) used to show that the route starts from and returns to SCC. Equations (10) and (11) state that each PCC is served exactly once on one route. Hereinafter, the vehicle's load capacity on a trip is the accumulation of the PCC capacity served, and its maximum capacity is not exceeded, as shown in Equation (12). This is because the supply is not large and the product volume is small, which allows the vehicle to carry the entire supply of smartphone consumers at once. Meanwhile, Equation (13) shows the vehicles going to the SCC to unload. However, the route completion time was calculated from the vehicle's total time plus the service time, which is loading–unloading time, and observed not to have exceeded the planning time horizon in a day, which is 7 h, as shown in Equations (14) and (15).

$$Min\ Z = \sum_{j(a)\in V} \sum_{j(b)\in V} \sum_{t\in T} \sum_{c\in C} C_{j(a)j(b)} X^t_{j(a)j(b)} \quad (7)$$

$$\sum_{j(b)\in P} X^t_{0j(b)c} = 1 \forall c \in C \quad (8)$$

$$\sum_{j(b)\in P} X^{t}_{j(a)0c} = 1 \forall c \in C \tag{9}$$

$$\sum_{j(b)\in P} \sum_{t\in T} \sum_{c\in C} X^{t}_{j(a)j(b)c} \forall j(a) \in P, j(b) \neq j(a) \tag{10}$$

$$\sum_{j(b)\in P} \sum_{t\in T} \sum_{c\in C} X^{t}_{j(a)j(b)c} \forall j(b) \in P, j(a) \neq j(b) \tag{11}$$

$$Q = \sum_{j(a)\in P} d_{j(a)} Y^{t}_{j(a)c} \forall t \in T, \forall c \in C \tag{12}$$

$$\sum_{j(a)\in P} X^{t}_{j(a)0c} = 1 \forall t \in T, \forall c \in C \tag{13}$$

$$CT = \sum_{j(a)\in V} \sum_{j(b)\in V} \sum_{t\in T} T_{j(a)j(b)} X^{t}_{j(a)j(b)} + \sum_{t\in T} S^{t}_{c}, \forall c \in C \tag{14}$$

$$CT \leq 7 \tag{15}$$

The steps to determine the initial solution using the nearest neighbor method [72] are as follows:

a. Select the center point as the starting point of transport, which is the SCC in this study.
b. Determine the point with the smallest distance from SCC and move to the PCC point.
c. The last point visited is the starting point; therefore, determine the point with the closest distance from the point.
d. Repeat the process until the vehicle does not have sufficient capacity for transportation; but because there is always enough capacity of the vehicle used in this research, the repetition is conducted until it meets the planning time horizon for a day but does not exceed it.
e. Drag this point to a line which is called a route with the working hours used as a constraint to form a freight route.

The Tabu search algorithm used in this study is based on [77,78] and includes the following steps:

a. Determine solution representation. This is a sequence of nodes where each is only visible once in the sequence. These nodes represent PCC and SCC.
b. Formulate initial solution formation, S.
c. Determine the neighborhood solution. This is an alternative solution obtained by moving the nodes such that each move produces a neighborhood solution and the number of solutions is calculated using the following Equation (16):

$$C_{(n,2)} = \frac{n!}{2!(n-2)!} \tag{16}$$

where n is the number of PCCs visited in a route.

d. Create a tabu list. This list contains the moved attribute previously found, and its length increases with the size of the issue and also corresponds to the number of PCCs to be visited.
e. Find the best solution, S*.
f. Fix the tabu list.
g. Determine aspiration criteria. This is a method of overturning the tabu status.
h. Determine termination criteria. These are used after all predetermined iterations have been fulfilled. The number of iterations selected is the same as the number of points visited because the maximum number of iterations is the same as the length of the tabu list [79].

3. Results

3.1. Number and Location of Collection Centers

The number and location of the PCCs were determined using the e-MCLP method. The solver software was used to determine the optimal solution. The calculations showed that 30 PCCs are to be built as shown in Figures 1 and 2 with a distribution of 1 unit in Yogyakarta city (Y6), 13 units in Gunung Kidul Regency (G1, G2, G4, G5, G6, G7, G8, G9,

G13, G14, G15, G16, and G17), 6 units in Bantul Regency (B4, B10, B12, B13, B15, and B16), 6 units in Sleman Regency (S2, S6, S11, S12, S13, and S17), and 4 units in Kulon Progo Regency (K5, K8, K10, and K12). The selected PCC numbers and locations are shown in Appendix C.

Y	Y1	Y2	Y3	Y4	Y5	Y6	Y7	Y8	Y9	Y10	Y11	Y12	Y13	Y14	Y15		
G	G1	G2	G3	G4	G5	G6	G7	G8	G9	G10	G11	G12	G13	G14	G15	G16	G17	G18
B	B1	B2	B3	B4	B5	B6	B7	B8	B9	B10	B11	B12	B13	B14	B15	B16	B17	
S	S1	S2	S3	S4	S5	S6	S7	S8	S9	S10	S11	S12	S13	S14	S15	S16	S17	
K	K1	K2	K3	K4	K5	K6	K7	K8	K9	K10	K11	K12						

Legend
Selected
Not selected

Figure 1. The output of the solver.

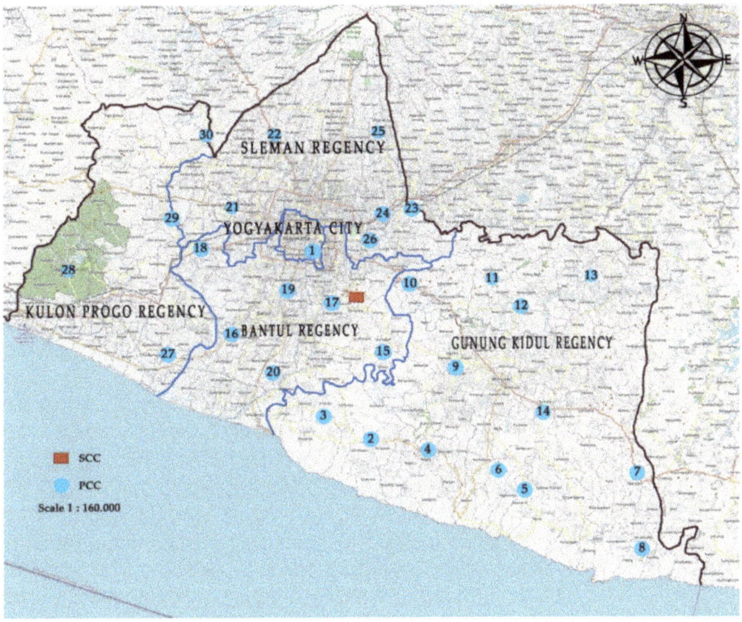

Figure 2. Locatio ns of selected PCCs.

The location chosen for the PCC construction in the city of Yogyakarta is Kotagede district. Those selected in Gunung Kidul Regency include the districts of Panggang, Purwosari, Sapto-sari, Tepus, Tanjungsari, Rongkop, Girisubo, Playen, Patuk, Gedangsari, Nglipar, Ngawen, and Semanu. Locations in Bantul Regency include the districts of Dlingo, Pandak, Pleret, Sedayu, Sewon, and Pundon while those in Sleman Regency are Godean, Sleman, Pambanan, Kalasan, Cangkringan, and Berbah districts. Those selected in Kulon Progo Regency included Lendah, Kokap, Nanggulan, and Kalibawang. The total accumulated distance from 30 selected PCCs to SCC is 948.7 km with a transportation cost of USD 126.31 for the scenario of collecting each pick-up from CCS to only one PCC. Since there are 30 CCPs established, the required annual investment cost is USD 2102.2. The total population of Yogyakarta in 2020 was 3,842,932 people; if 63.53% are smartphone users, the smartphone lifetime is 4.7 years, and the average weight of a smartphone is 0.5 kg, then the average smartphone waste supply in Yogyakarta per year is approximately 305,176 kg. Considering the capacity of the dropbox, pick-up should be done once a week, If the waste collection is done once a week, then the total cost required is USD 8165.07 per year

3.2. The Results of Scheduling the Transportation Route

The nearest neighbor method's search for initial solutions started with the 7 h obtained for planning horizon time, a loading time of 10 min for each PCC, and 30 min of unloading and administration time at SCC. This was followed by the determination of the depot as the starting location, which is the SCC. The vehicle has the capacity to accommodate the entire PCC because the supply is not large and the product volume is small; therefore, the planning time horizon was considered. The next step was the determination of the PCC with the closest distance, and this was discovered to be Pleret PCC, which has a distance of 4.3 km from the SCC. The distance matrices between the selected PCCs and from the selected PCC to the SCC are shown in Appendix D. It is important to note that the retrieval process was continued to the next PCC when the completion time (CT) was less than or equal to the planning time horizon but canceled when the completion time was greater than the planning time horizon. Furthermore, the next PCC was determined based on the closest distance with the initial steps implemented when it was discovered not to have been served. It is also important to point out that just one type of vehicle was used.

The number of trips or tours required to make the collection was calculated to be 3 with a total distance of 659.1 km, travel time of 14.65 h, and a completion time of 21.16 h, as shown in the sequence presented in Table 1. Route 1 had 13 PCCs with a total distance of 193.9 km, travel time of 4.31 h, and completion time of 6.98 h. Route 2 had 10 PCCs with a total distance of 198.4 km, travel time of 4.41 h, and completion time of 6.58 h. Route 3 had 7 PCCs with a total distance of 266.8 km, travel time of 5.93 h, and completion time of 7.6 h. It was discovered that Route 3 has a longer travel time than the planning time horizon, and it was used as an initial solution in the tabu search method with the expectation that it will improve and provide shorter distances and times for the optimal solution.

Table 1. The initial solution results using nearest neighbor.

Route	Picking Sequence
1	$SCC \rightarrow PCC_{17} \rightarrow PCC_{1} \rightarrow PCC_{26} \rightarrow PCC_{24} \rightarrow PCC_{23} \rightarrow PCC_{10} \rightarrow PCC_{9} \rightarrow PCC_{15} \rightarrow PCC_{19} \rightarrow PCC_{16} \rightarrow PCC_{27} \rightarrow PCC_{18} \rightarrow PCC_{29} \rightarrow SCC$
2	$SCC \rightarrow PCC_{20} \rightarrow PCC_{3} \rightarrow PCC_{2} \rightarrow PCC_{4} \rightarrow PCC_{6} \rightarrow PCC_{5} \rightarrow PCC_{8} \rightarrow PCC_{7} \rightarrow PCC_{14} \rightarrow PCC_{12} \rightarrow SCC$
3	$SCC \rightarrow PCC_{21} \rightarrow PCC_{22} \rightarrow PCC_{25} \rightarrow PCC_{11} \rightarrow PCC_{13} \rightarrow PCC_{30} \rightarrow PCC_{28} \rightarrow SCC$

The tabu search method was applied using the initial solution calculated from the nearest neighbor method. Route 1 was found to be SCC–Pleret PCC–Kotagede PCC–Sewon PCC–Pandak PCC–Bambanglipuro PCC–Sedayu PCC–Dlingo PCC–Playen PCC–Patuk PCC–Ngglipar PCC–Ngawen PCC–SCC. This was followed by the input of the number of elements to be searched, which was found to be in accordance with the points to be visited, i.e., 11 PCCs. The number of neighborhood solutions was later determined using Equation (14), and 55 lines were recorded. Furthermore, the tabu list length was discovered to be in line with the number of PCCs to be visited, which was 11 customer locations. This was followed by the maximum number of iterations, which was recorded to be 11 iterations in line with the number of PCCs. These steps were repeated for the other routes, and the determination of the best route produced three routes with a total distance of 602.2 km, a travel time of 13.4 h, and a completion time of 19.89 h. The time for a shipment was found to be 3 days. Furthermore, the best sequences for Routes 1, 2, and 3 had total distances of 178.3, 198.3, and 224.5 km; travel times of 3.98, 4.41, and 5.01 h; and completion times of 6.63, 6.58, and 6.68 h, respectively, as shown in Figure 3 and Table 2.

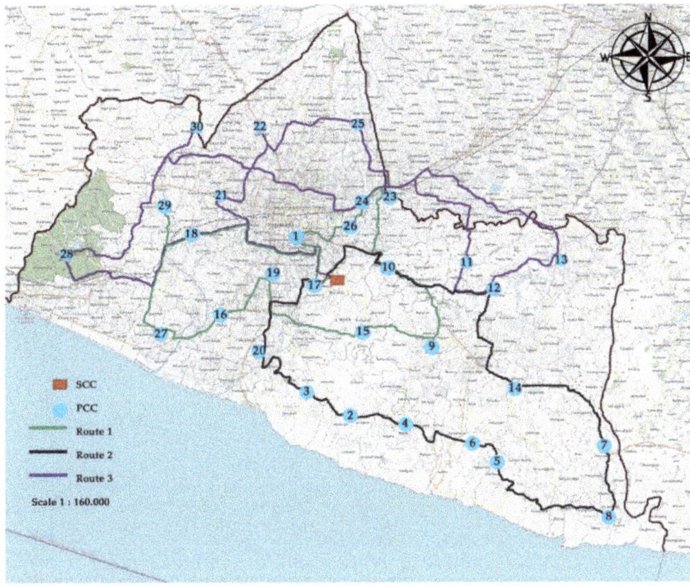

Figure 3. Collection routes.

Table 2. The final solution results using tabu search.

Route	Picking Sequence
1	SCC→PCC_1→PCC_{19}→PCC_{18}→PCC_{29}→PCC_{27}→PCC_{16}→PCC_{17}→PCC_{15}→PCC_9→PCC_{10}→PCC_{23}→PCC_{24}→PCC_{26}→SCC
2	SCC→PCC_{12}→PCC_{14}→PCC_7→PCC_8→PCC_5→PCC_6→PCC_4→PCC_2→PCC_3→PCC_{20}→SCC
3	SCC→PCC_{13}→PCC_{11}→PCC_{25}→PCC_{22}→PCC_{21}→PCC_{30}→PCC_{28}→SCC

4. Discussion

The results showed that the city/regency with the fewest PCCs is Yogyakarta city due to the short distance between its districts, with the one PCC established in Kota Gede district being found to have the ability to reach 13 other districts. The farthest is the Tegalrejo district, which is 9.5 km away, and this is also considered to be within the distance desired by the consumers. Meanwhile, most of the PCCs were built in Gunung Kidul Regency due to its large area relative to the other cities and regencies, and this caused quite a long distance between the districts. The area is 47% of the total area of Yogyakarta Province, as shown in Figure 2. Therefore, there is a need to build 13 PCCs in the existing 18 districts to cover all consumers, and the remaining 5 will be accessible because they are less than 10 km from the built locations. For example, Playen PCC covers Paliyan District while Semanu PCC covers Ponjong and Karangmojo districts. It is also possible for the waste from Wonosari District to be transported to Playen or Semanu PCC, while Ngawen PCC covers the Semin district.

This problem, if solved using MCLP as done by Church and Davis [70], Murray [17], Boonmee et al. [18], and Hartini et al. [26] to minimize the number of collection centers that must be built, results in the same number of PCCs that must be built as with e-MCLP, namely 30 PCCs; the number of PCCs established in each city/regency is the same, but there are several different locations. The comparison of selected CCP locations from each method is shown in Figure 4. The different locations are Kotagede, Semanu, Berbah, and Kalibawang when using the e-MCLP method. When using the MCLP method, the selected locations are Gondomanan, Ponjong, Minggir, and Samigaluh, as shown in Figure 4. The difference between the four locations will have implications for saving transportation costs from PCC to SCC because of the shorter distance, while the investment costs, in this case, are the same for each selected PCC. Comparison of the distance from PCC to SCC between the two methods is shown in Table 3

Figure 4. Comparison of the solver output between e-MCLP and MCLP methods.

Table 3. Distance comparison between e-MCLP and MCLP results.

MCLP		e-MCLP		The Difference in Distance (km)
Selected PCC	**Distance to SCC (km)**	**Selected PCC**	**Distance to SCC (km)**	
Gondomanan (Y4)	13.1	Kotagede (Y6)	9.0	4.1
Ponjong (G10)	45.8	Semanu (G9)	39.9	5.9
Minggir (S3)	34.0	Berbah (S13)	9.1	24.9
Samigaluh (K11)	46.6	Kalibawang (K12)	40.6	6.0
	Total of difference in distance			40.9

Table 4 shows that the distance between the selected PCCs and SCC is shorter in e-MCLP than MCLP. This indicates the numbers and locations calculated using the two approaches were able to accommodate the range expected by the consumers, but e-MCLP considered the investment costs and the distance between the PCC and SCC, unlike the MCLP. Therefore, MCLP provided a greater total PCC to SCC distance, which is indicated by 989.6 km with transportation costs of USD 131.76 when the collection is at only one PCC and the total cost required per year is USD 8426.27. Meanwhile, e-MCLP provided a shorter total distance of 40.9 km with 4.13% savings in transportation costs at USD 261.6 per year.

Table 4. Comparison of nearest neighbor and tabu search results.

Route	Nearest Neighbor			Tabu Search		
	D (km)	**TT (hours)**	**CT (hours)**	**D (km)**	**TT (hours)**	**CT (hours)**
1	193.9	4.31	6.98	178.3	3.98	6.63
2	198.4	4.41	6.58	198.4	4.41	6.58
3	266.8	5.93	7.60	225.5	5.01	6.68
Total	659.1	14.65	21.16	602.2	13.40	19.89

e-MCLP is very suitable for PCCs with large waste volumes because vehicle capacity is filled faster when the volume of waste is large so that there are fewer pick-up points on one route. When there are fewer pick-up points in one route, the more routes there will be, and the development of this method is suitable for implementation. This model's savings in transportation costs will be felt when the number of routes increases because vehicles will depart and return to SCC more often. That is, the closer CCP distance to the SCC is very beneficial for the vehicle. In this study, the selected location does not affect the investment cost because each candidate location requires a procurement cost of the same amount. However, the developed model can accommodate each candidate location requiring a different investment cost. Later, the selected location will provide a minimum total cost, including investment and transportation costs.

The best route was determined using the tabu search method to improve the results of the nearest neighbor. This is in line with the opinion found in [77,80,81] that metaheuristics are popular optimization problem-solving techniques to overcome the weaknesses of the heuristic method due to their ability to avoid being trapped in a local optimum solution [82]. The first route was found to be better than the original solution due to its ability to reduce the distance traveled by 15.6 km and the travel time by 0.33 h, thereby reducing the distance and travel time by 8%. Meanwhile, the optimal solution in the second route is the same as the initial solution, but the route completion sequence is reversed such that the first PCC visited using the nearest neighbor was the last in the tabu search method. This shows the nearest neighbor method also has the ability to provide the best solution, and this is in accordance with the findings of [41] that the nearest neighbor method produces the shortest route compared to other heuristic methods. The NN algorithm was able to minimize distribution costs [83] and could easily and quickly resolve problems for several small cities [84]. Furthermore, the initial solution was observed to be infeasible for the third route because the completion time, which was recorded to be 7.6 h, exceeds the planning time horizon, which is 7 h. The continuation of the iteration using the tabu search method changed the initially infeasible solution to feasible as indicated by the shortening of the completion time (CT) to 6.68 h with a total distance (D) of 225.5 km and a travel time (TT) of 5.01 h, saving 8.53% of travel time. This means the tabu search was able to reduce the distance and travel time by 15% as indicated by the 41.3 km and 0.92 h results when compared to the nearest neighbor method, as shown in Table 4.

The tabu search method was generally able to provide better performance than the nearest neighbor method. The metaheuristic approach gives better performance results than the heuristic approach [70]. The results showed the possibility of collecting all the smartphone waste in Yogyakarta Province using three routes. This can be completed in a day through the use of three vehicles or in three days through the use of one vehicle. The total distance required to be covered is 602.2 km with a travel time of 13.4 h and a total completion time of 19.89 h. This means the tabu search method generally saved 56.9 km (8.6%) distance and 1.25 h (8.5%) travel time.

Determination of smartphone waste collection routes in the province of Yogyakarta with one route picking up at several PCC points managed to save a mileage of 346.5 km compared to one route only picking up at one PCC point and a total of 30 pick-up points. If the smartphone waste collection is done once a week, this shorter distance can provide transportation cost savings of USD 2214.39 per year. The area of Yogyakarta Province is only 0.16% of the territory of Indonesia; if this model is implemented nationally, the estimated transportation cost savings will be more than USD 1 million.

This research is expected to be the initial framework in formulating e-waste management policies for the national formal channel. If this proposal is successfully implemented in Yogyakarta, it is likely to be implemented in other provinces in Indonesia. The developed model can also be used for other solid waste collection scenarios. The proposed e-MCLP model is very suitable for large e-waste because there is no need to proceed to route determination, considering that the supply from PCC may already meet vehicle capacity. One trip only picks up from a PCC and then returns to the SCC again. However, to use the proposed model, it is necessary to consider whether the community is willing to bring their large size/volume e-waste to the provided PCC. This research is also the first step in electronic waste management, which will then be followed by the next stage of management, which includes separation, repair, recycling, remanufacturing, or disposal.

5. Conclusions

Smartphone waste has a high economic value and has great potential. The tendency of people to store and dispose of smartphone waste is due to the absence of waste collection facilities and government regulations that specifically regulate electronic waste management mechanisms. With the public's willingness to bring smartphone waste to a collection point with a maximum reach of 11.2 km and the benefits that will be obtained, this is a challenge and an opportunity for the government to design an optimal collection channel. The design of the collection channel involves consumers as suppliers of electronic waste, primary collection centers (PCCs), and Secondary Collection Centers (SCC). Due to the small area of Yogyakarta Province, 1 SCC is sufficient to accommodate the supply of smartphone waste from all selected PCCs. Based on the results of calculations using e-MCLP, as many as 30 PCCs should be built, with a distribution of 1 PCC in Yogyakarta City, 13 PCCs in Gunung Kidul Regency, 6 PCCs in Bantul Regency, 6 PCCs in Sleman Regency, and 4 PCCs in Kulon Progo Regency. e-MCLP can produce the minimum number of primary collection facilities required to cover all consumers with the shortest distance from secondary collection facilities to minimize total costs, including investment and transportation costs, with a total cost of USD 3617.92 per year.

The best transportation route from PCC to SCC was determined using the nearest neighbor and tabu search method (NN-TB). The pick-up route starts and ends at SCC, and the result shows three routes to use in smartphone waste collection. These routes take three days to complete by using one vehicle or one day using three vehicles with a total time required of 19.89 h and a distance of 602.2 km.

Further research can expand the study of e-waste with a large volume because the large volume will affect the willingness of consumers to bring their e-waste and the need to calculate the capacity of the collection center. This research is expected to be the initial framework in formulating e-waste management policies for a formal national channel. Research can also be continued with the design of management following the collection of e-waste in a final disposal site, such as separation, repair, recycling, remanufacturing, or disposal.

Author Contributions: Conceptualization, D.P.S., N.A.M. and A.M.S.A.; methodology, D.P.S., N.A.M. and A.M.S.A.; software, D.P.S.; validation and analysis, D.P.S., N.A.M. and A.M.S.A.; data curation, D.P.S.; writing—original draft preparation, D.P.S.; writing—review and editing, N.A.M. and A.M.S.A.; visualization, D.P.S.; supervision, N.A.M. and A.M.S.A. All authors have read and agreed to the published version of the manuscript.

Funding: This paper is part of a PhD study that was financially supported by BUDI DN Grant from the Lembaga Pengelola Dana Penelitian (LPDP), Ministry of Finance, and Ministry of Education and Culture, Republic of Indonesia, grant number KET-193/LPDP.4/2020. The APC was funded by RTA 2021 grant number 3190/UN1/DITLIT/DIT-LIT/PT/2021 from Universitas Gadjah Mada.

Institutional Review Board Statement: Not applicable.

Informed Consent Statement: Not applicable.

Data Availability Statement: The data used to support the findings of this study are available from the corresponding author on request.

Conflicts of Interest: The authors declare no conflict of interest.

Abbreviations

B	Bantul Regency
CC	Collection Center
CT	Completion Time
D	Total Distance
e-MCLP	Extended Maximal Covering Location Problem
e-waste	Electronic Waste

G	Gunung Kidul Regency
K	Kulon Progo Regency
LSCP	Location Set Covering Problem
MCLP	Maximal Covering Location Problem
NN	Nearest Neighbor
NN-TS	Nearest Neighbor and Tabu Search
PCC	Primary Collection Center
S	Sleman Regency
SCC	Secondary Collection Center
TPA	Tempat Pembuangan Akhir (Final Disposal Site)
TS	Tabu Search
TT	Travel Time
VRP	Vehicle Routing Problem
Y	Yogyakarta City

Appendix A

Table A1. Regencies/Cities and Districts in Yogyakarta Province.

No	City/Regency	District	Code
1	Yogyakarta City (Y)	Danurejan	Y1
		Gedongtengen	Y2
		Gondokusuman	Y3
		Gondomanan	Y4
		Jetis	Y5
		Kotagede	Y6
		Kraton	Y7
		Mantrijeron	Y8
		Mergangsan	Y9
		Ngampilan	Y10
		Pakualaman	Y11
		Tegalrejo	Y12
		Umbulharjo	Y13
		Wirobrajan	Y14
2	Gunung Kidul Regency (G)	Panggang	G1
		Purwosari	G2
		Paliyan	G3
		Saptosari	G4
		Tepus	G5
		Tanjungsari	G6
		Rongkop	G7
		Girisubo	G8
		Semanu	G9
		Ponjong	G10
		Karangmojo	G11
		Wonosari	G12
		Playen	G13
		Patuk	G14
		Gedangsari	G15
		Nglipar	G16
		Ngawen	G17
		Semin	G18

Table A1. *Cont.*

No	City/Regency	District	Code
3	Bantul Regency (B)	Bambanglipuro	B1
		Banguntapan	B2
		Bantul	B3
		Dlingo	B4
		Imogiri	B5
		Jetis	B6
		Kasihan	B7
		Kretek	B8
		Pajangan	B9
		Pandak	B10
		Piyungan	B11
		Pleret	B12
		Pundong	B13
		Sanden	B14
		Sedayu	B15
		Sewon	B16
		Srandakan	B17
4	Sleman Regency (S)	Moyudan	S1
		Godean	S2
		Minggir	S3
		Gamping	S4
		Seyegan	S5
		Sleman	S6
		Ngaglik	S7
		Mlati	S8
		Tempel	S9
		Turi	S10
		Prambanan	S11
		Kalasan	S12
		Berbah	S13
		Ngemplak	S14
		Pakem	S15
		Depok	S16
		Cangkringan	S17
5	Kulon Progo Regency (K)	Temon	K1
		Wates	K2
		Panjatan	K3
		Galur	K4
		Lendah	K5
		Sentolo	K6
		Pengasih	K7
		Kokap	K8
		Girimulyo	K9
		Nanggulan	K10
		Samigaluh	K11
		Kalibawang	K12

Appendix B

Table A2. The distance matrice between PCC candidates and from the PCC candidates to the SCC in Yogyakarta City.

Distances (km)	Y1	Y2	Y3	Y4	Y5	Y6	Y7	Y8	Y9	Y10	Y11	Y12	Y13	Y14
Y1	0	3	1.9	2.7	2.8	6.2	4.1	5.3	4.4	3.2	1.9	5.1	4.5	4.3
Y2	3	0	2.7	2.2	2.1	6.9	3	4	5	2.1	2.3	2.9	6	2.5
Y3	1.9	2.7	0	4.6	3	5.5	6.4	7.3	5	5.5	3.8	4.7	4.1	6.5
Y4	2.7	2.2	4.6	0	3.5	5.1	2.4	3.6	3.2	1.6	1.1	4.7	3.3	2.6

Table A2. *Cont.*

Distances (km)	Y1	Y2	Y3	Y4	Y5	Y6	Y7	Y8	Y9	Y10	Y11	Y12	Y13	Y14
Y5	2.8	2.1	3	3.5	0	7.9	5.1	4.9	6.6	3	4	1.8	6.3	3.3
Y6	6.2	6.9	5.5	5.1	7.9	0	5.2	5.2	2.8	5.8	4.3	9.5	2.1	6.6
Y7	4.1	3	6.4	2.4	5.1	5.2	0	2.4	3.5	2	2.9	5.1	4.3	2.2
Y8	5.3	4	7.3	3.6	4.9	5.2	2.4	0	3.5	2.7	4	5.9	4.3	2.8
Y9	4.4	5	5	3.2	6.6	2.8	3.5	3.5	0	4.1	2.9	7.6	2	4.9
Y10	3.2	2.1	5.5	1.6	3	5.8	2	2.7	4.1	0	2	3.9	4.2	1.3
Y11	1.9	2.3	3.8	1.1	4	4.3	2.9	4	2.9	2	0	5	2.7	3.1
Y12	5.1	2.9	4.7	4.7	1.8	9.5	5.1	5.9	7.6	3.9	5	0	7.3	3.2
Y13	4.5	6	4.1	3.3	6.3	2.1	4.3	4.3	2	4.2	2.7	7.3	0	5.7
Y14	4.3	2.5	6.5	2.6	3.3	6.6	2.2	2.8	4.9	1.3	3.1	3.2	5.7	0
SCC	14.4	15.5	16.2	13.1	17.2	9	14	13.1	11	14.8	12.5	18.6	10.6	15.5

Table A3. The distance matrice between PCC candidates and from the PCC candidates to the SCC in Gunung Kidul Regency.

Distances (km)	G1	G2	G3	G4	G5	G6	G7	G8	G9	G10	G11	G12	G13	G14	G15	G16	G17	G18
G1	0	11.1	19	12.7	35.7	23.9	55.9	51.2	34.2	42.3	38.5	27.9	28.8	35.7	49.4	40.3	50.4	51.6
G2	11.1	0	29	22.7	45.7	33.9	65.9	61.2	44.3	52.3	48.5	37.8	36.5	45.7	59.4	50.3	60.4	61.7
G3	19	29	0	11.8	28.4	17.5	39.5	41.6	17.8	26.8	22	11.4	10	19.3	32.9	23.9	33.9	35.2
G4	12.7	22.7	11.8	0	28.5	16.7	48.7	44	27.1	36.1	31.3	20.7	19.3	28.5	42.2	33.1	43.2	44.5
G5	35.7	45.7	28.4	28.5	0	15.2	22.7	16.3	16.8	25.5	25.3	23.4	31.3	38.2	51.9	34.7	37.2	37.2
G6	23.9	33.9	17.5	16.7	15.2	0	27.6	30.7	20.4	29.2	27.7	17.8	25.7	32.6	46.3	29.1	40.9	40.9
G7	55.9	65.9	39.5	48.7	22.7	27.6	0	15.9	21.7	18.8	26.2	29.2	38.3	45.2	53	41	38.1	38.1
G8	51.2	61.2	41.6	44	16.3	30.7	15.9	0	27.2	30.9	35.7	34.5	43.6	50.2	62.5	46.3	47.6	47.6
G9	34.2	44.3	17.8	27.1	16.8	20.4	21.7	27.2	0	9.1	8.5	7.5	16.6	23.5	35.4	19.3	20.4	20.1
G10	42.3	52.3	26.8	36.1	25.5	29.2	18.8	30.9	9.1	0	8.4	14.4	22.5	29.4	35.3	21.6	20.3	20.3
G11	38.5	48.5	22	31.3	25.3	27.7	26.2	35.7	8.5	8.4	0	10.6	18.7	22.7	28.6	13.3	13.7	13.7
G12	27.9	37.8	11.4	20.7	23.4	17.8	29.2	34.5	7.5	14.4	10.6	0	8.5	15.4	29.1	11.7	23.8	23.8
G13	28.8	36.5	10	19.3	31.3	25.7	38.3	43.6	16.6	22.5	18.7	8.5	0	11.1	24.8	15.7	25.7	31.9
G14	35.7	45.7	19.3	28.5	38.2	32.6	45.2	50.2	23.5	29.4	22.7	15.4	11.1	0	19.6	14.6	24.5	31.4
G15	49.4	59.4	32.9	42.2	51.9	46.3	53	62.5	35.4	35.3	28.6	29.1	24.8	19.6	0	22	16.6	24.3
G16	40.3	50.3	23.9	33.1	34.7	29.1	41	46.3	19.3	21.6	13.3	11.7	15.7	14.6	22	0	10.6	20.6
G17	50.4	60.4	33.9	43.2	37.2	40.9	38.1	47.6	20.4	20.3	13.7	23.8	25.7	24.5	16.6	10.6	0	8.6
G18	51.6	61.7	35.2	44.5	37.2	40.9	38.1	47.6	20.1	20.3	13.7	23.8	31.9	31.4	24.3	20.6	8.6	0

Table A4. The distance matrice between PCC candidates and from the PCC candidates to the SCC in Bantul Regency.

Distances (km)	B1	B2	B3	B4	B5	B6	B7	B8	B9	B10	B11	B12	B13	B14	B15	B16	B17
B1	0	25.7	12	24.9	12.7	12.5	17.2	6.5	13.7	5.8	31.5	21.4	4.5	8.1	22.6	15.8	10.4
B2	25.7	0	16.2	22.5	14.3	12.8	14.1	28.3	18.9	23.3	8.4	9.8	23.8	28.8	26.4	7.5	30.5
B3	12	16.2	0	19.3	7.1	4.4	10.2	14	7.3	8.3	23.1	13.4	10.8	13.4	18.9	7.1	15.4
B4	24.9	22.5	19.3	0	13.6	16.9	30	29.6	28.2	26.5	13.7	12.7	25.1	31.9	37.2	24.3	33.7
B5	12.7	14.3	7.1	13.6	0	3.5	16.7	16.2	14.9	13.2	18.7	9.6	11.8	18.5	23.8	11.2	20.3
B6	12.5	12.8	4.4	16.9	3.5	0	13.3	16	11.5	12.4	18.6	9	11.6	18.3	23.1	7.6	19.6
B7	17.2	14.1	10.2	30	16.7	13.3	0	22.1	6.6	14.8	20.8	16.8	20.3	19.3	15.1	6.5	20.9
B8	6.5	28.3	14	29.6	16.2	16	22.1	0	18.7	10.7	35	24.9	6.7	6.5	27.5	19.3	11
B9	13.7	18.9	7.3	28.2	14.9	11.5	6.6	18.7	0	9.2	27	17.6	17.3	13.8	13.2	11.1	15.4

Table A4. *Cont.*

Distances (km)	B1	B2	B3	B4	B5	B6	B7	B8	B9	B10	B11	B12	B13	B14	B15	B16	B17
B10	5.8	23.3	8.3	26.5	13.2	12.4	14.8	10.7	9.2	0	30	20.9	9.7	7.8	18.2	14.3	5.5
B11	31.5	8.4	23.1	13.7	18.7	18.6	20.8	35	27	30	0	9.6	30	35.5	33	14.2	37.1
B12	21.4	9.8	13.4	12.7	9.6	9	16.8	24.9	17.6	20.9	9.6	0	20.5	27.3	28.7	10.2	29
B13	4.5	23.8	10.8	25.1	11.8	11.6	20.3	6.7	17.3	9.7	30	20.5	0	10.3	25.8	14.9	11.8
B14	8.1	28.8	13.4	31.9	18.5	18.3	19.3	6.5	13.8	7.8	35.5	27.3	10.3	0	22.1	19.9	5.3
B15	22.6	26.4	18.9	37.2	23.8	23.1	15.1	27.5	13.2	18.2	33	28.7	25.8	22.1	0	18.8	24.1
B16	15.8	7.5	7.1	24.3	11.2	7.6	6.5	19.3	11.1	14.3	14.2	10.2	14.9	19.9	18.8	0	21.4
B17	10.4	30.5	15.4	33.7	20.3	19.6	20.9	11	15.4	5.5	37.1	29	11.8	5.3	24.1	21.4	0
SCC	23.6	9.2	15.6	15.8	11.6	11.1	16.8	27.1	17.7	23.5	10.1	4.8	22.6	29.4	29.1	10.6	30.7

Table A5. The distance matrice between PCC candidates and from the PCC candidates to the SCC in Sleman Regency.

Distances (km)	S1	S2	S3	S4	S5	S6	S7	S8	S9	S10	S11	S12	S13	S14	S15	S16	S17
S1	0	6.0	6.9	10.6	11.6	19.3	23.8	16.2	16.5	26.8	38.4	31.1	29.9	31.5	30.5	22.6	37.8
S2	6.0	0	8.5	6.3	6.6	13.1	17.8	8.3	14.3	21.9	32.4	25.7	25.6	25.5	24.6	16.6	31.9
S3	6.9	8.5	0	14.9	8.4	16.1	21.9	13	13.3	23.6	38.7	30.9	32.3	30.4	27.3	0.92	34.6
S4	10.6	6.3	14.9	0	12.9	15.8	16.2	9.3	19.9	25	30.9	23.6	19.3	24	25	15.1	29.8
S5	11.6	6.6	8.4	12.9	0	8.6	14.1	5.2	8.8	19.4	31.5	23.6	24.5	22.6	20.9	15.1	27.5
S6	19.3	13.1	16.1	15.8	8.6	0	8.2	2.7	7.5	9.4	29.3	18.9	22.9	16	12.1	13.5	19.4
S7	23.8	17.8	21.9	16.2	14.1	8.2	0	9.3	14.5	13.9	23.1	12.8	16.7	8.9	11.8	7.7	14.8
S8	16.2	8.3	13.0	9.3	5.2	2.7	9.3	0	11.5	15.3	26.3	19	18.9	19.4	18	10.5	25.3
S9	16.5	14.3	13.3	19.9	8.8	7.5	14.5	11.5	0	11.4	35	24.1	27.6	23.2	15.2	19.3	22.5
S10	26.8	21.9	23.6	25	19.4	9.4	13.9	15.3	11.4	0	32.4	20.8	29.8	16.8	10	21.6	12.9
S11	38.4	32.4	38.7	30.9	31.5	29.3	23.1	26.3	35	32.4	0	11.8	10.6	18.9	28.2	15.7	25.9
S12	31.1	25.7	30.9	23.6	23.6	18.9	12.8	19	24.1	20.8	11.8	0	9	7.2	16.4	12.2	14.3
S13	29.9	25.6	32.3	19.3	24.5	22.9	16.7	18.9	27.6	29.8	10.6	9	0	16.2	26.9	8.8	23.2
S14	31.5	25.5	30.4	24	22.6	16	8.9	19.4	23.2	16.8	18.9	7.2	16.2	0	11.9	11.4	10.2
S15	30.5	24.6	27.3	25	20.9	12.1	11.8	18	15.2	10	28.2	16.4	26.9	11.9	0	16.7	7.8
S16	22.6	16.6	0.92	15.1	15.1	13.5	7.7	10.5	19.3	21.6	15.7	12.2	8.8	11.4	16.7	0	21.4
S17	37.8	31.9	34.6	29.8	27.5	19.4	14.8	25.3	22.5	12.9	25.9	14.3	23.2	10.2	7.8	21.4	0
SCC	29.8	25.5	34.0	19.1	30.7	29.8	25.8	27.1	33.8	38.7	17.7	17.8	9.1	23.9	34.6	20.1	30.9

Table A6. The distance matrice between PCC candidates and from the PCC candidates to the SCC in Kulon Progo Regency.

Distances (km)	K1	K2	K3	K4	K5	K6	K7	K8	K9	K10	K11	K12
K1	0.0	7.6	11.3	19	16.8	19.7	15.8	9.1	27.6	23.1	47.9	41.9
K2	7.6	0	3.7	12.1	9.2	13.4	11.4	13.4	29.1	18.3	41.6	35.6
K3	11.3	3.7	0	10.9	8.8	15.4	11.6	17.1	31.1	20.3	43.5	37.5
K4	19.0	12.1	10.9	0	6	13.6	19.8	24.8	29.3	22.6	42	35.8
K5	16.8	9.2	8.8	6	0	8.5	14.1	19.2	24.2	17.5	36.7	30.7
K6	19.7	13.4	15.4	13.6	8.5	0	9.5	17	19.1	12.4	31.5	25.5
K7	15.8	11.4	11.6	19.8	14.1	9.5	0	9.4	13.7	10.8	32.4	26.4
K8	9.1	13.4	17.1	24.8	19.2	17	9.4	0	23	19.2	37.9	31.8
K9	27.6	29.1	31.1	29.3	24.2	19.1	13.7	23	0	6.4	20.5	14.5
K10	23.1	18.3	20.3	22.6	17.5	12.4	10.8	19.2	6.4	0	21.3	17.7
K11	47.9	41.6	43.5	42	36.7	31.5	32.4	37.9	20.5	21.3	0	8.5
K12	41.9	35.6	37.5	35.8	30.7	25.5	26.4	31.8	14.5	17.7	8.5	0
SCC	48.8	42.2	41.8	33.8	35.5	35.8	40.8	49.2	42.2	35.5	46.6	40.6

Appendix C

Table A7. The selected PCC numbers and locations.

City/Regency	PCC Number	Location
Yogyakarta City	1	Kotagede
Gunung Kidul Regency	2	Panggang
	3	Purwosari
	4	Saptosari
	5	Tepus
	6	Tanjungsari
	7	Rongkop
	8	Girisubo
	9	Playen
	10	Patuk
	11	Gedangsari
	12	Ngglipar
	13	Ngawen
	14	Semanu
Bantul Regency	15	Dlingo
	16	Pandak
	17	Pleret
	18	Sedayu
	19	Sewon
	20	Pundon
Sleman Regency	21	Godean
	22	Sleman
	23	Prambanan
	24	Kalasan
	25	Cangkringan
	26	Berbah
Kulon Progo Regency	27	Lendah
	28	Kokap
	29	Nanggulan
	30	Kalibawang

Appendix D

Table A8. The distance matrice between the selected PCCs and from the selected PCC to the SCC.

Distances (km)		SCC	PCC														
			1	2	3	4	5	6	7	8	9	10	11	12	13	14	15
SCC		0	9	30.6	31	42.6	54.6	47.4	61.6	67	39.9	25.6	23.3	38.4	33.4	43.2	22.6
PCC	1	9	0	31.2	31.7	43.9	57.4	50.2	64.3	69.8	42.7	30.2	22.7	38.7	33.8	43.7	22.9
	2	30.6	31.2	0	11.1	12.7	35.7	23.9	55.9	51.2	34.2	28.8	35.7	49.4	40.3	50.4	22.5
	3	31	31.7	11.1	0	22.7	45.7	33.9	65.9	61.2	44.3	36.5	45.7	59.4	50.3	60.4	13.5
	4	42.6	43.9	12.7	22.7	0	28.5	16.7	48.7	44	27.1	19.3	28.5	42.2	33.1	43.2	35.2
	5	54.6	57.4	35.7	45.7	28.5	0	15.2	22.7	16.3	16.8	31.3	38.2	51.9	34.7	37.2	59
	6	47.4	50.2	23.9	33.9	16.7	15.2	0	27.6	30.7	20.4	25.7	32.6	46.3	29.1	40.9	46.4
	7	61.6	64.3	55.9	65.9	48.7	22.7	27.6	0	15.9	21.7	38.3	45.2	53	41	38.1	70.1
	8	67	69.8	51.2	61.2	44	16.3	30.7	15.9	0	27.2	43.6	50.2	62.5	46.3	47.6	72.3
	9	39.9	42.7	34.2	44.3	27.1	16.8	20.4	21.7	27.2	0	16.6	23.5	35.4	19.3	20.4	48.4
	10	25.6	30.2	28.8	36.5	19.3	31.3	25.7	38.3	43.6	16.6	0	11.1	24.8	15.7	25.7	36.7
	11	23.3	22.7	35.7	45.7	28.5	38.2	32.6	45.2	50.2	23.5	11.1	0	19.6	14.6	24.5	45.9
	12	38.4	38.7	49.4	59.4	42.2	51.9	46.3	53	62.5	35.4	24.8	19.6	0	22	16.6	59.6

Table A8. *Cont.*

Distances (km)		SCC	PCC														
			1	**2**	**3**	**4**	**5**	**6**	**7**	**8**	**9**	**10**	**11**	**12**	**13**	**14**	**15**
PCC	13	33.4	33.8	40.3	50.3	33.1	34.7	29.1	41	46.3	19.3	15.7	14.6	22	0	10.6	50.5
	14	43.2	43.7	50.4	60.4	43.2	37.2	40.9	38.1	47.6	20.4	25.7	24.5	16.6	10.6	0	60.6
	15	22.6	22.9	22.5	13.5	35.2	59	46.4	70.1	72.3	48.4	36.7	45.9	59.6	50.5	60.6	0
	16	15.8	22.7	35.7	31.5	28.6	40.6	35	47.5	52.9	25.9	11.6	20.6	34.3	25.1	35.2	25.1
	17	23.5	21.8	30.7	25.7	43.2	65.4	54.4	72.4	77.7	50.7	36.4	43.5	61	50	60.1	9.7
	18	4.8	10	28	28.4	39.5	51.6	45.9	58.5	63.8	36.9	22.5	20.6	36.7	31.7	41.5	20.5
	19	29.1	24.9	41.3	40.4	53.9	76.1	65.1	83.1	88.4	61.4	47.1	45.6	61.7	56.7	66.6	25.8
	20	10.6	8.5	28.8	32.7	41.4	61.5	55.9	68.4	73.8	46.8	34.4	26.8	45.3	37.9	47.8	14.9
	21	25.50	16.2	44	48	56.6	76.7	71.1	83.7	89	62	49.6	42	53.2	53.1	63	32.1
	22	29.80	22.4	52.2	56.2	67.3	77	71.3	83.9	89.3	62.3	49.8	40.3	51.3	53.4	59.3	26.4
	23	17.70	18.10	46.10	46.5	45.8	55.5	49.9	62.5	67.8	40.8	28.4	13.8	24.8	31.9	41.8	40
	24	17.80	15.3	46.2	46.6	50.8	60.5	54.8	67.4	72.8	45.8	33.3	22.1	30.2	36.9	38.6	32.9
	25	9.10	7.5	35.3	35.7	45	54.7	49.1	61.7	67	40	27.6	20	33.4	31.1	41	27
	26	30.90	28.8	59.6	60.1	63.2	72.9	67.2	79.8	85.2	58.2	45.7	34.5	37.1	49.3	45.6	45
	27	35.50	33.8	42.7	35	55.2	77.4	66.4	84.4	89.7	62.7	48.4	54.6	68.3	62	72.1	20.6
	28	49.20	40.9	61.5	53.8	74	96.2	85.2	103.2	108.5	81.5	67.2	65.7	81.9	76.8	86.7	24
	29	35.50	31.2	50.1	51.6	62.7	86.7	73.9	93.6	98.9	72	59.5	52	68.1	63.1	73	10.3
	30	40.60	36.4	60	59	72.5	91.8	83.8	98.8	104.1	77.1	64.7	57.1	65.7	68.2	74.1	44.4

Table A9. The distance matrice between the selected PCCs and from the selected PCC to the SCC.

Distances (km)		PCC														
		16	**17**	**18**	**19**	**20**	**21**	**22**	**23**	**24**	**25**	**26**	**27**	**28**	**29**	**30**
SCC		15.8	23.5	4.8	29.1	10.6	25.5	29.8	17.7	17.8	9.1	30.9	35.5	49.2	35.5	40.6
PCC	1	22.7	21.8	10	24.9	8.5	16.2	22.4	18.1	15.3	7.5	28.8	33.8	40.9	31.2	36.4
	2	35.7	30.7	28	41.3	28.8	44	52.2	46.1	46.2	35.3	59.6	42.7	61.5	50.1	60
	3	31.5	25.7	28.4	40.4	32.7	48	56.2	46.5	46.6	35.7	60.1	35	53.8	51.6	59
	4	28.6	43.2	39.5	53.9	41.4	56.6	67.3	45.8	50.8	45	63.2	55.2	74	62.7	72.5
	5	40.6	65.4	51.6	76.1	61.5	76.7	77	55.5	60.5	54.7	72.9	77.4	96.2	86.7	91.8
	6	35	54.4	45.9	65.1	55.9	71.1	71.3	49.9	54.8	49.1	67.2	66.4	85.2	73.9	83.8
	7	47.5	72.4	58.5	83.1	68.4	83.7	83.9	62.5	67.4	61.7	79.8	84.4	103.2	93.6	98.8
	8	52.9	77.7	63.8	88.4	73.8	89	89.3	67.8	72.8	67	85.2	89.7	108.5	98.9	104.1
	9	25.9	50.7	36.9	61.4	46.8	62	62.3	40.8	45.8	40	58.2	62.7	81.5	72	77.1
	10	11.6	36.4	22.5	47.1	34.4	49.6	49.8	28.4	33.3	27.6	45.7	48.4	67.2	59.5	64.7
	11	20.6	43.5	20.6	45.6	26.8	42	40.3	13.8	22.1	20	34.5	54.6	65.7	52	57.1
	12	34.3	61	36.7	61.7	45.3	53.2	51.3	24.8	30.2	33.4	37.1	68.3	81.9	68.1	65.7
	13	25.1	50	31.7	56.7	37.9	53.1	53.4	31.9	36.9	31.1	49.3	62	76.8	63.1	68.2
	14	35.2	60.1	41.5	66.6	47.8	63	59.3	41.8	38.6	41	45.6	72.1	86.7	73	74.1
	15	25.1	9.7	20.5	25.8	14.9	32.1	26.4	40	32.9	27	45	20.6	24	10.3	44.4
	16	0	26.5	12.7	37.2	24.3	37.8	46	25.1	30	21.1	42.4	38.5	57.3	46	55.8
	17	26.5	0	20.9	18.2	14.3	25.5	34.2	36.9	36.6	26.9	50.1	14.2	33	27	36.8
	18	12.7	20.9	0	28.7	10.2	25.1	33.3	19.6	19.7	11	32.9	33.9	52.7	35	40.2
	19	37.2	18.2	28.7	0	18.8	10.8	26.4	40	32.9	30	45	15.7	24	10.3	20.7
	20	24.3	14.3	10.2	18.8	0	17.9	26.1	23.9	23.6	13.9	37	26.4	41.6	27.9	33
	21	37.8	25.5	25.1	10.8	17.9	0	13.1	32.4	25.7	25.6	31.9	22.6	31.4	14.8	15.1

Table A9. *Cont.*

Distances (km)		PCC														
		16	**17**	**18**	**19**	**20**	**21**	**22**	**23**	**24**	**25**	**26**	**27**	**28**	**29**	**30**
PCC	22	46	34.2	33.3	26.4	26.1	13.10	0	29.3	18.9	22.9	19.4	38	45.3	28.7	22.8
	23	25.1	36.9	19.6	40	23.9	32.40	29.30	0	11.8	10.6	25.9	49.7	62.1	47.1	46
	24	30	36.6	19.7	32.9	23.6	25.70	18.90	11.80	0	9	14.3	48.7	58.6	44.8	37.2
	25	21.1	26.9	11	30	13.9	25.60	22.90	10.60	9	0	23.2	38.7	50.1	36.3	38
	26	42.4	50.1	32.9	45	37	31.9	19.4	25.9	14.3	23.2	0	57.8	63.2	46.6	40.8
	27	38.5	14.2	33.9	15.7	26.4	22.6	38	49.7	48.7	38.7	57.8	0	19.2	17.5	30.7
	28	57.3	33	52.7	24	41.6	31.4	45.3	62.1	58.6	50.1	63.2	19.2	0	19.2	31.8
	29	46	27	35	10.3	27.9	14.8	28.7	47.1	44.8	36.3	46.6	17.5	19.2	0	17.7
	30	55.8	36.8	40.2	20.7	33	15.1	22.8	46	37.2	38	40.8	30.7	31.8	17.7	0

References

1. Kumar, A.; Dixit, G. An analysis of barriers affecting the implementation of e-waste management practices in India: A novel ISM-DEMATEL approach. *Sustain. Prod. Consum.* **2018**, *14*, 36–52. [CrossRef]
2. Andarani, P.; Goto, N. Potential e-waste generated from households in Indonesia using material flow analysis. *J. Mater. Cycles Waste Manag.* **2014**, *16*, 306–320. [CrossRef]
3. Increased Media Device Usage Due to the Coronavirus Outbreak among Internet Users Worldwide as of March 2020, by Country. Available online: https://www.statista.com/statistics/1106607/device-usage-coronavirus-worldwide-by-country/ (accessed on 20 July 2020).
4. Baldé, C.P.; Forti, V.; Gray, V.; Kuehr, R.; Stegmann, P. *The Global E-Waste Monitor 2017 Quantities, Flows, and Resources*; United Nations University (UNU): Bonn, Germany; International Telecommunication Union (ITU): Geneva, Switzerland; International Solid Waste Association (ISWA): Vienna, Austria, 2017.
5. Sutanto, A.; Yuliandra, B.; Pratama, W. Manufaktur yang berkelanjutan pada sampah elektronik (e-waste) di kota Padang: Tinjauan kasus sampah kulkas. *J. Optimasi Sist. Ind.* **2017**, *16*, 1–25. [CrossRef]
6. Hanisah, K.; Kumar, S.; Tajul, A. The management of waste cooking oil: A preliminary survey. *Health Environ. J.* **2013**, *4*, 76–81.
7. Moltó, J.; Egea, S.; Conesa, J.A.; Font, R. Thermal decomposition of electronic wastes: Mobile phone case and other parts. *Waste Manag.* **2011**, *31*, 2546–2552. [CrossRef]
8. PACE. *A New Circular Vision for Electronics Time for a Global Reboot*; World Economic Forum: Cologny, Switzerland, 2019; pp. 1–24.
9. Pharino, C. E-waste Management in Thailand (Case Studies). In *Challenges for Sustainable Solid Waste Management*; Springer: Singapore, 2017; pp. 117–139.
10. Golev, A.; Schmeda-Lopez, D.R.; Smart, S.K.; Corder, G.D.; McFarland, E.W. Where next on e-waste in Australia? *Waste Manag.* **2016**, *58*, 348–358. [CrossRef]
11. Yong, K.; Heijungs, R.; Smit, R. *Comparative Life Cycle Assessment of the Informal and Formal Recycling Procedures of Mobile Phones*; Vrije Universiteit Amsterdam: Amsterdam, The Netherlands, 2020; pp. 9–13.
12. Penduduk, Laju Pertumbuhan Penduduk, Distribusi Persentase Penduduk, Kepadatan Penduduk, dan Rasio Jenis Kelamin Penduduk Menurut Provinsi. Available online: https://www.bps.go.id/indikator/indikator/view_data_pub/0000/api_pub/50/da_03/1 (accessed on 17 July 2021).
13. Nurhayati-Wolff, H. Share of Population Owning a Mobile Phone in Indonesia from 2010 to 2019. Available online: https://www.statista.com/statistics/1084069/indonesia-share-of-population-owning-a-mobile-phone/ (accessed on 17 July 2021).
14. Your Phone's Lifespan Is Five Years—Here's How to Keep It Going! Available online: https://computercare.net/2019/12/your-phones-lifespan-is-five-years-heres-how-to-keep-it-going/ (accessed on 17 July 2021).
15. Schluep, M.; Forum, W.R.; Kuehr, R.; Magalini, F. *Recycling—From E-Waste to Resources*; United Nations Environmental Program (UNEP): Nairobi, Kenya; United Nations University: Tokyo, Japan, 2015.
16. Sari, D.P.; Masruroh, N.A.; Asih, A.M.S. Consumer intention to participate in e-waste collection programs: A study of smartphone waste in Indonesia. *Sustainability* **2021**, *13*, 2759. [CrossRef]
17. Rimantho, D.; Noor, E.; Effendi, H. Risk assessment on failure factors of e-waste management process using FMEA method. *Int. J. Adv. Sci. Eng. Inf. Technol.* **2020**, *10*, 2504–2511. [CrossRef]
18. Cordeau, J.F.; Furini, F.; Ljubić, I. Benders decomposition for very large scale partial set covering and maximal covering location problems. *Eur. J. Oper. Res.* **2019**, *275*, 882–896. [CrossRef]
19. Zarandi, M.H.F.; Davari, S.; Sisakht, S.A.H. The large scale maximal covering location problem. *Sci. Iran.* **2011**, *18*, 1564–1570. [CrossRef]
20. Murray, A.T. Maximal coverage location problem: Impacts, significance, and evolution. *Int. Reg. Sci. Rev.* **2016**, *39*, 5–27. [CrossRef]

21. Boonmee, C.; Arimura, M.; Asada, T. Facility location optimization model for emergency humanitarian logistics. *Int. J. Disaster Risk Reduct.* **2017**, *24*, 485–498. [CrossRef]
22. Berman, O.; Kalcsics, J.; Krass, D. On covering location problems on networks with edge demand. *Comput. Oper. Res.* **2016**, *74*, 214–227. [CrossRef]
23. Azhmyakov, V.; Fernández-Gutiérrez, J.P.; Gadi, S.K.; Pickl, S. A novel numerical approach to the MCLP based resilent supply chain optimization. *IFAC-Papers* **2016**, *49*, 137–142. [CrossRef]
24. Eaton, D.J.; Héctor, M.L.S.U.; Lantigua, R.R.; Morgan, J. Determining ambulance deployment in Santo Domingo, Dominican Republic. *J. Oper. Res. Soc.* **1986**, *37*, 113–126. [CrossRef]
25. Current, J.; O'Kelly, M. Locating emergency warning sirens. *Decis. Sci.* **1992**, *23*, 221–234. [CrossRef]
26. Jia, H.; Ordóñez, F.; Dessouky, M.M. Solution approaches for facility location of medical supplies for large-scale emergencies. *Comput. Ind. Eng.* **2007**, *52*, 257–276. [CrossRef]
27. Murali, P.; Ordóñez, F.; Dessouky, M.M. Facility location under demand uncertainty: Response to a large-scale bio-terror attack. *Socioecon. Plann. Sci.* **2012**, *46*, 78–87. [CrossRef]
28. Chanta, S.; Sangsawang, O. Shelter-site selection during flood disaster. *Lect. Notes Magement Sci.* **2012**, *4*, 282–288.
29. Hartini, S.; Sari, D.P.; Utami, A.A. *The Use of Consumer Behavior to Identify the Flow Mapping of Waste Cooking Oil: A Finding from Semarang, Indonesia*; IOP Conference Series: Materials Science and Engineering: Bristol, UK, 2019; pp. 12–25.
30. Pastor, J.T. Bicriterion programs and managerial location decisions: Application to the banking sector. *J. Oper. Res. Soc.* **1994**, *45*, 1351–1362. [CrossRef]
31. Davari, S.; Zarandi, M.H.F.; Hemmati, A. Maximal covering location problem (MCLP) with fuzzy travel times. *Expert Syst. Appl.* **2011**, *38*, 14535–14541. [CrossRef]
32. Arana-Jiménez, M.; Blanco, V.; Fernández, E. On the fuzzy maximal covering location problem. *Eur. J. Oper. Res.* **2020**, *283*, 692–705. [CrossRef]
33. Vatsa, A.K.; Jayaswal, S. A new formulation and Benders decomposition for the multi-period maximal covering facility location problem with server uncertainty. *Eur. J. Oper. Res.* **2016**, *251*, 404–418. [CrossRef]
34. Vatsa, A.K.; Jayaswal, S. Capacitated multi-period maximal covering location problem with server uncertainty. *Eur. J. Oper. Res.* **2020**, *289*, 1107–1126. [CrossRef]
35. Yang, P.; Xiao, Y.; Zhang, Y.; Zhou, S.; Yang, J.; Xu, Y. The continuous maximal covering location problem in large-scale natural disaster rescue scenes. *Comput. Ind. Eng.* **2020**, *146*, 106608. [CrossRef]
36. ReVelle, C.; Scholssberg, M.; Williams, J. Solving the maximal covering location problem with heuristic concentration. *Comput. Oper. Res.* **2008**, *35*, 427–435. [CrossRef]
37. Ibarra-Rojas, O.J.; Ozuna, L.; López-Piñón, D. The maximal covering location problem with accessibility indicators. *Socioecon. Plann. Sci.* **2020**, *71*, 100758. [CrossRef]
38. Alizadeh, R.; Nishi, T. Hybrid set covering and dynamic modular covering location problem: Application to an emergency humanitarian logistics problem. *Appl. Sci.* **2020**, *10*, 7110. [CrossRef]
39. Alizadeh, R.; Nishi, T.; Bagherinejad, J.; Bashiri, M. Multi-period maximal covering location problem with capacitated facilities and modules for natural disaster relief services. *Appl. Sci.* **2021**, *11*, 397. [CrossRef]
40. Zhang, M.; Zhang, Y.; Qiu, Z.; Wu, H. Two-stage covering location model for air-ground medical rescue system. *Sustainability* **2019**, *11*, 3242. [CrossRef]
41. Sarjono, H. Determination of best route to minimize transportation costs using nearest neighbor procedure. *Appl. Math. Sci.* **2014**, *8*, 3063–3074. [CrossRef]
42. Dantzig, B.G.; Ramser, R. The truck dispatching problem. *Manag. Sci.* **1959**, *1*, 80. [CrossRef]
43. Agárdi, A.; Kovács, L.; Bányai, T. An attraction map framework of a complex multi-echelon vehicle routing problem with random walk analysis. *Appl. Sci.* **2021**, *11*, 2100. [CrossRef]
44. Sanz, F.T.; Gómez, E.N.E. The vehicle routing problem with limited vehicle capacities. *Int. J. Traffic Transp. Eng.* **2013**, *3*, 260–268. [CrossRef]
45. Toth, P.; Vigo, D. *The Vehicle Routing Problem*; SIAM: Philadelphia, PA, USA, 2002; pp. 1–26.
46. Asteria, C. Penentuan Rute Distribusi Dengan Algoritma Tabu Search Untuk VRP Dengan Time Windows: Studi Kasus di PT X. Ph.D. Thesis, Universitas Indonesia, Jakarta, Indonesia, 2008.
47. Solomon, M.M. Algorithms for and scheduling problems the vehicle routing with time window constraints. *Oper. Res.* **1987**, *35*, 254–265. [CrossRef]
48. Zheng, W.; Liao, Z.; Qin, J. Using a four-step heuristic algorithm to design personalized day tour route within a tourist attraction. *Tour. Manag.* **2017**, *62*, 335–349. [CrossRef]
49. Kim, S.; Moon, I. Traveling salesman problem with a drone station. *IEEE Trans. Syst. Man Cybern. Syst.* **2018**, *49*, 42–52. [CrossRef]
50. Arora, K.; Agarwal, S.; Tanwar, R. Solving TSP using genetic algorithm and nearest neighbour algorithm and their comparison. *Int. J. Sci. Eng. Res.* **2016**, *7*, 1014–1018.
51. Hermawan, F. Optimization of transportation of municipal solid waste from resource to intermediate treatment facility with nearest neighbour method (study on six sub sub district in DKI Jakarta province). *J. Environ. Sci. Sustain. Dev.* **2018**, *1*, 8. [CrossRef]

52. Zamar, D.S.; Gopaluni, B.; Sokhansanj, S. A constrained k-means and nearest neighbor approach for route optimization in the bale collection problem. *IFAC-PapersOnLine* **2017**, *50*, 12125–12130. [CrossRef]
53. Kurniawan, I.S.; Susanty, S.; Adianto, H. Usulan rute pendistribusian air mineral dalam kemasan menggunakan metode nearest neighbour dan clarke & wright savings. *Reka Integra.* **2014**, *1*, 125–316.
54. Glover, F. Tabu search—Part II. *ORSA J. Comput.* **1990**, *2*, 4–32. [CrossRef]
55. Sengupta, L.; Mariescu-Istodor, R.; Fränti, P. Which local search operator works best for the open-loop TSP? *Appl. Sci.* **2019**, *9*, 3985. [CrossRef]
56. Jia, H.; Li, Y.; Dong, B.; Ya, H. An improved tabu search approach to vehicle routing problem. *Procedia-Soc. Behav. Sci.* **2013**, *96*, 1208–1217. [CrossRef]
57. Cordeau, J.F.; Maischberger, M. A parallel iterated tabu search heuristic for vehicle routing problems. *Comput. Oper. Res.* **2012**, *39*, 2033–2050. [CrossRef]
58. Meliani, Y.; Hani, Y.; Elhaq, S.L.; El Mhamedi, A. A developed tabu search algorithm for heterogeneous fleet vehicle routing problem. *IFAC-PapersOnLine* **2019**, *52*, 1051–1056. [CrossRef]
59. Qiu, M.; Fu, Z.; Eglese, R.; Tang, Q. A tabu search algorithm for the vehicle routing problem with discrete split deliveries and pickups. *Comput. Oper. Res.* **2018**, *100*, 102–116. [CrossRef]
60. Silvestrin, P.V.; Ritt, M. An iterated tabu search for the multi-compartment vehicle routing problem. *Comput. Oper. Res.* **2017**, *81*, 192–202. [CrossRef]
61. Wang, Z.; Li, Y.; Hu, X. A heuristic approach and a tabu search for the heterogeneous multi-type fleet vehicle routing problem with time windows and an incompatible loading constraint. *Comput. Ind. Eng.* **2015**, *89*, 162–176. [CrossRef]
62. Soto, M.; Sevaux, M.; Rossi, A.; Reinholz, A. Multiple neighborhood search, tabu search and ejection chains for the multi-depot open vehicle routing problem. *Comput. Ind. Eng.* **2017**, *107*, 211–222. [CrossRef]
63. Wang, J.; Ranganathan Jagannathan, A.K.; Zuo, X.; Murray, C.C. Two-layer simulated annealing and tabu search heuristics for a vehicle routing problem with cross docks and split deliveries. *Comput. Ind. Eng.* **2017**, *112*, 84–98. [CrossRef]
64. Gmira, M.; Gendreau, M.; Lodi, A.; Potvin, J.Y. Tabu search for the time-dependent vehicle routing problem with time windows on a road network. *Eur. J. Oper. Res.* **2021**, *288*, 129–140. [CrossRef]
65. Tarantilis, C.D.; Stavropoulou, F.; Repoussis, P.P. A template-based tabu search algorithm for the consistent vehicle routing problem. *Expert Syst. Appl.* **2012**, *39*, 4233–4239. [CrossRef]
66. Lai, D.S.W.; Caliskan Demirag, O.; Leung, J.M.Y. A tabu search heuristic for the heterogeneous vehicle routing problem on a multigraph. *Transp. Res. Part E Logist. Transp. Rev.* **2016**, *86*, 32–52. [CrossRef]
67. Shi, Y.; Lv, L.; Hu, F.; Han, Q. A heuristic solution method for multi-depot vehicle routing-based waste collection problems. *Appl. Sci.* **2020**, *10*, 2403. [CrossRef]
68. Khan, A.S.; Khalid, Q.S.; Naeem, K.; Ahmad, R.; Khan, R.; Saleem, W. Application of exact and multi-heuristic approaches to a sustainable closed loop supply chain network design. *Sustainability* **2021**, *13*, 2433. [CrossRef]
69. Tebaldi, L.; Murino, T.; Bottani, E. An adapted version of the water wave optimization algorithm for the capacitated vehicle routing problem with time windows with application to a real case using probe data. *Sustainability* **2020**, *12*, 3666. [CrossRef]
70. Daerah Istimewa Yogyakarta Province in Figure. 2020. Available online: http://library1.nida.ac.th/termpaper6/sd/2554/19755.pdf. (accessed on 12 January 2021).
71. Sekilas Info TPST Piyungan | Dinas Lingkungan Hidup dan Kehutanan DIY. Available online: http://dlhk.jogjaprov.go.id/sekilas-info-tpst-piyungan. (accessed on 20 December 2020).
72. Pujawan, I.N. *Supply Chain Management*, 1st ed.; Guna Widya: Surabaya, Indonesia, 2005; pp. 173–188.
73. Herawati, C.; Adianto, R.H.; Mustofa, F.H. Usulan rute distribusi tabung gas 12 kg menggunakan algoritma nearest neighbour dan algoritma tabu search di PT.X Bandung. *Reka Integra.* **2015**, *3*, 209–220.
74. Saputro, W.P. Vehicle Routing Problem with Simultaneous Pickup and Delivery Menggunakan Algoritma Tabu Search. Bachelor's Thesis, University of Muhammadiyah Malang, Malang, Indonesia, 2016.
75. Church, R.L.; Davis, R.R. The fixed charge maximal covering location problem. *Pap. Reg. Sci.* **1992**, *71*, 199–215. [CrossRef]
76. Ryu, M.; Ahn, K.I.; Lee, K. Finding effective item assignment plans with weighted item associations using a hybrid genetic algorithm. *Appl. Sci.* **2021**, *11*, 2209. [CrossRef]
77. Santosa, B.; Ai, T.J. *Pengantar Metaheuristik Implementasi Dengan MATLAB*, 1st ed.; ITS Tekno Sains: Surabaya, Indonesia, 2017; pp. 107–116.
78. Sulistiono, S.; Mussafi, N.S.M. Rancang bangun vehicle routing problem menggunakan algoritma tabu search. *J. Fourier* **2015**, *4*, 113. [CrossRef]
79. Pradhana, F.E. Penerapan Algoritma Tabu Search untuk Menyelesaikan Vehicle Routing Problem. Bachelor's Thesis, Universitas Negeri Semarang, Semarang, Indonesia, 2011.
80. Talbi, E.G. *Metaheuristics: From Design to Implementation*; John Wiley & Sons: Hoboken, NJ, USA, 2009.
81. Hillier, J.; Lieberman, G.J. *Introduction to Operations Research*; McGraw-Hill International Edition: New York, NY, USA, 2010.
82. Mirjalili, S.; Mirjalili, S.M.; Lewis, A. Grey wolf optimizer. *Adv. Eng. Softw.* **2014**, *69*, 46–61. [CrossRef]
83. Prasetyo, W.; Tamyiz, M. Vehicle routing problem dengan aplikasi metode nearest neighbor. *J. Res. Technol.* **2017**, *3*, 88–199.
84. Khan, A.A.; Agrawala, H.A. Comparitive study of nearest neighbour algorithm and genetic algorithm in solving travelling salesman problem. *Int. Res. J. Eng. Technol.* **2016**, *3*, 234.

sustainability

Article

Predicting the Degree of Dissolved Oxygen Using Three Types of Multi-Layer Perceptron-Based Artificial Neural Networks

Fen Yang [1], Hossein Moayedi [2,3,*] and Amir Mosavi [4,*]

[1] School of Economics and Management, Beijing University of Technology, Beijing 100124, China; yappley666@126.com
[2] Institute of Research and Development, Duy Tan University, Da Nang 550000, Vietnam
[3] Faculty of Civil Engineering, Duy Tan University, Da Nang 550000, Vietnam
[4] John von Neumann Faculty of Informatics, Obuda University, 1034 Budapest, Hungary
* Correspondence: hosseinmoayedi@duytan.edu.vn (H.M.); amir.mosavi@mailbox.tu-dresden.de (A.M.)

Abstract: Predicting the level of dissolved oxygen (DO) is an important issue ensuring the sustainability of the inhabitants of a river. A prediction model can predict the DO level using a historical dataset with regard to water temperature, pH, and specific conductance for a given river. The model can be built using sophisticated computational procedures such as multi-layer perceptron-based artificial neural networks. Different types of networks can be constructed for this purpose. In this study, the authors constructed three networks, namely, multi-verse optimizer (MVO), black hole algorithm (BHA), and shuffled complex evolution (SCE). The networks were trained using the datasets collected from the Klamath River Station, Oregon, USA, for the period 2015–2018. We found that the trained networks could predict the DO level of 2019. We also found that both BHA- and SCE-based networks could predict the level of DO using a relatively simple configuration compared to that of MVO. From the viewpoints of absolute errors and Pearson's correlation coefficient, MVO- and SCE-based networks performed better than BHA-based networks. In synopsis, the authors recommend MVO- and MLP-based artificial neural networks for predicting the DO level of a river.

Keywords: water quality; dissolved oxygen; neural network; machine learning; artificial intelligence; deep learning; big data; data science; hydrological model; water treatment

check for updates

Citation: Yang, F.; Moayedi, H.; Mosavi, A. Predicting the Degree of Dissolved Oxygen Using Three Types of Multi-Layer Perceptron-Based Artificial Neural Networks. *Sustainability* **2021**, *13*, 9898. https://doi.org/10.3390/su13179898

Academic Editors: Dragan Pamucar and Željko Stević

Received: 30 April 2021
Accepted: 23 June 2021
Published: 3 September 2021

Publisher's Note: MDPI stays neutral with regard to jurisdictional claims in published maps and institutional affiliations.

Copyright: © 2021 by the authors. Licensee MDPI, Basel, Switzerland. This article is an open access article distributed under the terms and conditions of the Creative Commons Attribution (CC BY) license (https://creativecommons.org/licenses/by/4.0/).

1. Introduction

1.1. Problem Statement and Background

As is known, acquiring an appropriate forecast for water quality parameters such as dissolved oxygen (DO) is an important task due to their effects on aquatic health maintenance and reservoir management [1]. Constraints such as the influence of various environmental factors on the DO concentration [2] have driven many scholars to replace conventional models with sophisticated artificial intelligent techniques [3–6].

From a general point of view, the objective of many recent developments lies in facilitating complex analysis [7–9]. These efforts have resulted in finding promising approaches for various analyses, such as water treatment [10,11], energy [12,13], construction and waste reduction [14–16], optics-related simulations [17,18], environmental modeling [19–21], and remote sensing [22]. In this sense, different intelligent methods have been devised by experts to give a reliable simulation of engineering problems [23–25].

Going beyond typical soft computing, metaheuristic algorithms have been designed for optimization purposes [26–29]. Most of these techniques rely on the herding behavior of animals [30–33] (e.g., Harris hawks [34,35] and grey wolf [36,37]). When incorporated with conventional soft computing, metaheuristic algorithms aim to optimize the hyperparameters of a base model to achieve optimum configuration [38–41]. Some examples of these algorithms can be found in research by Hu et al. [42], Shen et al. [43], Wang and Chen [44], and Li et al. [45], dedicated to medical diagnosis.

1.2. Similar Works

Through applying support vector regression (SVR), Liu et al. [46] showed the efficiency of the maximal information coefficient technique used for feature selection in the estimation of DO concentration. The results of the optimized dataset were much more reliable (28.65% in terms of root mean square error, RMSE) than the original input configuration. Csábrági et al. [47] showed the appropriate efficiency of three conventional notions of artificial neural networks (ANNs), namely, multi-layer perceptron (MLP), radial basis function (RBF), and general regression neural network (GRNN), for this purpose. Similar efforts can be found in [48,49]. Heddam [50] introduced a new ANN-based model, namely, evolving fuzzy neural network, as a capable approach for DO simulation in a river ecosystem. The suitability of fuzzy-based models has been investigated in many studies [51]. Adaptive neuro-fuzzy inference system (ANFIS) is another potent data mining technique that has been discussed in many studies [52–54]. More attempts regarding the employment of machine learning tools can be found in [55–58].

Ouma et al. [59] compared the performance of a feed-forward ANN with multiple linear regression (MLR) in simulating DO in Nyando River, Kenya. It was shown that the correlation of the ANN is considerably greater than the MLR (i.e., 0.8546 vs. 0.6199). Zhang et al. [60] combined a recurrent neural network (RNN) with kernal principal component analysis to predict hourly DO concentration. Their suggested model was found to be more accurate than regular data mining techniques, including feed-forward ANN, SVR, and GRNN, by around 8%, 17%, and 12%. Additionally, the largest accuracy (the coefficient of determination $R^2 = 0.908$) was obtained for DO in the upcoming hour. Ali et al. [61] combined a so-called denoising method, namely, "complete ensemble empirical mode decomposition with adaptive noise", with two popular machine learning models, namely, random forest (RF) and extreme gradient boosting, to analyze various water quality parameters. It was shown that the RF-based ensemble is a more accurate approach for the simulation of DO, temperature, and specific conductance. They also proved the viability of the proposed approaches by comparing them with some benchmark tools. Likewise, Ahmed [62] showed the superiority of RF over MLR for DO modeling. He also revealed that water temperature and pH play the most significant roles in this process. Ay and Kişi [63] conducted a comparison among MLP, RBF, ANFIS (sub-clustering), and ANFIS (grid partitioning). The respective R^2 values of 0.98, 0.96, 0.95, and 0.86 for one station (Number: 02156500) revealed that the outcomes of MLP are better correlated with the observed DOs.

Synthesizing conventional approaches with auxiliary techniques has led to novel hybrid tools for various hydrological parameters [64–66]. Ravansalar et al. [67] showed that linking the ANN with discrete wavelet transform results in an improvement of accuracy (i.e., Nash–Sutcliffe coefficient) from 0.740 to 0.998. A similar improvement was achieved for the SVR applied to estimate biochemical oxygen demand in Karun River, Western Iran. Antanasijević et al. [68] presented a combination of Ward neural networks and a local similarity index for predicting DO in the Danube River. They noted the better performance of the proposed model compared to the multi-site DO evaluative approaches presented in the literature.

1.3. Novelty and Objective

Metaheuristic search methods such as teaching–learning based optimization [69] have provided suitable approaches for intricate problems. Ahmed and Shah [52] suggested three optimized versions of ANFIS using differential evolution, genetic algorithm (GA), and ant colony optimization for predicting water quality parameters, including electrical conductivity, sodium absorption ratio, and total hardness. In similar research, Mahmoudi et al. [70] coupled SVR with the shuffled frog leaping algorithm (SFLA) for the same objective. Zhu et al. [71] compared the efficiency of the fruit fly optimization algorithm (FOA) with the GA and particle swarm optimization (PSO) for optimizing a least-squares SVR for forecasting the trend of DO. Referring to the obtained mean absolute percentage errors of 0.35%, 1.3%, 2.03%, and 1.33%, the proposed model (i.e., FOA-LSSVR) surpassed the benchmark techniques. In this work, three stochastic search techniques of multi-verse optimizer (MVO), black hole algorithm (BHA), and shuffled complex evolution (SCE) are used to optimize an MLP neural network for predicting DO using recent data collected from the Klamath River Station. According to Sullivan et al. [72], the reach of interest is classified as very poor water quality based on the Oregon Water Quality Index. Additionally, the reach of Keno dam (downstream of the river) is labeled as "water quality limited" for ammonia and dissolved oxygen year-round, as well as pH and chlorophyll *a* in summer. It clearly highlights the importance of water quality assessments in this area. To the best of the authors' knowledge, up to now, few metaheuristic algorithms have been used for training the ANN in the field of DO modeling (e.g., firefly algorithm [73] and PSO [74]). Therefore, the models suggested in this study are deemed as innovative hybrids for this purpose.

2. Methodology

The steps of this research are shown in Figure 1. After providing the appropriate dataset, the MLP is submitted to MVO, BHA, and SCE algorithms to adjust its parameters through metaheuristic schemes. During an iterative process, the MLP is optimized to present the best possible prediction of the DO.

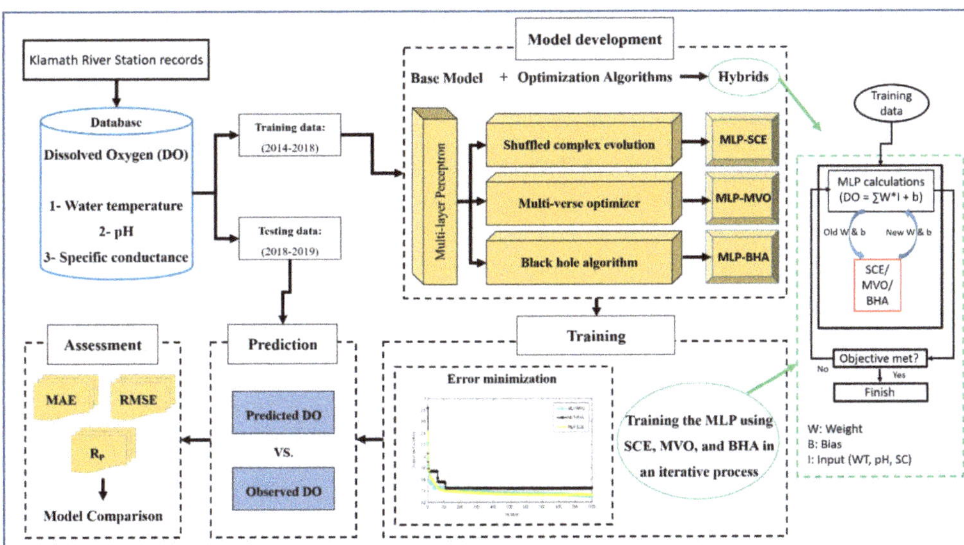

Figure 1. The steps taken in this research for predicting DO.

2.1. The MVO

As is implied by its name, the MVO is obtained from multi-verse theory in physics [75]. According to this theory, there is more than one big bang event, each of which has initiated a separate universe. The algorithm was introduced by Mirjalili et al. [76]. The main components of the MVO are wormholes, black holes, and white holes. The concepts of black and white holes run the exploration phase, while the wormhole concept is dedicated to the exploitation procedure. The pseudo code of the MVO is presented as Algorithm 1.

Algorithm 1. Pseudo code of the MVO [77]

```
Initialize the parameters (population size, iterations, wormhole existence probability (WEP),
travelling distance rate (TDR))
While maximum iteration not reached
Compute the fitness for each universe
AU = Sort the population
BI = Normalize the fitness values
for i = 2:N
Black hole = i
for j = 1:size(i)
Generate r1 as a random value
if r1 < BI(U_i)
white hole = Roulette Wheel Selection(-B)
U(black hole, j) = AU(white hole, j)
end if
Generate r2 as a random value
if r2 < WEP
Generate r3 and r4 as a random value
if r3 < 0.5
Update the position of the universe (Equation (4))
else
Update the position of the universe (Equation (4))
end if
end if
end for
end for
end while
Response = Best solution.
```

In the MVO, the so-called parameter of "rate of inflation" (ROI) is defined for each universe. The objects are transferred from the universes with larger ROIs to those with lower values to improve the average ROI of the whole cosmos. During an iteration, the organization of the universes is carried out with respect to their ROIs, and after a roulette wheel selection (RWS), one of them is deemed the white hole. In this relation, a set of universes can be defined as:

$$U = \begin{bmatrix} x_1^1 & x_1^2 & \cdots & x_1^g \\ x_2^1 & x_2^2 & \cdots & x_2^g \\ \vdots & \vdots & \vdots & \vdots \\ x_k^1 & x_k^2 & \cdots & x_k^g \end{bmatrix} \quad (1)$$

where g symbolizes the number of objects and k stands for the number of universes. The jth objective in the ith solution is generated according to the below equation:

$$\mathrm{x_i^j = lb_j + rand()\left(\left(ub_j - lb_j\right) + 1\right) \forall i \in (1, 2, \ldots, k)\hat{}\forall j \in (1, 2, \ldots, g)} \quad (2)$$

where $\mathrm{ub_j}$ and $\mathrm{lb_j}$ denote upper and lower bounds, and the function $\mathrm{rand}()$ produces a discrete randomly distributed number.

In each repetition, there are two options for the x_i^j: (i) it is selected from earlier solutions using RWS (e.g., $x_i^j \in (x_i^1, x_i^2, \ldots, x_i^{j-1})$ and (ii) it does not change. It can be written as follows:

$$x_i^j \begin{cases} x_m^j & rand_1 < Norm(U_i) \\ x_i^j & rand_1 \geq Norm(U_i) \end{cases} \quad (3)$$

In the above equation, U_i stands for the ith universe, $Norm(U_i)$ gives the corresponding normalized ROI, and $rand_1$ is a random value in [0, 1].

Equation (4) expresses the measures considered to deliver the variations of the whole universe. In this sense, the wormholes are supposed to enhance the ROI.

$$x_i^j = \begin{cases} \begin{cases} x_j + TDR \times \left(\left(ub_j - lb_j \right) \times r_4 + lb_j \right) & if\ r_3 < 0.5 \\ x_j - TDR \times \left(\left(ub_j - lb_j \right) \times r_4 + lb_j \right) & if\ r_3 \geq 0.5 \end{cases} & \begin{matrix} if\ r_2 < WEP \\ if\ r_2 \geq WEP \end{matrix} \\ x_i^j & \end{cases} \quad (4)$$

where x_j signifies the jth best-fitted universe obtained so far, and r_2, r_3, and r_4 are random values in [0, 1]. Moreover, the two parameters of WEP and TDR stand for the wormhole existence probability and the traveling distance rate, respectively. Given *Iter* as the running iteration, and $Iter_{max}$ as the maximum number of *Iter*s, these parameters can be calculated as follows:

$$WEP = a + Iter \times \left(\frac{b - a}{Iter_{max}} \right) \quad (5)$$

$$TDR = 1 - \frac{Iter^{1/q}}{{Iter_{max}}^{1/q}} \quad (6)$$

where q is the accuracy of exploitation, and a and b are constant pre-defined values [78,79].

2.2. The BHA

Inspired by black hole incidents in space, Hatamlou [80] proposed the BHA in 2013. Emerging after the collapse of massive stars, a black hole is distinguished by a huge gravitational power. The stars move toward this mass, and it explains the pivotal strategy of the BHA for achieving an optimum response. A randomly generated constellation of stars represents the initial population. Based on the fitness of these stars, the most powerful one is deemed as the black hole that will absorb the surrounding ones. The pseudo code of the BHA is presented as Algorithm 2.

Algorithm 2. Pseudo code of the BHA [81]

Initialize the stars x_i
Initialize the parameters (fitness function and iterations (*Iter*))
Select the best-fitted star (x_b) as the black hole (BH)
While maximum *Iter* not reached
for each x_i
Compute the fitness
if fitness (x_i) > fitness (x_b)
$x_b = x_i$
end if
Update fitness and compute Equation (8)
if $\sqrt{(x_b - x_i)^2} < R_{BH}$
Replace x_i with a new star
end if
end for
Iter = *Iter* + 1
end while
Response = Best solution.

In this procedure, the positions change according to the relationship below:

$$x_i^{Iter+1} = x_i^{Iter} + \text{rand} \times \left(x_{BH} - x_i^{Iter} \right) \quad \text{i} = 1,\ 2,\ \ldots,\ \text{Z} \quad (7)$$

where rand is a random number in [0, 1], x_{BH} is the black hole's position, Z is the total number of stars, and *Iter* symbolizes the iteration number.

Once the fitness of a star surpasses that of the black hole, they exchange their positions. In this regard, Equation (8) calculates the radius of the event horizon for the black hole.

$$R_S = F_{BH} / \sum_{i=1}^{Z} F_i \tag{8}$$

where F_i is the fitness of the ith star, and F_{BH} is the value for the black hole [82].

2.3. The SCE

Originally proposed by Duan et al. [82], the SCE has been efficiently used for dealing with optimization problems with high dimensions. The SCE can be defined as a hybrid of complex shuffling and competitive evolution concepts with the strengths of the controlled random search strategy. This algorithm (i.e., the SCE) benefits from a deterministic strategy to guide the search. Additionally, utilizing random elements has resulted in a flexible and robust algorithm. The pseudo code of the SCE is presented as Algorithm 3.

Algorithm 3. Pseudo code of the SCE [83]

Initialize the population (s)
Sample the population $\{x_1, x_2, \ldots, x_s\}$
Compute the fitness (f) for each member
Sort based on the obtained f values
Create $D^0 = \{x_i, f_i,$ *where* $i = 1, 2, \ldots, s\}$
Create complexes W.R.T Equation (9)
While maximum *It* not reached
for $q = 1{:}i$
Evolve C^q *using Competitive Complex Evolution*
end for
$D^i = D^{i+1}$
end while
Response = Best solution.

The SCE is implemented in seven steps. Assuming N_C as the number of complexes and N_P as the number of points existing in one complex, the sample size of the algorithm is generated as $S = N_C \times N_P$. In this sense, $N_C \geq 1$ and $N_P \geq 1$ + the number of design variables. Next, the samples $x_1, x_2, \ldots, x_s$ are created in the viable space (i.e., within the bounds). The fitness values are also calculated using sampling distribution. In the third step, these samples are arranged with reference to their fitness. An array-like $D = \{x_i, f_i,$ *where* $i = 1, 2, \ldots, s\}$ can be considered for storing them. This array is then divided into N_C complexes ($C^1, C^2, \ldots, C^{N_C}$), each of which contains N_P samples (Equation (9))

$$C^q = \{x_j^q, f_j^q \,|\, x_j^q = x_{q+N_C(j-1)}, f_j^q = f_{q+N_C(j-1)}, j = 1, 2, \ldots, N_p\} \tag{9}$$

In the fifth step, each complex is evolved by the competitive complex evolution algorithm. Later, in a process named the shuffling of the complexes, all complexes are replaced in array D. This array is then sorted based on the fitness values. Lastly, the algorithm checks for stopping criteria that terminate the process [84].

2.4. Accuracy Criteria

The quality of the results is lastly evaluated using Pearson's correlation coefficient (R_P) along with mean absolute error (MAE) and $RMSE$. They analyze the agreement and difference between the observed and predicted values of a target parameter. In the present work, given $DO_{i_{predicted}}$ and $DO_{i_{observed}}$ as the predicted and observed DOs, the R_P, MAE, and $RMSE$ are expressed by the following equations:

$$R_P = \frac{\sum_{i=1}^{K}(DO_{i_{predicted}} - \overline{DO}_{predicted})(DO_{i_{observed}} - \overline{DO}_{observed})}{\sqrt{\sum_{i=1}^{K}(DO_{i_{predicted}} - \overline{DO}_{predicted})^2}\sqrt{\sum_{i=1}^{K}(DO_{i_{observed}} - \overline{DO}_{observed})^2}} \quad (10)$$

$$MAE = \frac{1}{K}\sum_{i=1}^{K}\left|DO_{i_{observed}} - DO_{i_{predicted}}\right| \quad (11)$$

$$RMSE = \sqrt{\frac{1}{K}\sum_{i=1}^{K}[(DO_{i_{observed}} - DO_{i_{predicted}})]^2} \quad (12)$$

where K signifies the number of compared pairs.

3. Data

As a matter of fact, intelligent models should first learn the pattern of the intended parameter in order to predict it. This learning process is carried out by analyzing the dependence of the target parameter on some independent factors. In this work, DO is the target parameter for water temperature (WT), pH, and specific conductance (SC). This study uses the data belonging to a US Geological Survey (USGS) station, namely, the Klamath River (station number: 11509370). As Figure 2 illustrates, this station is located in Klamath County, Oregon State.

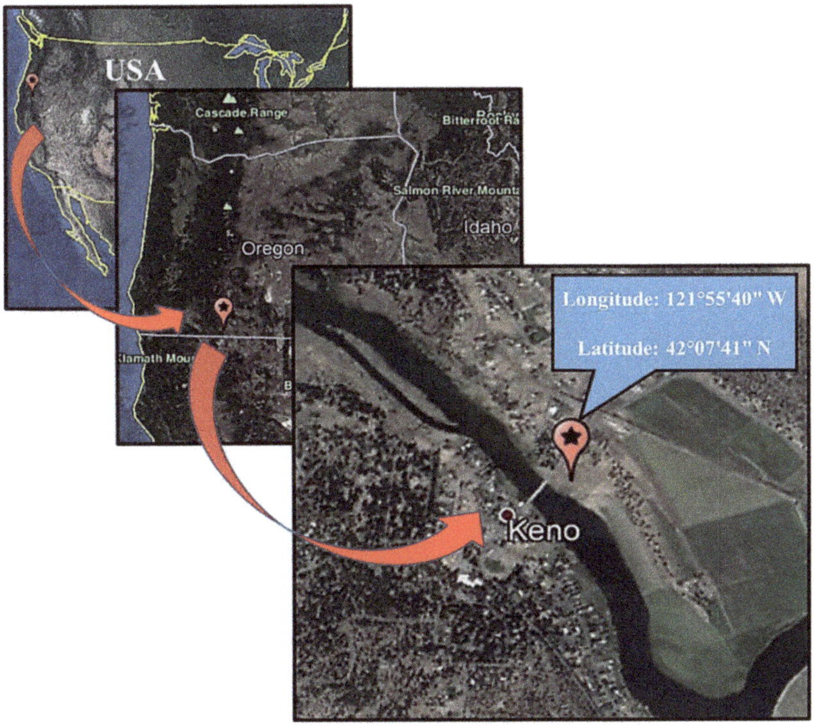

Figure 2. Location of the studied USGS station.

Pattern recognition is fulfilled by means of data obtained between 1 October 2014 and 30 September 2018. After training the models, the DO for the subsequent year (i.e., from 1 October 2018 to 30 September 2019) is predicted. Since the models do not know this data, the accuracy of this process will reflect their capability to predict DO in unseen conditions. Hereafter, these two groups are categorized as training data and testing data, respectively. Figure 3 depicts DO vs. WT, PH, and SC for the (a, c, and e) training and (b, d, and f) testing data. Based on the available data for the mentioned periods, the training and testing groups contain 1430 and 352 records, respectively. Moreover, the statistical description of these datasets is presented in Table 1

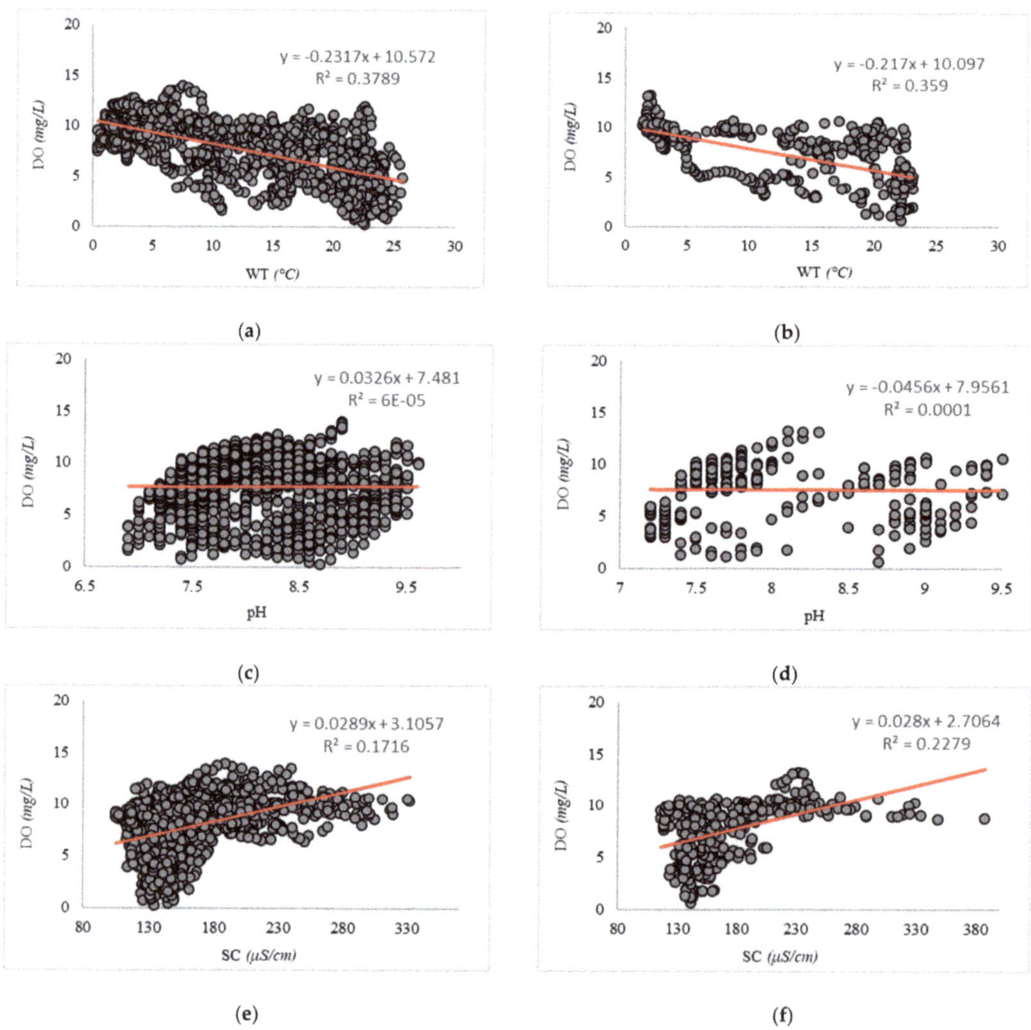

Figure 3. Scatterplots showing DO vs. (**a**) WT training, (**b**) WT testing, (**c**) pH training, (**d**) pH testing, (**e**) SC training, and (**f**) SC testing.

Table 1. Statistical indicators of DO and independent factors.

Indicator	Train Data				Test Data			
	WT (°C)	pH	SC (μS/cm)	DO (mg/L)	WT (°C)	pH	SC (μS/cm)	DO (mg/L)
Average	12.20	8.05	160.46	7.74	11.54	7.97	174.20	7.59
Standard Deviation	7.38	0.64	39.81	2.78	7.43	0.65	45.80	2.69
Sample Variance	54.43	0.41	1584.75	7.72	55.23	0.42	2097.97	7.24
Skewness	0.07	0.48	1.54	−0.54	0.11	0.96	1.49	−0.57
Minimum	0.40	6.90	105.00	0.30	1.40	7.20	116.00	0.60
Maximum	25.70	9.60	332.00	14.00	23.10	9.60	387.00	13.20

Moreover, the effect of the inputs on DO is investigated using a tree-based ensemble method. To do this, a bagged ensemble of 200 regression trees is implemented, and the outcome is reported as a value of importance. Figure 4 shows the results. As can be seen, SC is smaller than WT and the effect of WT smaller than pH. In other words, pH has the greatest impact on DO concentration in this case.

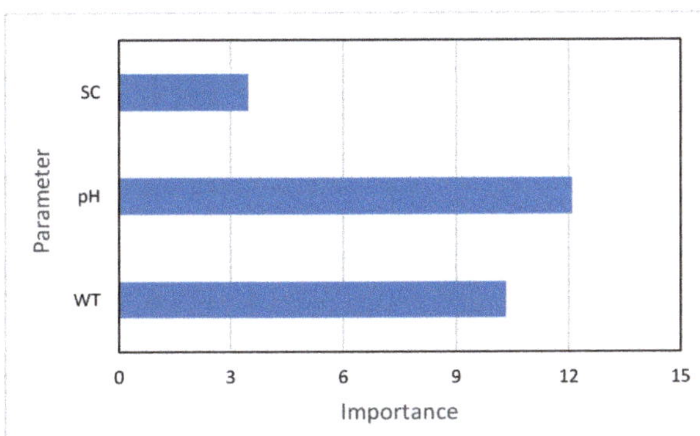

Figure 4. Importance assessment of the inputs.

4. Results and Discussion

4.1. Optimization and Weight Adjustment

As explained, the proposed hybrid models are designed in the way that MVO, BHA, and SCE algorithms are responsible for adjusting the weights and biases of the MLP. To do this, a raw MLP structure (that wants to predict DO from WT, pH, and SC) should be given as the optimization problem to the mentioned algorithms. In this work, a three-layered MLP with three neurons in the input layer (each for one input), six neurons in the middle layer, and one neuron (for DO) in the last layer was considered, noting that the value of 6 was determined after a trial and error process; this is schematically shown in Figure 5.

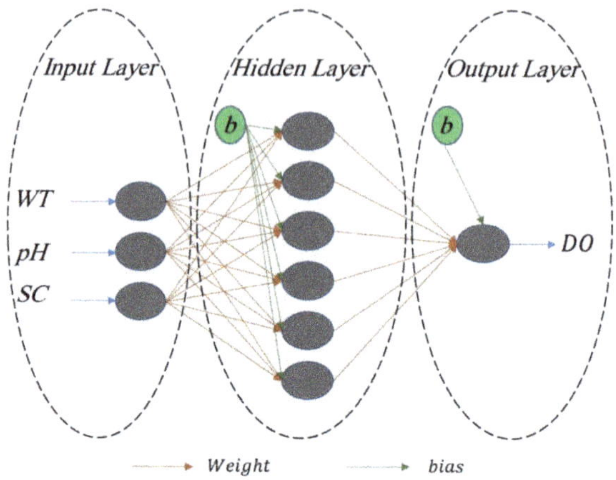

Figure 5. The used MLP structure.

The overall formulation of a neuron can be expressed as follows:

$$R_N = f(I_N \times W + b) \quad (13)$$

where $f(x)$ is the activation function used by the neurons in a layer; additionally, R_N and I_N denote the response and input of the neuron N, respectively. With a large number of these equations, each algorithm first suggests a stochastic response for the W and b values. In the next iterations, the algorithms improve this response in order to build a more accurate MLP. The formulation of a trained MLP network is presented at the end of the study to better illustrate this concept.

The created hybrids are implemented with different population sizes (N_{Pop}s) of the trainer algorithm to achieve the best results. Figure 6 shows the values of the objective function obtained for the N_{Pop}s of 10, 25, 50, 75, 100, 200, 300, 400, and 500. In the case of this study, the objective function is reported by the RMSE criterion. Figure 6 shows that unlike the SCE, which gives more quality training with small N_{Pop}s, the MVO performs better with the three largest N_{Pop}s. The BHA, however, did not show any specific behavior. Overall, the MVO, BHA, and SCE with the N_{Pop}s of 300, 50, and 10, respectively, could adjust the MLP parameters with the lowest error.

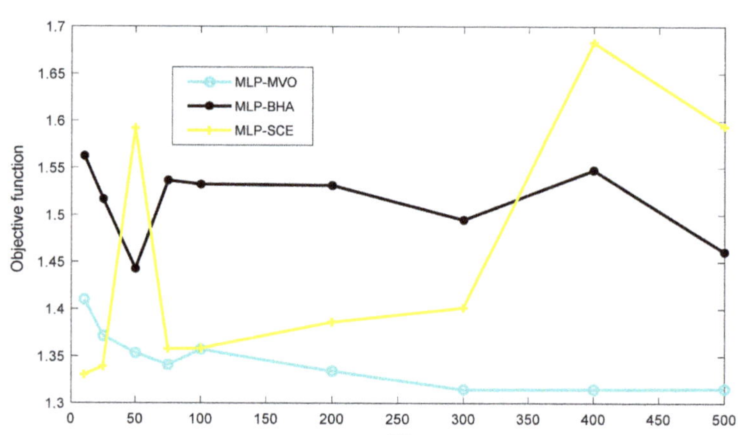

Figure 6. The quality of training for different configurations of the MVO, BHA, and SCE.

As stated, metaheuristic algorithms minimize the errors in an iterative process. Figure 7 shows the convergence curve plotted for the selected configurations of the MLP-MVO, BHA-MVO, and SCE-MVO. To this end, the training RMSE is calculated for a total of 1000 iterations. According to Figure 7, the optimum values of the objective function are 1.314816444, 1.442582978, and 1.33041779 for the MLP-MVO, BHA-MVO, and SCE-MVO, respectively. These configurations are applied in the next section to predict DO. Their results are then evaluated for accuracy assessment.

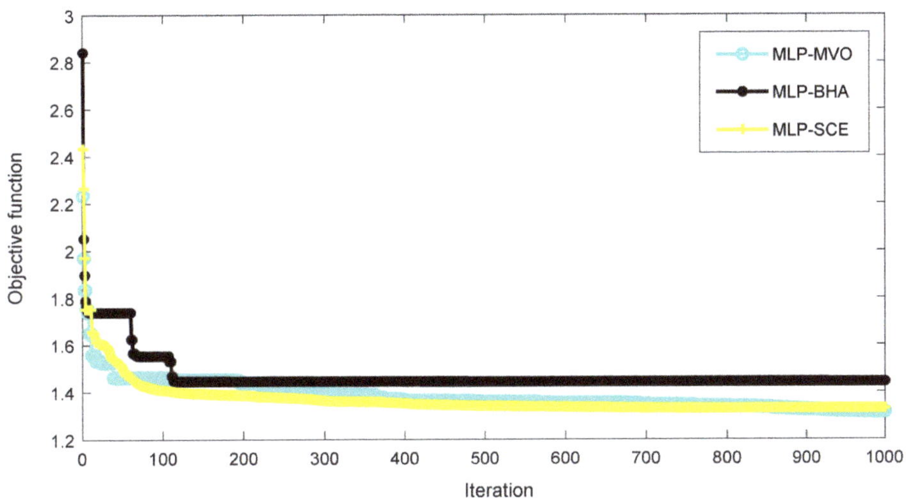

Figure 7. Error reduction carried out by the selected configurations of the MVO, BHA, and SCE.

4.2. Assessment of the Results

Figure 8 shows a comparison between the observed DO rates and those predicted by the MLP-MVO, MLP-BHA, and MLP-SCE for the whole five years. At a glance, all three models could properly capture DO behavior. It indicates that the algorithms have designated appropriate weights for each input parameter (WT, PH, and SC). The results of the training and testing datasets are presented in detail in the following figure.

Focusing on the training results, an acceptable level of accuracy is reflected by the RMSEs of 1.3148, 1.4426, and 1.3304 for the MLP-MVO, MLP-BHA, and MLP-SCE. In this sense, the values of the MAE (0.9687, 1.0931, and 0.9929) confirmed this statement and showed that all three models had understood the DO pattern with good accuracy. By comparison, it can be deduced that both error values of the MLP-MVO are lower than the two other models. Based on the same reason, the MLP-SCE outperformed the MLP-BHA.

Figure 9 depicts the errors obtained for the training data. This value is calculated as the difference between the predicted and observed DOs. The errors of the MLP-MVO, MLP-BHA, and MLP-SCE have the following ranges: [−4.6396, 4.7003], [−4.4964, 4.9537], and [−4.5585, 4.5653].

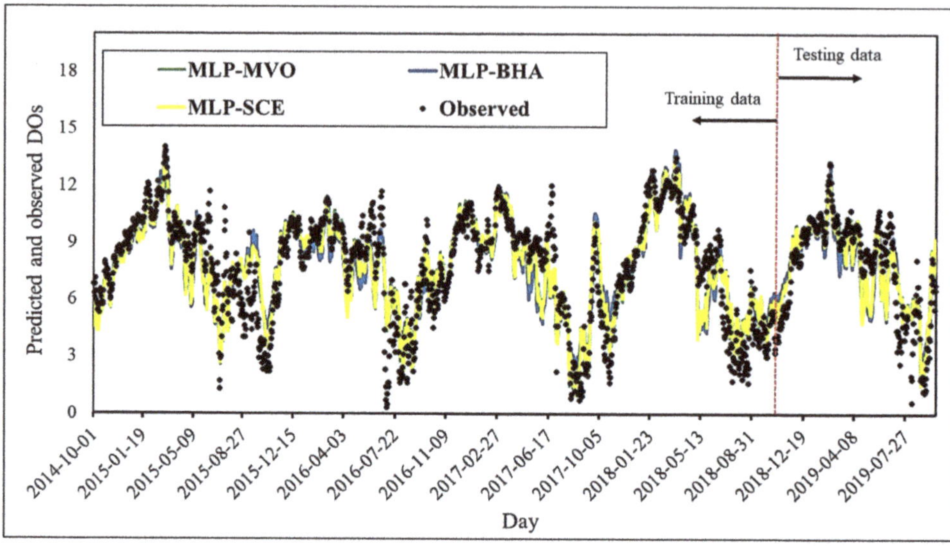

Figure 8. The predicted and observed DO rates from 1 October 2014 to 30 September 2019.

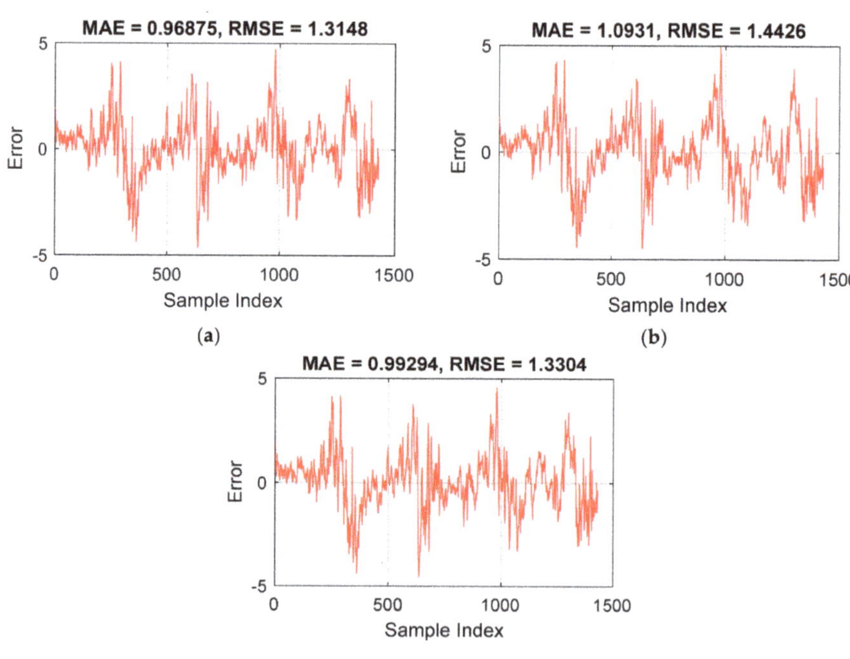

Figure 9. Training errors for the (**a**) MLP-MVO, (**b**) MLP-BHA, and (**c**) MLP-SCE.

As stated previously, the quality of the testing results shows how successful a trained model can be in confronting new conditions. The data of the fifth year were considered as these conditions in this study. Figure 10 depicts the histogram of the testing errors. In these charts, μ stands for the mean error, and σ represents the standard error. In this phase, the RMSEs of 1.3187, 1.4647, and 1.3085, along with the MEAs of 1.0161, 1.1997, and 1.0122, imply the power of the used models for dealing with strange data. It means that the weights (and biases) determined in the previous section have successfully mapped the relationship between DO and WT, PH, and SC for the second phase. From the comparison point of view, unlike the training phase, the SCE-based hybrid outperformed the MLP-MVO. The MLP-BHA, however, presented the poorest prediction of DO, again.

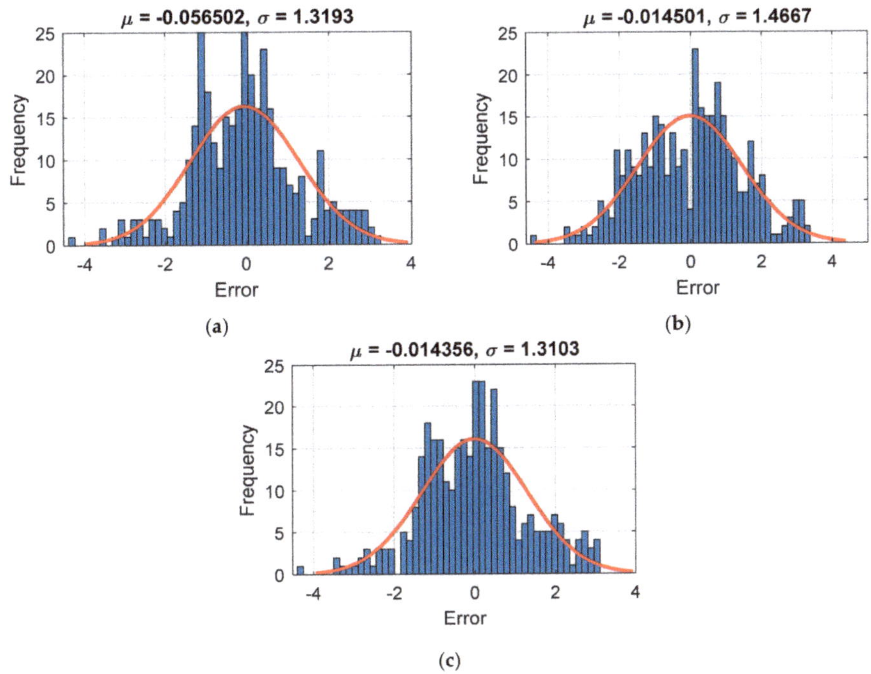

Figure 10. The histogram of the testing errors for the (**a**) MLP-MVO, (**b**) MLP-BHA, and (**c**) MLP-SCE.

The third accuracy indicator (i.e., the R_P) shows the agreement between the predicted and observed DO rates. This index can range within [−1, +1], where −1 (+1) indicates a totally negative (positive) correlation, and 0 means no correlation. Figure 11 shows a scatterplot for each model containing both training and testing results. As can be seen, all outputs are positively aggregated around the best-fit line (i.e., the black line). For the training results, the R_Ps of 0.8808, 0.8545, and 0.8778 indicate the higher consistency of the MLP-MVO results, while the values of 0.8741, 0.8453, and 0.8775 demonstrate the superiority of the MLP-SCE in the testing phase.

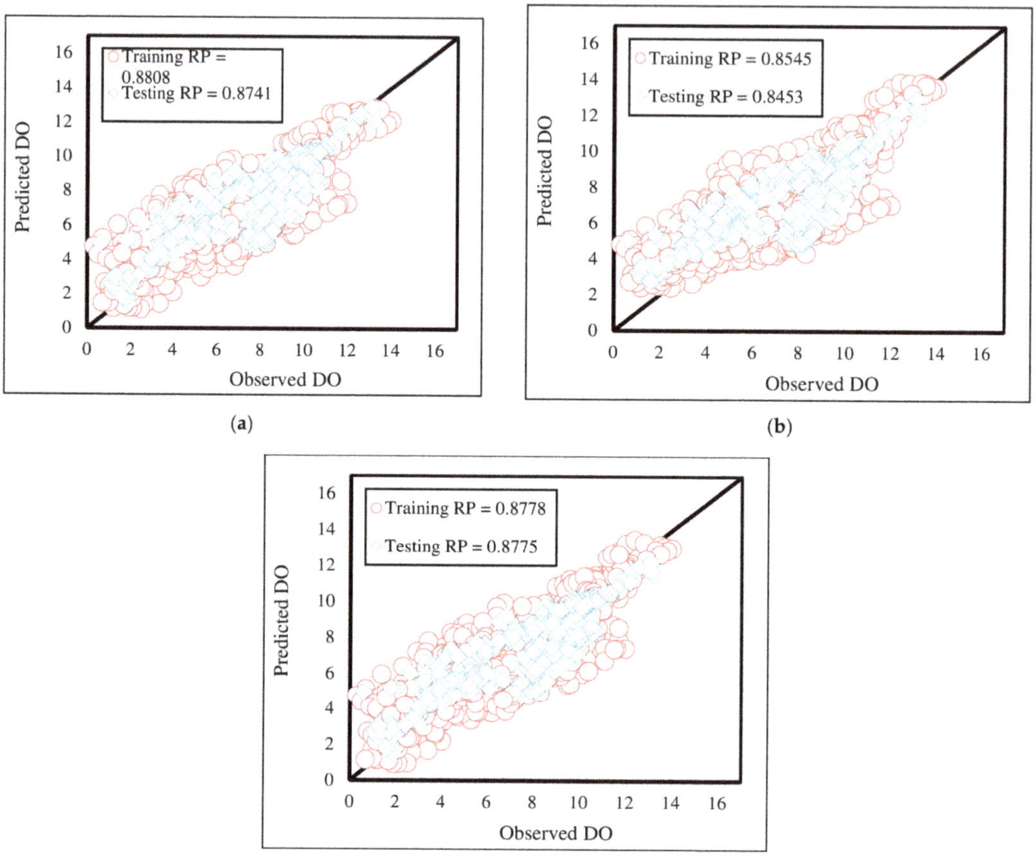

Figure 11. The error line and scatterplot plotted for the testing data of (**a**) MLP-MVO, (**b**) MLP-BHA, and (**c**) MLP-SCE.

4.3. Cross-Validation

In order to further verify the potential of the implemented models, the trained models (i.e., the MVO, BHA, and SCE, with the N_{Pop}s of 300, 50, and 10, respectively) are applied to another station called Fanno Creek (station number: 14206950). Similar to Klamath River, it is located in the Oregon State but in the northern part (longitude: 122°45′13″ W and latitude: 45°24′13″ N). The models predicted the DO for the water year 2019 using the daily records of WT, SC, and pH. Notably, the measured DO ranged from 5.40 to 13.20 mg/L in this site.

The results are shown in Table 2. As can be seen, the performance of all three models is satisfying, and the predicted values are well-correlated with the observed DO rates. It professes that the models can be used for new prediction cases. However, the combination of inputs (i.e., taking WT, pH, and SC into account) should be respected for all study cases.

Table 2. The results of cross-validation.

Model	RMSE	MAE	RP
MLP-MVO	0.7090	0.5767	0.9413
MLP-BHA	0.6980	0.5686	0.9430
MLP-SCE	0.6989	0.5686	0.9431

4.4. Further Discussion

Due to the water quality situation in the Klamath River, this study was dedicated to suggesting novel predictive tools for analyzing the DO concentration in this reach. Notably, this river is important for irrigation, flows, and hydropower generation [72]. It was discussed that DO should be evaluated in response to different key parameters. Another difficulty is that the DO does not follow a certain pattern over time. Hence, it is necessary to tackle these complexities by selecting appropriate models.

Many scholars have stated their tendency regarding the use of hybrid methods for similar issues (e.g., sediment concentration [85] and salinity [86] predictions). The reason for the development of such models can be the use of an optimization technique in the position of a trainer algorithm. In the case of DO modeling in this study, each of the MVO, BHA, and SCE algorithms drove an MLP neural network. In other words, their specific solutions were employed to analyze the relationship between DO and influential parameters through a neural framework.

The high level of accuracy achieved shows that DO can be promisingly predicted. However, there are some suggestions that should be applied for even more efficient solutions. The first part is related to methodology. Optimization using metaheuristic techniques needs to be carefully monitored to select appropriate parameters. For example, the number of iterations and the complexity of the population are better optimized vs. time, where time is as important as accuracy. However, the focus of this study was mostly on the accuracy of prediction (Figure 6). Another idea in this regard can be to test different metaheuristic techniques to find the quickest [87]. Another potential idea would be conducting comparisons with benchmark machine learning solutions such as the ANFIS and SVR as well as their hybridized versions.

The dataset is also a key item. The factors that can be regarded are optimizing the number of input parameters, selecting the appropriate time period, and the pre-processing and purification of misleading samples, among others. In this work, a valid 5-year dataset with three inputs was used, and the results showed that it could provide sufficient samples to be analyzed by the algorithms. Since the models could predict the DO of the testing period without prior knowledge of it, they can be used for further unseen events. Additionally, these results were achieved with the effect of only three influential parameters. Such simplicity is effective for avoiding complicated simulations and reducing the cost of computations.

4.5. SCE-Based Formula

In this section, the formulation of the MLP-SCE model is presented due to its superiority in the prediction phase as well as a simpler configuration compared to the MVO and BHA. Moreover, considering time efficiency, the SCE could optimize the ANN in a meaningfully shorter time. The elapsed times were approximately 5980, 675, and 531 s for the selected configurations of the MLP-MVO, MLP-BHA, and MLP-SCE, respectively (under a core i7 (at 1.8 GHz) operating system with 16 gigs of RAM).

This formula can serve as a predictive equation that estimates DO for given values of WT, pH, and SC. This model predicts DO based on Equation (14), in which Input stands for three inputs, and *IW* and *b1* are their corresponding weights and bias vectors, respectively. *LW* and *b2* symbolize the same values but for the output layer (Figure 5). These numbers are optimally tuned by the SCE for the MLP so that the lowest training error is achieved (Figure 7). Additionally, $f(x)$ (i.e., the activation function) is presented in Equation (20). It is worth noting that due to the neural network mechanism, this formula must be fed with normalized data.

$$DO = [LW] \times (f\ (([IW] \times [Input]) + [b1])) + [b2] \tag{14}$$

$$LW = \begin{bmatrix} -0.0317 & 0.4963 & 0.6755 & 0.4120 & 0.5135 & 0.0709 \end{bmatrix} \tag{15}$$

$$IW = \begin{bmatrix} -1.9543 & 0.0595 & -1.6276 \\ -0.6330 & 2.2098 & 1.0898 \\ -1.0687 & -1.5362 & 1.7233 \\ -1.5501 & -0.3963 & 1.9778 \\ -2.1836 & -0.9834 & 0.8583 \\ -0.7249 & 1.8623 & -1.5742 \end{bmatrix} \quad (16)$$

$$Input = \begin{bmatrix} WT \\ pH \\ SC \end{bmatrix} \quad (17)$$

$$b1 = \begin{bmatrix} 2.5440 \\ 1.5264 \\ 0.5088 \\ -0.5088 \\ -1.5264 \\ -2.5440 \end{bmatrix} \quad (18)$$

$$b2 = \begin{bmatrix} -0.5142 \end{bmatrix} \quad (19)$$

$$f(x) = \frac{2}{1 + e^{-2x}} - 1 \quad (20)$$

5. Conclusions

This research points out the suitability of metaheuristic strategies for analyzing the relationship between DO and three influential factors (WT, PH, and SC) through the principles of a multi-layer perceptron network. The algorithms used were multi-verse optimizer, black hole algorithm, and shuffled complex evolution, which showed high applicability for optimization objectives. A finding of this study was that while the MVO needs N_{Pop} = 300 to give proper training to the MLP, two other algorithms can do this with smaller populations (N_{Pop}s of 50 and 10). According to the findings of the training phase, the MVO can achieve a more profound understanding of the mentioned relationship. The RMSE of this model was 1.3148, which was found to be smaller than MLP-BHA (1.4426) and MLP-SCE (1.3304). However, different results were observed in the testing phase. The SCE-based model came up with the largest accuracy (the R_Ps were 0.8741, 0.8453, and 0.8775). All in all, the authors believe that the tested models can serve as promising ways for predicting DO. However, assessing other metaheuristic techniques and other hybridization strategies is recommended for future studies.

Author Contributions: F.Y., writing—review and editing, visualization, supervision; H.M., methodology, software validation, writing—original draft preparation; A.M., writing—review and editing, visualization, supervision, project administration. All authors have read and agreed to the published version of the manuscript.

Funding: This research received no external funding.

Institutional Review Board Statement: Not applicable.

Informed Consent Statement: Not applicable.

Data Availability Statement: Not applicable.

Conflicts of Interest: The authors declare no conflict of interest.

Abbreviations

DO	Dissolved oxygen
SVR	Support vector regression
ANN	Artificial neural networks
MLP	Multi-layer perceptron
RBF	Radial basis function
GRNN	General regression neural network
ANFIS	Adaptive neuro-fuzzy inference system
GA	Genetic algorithm
SFLA	Shuffled frog leaping algorithm
FOA	Fruit fly optimization
PSO	Particle swarm optimization
MVO	Multi-verse optimizer
BHA	Black hole algorithm
SCE	Shuffled complex evolution
MLR	Multiple linear regression
RNN	Recurrent neural network
ROI	Rate of inflation
RWS	Roulette wheel selection
R_P	Pearson's correlation coefficient
MAE	Mean absolute error
RMSE	Root mean square error
WT	Water temperature
SC	Specific conductance

References

1. Kisi, O.; Alizamir, M.; Gorgij, A.D. Dissolved oxygen prediction using a new ensemble method. *Environ. Sci. Pollut. Res.* **2020**, *27*, 9589–9603. [CrossRef] [PubMed]
2. Liu, S.; Yan, M.; Tai, H.; Xu, L.; Li, D. Prediction of dissolved oxygen content in aquaculture of *Hyriopsis cumingii* using Elman neural network. In Proceedings of the International Conference on Computer and Computing Technologies in Agriculture, Beijing, China, 29–31 October 2011; Springer: Berlin/Heidelberg, Germany, 2011; pp. 508–518.
3. Heddam, S.; Kisi, O. Extreme learning machines: A new approach for modeling dissolved oxygen (DO) concentration with and without water quality variables as predictors. *Environ. Sci. Pollut. Res.* **2017**, *24*, 16702–16724. [CrossRef]
4. Li, C.; Li, Z.; Wu, J.; Zhu, L.; Yue, J. A hybrid model for dissolved oxygen prediction in aquaculture based on multi-scale features. *Inf. Process. Agric.* **2018**, *5*, 11–20. [CrossRef]
5. Ay, M.; Kisi, O. Modeling of Dissolved Oxygen Concentration Using Different Neural Network Techniques in Foundation Creek, El Paso County, Colorado. *J. Environ. Eng.* **2012**, *138*, 654–662. [CrossRef]
6. Elkiran, G.; Nourani, V.; Abba, S.; Abdullahi, J. Artificial intelligence-based approaches for multi-station modelling of dissolve oxygen in river. *Glob. J. Environ. Sci. Manag.* **2018**, *4*, 439–450.
7. Zhang, M.; Zhang, L.; Tian, S.; Zhang, X.; Guo, J.; Guan, X.; Xu, P. Effects of graphite particles/Fe^{3+} on the properties of anoxic activated sludge. *Chemosphere* **2020**, *253*, 126638. [CrossRef] [PubMed]
8. Sun, M.; Hou, B.; Wang, S.; Zhao, Q.; Zhang, L.; Song, L.; Zhang, H. Effects of NaClO shock on MBR performance under continuous operating conditions. *Environ. Sci. Water Res. Technol.* **2021**, *7*, 396–404. [CrossRef]
9. Zhao, C.; Li, J. Equilibrium Selection under the Bayes-Based Strategy Updating Rules. *Symmetry* **2020**, *12*, 739. [CrossRef]
10. Liu, M.; Xue, Z.; Zhang, H.; Li, Y. Dual-channel membrane capacitive deionization based on asymmetric ion adsorption for continuous water desalination. *Electrochem. Commun.* **2021**, *125*, 106974. [CrossRef]
11. Yang, Y.; Li, Y.; Yao, J.; Iglauer, S.; Luquot, L.; Zhang, K.; Sun, H.; Zhang, L.; Song, W.; Wang, Z. Dynamic Pore-Scale Dissolution by CO_2-Saturated Brine in Carbonates: Impact of Homogeneous Versus Fractured Versus Vuggy Pore Structure. *Water Resour. Res.* **2020**, *56*, 26112. [CrossRef]
12. Zhao, X.; Gu, B.; Gao, F.; Chen, S. Matching Model of Energy Supply and Demand of the Integrated Energy System in Coastal Areas. *J. Coast. Res.* **2020**, *103*, 983–989. [CrossRef]
13. Zuo, X.; Dong, M.; Gao, F.; Tian, S. The Modeling of the Electric Heating and Cooling System of the Integrated Energy System in the Coastal Area. *J. Coast. Res.* **2020**, *103*, 1022–1029. [CrossRef]
14. Liu, J.; Yi, Y.; Wang, X. Exploring factors influencing construction waste reduction: A structural equation modeling approach. *J. Clean. Prod.* **2020**, *276*, 123185. [CrossRef]
15. Gao, N.; Wang, B.; Lu, K.; Hou, H. Complex band structure and evanescent Bloch wave propagation of periodic nested acoustic black hole phononic structure. *Appl. Acoust.* **2021**, *177*, 107906. [CrossRef]

16. Liu, J.; Liu, Y.; Wang, X. An environmental assessment model of construction and demolition waste based on system dynamics: A case study in Guangzhou. *Environ. Sci. Pollut. Res.* **2019**, *27*, 37237–37259. [CrossRef] [PubMed]
17. Zuo, C.; Chen, Q.; Tian, L.; Waller, L.; Asundi, A. Transport of intensity phase retrieval and computational imaging for partially coherent fields: The phase space perspective. *Opt. Lasers Eng.* **2015**, *71*, 20–32. [CrossRef]
18. Zuo, C.; Sun, J.; Jialin, Z.; Zhang, J.; Asundi, A.; Chen, Q. High-resolution transport-of-intensity quantitative phase microscopy with annular illumination. *Sci. Rep.* **2017**, *7*, 1–22. [CrossRef]
19. Zhang, L.; Zheng, J.; Tian, S.; Zhang, H.; Guan, X.; Zhu, S.; Zhang, X.; Bai, Y.; Xu, P.; Zhang, J.; et al. Effects of Al^{3+} on the microstructure and bioflocculation of anoxic sludge. *J. Environ. Sci.* **2020**, *91*, 212–221. [CrossRef]
20. Hong, X.-C.; Wang, G.-Y.; Liu, J.; Song, L.; Wu, E.T. Modeling the impact of soundscape drivers on perceived birdsongs in urban forests. *J. Clean. Prod.* **2021**, *292*, 125315. [CrossRef]
21. Zhang, W.; Hu, Y.; Liu, J.; Wang, H.; Wei, J.; Sun, P.; Wu, L.; Zheng, H. Progress of ethylene action mechanism and its application on plant type formation in crops. *Saudi J. Biol. Sci.* **2020**, *27*, 1667–1673. [CrossRef]
22. Han, C.; Zhang, B.; Chen, H.; Wei, Z.; Liu, Y. Spatially distributed crop model based on remote sensing. *Agric. Water Manag.* **2019**, *218*, 165–173. [CrossRef]
23. Qiao, W.; Wang, Y.; Zhang, J.; Tian, W.; Tian, Y.; Yang, Q. An innovative coupled model in view of wavelet transform for predicting short-term PM10 concentration. *J. Environ. Manag.* **2021**, *289*, 112438. [CrossRef]
24. Seyedashraf, O.; Mehrabi, M.; Akhtari, A.A. Novel approach for dam break flow modeling using computational intelligence. *J. Hydrol.* **2018**, *559*, 1028–1038. [CrossRef]
25. Moayedi, H.; Mehrabi, M.; Mosallanezhad, M.; Rashid, A.S.A.; Pradhan, B. Modification of landslide susceptibility mapping using optimized PSO-ANN technique. *Eng. Comput.* **2019**, *35*, 967–984. [CrossRef]
26. Xu, Y.; Chen, H.; Luo, J.; Zhang, Q.; Jiao, S.; Zhang, X. Enhanced Moth-flame optimizer with mutation strategy for global optimization. *Inf. Sci.* **2019**, *492*, 181–203. [CrossRef]
27. Xia, J.; Chen, H.; Li, Q.; Zhou, M.; Chen, L.; Cai, Z.; Fang, Y.; Zhou, H. Ultrasound-based differentiation of malignant and benign thyroid Nodules: An extreme learning machine approach. *Comput. Methods Programs Biomed.* **2017**, *147*, 37–49. [CrossRef]
28. Wang, M.; Chen, H.; Yang, B.; Zhao, X.; Hu, L.; Cai, Z.; Huang, H.; Tong, C. Toward an optimal kernel extreme learning machine using a chaotic moth-flame optimization strategy with applications in medical diagnoses. *Neurocomputing* **2017**, *267*, 69–84. [CrossRef]
29. Chen, H.-L.; Wang, G.; Ma, C.; Cai, Z.-N.; Liu, W.-B.; Wang, S.-J. An efficient hybrid kernel extreme learning machine approach for early diagnosis of Parkinson's disease. *Neurocomputing* **2016**, *184*, 131–144. [CrossRef]
30. Xu, X.; Chen, H.-L. Adaptive computational chemotaxis based on field in bacterial foraging optimization. *Soft Comput.* **2013**, *18*, 797–807. [CrossRef]
31. Zhao, D.; Liu, L.; Yu, F.; Heidari, A.A.; Wang, M.; Liang, G.; Muhammad, K.; Chen, H. Chaotic random spare ant colony optimization for multi-threshold image segmentation of 2D Kapur entropy. *Knowl.-Based Syst.* **2021**, *216*, 106510. [CrossRef]
32. Zhao, X.; Li, D.; Yang, B.; Ma, C.; Zhu, Y.; Chen, H. Feature selection based on improved ant colony optimization for online detection of foreign fiber in cotton. *Appl. Soft Comput.* **2014**, *24*, 585–596. [CrossRef]
33. Moayedi, H.; Mehrabi, M.; Kalantar, B.; Mu'Azu, M.A.; Rashid, A.S.A.; Foong, L.K.; Nguyen, H. Novel hybrids of adaptive neuro-fuzzy inference system (ANFIS) with several metaheuristic algorithms for spatial susceptibility assessment of seismic-induced landslide. *Geomat. Nat. Hazards Risk* **2019**, *10*, 1879–1911. [CrossRef]
34. Chen, H.; Heidari, A.A.; Chen, H.; Wang, M.; Pan, Z.; Gandomi, A.H. Multi-population differential evolution-assisted Harris hawks optimization: Framework and case studies. *Future Gener. Comput. Syst.* **2020**, *111*, 175–198. [CrossRef]
35. Zhang, Y.; Liu, R.; Wang, X.; Chen, H.; Li, C. Boosted binary Harris hawks optimizer and feature selection. *Eng. Comput.* **2020**, *25*, 1–30. [CrossRef]
36. Zhao, X.; Zhang, X.; Cai, Z.; Tian, X.; Wang, X.; Huang, Y.; Chen, H.; Hu, L. Chaos enhanced grey wolf optimization wrapped ELM for diagnosis of paraquat-poisoned patients. *Comput. Biol. Chem.* **2019**, *78*, 481–490. [CrossRef] [PubMed]
37. Hu, J.; Chen, H.; Heidari, A.A.; Wang, M.; Zhang, X.; Chen, Y.; Pan, Z. Orthogonal learning covariance matrix for defects of grey wolf optimizer: Insights, balance, diversity, and feature selection. *Knowl.-Based Syst.* **2021**, *213*, 106684. [CrossRef]
38. Zhang, Y.; Liu, R.; Heidari, A.A.; Wang, X.; Chen, Y.; Wang, M.; Chen, H. Towards augmented kernel extreme learning models for bankruptcy prediction: Algorithmic behavior and comprehensive analysis. *Neurocomputing* **2021**, *430*, 185–212. [CrossRef]
39. Tu, J.; Chen, H.; Liu, J.; Heidari, A.A.; Zhang, X.; Wang, M.; Ruby, R.; Pham, Q.-V. Evolutionary biogeography-based whale optimization methods with communication structure: Towards measuring the balance. *Knowl.-Based Syst.* **2021**, *212*, 106642. [CrossRef]
40. Shan, W.; Qiao, Z.; Heidari, A.A.; Chen, H.; Turabieh, H.; Teng, Y. Double adaptive weights for stabilization of moth flame optimizer: Balance analysis, engineering cases, and medical diagnosis. *Knowl.-Based Syst.* **2021**, *214*, 106728. [CrossRef]
41. Yu, H.; Li, W.; Chen, C.; Liang, J.; Gui, W.; Wang, M.; Chen, H. Dynamic Gaussian bare-bones fruit fly optimizers with abandonment mechanism: Method and analysis. *Eng. Comput.* **2020**, 1–29. [CrossRef]
42. Hu, L.; Hong, G.; Ma, J.; Wang, X.; Chen, H. An efficient machine learning approach for diagnosis of paraquat-poisoned patients. *Comput. Biol. Med.* **2015**, *59*, 116–124. [CrossRef]
43. Shen, L.; Chen, H.; Yu, Z.; Kang, W.; Zhang, B.; Li, H.; Yang, B.; Liu, D. Evolving support vector machines using fruit fly optimization for medical data classification. *Knowl.-Based Syst.* **2016**, *96*, 61–75. [CrossRef]

44. Wang, M.; Chen, H. Chaotic multi-swarm whale optimizer boosted support vector machine for medical diagnosis. *Appl. Soft Comput.* **2020**, *88*, 105946. [CrossRef]
45. Li, C.; Hou, L.; Sharma, B.Y.; Li, H.; Chen, C.; Li, Y.; Zhao, X.; Huang, H.; Cai, Z.; Chen, H. Developing a new intelligent system for the diagnosis of tuberculous pleural effusion. *Comput. Methods Programs Biomed.* **2018**, *153*, 211–225. [CrossRef] [PubMed]
46. Liu, Z.; Gilbert, G.; Cepeda, J.M.; Lysdahl, A.O.K.; Piciullo, L.; Hefre, H.; Lacasse, S. Modelling of shallow landslides with machine learning algorithms. *Geosci. Front.* **2021**, *12*, 385–393. [CrossRef]
47. Csábrági, A.; Molnár, S.; Tanos, P.; Kovács, J. Application of artificial neural networks to the forecasting of dissolved oxygen content in the Hungarian section of the river Danube. *Ecol. Eng.* **2017**, *100*, 63–72. [CrossRef]
48. Csábrági, A.; Molnár, S.; Tanos, P.; Kovács, J.; Molnár, M.; Szabó, I.; Hatvani, I.G. Estimation of dissolved oxygen in riverine ecosystems: Comparison of differently optimized neural networks. *Ecol. Eng.* **2019**, *138*, 298–309. [CrossRef]
49. Heddam, S. Simultaneous modelling and forecasting of hourly dissolved oxygen concentration (DO) using radial basis function neural network (RBFNN) based approach: A case study from the Klamath River, Oregon, USA. *Model. Earth Syst. Environ.* **2016**, *2*, 1–18. [CrossRef]
50. Heddam, S. Fuzzy Neural Network (EFuNN) for Modelling Dissolved Oxygen Concentration (DO). In *Intelligence Systems in Environmental Management: Theory and Applications*; Springer: Berlin/Heidelberg, Germany, 2017; pp. 231–253.
51. Khan, U.T.; Valeo, C. Dissolved oxygen prediction using a possibility theory based fuzzy neural network. *Hydrol. Earth Syst. Sci.* **2016**, *20*, 2267–2293. [CrossRef]
52. Ahmed, A.M.; Shah, S.M.A. Application of adaptive neuro-fuzzy inference system (ANFIS) to estimate the biochemical oxygen demand (BOD) of Surma River. *J. King Saud Univ. Eng. Sci.* **2017**, *29*, 237–243. [CrossRef]
53. Heddam, S. Modeling hourly dissolved oxygen concentration (DO) using two different adaptive neuro-fuzzy inference systems (ANFIS): A comparative study. *Environ. Monit. Assess.* **2014**, *186*, 597–619. [CrossRef] [PubMed]
54. Ranković, V.; Radulović, J.; Radojevic, I.; Ostojić, A.; Čomić, L. Prediction of dissolved oxygen in reservoirs using adaptive network-based fuzzy inference system. *J. Hydroinform.* **2011**, *14*, 167–179. [CrossRef]
55. Heddam, S.; Kisi, O. Modelling daily dissolved oxygen concentration using least square support vector machine, multivariate adaptive regression splines and M5 model tree. *J. Hydrol.* **2018**, *559*, 499–509. [CrossRef]
56. Kisi, O.; Akbari, N.; Sanatipour, M.; Hashemi, A.; Teimourzadeh, K.; Shiri, J. Modeling of Dissolved Oxygen in River Water Using Artificial Intelligence Techniques. *J. Environ. Inform.* **2013**, *22*. [CrossRef]
57. Li, X.; Sha, J.; Wang, Z.-L. A comparative study of multiple linear regression, artificial neural network and support vector machine for the prediction of dissolved oxygen. *Hydrol. Res.* **2016**, *48*, 1214–1225. [CrossRef]
58. Antanasijević, D.; Pocajt, V.; Pericgrujic, A.A.; Ristić, M. Modelling of dissolved oxygen in the Danube River using artificial neural networks and Monte Carlo Simulation uncertainty analysis. *J. Hydrol.* **2014**, *519*, 1895–1907. [CrossRef]
59. Ouma, Y.O.; Okuku, C.O.; Njau, E.N. Use of Artificial Neural Networks and Multiple Linear Regression Model for the Prediction of Dissolved Oxygen in Rivers: Case Study of Hydrographic Basin of River Nyando, Kenya. *Complexity* **2020**, *2020*, 9570789. [CrossRef]
60. Zhang, Y.-F.; Fitch, P.; Thorburn, P.J. Predicting the Trend of Dissolved Oxygen Based on the kPCA-RNN Model. *Water* **2020**, *12*, 585. [CrossRef]
61. Ali, S.A.; Parvin, F.; Vojteková, J.; Costache, R.; Linh, N.T.T.; Pham, Q.B.; Vojtek, M.; Gigović, L.; Ahmad, A.; Ghorbani, M.A. GIS-based landslide susceptibility modeling: A comparison between fuzzy multi-criteria and machine learning algorithms. *Geosci. Front.* **2021**, *12*, 857–876. [CrossRef]
62. Ahmed, M.H. Prediction of the Concentration of Dissolved Oxygen in Running Water by Employing A Random Forest Machine Learning Technique. *J. Hydrol.* **2021**. [CrossRef]
63. Ay, M.; Kişi, Ö. Estimation of dissolved oxygen by using neural networks and neuro fuzzy computing techniques. *J. Civil Eng.* **2017**, *21*, 1631–1639. [CrossRef]
64. Rajaee, T.; Khani, S.; Ravansalar, M. Artificial intelligence-based single and hybrid models for prediction of water quality in rivers: A review. *Chemom. Intell. Lab. Syst.* **2020**, *200*, 103978. [CrossRef]
65. Faruk, D.Ö. A hybrid neural network and ARIMA model for water quality time series prediction. *Eng. Appl. Artif. Intell.* **2010**, *23*, 586–594. [CrossRef]
66. Bui, D.T.; Khosravi, K.; Tiefenbacher, J.; Nguyen, H.; Kazakis, N. Improving prediction of water quality indices using novel hybrid machine-learning algorithms. *Sci. Total Environ.* **2020**, *721*, 137612. [CrossRef] [PubMed]
67. Ravansalar, M.; Rajaee, T.; Ergil, M. Prediction of dissolved oxygen in River Calder by noise elimination time series using wavelet transform. *J. Exp. Theor. Artif. Intell.* **2016**, *28*, 689–706. [CrossRef]
68. Antanasijević, D.; Pocajt, V.; Perić-Grujić, A.; Ristić, M. Multilevel split of high-dimensional water quality data using artificial neural networks for the prediction of dissolved oxygen in the Danube River. *Neural Comput. Appl.* **2020**, *32*, 3957–3966. [CrossRef]
69. Nacar, S.; Bayram, A.; Baki, O.T.; Kankal, M.; Aras, E. Spatial Forecasting of Dissolved Oxygen Concentration in the Eastern Black Sea Basin, Turkey. *Water* **2020**, *12*, 1041. [CrossRef]
70. Mahmoudi, N.; Orouji, H.; Fallah-Mehdipour, E. Integration of Shuffled Frog Leaping Algorithm and Support Vector Regression for Prediction of Water Quality Parameters. *Water Resour. Manag.* **2016**, *30*, 2195–2211. [CrossRef]
71. Zhu, C.; Liu, X.; Ding, W. Prediction model of dissolved oxygen based on FOA-LSSVR. In Proceedings of the 36th Chinese Control Conference (CCC), Dalian, China, 26–28 July 2017; IEEE: Piscataway, NJ, USA, 2017; pp. 9819–9823.

72. Sullivan, A.B.; Deas, M.L.; Asbill, J.; Kirshtein, J.D.; Butler, K.D.; Stewart, M.A.; Wellman, R.W.; Vaughn, J. Klamath River Water Quality and Acoustic Doppler Current Profiler Data from Link River Dam to Keno Dam, 2007. U.S. Geological Survey Open-File Report 2008-1185. Available online: https://pubs.usgs.gov/of/2008/1185/ (accessed on 19 July 2021).
73. Raheli, B.; Aalami, M.T.; El-Shafie, A.; Ghorbani, M.A.; Deo, R.C. Uncertainty assessment of the multilayer perceptron (MLP) neural network model with implementation of the novel hybrid MLP-FFA method for prediction of biochemical oxygen demand and dissolved oxygen: A case study of Langat River. *Environ. Earth Sci.* **2017**, *76*, 503. [CrossRef]
74. Deng, C.; Wei, X.; Guo, L. Application of Neural Network Based on PSO Algorithm in Prediction Model for Dissolved Oxygen in Fishpond. In Proceedings of the 6th World Congress on Intelligent Control and Automation, Dalian, China, 21–23 June 2006; IEEE: Piscataway, NJ, USA, 2006.
75. Barrow, J.D.; Davies, P.C.; Harper, C.L., Jr. *Science and Ultimate Reality: Quantum Theory, Cosmology, and Complexity*; Cambridge University Press: Cambridge, UK, 2004.
76. Mirjalili, S.; Mirjalili, S.M.; Hatamlou, A. Multi-verse optimizer: A nature-inspired algorithm for global optimization. *Neural Comput. Appl.* **2016**, *27*, 495–513. [CrossRef]
77. Chen, L.; Li, L.; Kuang, W. A hybrid multiverse optimisation algorithm based on differential evolution and adaptive mutation. *J. Exp. Theor. Artif. Intell.* **2021**, *33*, 239–261. [CrossRef]
78. Abasi, A.K.; Khader, A.T.; Al-Betar, M.A.; Naim, S.; Makhadmeh, S.N.; Alyasseri, Z.A.A. Link-based multi-verse optimizer for text documents clustering. *Appl. Soft Comput.* **2020**, *87*, 106002. [CrossRef]
79. Faris, H.; Hassonah, M.A.; Al-Zoubi, A.M.; Mirjalili, S.; Aljarah, I. A multi-verse optimizer approach for feature selection and optimizing SVM parameters based on a robust system architecture. *Neural Comput. Appl.* **2018**, *30*, 2355–2369. [CrossRef]
80. Hatamlou, A. Black hole: A new heuristic optimization approach for data clustering. *Inf. Sci.* **2013**, *222*, 175–184. [CrossRef]
81. Qasim, O.S.; Al-Thanoon, N.A.; Algamal, Z.Y. Feature selection based on chaotic binary black hole algorithm for data classification. *Chemom. Intell. Lab. Syst.* **2020**, *204*, 104104. [CrossRef]
82. Duan, Q.; Gupta, V.K.; Sorooshian, S. Shuffled complex evolution approach for effective and efficient global minimization. *J. Optim. Theory Appl.* **1993**, *76*, 501–521. [CrossRef]
83. Naeini, M.R.; Analui, B.; Gupta, H.V.; Duan, Q.; Sorooshian, S. Three decades of the Shuffled Complex Evolution (SCE-UA) optimization algorithm: Review and applications. *Sci. Iran.* **2019**, *26*, 2015–2031.
84. Majeed, K.; Qyyum, M.A.; Nawaz, A.; Ahmad, A.; Naqvi, M.; He, T.; Lee, M. Shuffled Complex Evolution-Based Performance Enhancement and Analysis of Cascade Liquefaction Process for Large-Scale LNG Production. *Energies* **2020**, *13*, 2511. [CrossRef]
85. Ehteram, M.; Ahmed, A.N.; Latif, S.D.; Huang, Y.F.; Alizamir, M.; Kisi, O.; Mert, C.; El-Shafie, A. Design of a hybrid ANN multi-objective whale algorithm for suspended sediment load prediction. *Environ. Sci. Pollut. Res.* **2021**, *28*, 1596–1611. [CrossRef] [PubMed]
86. Nazari, H.; Taghavi, B.; Hajizadeh, F. Groundwater salinity prediction using adaptive neuro-fuzzy inference system methods: A case study in Azarshahr, Ajabshir and Maragheh plains, Iran. *Environ. Earth Sci.* **2021**, *80*, 1–10. [CrossRef]
87. Moayedi, H.; Ghareh, S.; Foong, L.K. Quick integrative optimizers for minimizing the error of neural computing in pan evaporation modeling. *Eng. Comput.* **2021**, 1–17. [CrossRef]